SIXTH EDITION

Fundamentals of Physics

VOLUME 1

Mathematical Formulas*

Quadratic Formula

If $ax^2 + bx + c = 0$, then $x = \dfrac{-b \pm \sqrt{b^2 - 4ac}}{2a}$

Binomial Theorem

$$(1 + x)^n = 1 + \frac{nx}{1!} + \frac{n(n-1)x^2}{2!} + \ldots \qquad (x^2 < 1)$$

Products of Vectors

Let θ be the smaller of the two angles between $\vec{a}$ and $\vec{b}$. Then

$$\vec{a} \cdot \vec{b} = \vec{b} \cdot \vec{a} = a_x b_x + a_y b_y + a_z b_z = ab \cos \theta$$

$$\vec{a} \times \vec{b} = -\vec{b} \times \vec{a} = \begin{vmatrix} \hat{i} & \hat{j} & \hat{k} \\ a_x & a_y & a_z \\ b_x & b_y & b_z \end{vmatrix}$$

$$= \hat{i} \begin{vmatrix} a_y & a_z \\ b_y & b_z \end{vmatrix} - \hat{j} \begin{vmatrix} a_x & a_z \\ b_x & b_z \end{vmatrix} + \hat{k} \begin{vmatrix} a_x & a_y \\ b_x & b_y \end{vmatrix}$$

$$= (a_y b_z - b_y a_z)\hat{i} + (a_z b_x - b_z a_x)\hat{j} + (a_x b_y - b_x a_y)\hat{k}$$

$$|\vec{a} \times \vec{b}| = ab \sin \theta$$

Trigonometric Identities

$$\sin \alpha \pm \sin \beta = 2 \sin \tfrac{1}{2}(\alpha \pm \beta) \cos \tfrac{1}{2}(\alpha \mp \beta)$$
$$\cos \alpha + \cos \beta = 2 \cos \tfrac{1}{2}(\alpha + \beta) \cos \tfrac{1}{2}(\alpha - \beta)$$

* See Appendix E for a more complete list.

Derivatives and Integrals

$$\frac{d}{dx} \sin x = \cos x \qquad \int \sin x \, dx = -\cos x$$

$$\frac{d}{dx} \cos x = -\sin x \qquad \int \cos x \, dx = \sin x$$

$$\frac{d}{dx} e^x = e^x \qquad \int e^x \, dx = e^x$$

$$\int \frac{dx}{\sqrt{x^2 + a^2}} = \ln(x + \sqrt{x^2 + a^2})$$

$$\int \frac{x \, dx}{(x^2 + a^2)^{3/2}} = -\frac{1}{(x^2 + a^2)^{1/2}}$$

$$\int \frac{dx}{(x^2 + a^2)^{3/2}} = \frac{x}{a^2(x^2 + a^2)^{1/2}}$$

Cramer's Rule

Two simultaneous equations in unknowns x and y,

$$a_1 x + b_1 y = c_1 \qquad \text{and} \qquad a_2 x + b_2 y = c_2,$$

have the solutions

$$x = \frac{\begin{vmatrix} c_1 & b_1 \\ c_2 & b_2 \end{vmatrix}}{\begin{vmatrix} a_1 & b_1 \\ a_2 & b_2 \end{vmatrix}} = \frac{c_1 b_2 - c_2 b_1}{a_1 b_2 - a_2 b_1}$$

and

$$y = \frac{\begin{vmatrix} a_1 & c_1 \\ a_2 & c_2 \end{vmatrix}}{\begin{vmatrix} a_1 & b_1 \\ a_2 & b_2 \end{vmatrix}} = \frac{a_1 c_2 - a_2 c_1}{a_1 b_2 - a_2 b_1}.$$

The Greek Alphabet

Alpha	A	α	Iota	I	ι	Rho	P	ρ
Beta	B	β	Kappa	K	κ	Sigma	Σ	σ
Gamma	Γ	γ	Lambda	Λ	λ	Tau	T	τ
Delta	Δ	δ	Mu	M	μ	Upsilon	Y	υ
Epsilon	E	ϵ	Nu	N	ν	Phi	Φ	ϕ, φ
Zeta	Z	ζ	Xi	Ξ	ξ	Chi	X	χ
Eta	H	η	Omicron	O	o	Psi	Ψ	ψ
Theta	Θ	θ	Pi	Π	π	Omega	Ω	ω

SIXTH EDITION

Fundamentals of Physics

VOLUME 1

David Halliday
University of Pittsburgh

Robert Resnick
Rensselaer Polytechnic Institute

Jearl Walker
Cleveland State University

John Wiley & Sons, Inc.

New York / Chichester / Weinheim / Brisbane
Singapore / Toronto

ACQUISITIONS EDITOR   Stuart Johnson
DEVELOPMENTAL EDITOR   Ellen Ford
MARKETING MANAGER   Sue Lyons
ASSOCIATE PRODUCTION DIRECTOR   Lucille Buonocore
SENIOR PRODUCTION EDITOR   Monique Calello
TEXT/COVER DESIGNER   Madelyn Lesure
COVER PHOTO   Tsuyoshi Nishiinoue/Orion Press
PHOTO MANAGER   Hilary Newman
PHOTO RESEARCHER   Jennifer Atkins
DUMMY DESIGNER   Lee Goldstein
ILLUSTRATION EDITORS   Edward Starr and Anna Melhorn
ILLUSTRATION   Radiant/Precision Graphics
COPYEDITOR   Helen Walden
PROOFREADER   Lilian Brady
TECHNICAL PROOFREADER   Georgia Mederer
INDEXER   Dorothy M. Jahoda

This book was set in 10/12 Times Roman by Progressive Information Technologies and printed and bound by Von Hoffmann Press, Inc. The cover was printed by Brady Palmer Printing Company.

This book is printed on acid-free paper.

The paper in this book was manufactured by a mill whose forest management programs include sustained yield harvesting of its timberlands. Sustained yield harvesting principles ensure that the number of trees cut each year does not exceed the amount of new growth.

Library of Congress Cataloging-in-Publication Data

Halliday, David
    Fundamentals of physics/David Halliday, Robert Resnick, Jearl Walker.—6th ed.
      p. cm.
    Includes index.
    ISBN 0-471-33235-6 (v. 1 : cloth : alk. paper)
    1. Physics.   I. Resnick, Robert.   II. Walker, Jearl.   III. Title.

QC21,2.H35 2001
530–dc21   00-027365

To order books or for customer service call 1-800-CALL-WILEY (225-5945).

ISBN 0-471-33235-6

Printed in the United States of America

10   9   8   7   6   5   4   3   2

BRIEF CONTENTS

v

TABLES

CONTENTS

CHAPTER 3

Vectors   31

How can vectors be used in cave exploring?

CHAPTER 4

Motion in Two and Three Dimensions   48

How can the placement of the net be determined for a human cannonball?

CHAPTER 5

Force and Motion—I 72

Can a man pull two railroad passenger cars with his teeth?

CHAPTER 6

Force and Motion—II 98

Why do cats sometimes survive long falls better than shorter ones?

CHAPTER 7

Kinetic Energy and Work 116

How much work is required in lifting great weights?

CHAPTER 8

Potential Energy and Conservation of Energy   139

Did the placement of the giant statues on Easter Island require a mysterious energy source?

CHAPTER 9

Systems of Particles   168

How does a ballerina seemingly "turn off" the gravitational force?

CHAPTER 10

Collisions   193

Is a board or a concrete block easier to break in karate?

CHAPTER 11

Rotation   215

What advantages does physics offer in judo throws?

CHAPTER 12

PART 2

CHAPTER 13

CHAPTER 17

Waves—I   370

How does a scorpion detect a beetle without using sight or sound?

CHAPTER 18

Waves—II   398

How does a bat detect a moth in total darkness?

CHAPTER 19

Temperature, Heat, and the First Law of Thermodynamics   425

What thermal protection do bees have against a hornet invasion?

CHAPTER 20

The Kinetic Theory of Gases   454

Why can a cold container of a carbonated drink form a fog when opened?

CHAPTER 21

Entropy and the Second Law of Thermodynamics   482

What in the world gives direction to time?

APPENDICES

ANSWERS TO CHECKPOINTS AND ODD-NUMBERED QUESTIONS, EXERCISES, AND PROBLEMS

INDEX

PREFACE

This sixth edition of *Fundamentals of Physics* contains a redesign and major rewrites of the widely used fifth edition, while maintaining many elements of the classic text first written by David Halliday and Robert Resnick. Nearly all the changes are based on suggestions from instructors and students using the fifth edition, from reviewers of the manuscripts for the sixth edition, and from research done on the process of learning. You can send suggestions, corrections, and positive or negative comments to John Wiley & Sons (http://www.wiley.com/college/hrw) or Jearl Walker (mail address: Physics Department, Cleveland State University, Cleveland OH 44115 USA; fax number: (USA) (216) 687-2424; or email address: physics@wiley.com). We may not be able to respond to all suggestions, but we keep and study each of them.

Design Changes

➤ *More open format.* Previous editions have been printed in a double-column format, which many students and instructors have found cluttered and distracting. In this edition, the narrative is presented in a single-column format with a wide margin for note-taking.

➤ *Streamlined presentation.* It is a common complaint of all texts that they cover too much material. As a response to this criticism, the sixth edition has been shortened in two ways.

1. Material regarding special relativity and quantum physics has been moved from the early chapters to the later chapters devoted to those subjects.

2. The essential sample problems have been retained in this book, but the more specialized sample problems have been shifted to the Problem Supplement that is automatically provided with this book. (The Problem Supplement is described on the next page.)

➤ *Vector notation.* Vectors are now presented with an overhead arrow (such as $\vec{F}$) instead of as a bold symbol (such as **F**).

➤ *Emphasis on metric units.* Except in Chapter 1 (in which various systems of units are employed) and certain problems involving baseball (in which English units are traditional), metric units are used almost exclusively.

➤ *Structured versus unstructured order of problems.* The homework problems in this book are still ordered approxi-

mately according to their difficulty and grouped under section titles corresponding to the narrative of the chapter. However, many of the homework problems of the fifth edition have been shifted, without order or grouping, to the Problem Supplement. (The total number of problems in this book and in the Problem Supplement exceeds what was available in the fifth edition.)

➤ *Icons for additional help.* When worked-out solutions are provided either in print or electronically for certain of the odd-numbered homework problems, the statements for those problems include a trailing icon to alert both student and instructor as to where the solutions are located. An icon guide is provided here and at the beginning of each set of homework problems:

ssm	Solution is in the Student Solutions Manual.
www	Solution is available on the World Wide Web at: http://www.wiley.com/college/hrw
ilw	Solution is available on the Interactive LearningWare.

These resources are described later in this preface.

Pedagogy Changes

➤ *Reasoning versus plug-and-chug.* The primary goal of this book is to teach students to reason through challenging situations, from basic principles to a solution. Although some plug-and-chug homework problems remain in this book, most homework problems emphasize reasoning.

➤ *Key Ideas in the sample problems.* The solutions to all 360 sample problems in this book and in the Problem Supplement have been rewritten to begin with one or more Key Ideas based on basic principles.

➤ *Lengthened solutions to sample problems.* Most of the solutions to the sample problems are now longer because they build step by step from the beginning Key Ideas to an answer, often repeating some of the important reasoning of the narrative preceding the sample problems. For example, see Sample Problem 8-3 on page 148 and Sample Problem 10-2 on pages 200–201.

➤ *Use of vector-capable calculators.* When vector calculations in a sample problem can be performed directly on-screen with a vector-capable calculator, the solution of the sample problem indicates that fact but still carries through the traditional component analysis. When vector calculations cannot be performed directly on-screen, the solution explains why.

➤ *Problems with applied physics,* based on published research, have been added in many places, either as sample problems or homework problems. For example, see Sample Problem 11-6 on page 229, homework problem 64 on page 71, and homework problem 56 on page 214. For an example of homework problems that build with a continuing story, see problems 4, 32, and 48 (on pages 112, 114, and 115, respectively) in Chapter 6.

Content Changes

➤ *Chapter 5 on force and motion* contains clearer explanations of the gravitational force, weight, and normal force (pages 80–82).

➤ *Chapter 7 on kinetic energy and work* begins with a rough definition of energy. It then defines kinetic energy, work, and the work–kinetic energy theorem in ways that are more closely tied to Newton's second law than in the fifth edition, while keeping those definitions consistent with thermodynamics (pages 117–120).

➤ *Chapter 8 on the conservation of energy* avoids the much criticized definition of work done by a nonconservative force by explaining, instead, the energy transfers that occur due to a nonconservative force (page 153). (The wording still allows an instructor to superimpose a definition of work done by a nonconservative force.)

➤ *Chapter 10 on collisions* now presents the general situation of inelastic one-dimensional collisions (pages 198–200) before the special situation of elastic one-dimensional collisions (pages 202–204).

➤ *Chapters 16, 17, and 18 on SHM and waves* have been rewritten to better ease a student into these difficult subjects.

➤ *Chapter 21 on entropy* now presents a Carnot engine as the ideal heat engine with the greatest efficiency.

Chapter Features

➤ *Opening puzzlers.* A curious puzzling situation opens each chapter and is explained somewhere within the chapter, to entice a student to read the chapter.

➤ *Checkpoints* are stopping points that effectively ask the student, "Can you answer this question with some reasoning based on the narrative or sample problem that you just read?" If not, then the student should go back over that previous material before traveling deeper into the chapter. For ex-

ample, see Checkpoint 3 on page 78 and Checkpoint 1 on page 101. **Answers to all checkpoints are in the back of the book.**

➤ *Sample problems* have been chosen to help the student organize the basic concepts of the narrative and to develop problem-solving skills. Each sample problem builds step by step from one or more Key Ideas to a solution.

➤ *Problem-solving tactics* contain helpful instructions to guide the beginning physics student as to how to solve problems and to avoid common errors.

➤ *Review & Summary* is a brief outline of the chapter contents that contains the essential concepts but which is not a substitute for reading the chapter.

➤ *Questions* are like the checkpoints and require reasoning and understanding rather than calculations. **Answers to the odd-number questions are in the back of the book.**

➤ *Exercises & Problems* are ordered approximately according to difficulty and grouped under section titles. **The odd-numbered ones are answered in the back of the book.** Worked-out solutions to the odd-numbered problems with trailing icons are available either in print or electronically. (See the icon guide at the beginning of the Exercises & Problems.) A problem number with a star indicates an especially challenging problem.

➤ *Additional Problems* appear at the end of the Exercises & Problems in certain chapters. They are not sorted according to section titles and many involve applied physics.

Problem Supplement

A problem supplement is automatically provided with this book. The *Problem Supplement #1* (green book) will be provided until May 15, 2002. Thereafter, the *Problem Supplement #2* (blue book) will be provided. The blue book will have a different set of questions and homework problems and will contain more sample problems. The features of both versions of the problem supplement are the following:

➤ *Additional sample problems* that were shifted from the main book, plus many new ones. All begin with the basic *Key Ideas* and then build step by step to a solution.

➤ *Questions* include:
  1. *Checkpoint-style questions,* as in the main book.
  2. *Organizing questions,* which request that equations be set up for common situations, as a warm-up for the homework problems.

3. *Discussion questions* from the fourth and earlier editions of the book (back by request).

➤ *Exercises & Problems.* More homework problems, including many shifted from the main book. These are *not* ordered according to difficulty, section titles, or appearance of the associated physics in the chapter. Some of the new problems involve applied physics. In some chapters the homework problems end with *Clustered Problems,* in which similar problems are grouped together. In the other chapters, the homework problems end with *Tutorial Problems,* in which solutions are worked out.

Versions of the Text

The sixth edition of *Fundamentals of Physics* is available in a number of different versions, to accommodate the individual needs of instructors and students. The Regular Edition consists of Chapters 1 through 38 (ISBN 0-471-32000-5). The Extended Edition contains seven additional chapters on quantum physics and cosmology (Chapters 1–45) (ISBN 0-471-33236-4). Both editions are available as single, hardcover books, or in the following alternative versions:

➤ *Volume 1—Chapters 1–21 (Mechanics/Thermodynamics), hardcover, 0-471-33235-6*

➤ *Volume 2—Chapters 22–45 (E&M and Modern Physics), hardcover, 0-471-36037-6*

➤ *Part 1—Chapters 1–12, paperback, 0-471-33234-8*

➤ *Part 2—Chapters 13–21, paperback, 0-471-36041-4*

➤ *Part 3—Chapters 22–33, paperback, 0-471-36040-6*

➤ *Part 4—Chapters 34–38, paperback, 0-471-36039-2*

➤ *Part 5—Chapters 39–45, paperback, 0-471-36038-4*

Supplements

The sixth edition of *Fundamentals of Physics* is supplemented by a comprehensive ancillary package carefully developed to help teachers teach and students learn.

Instructor's Supplements

➤ *Instructor's Manual* by J. RICHARD CHRISTMAN, U.S. Coast Guard Academy. This manual contains lecture notes outlining the most important topics of each chapter, demonstration experiments, laboratory and computer projects, film and video sources, answers to all Questions, Exercises & Problems, and Checkpoints, and a correlation guide to the Questions and Exercises & Problems in the previous edition.

➤ *Instructor's Solutions Manual* by JAMES WHITENTON, Southern Polytechnic University. This manual provides worked-out solutions for all the exercises and problems found at the end of each chapter within the text and in the Problem Supplement #1. *This supplement is available only to instructors.*

➤ *Test Bank* by J. RICHARD CHRISTMAN, U.S. Coast Guard Academy. More than 2200 multiple-choice questions are included in this manual. These items are also available in the Computerized Test Bank (see below).

➤ *Instructor's Resource CD.* This CD contains:
- All of the Instructor's Solutions Manual in both LaTex and PDF files.
- Computerized Test Bank in both IBM and Macintosh versions, with full editing features to help instructors customize tests.
- All text illustrations suitable for both classroom presentation and printing.

➤ *Transparencies.* More than 200 four-color illustrations from the text are provided in a form suitable for projection in the classroom.

➤ *On-line Course Management.*
- WebAssign, CAPA, and Wiley eGrade are on-line homework and quizzing programs that give instructors the ability to deliver and grade homework and quizzes over the Internet.
- Instructors will also have access to WebCT course materials. WebCT is a powerful Web site program that allows instructors to set up complete on-line courses with chat rooms, bulletin boards, quizzing, student tracking, etc. Please contact your local Wiley representative for more information.

Student's Supplements

➤ *A Student Companion* by J. RICHARD CHRISTMAN, U.S. Coast Guard Academy. This student study guide consists of a traditional print component and an accompanying Web site, which together provide a rich, interactive environment for review and study. The Student Companion Web site includes self-quizzes, simulation exercises, hints for solving end-of-chapter problems, the *Interactive LearningWare* program (see the next page), and links to other Web sites that offer physics tutorial help.

➤ *Student Solutions Manual* by J. RICHARD CHRISTMAN, U.S. Coast Guard Academy and EDWARD DERRINGH, Wentworth Institute. This manual provides students with complete worked-out solutions to 30 percent of the exercises and problems found at the end of each chapter within the text. These problems are indicated with an ssm icon in the text.

➤ *Interactive LearningWare.* This software guides students through solutions to 200 of the end-of-chapter problems. The solutions process is developed interactively, with appropriate feedback and access to error-specific help for the most common mistakes. These problems are indicated with an ilw icon in the text.

➤ *CD-Physics, 3.0.* This CD-ROM based version of *Fundamentals of Physics,* Sixth Edition, contains the complete, extended version of the text, *A Student's Companion,* the *Student's Solutions Manual,* the *Interactive LearningWare,* and numerous simulations all connected with extensive hyperlinking.

➤ *Take Note!* This bound notebook lets students take notes directly onto large, black-and-white versions of textbook illustrations. All of the illustrations from the transparency set are included. In-class time spent copying illustrations is substantially reduced by this supplement.

➤ *Physics Web Site.* This Web site, **http://www.wiley.com/college/hrw**, was developed specifically for *Fundamentals of Physics,* Sixth Edition, and is designed to further assist students in the study of physics and offers additional physics resources. The site also includes solutions to selected end-of-chapter problems. These problems are identified with a www icon in the text.

ACKNOWLEDGMENTS

A textbook contains far more contributions to the elucidation of a subject than those made by the authors alone. J. Richard Christman, of the U.S. Coast Guard Academy, has once again created many fine supplements for us; his knowledge of our book and his recommendations to students and faculty are invaluable. James Tanner, of Georgia Institute of Technology, and Gary Lewis, of Kennesaw State College, have provided us with innovative software, closely tied to the text exercises and problems. James Whitenton, of Southern Polytechnic State University, and Jerry Shi, of Pasadena City College, performed the Herculean task of working out solutions for every one of the Exercises & Problems in the text. We thank John Merrill, of Brigham Young University, and Edward Derringh, of the Wentworth Institute of Technology, for their many contributions in the past. We also thank George W. Hukle of Oxnard, California, and Frank G. Jacobs of Evanston, Illinois, for their check of the answers for the problems in the book.

At John Wiley publishers, we have been fortunate to receive strong coordination and support from our former editor, Cliff Mills. Cliff guided our efforts and encouraged us along the way. When Cliff moved on to other responsibilities at Wiley, we were ably guided to completion by his successor, Stuart Johnson. Ellen Ford has coordinated the developmental editing and multilayered preproduction process. Sue Lyons, our marketing manager, has been tireless in her efforts on behalf of this edition. Joan Kalkut has built a fine supporting package of ancillary materials. Thomas Hempstead managed the reviews of manuscript and the multiple administrative duties admirably.

We thank Lucille Buonocore, our production director, and Monique Calello, our production editor, for pulling all the pieces together and guiding us through the complex production process. We also thank Maddy Lesure, for her design; Helen Walden for her copyediting; Edward Starr and Anna Melhorn, for managing the illustration program; Georgia Mederer, Katrina Avery, and Lilian Brady, for their proofreading; and all other members of the production team.

Hilary Newman and her team of photo researchers were inspired in their search for unusual and interesting photographs that communicate physics principles beautifully. We also owe a debt of gratitude for the line art to the late John Balbalis, whose careful hand and understanding of physics can still be seen in every diagram.

We especially thank Edward Millman for his developmental work on the manuscript. With us, he has read every word, asking many questions from the point of view of a student. Many of his questions and suggested changes have added to the clarity of this volume.

We owe a particular debt of gratitude to the numerous students who used the previous editions of *Fundamentals of Physics* and took the time to fill out the response cards and return them to us. As the ultimate consumers of this text, students are extremely important to us. By sharing their opinions with us, your students help us ensure that we are providing the best possible product and the most value for their textbook dollars. We encourage the users of this book to contact us with their thoughts and concerns so that we can continue to improve this text in the years to come.

Finally, our external reviewers have been outstanding and we acknowledge here our debt to each member of that team:

Edward Adelson
Ohio State University

Mark Arnett
Kirkwood Community College

Arun Bansil
Northeastern University

J. Richard Christman
U.S. Coast Guard Academy

Robert N. Davie, Jr.
St. Petersburg Junior College

Cheryl K. Dellai
Glendale Community College

Eric R. Dietz
California State University at Chico

N. John DiNardo
Drexel University

Harold B. Hart
Western Illinois University

Rebecca Hartzler
Edmonds Community College

Joey Huston
Michigan State University

Hector Jimenez
University of Puerto Rico

Sudhakar B. Joshi
York University

Leonard M. Kahn
University of Rhode Island

Yuichi Kubota
Cornell University

Priscilla Laws
Dickinson College

Edbertho Leal
Polytechnic University of Puerto Rico

Dale Long
Virginia Tech

Andreas Mandelis
University of Toronto

Paul Marquard
Caspar College

James Napolitano
Rensselaer Polytechnic Institute

Des Penny
Southern Utah University

Joe Redish
University of Maryland

Timothy M. Ritter
University of North Carolina at Pembroke

Gerardo A. Rodriguez
Skidmore College

John Rosendahl
University of California at Irvine

Michael Schatz
Georgia Institute of Technology

Michael G. Strauss
University of Oklahoma

Dan Styer
Oberlin College

Marshall Thomsen
Eastern Michigan University

Fred F. Tomblin
New Jersey Institute of Technology

B. R. Weinberger
Trinity College

William M. Whelan
Ryerson Polytechnic University

William Zimmerman, Jr.
University of Minnesota

Reviewers of the Fifth and Previous Editions

Maris A. Abolins
Michigan State University

Barbara Andereck
Ohio Wesleyan University

Albert Bartlett
University of Colorado

Michael E. Browne
University of Idaho

Timothy J. Burns
Leeward Community College

Joseph Buschi
Manhattan College

Philip A. Casabella
Rensselaer Polytechnic Institute

Randall Caton
Christopher Newport College

J. Richard Christman
U.S. Coast Guard Academy

Roger Clapp
University of South Florida

W. R. Conkie
Queen's University

Peter Crooker
University of Hawaii at Manoa

William P. Crummett
*Montana College of Mineral Science
and Technology*

Eugene Dunnam
University of Florida

Robert Endorf
University of Cincinnati

F. Paul Esposito
University of Cincinnati

Jerry Finkelstein
San Jose State University

Alexander Firestone
Iowa State University

Alexander Gardner
Howard University

Andrew L. Gardner
Brigham Young University

John Gieniec
Central Missouri State University

John B. Gruber
San Jose State University

Ann Hanks
American River College

Samuel Harris
Purdue University

Emily Haught
Georgia Institute of Technology

Laurent Hodges
Iowa State University

John Hubisz
North Carolina State University

Joey Huston
Michigan State University

Darrell Huwe
Ohio University

Claude Kacser
University of Maryland

Leonard Kleinman
University of Texas at Austin

Earl Koller
Stevens Institute of Technology

Arthur Z. Kovacs
Rochester Institute of Technology

Kenneth Krane
Oregon State University

Sol Krasner
University of Illinois at Chicago

Peter Loly
University of Manitoba

Robert R. Marchini
Memphis State University

David Markowitz
University of Connecticut

Howard C. McAllister
University of Hawaii at Manoa

W. Scott McCullough
Oklahoma State University

James H. McGuire
Tulane University

David M. McKinstry
Eastern Washington University

Joe P. Meyer
Georgia Institute of Technology

Roy Middleton
University of Pennsylvania

Irvin A. Miller
Drexel University

Eugene Mosca
United States Naval Academy

Michael O'Shea
Kansas State University

Patrick Papin
San Diego State University

George Parker
North Carolina State University

Robert Pelcovits
Brown University

Oren P. Quist
South Dakota State University

Jonathan Reichart
SUNY—Buffalo

Manuel Schwartz
University of Louisville

Darrell Seeley
Milwaukee School of Engineering

Bruce Arne Sherwood
Carnegie Mellon University

John Spangler
St. Norbert College

Ross L. Spencer
Brigham Young University

Harold Stokes
Brigham Young University

Jay D. Strieb
Villanova University

David Toot
Alfred University

J. S. Turner
University of Texas at Austin

T. S. Venkataraman
Drexel University

Gianfranco Vidali
Syracuse University

Fred Wang
Prairie View A & M

Robert C. Webb
Texas A & M University

George Williams
University of Utah

David Wolfe
University of New Mexico

1 Measurement

You can watch the Sun set and disappear over a calm ocean once while you lie on a beach, and then once again if you stand up. Surprisingly, by measuring the time between the two sunsets, you can approximate Earth's radius.

How can such a simple observation be used to measure Earth?

The answer is in this chapter.

1-1 Measuring Things

Physics is based on measurement. We discover physics by learning how to measure the quantities that are involved in physics. Among these quantities are length, time, mass, temperature, pressure, and electric current.

We measure each physical quantity in its own units, by comparison with a **standard.** The **unit** is a unique name we assign to measures of that quantity—for example, meter (or m) for the quantity length. The standard corresponds to exactly 1.0 unit of the quantity. As you will see, the standard for length, which corresponds to exactly 1.0 m, is the distance traveled by light in a vacuum during a certain fraction of a second. We can define a unit and its standard in any way we care to. However, the important thing is to do so in such a way that scientists around the world will agree that our definitions are both sensible and practical.

Once we have set up a standard, say, for length, we must work out procedures by which any length whatever, be it the radius of a hydrogen atom, the wheelbase of a skateboard, or the distance to a star, can be expressed in terms of the standard. Rulers, which approximate our length standard, give us one such procedure for measuring length. However, many of our comparisons must be indirect. You cannot use a ruler, for example, to measure the radius of an atom or the distance to a star.

There are so many physical quantities that it is a problem to organize them. Fortunately, they are not all independent; for example, speed is the ratio of a length to a time. Thus, what we do is pick out—by international agreement—a small number of physical quantities, such as length and time, and assign standards to them alone. We then define all other physical quantities in terms of these *base quantities* and their standards (called *base standards*). Speed, for example, is defined in terms of the base quantities length and time and the associated base standards.

Base standards must be both accessible and invariable. If we define the length standard as the distance between one's nose and the index finger on an outstretched arm, we certainly have an accessible standard—but it will, of course, vary from person to person. The demand for precision in science and engineering pushes us to aim first for invariability. We then exert great effort to make duplicates of the base standards that are accessible to those who need them.

1-2 The International System of Units

In 1971, the 14th General Conference on Weights and Measures picked seven quantities as base quantities, thereby forming the basis of the International System of Units, abbreviated SI from its French name and popularly known as the *metric system*. Table 1-1 shows the units for the three base quantities—length, mass, and time—that we use in the early chapters of this book. These units were defined to be on a "human scale."

Many SI *derived units* are defined in terms of these base units. For example, the SI unit for power, called the **watt** (symbol: W), is defined in terms of the base units for mass, length, and time. Thus, as you will see in Chapter 7,

$$1 \text{ watt} = 1 \text{ W} = 1 \text{ kg} \cdot \text{m}^2/\text{s}^3, \tag{1-1}$$

where the last collection of unit symbols is read as kilogram−square meter per second−cubed.

To express the very large and very small quantities that we often run into in physics, we use *scientific notation*, which employs powers of 10. In this notation,

$$3\,560\,000\,000 \text{ m} = 3.56 \times 10^9 \text{ m} \tag{1-2}$$

and

$$0.000\,000\,492 \text{ s} = 4.92 \times 10^{-7} \text{ s}. \tag{1-3}$$

TABLE 1-1   Some SI Base Units

Quantity	Unit Name	Unit Symbol
Length	meter	m
Time	second	s
Mass	kilogram	kg

Factor	Prefix*a*	Symbol
10^{24}	yotta-	Y
10^{21}	zetta-	Z
10^{18}	exa-	E
10^{15}	peta-	P
10^{12}	tera-	T
10^{9}	**giga-**	**G**
10^{6}	**mega-**	**M**
10^{3}	**kilo-**	**k**
10^{2}	hecto-	h
10^{1}	deka-	da
10^{-1}	deci-	d
10^{-2}	**centi-**	**c**
10^{-3}	**milli-**	**m**
10^{-6}	**micro-**	**μ**
10^{-9}	**nano-**	**n**
10^{-12}	**pico-**	**p**
10^{-15}	femto-	f
10^{-18}	atto-	a
10^{-21}	zepto-	z
10^{-24}	yocto-	y

TABLE 1-2 Prefixes for SI Units

*a*The most commonly used prefixes are shown in bold type.

Scientific notation on computers sometimes takes on an even briefer look, as in 3.56 E9 and 4.92 E−7, where E stands for "exponent of ten." It is briefer still on some calculators, where E is replaced with an empty space.

As a further convenience when dealing with very large or very small measurements, we use the prefixes listed in Table 1-2. As you can see, each prefix represents a certain power of 10, as a factor. Attaching a prefix to an SI unit has the effect of multiplying by the associated factor. Thus, we can express a particular electric power as

$$1.27 \times 10^9 \text{ watts} = 1.27 \text{ gigawatts} = 1.27 \text{ GW} \tag{1-4}$$

or a particular time interval as

$$2.35 \times 10^{-9} \text{ s} = 2.35 \text{ nanoseconds} = 2.35 \text{ ns.} \tag{1-5}$$

Some prefixes, as used in milliliter, centimeter, kilogram, and megabyte, are probably familiar to you.

1-3 Changing Units

We often need to change the units in which a physical quantity is expressed. We do so by a method called *chain-link conversion*. In this method, we multiply the original measurement by a **conversion factor** (a ratio of units that is equal to unity). For example, because 1 min and 60 s are identical time intervals, we have

$$\frac{1 \text{ min}}{60 \text{ s}} = 1 \quad \text{and} \quad \frac{60 \text{ s}}{1 \text{ min}} = 1.$$

Thus, the ratios (1 min)/(60 s) and (60 s)/(1 min) can be used as conversion factors. This is *not* the same as writing $\frac{1}{60} = 1$ or $60 = 1$; each *number* and its *unit* must be treated together.

Because multiplying any quantity by unity leaves it unchanged, we can introduce such conversion factors wherever we find them useful. In chain-link conversion, we use the factors to cancel unwanted units. For example, to convert 2 min to seconds, we have

$$2 \text{ min} = (2 \text{ min})(1) = (2 \text{ min})\left(\frac{60 \text{ s}}{1 \text{ min}}\right) = 120 \text{ s.} \tag{1-6}$$

If you introduce a conversion factor in such a way that unwanted units do *not* cancel, invert the factor and try again. In conversions, the units obey the same algebraic rules as variables and numbers.

Appendix D and the inside back cover give conversion factors between SI and other systems of units, including non-SI units still used in the United States. However, the conversion factors are written in the style of "1 min = 60 s" rather than as a ratio as used previously. The following sample problem gives an example of how to set up such ratios.

Sample Problem 1-1

When Pheidippides ran from Marathon to Athens in 490 B.C. to bring word of the Greek victory over the Persians, he probably ran at a speed of about 23 rides per hour (rides/h). The ride is an ancient Greek unit for length, as are the stadium and the plethron: 1 ride was defined to be 4 stadia, 1 stadium was defined to be 6 plethra, and, in terms of a modern unit, 1 plethron is 30.8 m. How fast did Pheidippides run in kilometers per second (km/s)?

SOLUTION: The Key Idea in chain-link conversions is to write the conversion factors as ratios that will eliminate unwanted units. Here

we write

$$23 \text{ rides/h} = \left(23 \frac{\text{rides}}{\text{h}}\right)\left(\frac{4 \text{ stadia}}{1 \text{ ride}}\right)\left(\frac{6 \text{ plethra}}{1 \text{ stadium}}\right)$$

$$\times \left(\frac{30.8 \text{ m}}{1 \text{ plethron}}\right)\left(\frac{1 \text{ km}}{1000 \text{ m}}\right)\left(\frac{1 \text{ h}}{3600 \text{ s}}\right)$$

$$= 4.7227 \times 10^{-3} \text{ km/s} \approx 4.7 \times 10^{-3} \text{ km/s.}$$

(Answer)

Sample Problem 1-2

The cran is a British volume unit for freshly caught herrings: 1 cran = 170.474 liters (L) of fish, about 750 herrings. Suppose that, to be cleared through customs in Saudia Arabia, a shipment of 1255 crans must be declared in terms of cubic covidos, where the covido is an Arabic unit of length: 1 covido = 48.26 cm. What is the required declaration?

SOLUTION: From Appendix D we see that 1 L is equivalent to 1000 cm³. A Key Idea then helps: To convert from *cubic* centimeters to *cubic* covidos, we must *cube* the conversion ratio between centimeters and covidos. Thus, we write the following chain-link conversion:

1255 crans

$$= (1255 \text{ crans})\left(\frac{170.474 \text{ L}}{1 \text{ cran}}\right)\left(\frac{1000 \text{ cm}^3}{1 \text{ L}}\right)\left(\frac{1 \text{ covido}}{48.26 \text{ cm}}\right)^3$$

$$= 1.903 \times 10^3 \text{ covidos}^3. \qquad \text{(Answer)}$$

PROBLEM-SOLVING TACTICS

Tactic 1: *Significant Figures and Decimal Places*

If you calculated the answer to Sample Problem 1-1 without your calculator automatically rounding it off, the number 4.722 666 666 67 × 10⁻³ might have appeared in the display. The precision implied by this number is meaningless. We rounded the answer to 4.7 × 10⁻³ km/s so as not to imply that it is more precise than the given data. The given speed of 23 rides/h consists of two digits, called **significant figures.** Thus, we rounded the answer to two significant figures. In this book, final results of calculations are often rounded to match the least number of significant figures in the given data. (However, sometimes an extra significant figure is kept.) When the leftmost of the digits to be discarded is 5 or more, the last remaining digit is rounded up; otherwise it is retained as is. For example, 11.3516 is rounded to three significant figures as 11.4 and 11.3279 is rounded to three significant figures as 11.3. (The answers to sample problems in this book are usually presented with the symbol = instead of ≈ even if rounding off is involved.)

When a number such as 3.15 or 3.15 × 10³ is provided in a problem, the number of significant figures is apparent, but how about the number 3000? Is it known to only one significant figure (could it be written as 3 × 10³)? Or is it known to as many as four significant figures (could it be written as 3.000 × 10³)? In this book, we assume that all the zeros in such given numbers as 3000 are significant, but you had better not make that assumption elsewhere.

Don't confuse *significant figures* with *decimal places*. Consider the lengths 35.6 mm, 3.56 m, and 0.00356 m. They all have three significant figures but they have one, two, and five decimal places, respectively.

1-4   Length

In 1792, the newborn Republic of France established a new system of weights and measures. Its cornerstone was the meter, defined to be one ten-millionth of the distance from the north pole to the equator. Later, for practical reasons, this Earth standard was abandoned and the meter came to be defined as the distance between two fine lines engraved near the ends of a platinum–iridium bar, the **standard meter bar,** which was kept at the International Bureau of Weights and Measures near Paris. Accurate copies of the bar were sent to standardizing laboratories throughout the world. These **secondary standards** were used to produce other, still more accessible standards so that ultimately every measuring device derived its authority from the standard meter bar through a complicated chain of comparisons.

Eventually, modern science and technology required a more precise standard than the distance between two fine scratches on a metal bar. In 1960, a new standard for the meter, based on the wavelength of light, was adopted. Specifically, the standard for the meter was redefined to be 1 650 763.73 wavelengths of a particular orange-red light emitted by atoms of krypton-86 (a particular isotope, or type, of krypton) in a gas discharge tube. This awkward number of wavelengths was chosen so that the new standard would be close to the old meter-bar standard.

By 1983, however, the demand for higher precision had reached such a point that even the krypton-86 standard could not meet it, and in that year a bold step was taken. The meter was redefined as the distance traveled by light in a specified time interval. In the words of the 17th General Conference on Weights and Measures:

TABLE 1-3   Some Approximate Lengths

Measurement	Length in Meters
Distance to the first galaxies formed	2×10^{26}
Distance to the Andromeda galaxy	2×10^{22}
Distance to the nearest star (Proxima Centauri)	4×10^{16}
Distance to Pluto	6×10^{12}
Radius of Earth	6×10^{6}
Height of Mt. Everest	9×10^{3}
Thickness of this page	1×10^{-4}
Length of a typical virus	1×10^{-8}
Radius of a hydrogen atom	5×10^{-11}
Radius of a proton	1×10^{-15}

> The meter is the length of the path traveled by light in a vacuum during a time interval of 1/299 792 458 of a second.

This time interval was chosen so that the speed of light c is exactly

$$c = 299\ 792\ 458 \text{ m/s.}$$

Measurements of the speed of light had become extremely precise, so it made sense to adopt the speed of light as a defined quantity and to use it to redefine the meter.

Table 1-3 shows a wide range of lengths, from that of the universe to those of some very small objects.

PROBLEM-SOLVING TACTICS

Tactic 2: *Order of Magnitude*
The *order of magnitude* of a number is the power of ten when the number is expressed in scientific notation. For example, if $A = 2.3 \times 10^4$ and $B = 7.8 \times 10^4$, then the orders of magnitude of both A and B are 4.

Often, engineering and science professionals will estimate the result of a calculation to the *nearest* order of magnitude. For our example, the nearest order of magnitude is 4 for A and 5 for B. Such estimation is common when detailed or precise data required in the calculation are not known or easily found. Sample Problem 1-3 gives an example.

Sample Problem 1-3

The world's largest ball of string is about 2 m in radius. To the nearest order of magnitude, what is the total length L of the string in the ball?

SOLUTION: We could, of course, take the ball apart and measure the total length L, but that would take great effort and make the ball's builder most unhappy. A Key Idea here is that, because we want only the nearest order of magnitude, we can estimate any quantities required in the calculation.

Let us assume the ball is spherical with radius $R = 2$ m. The string in the ball is not closely packed (there are uncountable gaps between nearby sections of string). To allow for these gaps, let us somewhat overestimate the cross-sectional area of the string by assuming the cross section is square, with an edge length $d =$ 4 mm. Then, with a cross-sectional area of d^2 and a length L, the string occupies a total volume of

$$V = (\text{cross-sectional area})(\text{length}) = d^2 L.$$

This is approximately equal to the volume of the ball, given by $\frac{4}{3}\pi R^3$, which is about $4R^3$ because π is about 3. Thus, we have

$$d^2 L = 4R^3,$$

or $\quad L = \dfrac{4R^3}{d^2} = \dfrac{4(2 \text{ m})^3}{(4 \times 10^{-3} \text{ m})^2}$

$$= 2 \times 10^6 \text{ m} \approx 10^6 \text{ m} = 10^3 \text{ km.} \quad \text{(Answer)}$$

(Note that you do not need a calculator for such a simplified calculation.) Thus, to the nearest order of magnitude, the ball contains about 1000 km of string!

TABLE 1-4   Some Approximate Time Intervals

Measurement	Time Interval in Seconds
Lifetime of the proton (predicted)	1×10^{39}
Age of the universe	5×10^{17}
Age of the pyramid of Cheops	1×10^{11}
Human life expectancy	2×10^{9}
Length of a day	9×10^{4}
Time between human heartbeats	8×10^{-1}
Lifetime of the muon	2×10^{-6}
Shortest lab light pulse	6×10^{-15}
Lifetime of the most unstable particle	1×10^{-23}
The Planck time[a]	1×10^{-43}

[a]This is the earliest time after the big bang at which the laws of physics as we know them can be applied.

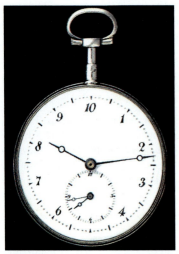

Fig. 1-1 When the metric system was proposed in 1792, the hour was redefined to provide a 10-hour day. The idea did not catch on. The maker of this 10-hour watch wisely provided a small dial that kept conventional 12-hour time. Do the two dials indicate the same time?

1-5 Time

Time has two aspects. For civil and some scientific purposes, we want to know the time of day so that we can order events in sequence. In much scientific work, we want to know how long an event lasts. Thus, any time standard must be able to answer two questions: "*When* did it happen?" and "What is its *duration*?" Table 1-4 shows some time intervals.

Any phenomenon that repeats itself is a possible time standard. Earth's rotation, which determines the length of the day, has been used in this way for centuries; Fig. 1-1 shows one novel example of a watch based on that rotation. A quartz clock, in which a quartz ring is made to vibrate continuously, can be calibrated against Earth's rotation via astronomical observations and used to measure time intervals in the laboratory. However, the calibration cannot be carried out with the accuracy called for by modern scientific and engineering technology.

To meet the need for a better time standard, atomic clocks have been developed. An atomic clock at the National Institute of Standards and Technology (NIST) in Boulder, Colorado, is the standard for Coordinated Universal Time (UTC) in the United States. Its time signals are available by shortwave radio (stations WWV and WWVH) and by telephone (303-499-7111). Time signals (and related information) are also available from the United States Naval Observatory at Web site http://tycho.usno.navy.mil/time.html. (To set a clock extremely accurately at your particular location, you would have to account for the travel time that is required for these signals to reach you.)

Figure 1-2 shows variations in the length of one day on Earth over a 4-year period, as determined by comparison with a cesium (atomic) clock. Because the variation displayed by Fig. 1-2 is seasonal and repetitious, we suspect the rotating Earth when there is a difference between Earth and atom as timekeepers. The variation is probably due to tidal effects caused by the Moon and to large-scale winds.

The 13th General Conference on Weights and Measures in 1967 adopted a standard second based on the cesium clock:

> One second is the time taken by 9 192 631 770 oscillations of the light (of a specified wavelength) emitted by a cesium-133 atom.

Atomic clocks are so consistent that, in principle, two cesium clocks would have to run for 6000 years before their readings would differ by more than 1 s. Even such accuracy pales in comparison to that of clocks currently being developed; their precision may be 1 part in 10^{18}—that is, 1 s in 1×10^{18} s (about 3×10^{10} y).

Fig. 1-2 Variations in the length of the day over a 4-year period. Note that the entire vertical scale amounts to only 3 ms (3 milliseconds = 0.003 s).

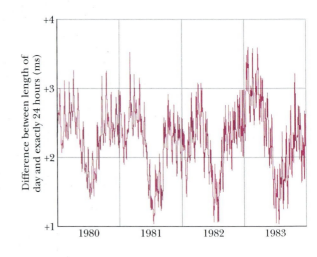

Sample Problem 1-4*

Suppose that while lying on a beach watching the Sun set over a calm ocean, you start a stopwatch just as the top of the Sun disappears. You then stand, elevating your eyes by a height $h = 1.70$ m, and stop the watch when the top of the Sun again disappears. If the elapsed time on the watch is $t = 11.1$ s, what is the radius r of Earth?

SOLUTION: A Key Idea here is that just as the Sun disappears, your line of sight to the top of the Sun is tangent to Earth's surface. Two such lines of sight are shown in Fig. 1-3. There your eyes are located at point A while you are lying, and at height h above point A while you are standing. For the latter situation, the line of sight is tangent to Earth's surface at point B. Let d represent the distance between point B and the location of your eyes when you are standing, and draw radii r as shown in Fig. 1-3. From the Pythagorean theorem, we then have

$$d^2 + r^2 = (r + h)^2 = r^2 + 2rh + h^2,$$

or
$$d^2 = 2rh + h^2. \qquad (1\text{-}7)$$

Because the height h is so much smaller than Earth's radius r, the term h^2 is negligible compared to the term $2rh$, and we can rewrite Eq. 1-7 as

$$d^2 = 2rh. \qquad (1\text{-}8)$$

In Fig. 1-3, the angle between the radii to the two tangent points A and B is θ, which is also the angle through which the Sun moves about Earth during the measured time $t = 11.1$ s. During a full day, which is approximately 24 h, the Sun moves through an angle of 360° about Earth. This allows us to write

$$\frac{\theta}{360°} = \frac{t}{24 \text{ h}},$$

*Adapted from "Doubling Your Sunsets, or How Anyone Can Measure the Earth's Size with a Wristwatch and Meter Stick," by Dennis Rawlins, *American Journal of Physics,* Feb. 1979, Vol. 47, pp. 126–128. This technique works best at the equator.

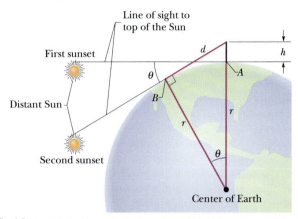

Fig. 1-3 Sample Problem 1-4. Your line of sight to the top of the setting Sun rotates through the angle θ when you stand up at point A and elevate your eyes by a distance h. (Angle θ and distance h are exaggerated here for clarity.)

which, with $t = 11.1$ s, gives us

$$\theta = \frac{(360°)(11.1 \text{ s})}{(24 \text{ h})(60 \text{ min/h})(60 \text{ s/min})} = 0.04625°.$$

Again in Fig. 1-3, we see that $d = r \tan \theta$. Substituting this for d in Eq. 1-8 gives us

$$r^2 \tan^2\theta = 2rh,$$

or
$$r = \frac{2h}{\tan^2\theta}.$$

Substituting $\theta = 0.04625°$ and $h = 1.70$ m, we find

$$r = \frac{(2)(1.70 \text{ m})}{\tan^2 0.04625°} = 5.22 \times 10^6 \text{ m}, \qquad \text{(Answer)}$$

which is within 20% of the accepted value (6.37×10^6 m) for the (mean) radius of Earth.

1-6 Mass

The Standard Kilogram

The SI standard of mass is a platinum–iridium cylinder (Fig. 1-4) kept at the International Bureau of Weights and Measures near Paris and assigned, by international agreement, a mass of 1 kilogram. Accurate copies have been sent to standardizing laboratories in other countries, and the masses of other bodies can be determined by balancing them against a copy. Table 1-5 shows some masses expressed in kilograms, ranging over about 83 orders of magnitude.

Fig. 1-4 The international 1 kg standard of mass, a platinum–iridium cylinder 3.9 cm in height and in diameter.

TABLE 1-5 Some Approximate Masses

Object	Mass in Kilograms
Known universe	1×10^{53}
Our galaxy	2×10^{41}
Sun	2×10^{30}
Moon	7×10^{22}
Asteroid Eros	5×10^{15}
Small mountain	1×10^{12}
Ocean liner	7×10^{7}
Elephant	5×10^{3}
Grape	3×10^{-3}
Speck of dust	7×10^{-10}
Penicillin molecule	5×10^{-17}
Uranium atom	4×10^{-25}
Proton	2×10^{-27}
Electron	9×10^{-31}

The U.S. copy of the standard kilogram is housed in a vault at NIST. It is removed, no more than once a year, for the purpose of checking duplicate copies that are used elsewhere. Since 1889, it has been taken to France twice for recomparison with the primary standard.

A Second Mass Standard

The masses of atoms can be compared with each other more precisely than they can be compared with the standard kilogram. For this reason, we have a second mass standard. It is the carbon-12 atom, which, by international agreement, has been assigned a mass of 12 **atomic mass units** (u). The relation between the two units is

$$1 \text{ u} = 1.6605402 \times 10^{-27} \text{ kg}, \tag{1-9}$$

with an uncertainty of ± 10 in the last two decimal places. Scientists can, with reasonable precision, experimentally determine the masses of other atoms relative to the mass of carbon-12. What we presently lack is a reliable means of extending that precision to more common units of mass, such as a kilogram.

REVIEW & SUMMARY

Measurement in Physics Physics is based on measurement of physical quantities. Certain physical quantities have been chosen as **base quantities** (such as length, time, and mass); each has been defined in terms of a **standard** and given a **unit** of measure (such as meter, second, and kilogram). Other physical quantities are defined in terms of the base quantities and their standards and units.

SI Units The unit system emphasized in this book is the International System of Units (SI). The three physical quantities displayed in Table 1-1 are used in the early chapters. Standards, which must be both accessible and invariable, have been established for these base quantities by international agreement. These standards are used in all physical measurement, for both the base quantities and the quantities derived from them. Scientific notation and the prefixes of Table 1-2 can be used to simplify measurement notation in many cases.

Changing Units Conversion of units from one system to another (for example, from miles per hour to kilometers per second) may be performed by using *chain-link conversions* in which the original data are multiplied successively by conversion factors written as unity and the units are manipulated like algebraic quantities until only the desired units remain.

Length The unit of length—the meter—is defined as the distance traveled by light during a precisely specified time interval.

Time The unit of time—the second—was formerly defined in terms of the rotation of Earth. It is now defined in terms of the oscillations of light emitted by an atomic (cesium-133) source. Accurate time signals are sent worldwide by radio signals keyed to atomic clocks in standardizing laboratories.

Mass The unit of mass—the kilogram—is defined in terms of a particular platinum–iridium prototype kept near Paris, France. For measurements on an atomic scale, the atomic mass unit, defined in terms of the atom carbon-12, is usually used.

EXERCISES & PROBLEMS

ssm Solution is in the Student Solutions Manual.
www Solution is available on the World Wide Web at:
          http://www.wiley.com/college/hrw
ilw Solution is available on the Interactive LearningWare.

SEC. 1-4   Length

1E. The micrometer (1 μm) is often called the *micron*. (a) How many microns make up 1.0 km? (b) What fraction of a centimeter equals 1.0 μm? (c) How many microns are in 1.0 yd? ssm

2E. Two types of *barrel* units were in use in the 1920s in the United States. The apple barrel had a legally set volume of 7056 cubic inches; the cranberry barrel, 5826 cubic inches. If a merchant sells 20 cranberry barrels of goods to a customer who thinks he is receiving apple barrels, what is the discrepancy in the shipment volume in liters?

3F. Horses are to race over a certain English meadow for a distance of 4.0 furlongs. What is the race distance in units of (a) rods and (b) chains? (1 furlong = 201.168 m, 1 rod = 5.0292 m, and 1 chain = 20.117 m.) ssm

4E. Spacing in this book was generally done in units of points and picas: 12 points = 1 pica, and 6 picas = 1 inch. If a figure was misplaced in the page proofs by 0.80 cm, what was the misplacement in (a) points and (b) picas?

5E. Earth is approximately a sphere of radius 6.37×10^6 m. What are (a) its circumference in kilometers, (b) its surface area in square kilometers, and (c) its volume in cubic kilometers? ssm

6E. An old manuscript reveals that a landowner in the time of King Arthur held 3.00 acres of plowed land plus a livestock area of 25.0 perches by 4.00 perches. What was the total area in (a) the old unit of roods and (b) the more modern unit of square meters? Here, 1 acre is an area of 40 perches by 4 perches, 1 rood is 40 perches by 1 perch, and 1 perch is 16.5 ft.

7P. Antarctica is roughly semicircular, with a radius of 2000 km (Fig. 1-5). The average thickness of its ice cover is 3000 m. How many cubic centimeters of ice does Antarctica contain? (Ignore the curvature of Earth.) ssm

Fig. 1-5 Problem 7.

8P. In the United States, a doll house has the scale of 1:12 of a real house (that is, each length of the doll house is $\frac{1}{12}$ that of the real house) and a miniature house (a doll house to fit within a doll house) has the scale of 1:144 of a real house. Suppose a real house (Fig. 1-6) has a front length of 20 m, a depth of 12 m, a height of 6.0 m, and a standard sloped roof (vertical triangular faces on the ends) of height 3.0 m. In cubic meters, what are the volumes of the corresponding (a) doll house and (b) miniature house?

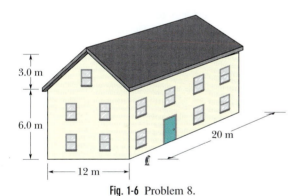

Fig. 1-6 Problem 8.

9P. Hydraulic engineers in the United States often use, as a unit of volume of water, the *acre-foot,* defined as the volume of water that will cover 1 acre of land to a depth of 1 ft. A severe thunderstorm dumped 2.0 in. of rain in 30 min on a town of area 26 km². What volume of water, in acre-feet, fell on the town? ssm ilw www

SEC. 1-5 Time

10E. Physicist Enrico Fermi once pointed out that a standard lecture period (50 min) is close to 1 microcentury. (a) How long is a microcentury in minutes? (b) Using

$$\text{percentage difference} = \left(\frac{\text{actual} - \text{approximation}}{\text{actual}} \right) 100,$$

find the percentage difference from Fermi's approximation.

11E. Express the speed of light, 3.0×10^8 m/s, in (a) feet per nanosecond and (b) millimeters per picosecond. ssm

12E. A unit of time sometimes used in microscopic physics is the *shake.* One shake equals 10^{-8} s. (a) Are there more shakes in a second than there are seconds in a year? (b) Humans have existed for about 10^6 years, whereas the universe is about 10^{10} years old. If the age of the universe now is taken to be 1 "universe day," for how many "universe seconds" have humans existed?

13P. Five clocks are being tested in a laboratory. Exactly at noon, as determined by the WWV time signal, on successive days of a week the clocks read as in the following table. Rank the five clocks according to their relative value as good timekeepers, best to worst. Justify your choice. ssm

Clock	Sun.	Mon.	Tues.	Wed.	Thurs.	Fri.	Sat.
A	12:36:40	12:36:56	12:37:12	12:37:27	12:37:44	12:37:59	12:38:14
B	11:59:59	12:00:02	11:59:57	12:00:07	12:00:02	11:59:56	12:00:03
C	15:50:45	15:51:43	15:52:41	15:53:39	15:54:37	15:55:35	15:56:33
D	12:03:59	12:02:52	12:01:45	12:00:38	11:59:31	11:58:24	11:57:17
E	12:03:59	12:02:49	12:01:54	12:01:52	12:01:32	12:01:22	12:01:12

14P. Three digital clocks A, B, and C run at different rates and do not have simultaneous readings of zero. Figure 1-7 shows simultaneous readings on pairs of the clocks for four occasions. (At the earliest occasion, for example, B reads 25.0 s and C reads 92.0 s.) If two events are 600 s apart on clock A, how far apart are they on (a) clock B and (b) clock C? (c) When clock A reads 400 s, what does clock B read? (d) When clock C reads 15.0 s, what does clock B read? (Assume negative readings for prezero times.)

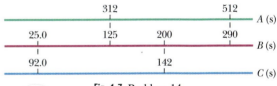

Fig. 1-7 Problem 14.

15P. An astronomical unit (AU) is the average distance of Earth from the Sun, approximately 1.50×10^8 km. The speed of light is about 3.0×10^8 m/s. Express the speed of light in terms of astronomical units per minute. ssm

16P. Until 1883, every city and town in the United States kept its own local time. Today, travelers reset their watches only when the time change equals 1.0 h. How far, on the average, must you travel in degrees of longitude until your watch must be reset by 1.0 h? (*Hint:* Earth rotates 360° in about 24 h.)

17P. Assuming the length of the day uniformly increases by 0.0010 s per century, calculate the cumulative effect on the measure of time over 20 centuries. (Such slowing of Earth's rotation is indicated by observations of the occurrences of solar eclipses during this period.) ssm www

18P. Time standards are now based on atomic clocks. A promising second standard is based on *pulsars,* which are rotating neutron stars (highly compact stars consisting only of neutrons). Some rotate at a rate that is highly stable, sending out a radio beacon that sweeps briefly across Earth once with each rotation, like a lighthouse beacon. Pulsar PSR 1937+21 is an example; it rotates once every 1.557 806 448 872 75 ± 3 ms, where the trailing ±3 indicates the uncertainty in the last decimal place (it does *not* mean ±3 ms). (a) How many times does PSR 1937+21 rotate in 7.00 days? (b) How much time does the pulsar take to rotate 1.0×10^6 times and (c) what is the associated uncertainty?

SEC. 1-6   Mass

19E. Earth has a mass of 5.98×10^{24} kg. The average mass of the atoms that make up Earth is 40 u. How many atoms are there in Earth? ssm

20P. Gold, which has a mass of 19.32 g for each cubic centimeter of volume, is the most ductile metal and can be pressed into a thin leaf or drawn out into a long fiber. (a) If 1.000 oz of gold, with a mass of 27.63 g, is pressed into a leaf of 1.000 μm thickness, what is the area of the leaf? (b) If, instead, the gold is drawn out into a cylindrical fiber of radius 2.500 μm, what is the length of the fiber?

21P. (a) Assuming that each cubic centimeter of water has a mass of exactly 1 g, find the mass of one cubic meter of water in kilograms. (b) Suppose that it takes 10.0 h to drain a container of 5700 m^3 of water. What is the "mass flow rate," in kilograms per second, of water from the container? ssm

22P. What mass of water fell on the town in Problem 9 during the thunderstorm? One cubic meter of water has a mass of 10^3 kg.

23P. Iron has a mass of 7.87 g per cubic centimeter of volume, and the mass of an iron atom is 9.27×10^{-26} kg. If the atoms are spherical and tightly packed, (a) what is the volume of an iron atom and (b) what is the distance between the centers of adjacent atoms? ssm

24P. Grains of fine California beach sand are approximately spheres with an average radius of 50 μm and are made of silicon dioxide. A solid cube of silicon dioxide with a volume of 1.00 m^3 has a mass of 2600 kg. What mass of sand grains would have a total surface area (the total area of all the individual spheres) equal to the surface area of a cube 1 m on an edge?

Additional Problems

25. Harvard Bridge, which connects MIT with its fraternities across the Charles River, has a length of 364.4 Smoots plus one ear. The unit of one Smoot is based on the length of Oliver Reed Smoot, Jr., class of 1962, who was carried or dragged length by length across the bridge so that other pledge members of the Lambda Chi Alpha fraternity could mark off (with paint) 1-Smoot lengths along the bridge. The marks have been repainted biannually by fraternity pledges since the initial measurement, usually during times of traf-

fic congestion so that the police could not easily interfere. (Presumably, the police were originally upset because a Smoot is not an SI base unit, but these days they seem to have accepted the unit.) Figure 1-8 shows three parallel paths, measured in Smoots (S), Willies (W), and Zeldas (Z). What is the length of 50.0 Smoots in (a) Willies and (b) Zeldas?

26. An old English children's rhyme states, "Little Miss Muffet sat on her tuffet, eating her curds and whey, when along came a spider who sat down beside her. . . ." The spider sat down not because of the curds and whey but because Miss Muffet had a stash of 11 tuffets of dried flies. The volume measure of a tuffet is given by 1 tuffet = 2 pecks = 0.50 bushel, where 1 Imperial (British) bushel = 36.3687 liters (L). What was Miss Muffet's stash in (a) pecks, (b) bushels, and (c) liters?

27. During the summers at high latitudes, ghostly, silver-blue clouds occasionally appear after sunset when common clouds are in Earth's shadow and are no longer visible. The ghostly clouds have been called *noctilucent clouds* (NLC), which means "luminous night clouds," but now are often called *mesospheric clouds,* after the *mesosphere,* the name of the atmosphere at the altitude of the clouds.

These clouds were first seen in June 1885, after dust and water from the massive 1883 volcanic explosion of Krakatoa Island (near Java in the Southeast Pacific) reached the high altitudes in the Northern Hemisphere. In the low temperatures of the mesosphere, the water collected and froze on the volcanic dust (and perhaps on comet and meteor dust already present there) to form the particles that made up the first clouds. Since then, mesospheric clouds have generally increased in occurrence and brightness, probably because of the increased production of methane by industries, rice paddies, landfills, and livestock flatulence. The methane works its way into the upper atmosphere, undergoes chemical changes, and results in an increase of water molecules there, and also in bits of ice for the mesospheric clouds.

If mesospheric clouds are spotted 38 min after sunset and then quickly dim, what is their altitude if they are directly over the observer? (*Hint:* See Sample Problem 1-4.)

28. A standard interior staircase has steps each with a rise (height) of 19 cm and a run (horizontal depth) of 23 cm. Research suggests that the stairs would be safer for descent if the run were, instead, 28 cm. For a particular staircase of total height 4.57 m, how much farther would the staircase extend into the room at the foot of the stairs if this change in run were made?

29. As a contrast between the old and the modern and between the large and the small, consider the following: In old rural England 1 hide (between 100 and 120 acres) was the area of land needed to sustain one family with a single plough for one year. (An area of 1 acre is equal to 4047 m^2.) Also, 1 wapentake was the area of land needed by 100 such families. In quantum physics, the cross-sectional area of a nucleus (defined in terms of the chance of a particle hitting and being absorbed by it) is measured in units of barns, where 1 barn is 1×10^{-28} m^2. (In nuclear physics jargon, if a nucleus is "large," then shooting a particle at it is like shooting a bullet at a barn door, which can hardly be missed.) What is the ratio of 25 wapentakes to 11 barns?

Fig. 1-8 Problem 25.

2 Motion Along a Straight Line

On September 26, 1993, Dave Munday, a diesel mechanic by trade, went over the Canadian edge of Niagara Falls for the second time, freely falling 48 m to the water (and rocks) below. On this attempt, he rode in a steel ball with a hole for air. Munday, keen on surviving this plunge that had killed four other stuntmen, had done considerable research on the physics and engineering aspects of the plunge.

How long would he fall, and at what speed would he hit the turbulent water at the bottom?

The answer is in this chapter.

2-1 Motion

The world, and everything in it, moves. Even seemingly stationary things, such as a roadway, move with Earth's rotation, Earth's orbit around the Sun, the Sun's orbit around the center of the Milky Way galaxy, and that galaxy's migration relative to other galaxies. The classification and comparison of motions (called **kinematics**) is often challenging. What exactly do you measure, and how do you compare?

Before we attempt an answer, we shall examine some general properties of motion that is restricted in three ways.

1. The motion is along a straight line only. The line may be vertical (that of a falling stone), horizontal (that of a car on a level highway), or slanted, but it must be straight.

2. Forces (pushes and pulls) cause motion, but forces will not be discussed until Chapter 5. In this chapter you study only the motion itself, and changes in the motion. Does the moving object speed up, slow down, stop, or reverse direction? If the motion does change, how is time involved in the change?

3. The moving object is either a **particle** (by which we mean a pointlike object such as an electron) or an object that moves like a particle (such that every portion moves in the same direction and at the same rate). A stiff pig slipping down a straight playground slide might be considered to be moving like a particle; however, a tumbling tumbleweed would not, because different points inside it move in different directions.

2-2 Position and Displacement

To locate an object means to find its position relative to some reference point, often the **origin** (or zero point) of an axis such as the x axis in Fig. 2-1. The **positive direction** of the axis is in the direction of increasing numbers (coordinates), which is toward the right in Fig. 2-1. The opposite direction is the **negative direction.**

For example, a particle might be located at $x = 5$ m, which means that it is 5 m in the positive direction from the origin. If it were at $x = -5$ m, it would be just as far from the origin but in the opposite direction. On the axis, a coordinate of -5 m is less than one of -1 m, and both coordinates are less than a coordinate of $+5$ m. A plus sign for a coordinate need not be shown, but a minus sign must always be shown.

A change from one position x_1 to another position x_2 is called a **displacement** Δx, where

$$\Delta x = x_2 - x_1. \tag{2-1}$$

(The symbol Δ, the Greek uppercase delta, represents a change in a quantity, and it means the final value of that quantity minus the initial value.) When numbers are inserted for the position values x_1 and x_2, a displacement in the positive direction (toward the right in Fig. 2-1) always comes out positive, and one in the opposite direction (left in the figure), negative. For example, if the particle moves from $x_1 = 5$ m to $x_2 = 12$ m, then $\Delta x = (12\text{ m}) - (5\text{ m}) = +7$ m. The positive result indicates that the motion is in the positive direction. If the particle then returns to $x = 5$ m, the displacement for the full trip is zero. The actual number of meters covered for the full trip is irrelevant; displacement involves only the original and final positions.

A plus sign for a displacement need not be shown, but a minus sign must always be shown. If we ignore the sign (and thus the direction) of a displacement, we are

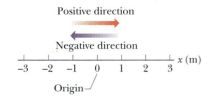

Fig. 2-1 Position is determined on an axis that is marked in units of length (here meters) and that extends indefinitely in opposite directions. The axis label, here x, is always on the positive side of the origin.

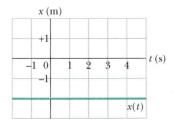

Fig. 2-2 The graph of $x(t)$ for an armadillo that is stationary at $x = -2$ m. The value of x is -2 m for all times t.

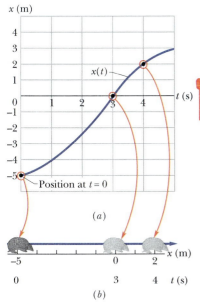

(a)

(b)

Fig. 2-3 (a) The graph of $x(t)$ for a moving armadillo. (b) The path associated with the graph. The scale below the x axis shows the times at which the armadillo reaches various x values.

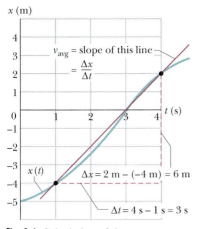

Fig. 2-4 Calculation of the average velocity between $t = 1$ s and $t = 4$ s as the slope of the line that connects the points (on the $x(t)$ curve) representing those times.

left with the **magnitude** (or absolute value) of the displacement. In the previous example, the magnitude of Δx is 7 m.

Displacement is an example of a **vector quantity,** which is a quantity that has both a direction and a magnitude. We explore vectors more fully in Chapter 3 (in fact, some of you may have already read that chapter), but here all we need is the idea that displacement has two features: (1) Its *magnitude* is the distance (such as the number of meters) between the original and final positions. (2) Its *direction*, from an original position to a final position, can be represented by a plus sign or a minus sign if the motion is along a single axis.

What follows is the first of many checkpoints that you will see in this book. Each consists of one or more questions whose answers require some reasoning or a mental calculation, and each gives you a quick check of your understanding. The answers are listed in the back of the book.

✓ CHECKPOINT 1: Here are three pairs of initial and final positions, respectively, along an x axis. Which pairs give a negative displacement: (a) -3 m, $+5$ m; (b) -3 m, -7 m; (c) 7 m, -3 m?

2-3 Average Velocity and Average Speed

A compact way to describe position is with a graph of position x plotted as a function of time t—a graph of $x(t)$. (The notation $x(t)$ represents a function x of t, not the product x times t.) As a simple example, Fig. 2-2 shows the position function $x(t)$ for a stationary armadillo (which we treat as a particle) at $x = -2$ m.

Figure 2-3a, also for an armadillo, is more interesting, because it involves motion. The armadillo is apparently first noticed at $t = 0$ when it is at the position $x = -5$ m. It moves toward $x = 0$, passes through that point at $t = 3$ s, and then moves on to increasingly larger positive values of x.

Figure 2-3b depicts the actual straight-line motion of the armadillo and is something like what you would see. The graph in Fig. 2-3a is more abstract and quite unlike what you would see, but it is richer in information. It also reveals how fast the armadillo moves.

Actually, several quantities are associated with the phrase "how fast." One of them is the **average velocity** v_{avg}, which is the ratio of the displacement Δx that occurs during a particular time interval Δt to that interval:

$$v_{avg} = \frac{\Delta x}{\Delta t} = \frac{x_2 - x_1}{t_2 - t_1}. \qquad (2\text{-}2)$$

The notation means that the position is x_1 at time t_1 and then x_2 at time t_2. A common unit for v_{avg} is the meter per second (m/s). You may see other units in the problems, but they are always in the form of length/time.

On a graph of x versus t, v_{avg} is the **slope** of the straight line that connects two particular points on the $x(t)$ curve: one is the point that corresponds to x_2 and t_2, and the other is the point that corresponds to x_1 and t_1. Like displacement, v_{avg} has both magnitude and direction (it is another vector quantity). Its magnitude is the magnitude of the line's slope. A positive v_{avg} (and slope) tells us that the line slants upward to the right; a negative v_{avg} (and slope), that the line slants downward to the right. The average velocity v_{avg} always has the same sign as the displacement Δx because Δt in Eq. 2-2 is always positive.

Figure 2-4 shows how to find v_{avg} for the armadillo of Fig. 2-3, for the time interval $t = 1$ s to $t = 4$ s. We draw the straight line that connects the point on the position curve at the beginning of the interval and the point on the curve at the end of the interval. Then we find the slope $\Delta x / \Delta t$ of the straight line. For the given time interval, the average velocity is

$$v_{avg} = \frac{6 \text{ m}}{3 \text{ s}} = 2 \text{ m/s}.$$

Average speed s_{avg} is a different way of describing "how fast" a particle moves. Whereas the average velocity involves the particle's displacement Δx, the average speed involves the total distance covered (for example, the number of meters moved), independent of direction; that is,

$$s_{avg} = \frac{\text{total distance}}{\Delta t}. \qquad (2\text{-}3)$$

Because average speed does *not* include direction, it lacks any algebraic sign. Sometimes s_{avg} is the same (except for the absence of a sign) as v_{avg}. However, as is demonstrated in Sample Problem 2-1, when an object doubles back on its path, the two can be quite different.

Sample Problem 2-1

You drive a beat-up pickup truck along a straight road for 8.4 km at 70 km/h, at which point the truck runs out of gasoline and stops. Over the next 30 min, you walk another 2.0 km farther along the road to a gasoline station.

(a) What is your overall displacement from the beginning of your drive to your arrival at the station?

SOLUTION: Assume, for convenience, that you move in the positive direction of an x axis, from a first position of $x_1 = 0$ to a second position of x_2 at the station. That second position must be at $x_2 = 8.4$ km $+ 2.0$ km $= 10.4$ km. Then the Key Idea here is that your displacement Δx along the x axis is the second position minus the first position. From Eq. 2-1, we have

$$\Delta x = x_2 - x_1 = 10.4 \text{ km} - 0 = 10.4 \text{ km}. \quad \text{(Answer)}$$

Thus, your overall displacement is 10.4 km in the positive direction of the x axis.

(b) What is the time interval Δt from the beginning of your drive to your arrival at the station?

SOLUTION: We already know the time interval Δt_{wlk} ($= 0.50$ h) for the walk, but we lack the time interval Δt_{dr} for the drive. However, we know that for the drive the displacement Δx_{dr} is 8.4 km and the average velocity $v_{avg,dr}$ is 70 km/h. A Key Idea to use here comes from Eq. 2-2: This average velocity is the ratio of the displacement for the drive to the time interval for the drive:

$$v_{avg,dr} = \frac{\Delta x_{dr}}{\Delta t_{dr}}.$$

Rearranging and substituting data then give us

$$\Delta t_{dr} = \frac{\Delta x_{dr}}{v_{avg,dr}} = \frac{8.4 \text{ km}}{70 \text{ km/h}} = 0.12 \text{ h}.$$

So,

$$\Delta t = \Delta t_{dr} + \Delta t_{wlk}$$
$$= 0.12 \text{ h} + 0.50 \text{ h} = 0.62 \text{ h}. \quad \text{(Answer)}$$

(c) What is your average velocity v_{avg} from the beginning of your drive to your arrival at the station? Find it both numerically and graphically.

SOLUTION: The Key Idea here again comes from Eq. 2-2: v_{avg} *for the entire trip* is the ratio of the displacement of 10.4 km *for the entire trip* to the time interval of 0.62 h *for the entire trip*. With Eq. 2-2, we find it is

$$v_{avg} = \frac{\Delta x}{\Delta t} = \frac{10.4 \text{ km}}{0.62 \text{ h}}$$
$$= 16.8 \text{ km/h} \approx 17 \text{ km/h}. \quad \text{(Answer)}$$

To find v_{avg} graphically, first we graph $x(t)$ as shown in Fig. 2-5, where the beginning and arrival points on the graph are the origin and the point labeled as "Station." The Key Idea here is that your average velocity is the slope of the straight line connecting those

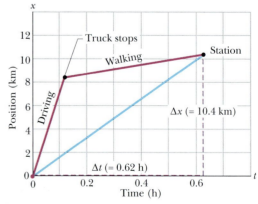

Fig. 2-5 Sample Problem 2-1. The lines marked "Driving" and "Walking" are the position–time plots for the driving and walking stages. (The plot for the walking stage assumes a constant rate of walking.) The slope of the straight line joining the origin and the point labeled "Station" is the average velocity for the trip, from the beginning to the station.

points; that is, it is the ratio of the *rise* (Δx = 10.4 km) to the *run* (Δt = 0.62 h), which gives us v_{avg} = 16.8 km/h.

(d) Suppose that to pump the gasoline, pay for it, and walk back to the truck takes you another 45 min. What is your average speed from the beginning of your drive to your return to the truck with the gasoline?

SOLUTION: The Key Idea here is that your average speed is the ratio of the total distance you move to the total time interval you take

to make that move. The total distance is 8.4 km + 2.0 km + 2.0 km = 12.4 km. The total time interval is 0.12 h + 0.50 h + 0.75 h = 1.37 h. Thus, Eq. 2-3 gives us

$$ s_{avg} = \frac{12.4 \text{ km}}{1.37 \text{ h}} = 9.1 \text{ km/h}. \qquad \text{(Answer)} $$

✔CHECKPOINT 2:   In this sample problem, suppose that right after refueling the truck, you drive back to x_1 at 35 km/h. What is your average velocity for your entire trip?

PROBLEM-SOLVING TACTICS

Tactic 1: *Do You Understand the Problem?*
For beginning problem solvers, no difficulty is more common than simply not understanding the problem. The best test of understanding is this: Can you explain the problem in your own words?

Write down the given data, with units, using the symbols of the chapter. (In Sample Problem 2-1, the given data allow you to find your net displacement Δx in part (a) and the corresponding time interval Δt in part (b).) Identify the unknown and its symbol. (In the sample problem, the unknown in part (c) is your average velocity v_{avg}.) Then find the connection between the unknown and the data. (The connection is Eq. 2-2, the definition of average velocity.)

Tactic 2: *Are the Units OK?*
Be sure to use a consistent set of units when putting numbers into the equations. In Sample Problem 2-1, the logical units in terms of the given data are kilometers for distances, hours for time intervals, and kilometers per hour for velocities. You may sometimes need to make conversions.

Tactic 3: *Is Your Answer Reasonable?*
Does your answer make sense? Is it far too large or far too small? Is the sign correct? Are the units appropriate? In part (c) of Sample Problem 2-1, for example, the correct answer is 17 km/h. If you find 0.00017 km/h, −17 km/h, 17 km/s, or 17,000 km/h, you should realize at once that you have done something wrong. The error may lie in your method, in your algebra, or in your keystroking of numbers on a calculator.

Tactic 4: *Reading a Graph*
Figures 2-2, 2-3a, 2-4, and 2-5 are graphs that you should be able to read easily. In each graph, the variable on the horizontal axis is the time t, with the direction of increasing time to the right. In each, the variable on the vertical axis is the position x of the moving particle with respect to the origin, with the positive direction of x upward. Always note the units (seconds or minutes; meters or kilometers) in which the variables are expressed.

2-4 Instantaneous Velocity and Speed

You have now seen two ways to describe how fast something moves: average velocity and average speed, both of which are measured over a time interval Δt. However, the phrase "how fast" more commonly refers to how fast a particle is moving at a given instant—and that is its **instantaneous velocity** (or simply **velocity**) v.

The velocity at any instant is obtained from the average velocity by shrinking the time interval Δt closer and closer to 0. As Δt dwindles, the average velocity approaches a limiting value, which is the velocity at that instant:

$$ v = \lim_{\Delta t \to 0} \frac{\Delta x}{\Delta t} = \frac{dx}{dt}. \qquad (2\text{-}4) $$

This equation displays two features of the instantaneous velocity v. First, v is the rate at which the particle's position x is changing with time at a given instant; that is, v is the derivative of x with respect to t. Second, v at any instant is the slope of the particle's position–time curve at the point representing that instant. Velocity is another vector quantity and thus has an associated direction.

Speed is the magnitude of velocity; that is, speed is velocity that has been stripped of any indication of direction, either in words or via an algebraic sign. (*Caution:* Speed and average speed can be quite different.) A velocity of +5 m/s and one of −5 m/s both have an associated speed of 5 m/s. The speedometer in a car measures the speed, not the velocity, because it cannot determine the direction.

Sample Problem 2-2

Figure 2-6a is an $x(t)$ plot for an elevator that is initially stationary, then moves upward (which we take to be the positive direction of x), and then stops. Plot v as a function of time.

SOLUTION: The Key Idea here is that we can find the velocity at any time from the slope of the curve of $x(t)$ at that time. The slope of $x(t)$, and so also the velocity, is zero in the intervals from 0 to 1 s and from 9 s on, so then the cab is stationary. During the interval bc the slope is constant and nonzero; so then the cab moves with constant velocity. We calculate the slope of $x(t)$ then as

$$\frac{\Delta x}{\Delta t} = v = \frac{24 \text{ m} - 4.0 \text{ m}}{8.0 \text{ s} - 3.0 \text{ s}} = +4.0 \text{ m/s}.$$

The plus sign indicates that the cab is moving in the positive x direction. These intervals (where $v = 0$ and $v = 4$ m/s) are plotted in Fig. 2-6b. In addition, as the cab initially begins to move and then later slows to a stop, v varies as indicated in the intervals 1 s to 3 s and 8 s to 9 s. Thus, Fig. 2-6b is the required plot. (Figure 2-6c is considered in Section 2-5.)

Given a $v(t)$ graph such as Fig. 2-6b, we could "work backward" to produce the shape of the associated $x(t)$ graph (Fig. 2-6a). However, we would not know the actual values for x at various times, because the $v(t)$ graph indicates only *changes* in x. To find such a change in x during any interval, we must, in the language of calculus, calculate the area "under the curve" on the $v(t)$ graph for that interval. For example, during the interval 3 s to 8 s in which the cab has a velocity of 4.0 m/s, the change in x is

$$\Delta x = (4.0 \text{ m/s})(8.0 \text{ s} - 3.0 \text{ s}) = +20 \text{ m}.$$

(This area is positive because the $v(t)$ curve is above the t axis.) Figure 2-6a shows that x does indeed increase by 20 m in that interval. However, Fig. 2-6b does not tell us the *values* of x at the beginning and end of the interval. For that we need additional information such as the value of x for some given instant.

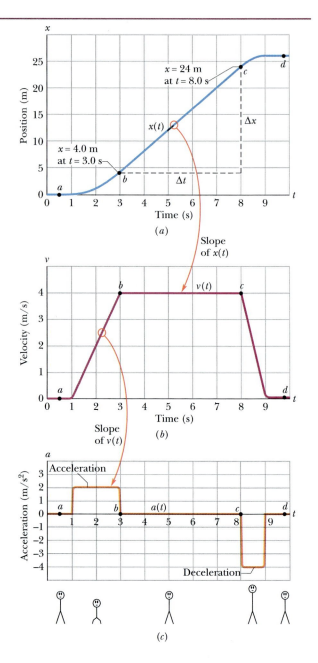

Fig. 2-6 Sample Problem 2-2. (a) The $x(t)$ curve for an elevator cab that moves upward along an x axis. (b) The $v(t)$ curve for the cab. Note that it is the derivative of the $x(t)$ curve ($v = dx/dt$). (c) The $a(t)$ curve for the cab. It is the derivative of the $v(t)$ curve ($a = dv/dt$). The stick figures along the bottom suggest how a passenger's body might feel during the accelerations.

Sample Problem 2-3

The position of a particle moving on an x axis is given by

$$x = 7.8 + 9.2t - 2.1t^3, \qquad (2\text{-}5)$$

with x in meters and t in seconds. What is its velocity at $t = 3.5$ s? Is the velocity constant, or is it continuously changing?

SOLUTION: For simplicity, the units have been omitted from Eq. 2-5, but you can insert them if you like by changing the coefficients to 7.8 m, 9.2 m/s, and -2.1 m/s^3. The Key Idea here is that velocity is the first derivative (with respect to time) of the position function $x(t)$. Thus, we write

$$v = \frac{dx}{dt} = \frac{d}{dt}(7.8 + 9.2t - 2.1t^3),$$

which becomes

$$v = 0 + 9.2 - (3)(2.1)t^2 = 9.2 - 6.3t^2. \qquad (2\text{-}6)$$

At $t = 3.5$ s,

$$v = 9.2 - (6.3)(3.5)^2 = -68 \text{ m/s}. \qquad \text{(Answer)}$$

At $t = 3.5$ s, the particle is moving in the negative direction of x (note the minus sign) with a speed of 68 m/s. Since the quantity t appears in Eq. 2-6, the velocity v depends on t and so is continuously changing.

✔**CHECKPOINT 3:**   The following equations give the position $x(t)$ of a particle in four situations (in each equation, x is in meters, t is in seconds, and $t > 0$): (1) $x = 3t - 2$; (2) $x = -4t^2 - 2$; (3) $x = 2/t^2$; and (4) $x = -2$. (a) In which situation is the velocity v of the particle constant? (b) In which is v in the negative x direction?

2-5 Acceleration

When a particle's velocity changes, the particle is said to undergo **acceleration** (or to accelerate). For motion along an axis, the **average acceleration** a_{avg} over a time interval Δt is

$$a_{avg} = \frac{v_2 - v_1}{t_2 - t_1} = \frac{\Delta v}{\Delta t}, \tag{2-7}$$

where the particle has velocity v_1 at time t_1 and then velocity v_2 at time t_2. The **instantaneous acceleration** (or simply **acceleration**) is the derivative of the velocity with respect to time:

$$a = \frac{dv}{dt}. \tag{2-8}$$

In words, the acceleration of a particle at any instant is the rate at which its velocity is changing at that instant. Graphically, the acceleration at any point is the slope of the curve of $v(t)$ at that point.

We can combine Eq. 2-8 with Eq. 2-4 to write

$$a = \frac{dv}{dt} = \frac{d}{dt}\left(\frac{dx}{dt}\right) = \frac{d^2x}{dt^2}. \tag{2-9}$$

In words, the acceleration of a particle at any instant is the second derivative of its position $x(t)$ with respect to time.

A common unit of acceleration is the meter per second per second: m/(s · s) or m/s². You will see other units in the problems, but they will each be in the form of length/(time · time) or length/time². Acceleration has both magnitude and direction (it is yet another vector quantity). Its algebraic sign represents its direction on an axis just as for displacement and velocity; that is, acceleration with a positive value is in the positive direction of an axis, and acceleration with a negative value is in the negative direction.

Figure 2-6c is a plot of the acceleration of the elevator cab discussed in Sample Problem 2-2. Compare this $a(t)$ curve with the $v(t)$ curve—each point on the $a(t)$ curve shows the derivative (slope) of the $v(t)$ curve at the corresponding time. When v is constant (at either 0 or 4 m/s), the derivative is zero and so also is the acceleration. When the cab first begins to move, the $v(t)$ curve has a positive derivative (the slope is positive), which means that $a(t)$ is positive. When the cab slows to a stop, the derivative and slope of the $v(t)$ curve are negative; that is, $a(t)$ is negative.

Next compare the slopes of the $v(t)$ curve during the two acceleration periods. The slope associated with the cab's slowing down (commonly called "deceleration") is steeper, because the cab stops in half the time it took to get up to speed. The steeper slope means that the magnitude of the deceleration is larger than that of the acceleration, as indicated in Fig. 2-6c.

The sensations you would feel while riding in the cab of Fig. 2-6 are indicated by the sketched figures. When the cab first accelerates, you feel as though you are pressed downward; when later the cab is braked to a stop, you seem to be stretched upward. In between, you feel nothing special. Your body reacts to accelerations (it is an accelerometer) but not to velocities (it is not a speedometer). When you are in

Fig. 2-7 Colonel J. P. Stapp in a rocket sled as it is brought up to high speed (acceleration out of the page) and then very rapidly braked (acceleration into the page).

a car traveling at 90 km/h or an airplane traveling at 900 km/h, you have no bodily awareness of the motion. However, if the car or plane quickly changes velocity, you may become keenly aware of the change, perhaps even frightened by it. Part of the thrill of an amusement park ride is due to the quick changes of velocity that you undergo (you pay for the accelerations, not for the speed). A more extreme example is shown in the photographs of Fig. 2-7, which were taken while a rocket sled was rapidly accelerated along a track and then rapidly braked to a stop.

Large accelerations are sometimes expressed in terms of g units, with

$$1g = 9.8 \text{ m/s}^2 \qquad (g \text{ unit}). \qquad (2\text{-}10)$$

(As we shall discuss in Section 2-8, g is the magnitude of the acceleration of a falling object near Earth's surface.) On a roller coaster, you may experience brief accelerations up to $3g$, which is $(3)(9.8 \text{ m/s}^2)$ or about 29 m/s², more than enough to justify the cost of the ride.

PROBLEM-SOLVING TACTICS

Tactic 5: *An Acceleration's Sign*

In common language, the sign of an acceleration has a nonscientific meaning: positive acceleration means that the speed of an object is increasing, and negative acceleration means that the speed is decreasing (the object is decelerating). In this book, however, the sign of an acceleration indicates a direction, not whether an object's speed is increasing or decreasing.

For example, if a car with an initial velocity $v = -25$ m/s is braked to a stop in 5.0 s, then $a_{\text{avg}} = +5.0$ m/s². The acceleration is *positive,* but the car's speed has decreased. The reason is the difference in signs: the direction of the acceleration is opposite that of the velocity.

Here then is the proper way to interpret the signs:

▶ If the signs of the velocity and acceleration of a particle are the same, the speed of the particle increases. If the signs are opposite, the speed decreases.

✔**CHECKPOINT 4:** A wombat moves along an x axis. What is the sign of its acceleration if it is moving (a) in the positive direction with increasing speed, (b) in the positive direction with decreasing speed, (c) in the negative direction with increasing speed, and (d) in the negative direction with decreasing speed?

Sample Problem 2-4

A particle's position on the x axis of Fig. 2-1 is given by

$$x = 4 - 27t + t^3,$$

with x in meters and t in seconds.

(a) Find the particle's velocity function $v(t)$ and acceleration function $a(t)$.

SOLUTION: One **Key Idea** is that to get the velocity function $v(t)$, we differentiate the position function $x(t)$ with respect to time. Here we find

$$v = -27 + 3t^2, \qquad \text{(Answer)}$$

with v in meters per second.

Another **Key Idea** is that to get the acceleration function $a(t)$,

Fig. 2-7 *Continued*

we differentiate the velocity function $v(t)$ with respect to time. This gives us

$$a = +6t, \qquad \text{(Answer)}$$

with a in meters per second squared.

(b) Is there ever a time when $v = 0$?

SOLUTION: Setting $v(t) = 0$ yields

$$0 = -27 + 3t^2,$$

which has the solution

$$t = \pm 3 \text{ s.} \qquad \text{(Answer)}$$

Thus, the velocity is zero both 3 s before and 3 s after the clock reads 0.

(c) Describe the particle's motion for $t \geq 0$.

SOLUTION: The Key Idea is to examine the expressions for $x(t)$, $v(t)$, and $a(t)$.

At $t = 0$, the particle is at $x(0) = +4$ m and is moving with a velocity of $v(0) = -27$ m/s—that is, in the negative direction of the x axis. Its acceleration is $a(0) = 0$, because just then the particle's velocity is not changing.

For $0 < t < 3$ s, the particle still has a negative velocity, so it continues to move in the negative direction. However, its acceleration is no longer 0 but is increasing and positive. Because the signs of the velocity and the acceleration are opposite, the particle must be slowing.

Indeed, we already know that it stops momentarily at $t = 3$ s. Just then the particle is as far to the left of the origin in Fig. 2-1 as it will ever get. Substituting $t = 3$ s into the expression for $x(t)$, we find that the particle's position just then is $x = -50$ m. Its acceleration is still positive.

For $t > 3$ s, the particle moves to the right on the axis. Its acceleration remains positive and grows progressively larger in magnitude. The velocity is now positive, and it too grows progressively larger in magnitude.

2-6 Constant Acceleration: A Special Case

In many types of motion, the acceleration is either constant or approximately so. For example, you might accelerate a car at an approximately constant rate when a traffic light turns from red to green. Then graphs of your position, velocity, and acceleration would resemble those in Fig. 2-8. (Note that $a(t)$ in Fig. 2-8c is constant, which requires that $v(t)$ in Fig. 2-8b have a constant slope.) Later when you brake the car to a stop, the deceleration might also be approximately constant.

Such cases are so common that a special set of equations has been derived for dealing with them. One approach to the derivation of these equations is given in this section. A second approach is given in the next section. Throughout both sections and later when you work on the homework problems, keep in mind that *these equations are valid only for constant acceleration (or situations in which you can approximate the acceleration as being constant)*.

When the acceleration is constant, the average acceleration and instantaneous acceleration are equal and we can write Eq. 2-7, with some changes in notation, as

$$a = a_{\text{avg}} = \frac{v - v_0}{t - 0}.$$

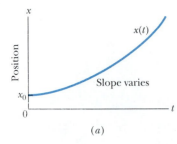

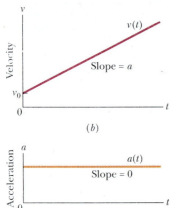

Fig. 2-8 (*a*) The position $x(t)$ of a particle moving with constant acceleration. (*b*) Its velocity $v(t)$, given at each point by the slope of the curve in (*a*). (*c*) Its (constant) acceleration, equal to the (constant) slope of the curve of $v(t)$.

Here v_0 is the velocity at time $t = 0$, and v is the velocity at any later time t. We can recast this equation as

$$v = v_0 + at. \tag{2-11}$$

As a check, note that this equation reduces to $v = v_0$ for $t = 0$, as it must. As a further check, take the derivative of Eq. 2-11. Doing so yields $dv/dt = a$, which is the definition of a. Figure 2-8*b* shows a plot of Eq. 2-11, the $v(t)$ function; the function is linear and thus the plot is a straight line.

In a similar manner we can rewrite Eq. 2-2 (with a few changes in notation) as

$$v_{\text{avg}} = \frac{x - x_0}{t - 0}$$

and then as

$$x = x_0 + v_{\text{avg}}t, \tag{2-12}$$

in which x_0 is the position of the particle at $t = 0$, and v_{avg} is the average velocity between $t = 0$ and a later time t.

For the linear velocity function in Eq. 2-11, the *average* velocity over any time interval (say, from $t = 0$ to a later time t) is the average of the velocity at the beginning of the interval ($= v_0$) and the velocity at the end of the interval ($= v$). For the interval from $t = 0$ to the later time t then, the average velocity is

$$v_{\text{avg}} = \tfrac{1}{2}(v_0 + v). \tag{2-13}$$

Substituting the right side of Eq. 2-11 for v yields, after a little rearrangement,

$$v_{\text{avg}} = v_0 + \tfrac{1}{2}at. \tag{2-14}$$

Finally, substituting Eq. 2-14 into Eq. 2-12 yields

$$x - x_0 = v_0 t + \tfrac{1}{2}at^2. \tag{2-15}$$

As a check, note that putting $t = 0$ yields $x = x_0$, as it must. As a further check, taking the derivative of Eq. 2-15 yields Eq. 2-11, again as it must. Figure 2-8*a* shows a plot of Eq. 2-15; the function is quadratic and thus the plot is curved.

Equations 2-11 and 2-15 are the *basic equations for constant acceleration*; they can be used to solve any constant acceleration problem in this book. However, we can derive other equations that might prove useful in certain specific situations. First, note that five quantities can possibly be involved in any problem regarding constant acceleration, namely, $x - x_0$, v, t, a, and v_0. Usually, one of these quantities is *not* involved in the problem, *either as a given or as an unknown*. We are then presented with three of the remaining quantities and asked to find the fourth.

Equations 2-11 and 2-15 each contain four of these quantities, but not the same four. In Eq. 2-11, the "missing ingredient" is the displacement, $x - x_0$. In Eq. 2-15, it is the velocity v. These two equations can also be combined in three ways to yield three additional equations, each of which involves a different "missing variable." First, we can eliminate t to obtain

$$v^2 = v_0^2 + 2a(x - x_0). \tag{2-16}$$

This equation is useful if we do not know t and are not required to find it. Second, we can eliminate the acceleration a between Eqs. 2-11 and 2-15 to produce an equation in which a does not appear:

$$x - x_0 = \tfrac{1}{2}(v_0 + v)t. \tag{2-17}$$

TABLE 2-1 Equations for Motion with Constant Acceleration[a]

Equation Number	Equation	Missing Quantity
2-11	$v = v_0 + at$	$x - x_0$
2-15	$x - x_0 = v_0 t + \frac{1}{2}at^2$	v
2-16	$v^2 = v_0^2 + 2a(x - x_0)$	t
2-17	$x - x_0 = \frac{1}{2}(v_0 + v)t$	a
2-18	$x - x_0 = vt - \frac{1}{2}at^2$	v_0

[a] Make sure that the acceleration is indeed constant before using the equations in this table.

Finally, we can eliminate v_0, obtaining

$$x - x_0 = vt - \tfrac{1}{2}at^2. \qquad (2\text{-}18)$$

Note the subtle difference between this equation and Eq. 2-15. One involves the initial velocity v_0; the other involves the velocity v at time t.

Table 2-1 lists the basic constant acceleration equations (Eqs. 2-11 and 2-15) as well as the specialized equations that we have derived. To solve a simple constant acceleration problem, you can usually use an equation from this list (*if* you have the list). Choose an equation for which the only unknown variable is the variable requested in the problem. A simpler plan is to remember only Eqs. 2-11 and 2-15, and then solve them as simultaneous equations whenever needed. An example is given in Sample Problem 2-5.

✓CHECKPOINT 5: The following equations give the position $x(t)$ of a particle in four situations: (1) $x = 3t - 4$; (2) $x = -5t^3 + 4t^2 + 6$; (3) $x = 2/t^2 - 4/t$; (4) $x = 5t^2 - 3$. To which of these situations do the equations of Table 2-1 apply?

Sample Problem 2-5

Spotting a police car, you brake a Porsche from a speed of 100 km/h to a speed of 80.0 km/h during a displacement of 88.0 m, at a constant acceleration.

(a) What is that acceleration?

SOLUTION: Assume that the motion is along the positive direction of an x axis. For simplicity, let us take the beginning of the braking to be at time $t = 0$, at position x_0. The **Key Idea** here is that, with the acceleration constant, we can relate the car's acceleration to its velocity and displacement via the basic constant acceleration equations (Eqs. 2-11 and 2-15). The initial velocity is $v_0 = 100$ km/h $= 27.78$ m/s, the displacement is $x - x_0 = 88.0$ m, and the velocity at the end of that displacement is $v = 80.0$ km/h $= 22.22$ m/s. However, we do not know the acceleration a and time t, which appear in both basic equations. So, we must solve those equations simultaneously.

To eliminate the unknown t, we use Eq. 2-11 to write

$$t = \frac{v - v_0}{a}, \qquad (2\text{-}19)$$

and then we substitute this expression into Eq. 2-15 to write

$$x - x_0 = v_0 \left(\frac{v - v_0}{a} \right) + \tfrac{1}{2}a \left(\frac{v - v_0}{a} \right)^2.$$

Solving for a and substituting known data then yield

$$a = \frac{v^2 - v_0^2}{2(x - x_0)} = \frac{(22.22 \text{ m/s})^2 - (27.78 \text{ m/s})^2}{2(88.0 \text{ m})}$$

$$= -1.58 \text{ m/s}^2. \qquad \text{(Answer)}$$

Note that we could have used Eq. 2-16 instead to solve for a because the unknown t is the missing variable in that equation.

(b) How much time is required for the given decrease in speed?

SOLUTION: Now that we know a, we can use Eq. 2-19 to solve for t:

$$t = \frac{v - v_0}{a} = \frac{22.22 \text{ m/s} - 27.78 \text{ m/s}}{-1.58 \text{ m/s}^2}$$

$$= 3.519 \text{ s} \approx 3.52 \text{ s}. \qquad \text{(Answer)}$$

If you are initially speeding and trying to slow to the speed limit, this is plenty of time for the police officer to measure your speed.

Tactic 6: *Check the Dimensions*

The dimension of a velocity is L/T—that is, length L divided by time T—and the dimension of an acceleration is L/T^2. In any equation, the dimensions of all terms must be the same. If you are in doubt about an equation, check its dimensions.

To check the dimensions of Eq. 2-15 ($x - x_0 = v_0 t + \frac{1}{2}at^2$), we note that every term must be a length, because that is the dimension of x and of x_0. The dimension of the term $v_0 t$ is $(L/T)(T)$, which is L. The dimension of $\frac{1}{2}at^2$ is $(L/T^2)(T^2)$, which is also L. Thus, this equation checks out.

2-7 Another Look at Constant Acceleration*

The first two equations in Table 2-1 are the basic equations from which the others are derived. Those two can be obtained by integration of the acceleration with the condition that a is constant. To find Eq. 2-11, we rewrite the definition of acceleration (Eq. 2-8) as

$$dv = a\, dt.$$

We next write the *indefinite integral* (or *antiderivative*) of both sides:

$$\int dv = \int a\, dt.$$

Since acceleration a is a constant, it can be taken outside the integration. Then we obtain

$$\int dv = a \int dt$$

or

$$v = at + C. \tag{2-20}$$

To evaluate the constant of integration C, we let $t = 0$, at which time $v = v_0$. Substituting these values into Eq. 2-20 (which must hold for all values of t, including $t = 0$) yields

$$v_0 = (a)(0) + C = C.$$

Substituting this into Eq. 2-20 gives us Eq. 2-11.

To derive Eq. 2-15, we rewrite the definition of velocity (Eq. 2-4) as

$$dx = v\, dt$$

and then take the indefinite integral of both sides to obtain

$$\int dx = \int v\, dt.$$

Generally v is not constant, so we cannot move it outside the integration. However, we can substitute for v with Eq. 2-11:

$$\int dx = \int (v_0 + at)\, dt.$$

Since v_0 is a constant, as is the acceleration a, this can be rewritten as

$$\int dx = v_0 \int dt + a \int t\, dt.$$

Integration now yields

$$x = v_0 t + \tfrac{1}{2}at^2 + C', \tag{2-21}$$

where C' is another constant of integration. At time $t = 0$, we have $x = x_0$. Substituting these values in Eq. 2-21 yields $x_0 = C'$. Replacing C' with x_0 in Eq. 2-21 gives us Eq. 2-15.

---

*This section is intended for students who have had integral calculus.

2-8 Free-Fall Acceleration

If you tossed an object either up or down and could somehow eliminate the effects of air on its flight, you would find that the object accelerates downward at a certain constant rate. That rate is called the **free-fall acceleration,** and its magnitude is represented by g. The acceleration is independent of the object's characteristics, such as mass, density, or shape; it is the same for all objects.

Two examples of free-fall acceleration are shown in Fig. 2-9, which is a series of stroboscopic photos of a feather and an apple. As these objects fall, they accelerate downward—both at the same rate g. Their speeds increase together.

The value of g varies slightly with latitude and with elevation. At sea level in Earth's midlatitudes the value is 9.8 m/s^2 (or 32 ft/s^2), which is what you should use for the problems in this chapter.

The equations of motion in Table 2-1 for constant acceleration also apply to free fall near Earth's surface; that is, they apply to an object in vertical flight, either up or down, when the effects of the air can be neglected. However, note that for free fall: (1) The directions of motion are now along a vertical y axis instead of the x axis, with the positive direction of y upward. (This is important for later chapters when combined horizontal and vertical motions are examined.) (2) The free-fall acceleration is now negative—that is, downward on the y axis, toward Earth's center—and so it has the value $-g$ in the equations.

Fig. 2-9 A feather and an apple, undergoing free fall in a vacuum, move downward with the same magnitude of acceleration g. The acceleration causes the increase in distance between successive images during the fall, but note that, in the absence of air, the feather and apple fall the same distance each time.

> ► The free-fall acceleration near Earth's surface is $a = -g = -9.8$ m/s^2, and the *magnitude* of the acceleration is $g = 9.8$ m/s^2. Do not substitute -9.8 m/s^2 for g.

Suppose that you toss a tomato directly upward with an initial (positive) velocity v_0 and then catch it when it returns to the release level. During its *free-fall flight* (just after its release and just before it is caught), the equations of Table 2-1 apply to its motion. The acceleration is always $a = -g = -9.8$ m/s^2, negative and thus downward. The velocity, however, changes, as indicated by Eqs. 2-11 and 2-16: during the ascent, the magnitude of the positive velocity decreases, until it momentarily becomes zero. Because the tomato has then stopped, it is at its maximum height. During the descent, the magnitude of the (now negative) velocity increases.

Sample Problem 2-6

Let us return to the opening story about Dave Munday's Niagara free fall in a steel ball. He fell 48 m. Let us assume that his initial velocity was zero and neglect the effect of the air on the ball during the fall.

(a) How long did Munday fall to reach the water surface below the falls?

SOLUTION: The **Key Idea** here is that, because Munday's fall was a free fall, the equations of Table 2-1 apply. Let us place a y axis along the path of Munday's fall, with $y = 0$ at his starting point and the positive direction upward along the axis (Fig. 2-10). Then the acceleration is $a = -g$ along that axis and the water level is at $y = -48$ m (negative because it is below $y = 0$). Let the fall begin at time $t = 0$, with initial velocity $v_0 = 0$.

Then from Table 2-1 we choose Eq. 2-15 (but in y notation) because it contains the requested time t and all the other variables

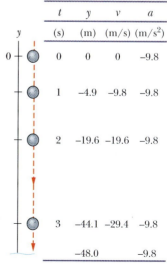

	t	y	v	a
	(s)	(m)	(m/s)	(m/s^2)
0	0	0	0	−9.8
	1	−4.9	−9.8	−9.8
	2	−19.6	−19.6	−9.8
	3	−44.1	−29.4	−9.8
		−48.0		−9.8

Fig. 2-10 Sample Problem 2-6. The position, velocity, and acceleration of a freely falling object, here the steel ball ridden by Dave Munday over Niagara Falls.

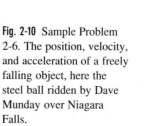

have known values. We find

$$y - y_0 = v_0 t - \tfrac{1}{2}gt^2,$$

$$-48 \text{ m} - 0 = 0t - \tfrac{1}{2}(9.8 \text{ m/s}^2)t^2,$$

$$t^2 = 48/4.9,$$

and $t = 3.1$ s.   (Answer)

Note that Munday's displacement $y - y_0$ is a negative quantity— Munday fell down, in the *negative direction* of the y axis (he did not fall up!). Also note that 48/4.9 has two square roots: 3.1 and −3.1. Here we choose the positive root because Munday obviously reaches the water surface *after* he begins to fall at $t = 0$.

(b) Munday could count off the three seconds of free fall but could not see how far he had fallen with each count. Determine his position at each full second.

SOLUTION: We again use Eq. 2-15 but now we substitute, in turn, the values $t = 1.0$ s, 2.0 s, and 3.0 s, and solve for Munday's position y. The results are shown in Fig. 2-10.

(c) What was Munday's velocity as he reached the water surface?

SOLUTION: To find the velocity from the original data without using the time of fall from (a), we rewrite Eq. 2-16 in y notation and

then substitute known data:

$$v^2 = v_0^2 - 2g(y - y_0) = 0 - (2)(9.8 \text{ m/s}^2)(-48 \text{ m}),$$

so $v = -30.67$ m/s ≈ -31 m/s $= -110$ km/h.   (Answer)

We chose the negative root here, because the velocity was in the negative direction.

(d) What was Munday's velocity at each count of one full second? Was Munday aware of this increasing speed?

SOLUTION: To find the velocities from the original data without using the positions from (b), we let $a = -g$ in Eq. 2-11 and then substitute, in turn, the values $t = 1.0$ s, 2.0 s, and 3.0 s. Here is an example:

$$v = v_0 - gt$$
$$= 0 - (9.8 \text{ m/s}^2)(1.0 \text{ s}) = -9.8 \text{ m/s}.$$ (Answer)

The other results are shown in Fig. 2-10.

Once he was in free fall, Munday was unaware of the increasing speed because the acceleration during the fall was always -9.8 m/s^2, as noted in the last column of Fig. 2-10. He was, of course, sharply aware of hitting the water because then the acceleration abruptly changed. (Munday survived the fall but then faced stiff legal fines for his daredevil action.)

Sample Problem 2-7

In Fig. 2-11, a pitcher tosses a baseball up along a y axis, with an initial speed of 12 m/s.

(a) How long does the ball take to reach its maximum height?

SOLUTION: One Key Idea here is that once the ball leaves the pitcher and before it returns, its acceleration is the free-fall acceleration

Ball
$v = 0$ at highest point

During ascent, $a = -g$, speed decreases, and velocity becomes less positive

During descent, $a = -g$, speed increases, and velocity becomes more negative

$y = 0$

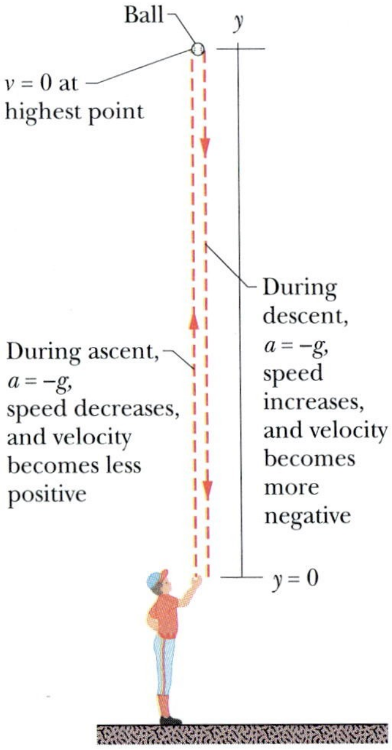

Fig. 2-11 Sample Problem 2-7. A pitcher tosses a baseball straight up into the air. The equations of free fall apply for rising as well as for falling objects, provided any effects from the air can be neglected.

$a = -g$. Because this is constant, Table 2-1 applies to the motion. A second Key Idea is that the velocity v at the maxium height must be 0. So, knowing v, a, and the initial velocity $v_0 = 12$ m/s, and seeking t, we solve Eq. 2-11, which contains those four variables. Rearranging yields

$$t = \frac{v - v_0}{a} = \frac{0 - 12 \text{ m/s}}{-9.8 \text{ m/s}^2} = 1.2 \text{ s}.$$ (Answer)

(b) What is the ball's maximum height above its release point?

SOLUTION: We can take the ball's release point to be $y_0 = 0$. We can then write Eq. 2-16 in y notation, set $y - y_0 = y$ and $v = 0$ (at the maximum height), and solve for y. We get

$$y = \frac{v^2 - v_0^2}{2a} = \frac{0 - (12 \text{ m/s})^2}{2(-9.8 \text{ m/s}^2)} = 7.3 \text{ m}.$$ (Answer)

(c) How long does the ball take to reach a point 5.0 m above its release point?

SOLUTION: We know v_0, $a = -g$, and the displacement $y - y_0 = 5.0$ m, and we want t, so we choose Eq. 2-15. Rewriting it for y and setting $y_0 = 0$ give us

$$y = v_0 t - \tfrac{1}{2}gt^2,$$

or $5.0 \text{ m} = (12 \text{ m/s})t - (\tfrac{1}{2})(9.8 \text{ m/s}^2)t^2.$

If we temporarily omit the units (having noted that they are consistent), we can rewrite this as

$$4.9t^2 - 12t + 5.0 = 0.$$

Solving this quadratic equation for t yields

$$t = 0.53 \text{ s} \quad \text{and} \quad t = 1.9 \text{ s.} \qquad \text{(Answer)}$$

There are two such times! This is not really surprising because the ball passes twice through $y = 5.0$ m, once on the way up and once on the way down.

✓CHECKPOINT 6: (a) In this sample problem, what is the sign of the ball's displacement for the ascent, from the release point to the highest point? (b) What is it for the descent, from the highest point back to the release point? (c) What is the ball's acceleration at its highest point?

PROBLEM-SOLVING TACTICS

Tactic 7: *Meanings of Minus Signs*
In Sample Problems 2-6 and 2-7, many answers emerged automatically with minus signs. It is important to know what these signs mean. For these two falling-body problems, we established a vertical axis (the y axis) and we chose—quite arbitrarily—its upward direction to be positive.

We then chose the origin of the y axis (that is, the $y = 0$ position) to suit the problem. In Sample Problem 2-6, the origin was at the top of the falls, and in Sample Problem 2-7 it was at the pitcher's hand. A negative value of y then means that the body is below the chosen origin. A negative velocity means that the body is moving in the negative direction of the y axis—that is, downward. This is true no matter where the body is located.

We take the acceleration to be negative (-9.8 m/s^2) in all problems dealing with falling bodies. A negative acceleration

means that, as time goes on, the velocity of the body becomes either less positive or more negative. This is true no matter where the body is located and no matter how fast or in what direction it is moving. In Sample Problem 2-7, the acceleration of the ball is negative (downward) throughout its flight, whether the ball is rising or falling.

Tactic 8: *Unexpected Answers*
Mathematics often generates answers that you might not have thought of as possibilities, as in Sample Problem 2-7c. If you get more answers than you expect, do not automatically discard the ones that do not seem to fit. Examine them carefully for physical meaning. If time is your variable, even a negative value can mean something; negative time simply refers to time before $t = 0$, the (arbitrary) time at which you decided to start your stopwatch.

REVIEW & SUMMARY

Position   The *position x* of a particle on an x axis locates the particle with respect to the **origin,** or zero point, of the axis. The position is either positive or negative, according to which side of the origin the particle is on, or zero if the particle is at the origin. The **positive direction** on an axis is the direction of increasing positive numbers; the opposite direction is the **negative direction**.

Displacement   The *displacement* Δx of a particle is the change in its position:

$$\Delta x = x_2 - x_1. \qquad (2\text{-}1)$$

Displacement is a vector quantity. It is positive if the particle has moved in the positive direction of the x axis, and negative if the particle has moved in the negative direction.

Average Velocity   When a particle has moved from position x_1 to x_2 during a time interval $\Delta t = t_2 - t_1$, its *average velocity* during that interval is

$$v_{\text{avg}} = \frac{\Delta x}{\Delta t} = \frac{x_2 - x_1}{t_2 - t_1}. \qquad (2\text{-}2)$$

The algebraic sign of v_{avg} indicates the direction of motion (v_{avg} is a vector quantity). Average velocity does not depend on the actual distance a particle moves, but instead depends on its original and final positions.

On a graph of x versus t, the average velocity for a time interval Δt is the slope of the straight line connecting the points on the curve that represent the ends of the interval.

Average Speed   The *average speed* s_{avg} of a particle during a time interval Δt depends on the total distance the particle moves in that time interval:

$$s_{\text{avg}} = \frac{\text{total distance}}{\Delta t}. \qquad (2\text{-}3)$$

Instantaneous Velocity   The *instantaneous velocity* (or simply **velocity**) v of a moving particle is

$$v = \lim_{\Delta t \to 0} \frac{\Delta x}{\Delta t} = \frac{dx}{dt}, \qquad (2\text{-}4)$$

where Δx and Δt are defined by Eq. 2-2. The instantaneous velocity (at a particular time) may be found as the slope (at that particular time) of the graph of x versus t. **Speed** is the magnitude of instantaneous velocity.

Average Acceleration   *Average acceleration* is the ratio of a change in velocity Δv to the time interval Δt in which the change occurs:

$$a_{\text{avg}} = \frac{\Delta v}{\Delta t}. \qquad (2\text{-}7)$$

The algebraic sign indicates the direction of a_{avg}.

Instantaneous Acceleration   *Instantaneous acceleration* (or simply **acceleration**) a is the rate of change of velocity with time and the second derivative of position $x(t)$ with respect to time:

$$a = \frac{dv}{dt} = \frac{d^2x}{dt^2}. \qquad (2\text{-}8, 2\text{-}9)$$

On a graph of v versus t, the acceleration a at any time t is the slope of the curve at the point that represents t.

Constant Acceleration   The five equations in Table 2-1 describe the motion of a particle with constant acceleration:

$$v = v_0 + at, \qquad (2\text{-}11)$$

$$x - x_0 = v_0 t + \tfrac{1}{2}at^2, \qquad (2\text{-}15)$$

$$v^2 = v_0^2 + 2a(x - x_0), \qquad (2\text{-}16)$$

$$x - x_0 = \tfrac{1}{2}(v_0 + v)t, \qquad (2\text{-}17)$$

$$x - x_0 = vt - \tfrac{1}{2}at^2. \qquad (2\text{-}18)$$

These are *not* valid when the acceleration is not constant.

Free-Fall Acceleration   An important example of straight-line motion with constant acceleration is that of an object rising or falling freely near Earth's surface. The constant acceleration equations describe this motion, but we make two changes in notation: (1) we refer the motion to the vertical y axis with $+y$ vertically *up*; (2) we replace a with $-g$, where g is the magnitude of the free-fall acceleration. Near Earth's surface, $g = 9.8$ m/s^2 ($= 32$ ft/s^2).

QUESTIONS

1. Figure 2-12 shows four paths along which objects move from a starting point to a final point, all in the same time. The paths pass over a grid of equally spaced straight lines. Rank the paths according to (a) the average velocity of the objects and (b) the average speed of the objects, greatest first.

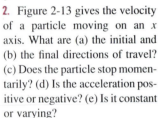

Fig. 2-12 Question 1.

2. Figure 2-13 gives the velocity of a particle moving on an x axis. What are (a) the initial and (b) the final directions of travel? (c) Does the particle stop momentarily? (d) Is the acceleration positive or negative? (e) Is it constant or varying?

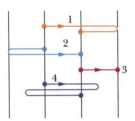

Fig. 2-13 Question 2.

3. Figure 2-14 gives the acceleration $a(t)$ of a Chihuahua as it chases a German shepherd along an axis. In which of the time periods indicated does the Chihuahua move at constant speed?

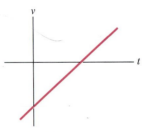

Fig. 2-14 Question 3.

4. At $t = 0$, a particle moving along an x axis is at position $x_0 = -20$ m. The signs of the particle's initial velocity v_0 (at time t_0) and constant acceleration a are, respectively, for four situations: (1) +, +; (2) +, −; (3) −, +; (4) −, −. In which situation will the particle (a) undergo a momentary stop, (b) definitely pass through the origin (given enough time), and (c) definitely not pass through the origin?

5. The following equations give the velocity $v(t)$ of a particle in four situations: (a) $v = 3$; (b) $v = 4t^2 + 2t - 6$; (c) $v = 3t - 4$; (d) $v = 5t^2 - 3$. To which of these situations do the equations of Table 2-1 apply?

6. The driver of a blue car, moving at a speed of 80 km/h, suddenly realizes that she is about to rear-end a red car, moving at a speed of 60 km/h. To avoid a collision, what is the maximum speed the blue car can have just as it reaches the red car? (Warm-up for Problem 38)

7. At $t = 0$ and $x = 0$, an initially stationary blue car begins to accelerate at the constant rate of 2.0 m/s^2 in the positive direction of the x axis. At $t = 2$ s, a red car traveling in an adjacent lane and in the same direction, passes $x = 0$ with a speed of 8.0 m/s and a constant acceleration of 3.0 m/s^2. What pair of simultaneous equations should be solved to find when the red car passes the blue car? (Warm-up for Problem 36)

8. In Fig. 2-15, a cream tangerine is thrown directly upward past three evenly spaced windows of equal heights. Rank the windows according to (a) the average speed of the cream tangerine while passing them, (b) the time the cream tangerine takes to pass them, (c) the magnitude of the acceleration of the cream tangerine while passing them, and (d) the change Δv in the speed of the cream tangerine during the passage, greatest first.

9. You throw a ball straight up from the edge of a cliff, and it lands on the ground below the cliff. If you had, instead, thrown the ball down from the cliff edge with the same speed, would the ball's speed just before landing be larger than, smaller than, or the same as previously? (*Hint:* Consider Eq. 2-16.)

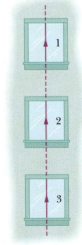

Fig. 2-15 Question 8.

EXERCISES & PROBLEMS

In several of the problems that follow you are asked to graph position, velocity, and acceleration versus time. Usually a sketch will suffice, appropriately labeled and with straight and curved portions apparent. If you have a computer or graphing calculator, you might use it to produce the graph.

SEC. 2-3   Average Velocity and Average Speed

1E. If a baseball pitcher throws a fastball at a horizontal speed of 160 km/h, how long does the ball take to reach home plate 18.4 m away? ssm

2E. A world speed record for bicycles was set in 1992 by Chris Huber riding Cheetah, a high-tech bicycle built by three mechanical engineering graduates. The record (average) speed was 110.6 km/h through a measured length of 200.0 m on a desert road. At the end of the run, Huber commented, "Cogito ergo zoom!" (I think, therefore I go fast!) What was Huber's elapsed time through the 200.0 m?

3E. An automobile travels on a straight road for 40 km at 30 km/h. It then continues in the same direction for another 40 km at 60 km/h. (a) What is the average velocity of the car during this 80 km trip? (Assume that it moves in the positive x direction.) (b) What is the average speed? (c) Graph x versus t and indicate how the average velocity is found on the graph. ssm

4P. A top-gun pilot, practicing radar avoidance maneuvers, is manually flying horizontally at 1300 km/h, just 35 m above the level ground. Suddenly, the plane encounters terrain that slopes gently upward at 4.3°, an amount difficult to detect visually (Fig. 2-16). How much times does the pilot have to make a correction to avoid flying into the ground?

4.3°

35 m

Fig. 2-16 Problem 4.

5P. You drive on Interstate 10 from San Antonio to Houston, half the *time* at 55 km/h and the other half at 90 km/h. On the way back you travel half the *distance* at 55 km/h and the other half at 90 km/h. What is your average speed (a) from San Antonio to Houston, (b) from Houston back to San Antonio, and (c) for the entire trip? (d) What is your average velocity for the entire trip? (e) Sketch x versus t for (a), assuming the motion is all in the positive x direction. Indicate how the average velocity can be found on the sketch. ilw

6P. Compute your average velocity in the following two cases: (a) You walk 73.2 m at a speed of 1.22 m/s and then run 73.2 m at a speed of 3.05 m/s along a straight track. (b) You walk for 1.00 min at a speed of 1.22 m/s and then run for 1.00 min at 3.05 m/s

along a straight track. (c) Graph x versus t for both cases and indicate how the average velocity is found on the graph.

7P. The position of an object moving along an x axis is given by $x = 3t - 4t^2 + t^3$, where x is in meters and t in seconds. (a) What is the position of the object at $t = 1, 2, 3,$ and 4 s? (b) What is the object's displacement between $t = 0$ and $t = 4$ s? (c) What is its average velocity for the time interval from $t = 2$ s to $t = 4$ s? (d) Graph x versus t for $0 \le t \le 4$ s and indicate how the answer for (c) can be found on the graph. ssm www

8P. Two trains, each having a speed of 30 km/h, are headed at each other on the same straight track. A bird that can fly 60 km/h flies off the front of one train when they are 60 km apart and heads directly for the other train. On reaching the other train it flies directly back to the first train, and so forth. (We have no idea *why* a bird would behave in this way.) What is the total distance the bird travels?

9P. On two *different* tracks, the winners of the one-kilometer race ran their races in 2 min, 27.95 s and 2 min, 28.15 s. In order to conclude that the runner with the shorter time was indeed faster, how much longer can the other track be in *actual* length? ilw

SEC. 2-4   Instantaneous Velocity and Speed

10E. The graph in Fig. 2-17 is for an armadillo that scampers left (negative direction of x) and right along an x axis. (a) When, if ever, is the animal to the left of the origin on the axis? When, if ever, is its velocity (b) negative, (c) positive, or (d) zero?

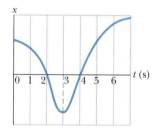

Fig. 2-17 Exercise 10.

11E. (a) If a particle's position is given by $x = 4 - 12t + 3t^2$ (where t is in seconds and x is in meters), what is its velocity at $t = 1$ s? (b) Is it moving in the positive or negative direction of x just then? (c) What is its speed just then? (d) Is the speed larger or smaller at later times? (Try answering the next two questions without further calculation.) (e) Is there ever an instant when the velocity is zero? (f) Is there a time after $t = 3$ s when the particle is moving in the negative direction of x?

12P. The position of a particle moving along the x axis is given in centimeters by $x = 9.75 + 1.50t^3$, where t is in seconds. Calculate (a) the average velocity during the time interval $t = 2.00$ s to $t = 3.00$ s; (b) the instantaneous velocity at $t = 2.00$ s; (c) the instantaneous velocity at $t = 3.00$ s; (d) the instantaneous velocity at $t = 2.50$ s; and (e) the instantaneous velocity when the particle is midway between its positions at $t = 2.00$ s and $t = 3.00$ s. (f) Graph x versus t and indicate your answers graphically.

13P. How far does the runner whose velocity–time graph is shown in Fig. 2-18 travel in 16 s? **itw**

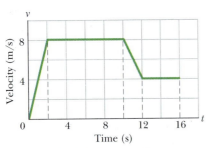

Fig. 2-18 Problem 13.

SEC. 2-5   Acceleration

14E. Sketch a graph that is a possible description of position as a function of time for a particle that moves along the x axis and, at $t = 1$ s, has (a) zero velocity and positive acceleration; (b) zero velocity and negative acceleration; (c) negative velocity and positive acceleration; (d) negative velocity and negative acceleration. (e) For which of these situations is the speed of the particle increasing at $t = 1$ s?

15E. What do the quantities (a) $(dx/dt)^2$ and (b) d^2x/dt^2 represent? (c) What are their SI units?

16E. A frightened ostrich moves in a straight line with velocity described by the velocity–time graph of Fig. 2-19. Sketch acceleration versus time.

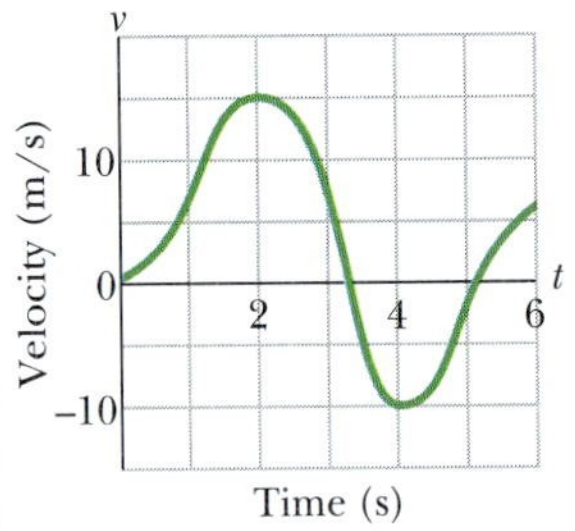

Fig. 2-19 Exercise 16.

17E. A particle had a speed of 18 m/s at a certain time, and 2.4 s later its speed was 30 m/s in the opposite direction. What were the magnitude and direction of the average acceleration of the particle during this 2.4 s interval? **ssm**

18P. From $t = 0$ to $t = 5.00$ min, a man stands still, and from $t = 5.00$ min to $t = 10.0$ min, he walks briskly in a straight line at a constant speed of 2.20 m/s. What are (a) his average velocity v_{avg} and (b) his average acceleration a_{avg} in the time interval 2.00 min to 8.00 min? What are (c) v_{avg} and (d) a_{avg} in the time interval 3.00 min to 9.00 min? (e) Sketch x versus t and v versus t, and indicate how the answers to (a) through (d) can be obtained from the graphs.

19P. A proton moves along the x axis according to the equation $x = 50t + 10t^2$, where x is in meters and t is in seconds. Calculate (a) the average velocity of the proton during the first 3.0 s of its motion, (b) the instantaneous velocity of the proton at $t = 3.0$ s, and (c) the instantaneous acceleration of the proton at $t = 3.0$ s. (d) Graph x versus t and indicate how the answer to (a) can be obtained from the plot. (e) Indicate the answer to (b) on the graph. (f) Plot v versus t and indicate on it the answer to (c). **ssm**

20P. A electron moving along the x axis has a position given by $x = 16te^{-t}$ m, where t is in seconds. How far is the electron from the origin when it momentarily stops?

21P. The position of a particle moving along the x axis depends on the time according to the equation $x = ct^2 - bt^3$, where x is in meters and t in seconds. (a) What units must c and b have? Let their numerical values be 3.0 and 2.0, respectively. (b) At what time does the particle reach its maximum positive x position? From $t = 0.0$ s to $t = 4.0$ s, (c) what distance does the particle move and (d) what is its displacement? At $t = 1.0$, 2.0, 3.0, and 4.0 s, what are (e) its velocities and (f) its accelerations? **ssm**

SEC. 2-6   Constant Acceleration: A Special Case

22E. An automobile driver increases the speed at a constant rate from 25 km/h to 55 km/h in 0.50 min. A bicycle rider speeds up at a constant rate from rest to 30 km/h in 0.50 min. Calculate their accelerations.

23E. A muon (an elementary particle) enters a region with a speed of 5.00×10^6 m/s and then is slowed at the rate of 1.25×10^{14} m/s². (a) How far does the muon take to stop? (b) Graph x versus t and v versus t for the muon. **ssm**

24E. The head of a rattlesnake can accelerate at 50 m/s² in striking a victim. If a car could do as well, how long would it take to reach a speed of 100 km/h from rest?

25E. An electron has a constant acceleration of +3.2 m/s². At a certain instant its velocity is +9.6 m/s. What is its velocity (a) 2.5 s earlier and (b) 2.5 s later? **ssm**

26E. The speed of a bullet is measured to be 640 m/s as the bullet emerges from a barrel of length 1.20 m. Assuming constant acceleration, find the time that the bullet spends in the barrel after it is fired.

27E. Suppose a rocket ship in deep space moves with constant acceleration equal to 9.8 m/s², which gives the illusion of normal gravity during the flight. (a) If it starts from rest, how long will it take to acquire a speed one-tenth that of light, which travels at 3.0×10^8 m/s? (b) How far will it travel in so doing? **ssm**

28E. A jumbo jet must reach a speed of 360 km/h on the runway for takeoff. What is the least constant acceleration needed for takeoff from a 1.80 km runway?

29E. An electron with initial velocity $v_0 = 1.50 \times 10^5$ m/s enters a region 1.0 cm long where it is electrically accelerated (Fig. 2-20). It emerges with velocity $v = 5.70 \times 10^6$ m/s. What is its acceleration, assumed constant? (Such a process occurs in conventional television sets.) **ssm**

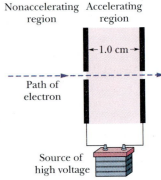

Nonaccelerating region   Accelerating region

←1.0 cm→

Path of electron

Source of high voltage

Fig. 2-20 Exercise 29.

30E. A world's land speed record was set by Colonel John P. Stapp when in March 1954 he rode a rocket-propelled sled that moved along a track at 1020 km/h. He and the sled were brought to a stop in 1.4 s. (See Fig. 2-7.) In g units, what acceleration did he experience while stopping?

31E. The brakes on your automobile are capable of creating a deceleration of 5.2 m/s². (a) If you are going 137 km/h and suddenly see a state trooper, what is the minimum time in which you can get your car under the 90 km/h speed limit? (The answer reveals the futility of braking to keep your high speed from being detected with a radar or laser gun.) (b) Graph x versus t and v versus t for such a deceleration. ssm www

32E. Figure 2-21 depicts the motion of a particle moving along an x axis with a constant acceleration. What are the magnitude and direction of the particle's acceleration?

33P. A car traveling 56.0 km/h is 24.0 m from a barrier when the driver slams on the brakes. The car hits the barrier 2.00 s later. (a) What is the car's constant deceleration before impact? (b) How fast is the car traveling at impact? ssm ilw

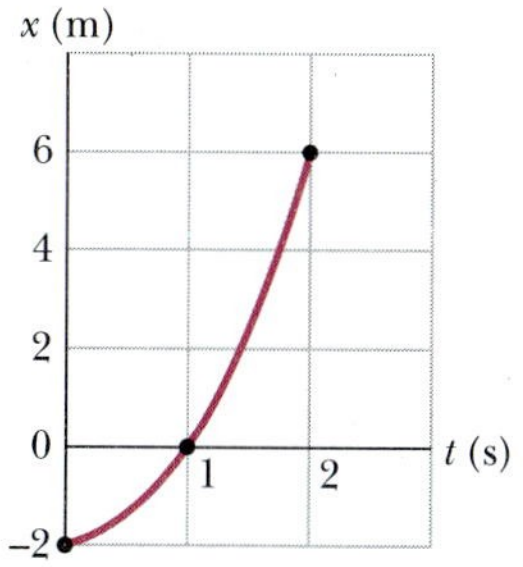

Fig. 2-21 Exercise 32.

34P. A red train traveling at 72 km/h and a green train traveling at 144 km/h are headed toward one another along a straight, level track. When they are 950 m apart, each engineer sees the other's train and applies the brakes. The brakes decelerate each train at the rate of 1.0 m/s². Is there a collision? If so, what is the speed of each train at impact? If not, what is the separation between the trains when they stop?

35P. A car moving with constant acceleration covered the distance between two points 60.0 m apart in 6.00 s. Its speed as it passes the second point was 15.0 m/s. (a) What was the speed at the first point? (b) What was the acceleration? (c) At what prior distance from the first point was the car at rest? (d) Graph x versus t and v versus t for the car, from rest ($t = 0$). ssm www

36P. At the instant the traffic light turns green, an automobile starts with a constant acceleration a of 2.2 m/s². At the same instant a truck, traveling with a constant speed of 9.5 m/s, overtakes and passes the automobile. (a) How far beyond the traffic signal will the automobile overtake the truck? (b) How fast will the car be traveling at that instant?

37P. To stop a car, first you require a certain reaction time to begin braking; then the car slows under the constant braking deceleration. Suppose that the total distance moved by your car during these two phases is 56.7 m when its initial speed is 80.5 km/h, and 24.4 m when its initial speed is 48.3 km/h. What are (a) your reaction time and (b) the magnitude of the deceleration? ssm

38P. When a high-speed passenger train traveling at 161 km/h rounds a bend, the engineer is shocked to see that a locomotive has improperly entered onto the track from a siding and is a distance $D = 676$ m ahead (Fig. 2-22). The locomotive is moving at 29.0 km/h. The engineer of the high-speed train immediately applies the brakes. (a) What must be the magnitude of the resulting constant deceleration if a collision is to be just avoided? (b) Assume that the engineer is at $x = 0$ when, at $t = 0$, he first spots the locomotive. Sketch the $x(t)$ curves representing the locomotive and high-speed train for the situations in which a collision is just avoided and is not quite avoided.

39P. An elevator cab in the New York Marquis Marriott has a total run of 190 m. Its maximum speed is 305 m/min. Its acceleration and deceleration both have a magnitude of 1.22 m/s². (a) How far does the cab move while accelerating to full speed from rest? (b) How long does it take to make the nonstop 190 m run, starting and ending at rest? ilw

SEC. 2-8   Free-Fall Acceleration

40E. Raindrops fall 1700 m from a cloud to the ground. (a) If they were not slowed by air resistance, how fast would the drops be moving when they struck the ground? (b) Would it be safe to walk outside during a rainstorm?

41E. At a construction site a pipe wrench struck the ground with a speed of 24 m/s. (a) From what height was it inadvertently dropped? (b) How long was it falling? (c) Sketch graphs of y, v, and a versus t for the wrench. ssm

42E. A hoodlum throws a stone vertically downward with an initial speed of 12.0 m/s from the roof of a building, 30.0 m above the ground. (a) How long does it take the stone to reach the ground? (b) What is the speed of the stone at impact?

43E. (a) With what speed must a ball be thrown vertically from ground level to rise to a maximum height of 50 m? (b) How long will it be in the air? (c) Sketch graphs of y, v, and a versus t for the ball. On the first two graphs, indicate the time at which 50 m is reached. ssm

44E. The Zero Gravity Research Facility at the NASA Lewis Research Center includes a 145 m drop tower. This is an evacuated vertical tower through which, among other possibilities, a 1 m diameter sphere containing an experimental package can be dropped. (a) How long is the sphere in free fall? (b) What is its speed just as it reaches a catching device at the bottom of the tower? (c) When caught, the sphere experiences an average deceleration of 25g as its speed is reduced to zero. Through what distance does it travel during the deceleration?

45E. A rock is dropped from a 100-m-high cliff. How long does it take to fall (a) the first 50 m and (b) the second 50 m? ssm

46P. A ball is thrown *down* vertically with an initial *speed* of v_0 from a height of h. (a) What is its speed just before it strikes the ground? (b) How long does the ball take to reach the ground? What would be the answers to (c) part a and (d) part b if the ball were thrown *upward* from the same height and with the same initial speed? Before solving any equations, decide whether the answers to (c) and (d) should be greater than, less than, or the same as in (a) and (b).

47P. A startled armadillo leaps upward, rising 0.544 m in the first 0.200 s. (a) What is its initial speed as it leaves the ground? (b) What is its speed at the height of 0.544 m? (c) How much higher does it go? ssm www

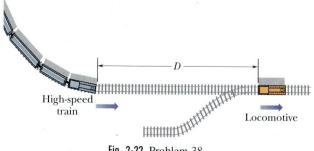

Fig. 2-22 Problem 38.

48P. A rock is dropped (from rest) from the top of a 60-m-tall building. How far above the ground is the rock 1.2 s before it reaches the ground?

49P. A key falls from a bridge that is 45 m above the water. It falls directly into a model boat, moving with constant velocity, that is 12 m from the point of impact when the key is released. What is the speed of the boat? ssm ilw

50P. A ball is thrown vertically downward from the top of a 36.6-m-tall building. The ball passes the top of a window that is 12.2 m above the ground 2.00 s after being thrown. What is the speed of the ball as it passes the top of the window?

51P. A ball of moist clay falls 15.0 m to the ground. It is in contact with the ground for 20.0 ms before stopping. What is the average acceleration of the ball during the time it is in contact with the ground? (Treat the ball as a particle.) ssm

52P. A model rocket fired vertically from the ground ascends with a constant vertical acceleration of 4.00 m/s² for 6.00 s. Its fuel is then exhausted, so it continues upward as a free-fall particle and then falls back down. (a) What is the maximum altitude reached? (b) What is the total time elapsed from takeoff until the rocket strikes the ground?

53P. To test the quality of a tennis ball, you drop it onto the floor from a height of 4.00 m. It rebounds to a height of 2.00 m. If the ball is in contact with the floor for 12.0 ms, what is its average acceleration during that contact? ssm

54P. A basketball player, standing near the basket to grab a rebound, jumps 76.0 cm vertically. How much (total) time does the player spend (a) in the top 15.0 cm of this jump and (b) in the bottom 15.0 cm? Does this help explain why such players seem to hang in the air at the tops of their jumps?

55P. Water drips from the nozzle of a shower onto the floor 200 cm below. The drops fall at regular (equal) intervals of time, the first drop striking the floor at the instant the fourth drop begins to fall. Find the locations of the second and third drops when the first strikes the floor.

56P. A ball is shot vertically upward from the surface of a planet in a distant solar system. A plot of y versus t for the ball is shown in Fig. 2-23, where y is the height of the ball above its starting point and t = 0 at the instant the ball is shot. What are the magnitudes of (a) the free-fall acceleration on the planet and (b) the initial velocity of the ball?

57P. Two diamonds begin a free fall from rest from the same height, 1.0 s apart. How long after the first diamond begins to fall will the two diamonds be 10 m apart? ssm

58P. A certain juggler usually tosses balls vertically to a height H. To what height must they be tossed if they are to spend twice as much time in the air?

59P. A hot-air balloon is ascending at the rate of 12 m/s and is 80 m above the ground when a package is dropped over the side. (a) How long does the package take to reach the ground? (b) With what speed does it hit the ground? ssm

60P. A stone is dropped into a river from a bridge 43.9 m above the water. Another stone is thrown vertically down 1.00 s after the first is dropped. Both stones strike the water at the same time. (a) What is the initial speed of the second stone? (b) Plot velocity versus time on a graph for each stone, taking zero time as the instant the first stone is released.

61P. An elevator without a ceiling is ascending with a constant speed of 10 m/s. A boy on the elevator shoots a ball directly upward, from a height of 2.0 m above the elevator floor, just as the elevator floor is 28 m above the ground. The initial speed of the ball with respect to the elevator is 20 m/s. (a) What maximum height above the ground does the ball reach? (b) How long does the ball take to return to the elevator floor? ssm

62P. A stone is thrown vertically upward. On its way up it passes point A with speed v, and point B, 3.00 m higher than A, with speed $\frac{1}{2}v$. Calculate (a) the speed v and (b) the maximum height reached by the stone above point B.

63P. Figure 2-24 shows a simple device for measuring your reaction time. It consists of a cardboard strip marked with a scale and two large dots. A friend holds the strip *vertically,* with thumb and forefinger at the dot on the right in Fig. 2-24. You then position your thumb and forefinger at the other dot (on the left in Fig. 2-24), being careful not to touch the strip. Your friend releases the strip, and you try to pinch it as soon as possible after you see it begin to fall. The mark at the place where you pinch the strip gives your reaction time. (a) How far from the lower dot should you place the 50.0 ms mark? (b) How much higher should the marks for 100, 150, 200, and 250 ms be? (For example, should the 100 ms marker be two times as far from the dot as the 50 ms marker? Can you find any pattern in the answers?)

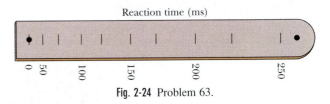

Reaction time (ms)

Fig. 2-24 Problem 63.

64P. A parachutist bails out and freely falls 50 m. Then the parachute opens, and thereafter she decelerates at 2.0 m/s². She reaches the ground with a speed of 3.0 m/s. (a) How long is the parachutist in the air? (b) At what height does the fall begin?

65P. A drowsy cat spots a flowerpot that sails first up and then down past an open window. The pot is in view for a total of 0.50 s, and the top-to-bottom height of the window is 2.00 m. How high above the window top does the flowerpot go?

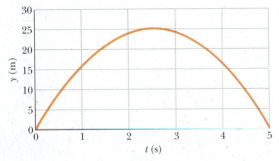

Fig. 2-23 Problem 56.

3 Vectors

For two decades spelunking teams crawled, climbed, and squirmed through 200 km of Mammoth Cave and the Flint Ridge cave system, seeking a connection. The photograph shows Richard Zopf pushing his pack through the Tight Tube, far inside the Flint Ridge system. After 12 hours of "caving" along a labyrinthine route, Zopf and six others waded through a stretch of chilling water and found themselves in Mammoth Cave. Their breakthrough established the Mammoth—Flint cave system as the longest cave in the world.

How does their final point relate to their initial point other than in terms of the actual route they covered?

The answer is in this chapter.

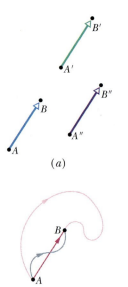

(a)

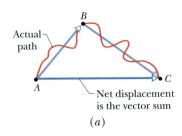

(b)

Fig. 3-1 *(a)* All three arrows have the same magnitude and direction and thus represent the same displacement. *(b)* All three paths connecting the two points correspond to the same displacement vector.

3-1 Vectors and Scalars

A particle moving along a straight line can move in only two directions. We can take its motion to be positive in one of these directions and negative in the other. For a particle moving in three dimensions, however, a plus sign or minus sign is no longer enough to indicate the direction of the motion. Instead, we must use a *vector*.

A **vector** has magnitude as well as direction, and vectors follow certain (vector) rules of combination, which we examine in this chapter. A **vector quantity** is a quantity that has both a magnitude and a direction and thus can be represented with a vector. Some physical quantities that are vector quantities are displacement, velocity, and acceleration. You will see many more throughout this book, so learning the rules of vector combination now will help you greatly in later chapters.

Not all physical quantities involve a direction. Temperature, pressure, energy, mass, and time, for example, do not "point" in the spatial sense. We call such quantities **scalars,** and we deal with them by the rules of ordinary algebra. A single value, with a sign (as in a temperature of $-40°F$), specifies a scalar.

The simplest vector quantity is displacement, or change of position. A vector that represents a displacement is called, reasonably, a **displacement vector.** (Similarly, we have velocity vectors and acceleration vectors.) If a particle changes its position by moving from *A* to *B* in Fig. 3-1*a*, we say that it undergoes a displacement from *A* to *B*, which we represent with an arrow pointing from *A* to *B*. The arrow specifies the vector graphically. To distinguish vector symbols from other kinds of arrows in this book, we use the outline of a triangle as the arrowhead.

In Fig. 3-1*a*, the arrows from *A* to *B*, from *A′* to *B′*, and from *A″* to *B″* have the same magnitude and direction. Thus, they specify identical displacement vectors and represent the same *change of position* for the particle. A vector can be shifted without changing its value, *if* its magnitude (length) and direction are not changed.

The displacement vector tells us nothing about the actual path that the particle takes. In Fig. 3-1*b*, for example, all three paths connecting points *A* and *B* correspond to the same displacement vector, that of Fig. 3-1*a*. Displacement vectors represent only the overall effect of the motion, not the motion itself.

3-2 Adding Vectors Geometrically

Suppose that, as in the vector diagram of Fig. 3-2*a*, a particle moves from *A* to *B* and then later from *B* to *C*. We can represent its overall displacement (no matter what its actual path) with two successive displacement vectors, *AB* and *BC*. The *net displacement* of these two displacements is a single displacement from *A* to *C*. We call *AC* the **vector sum** (or **resultant**) of the vectors *AB* and *BC*. This sum is not the usual algebraic sum.

In Fig. 3-2*b*, we redraw the vectors of Fig. 3-2*a* and relabel them in the way that we shall use from now on, namely, with an arrow over an italic symbol, as in $\vec{a}$. If we want to indicate only the magnitude of the vector (a quantity that lacks a sign or direction), we shall use the italic symbol, as in *a*, *b*, and *s*. (You can use just a handwritten symbol.) A symbol with an overhead arrow always implies both properties of a vector, magnitude and direction.

We can represent the relation among the three vectors in Fig. 3-2*b* with the *vector equation*

$$\vec{s} - \vec{a} + \vec{b}, \qquad (3\text{-}1)$$

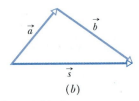

(a)

(b)

Fig. 3-2 *(a) AC* is the vector sum of the vectors *AB* and *BC*. *(b)* The same vectors relabeled.

which says that the vector $\vec{s}$ is the vector sum of vectors $\vec{a}$ and $\vec{b}$. The symbol + in Eq. 3-1 and the words "sum" and "add" have different meanings for vectors than they do in the usual algebra because they involve both magnitude *and* direction.

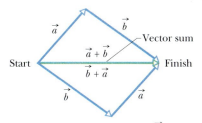

Fig. 3-3 The two vectors $\vec{a}$ and $\vec{b}$ can be added in either order; see Eq. 3-2.

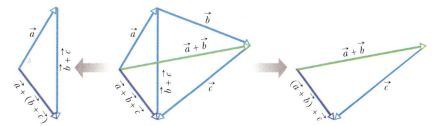

Fig. 3-4 The three vectors $\vec{a}$, $\vec{b}$, and $\vec{c}$ can be grouped in any way as they are added; see Eq. 3-3.

Figure 3-2 suggests a procedure for adding two-dimensional vectors $\vec{a}$ and $\vec{b}$ geometrically. (1) On paper, sketch vector $\vec{a}$ to some convenient scale and at the proper angle. (2) Sketch vector $\vec{b}$ to the same scale, with its tail at the head of vector $\vec{a}$, again at the proper angle. (3) The vector sum $\vec{s}$ is the vector that extends from the tail of $\vec{a}$ to the head of $\vec{b}$.

Vector addition, defined in this way, has two important properties. First, the order of addition does not matter. Adding $\vec{a}$ to $\vec{b}$ gives the same result as adding $\vec{b}$ to $\vec{a}$ (Fig. 3-3); that is,

$$\vec{a} + \vec{b} = \vec{b} + \vec{a} \qquad \text{(commutative law).} \qquad (3\text{-}2)$$

Second, when there are more than two vectors, we can group them in any order as we add them. Thus, if we want to add vectors $\vec{a}$, $\vec{b}$, and $\vec{c}$, we can add $\vec{a}$ and $\vec{b}$ first and then add their vector sum to $\vec{c}$. We can also add $\vec{b}$ and $\vec{c}$ first and then add *that* sum to $\vec{a}$. We get the same result either way, as shown in Fig. 3-4. That is,

$$(\vec{a} + \vec{b}) + \vec{c} = \vec{a} + (\vec{b} + \vec{c}) \qquad \text{(associative law).} \qquad (3\text{-}3)$$

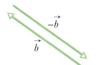

Fig. 3-5 The vectors $\vec{b}$ and $-\vec{b}$ have the same magnitude and opposite directions.

The vector $-\vec{b}$ is a vector with the same magnitude as $\vec{b}$ but the opposite direction (see Fig. 3-5). Adding the two vectors in Fig. 3-5 would yield

$$\vec{b} + (-\vec{b}) = 0.$$

Thus, adding $-\vec{b}$ has the effect of subtracting $\vec{b}$. We use this property to define the difference between two vectors: let $\vec{d} = \vec{a} - \vec{b}$. Then

$$\vec{d} = \vec{a} - \vec{b} = \vec{a} + (-\vec{b}) \qquad \text{(vector subtraction);} \qquad (3\text{-}4)$$

that is, we find the difference vector $\vec{d}$ by adding the vector $-\vec{b}$ to the vector $\vec{a}$. Figure 3-6 shows how this is done geometrically.

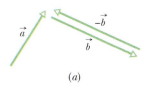

(a)

As in the usual algebra, we can move a term that includes a vector symbol from one side of a vector equation to the other, but we must change its sign. For example, if we are given Eq. 3-4 and need to solve for $\vec{a}$, we can rearrange the equation as

$$\vec{d} + \vec{b} = \vec{a} \quad \text{or} \quad \vec{a} = \vec{d} + \vec{b}.$$

Remember, although we have used displacement vectors here, the rules for addition and subtraction hold for vectors of all kinds, whether they represent velocities, accelerations, or any other vector quantity. However, we can add only vectors of the same kind. For example, we can add two displacements, or two velocities, but adding a displacement and a velocity makes no sense. In the arithmetic of scalars, that would be like trying to add 21 s and 12 m.

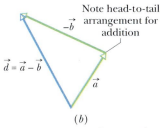

(b)

Fig. 3-6 (a) Vectors $\vec{a}$, $\vec{b}$, and $-\vec{b}$. (b) To subtract vector $\vec{b}$ from vector $\vec{a}$, add vector $-\vec{b}$ to vector $\vec{a}$.

✓**CHECKPOINT 1:** The magnitudes of displacements $\vec{a}$ and $\vec{b}$ are 3 m and 4 m, respectively, and $\vec{c} = \vec{a} + \vec{b}$. Considering various orientations of $\vec{a}$ and $\vec{b}$, what is (a) the maximum possible magnitude for $\vec{c}$ and (b) the minimum possible magnitude?

Sample Problem 3-1

In an orienteering class, you have the goal of moving as far (straight-line distance) from base camp as possible by making three straight-line moves. You may use the following displacements in any order: (a) $\vec{a}$, 2.0 km due east (directly toward the east); (b) $\vec{b}$, 2.0 km 30° north of east (at an angle of 30° toward the north from due east); (c) $\vec{c}$, 1.0 km due west. Alternatively, you may substitute either $-\vec{b}$ for $\vec{b}$ or $-\vec{c}$ for $\vec{c}$. What is the greatest distance you can be from base camp at the end of the third displacement?

SOLUTION: Using a convenient scale, we draw vectors $\vec{a}, \vec{b}, \vec{c}, -\vec{b}$, and $-\vec{c}$ as in Fig. 3-7a. We then mentally slide the vectors over the page, connecting three of them at a time in head-to-tail arrangements to find their vector sum $\vec{d}$. The tail of the first vector represents base camp. The head of the third vector represents the point at which you stop. The vector sum $\vec{d}$ extends from the tail of the first vector to the head of the third vector. Its magnitude d is your distance from base camp.

We find that distance d is greatest for a head-to-tail arrangement of vectors $\vec{a}, \vec{b}$, and $-\vec{c}$. They can be in any order, because their vector sum is the same for any order. The order shown in Fig.

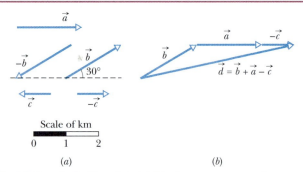

Scale of km

(a)   (b)

Fig. 3-7 Sample Problem 3-1. (a) Displacement vectors; three are to be used. (b) Your distance from base camp is greatest if you undergo displacements $\vec{a}, \vec{b}$, and $-\vec{c}$, in any order. One choice of order is shown; it gives vector sum $\vec{d} = \vec{b} + \vec{a} - \vec{c}$.

3-7b is for the vector sum

$$\vec{d} = \vec{b} + \vec{a} + (-\vec{c}).$$

Using the scale given in Fig. 3-7a, we measure the length d of this vector sum, finding

$$d = 4.8 \text{ m.} \qquad \text{(Answer)}$$

3-3   Components of Vectors

Adding vectors geometrically can be tedious. A neater and easier technique involves algebra but requires that the vectors be placed on a rectangular coordinate system. The x and y axes are usually drawn in the plane of the page, as in Fig. 3-8a. The z axis comes directly out of the page at the origin; we ignore it for now and deal only with two-dimensional vectors.

A **component** of a vector is the projection of the vector on an axis. In Fig. 3-8a, for example, a_x is the component of vector $\vec{a}$ on (or along) the x axis and a_y is the component along the y axis. To find the projection of a vector along an axis, we draw perpendicular lines from the two ends of the vector to the axis, as shown. The projection of a vector on an x axis is its x component, and similarly the projection on the y axis is the y component. The process of finding the components of a vector is called **resolving the vector.**

A component of a vector has the same direction (along an axis) as the vector. In Fig. 3-8, a_x and a_y are both positive because $\vec{a}$ extends in the positive direction of both axes. (Note the small arrowheads on the components, to indicate their direction.) If we were to reverse vector $\vec{a}$, then both components would be negative and their arrowheads would point toward negative x and y. Resolving vector $\vec{b}$ in Fig. 3-9 yields a positive component b_x and a negative component b_y.

In general, a vector has three components, although for the case of Fig. 3-8a the component along the z axis is zero. As Figs. 3-8a and b show, if you shift a vector without changing its direction, its components do not change.

We can find the components of $\vec{a}$ in Fig. 3-8a geometrically from the right triangle there:

$$a_x = a \cos \theta \quad \text{and} \quad a_y = a \sin \theta, \qquad (3\text{-}5)$$

where θ is the angle that the vector $\vec{a}$ makes with the positive direction of the x axis, and a is the magnitude of $\vec{a}$. Figure 3-8c shows that $\vec{a}$ and its x and y components

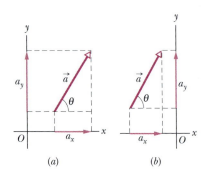

(a)   (b)

(c)

Fig. 3-8 (a) The components a_x and a_y of vector $\vec{a}$. (b) The components are unchanged if the vector is shifted, as long as the magnitude and orientation are maintained. (c) The components form the legs of a right triangle whose hypotenuse is the magnitude of the vector.

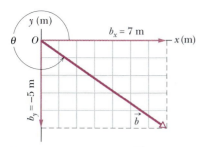

Fig. 3-9 The component of $\vec{b}$ on the x axis is positive, and that on the y axis is negative.

form a right triangle. It also shows how we can reconstruct a vector from its components: we arrange those components *head to tail*. Then we complete a right triangle with the vector forming the hypotenuse, from the tail of one component to the head of the other component.

Once a vector has been resolved into its components along a set of axes, the components themselves can be used in place of the vector. For example, $\vec{a}$ in Fig. 3-8*a* is given (completely determined) by a and θ. It can also be given by its components a_x and a_y. Both pairs of values contain the same information. If we know a vector in *component notation* (a_x and a_y) and want it in *magnitude-angle notation* (a and θ), we can use the equations

$$a = \sqrt{a_x^2 + a_y^2} \quad \text{and} \quad \tan \theta = \frac{a_y}{a_x} \tag{3-6}$$

to transform it.

In the more general three-dimensional case, we need a magnitude and two angles (say, a, θ, and ϕ) or three components (a_x, a_y, and a_z) to specify a vector.

✔**CHECKPOINT 2:** In the figure, which of the indicated methods for combining the x and y components of vector $\vec{a}$ are proper to determine that vector?

Sample Problem 3-2

A small airplane leaves an airport on an overcast day and is later sighted 215 km away, in a direction making an angle of 22° east of north. How far east and north is the airplane from the airport when sighted?

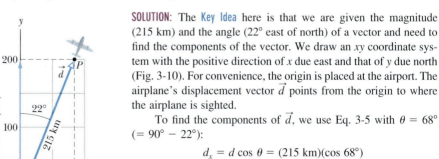

Fig. 3-10 Sample Problem 3-2. A plane takes off from an airport at the origin and is later sighted at P.

SOLUTION: The **Key Idea** here is that we are given the magnitude (215 km) and the angle (22° east of north) of a vector and need to find the components of the vector. We draw an xy coordinate system with the positive direction of x due east and that of y due north (Fig. 3-10). For convenience, the origin is placed at the airport. The airplane's displacement vector $\vec{d}$ points from the origin to where the airplane is sighted.

To find the components of $\vec{d}$, we use Eq. 3-5 with $\theta = 68°$ ($= 90° - 22°$):

$$d_x = d \cos \theta = (215 \text{ km})(\cos 68°)$$
$$= 81 \text{ km} \quad \text{(Answer)}$$
$$d_y = d \sin \theta = (215 \text{ km})(\sin 68°)$$
$$= 199 \text{ km}. \quad \text{(Answer)}$$

Thus, the airplane is 81 km east and 199 km north of the airport.

Sample Problem 3-3

The 1972 team that connected the Mammoth–Flint cave system went from Austin Entrance in the Flint Ridge system to Echo River in Mammoth Cave (Fig. 3-11a), traveling a net 2.6 km westward, 3.9 km southward, and 25 m upward. What was their displacement vector from start to finish?

SOLUTION: The Key Idea here is that we have the components of a three-dimensional vector, and we need to find the vector's magnitude and two angles to specify the vector's direction. We first draw the components as in Fig. 3-11b. The horizontal components (2.6 km west and 3.9 km south) form the legs of a horizontal right triangle. The team's horizontal displacement forms the hypotenuse of the triangle, and its magnitude d_h is given by the Pythagorean theorem:

$$d_h = \sqrt{(2.6 \text{ km})^2 + (3.9 \text{ km})^2} = 4.69 \text{ km}.$$

Also from the horizontal triangle in Fig. 3-11b, we see that this horizontal displacement is directed south of due west by an angle θ_h given by

$$\tan \theta_h = \frac{3.9 \text{ km}}{2.6 \text{ km}},$$

so

$$\theta_h = \tan^{-1} \frac{3.9 \text{ km}}{2.6 \text{ km}} = 56°, \qquad \text{(Answer)}$$

which is one of the two angles we need to specify the direction of the overall displacement.

To include the vertical component (25 m = 0.025 km), we now take a side view of Fig. 3-11b, looking northwest. We get Fig. 3-11c, where the vertical component and the horizontal displacement d_h form the legs of another right triangle. Now the team's overall displacement forms the hypotenuse of that triangle, with a magnitude d given by

$$d = \sqrt{(4.69 \text{ km})^2 + (0.025 \text{ km})^2} = 4.69 \text{ km}$$

$$\approx 4.7 \text{ km}. \qquad \text{(Answer)}$$

This displacement is directed upward from the horizontal displacement by the angle

$$\theta_v = \tan^{-1} \frac{0.025 \text{ km}}{4.69 \text{ km}} = 0.3°. \qquad \text{(Answer)}$$

Thus, the team's displacement vector had a magnitude of 4.7 km and was at an angle of 56° south of west and at an angle of 0.3° upward. The net vertical motion was, of course, insignificant compared to the horizontal motion. However, that fact would have been no comfort to the team, which had to climb up and down countless times to get through the cave. The route they actually covered was quite different from the displacement vector, which merely points in a straight line from start to finish.

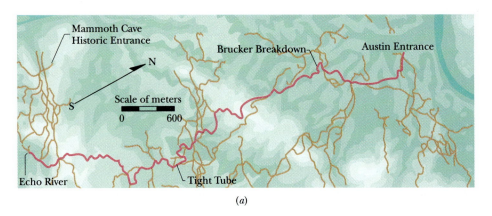

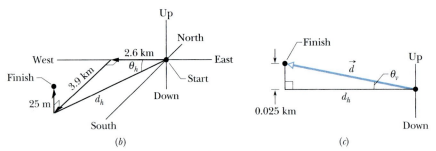

Fig. 3-11 Sample Problem 3-3. (a) Part of the Mammoth–Flint cave system, with the spelunking team's route from Austin Entrance to Echo River indicated in red. (b) The components of the team's overall displacement and their horizontal displacement d_h. (c) A side view showing d_h and the team's overall displacement vector $\vec{d}$. (Adapted from map by Cave Research Foundation.)

PROBLEM-SOLVING TACTICS

Tactic 1: *Angles—Degrees and Radians*
Angles that are measured relative to the positive direction of the x axis are positive if they are measured in the counterclockwise direction, and negative if measured clockwise. For example, $210°$ and $-150°$ are the same angle.

Angles may be measured in degrees or radians (rad). You can relate the two measures by remembering that one full circle is equivalent to $360°$ and to 2π rad. If you needed to convert, say, $40°$ to radians, you would write

$$40° \frac{2\pi \text{ rad}}{360°} = 0.70 \text{ rad}.$$

Tactic 2: *Trig Functions*
You need to know the definitions of the common trigonometric functions—sine, cosine, and tangent—because they are part of the language of science and engineering. They are given in Fig. 3-12 in a form that does not depend on how the triangle is labeled.

You should also be able to sketch how the trig functions vary with angle, as in Fig. 3-13, in order to be able to judge whether a calculator result is reasonable. Even knowing the signs of the functions in the various quadrants can be of help.

Tactic 3: *Inverse Trig Functions*
When the inverse trig functions $\sin^{-1}$, $\cos^{-1}$, and $\tan^{-1}$ are taken on a calculator, you must consider the reasonableness of the answer you get, because there is usually another possible answer that the calculator does not give. The range of operation for a calculator in taking each inverse trig function is indicated in Fig. 3-13. As an example, $\sin^{-1} 0.5$ has associated angles of $30°$ (which is displayed by the calculator, since $30°$ falls within its range of operation) and $150°$. To see both values, draw a horizontal line through 0.5 in Fig. 3-13a and note where it cuts the sine curve.

How do you distinguish a correct answer? It is the one that seems more reasonable for the given situation. As an example, reconsider the calculation of θ_h in Sample Problem 3-3, where $\tan \theta_h = 3.9/2.6 = 1.5$. Taking $\tan^{-1} 1.5$ on your calculator tells you that $\theta_h = 56°$, but $\theta_h = 236° (= 180° + 56°)$ also has a tangent

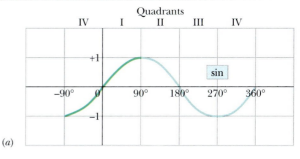

(a)

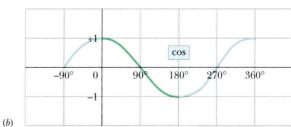

(b)

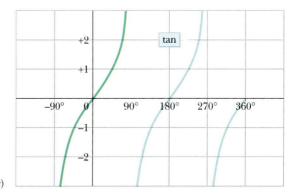

(c)

Fig. 3-13 Three useful curves to remember. A calculator's range of operation for taking *inverse* trig functions is indicated by the darker portions of the colored curves.

of 1.5. Which is correct? From the physical situation (Fig. 3-11b), $56°$ is reasonable and $236°$ is clearly not.

Tactic 4: *Measuring Vector Angles*
The equations for $\cos \theta$ and $\sin \theta$ in Eq. 3-5 and the equation for $\tan \theta$ in Eq. 3-6 are valid only if the angle is measured relative to the positive direction of the x axis. If it is measured relative to some other direction, then the trig functions in Eq. 3-5 may have to be interchanged, and the ratio in Eq. 3-6 may have to be inverted. A safer method is to convert the given angle into one that is measured from the positive direction of the x axis.

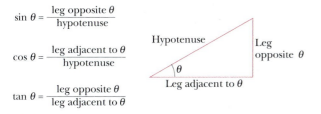

$$\sin \theta = \frac{\text{leg opposite } \theta}{\text{hypotenuse}}$$

$$\cos \theta = \frac{\text{leg adjacent to } \theta}{\text{hypotenuse}}$$

$$\tan \theta = \frac{\text{leg opposite } \theta}{\text{leg adjacent to } \theta}$$

Fig. 3-12 A triangle used to define the trigonometric functions. See also Appendix E.

3-4 Unit Vectors

A **unit vector** is a vector that has a magnitude of exactly 1 and points in a particular direction. It lacks both dimension and unit. Its sole purpose is to point—that is, to specify a direction. The unit vectors in the positive directions of the x, y, and z axes are labeled $\hat{i}$, $\hat{j}$, and $\hat{k}$, where the hat ˆ is used instead of an overhead arrow as for

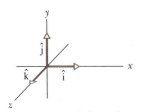

Fig. 3-14 Unit vectors $\hat{i}$, $\hat{j}$, and $\hat{k}$ define the directions of a right-handed coordinate system.

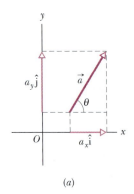

(a)

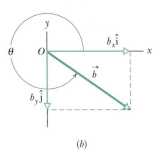

(b)

Fig. 3-15 (a) The vector components of vector $\vec{a}$. (b) The vector components of vector $\vec{b}$.

other vectors (Fig. 3-14). The arrangement of axes in Fig. 3-14 is said to be a **right-handed coordinate system.** The system remains right-handed if it is rotated rigidly to a new orientation. We use such coordinate systems exclusively in this book.

Unit vectors are very useful for expressing other vectors; for example, we can express $\vec{a}$ and $\vec{b}$ of Figs. 3-8 and 3-9 as

$$\vec{a} = a_x\hat{i} + a_y\hat{j} \tag{3-7}$$

and

$$\vec{b} = b_x\hat{i} + b_y\hat{j}. \tag{3-8}$$

These two equations are illustrated in Fig. 3-15. The quantities $a_x\hat{i}$ and $a_y\hat{j}$ are vectors and are called the **vector components** of $\vec{a}$. The quantities a_x and a_y are scalars and are called the **scalar components** of $\vec{a}$ (or, as before, simply its **components**).

As an example, let us write the displacement $\vec{d}$ of the spelunking team of Sample Problem 3-3 in terms of unit vectors. First, superimpose the coordinate system of Fig. 3-14 on the one shown in Fig. 3-11b. Then the direction of $\hat{i}$, $\hat{j}$, and $\hat{k}$ are toward the east, up, and toward the south, respectively. Thus, displacement $\vec{d}$ from start to finish is neatly expressed in unit-vector notation as

$$\vec{d} = -(2.6 \text{ km})\hat{i} + (0.025 \text{ km})\hat{j} + (3.9 \text{ km})\hat{k}. \tag{3-9}$$

Here $-(2.6 \text{ km})\hat{i}$ is the vector component $d_x\hat{i}$ along the x axis, and $-(2.6 \text{ km})$ is the x component d_x.

3-5   Adding Vectors by Components

Using a sketch, we can add vectors geometrically. On a vector-capable calculator, we can add them directly on the screen. A third way to add vectors is to combine their components, axis by axis.

To start, consider the statement

$$\vec{r} = \vec{a} + \vec{b}, \tag{3-10}$$

which says that the vector $\vec{r}$ is the same as the vector $(\vec{a} + \vec{b})$. If that is so, then each component of $\vec{r}$ must be the same as the corresponding component of $(\vec{a} + \vec{b})$:

$$r_x = a_x + b_x \tag{3-11}$$
$$r_y = a_y + b_y \tag{3-12}$$
$$r_z = a_z + b_z. \tag{3-13}$$

In other words, two vectors must be equal if their corresponding components are equal. Equations 3-10 to 3-13 tell us that to add vectors $\vec{a}$ and $\vec{b}$, we must (1) resolve the vectors into their scalar components; (2) combine these scalar components, axis by axis, to get the components of the sum $\vec{r}$; and (3) combine the components of $\vec{r}$ to get $\vec{r}$ itself. We have a choice in step 3. We can express $\vec{r}$ in unit-vector notation (as in Eq. 3-9) or in magnitude-angle notation (as in the answer to Sample Problem 3-3).

This procedure for adding vectors by components also applies to vector subtractions. Recall that a subtraction such as $\vec{d} = \vec{a} - \vec{b}$ can be rewritten as an addition $\vec{d} = \vec{a} + (-\vec{b})$. To subtract we simply add $\vec{a}$ and $-\vec{b}$ by components, to get

$$d_x = a_x - b_x, \quad d_y = a_y - b_y, \quad \text{and} \quad d_z = a_z - b_z,$$

where

$$\vec{d} = d_x\hat{i} + d_y\hat{j} + d_z\hat{k}.$$

CHECKPOINT 3:   (a) In the figure here, what are the signs of the x components of $\vec{d_1}$ and $\vec{d_2}$? (b) What are the signs of the y components of $\vec{d_1}$ and $\vec{d_2}$? (c) What are the signs of the x and y components of $\vec{d_1} + \vec{d_2}$?

Sample Problem 3-4

Figure 3-16a shows the following three vectors:

$$\vec{a} = (4.2 \text{ m})\hat{i} - (1.5 \text{ m})\hat{j},$$

$$\vec{b} = (-1.6 \text{ m})\hat{i} + (2.9 \text{ m})\hat{j},$$

and $\vec{c} = (-3.7 \text{ m})\hat{j}.$

What is their vector sum $\vec{r}$, which is also shown?

SOLUTION: The Key Idea here is that we can add the three vectors by components, axis by axis. For the x axis, we add the x components of $\vec{a}$, $\vec{b}$, and $\vec{c}$ to get the x component of $\vec{r}$:

$$r_x = a_x + b_x + c_x$$
$$= 4.2 \text{ m} - 1.6 \text{ m} + 0 = 2.6 \text{ m}.$$

Similarly, for the y axis,

$$r_y = a_y + b_y + c_y$$
$$= -1.5 \text{ m} + 2.9 \text{ m} - 3.7 \text{ m} = -2.3 \text{ m}.$$

Another Key Idea is that we can combine these components of $\vec{r}$ to write the vector in unit-vector notation:

$$\vec{r} = (2.6 \text{ m})\hat{i} - (2.3 \text{ m})\hat{j}, \qquad \text{(Answer)}$$

where $(2.6 \text{ m})\hat{i}$ is the vector component of $\vec{r}$ along the x axis and $-(2.3 \text{ m})\hat{j}$ is that along the y axis. Figure 3-16b shows one way to arrange these vector components to form $\vec{r}$. (Can you sketch the other way?)

A third Key Idea is that we can also answer the question by giving the magnitude and an angle for $\vec{r}$. From Eq. 3-6, the magnitude is

$$r = \sqrt{(2.6 \text{ m})^2 + (-2.3 \text{ m})^2} \approx 3.5 \text{ m} \qquad \text{(Answer)}$$

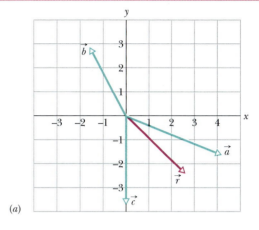

(a)

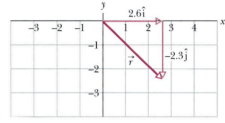

(b)

Fig. 3-16 Sample Problem 3-4. Vector $\vec{r}$ is the vector sum of the other three vectors.

and the angle (measured from the positive direction of x) is

$$\theta = \tan^{-1}\left(\frac{-2.3 \text{ m}}{2.6 \text{ m}}\right) = -41°, \qquad \text{(Answer)}$$

where the minus sign means that the angle is measured clockwise.

Sample Problem 3-5

Figure 3-17 gives an incomplete map of a road rally. From the starting point (at the origin), you must use available roads to go through the following displacements:

(1) $\vec{a}$ to checkpoint Able, magnitude 36 km, due east,
(2) $\vec{b}$ to checkpoint Baker, due north,
(3) $\vec{c}$ to checkpoint Charlie, magnitude 25 km, at the angle shown.

Your net displacement $\vec{d}$ from the starting point is 62.0 km. What is the magnitude b of $\vec{b}$?

SOLUTION: The Key Idea here is that the net displacement $\vec{d}$ is the vector sum of the three individual displacements, so we can write

$$\vec{d} = \vec{a} + \vec{b} + \vec{c},$$

which gives us

$$\vec{b} = \vec{d} - \vec{a} - \vec{c}. \qquad \text{(3-14)}$$

Although we know both magnitude and direction for $\vec{a}$ and $\vec{c}$, we do not know both for $\vec{d}$, so we cannot directly solve for $\vec{b}$

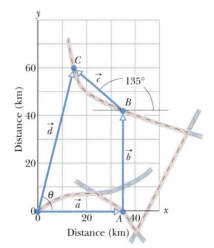

Fig. 3-17 Sample Problem 3-5. A rally route, showing the origin, checkpoints Able (A), Baker (B), and Charlie (C), and the roads.

on a vector-capable calculator. However, we can write Eq. 3-14 in terms of components along both the x axis and the y axis. Since $\vec{b}$ points parallel to the y axis, choosing that axis may give us the magnitude of $\vec{b}$. We thus write

$$b_y = d_y - a_y - c_y. \tag{3-15}$$

Following Eq. 3-5, inserting known data, and realizing that $b = b_y$, we have

$$b = (62 \text{ km}) \sin \theta - 0 - (25 \text{ km}) \sin 135°. \tag{3-16}$$

Unfortunately, we do not know θ. To find it, we write Eq. 3-14 for components along the x axis:

$$b_x = d_x - a_x - c_x, \tag{3-17}$$

which gives us

$$0 = (62 \text{ km}) \cos \theta - 36 \text{ km} - (25 \text{ km}) \cos 135°$$

and

$$\theta = \cos^{-1} \frac{36 + (25)(\cos 135°)}{62} = 72.81°.$$

Inserting this into Eq. 3-16, we find

$$b \approx 42 \text{ km}. \tag{Answer}$$

3-6   Vectors and the Laws of Physics

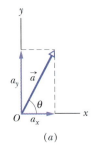

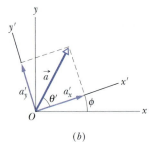

Fig. 3-18 (a) The vector $\vec{a}$ and its components. (b) The same vector, with the axes of the coordinate system rotated through an angle ϕ.

So far, in every figure that includes a coordinate system, the x and y axes are parallel to the edges of the book page. Thus, when a vector $\vec{a}$ is included, its components a_x and a_y are also parallel to the edges (as in Fig. 3-18a). The only reason for that orientation of the axes is that it looks "proper": there is no deeper reason. We could, instead, rotate the axes (but not the vector $\vec{a}$) through an angle ϕ as in Fig. 3-18b, in which case the components would have new values, call them a'_x and a'_y. Since there are an infinite number of choices of ϕ, there are an infinite number of different pairs of components for $\vec{a}$.

Which then is the "right" pair of components? The answer is that they are all equally valid because each pair (with its axes) just gives us a different way of describing the same vector $\vec{a}$; all produce the same magnitude and direction for the vector. In Fig. 3-18 we have

$$a = \sqrt{a_x^2 + a_y^2} = \sqrt{a_x'^2 + a_y'^2} \tag{3-18}$$

and

$$\theta = \theta' + \phi. \tag{3-19}$$

The point is that we have great freedom in choosing a coordinate system, because the relations among vectors (including, for example, the vector addition of Eq. 3-1) do not depend on the location of the origin of the coordinate system or on the orientation of the axes. This is also true of the relations of physics; they are all independent of the choice of coordinate system. Add to that the simplicity and richness of the language of vectors and you can see why the laws of physics are almost always presented in that language: one equation, like Eq. 3-10, can represent three (or even more) relations, like Eqs. 3-11, 3-12, and 3-13.

3-7   Multiplying Vectors*

There are three ways in which vectors can be multiplied, but none is exactly like the usual algebraic multiplication. As you read this section, keep in mind that a vector-capable calculator will help you multiply vectors only if you understand the basic rules of that multiplication.

Multiplying a Vector by a Scalar

If we multiply a vector $\vec{a}$ by a scalar s, we get a new vector. Its magnitude is the product of the magnitude of $\vec{a}$ and the absolute value of s. Its direction is the direction

---

* This material will not be employed until later (Chapter 7 for scalar products and Chapter 12 for vector products), and so your instructor may wish to postpone assignment of the section.

of $\vec{a}$ if s is positive, but the opposite direction if s is negative. To divide $\vec{a}$ by s, we multiply $\vec{a}$ by $1/s$.

Multiplying a Vector by a Vector

There are two ways to multiply a vector by a vector: one way produces a scalar (called the *scalar product*), and the other produces a new vector (called the *vector product*). Students commonly confuse the two ways, and so starting now, you should carefully distinguish between them.

The Scalar Product

The **scalar product** of the vectors $\vec{a}$ and $\vec{b}$ in Fig. 3-19a is written as $\vec{a} \cdot \vec{b}$ and defined to be

$$\vec{a} \cdot \vec{b} = ab \cos \phi, \tag{3-20}$$

where a is the magnitude of $\vec{a}$, b is the magnitude of $\vec{b}$, and ϕ is the angle between $\vec{a}$ and $\vec{b}$ (or, more properly, between the directions of $\vec{a}$ and $\vec{b}$). There are actually two such angles: ϕ and $360° - \phi$. Either can be used in Eq. 3-20, because their cosines are the same.

Note that there are only scalars on the right side of Eq. 3-20 (including the value of $\cos \phi$). Thus $\vec{a} \cdot \vec{b}$ on the left side represents a *scalar* quantity. Because of the notation, $\vec{a} \cdot \vec{b}$ is also known as the **dot product** and is spoken as "a dot b."

A dot product can be regarded as the product of two quantities: (1) the magnitude of one of the vectors and (2) the scalar component of the second vector along the direction of the first vector. For example, in Fig. 3-19b, $\vec{a}$ has a scalar component $a \cos \phi$ along the direction of $\vec{b}$; note that a perpendicular dropped from the head of $\vec{a}$ to $\vec{b}$ determines that component. Similarly, $\vec{b}$ has a scalar component $b \cos \phi$ along the direction of $\vec{a}$.

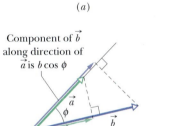

Component of $\vec{b}$
along direction of
$\vec{a}$ is $b \cos \phi$

Component of $\vec{a}$
along direction of
$\vec{b}$ is $a \cos \phi$

(b)

Fig. 3-19 (a) Two vectors $\vec{a}$ and $\vec{b}$, with an angle ϕ between them. (b) Each vector has a component along the direction of the other vector.

> If the angle ϕ between two vectors is 0°, the component of one vector along the other is maximum, and so also is the dot product of the vectors. If, instead, ϕ is 90°, the component of one vector along the other is zero, and so is the dot product.

Equation 3-20 can be rewritten as follows to emphasize the components:

$$\vec{a} \cdot \vec{b} = (a \cos \phi)(b) = (a)(b \cos \phi). \tag{3-21}$$

The commutative law applies to a scalar product, so we can write

$$\vec{a} \cdot \vec{b} = \vec{b} \cdot \vec{a}.$$

When two vectors are in unit-vector notation, we write their dot product as

$$\vec{a} \cdot \vec{b} = (a_x \hat{i} + a_y \hat{j} + a_z \hat{k}) \cdot (b_x \hat{i} + b_y \hat{j} + b_z \hat{k}), \tag{3-22}$$

which we can expand according to the distributive law: Each vector component of the first vector is to be dotted with each vector component of the second vector. By doing so, we can show that

$$\vec{a} \cdot \vec{b} = a_x b_x + a_y b_y + a_z b_z. \tag{3-23}$$

✓**CHECKPOINT 4:** Vectors $\vec{C}$ and $\vec{D}$ have magnitudes of 3 units and 4 units, respectively. What is the angle between the directions of $\vec{C}$ and $\vec{D}$ if $\vec{C} \cdot \vec{D}$ equals (a) zero, (b) 12 units, and (c) -12 units?

Sample Problem 3-6

What is the angle ϕ between $\vec{a} = 3.0\hat{i} - 4.0\hat{j}$ and $\vec{b} = -2.0\hat{i} + 3.0\hat{k}$?

SOLUTION: First, a caution: Although many of the following steps can be bypassed with a vector-capable calculator, you will learn more about scalar products if, at least here, you use these steps.

One **Key Idea** here is that the angle between the directions of two vectors is included in the definition of their scalar product (Eq. 3-20):

$$\vec{a} \cdot \vec{b} = ab \cos \phi. \qquad (3\text{-}24)$$

In this equation, a is the magnitude of $\vec{a}$, or

$$a = \sqrt{3.0^2 + (-4.0)^2} = 5.00, \qquad (3\text{-}25)$$

and b is the magnitude of $\vec{b}$, or

$$b = \sqrt{(-2.0)^2 + 3.0^2} = 3.61. \qquad (3\text{-}26)$$

A second **Key Idea** is that we can separately evaluate the left side of

Eq. 3-24 by writing the vectors in unit-vector notation and using the distributive law:

$$\vec{a} \cdot \vec{b} = (3.0\hat{i} - 4.0\hat{j}) \cdot (-2.0\hat{i} + 3.0\hat{k})$$
$$= (3.0\hat{i}) \cdot (-2.0\hat{i}) + (3.0\hat{i}) \cdot (3.0\hat{k})$$
$$+ (-4.0\hat{j}) \cdot (-2.0\hat{i}) + (-4.0\hat{j}) \cdot (3.0\hat{k}).$$

We next apply Eq. 3-20 to each term in this last expression. The angle between the vectors in the first term ($3.0\hat{i}$ and $-2.0\hat{i}$) is $0°$, and in the other terms it is $90°$. We then have

$$\vec{a} \cdot \vec{b} = -(6.0)(1) + (9.0)(0) + (8.0)(0) - (12)(0) = -6.0.$$

Substituting this and the results of Eqs. 3-25 and 3-26 into Eq. 3-24 yields

$$-6.0 = (5.00)(3.61) \cos \phi,$$

so

$$\phi = \cos^{-1} \frac{-6.0}{(5.00)(3.61)} = 109° \approx 110°. \qquad \text{(Answer)}$$

The Vector Product

The **vector product** of $\vec{a}$ and $\vec{b}$, written $\vec{a} \times \vec{b}$, produces a third vector $\vec{c}$ whose magnitude is

$$c = ab \sin \phi, \qquad (3\text{-}27)$$

where ϕ is the *smaller* of the two angles between $\vec{a}$ and $\vec{b}$. (You must use the smaller of the two angles between the vectors because $\sin \phi$ and $\sin(360° - \phi)$ differ in algebraic sign.) Because of the notation, $\vec{a} \times \vec{b}$ is also known as the **cross product,** and in speech it is "a cross b."

▶ If $\vec{a}$ and $\vec{b}$ are parallel or antiparallel, $\vec{a} \times \vec{b} = 0$. The magnitude of $\vec{a} \times \vec{b}$, which can be written as $|\vec{a} \times \vec{b}|$, is maximum when $\vec{a}$ and $\vec{b}$ are perpendicular to each other.

The direction of $\vec{c}$ is perpendicular to the plane that contains $\vec{a}$ and $\vec{b}$. Figure 3-20a shows how to determine the direction of $\vec{c} = \vec{a} \times \vec{b}$ with what is known as the **right-hand rule.** Place the vectors $\vec{a}$ and $\vec{b}$ tail to tail without altering their orientations, and imagine a line that is perpendicular to their plane where they meet. Pretend to place your *right* hand around that line in such a way that your fingers would sweep $\vec{a}$ into $\vec{b}$ through the smaller angle between them. Your outstretched thumb points in the direction of $\vec{c}$.

The order of the vector multiplication is important. In Fig. 3-20b, we are determining the direction of $\vec{c}' = \vec{b} \times \vec{a}$, so the fingers are placed to sweep $\vec{b}$ into $\vec{a}$ through the smaller angle. The thumb ends up in the opposite direction from previously, and so it must be that $\vec{c}' = -\vec{c}$; that is,

$$\vec{b} \times \vec{a} = -(\vec{a} \times \vec{b}). \qquad (3\text{-}28)$$

In other words, the commutative law does not apply to a vector product.

In unit-vector notation, we write

$$\vec{a} \times \vec{b} = (a_x\hat{i} + a_y\hat{j} + a_z\hat{k}) \times (b_x\hat{i} + b_y\hat{j} + b_z\hat{k}), \qquad (3\text{-}29)$$

which can be expanded according to the distributive law; that is, each component of the first vector is to be crossed with each component of the second vector. The cross products of unit vectors are given in Appendix E (see "Products of Vectors").

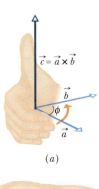

(a)

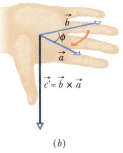

(b)

Fig. 3-20 Illustration of the right-hand rule for vector products. (a) Sweep vector $\vec{a}$ into vector $\vec{b}$ with the fingers of your right hand. Your outstretched thumb shows the direction of vector $\vec{c} = \vec{a} \times \vec{b}$. (b) Showing that $\vec{a} \times \vec{b}$ is the reverse of $\vec{b} \times \vec{a}$.

For example, in the expansion of Eq. 3-29, we have

$$a_x\hat{i} \times b_x\hat{i} = a_xb_x(\hat{i} \times \hat{i}) = 0,$$

because the two unit vectors $\hat{i}$ and $\hat{i}$ are parallel and thus have a zero cross product. Similarly, we have

$$a_x\hat{i} \times b_y\hat{j} = a_xb_y(\hat{i} \times \hat{j}) = a_xb_y\hat{k}.$$

In the last step we used Eq. 3-27 to evaluate the magnitude of $\hat{i} \times \hat{j}$ as unity. (The vectors $\hat{i}$ and $\hat{j}$ each have a magnitude of unity, and the angle between them is 90°.) Also, we used the right-hand rule to get the direction of $\hat{i} \times \hat{j}$ as being in the positive direction of the z axis (thus in the direction of $\hat{k}$).

Continuing to expand Eq. 3-29, you can show that

$$\vec{a} \times \vec{b} = (a_yb_z - b_ya_z)\hat{i} + (a_zb_x - b_za_x)\hat{j} + (a_xb_y - b_xa_y)\hat{k}. \quad (3\text{-}30)$$

You can also evaluate a cross product by setting up and evaluating a determinant (as shown in Appendix E) or by using a vector-capable calculator.

To check whether any xyz coordinate system is a right-handed coordinate system, use the right-hand rule for the cross product $\hat{i} \times \hat{j} = \hat{k}$ with that system. If your fingers sweep $\hat{i}$ (positive direction of x) into $\hat{j}$ (positive direction of y) with the outstretched thumb pointing in the positive direction of z, then the system is right-handed.

✓**CHECKPOINT 5:** Vectors $\vec{C}$ and $\vec{D}$ have magnitudes of 3 units and 4 units, respectively. What is the angle between the directions of $\vec{C}$ and $\vec{D}$ if the magnitude of the vector product $\vec{C} \times \vec{D}$ is (a) zero and (b) 12 units?

Sample Problem 3-7

In Fig. 3-21, vector $\vec{a}$ lies in the xy plane, has a magnitude of 18 units, and points in a direction 250° from the positive direction of x. Also, vector $\vec{b}$ has a magnitude of 12 units and points along the positive direction of z. What is the vector product $\vec{c} = \vec{a} \times \vec{b}$?

SOLUTION: One **Key Idea** is that when we have two vectors in magnitude-angle notation, we find the magnitude of their cross product (that is, the vector that results from taking their cross product) with Eq. 3-27. Here that means the magnitude of $\vec{c}$ is

$$c = ab \sin\phi = (18)(12)(\sin 90°) = 216. \quad \text{(Answer)}$$

A second **Key Idea** is that with two vectors in magnitude-angle notation, we find the direction of their cross product with the right-hand rule of Fig. 3-20. In Fig. 3-21, imagine placing the fingers of your right hand around a line perpendicular to the plane of $\vec{a}$ and $\vec{b}$ (the line on which $\vec{c}$ is shown) such that your fingers sweep $\vec{a}$

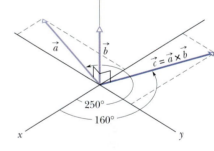

Fig. 3-21 Sample Problem 3-7. Vector $\vec{c}$ (in the xy plane) is the vector (or cross) product of vectors $\vec{a}$ and $\vec{b}$.

into $\vec{b}$. Your outstretched thumb then gives the direction of $\vec{c}$. Thus, as shown in Fig. 3-21, $\vec{c}$ lies in the xy plane. Because its direction is perpendicular to the direction of $\vec{a}$, it is at an angle of

$$250° - 90° = 160° \quad \text{(Answer)}$$

from the positive direction of x.

Sample Problem 3-8

If $\vec{a} = 3\hat{i} - 4\hat{j}$ and $\vec{b} = -2\hat{i} + 3\hat{k}$, what is $\vec{c} = \vec{a} \times \vec{b}$?

SOLUTION: The **Key Idea** is that when two vectors are in unit-vector notation, we can find their cross product by using the distributive law. Here that means we can write

$$\vec{c} = (3\hat{i} - 4\hat{j}) \times (-2\hat{i} + 3\hat{k})$$
$$= 3\hat{i} \times (-2\hat{i}) + 3\hat{i} \times 3\hat{k} + (-4\hat{j}) \times (-2\hat{i}) + (-4\hat{j}) \times 3\hat{k}.$$

We next evaluate each term with Eq. 3-27, determining the direc-

tion with the right-hand rule. For the first term here, the angle ϕ between the two vectors being crossed is 0. For the other terms, ϕ is 90°. We find

$$\vec{c} = 6(0) + 9(-\hat{j}) + 8(-\hat{k}) - 12\hat{i}$$
$$= -12\hat{i} - 9\hat{j} - 8\hat{k}. \quad \text{(Answer)}$$

This vector $\vec{c}$ is perpendicular to both $\vec{a}$ and $\vec{b}$, a fact you can check by showing that $\vec{c} \cdot \vec{a} = 0$ and $\vec{c} \cdot \vec{b} = 0$; that is, there is no component of $\vec{c}$ along the direction of either $\vec{a}$ or $\vec{b}$.

Tactic 5: *Common Errors with Cross Products*
Several errors are common in finding a cross product. (1) Failure to arrange vectors tail to tail is tempting when an illustration presents them head to tail: you must mentally shift (or better, redraw) one vector to the proper arrangement without changing its orientation. (2) Failing to use the right hand in applying the right-hand rule is easy when the right hand is occupied with a calculator or

pencil. (3) Failure to sweep the first vector of the product into the second vector can occur when the orientations of the vectors require an awkward twisting of your hand to apply the right-hand rule. Sometimes that happens when you try to make the sweep mentally rather than actually using your hand. (4) Failure to work with a right-handed coordinate system results when you forget how to draw such a system (see Fig. 3-14).

REVIEW & SUMMARY

Scalars and Vectors   *Scalars,* such as temperature, have magnitude only. They are specified by a number with a unit (10°C) and obey the rules of arithmetic and ordinary algebra. *Vectors,* such as displacement, have both magnitude and direction (5 m, north) and obey the special rules of vector algebra.

Adding Vectors Geometrically   Two vectors $\vec{a}$ and $\vec{b}$ may be added geometrically by drawing them to a common scale and placing them head to tail. The vector connecting the tail of the first to the head of the second is the vector sum $\vec{s}$. To subtract $\vec{b}$ from $\vec{a}$, reverse the direction of $\vec{b}$ to get $-\vec{b}$; then add $-\vec{b}$ to $\vec{a}$. Vector addition is commutative and obeys the associative law.

Components of a Vector   The (scalar) *components* a_x and a_y of any two-dimensional vector $\vec{a}$ along the coordinate axes are found by dropping perpendicular lines from the ends of $\vec{a}$ onto the coordinate axes. The components are given by

$$a_x = a \cos \theta \quad \text{and} \quad a_y = a \sin \theta, \quad (3\text{-}5)$$

where θ is the angle between the positive direction of the x axis and the direction of $\vec{a}$. The algebraic sign of a component indicates its direction along the associated axis. Given its components, we can find the magnitude and orientation of the vector $\vec{a}$ with

$$a = \sqrt{a_x^2 + a_y^2} \quad \text{and} \quad \tan \theta = \frac{a_y}{a_x}. \quad (3\text{-}6)$$

Unit-Vector Notation   *Unit vectors* $\hat{i}, \hat{j},$ and $\hat{k}$ have magnitudes of unity and are directed in the positive directions of the x, y, and z axes, respectively, in a right-handed coordinate system. We can write a vector $\vec{a}$ in terms of unit vectors as

$$\vec{a} = a_x\hat{i} + a_y\hat{j} + a_z\hat{k}, \quad (3\text{-}7)$$

in which $a_x\hat{i}$, $a_y\hat{j}$, and $a_z\hat{k}$ are the **vector components** of $\vec{a}$ and a_x, a_y, and a_z are its **scalar components.**

Adding Vectors in Component Form   To add vectors in component form, we use the rules

$$r_x = a_x + b_x \quad r_y = a_y + b_y \quad r_z = a_z + b_z. \quad (3\text{-}11 \text{ to } 3\text{-}13)$$

Here $\vec{a}$ and $\vec{b}$ are the vectors to be added, and $\vec{r}$ is the vector sum.

Vectors and Physical Laws   Any physical situation involving vectors can be described using many possible coordinate systems. We usually choose the one that most simplifies the situation. However, the relationship between the vector quantities does not depend on our choice of coordinates. The laws of physics are also independent of that choice.

Product of a Scalar and a Vector   The product of a scalar s and a vector $\vec{v}$ is a new vector whose magnitude is sv and whose direction is the same as that of $\vec{v}$ if s is positive, and opposite that of $\vec{v}$ if s is negative. To divide $\vec{v}$ by s, multiply $\vec{v}$ by $(1/s)$.

The Scalar Product   The **scalar** (or **dot**) **product** of two vectors $\vec{a}$ and $\vec{b}$ is written $\vec{a} \cdot \vec{b}$ and is the *scalar* quantity given by

$$\vec{a} \cdot \vec{b} = ab \cos \phi, \quad (3\text{-}20)$$

in which ϕ is the angle between the directions of $\vec{a}$ and $\vec{b}$. The scalar product may be positive, zero, or negative, depending on the value of ϕ. A scalar product is the product of the magnitude of one vector and the component of the second vector along the direction of the first vector.

In unit-vector notation,

$$\vec{a} \cdot \vec{b} = (a_x\hat{i} + a_y\hat{j} + a_z\hat{k}) \cdot (b_x\hat{i} + b_y\hat{j} + b_z\hat{k}), \quad (3\text{-}22)$$

which may be expanded according to the distributive law. Note that $\vec{a} \cdot \vec{b} = \vec{b} \cdot \vec{a}$.

The Vector Product   The **vector** (or **cross**) **product** of two vectors $\vec{a}$ and $\vec{b}$ is written $\vec{a} \times \vec{b}$ and is a *vector* $\vec{c}$ whose magnitude c is given by

$$c = ab \sin \phi, \quad (3\text{-}27)$$

in which ϕ is the smaller of the angles between the directions of $\vec{a}$ and $\vec{b}$. The direction of $\vec{c}$ is perpendicular to the plane defined by $\vec{a}$ and $\vec{b}$ and is given by a right-hand rule, as shown in Fig. 3-20. Note that $\vec{a} \times \vec{b} = -(\vec{b} \times \vec{a})$. In unit-vector notation,

$$\vec{a} \times \vec{b} = (a_x\hat{i} + a_y\hat{j} + a_z\hat{k}) \times (b_x\hat{i} + b_y\hat{j} + b_z\hat{k}), \quad (3\text{-}29)$$

which we may expand with the distributive law.

QUESTIONS

1. Displacement $\vec{D}$ points from coordinates (5 m, 3 m) to coordinates (7 m, 6 m) in the xy plane. Which of the following displacement vectors are equivalent to $\vec{D}$: vector $\vec{A}$, which points from (−6 m, −5 m) to (−4 m, −2 m); vector $\vec{B}$, which points from (−6 m, 1 m) to (−4 m, 4 m); and vector $\vec{C}$, which points from (−8 m, −6 m) to (−10 m, −9 m)?

2. Can the magnitude of the difference between two vectors ever be greater than (a) the magnitude of one of the vectors, (b) the magnitudes of both vectors, and (c) the magnitude of their sum?

3. Equation 3-2 shows that the addition of two vectors $\vec{a}$ and $\vec{b}$ is commutative. Does that mean subtraction is commutative, so that $\vec{a} - \vec{b} = \vec{b} - \vec{a}$?

4. If $\vec{d} = \vec{a} + \vec{b} + (-\vec{c})$, does (a) $\vec{a} + (-\vec{d}) = \vec{c} + (-\vec{b})$, (b) $\vec{a} = (-\vec{b}) + \vec{d} + \vec{c}$, and (c) $\vec{c} + (-\vec{d}) = \vec{a} + \vec{b}$?

5. Describe two vectors $\vec{a}$ and $\vec{b}$ such that
(a) $\vec{a} + \vec{b} = \vec{c}$ and $a + b = c$;
(b) $\vec{a} + \vec{b} = \vec{a} - \vec{b}$;
(c) $\vec{a} + \vec{b} = \vec{c}$ and $a^2 + b^2 = c^2$.

6. In Fig. 3-22, are (a) the x component and (b) the y component of vector $\vec{A}$ positive or negative? Are (c) the x component and (d) the y component of the vector combination $\vec{A} - \vec{B}$ positive or negative?

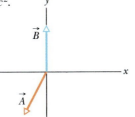

Fig. 3-22 Question 6.

7. Which of the arrangements of axes in Fig. 3-23 can be labeled "right-handed coordinate system"? As usual, each axis label indicates the positive side of the axis.

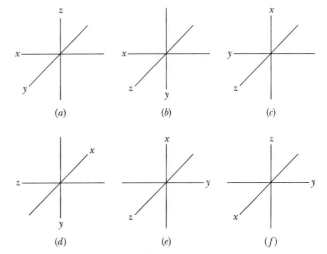

Fig. 3-23 Question 7.

8. If $\vec{a} \cdot \vec{b} = \vec{a} \cdot \vec{c}$, must $\vec{b}$ equal $\vec{c}$?

9. If $\vec{A} = 2\hat{i} + 4\hat{j}$, what is $\vec{A} \times \vec{B}$ when (a) $\vec{B} = 8\hat{i} + 16\hat{j}$ and (b) $\vec{B} = -8\hat{i} - 16\hat{j}$? (This question can be answered without computation.)

10. Figure 3-24 shows vector $\vec{A}$ and four other vectors that have the same magnitude but differ in orientation. (a) Which of those other four vectors have the same dot product with $\vec{A}$? (b) Which have a negative dot product with $\vec{A}$?

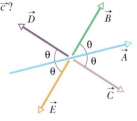

Fig. 3-24 Question 10.

EXERCISES & PROBLEMS

SEC. 3-2   Adding Vectors Geometrically

1E. Consider two displacements, one of magnitude 3 m and another of magnitude 4 m. Show how the displacement vectors may be combined to get a resultant displacement of magnitude (a) 7 m, (b) 1 m, and (c) 5 m.

2P. A bank in downtown Boston is robbed (see the map in Fig. 3-25). To elude police, the robbers escape by helicopter, making three successive flights described by the following displacements: 32 km, 45° south of east; 53 km, 26° north of west; 26 km, 18° east of south. At the end of the third flight they are captured. In what town are they apprehended? (Use the geometrical method to add these displacements on the map.)

SEC. 3-3   Components of Vectors

3E. What are (a) the x component and (b) the y component of a vector $\vec{a}$ in the xy plane if its direction is 250° counterclockwise

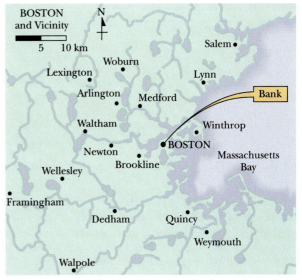

Fig. 3-25 Problem 2.

from the positive direction of the x axis and its magnitude is 7.3 m? ssm

4E. Express the following angles in radians: (a) 20.0°, (b) 50.0°, (c) 100°. Convert the following angles to degrees: (d) 0.330 rad, (e) 2.10 rad, (f) 7.70 rad.

5E. The x component of vector $\vec{A}$ is -25.0 m and the y component is $+40.0$ m. (a) What is the magnitude of $\vec{A}$? (b) What is the angle between the direction of $\vec{A}$ and the positive direction of x? ssm

6E. A displacement vector $\vec{r}$ in the xy plane is 15 m long and directed as shown in Fig. 3-26. Determine (a) the x component and (b) the y component of the vector.

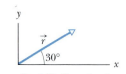

Fig. 3-26 Exercise 6.

7P. A wheel with a radius of 45.0 cm rolls without slipping along a horizontal floor (Fig. 3-27). At time t_1, the dot P painted on the rim of the wheel is at the point of contact between the wheel and the floor. At a later time t_2, the wheel has rolled through one-half of a revolution. What are (a) the magnitude and (b) the angle (relative to the floor) of the displacement of P during this interval? ssm

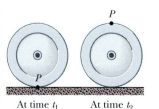

At time t_1 At time t_2

Fig. 3-27 Problem 7.

8P. Rock *faults* are ruptures along which opposite faces of rock have slid past each other. In Fig. 3-28, points A and B coincided before the rock in the foreground slid down to the right. The net displacement $\overrightarrow{AB}$ is along the plane of the fault. The horizontal component of $\overrightarrow{AB}$ is the *strike-slip AC*. The component of $\overrightarrow{AB}$ that is directly down the plane of the fault is the *dip-slip AD*. (a) What is the magnitude of the net displacement $\overrightarrow{AB}$ if the strike-slip is 22.0 m and the dip-slip is 17.0 m? (b) If the plane of the fault is inclined 52.0° to the horizontal, what is the vertical component of $\overrightarrow{AB}$?

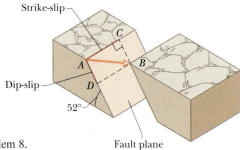

Strike-slip

Dip-slip

52°

Fig. 3-28 Problem 8.   Fault plane

9P. A room has dimensions 3.00 m (height) × 3.70 m × 4.30 m. A fly starting at one corner flies around, ending up at the diagonally opposite corner. (a) What is the magnitude of its displacement? (b) Could the length of its path be less than this magnitude? (c) Greater than this magnitude? (d) Equal to this magnitude? (e) Choose a suitable coordinate system and find the components of the displacement vector in that system. (f) If the fly walks rather than flies, what is the length of the shortest path it can take? (*Hint*: This can be answered without calculus. The room is like a box. Unfold its walls to flatten them into a plane.) ssm www

SEC. 3-5   Adding Vectors by Components

10E. A car is driven east for a distance of 50 km, then north for 30 km, and then in a direction 30° east of north for 25 km. Sketch the vector diagram and determine (a) the magnitude and (b) the angle of the car's total displacement from its starting point.

11E. A woman walks 250 m in the direction 30° east of north, then 175 m directly east. Find (a) the magnitude and (b) the angle of her final displacement from the starting point. (c) Find the distance she walks. (d) Which is greater, that distance or the magnitude of her displacement? ssm

12E. A person walks in the following pattern: 3.1 km north, then 2.4 km west, and finally 5.2 km south. (a) Sketch the vector diagram that represents this motion. (b) How far and (c) in what direction would a bird fly in a straight line from the same starting point to the same final point?

13E. (a) In unit-vector notation, what is the sum of

$$\vec{a} = (4.0 \text{ m})\hat{i} + (3.0 \text{ m})\hat{j} \quad \text{and} \quad \vec{b} = (-13.0 \text{ m})\hat{i} + (7.0 \text{ m})\hat{j}?$$

What are (b) the magnitude and (c) the direction of $\vec{a} + \vec{b}$ (relative to $\hat{i}$)? ssm

14E. Find the (a) x, (b) y, and (c) z components of the sum $\vec{r}$ of the displacements $\vec{c}$ and $\vec{d}$ whose components in meters along the three axes are $c_x = 7.4$, $c_y = -3.8$, $c_z = -6.1$; $d_x = 4.4$, $d_y = -2.0$, $d_z = 3.3$.

15E. Vector $\vec{a}$ has a magnitude of 5.0 m and is directed east. Vector $\vec{b}$ has a magnitude of 4.0 m and is directed 35° west of north. What are (a) the magnitude and (b) the direction of $\vec{a} + \vec{b}$? What are (c) the magnitude and (d) the direction of $\vec{b} - \vec{a}$? (e) Draw a vector diagram for each combination. ssm

16E. For the vectors

$$\vec{a} = (3.0 \text{ m})\hat{i} + (4.0 \text{ m})\hat{j} \quad \text{and} \quad \vec{b} = (5.0 \text{ m})\hat{i} + (-2.0 \text{ m})\hat{j},$$

give $\vec{a} + \vec{b}$ in (a) unit-vector notation, and as (b) a magnitude and (c) an angle (relative to $\hat{i}$). Now give $\vec{b} - \vec{a}$ in (d) unit-vector notation, and as (e) a magnitude and (f) an angle.

17E. Two vectors are given by

$$\vec{a} = (4.0 \text{ m})\hat{i} - (3.0 \text{ m})\hat{j} + (1.0 \text{ m})\hat{k}$$

and $$\vec{b} = (-1.0 \text{ m})\hat{i} + (1.0 \text{ m})\hat{j} + (4.0 \text{ m})\hat{k}.$$

In unit-vector notation, find (a) $\vec{a} + \vec{b}$, (b) $\vec{a} - \vec{b}$, and (c) a third vector $\vec{c}$ such that $\vec{a} - \vec{b} + \vec{c} = 0$. ssm

18P. Here are two vectors:

$$\vec{a} = (4.0 \text{ m})\hat{i} - (3.0 \text{ m})\hat{j} \quad \text{and} \quad \vec{b} = (6.0 \text{ m})\hat{i} + (8.0 \text{ m})\hat{j}.$$

What are (a) the magnitude and (b) the angle (relative to $\hat{i}$) of $\vec{a}$? What are (c) the magnitude and (d) the angle of $\vec{b}$? What are (e) the magnitude and (f) the angle of $\vec{a} + \vec{b}$; (g) the magnitude and (h) the angle of $\vec{b} - \vec{a}$; and (i) the magnitude and (j) the angle of $\vec{a} - \vec{b}$? (k) What is the angle between the directions of $\vec{b} - \vec{a}$ and $\vec{a} - \vec{b}$?

19P. Three vectors $\vec{a}$, $\vec{b}$, and $\vec{c}$ each have a magnitude of 50 m and lie in an xy plane. Their directions relative to the positive direction of the x axis are 30°, 195°, and 315°, respectively. What are (a) the magnitude and (b) the angle of the vector $\vec{a} + \vec{b} + \vec{c}$, and (c) the magnitude and (d) the angle of $\vec{a} - \vec{b} + \vec{c}$? What are (e) the magnitude and (f) the angle of a fourth vector $\vec{d}$ such that $(\vec{a} + \vec{b}) - (\vec{c} + \vec{d}) = 0$? ilw

20P. What is the sum of the following four vectors in (a) unit-vector notation and (b) magnitude-angle notation? For the latter, give the angle in both degrees and radians. Positive angles are counterclockwise from the positive direction of the *x* axis; negative angles are clockwise.

$\vec{E}$: 6.00 m at +0.900 rad $\vec{F}$: 5.00 m at −75.0°

$\vec{G}$: 4.00 m at +1.20 rad $\vec{H}$: 6.00 m at −210°

21P. The two vectors $\vec{a}$ and $\vec{b}$ in Fig. 3-29 have equal magnitudes of 10.0 m. Find (a) the *x* component and (b) the *y* component of their vector sum $\vec{r}$, (c) the magnitude of $\vec{r}$, and (d) the angle $\vec{r}$ makes with the positive direction of the *x* axis. ssm ilw www

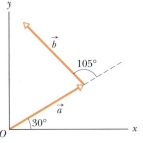

22P. In the sum $\vec{A} + \vec{B} = \vec{C}$, vector $\vec{A}$ has a magnitude of 12.0 m and is angled 40.0° counterclockwise from the +*x* direction, and vector $\vec{C}$ has a magnitude of 15.0 m and is angled 20.0° counterclockwise from the −*x* direction. What are (a) the magnitude and (b) the angle (relative to +*x*) of $\vec{B}$?

Fig. 3-29 Problem 21.

23P. Prove that two vectors must have equal magnitudes if their sum is perpendicular to their difference. ssm

24P. Find the sum of the following four vectors in (a) unit-vector notation, and as (b) a magnitude and (c) an angle relative to +*x*.

$\vec{P}$: 10.0 m, at 25.0° counterclockwise from +*x*

$\vec{Q}$: 12.0 m, at 10.0° counterclockwise from +*y*

$\vec{R}$: 8.00 m, at 20.0° clockwise from −*y*

$\vec{S}$: 9.00 m, at 40.0° counterclockwise from −*y*

25P. Two vectors of magnitudes *a* and *b* make an angle *θ* with each other when placed tail to tail. Prove, by taking components along two perpendicular axes, that

$$r = \sqrt{a^2 + b^2 + 2ab \cos \theta}$$

gives the magnitude of the sum $\vec{r}$ of the two vectors. ssm

26P. What is the sum of the following four vectors in (a) unit-vector notation, and as (b) a magnitude and (c) an angle? Positive angles are counterclockwise from the positive direction of the *x* axis; negative angles are clockwise.

$\vec{A} = (2.00 \text{ m})\hat{i} + (3.00 \text{ m})\hat{j}$ $\vec{B}$: 4.00 m, at +65.0°

$\vec{C} = (-4.00 \text{ m})\hat{i} - (6.00 \text{ m})\hat{j}$ $\vec{D}$: 5.00 m, at −235°

27P. (a) Using unit vectors, write expressions for the four body diagonals (the straight lines from one corner to another through the center) of a cube in terms of its edges, which have length *a*. (b) Determine the angles that the body diagonals make with the adjacent edges. (c) Determine the length of the body diagonals in terms of *a*. ssm

SEC. 3-6   Vectors and the Laws of Physics

28E. $\vec{A}$ has the magnitude 12.0 m and is angled 60.0° counterclockwise from the positive direction of the *x* axis of an *xy* coordinate system. Also, $\vec{B} = (12.0 \text{ m})\hat{i} + (8.00 \text{ m})\hat{j}$ on that same coordinate system. We now rotate the system counterclockwise about the origin by 20.0° to form an *x'y'* system. On this new system, what are (a) $\vec{A}$ and (b) $\vec{B}$, both in unit-vector notation?

SEC. 3-7   Multiplying Vectors

29E. A vector $\vec{a}$ of magnitude 10 units and another vector $\vec{b}$ of magnitude 6.0 units differ in directions by 60°. Find (a) the scalar product of the two vectors and (b) the magnitude of the vector product $\vec{a} \times \vec{b}$. ssm

30E. Derive Eq. 3-23 for a scalar product in unit-vector notation.

31P. Use the definition of scalar product, $\vec{a} \cdot \vec{b} = ab \cos \theta$, and the fact that $\vec{a} \cdot \vec{b} = a_x b_x + a_y b_y + a_z b_z$ (see Exercise 30) to calculate the angle between the two vectors given by $\vec{a} = 3.0\hat{i} + 3.0\hat{j} + 3.0\hat{k}$ and $\vec{b} = 2.0\hat{i} + 1.0\hat{j} + 3.0\hat{k}$. ssm ilw www

32P. Derive Eq. 3-30 for a vector product in unit-vector notation.

33P. Show that the area of the triangle contained between $\vec{a}$ and $\vec{b}$ and the red line in Fig. 3-30 is $\frac{1}{2}|\vec{a} \times \vec{b}|$. ssm

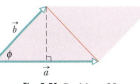

Fig. 3-30 Problem 33.

34P. In the product $\vec{F} = q\vec{v} \times \vec{B}$, take *q* = 2,

$$\vec{v} = 2.0\hat{i} + 4.0\hat{j} + 6.0\hat{k} \quad \text{and} \quad \vec{F} = 4.0\hat{i} - 20\hat{j} + 12\hat{k}.$$

What then is $\vec{B}$ in unit-vector notation if $B_x = B_y$?

35P. (a) Show that $\vec{a} \cdot (\vec{b} \times \vec{a})$ is zero for all vectors $\vec{a}$ and $\vec{b}$. (b) What is the magnitude of $\vec{a} \times (\vec{b} \times \vec{a})$ if there is an angle *φ* between the directions of $\vec{a}$ and $\vec{b}$? ssm

36P. For the following three vectors, what is $3\vec{C} \cdot (2\vec{A} \times \vec{B})$?

$$\vec{A} = 2.00\hat{i} + 3.00\hat{j} - 4.00\hat{k}$$

$$\vec{B} = -3.00\hat{i} + 4.00\hat{j} + 2.00\hat{k} \quad \vec{C} = 7.00\hat{i} - 8.00\hat{j}$$

37P. The three vectors in Fig. 3-31 have magnitudes *a* = 3.00 m, *b* = 4.00 m, and *c* = 10.0 m. What are (a) the *x* component and (b) the *y* component of $\vec{a}$; (c) the *x* component and (d) the *y* component of $\vec{b}$; and (e) the *x* component and (f) the *y* component of $\vec{c}$? If $\vec{c} = p\vec{a} + q\vec{b}$, what are the values of (g) *p* and (h) *q*? ilw

38P. Two vectors $\vec{a}$ and $\vec{b}$ have the components, in meters, $a_x = 3.2$, $a_y = 1.6$, $b_x = 0.50$, $b_y = 4.5$. (a) Find the angle between the directions of $\vec{a}$ and $\vec{b}$. There are two vectors in the *xy* plane that are perpendicular to $\vec{a}$ and have a magnitude of 5.0 m. One, vector $\vec{c}$, has a positive *x* component and the other, vector $\vec{d}$, a negative *x* component. What are (b) the *x* component and (c) the *y* component of $\vec{c}$, and (d) the *x* component and (e) the *y* component of vector $\vec{d}$?

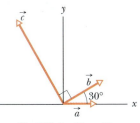

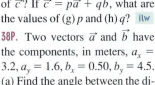

Fig. 3-31 Problem 37.

4 Motion in Two and Three Dimensions

In 1922, one of the Zacchinis, a famous family of circus performers, was the first human cannonball to be shot across an arena and into a net. To increase the excitement, the family gradually increased the height and distance of the flight until, in 1939 or 1940, Emanuel Zacchini soared over three Ferris wheels and through a horizontal distance of 69 m.

How could he know where to place the net, and how could he be certain he would clear the Ferris wheels?

The answer is in this chapter.

4-1   Moving in Two or Three Dimensions

This chapter extends the material of the preceding two chapters to two and three dimensions. Many of the ideas of Chapter 2, such as position, velocity, and acceleration, are used here, but they are now a little more complex because of the extra dimensions. To keep the notation manageable, we use the vector algebra of Chapter 3. As you read this chapter, you might want to thumb back to those previous chapters to refresh your memory.

4-2   Position and Displacement

One general way of locating a particle (or particle-like object) is with a **position vector** $\vec{r}$, which is a vector that extends from a reference point (usually the origin of a coordinate system) to the particle. In the unit-vector notation of Section 3-4, $\vec{r}$ can be written

$$\vec{r} = x\hat{i} + y\hat{j} + z\hat{k}, \tag{4-1}$$

where $x\hat{i}$, $y\hat{j}$, and $z\hat{k}$ are the vector components of $\vec{r}$, and the coefficients x, y, and z are its scalar components.

The coefficients x, y, and z give the particle's location along the coordinate axes and relative to the origin; that is, the particle has the rectangular coordinates (x, y, z). For instance, Fig. 4-1 shows a particle with position vector

$$\vec{r} = (-3\text{ m})\hat{i} + (2\text{ m})\hat{j} + (5\text{ m})\hat{k}$$

and rectangular coordinates $(-3\text{ m}, 2\text{ m}, 5\text{ m})$. Along the x axis the particle is 3 m from the origin, in the $-\hat{i}$ direction. Along the y axis it is 2 m from the origin, in the $+\hat{j}$ direction. Along the z axis it is 5 m from the origin, in the $+\hat{k}$ direction.

As a particle moves, its position vector changes in such a way that the vector always extends to the particle from the reference point (the origin). If the position vector changes, say, from $\vec{r}_1$ to $\vec{r}_2$ during a certain time interval, then the particle's **displacement** $\Delta\vec{r}$ during that time interval is

$$\Delta\vec{r} = \vec{r}_2 - \vec{r}_1. \tag{4-2}$$

Using the unit-vector notation of Eq. 4-1, we can rewrite this displacement as

$$\Delta\vec{r} = (x_2\hat{i} + y_2\hat{j} + z_2\hat{k}) - (x_1\hat{i} + y_1\hat{j} + z_1\hat{k})$$

or as

$$\Delta\vec{r} = (x_2 - x_1)\hat{i} + (y_2 - y_1)\hat{j} + (z_2 - z_1)\hat{k}, \tag{4-3}$$

where coordinates (x_1, y_1, z_1) correspond to position vector $\vec{r}_1$ and coordinates (x_2, y_2, z_2) correspond to position vector $\vec{r}_2$. We can also rewrite the displacement by substituting Δx for $(x_2 - x_1)$, Δy for $(y_2 - y_1)$, and Δz for $(z_2 - z_1)$:

$$\Delta\vec{r} = \Delta x\hat{i} + \Delta y\hat{j} + \Delta z\hat{k}. \tag{4-4}$$

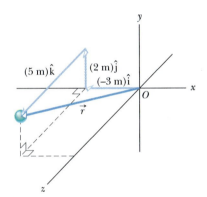

Fig. 4-1 The position vector $\vec{r}$ for a particle is the vector sum of its vector components.

Sample Problem 4-1

In Fig. 4-2, the position vector for a particle is initially

$$\vec{r}_1 = (-3.0\text{ m})\hat{i} + (2.0\text{ m})\hat{j} + (5.0\text{ m})\hat{k}$$

and then later is

$$\vec{r}_2 = (9.0\text{ m})\hat{i} + (2.0\text{ m})\hat{j} + (8.0\text{ m})\hat{k}.$$

What is the particle's displacement $\Delta\vec{r}$ from $\vec{r}_1$ to $\vec{r}_2$?

SOLUTION: The Key Idea is that the displacement $\Delta\vec{r}$ is obtained by subtracting the initial position vector $\vec{r}_1$ from the later position vector $\vec{r}_2$. That is most easily done by components:

$$\Delta\vec{r} = \vec{r}_2 - \vec{r}_1$$
$$= [9.0 - (-3.0)]\hat{i} + [2.0 - 2.0]\hat{j} + [8.0 - 5.0]\hat{k}$$
$$= (12\text{ m})\hat{i} + (3.0\text{ m})\hat{k}. \tag{Answer}$$

This displacement vector is parallel to the *xz* plane, because it lacks any *y* component, a fact that is easier to see in the numerical result than in Fig. 4-2.

✓ **CHECKPOINT 1:**   (a) If a wily bat flies from *xyz* coordinates (−2 m, 4 m, −3 m) to coordinates (6 m, −2 m, −3 m), what is its displacement $\Delta\vec{r}$ in unit-vector notation? (b) Is $\Delta\vec{r}$ parallel to one of the three coordinate planes? If so, which plane?

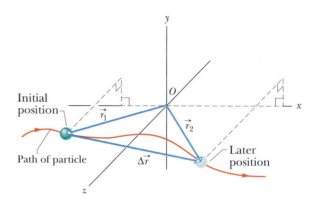

Fig. 4-2 Sample Problem 4-1. The displacement $\Delta\vec{r} = \vec{r}_2 - \vec{r}_1$ extends from the head of the initial position vector $\vec{r}_1$ to the head of the later position vector $\vec{r}_2$.

Sample Problem 4-2

A rabbit runs across a parking lot on which a set of coordinate axes has, strangely enough, been drawn. The coordinates of the rabbit's position as functions of time *t* are given by

$$x = -0.31t^2 + 7.2t + 28 \qquad (4\text{-}5)$$

and

$$y = 0.22t^2 - 9.1t + 30, \qquad (4\text{-}6)$$

with *t* in seconds and *x* and *y* in meters.

(a) At *t* = 15 s, what is the rabbit's position vector $\vec{r}$ in unit-vector notation and as a magnitude and an angle?

SOLUTION: The **Key Idea** here is that the *x* and *y* coordinates of the rabbit's position, as given by Eqs. 4-5 and 4-6, are the scalar components of the rabbit's position vector $\vec{r}$. Thus, we can write

$$\vec{r}(t) = x(t)\hat{i} + y(t)\hat{j}. \qquad (4\text{-}7)$$

(We write $\vec{r}(t)$ rather than $\vec{r}$ because the components are functions of *t*, and thus $\vec{r}$ is also.)

At *t* = 15 s, the scalar components are

$$x = (-0.31)(15)^2 + (7.2)(15) + 28 = 66 \text{ m}$$

and

$$y = (0.22)(15)^2 - (9.1)(15) + 30 = -57 \text{ m}.$$

Thus, at *t* = 15 s,

$$\vec{r} = (66 \text{ m})\hat{i} - (57 \text{ m})\hat{j}, \qquad \text{(Answer)}$$

which is drawn in Fig. 4-3*a*.

To get the magnitude and angle of $\vec{r}$, we can use a vector-capable calculator, or we can be guided by Eq. 3-6 to write

$$r = \sqrt{x^2 + y^2} = \sqrt{(66 \text{ m})^2 + (-57 \text{ m})^2}$$

$$= 87 \text{ m}, \qquad \text{(Answer)}$$

and

$$\theta = \tan^{-1}\frac{y}{x} = \tan^{-1}\left(\frac{-57 \text{ m}}{66 \text{ m}}\right) = -41°. \quad \text{(Answer)}$$

(Although $\theta = 139°$ has the same tangent as −41°, study of the signs of the components of $\vec{r}$ rules out 139°.)

(b) Graph the rabbit's path for *t* = 0 to *t* = 25 s.

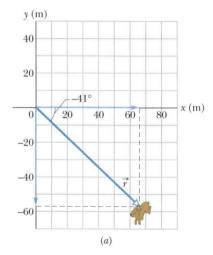

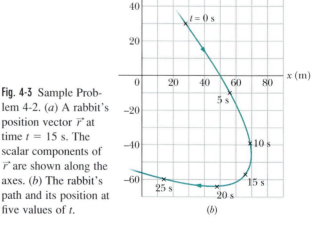

Fig. 4-3 Sample Problem 4-2. (*a*) A rabbit's position vector $\vec{r}$ at time *t* = 15 s. The scalar components of $\vec{r}$ are shown along the axes. (*b*) The rabbit's path and its position at five values of *t*.

SOLUTION: We can repeat part (a) for several values of *t* and then plot the results. Figure 4-3*b* shows the plots for five values of *t* and the path connecting them. We can also use a graphing calculator to make a *parametric graph;* that is, we would have the calculator plot *y* versus *x*, where these coordinates are given by Eqs. 4-5 and 4-6 as functions of time *t*.

4-3 Average Velocity and Instantaneous Velocity

If a particle moves through a displacement $\Delta \vec{r}$ in a time interval Δt, then its **average velocity** $\vec{v}_{avg}$ is

$$\text{average velocity} = \frac{\text{displacement}}{\text{time interval}},$$

or

$$\vec{v}_{avg} = \frac{\Delta \vec{r}}{\Delta t}. \tag{4-8}$$

This tells us the direction of $\vec{v}_{avg}$ (the vector on the left side of Eq. 4-8) must be the same as that of the displacement $\Delta \vec{r}$ (the vector on the right side). Using Eq. 4-4, we can write Eq. 4-8 in vector components as

$$\vec{v}_{avg} = \frac{\Delta x\hat{i} + \Delta y\hat{j} + \Delta z\hat{k}}{\Delta t} = \frac{\Delta x}{\Delta t}\hat{i} + \frac{\Delta y}{\Delta t}\hat{j} + \frac{\Delta z}{\Delta t}\hat{k}. \tag{4-9}$$

For example, if the particle in Sample Problem 4-1 moves from its initial position to its later position in 2.0 s, then its average velocity during that move is

$$\vec{v}_{avg} = \frac{\Delta \vec{r}}{\Delta t} = \frac{(12\text{ m})\hat{i} + (3.0\text{ m})\hat{k}}{2.0\text{ s}} = (6.0\text{ m/s})\hat{i} + (1.5\text{ m/s})\hat{k}.$$

When we speak of the **velocity** of a particle, we usually mean the particle's **instantaneous velocity** $\vec{v}$ at some instant. This $\vec{v}$ is the value that $\vec{v}_{avg}$ approaches in the limit as we shrink the time interval Δt to 0 about that instant. Using the language of calculus, we may write $\vec{v}$ as the derivative

$$\vec{v} = \frac{d\vec{r}}{dt}. \tag{4-10}$$

Figure 4-4 shows the path of a particle that is restricted to the xy plane. As the particle travels to the right along the curve, its position vector sweeps to the right. During time interval Δt, the position vector changes from $\vec{r}_1$ to $\vec{r}_2$, and the particle's displacement is $\Delta \vec{r}$.

To find the instantaneous velocity of the particle at, say, instant t_1 (when the particle is at position 1), we shrink interval Δt to 0 about t_1. Three things happen as we do so: (1) Position vector $\vec{r}_2$ in Fig. 4-4 moves toward $\vec{r}_1$ so that $\Delta \vec{r}$ shrinks toward zero. (2) The direction of $\Delta \vec{r}/\Delta t$ (thus of $\vec{v}_{avg}$) approaches the direction of the tangent line to the particle's path at position 1. (3) The average velocity $\vec{v}_{avg}$ approaches the instantaneous velocity $\vec{v}$ at t_1.

In the limit as $\Delta t \to 0$, we have $\vec{v}_{avg} \to \vec{v}$ and, most important here, $\vec{v}_{avg}$ takes on the direction of the tangent line. Thus, $\vec{v}$ has that direction as well:

▶ The direction of the instantaneous velocity $\vec{v}$ of a particle is always tangent to the particle's path at the particle's position.

The result is the same in three dimensions: $\vec{v}$ is always tangent to the particle's path.

To write Eq. 4-10 in unit-vector form, we substitute for $\vec{r}$ from Eq. 4-1:

$$\vec{v} = \frac{d}{dt}(x\hat{i} + y\hat{j} + z\hat{k}) = \frac{dx}{dt}\hat{i} + \frac{dy}{dt}\hat{j} + \frac{dz}{dt}\hat{k}.$$

This equation can be simplified somewhat by writing it as

$$\vec{v} = v_x\hat{i} + v_y\hat{j} + v_z\hat{k}, \tag{4-11}$$

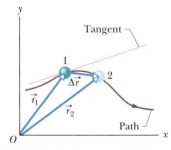

Fig. 4-4 The displacement $\Delta \vec{r}$ of a particle during a time interval Δt, from position 1 with position vector $\vec{r}_1$ at time t_1 to position 2 with position vector $\vec{r}_2$ at time t_2. The tangent to the particle's path at position 1 is shown.

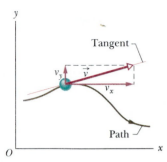

Fig. 4-5 The velocity $\vec{v}$ of a particle, along with the scalar components of $\vec{v}$.

where the scalar components of $\vec{v}$ are

$$v_x = \frac{dx}{dt}, \quad v_y = \frac{dy}{dt}, \quad \text{and} \quad v_z = \frac{dz}{dt}. \tag{4-12}$$

For example, dx/dt is the scalar component of $\vec{v}$ along the x axis. Thus, we can find the scalar components of $\vec{v}$ by differentiating the scalar components of $\vec{r}$.

Figure 4-5 shows a velocity vector $\vec{v}$ and its scalar x and y components. Note that $\vec{v}$ is tangent to the particle's path at the particle's position. *Caution:* When a position vector is drawn as in Figs. 4-1 through 4-4, it is an arrow that extends from one point (a "here") to another point (a "there"). However, when a velocity vector is drawn as in Fig. 4-5, it does *not* extend from one point to another. Rather, it shows the instantaneous direction of travel of a particle located at the tail, and its length (representing the velocity magnitude) can be drawn to any scale.

✔**CHECKPOINT 2:**   The figure shows a circular path taken by a particle. If the instantaneous velocity of the particle is $\vec{v} = (2 \text{ m/s})\hat{i} - (2 \text{ m/s})\hat{j}$, through which quadrant is the particle moving when it is traveling (a) clockwise and (b) counterclockwise around the circle? For both cases, draw $\vec{v}$ on the figure.

Sample Problem 4-3

For the rabbit in Sample Problem 4-2, find the velocity $\vec{v}$ at time $t = 15$ s, in unit-vector notation and as a magnitude and an angle.

SOLUTION: There are two **Key Ideas** here: (1) We can find the rabbit's velocity $\vec{v}$ by first finding the velocity components. (2) We can find those components by taking derivatives of the components of the rabbit's position vector. Applying the first of Eqs. 4-12 to Eq. 4-5,

we find the x component of $\vec{v}$ to be

$$v_x = \frac{dx}{dt} = \frac{d}{dt}(-0.31t^2 + 7.2t + 28)$$

$$= -0.62t + 7.2. \tag{4-13}$$

At $t = 15$ s, this gives $v_x = -2.1$ m/s. Similarly, applying the second of Eqs. 4-12 to Eq. 4-6, we find that the y component is

$$v_y = \frac{dy}{dt} = \frac{d}{dt}(0.22t^2 - 9.1t + 30)$$

$$= 0.44t - 9.1. \tag{4-14}$$

At $t = 15$ s, this gives $v_y = -2.5$ m/s. Equation 4-11 then yields

$$\vec{v} = -(2.1 \text{ m/s})\hat{i} - (2.5 \text{ m/s})\hat{j}, \qquad \text{(Answer)}$$

which is shown in Fig. 4-6, tangent to the rabbit's path and in the direction the rabbit is running at $t = 15$ s.

To get the magnitude and angle of $\vec{v}$, either we use a vector-capable calculator or we follow Eq. 3-6 to write

$$v = \sqrt{v_x^2 + v_y^2} = \sqrt{(-2.1 \text{ m/s})^2 + (-2.5 \text{ m/s})^2}$$

$$= 3.3 \text{ m/s} \qquad \text{(Answer)}$$

and

$$\theta = \tan^{-1} \frac{v_y}{v_x} = \tan^{-1}\left(\frac{-2.5 \text{ m/s}}{-2.1 \text{ m/s}}\right)$$

$$= \tan^{-1} 1.19 = -130°. \qquad \text{(Answer)}$$

(Although 50° has the same tangent as $-130°$, inspection of the signs of the velocity components indicates that the desired angle is in the third quadrant, given by $50° - 180° = -130°$.)

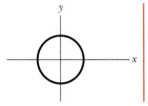

Fig. 4-6 Sample Problem 4-3. The rabbit's velocity $\vec{v}$ at $t = 15$ s. The velocity vector is tangent to the path at the rabbit's position at that instant. The scalar components of $\vec{v}$ are shown.

4-4 Average Acceleration and Instantaneous Acceleration

When a particle's velocity changes from $\vec{v}_1$ to $\vec{v}_2$ in a time interval Δt, its **average acceleration** $\vec{a}_{avg}$ during Δt is

$$\frac{\text{average}}{\text{acceleration}} = \frac{\text{change in velocity}}{\text{time interval}},$$

or $$\vec{a}_{avg} = \frac{\vec{v}_2 - \vec{v}_1}{\Delta t} = \frac{\Delta \vec{v}}{\Delta t}. \qquad (4\text{-}15)$$

If we shrink Δt to zero about some instant, then in the limit $\vec{a}_{avg}$ approaches the **instantaneous acceleration** (or **acceleration**) $\vec{a}$ at that instant; that is,

$$\vec{a} = \frac{d\vec{v}}{dt}. \qquad (4\text{-}16)$$

If the velocity changes in *either* magnitude *or* direction (or both), the particle must have an acceleration.

We can write Eq. 4-16 in unit-vector form by substituting for $\vec{v}$ from Eq. 4-11 to obtain

$$\vec{a} = \frac{d}{dt}(v_x\hat{i} + v_y\hat{j} + v_z\hat{k})$$

$$= \frac{dv_x}{dt}\hat{i} + \frac{dv_y}{dt}\hat{j} + \frac{dv_z}{dt}\hat{k}.$$

We can rewrite this as

$$\vec{a} = a_x\hat{i} + a_y\hat{j} + a_z\hat{k}, \qquad (4\text{-}17)$$

where the scalar components of $\vec{a}$ are

$$a_x = \frac{dv_x}{dt}, \quad a_y = \frac{dv_y}{dt}, \quad \text{and} \quad a_z = \frac{dv_z}{dt}. \qquad (4\text{-}18)$$

Thus, we can find the scalar components of $\vec{a}$ by differentiating the scalar components of $\vec{v}$.

Figure 4-7 shows an acceleration vector $\vec{a}$ and its scalar components for a particle moving in two dimensions. *Caution:* When an acceleration vector is drawn as in Fig. 4-7, it does *not* extend from one position to another. Rather, it shows the direction of acceleration for a particle located at its tail, and its length (representing the acceleration magnitude) can be drawn to any scale.

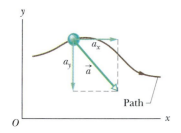

Fig. 4-7 The acceleration $\vec{a}$ of a particle and the scalar components of $\vec{a}$.

Sample Problem 4-4

For the rabbit in Sample Problems 4-2 and 4-3, find the acceleration $\vec{a}$ at time $t = 15$ s, in unit-vector notation and as a magnitude and an angle.

SOLUTION: There are two Key Ideas here: (1) We can find the rabbit's acceleration $\vec{a}$ by first finding the acceleration components. (2) We can find those components by taking derivatives of the rabbit's velocity components. Applying the first of Eqs. 4-18 to Eq. 4-13, we find the x component of $\vec{a}$ to be

$$a_x = \frac{dv_x}{dt} = \frac{d}{dt}(-0.62t + 7.2) = -0.62 \text{ m/s}^2.$$

Similarly, applying the second of Eqs. 4-18 to Eq. 4-14 yields the y component as

$$a_y = \frac{dv_y}{dt} = \frac{d}{dt}(0.44t - 9.1) = 0.44 \text{ m/s}^2.$$

We see that the acceleration does not vary with time (it is a constant) because the time variable t does not appear in the expression for either acceleration component. Equation 4-17 then yields

$$\vec{a} = (-0.62 \text{ m/s}^2)\hat{i} + (0.44 \text{ m/s}^2)\hat{j}, \qquad \text{(Answer)}$$

which is shown superimposed on the rabbit's path in Fig. 4-8.

To get the magnitude and angle of $\vec{a}$, either we use a vector-capable calculator or we follow Eq. 3-6. For the magnitude we have

$$a = \sqrt{a_x^2 + a_y^2} = \sqrt{(-0.62 \text{ m/s}^2)^2 + (0.44 \text{ m/s}^2)^2}$$

$$= 0.76 \text{ m/s}^2. \qquad \text{(Answer)}$$

For the angle we have

$$\theta = \tan^{-1} \frac{a_y}{a_x} = \tan^{-1}\left(\frac{0.44 \text{ m/s}^2}{-0.62 \text{ m/s}^2}\right) = -35°.$$

However, this last result, which is displayed on a calculator, indicates that $\vec{a}$ is directed to the right and downward in Fig. 4-8. Yet, we know from the components above that $\vec{a}$ must be directed to the left and upward. To find the other angle that has the same tangent as $-35°$, but which is not displayed on a calculator, we add 180°:

$$-35° + 180° = 145°. \qquad \text{(Answer)}$$

This *is* consistent with the components of $\vec{a}$. Note that $\vec{a}$ has the same magnitude and direction throughout the rabbit's run because, as we noted previously, the acceleration is constant.

✓**CHECKPOINT 3:**   Here are four descriptions of the position (in meters) of a hockey puck as it moves in the *xy* plane:

(1) $x = -3t^2 + 4t - 2$ and $y = 6t^2 - 4t$
(2) $x = -3t^3 - 4t$ and $y = -5t^2 + 6$
(3) $\vec{r} = 2t^2\hat{i} - (4t + 3)\hat{j}$
(4) $\vec{r} = (4t^3 - 2t)\hat{i} + 3\hat{j}$

For each description, determine whether the *x* and *y* components of the puck's acceleration are constant, and whether the acceleration $\vec{a}$ is constant.

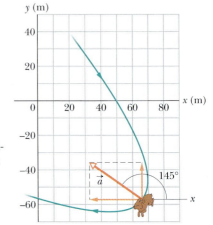

Fig. 4-8 Sample Problem 4-4. The acceleration $\vec{a}$ of the rabbit at $t = 15$ s. The rabbit happens to have this same acceleration at all points of its path.

Sample Problem 4-5

A particle with velocity $\vec{v}_0 = -2.0\hat{i} + 4.0\hat{j}$ (in meters per second) at $t = 0$ undergoes a constant acceleration $\vec{a}$ of magnitude $a = 3.0 \text{ m/s}^2$ at an angle $\theta = 130°$ from the positive direction of the *x* axis. What is the particle's velocity $\vec{v}$ at $t = 5.0$ s, in unit-vector notation and as a magnitude and an angle?

SOLUTION: We first note that this is two-dimensional motion, in the *xy* plane. Then there are two **Key Ideas** here. One is that, because the acceleration is constant, Eq. 2-11 ($v = v_0 + at$) applies. The second is that, because Eq. 2-11 applies only to straight-line motion, we must apply it separately for motion parallel to the *x* axis and motion parallel to the *y* axis; that is, we must find the velocity components v_x and v_y from the equations

$$v_x = v_{0x} + a_x t \quad \text{and} \quad v_y = v_{0y} + a_y t.$$

In these equations, v_{0x} ($= -2.0$ m/s) and v_{0y} ($= 4.0$ m/s) are the *x* and *y* components of $\vec{v}_0$, and a_x and a_y are the *x* and *y* components of $\vec{a}$. To find a_x and a_y, we resolve $\vec{a}$ either with a vector-capable calculator or with Eqs. 3-5:

$$a_x = a \cos \theta = (3.0 \text{ m/s}^2)(\cos 130°) = -1.93 \text{ m/s}^2,$$

$$a_y = a \sin \theta = (3.0 \text{ m/s}^2)(\sin 130°) = +2.30 \text{ m/s}^2.$$

When these values are inserted into the equations for v_x and v_y, we find that, at time $t = 5.0$ s,

$$v_x = -2.0 \text{ m/s} + (-1.93 \text{ m/s}^2)(5.0 \text{ s}) = -11.65 \text{ m/s},$$

$$v_y = 4.0 \text{ m/s} + (2.30 \text{ m/s}^2)(5.0 \text{ s}) = 15.50 \text{ m/s}.$$

Thus, at $t = 5.0$ s, we have, after rounding,

$$\vec{v} = (-12 \text{ m/s})\hat{i} + (16 \text{ m/s})\hat{j}. \qquad \text{(Answer)}$$

Either using a vector-capable calculator or following Eq. 3-6, we find that the magnitude and angle of $\vec{v}$ are

$$v = \sqrt{v_x^2 + v_y^2} = 19.4 \approx 19 \text{ m/s} \qquad \text{(Answer)}$$

and

$$\theta = \tan^{-1} \frac{v_y}{v_x} = 127° \approx 130°. \qquad \text{(Answer)}$$

Check the last line with your calculator. Does 127° appear on the display, or does $-53°$ appear? Now sketch the vector $\vec{v}$ with its components to see which angle is reasonable.

✓**CHECKPOINT 4:**   If the position of a hobo's marble is given by $\vec{r} = (4t^3 - 2t)\hat{i} + 3\hat{j}$, with $\vec{r}$ in meters and t in seconds, what must be the units of the coefficients 4, -2, and 3?

4-5 Projectile Motion

We next consider a special case of two-dimensional motion: A particle moves in a vertical plane with some initial velocity $\vec{v}_0$ but its acceleration is always the free-fall acceleration $\vec{g}$, which is downward. Such a particle is called a **projectile** (meaning that it is projected or launched) and its motion is called **projectile motion.** A projectile might be a golf ball (Fig. 4-9) or a baseball in flight, but it is not an airplane or a duck in flight. Our goal here is to analyze projectile motion using the tools for two-dimensional motion in Sections 4-2 through 4-4 and making the assumption that air has no effect on the projectile.

Figure 4-10, which is analyzed in the next section, shows the path followed by a projectile when the air has no effect. The projectile is launched with an initial velocity $\vec{v}_0$ that can be written as

$$\vec{v}_0 = v_{0x}\hat{i} + v_{0y}\hat{j}. \tag{4-19}$$

The components v_{0x} and v_{0y} can then be found if we know the angle θ_0 between $\vec{v}_0$ and the positive x direction:

$$v_{0x} = v_0 \cos\theta_0 \quad \text{and} \quad v_{0y} = v_0 \sin\theta_0. \tag{4-20}$$

During its two-dimensional motion, the projectile's position vector $\vec{r}$ and velocity vector $\vec{v}$ change continuously, but its acceleration vector $\vec{a}$ is constant and *always* directed vertically downward. The projectile has *no* horizontal acceleration.

Projectile motion, like that in Figs. 4-9 and 4-10, looks complicated, but we have the following simplifying feature (known from experiment):

▶ In projectile motion, the horizontal motion and the vertical motion are independent of each other; that is, neither motion affects the other.

This feature allows us to break up a problem involving two-dimensional motion into two separate and easier one-dimensional problems, one for the horizontal motion (with *zero acceleration*) and one for the vertical motion (with *constant downward acceleration*). Here are two experiments that show that the horizontal motion and the vertical motion are independent.

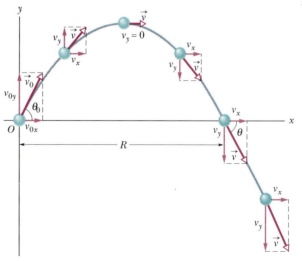

Fig. 4-10 The path of a projectile that is launched at $x_0 = 0$ and $y_0 = 0$, with an initial velocity $\vec{v}_0$. The initial velocity and the velocities at various points along its path are shown, along with their components. Note that the horizontal velocity component remains constant but the vertical velocity component changes continuously. The *range R* is the horizontal distance the projectile has traveled *when it returns to its launch height.*

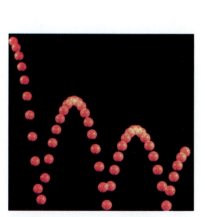

Fig. 4-9 A stroboscopic photograph of an orange golf ball bouncing off a hard surface. Between impacts, the ball has projectile motion.

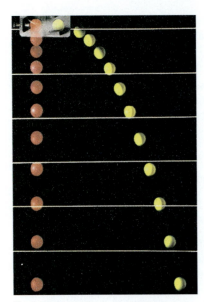

Fig. 4-11 One ball is released from rest at the same instant that another ball is shot horizontally to the right. Their vertical motions are identical.

Two Golf Balls

Figure 4-11 is a stroboscopic photograph of two golf balls, one simply released and the other shot horizontally by a spring. The golf balls have the same vertical motion, both falling through the same vertical distance in the same interval of time. *The fact that one ball is moving horizontally while it is falling has no effect on its vertical motion;* that is, the horizontal and vertical motions are independent.

A Great Student Rouser

Figure 4-12 shows a demonstration that has enlivened many a physics lecture. It involves a blow gun G, using a ball as a projectile. The target is a can suspended from a magnet M, and the tube of the blow gun is aimed directly at the can. The experiment is arranged so that the magnet releases the can just as the ball leaves the blow gun.

If g (the magnitude of the free-fall acceleration) were zero, the ball would follow the straight line shown in Fig. 4-12 and the can would float in place after the magnet released it. The ball would certainly hit the can.

However, g is *not* zero. The ball *still* hits the can! As Fig. 4-12 shows, during the time of flight of the ball, both ball and can fall the same distance h from their zero-g locations. The harder the demonstrator blows, the greater the ball's initial speed, the shorter the time of flight, and the smaller the value of h.

4-6 Projectile Motion Analyzed

Now we are ready to analyze projectile motion, horizontally and vertically.

The Horizontal Motion

Because there is *no acceleration* in the horizontal direction, the horizontal component v_x of the projectile's velocity remains unchanged from its initial value v_{0x} throughout the motion, as demonstrated in Fig. 4-13. At any time t, the projectile's horizontal displacement $x - x_0$ from an initial position x_0 is given by Eq. 2-15 with $a = 0$, which we write as

$$x - x_0 = v_{0x}t.$$

Because $v_{0x} = v_0 \cos \theta_0$, this becomes

$$x - x_0 = (v_0 \cos \theta_0)t. \tag{4-21}$$

The Vertical Motion

The vertical motion is the motion we discussed in Section 2-8 for a particle in free fall. Most important is that the acceleration is constant. Thus, the equations of Table 2-1 apply, provided we substitute $-g$ for a and switch to y notation. Then, for example, Eq. 2-15 becomes

$$y - y_0 = v_{0y}t - \tfrac{1}{2}gt^2$$
$$= (v_0 \sin \theta_0)t - \tfrac{1}{2}gt^2, \tag{4-22}$$

where the initial vertical velocity component v_{0y} is replaced with the equivalent $v_0 \sin \theta_0$. Similarly, Eqs. 2-11 and 2-16 become

$$v_y = v_0 \sin \theta_0 - gt \tag{4-23}$$

and

$$v_y^2 = (v_0 \sin \theta_0)^2 - 2g(y - y_0). \tag{4-24}$$

As is illustrated in Fig. 4-10 and Eq. 4-23, the vertical velocity component behaves just as for a ball thrown vertically upward. It is directed upward initially

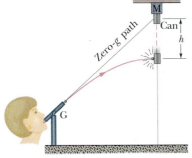

Fig. 4-12 The projectile ball always hits the falling can. Each falls a distance h from where it would be were there no free-fall acceleration.

Fig. 4-13 The vertical component of this skateboarder's velocity is changing, but not the horizontal component, which matches the skateboard's velocity. As a result, the skateboard stays underneath him, allowing him to land on it.

and its magnitude steadily decreases to zero, *which marks the maximum height of the path.* The vertical velocity component then reverses direction, and its magnitude becomes larger with time.

The Equation of the Path

We can find the equation of the projectile's path (its **trajectory**) by eliminating t between Eqs. 4-21 and 4-22. Solving Eq. 4-21 for t and substituting into Eq. 4-22, we obtain, after a little rearrangement,

$$y = (\tan \theta_0)x - \frac{gx^2}{2(v_0 \cos \theta_0)^2} \qquad \text{(trajectory).} \qquad (4\text{-}25)$$

This is the equation of the path shown in Fig. 4-10. In deriving it, for simplicity we let $x_0 = 0$ and $y_0 = 0$ in Eqs. 4-21 and 4-22, respectively. Because g, θ_0, and v_0 are constants, Eq. 4-25 is of the form $y = ax + bx^2$, in which a and b are constants. This is the equation of a parabola, so the path is *parabolic*.

The Horizontal Range

The *horizontal range R* of the projectile, as Fig. 4-10 shows, is the *horizontal* distance the projectile has traveled when it returns to its initial (launch) height. To find range R, let us put $x - x_0 = R$ in Eq. 4-21 and $y - y_0 = 0$ in Eq. 4-22, obtaining

$$R = (v_0 \cos \theta_0)t$$

and

$$0 = (v_0 \sin \theta_0)t - \tfrac{1}{2}gt^2.$$

Eliminating t between these two equations yields

$$R = \frac{2v_0^2}{g} \sin \theta_0 \cos \theta_0.$$

Using the identity $\sin 2\theta_0 = 2 \sin \theta_0 \cos \theta_0$ (see Appendix E), we obtain

$$R = \frac{v_0^2}{g} \sin 2\theta_0. \qquad (4\text{-}26)$$

Caution: This equation does *not* give the horizontal distance traveled by a projectile when the final height is not the launch height.

Note that R in Eq. 4-26 has its maximum value when $\sin 2\theta_0 = 1$, which corresponds to $2\theta_0 = 90°$ or $\theta_0 = 45°$.

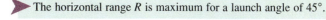

▶ The horizontal range R is maximum for a launch angle of 45°.

The Effects of the Air

We have assumed that the air through which the projectile moves has no effect on its motion. However, in many situations, the disagreement between our calculations and the actual motion of the projectile can be large because the air resists (or opposes) the motion. Figure 4-14, for example, shows two paths for a fly ball that leaves the

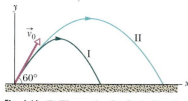

Fig. 4-14 (I) The path of a fly ball, calculated by taking air resistance into account. (II) The path the ball would follow in a vacuum, calculated by the methods of this chapter. See Table 4-1 for corresponding data. (Adapted from "The Trajectory of a Fly Ball," by Peter J. Brancazio, *The Physics Teacher*, January 1985.)

TABLE 4-1 Two Fly Balls[a]

	Path I (Air)	Path II (Vacuum)
Range	98.5 m	177 m
Maximum height	53.0 m	76.8 m
Time of flight	6.6 s	7.9 s

[a]See Fig. 4-14. The launch angle is 60° and the launch speed is 44.7 m/s.

bat at an angle of 60° with the horizontal and an initial speed of 44.7 m/s. Path I (the baseball player's fly ball) is a calculated path that approximates normal conditions of play, in air. Path II (the physics professor's fly ball) is the path that the ball would follow in a vacuum.

✔**CHECKPOINT 5:**   A fly ball is hit to the outfield. During its flight (ignore the effects of the air), what happens to its (a) horizontal and (b) vertical components of velocity? What are the (c) horizontal and (d) vertical components of its acceleration during its ascent and its descent, and at the topmost point of its flight?

Sample Problem 4-6

In Fig. 4-15, a rescue plane flies at 198 km/h (= 55.0 m/s) and a constant elevation of 500 m toward a point directly over a boating accident victim struggling in the water. The pilot wants to release a rescue capsule so that it hits the water very close to the victim. (a) What should be the angle ϕ of the pilot's line of sight to the victim when the release is made?

SOLUTION: The **Key Idea** here is that, once released, the capsule is a projectile, so its horizontal and vertical motions are independent and can be considered separately (we need not consider the actual curved path of the capsule). Figure 4-15 includes a coordinate system with its origin at the point of release, and we see there that ϕ is given by

$$\phi = \tan^{-1}\frac{x}{h}, \qquad (4\text{-}27)$$

where x is the horizontal coordinate of the victim at release (and of the capsule when it hits the water), and h is the height of the plane. That height is 500 m, so we need only x in order to find ϕ. We should be able to find x with Eq. 4-21:

$$x - x_0 = (v_0 \cos \theta_0)t. \qquad (4\text{-}28)$$

Here we know $x_0 = 0$ because the origin is placed at the point of release. Because the capsule is *released* and not shot from the plane, its initial velocity $\vec{v}_0$ is equal to the plane's velocity. Thus, we know also that the initial velocity has magnitude $v_0 = 55.0$ m/s and angle $\theta_0 = 0°$ (measured relative to the positive direction of the x axis). However, we do not know the time t the capsule takes to move from the plane to the victim.

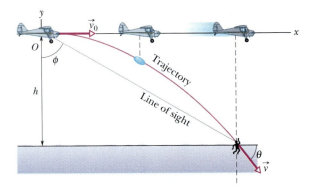

Fig. 4-15 Sample Problem 4-6. A plane drops a rescue capsule while moving at constant velocity in level flight. While the capsule is falling, its horizontal velocity component remains equal to the velocity of the plane.

To find t, we next consider the vertical motion and specifically Eq. 4-22:

$$y - y_0 = (v_0 \sin \theta_0)t - \tfrac{1}{2}gt^2. \qquad (4\text{-}29)$$

Here the vertical displacement $y - y_0$ of the capsule is -500 m (the negative value indicates that the capsule moves *downward*). Putting this and other known values into Eq. 4-29 gives us

$$-500 \text{ m} = (55.0 \text{ m/s})(\sin 0°)t - \tfrac{1}{2}(9.8 \text{ m/s}^2)t^2.$$

Solving for t, we find $t = 10.1$ s. Using that value in Eq. 4-28 yields

$$x - 0 = (55.0 \text{ m/s})(\cos 0°)(10.1 \text{ s}),$$

or

$$x = 555.5 \text{ m}.$$

Then Eq. 4-27 gives us

$$\phi = \tan^{-1}\frac{555.5 \text{ m}}{500 \text{ m}} = 48°. \qquad \text{(Answer)}$$

(b) As the capsule reaches the water, what is its velocity $\vec{v}$ in unit-vector notation and as a magnitude and an angle?

SOLUTION: Again, we need the **Key Idea** that the horizontal and vertical motions of the capsule are independent during the capsule's flight. In particular, the horizontal and vertical components of the capsule's velocity are independent of each other.

A second **Key Idea** is that the horizontal component of velocity v_x does not change from its initial value $v_{0x} = v_0 \cos \theta_0$ because there is no horizontal acceleration. Thus, when the capsule reaches the water,

$$v_x = v_0 \cos \theta_0 = (55.0 \text{ m/s})(\cos 0°) = 55.0 \text{ m/s}.$$

A third **Key Idea** is that the vertical component of velocity v_y changes from its initial value $v_{0y} = v_0 \sin \theta_0$ because there is a vertical acceleration. Using Eq. 4-23 and the capsule's time of fall $t = 10.1$ s, we find that when the capsule reaches the water,

$$v_y = v_0 \sin \theta_0 - gt$$
$$= (55.0 \text{ m/s})(\sin 0°) - (9.8 \text{ m/s}^2)(10.1 \text{ s})$$
$$= -99.0 \text{ m/s}.$$

Thus, when the capsule reaches the water it has the velocity

$$\vec{v} = (55.0 \text{ m/s})\hat{i} - (99.0 \text{ m/s})\hat{j}. \qquad \text{(Answer)}$$

Using either Eq. 3-6 as a guide or a vector-capable calculator, we find that the magnitude and the angle of $\vec{v}$ are

$$v = 113 \text{ m/s} \quad \text{and} \quad \theta = -61°. \qquad \text{(Answer)}$$

Sample Problem 4-7

Figure 4-16 shows a pirate ship 560 m from a fort defending the harbor entrance of an island. A defense cannon, located at sea level, fires balls at initial speed $v_0 = 82$ m/s.

(a) At what angle θ_0 from the horizontal must a ball be fired to hit the ship?

SOLUTION: The **Key Idea** here is obvious: A fired cannonball is a projectile and the projectile equations apply. We want an equation that relates the launch angle θ_0 to the horizontal displacement of the ball from the cannon to the ship.

A second **Key Idea** is that, because the cannon and the ship are at the same height, the horizontal displacement is the range. We can then relate the launch angle θ_0 to the range R with Eq. 4-26,

$$R = \frac{v_0^2}{g} \sin 2\theta_0, \tag{4-30}$$

which gives us

$$2\theta_0 = \sin^{-1} \frac{gR}{v_0^2} = \sin^{-1} \frac{(9.8 \text{ m/s}^2)(560 \text{ m})}{(82 \text{ m/s})^2}$$

$$= \sin^{-1} 0.816. \tag{4-31}$$

The inverse function $\sin^{-1}$ always has two possible solutions. One solution (here, 54.7°) is displayed by a calculator; we subtract it from 180° to get the other solution (here, 125.3°). Thus, Eq. 4-31 gives us

$$\theta_0 = \tfrac{1}{2}(54.7°) \approx 27° \qquad \text{(Answer)}$$

and

$$\theta_0 = \tfrac{1}{2}(125.3°) \approx 63°. \qquad \text{(Answer)}$$

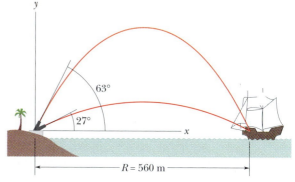

Fig. 4-16 Sample Problem 4-7. At this range, the cannon can hit the pirate ship at two elevation angles of the cannon.

The commandant of the fort can elevate the cannon to either of these two angles and (if only there were no intervening air!) hit the pirate ship.

(b) How far should the pirate ship be from the cannon if it is to be beyond the maximum range of the cannonballs?

SOLUTION: We have seen that maximum range corresponds to an elevation angle θ_0 of 45°. Thus, from Eq. 4-30 with $\theta_0 = 45°$,

$$R = \frac{v_0^2}{g} \sin 2\theta_0 = \frac{(82 \text{ m/s})^2}{9.8 \text{ m/s}^2} \sin (2 \times 45°)$$

$$= 686 \text{ m} \approx 690 \text{ m}. \qquad \text{(Answer)}$$

As the pirate ship sails away, the two elevation angles at which the ship can be hit draw together, eventually merging at $\theta_0 = 45°$ when the ship is 690 m away. Beyond that distance the ship is safe.

Sample Problem 4-8

Figure 4-17 illustrates the flight of Emanuel Zacchini over three Ferris wheels, located as shown and each 18 m high. Zacchini is launched with speed $v_0 = 26.5$ m/s, at an angle $\theta_0 = 53°$ up from the horizontal and with an initial height of 3.0 m above the ground. The net in which he is to land is at the same height.

(a) Does he clear the first Ferris wheel?

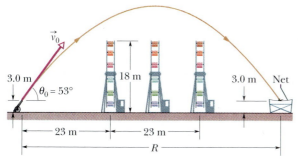

Fig. 4-17 Sample Problem 4-8. The flight of a human cannonball over three Ferris wheels and into a net.

SOLUTION: A **Key Idea** here is that Zacchini is a human projectile, so we can use the projectile equations. To do so, we place the origin of an xy coordinate system at the cannon muzzle. Then $x_0 = 0$ and $y_0 = 0$ and we want his height y when $x = 23$ m, but we do not know the time t when he reaches that height. To relate y to x without t, we use Eq. 4-25:

$$y = (\tan \theta_0)x - \frac{gx^2}{2(v_0 \cos \theta_0)^2}$$

$$= (\tan 53°)(23 \text{ m}) - \frac{(9.8 \text{ m/s}^2)(23 \text{ m})^2}{2(26.5 \text{ m/s})^2(\cos 53°)^2}$$

$$= 20.3 \text{ m}.$$

Since he begins 3.0 m off the ground, he clears the Ferris wheel by about 5.3 m.

(b) If he reaches his maximum height when he is over the middle Ferris wheel, what is his clearance above it?

SOLUTION: A **Key Idea** here is that the vertical component v_y of his velocity is zero when he reaches his maximum height. Since Eq. 4-24 relates v_y and his height y, we write

$$v_y^2 = (v_0 \sin \theta_0)^2 - 2gy = 0.$$

Solving for y gives us

$$y = \frac{(v_0 \sin \theta_0)^2}{2g} = \frac{(26.5 \text{ m/s})^2(\sin 53°)^2}{(2)(9.8 \text{ m/s}^2)} = 22.9 \text{ m},$$

which means that he clears the middle Ferris wheel by 7.9 m.

(c) How far from the cannon should the center of the net be positioned?

SOLUTION: The additional **Key Idea** here is that, because Zacchini's initial and landing heights are the same, the horizontal distance from cannon muzzle to net is his horizontal range for the flight. From Eq. 4-26, we find

$$R = \frac{v_0^2}{g} \sin 2\theta_0 = \frac{(26.5 \text{ m/s})^2}{9.8 \text{ m/s}^2} \sin 2(53°)$$

$$= 69 \text{ m}. \qquad \text{(Answer)}$$

We can now answer the questions that opened this chapter: How could Zacchini know where to place the net, and how could he be certain he would clear the Ferris wheels? He (or someone) did the calculations as we have here. Although he could not take into account the complicated effects of the air on his flight, Zacchini knew that the air would slow him and thus decrease his range from the calculated value, so he used a wide net and biased it toward the cannon. He was then relatively safe whether the effects of the air in a particular flight happened to slow him considerably or very little. Still, the variability of this factor of air effects must have played on his imagination before each flight.

Zacchini still faced a subtle danger: Even for shorter flights, his propulsion through the cannon was so severe that he underwent a momentary blackout. If he landed during the blackout, he could break his neck. To avoid this, he had trained himself to awake quickly. Indeed, not waking up in time presents the only real danger to a human cannonball in the short flights today.

PROBLEM-SOLVING TACTICS

Tactic 1: *Numbers Versus Algebra*
One way to avoid rounding errors and other numerical errors is to solve problems algebraically, substituting numbers only in the final step. That is easy to do in Sample Problems 4-6 to 4-8, and that is

the way experienced problem solvers operate. In these early chapters, however, we prefer to solve most problems in parts, to give you a firmer numerical grasp of what is going on. Later we shall stick to the algebra longer.

4-7   Uniform Circular Motion

A particle is in **uniform circular motion** if it travels around a circle or a circular arc at constant (*uniform*) speed. Although the speed does not vary, *the particle is accelerating*. That fact may be surprising because we often think of acceleration (a change in velocity) as an increase or decrease in speed. However, actually velocity is a vector, not a scalar. Thus, even if a velocity changes only in direction, there is still an acceleration, and that is what happens in uniform circular motion.

Figure 4-18 shows the relation between the velocity and acceleration vectors at various stages during uniform circular motion. Both vectors have constant magnitude as the motion progresses, but their directions change continuously. The velocity is always directed tangent to the circle in the direction of motion. The acceleration is always directed *radially inward*. Because of this, the acceleration associated with uniform circular motion is called a **centripetal** (meaning "center seeking") **acceleration**. As we prove next, the magnitude of this acceleration $\vec{a}$ is

$$a = \frac{v^2}{r} \qquad \text{(centripetal acceleration)}, \qquad (4\text{-}32)$$

where r is the radius of the circle and v is the speed of the particle.

In addition, during this acceleration at constant speed, the particle travels the circumference of the circle (a distance of $2\pi r$) in time

$$T = \frac{2\pi r}{v} \qquad \text{(period)}. \qquad (4\text{-}33)$$

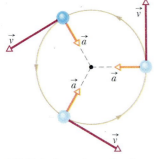

Fig. 4-18 Velocity and acceleration vectors for a particle in counterclockwise uniform circular motion. Both have constant magnitude but vary continuously in direction.

T is called the *period of revolution*, or simply the *period*, of the motion. It is, in general, the time for a particle to go around a closed path exactly once.

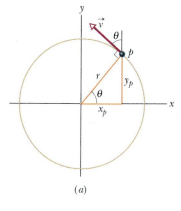

(a)

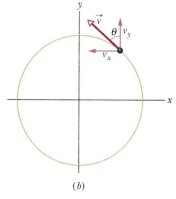

(b)

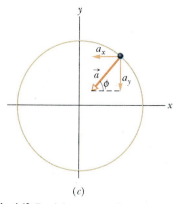

(c)

Fig. 4-19 Particle p moves in counterclockwise uniform circular motion. (a) Its position and velocity $\vec{v}$ at a certain instant. (b) Velocity $\vec{v}$ and its components. (c) The particle's acceleration $\vec{a}$ and its components.

Proof of Eq. 4-32

To find the magnitude and direction of the acceleration for uniform circular motion, we consider Fig. 4-19. In Fig. 4-19a, particle p moves at constant speed v around a circle of radius r. At the instant shown, p has coordinates x_p and y_p.

Recall from Section 4-3 that the velocity $\vec{v}$ of a moving particle is always tangent to the particle's path at the particle's position. In Fig. 4-19a, that means $\vec{v}$ is perpendicular to a radius r drawn to the particle's position. Then the angle θ that $\vec{v}$ makes with a vertical at p equals the angle θ that radius r makes with the x axis.

The scalar components of $\vec{v}$ are shown in Fig. 4-19b. With them, we can write the velocity $\vec{v}$ as

$$\vec{v} = v_x\hat{i} + v_y\hat{j} = (-v\sin\theta)\hat{i} + (v\cos\theta)\hat{j}. \qquad (4\text{-}34)$$

Now, using the right triangle in Fig. 4-19a, we can replace $\sin\theta$ with y_p/r and $\cos\theta$ with x_p/r to write

$$\vec{v} = \left(-\frac{vy_p}{r}\right)\hat{i} + \left(\frac{vx_p}{r}\right)\hat{j}. \qquad (4\text{-}35)$$

To find the acceleration $\vec{a}$ of particle p, we must take the time derivative of this equation. Noting that speed v and radius r do not change with time, we obtain

$$\vec{a} = \frac{d\vec{v}}{dt} = \left(-\frac{v}{r}\frac{dy_p}{dt}\right)\hat{i} + \left(\frac{v}{r}\frac{dx_p}{dt}\right)\hat{j}. \qquad (4\text{-}36)$$

Now note that the rate dy_p/dt at which y_p changes is equal to the velocity component v_y. Similarly, $dx_p/dt = v_x$, and, again from Fig. 4-19b, we see that $v_x = -v\sin\theta$ and $v_y = v\cos\theta$. Making these substitutions in Eq. 4-36, we find

$$\vec{a} = \left(-\frac{v^2}{r}\cos\theta\right)\hat{i} + \left(-\frac{v^2}{r}\sin\theta\right)\hat{j}. \qquad (4\text{-}37)$$

This vector and its components are shown in Fig. 4-19c. Following Eq. 3-6, we find that the magnitude of $\vec{a}$ is

$$a = \sqrt{a_x^2 + a_y^2} = \frac{v^2}{r}\sqrt{(\cos\theta)^2 + (\sin\theta)^2} = \frac{v^2}{r},$$

as we wanted to prove. To orient $\vec{a}$, we can find the angle ϕ shown in Fig. 4-19c:

$$\tan\phi = \frac{a_y}{a_x} = \frac{-(v^2/r)\sin\theta}{-(v^2/r)\cos\theta} = \tan\theta.$$

Thus, $\phi = \theta$, which means that $\vec{a}$ is directed along the radius r of Fig. 4-19a, toward the circle's center, as we wanted to prove.

✔**CHECKPOINT 6:** An object moves at constant speed along a circular path in a horizontal xy plane, with the center at the origin. When the object is at $x = -2$ m, its velocity is $-(4$ m/s$)\hat{j}$. Give the object's (a) velocity and (b) acceleration when it is at $y = 2$ m.

Sample Problem 4-9

"Top gun" pilots have long worried about taking a turn too tightly. As a pilot's body undergoes centripetal acceleration, with the head toward the center of curvature, the blood pressure in the brain decreases, leading to loss of brain function.

There are several warning signs to signal a pilot to ease up: when the centripetal acceleration is $2g$ or $3g$, the pilot feels heavy. At about $4g$, the pilot's vision switches to black and white and narrows to "tunnel vision." If that acceleration is sustained or increased, vision ceases and, soon after, the pilot is unconscious—a condition known as g-LOC for "g-induced loss of consciousness."

What is the centripetal acceleration, in g units, of a pilot flying an F-22 at speed $v = 2500$ km/h (694 m/s) through a circular arc with radius of curvature $r = 5.80$ km?

SOLUTION: The **Key Idea** here is that although the pilot's speed is constant, the circular path requires a (centripetal) acceleration, with

magnitude given by Eq. 4-32:

$$a = \frac{v^2}{r} = \frac{(694 \text{ m/s})^2}{5800 \text{ m}} = 83.0 \text{ m/s}^2 = 8.5g. \quad \text{(Answer)}$$

If an unwary pilot caught in a dogfight puts the aircraft into such a tight turn, the pilot goes into g-LOC almost immediately, with no warning signs to signal the danger.

4-8 Relative Motion in One Dimension

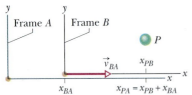

Fig. 4-20 Alex (frame A) and Barbara (frame B) watch car P, as both B and P move at different velocities along the common x axes of the two frames. At the instant shown, x_{BA} is the coordinate of B in the A frame. Also, P is at coordinate x_{PB} in the B frame and coordinate $x_{PA} = x_{PB} + x_{BA}$ in the A frame.

Suppose you see a duck flying north at, say, 30 km/h. To another duck flying alongside, the first duck seems to be stationary. In other words, the velocity of a particle depends on the **reference frame** of whoever is observing or measuring the velocity. For our purposes, a reference frame is the physical object to which we attach our coordinate system. In everyday life, that object is the ground. For example, the speed listed on a speeding ticket is always measured relative to the ground. The speed relative to the police officer would be different if the officer were moving while making the speed measurement.

Suppose that Alex (at the origin of frame A) is parked by the side of a highway, watching car P (the "particle") speed past. Barbara (at the origin of frame B) is driving along the highway at constant speed and is also watching car P. Suppose that, as in Fig. 4-20, they both measure the position of the car at a given moment. From the figure we see that

$$x_{PA} = x_{PB} + x_{BA}. \quad (4\text{-}38)$$

The equation is read: "The coordinate x_{PA} of P as measured by A *is equal to* the coordinate x_{PB} of P as measured by B *plus* the coordinate x_{BA} of B as measured by A." Note how this reading is supported by the sequence of the subscripts.

Taking the time derivative of Eq. 4-38, we obtain

$$\frac{d}{dt}(x_{PA}) = \frac{d}{dt}(x_{PB}) + \frac{d}{dt}(x_{BA}),$$

or (because $v = dx/dt$)

$$v_{PA} = v_{PB} + v_{BA}. \quad (4\text{-}39)$$

This equation is read: "The velocity v_{PA} of P as measured by A *is equal to* the velocity v_{PB} of P as measured by B *plus* the velocity v_{BA} of B as measured by A." The term v_{BA} is the velocity of frame B relative to frame A. (Because the motions are along a single axis, we can use components along that axis in Eq. 4-39 and omit overhead vector arrows.)

Here we consider only frames that move at constant velocity relative to each other. In our example, this means that Barbara (frame B) will drive always at constant velocity v_{BA} relative to Alex (frame A). Car P (the moving particle), however, may speed up, slow down, come to rest, or reverse direction (that is, it can accelerate).

To relate an acceleration of P as measured by Barbara and by Alex, we take the time derivative of Eq. 4-39:

$$\frac{d}{dt}(v_{PA}) = \frac{d}{dt}(v_{PB}) + \frac{d}{dt}(v_{BA}).$$

Because v_{BA} is constant, the last term is zero and we have

$$a_{PA} = a_{PB}. \qquad (4\text{-}40)$$

In other words,

> Observers on different frames of reference (that move at constant velocity relative to each other) will measure the same acceleration for a moving particle.

Situation	v_{BA}	v_{PA}	v_{PB}
(a)	+50	+50	
(b)	+30		+40
(c)		+60	−20

✔ **CHECKPOINT 7:**   The table here gives velocities (km/h) for Barbara and car *P* of Fig. 4-20 for three situations. For each, what is the missing value and how is the distance between Barbara and car *P* changing?

Sample Problem 4-10

For the situation of Fig. 4-20 and this section, Barbara's velocity relative to Alex is a constant $v_{BA} = 52$ km/h and car *P* is moving in the negative direction of the *x* axis.

(a) If Alex measures a constant velocity $v_{PA} = -78$ km/h for car *P*, what velocity v_{PB} will Barbara measure?

SOLUTION: The **Key Idea** here is that we can attach a frame of reference *A* to Alex and another frame of reference *B* to Barbara. Further, because these two frames move at constant velocity relative to each other, along a single axis, we can use Eq. 4-39 to relate v_{PB} and v_{PA}. We find

$$v_{PA} = v_{PB} + v_{BA}$$

or $-78 \text{ km/h} = v_{PB} + 52 \text{ km/h}.$

Thus, $v_{PB} = -130 \text{ km/h}.$ (Answer)

If car *P* were connected to Barbara's car by a cord wound on a spool, the cord would be unwinding at a speed of 130 km/h as the two cars separated.

(b) If car *P* brakes to a stop relative to Alex (and thus the ground) in time $t = 10$ s at constant acceleration, what is its acceleration a_{PA} relative to Alex?

SOLUTION: The **Key Idea** here is that, to calculate the acceleration of car *P relative to Alex*, we must use the car's velocities *relative to*

Alex. Because the acceleration is constant, we can use Eq. 2-11 ($v = v_0 + at$) to relate the acceleration to the initial and final velocities of *P*. The initial velocity of *P* relative to Alex is $v_{PA} = -78$ km/h and the final velocity is 0. Thus, Eq. 2-11 gives us

$$a_{PA} = \frac{v - v_0}{t} = \frac{0 - (-78 \text{ km/h})}{10 \text{ s}} \frac{1 \text{ m/s}}{3.6 \text{ km/h}}$$

$$= 2.2 \text{ m/s}^2. \qquad \text{(Answer)}$$

(c) What is the acceleration a_{PB} of car *P* relative to Barbara during the braking?

SOLUTION: The **Key Idea** is that now, to calculate the acceleration of car *P relative to Barbara*, we must use the car's velocities *relative to Barbara*. We know the initial velocity of *P* relative to Barbara from part (a) ($v_{PB} = -130$ km/h). The final velocity of *P* relative to Barbara is -52 km/h (this is the velocity of the stopped car relative to the moving Barbara). Thus, again from Eq. 2-11,

$$a_{PB} = \frac{v - v_0}{t} = \frac{-52 \text{ km/h} - (-130 \text{ km/h})}{10 \text{ s}} \frac{1 \text{ m/s}}{3.6 \text{ km/h}}$$

$$= 2.2 \text{ m/s}^2. \qquad \text{(Answer)}$$

We should have foreseen this result: Because Alex and Barbara have a constant relative velocity, they must measure the same acceleration for the car.

4-9 Relative Motion in Two Dimensions

Now we turn from relative motion in one dimension to relative motion in two (and, by extension, in three) dimensions. In Fig. 4-21, our two observers are again watching a moving particle *P* from the origins of reference frames *A* and *B*, while *B* moves at a constant velocity $\vec{v}_{BA}$ relative to *A*. (The corresponding axes of these two frames remain parallel.)

Figure 4-21 shows a certain instant during the motion. At that instant, the position vector of *B* relative to *A* is $\vec{r}_{BA}$. Also, the position vectors of particle *P* are

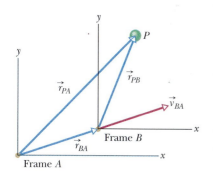

Fig. 4-21 Frame *B* has the constant two-dimensional velocity $\vec{v}_{BA}$ relative to frame *A*. The position vector of *B* relative to *A* is $\vec{r}_{BA}$. The position vectors of particle *P* are $\vec{r}_{PA}$ relative to *A* and $\vec{r}_{PB}$ relative to *B*.

$\vec{r}_{PA}$ relative to A and $\vec{r}_{PB}$ relative to B. From the arrangement of heads and tails of those three position vectors, we can relate the vectors with

$$\vec{r}_{PA} = \vec{r}_{PB} + \vec{r}_{BA}. \qquad (4\text{-}41)$$

By taking the time derivative of this equation, we can relate the velocities $\vec{v}_{PA}$ and $\vec{v}_{PB}$ of particle P relative to our observers. We get

$$\vec{v}_{PA} = \vec{v}_{PB} + \vec{v}_{BA}. \qquad (4\text{-}42)$$

By taking the time derivative of this relation, we can relate the accelerations $\vec{a}_{PA}$ and $\vec{a}_{PB}$ of the particle P relative to our observers. However, note that because $\vec{v}_{BA}$ is constant, its time derivative is zero. Thus, we get

$$\vec{a}_{PA} = \vec{a}_{PB}. \qquad (4\text{-}43)$$

As for one-dimensional motion, we have the following rule: Observers on different frames of reference that move at constant velocity relative to each other will measure the *same* acceleration for a moving particle.

Sample Problem 4-11

In Fig. 4-22a, a plane moves due east (directly toward the east) while the pilot points the plane somewhat south of east, toward a steady wind that blows to the northeast. The plane has velocity $\vec{v}_{PW}$ relative to the wind, with an airspeed (speed relative to the wind) of 215 km/h, directed at angle θ south of east. The wind has velocity $\vec{v}_{WG}$ relative to the ground, with a speed of 65.0 km/h, directed 20.0° east of north. What is the magnitude of the velocity $\vec{v}_{PG}$ of the plane relative to the ground, and what is θ?

SOLUTION: The Key Idea is that the situation is like the one in Fig. 4-21. Here the moving particle P is the plane, frame A is attached to the ground (call it G), and frame B is "attached" to the wind (call it W). We need to construct a vector diagram like that in Fig. 4-21 but this time using the three velocity vectors.

First construct a sentence that relates the three vectors:

velocity of plane   velocity of plane   velocity of wind
relative to ground = relative to wind   + relative to ground.
      (PG)   (PW)   (WG)

This relation can be drawn as in Fig. 4-22b and written in vector notation as

$$\vec{v}_{PG} = \vec{v}_{PW} + \vec{v}_{WG}. \qquad (4\text{-}44)$$

We want the magnitude of the first vector and the direction of the second vector. With unknowns in two vectors, we cannot solve Eq. 4-44 directly on a vector-capable calculator. Instead, we need to resolve the vectors into components on the coordinate system of Fig. 4-22b, and then solve Eq. 4-44 axis by axis (see Section 3-5). For the y components, we find

$$v_{PG,y} = v_{PW,y} + v_{WG,y}$$

or $0 = -(215 \text{ km/h}) \sin\theta + (65.0 \text{ km/h})(\cos 20.0°).$

Solving for θ gives us

$$\theta = \sin^{-1}\frac{(65.0 \text{ km/h})(\cos 20.0°)}{215 \text{ km/h}} = 16.5°. \qquad \text{(Answer)}$$

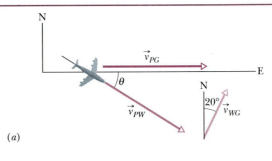

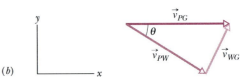

(a)

(b)

Fig. 4-22 Sample Problem 4-11. To travel due east, the plane must head somewhat into the wind.

Similarly, for the x components we find

$$v_{PG,x} = v_{PW,x} + v_{WG,x}.$$

Here, because $\vec{v}_{PG}$ is parallel to the x axis, the component $v_{PG,x}$ is equal to the magnitude v_{PG}. Substituting this and $\theta = 16.5°$, we find

$$v_{PG} = (215 \text{ km/h})(\cos 16.5°) + (65.0 \text{ km/h})(\sin 20.0°)$$
$$= 228 \text{ km/h}. \qquad \text{(Answer)}$$

✔CHECKPOINT 8: In this sample problem, suppose the pilot turns the plane to point it due east but without changing the air speed. Do the following magnitudes increase, decrease, or remain the same: (a) $v_{PG,y}$, (b) $v_{PG,x}$, and (c) v_{PG}? (You can answer without computation.)

REVIEW & SUMMARY

Position Vector   The location of a particle relative to the origin of a coordinate system is given by a *position vector* $\vec{r}$, which in unit-vector notation is

$$\vec{r} = x\hat{i} + y\hat{j} + z\hat{k}. \tag{4-1}$$

Here $x\hat{i}$, $y\hat{j}$, and $z\hat{k}$ are the *vector components* of position vector $\vec{r}$, and x, y, and z are its *scalar components* (as well as the coordinates of the particle). A position vector is described by a magnitude and one or two angles for orientation, or by its vector or scalar components.

Displacement   If a particle moves so that its position vector changes from $\vec{r}_1$ to $\vec{r}_2$, then the particle's *displacement* $\Delta\vec{r}$ is

$$\Delta\vec{r} = \vec{r}_2 - \vec{r}_1. \tag{4-2}$$

The displacement can also be written as

$$\Delta\vec{r} = (x_2 - x_1)\hat{i} + (y_2 - y_1)\hat{j} + (z_2 - z_1)\hat{k} \tag{4-3}$$

$$= \Delta x\hat{i} + \Delta y\hat{j} + \Delta z\hat{k}, \tag{4-4}$$

where coordinates (x_1, y_1, z_1) correspond to position vector $\vec{r}_1$ and coordinates (x_2, y_2, z_2) correspond to position vector $\vec{r}_2$.

Average Velocity and (Instantaneous) Velocity   If a particle undergoes a displacement $\Delta\vec{r}$ in time Δt, its *average velocity* $\vec{v}_{avg}$ for that time interval is

$$\vec{v}_{avg} = \frac{\Delta\vec{r}}{\Delta t}. \tag{4-8}$$

As Δt in Eq. 4-8 is shrunk to 0, $\vec{v}_{avg}$ reaches a limit called the *velocity* or *instantaneous velocity* $\vec{v}$:

$$\vec{v} = \frac{d\vec{r}}{dt}, \tag{4-10}$$

which can be rewritten in unit-vector notation as

$$\vec{v} = v_x\hat{i} + v_y\hat{j} + v_z\hat{k}, \tag{4-11}$$

where $v_x = dx/dt$, $v_y = dy/dt$, and $v_z = dz/dt$. The instantaneous velocity $\vec{v}$ of a particle is always directed along the tangent to the particle's path at the particle's position.

Average Acceleration and (Instantaneous) Acceleration   If a particle's velocity changes from $\vec{v}_1$ to $\vec{v}_2$ in time interval Δt, its *average acceleration* during Δt is

$$\vec{a}_{avg} = \frac{\vec{v}_2 - \vec{v}_1}{\Delta t} = \frac{\Delta\vec{v}}{\Delta t}. \tag{4-15}$$

As Δt in Eq. 4-15 is shrunk to 0, $\vec{a}_{avg}$ reaches a limiting value called the *acceleration* or *instantaneous acceleration* $\vec{a}$:

$$\vec{a} = \frac{d\vec{v}}{dt}. \tag{4-16}$$

In unit-vector notation,

$$\vec{a} = a_x\hat{i} + a_y\hat{j} + a_z\hat{k}, \tag{4-17}$$

where $a_x = dv_x/dt$, $a_y = dv_y/dt$, and $a_z = dv_z/dt$.

Projectile Motion   *Projectile motion* is the motion of a particle that is launched with an initial velocity $\vec{v}_0$. During its flight, the particle's horizontal acceleration is zero and its vertical acceleration is the free-fall acceleration $-g$. (Upward is taken to be a positive direction.) If $\vec{v}_0$ is expressed as a magnitude (the speed v_0) and an angle θ_0, the particle's equations of motion along the horizontal x axis and vertical y axis are

$$x - x_0 = (v_0 \cos \theta_0)t, \tag{4-21}$$

$$y - y_0 = (v_0 \sin \theta_0)t - \tfrac{1}{2}gt^2, \tag{4-22}$$

$$v_y = v_0 \sin \theta_0 - gt, \tag{4-23}$$

$$v_y^2 = (v_0 \sin \theta_0)^2 - 2g(y - y_0). \tag{4-24}$$

The **trajectory** (path) of a particle in projectile motion is parabolic and is given by

$$y = (\tan \theta_0)x - \frac{gx^2}{2(v_0 \cos \theta_0)^2}, \tag{4-25}$$

where the origin has been chosen so that x_0 and y_0 of Eqs. 4-21 to 4-24 are zero. The particle's **horizontal range** R, which is the horizontal distance from the launch point to the point at which the particle returns to the launch height, is

$$R = \frac{v_0^2}{g} \sin 2\theta_0. \tag{4-26}$$

Uniform Circular Motion   If a particle travels along a circle or circular arc with radius r at constant speed v, it is in *uniform circular motion* and has an acceleration $\vec{a}$ of magnitude

$$a = \frac{v^2}{r}. \tag{4-32}$$

The direction of $\vec{a}$ is toward the center of the circle or circular arc, and $\vec{a}$ is said to be *centripetal*. The time for the particle to complete a circle is

$$T = \frac{2\pi r}{v}. \tag{4-33}$$

T is called the *period of revolution*, or simply the *period*, of the motion.

Relative Motion   When two frames of reference A and B are moving relative to each other at constant velocity, the velocity of a particle P as measured by an observer in frame A usually differs from that measured from frame B. The two measured velocities are related by

$$\vec{v}_{PA} = \vec{v}_{PB} + \vec{v}_{BA}, \tag{4-42}$$

in which $\vec{v}_{BA}$ is the velocity of B with respect to A. Both observers measure the same acceleration for the particle; that is,

$$\vec{a}_{PA} = \vec{a}_{PB}. \tag{4-43}$$

QUESTIONS

1. Figure 4-23 shows the initial position i and the final position f of a particle. What are the particle's (a) initial position vector $\vec{r}_i$ and (b) final position vector $\vec{r}_f$, both in unit-vector notation? (c) What is the x component of the particle's displacement $\Delta\vec{r}$?

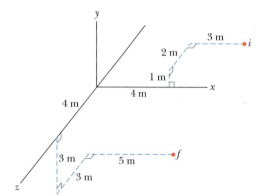

Fig. 4-23
Question 1.

2. Here are four descriptions for the velocity of a hockey puck in the xy plane, all in meters per second:
(1) $v_x = -3t^2 + 4t - 2$ and $v_y = 6t - 4$
(2) $v_x = -3$ and $v_y = -5t^2 + 6$
(3) $\vec{v} = 2t^2\hat{i} - (4t + 3)\hat{j}$
(4) $\vec{v} = -2t\hat{i} + 3\hat{j}$
(a) For each description, are the x and y components of the acceleration constant, and is the acceleration vector $\vec{a}$ constant? (b) In description (4), if $\vec{v}$ is in meters per second and t is in seconds, what must be the units of the coefficients -2 and 3?

3. Figure 4-24 shows three situations in which identical projectiles are launched from the ground (at the same level) at identical initial speeds and angles. The projectiles do not land on the same terrain, however. Rank the situations according to the final speeds of the projectiles just before they land, greatest first.

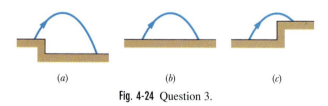

Fig. 4-24 Question 3.

4. At a certain instant, a fly ball has velocity $\vec{v} = 25\hat{i} - 4.9\hat{j}$ (the x axis is horizontal, the y axis is upward, and $\vec{v}$ is in meters per second). Has the ball passed the highest point of its trajectory?

5. You are to launch a rocket, from just above the ground, with one of the following initial velocity vectors: (1) $\vec{v}_0 = 20\hat{i} + 70\hat{j}$, (2) $\vec{v}_0 = -20\hat{i} + 70\hat{j}$, (3) $\vec{v}_0 = 20\hat{i} - 70\hat{j}$, (4) $\vec{v}_0 = -20\hat{i} - 70\hat{j}$. In your coordinate system, x runs along level ground and y increases upward. (a) Rank the vectors according to the launch speed of the projectile, greatest first. (b) Rank the vectors according to the time of flight of the projectile, greatest first.

6. A mud ball is launched 2 m above the ground with initial velocity $\vec{v}_0 = (2\hat{i} + 4\hat{j})$ m/s. What is its velocity just before it lands on a surface that is 2 m above the ground?

7. In Fig. 4-25, a cream tangerine is thrown up past windows 1, 2, and 3, which are identical in size and regularly spaced vertically. Rank those three windows according to (a) the time the cream tangerine takes to pass them and (b) the average speed of the cream tangerine during the passage, greatest first.

The cream tangerine then moves down past windows 4, 5, and 6, which are identical in size and irregularly spaced horizontally. Rank those three windows according to (c) the time the cream tangerine takes to pass them and (d) the average speed of the cream tangerine during the passage, greatest first.

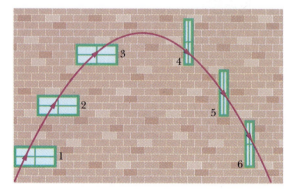

Fig. 4-25 Question 7.

8. An airplane flying horizontally at a constant speed of 350 km/h over level ground releases a bundle of food supplies. Ignore the effect of the air on the bundle. What are the bundle's initial (a) vertical and (b) horizontal components of velocity? (c) What is its horizontal component of velocity just before hitting the ground? (d) If the airplane's speed were, instead, 450 km/h, would the time of fall be larger, smaller, or the same?

9. You throw a ball with a launch velocity of $\vec{v}_i = (3 \text{ m/s})\hat{i} + (4 \text{ m/s})\hat{j}$ toward a wall, where it hits at height h_1 in time t_1 after the launch (Fig. 4-26). Suppose that the launch velocity were, instead, $\vec{v}_i = (5 \text{ m/s})\hat{i} + (4 \text{ m/s})\hat{j}$. (a) Would the time taken by the ball to reach the wall be greater than, less than, or equal to t_1, or is the question unanswerable without more information? (b) Would the height at which the ball hits be greater than, less than, or equal to h_1, or is the question unanswerable?

Suppose, instead, that the launch velocity were $\vec{v}_i = (3 \text{ m/s})\hat{i} + (5 \text{ m/s})\hat{j}$. (c) Would the time taken by the ball to reach the wall be greater than, less than, or equal to t_1, or is the question unanswerable? (d) Would the height at which the ball hits be greater than, less than, or equal to h_1, or is the question unanswerable?

Fig. 4-26 Question 9.

10. Figure 4-27 shows three paths for a football kicked from ground level. Ignoring the effects of air

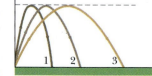

Fig. 4-27 Question 10.

on the flight, rank the paths according to (a) time of flight, (b) initial vertical velocity component, (c) initial horizontal velocity component, and (d) initial speed, greatest first.

11. Figure 4-28 shows the velocity and acceleration of a particle at a particular instant in three situations. In which situation is (a) the speed of the particle increasing, (b) the speed decreasing, (c) the speed not changing, (d) $\vec{v} \cdot \vec{a}$ positive, (e) $\vec{v} \cdot \vec{a}$ negative, and (f) $\vec{v} \cdot \vec{a} = 0$?

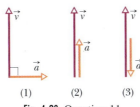

(1)   (2)   (3)

Fig. 4-28 Question 11.

12. Figure 4-29 shows four tracks (either half- or quarter-circles) that can be taken by a train, which moves at a constant speed. Rank the tracks according to the magnitude of a train's acceleration on the curved portion, greatest first.

13. (a) Is it possible to be accelerating while traveling at constant speed? Is it possible to round a curve with (b) zero acceleration and (c) a constant magnitude of acceleration?

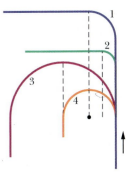

Fig. 4-29 Question 12.

EXERCISES & PROBLEMS

SEC. 4-2   Position and Displacement

1E. A watermelon seed has the following coordinates: $x = -5.0$ m, $y = 8.0$ m, and $z = 0$ m. Find its position vector (a) in unit-vector notation and as (b) a magnitude and (c) an angle relative to the positive direction of the x axis. (d) Sketch the vector on a right-handed coordinate system. If the seed is moved to the xyz coordinates (3.00 m, 0 m, 0 m), what is its displacement (e) in unit-vector notation and as (f) a magnitude and (g) an angle relative to the positive direction of the x axis?

2E. The position vector for an electron is $\vec{r} = (5.0$ m$)\hat{i} - (3.0$ m$)\hat{j} + (2.0$ m$)\hat{k}$. (a) Find the magnitude of $\vec{r}$. (b) Sketch the vector on a right-handed coordinate system.

3E. The position vector for a proton is initially $\vec{r} = 5.0\hat{i} - 6.0\hat{j} + 2.0\hat{k}$ and then later is $\vec{r} = -2.0\hat{i} + 6.0\hat{j} + 2.0\hat{k}$, all in meters. (a) What is the proton's displacement vector, and (b) to what plane is that vector parallel?

4P. A radar station detects an airplane approaching directly from the east. At first observation, the range to the plane is 360 m at 40° above the horizon. The airplane is tracked for another 123° in the vertical east–west plane, the range at final contact being 790 m. See Fig. 4-30. Find the displacement of the airplane during the period of observation.

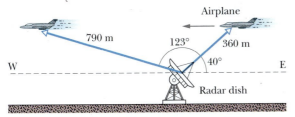

Fig. 4-30 Problem 4.

SEC. 4-3   Average Velocity and Instantaneous Velocity

5E. A train moving at a constant speed of 60.0 km/h moves east for 40.0 min, then in a direction 50.0° east of north for 20.0 min, and finally west for 50.0 min. What is the average velocity of the train during this trip? **ssm**

6E. An ion's position vector is initially $\vec{r} = 5.0\hat{i} - 6.0\hat{j} + 2.0\hat{k}$, and 10 s later it is $\vec{r} = -2.0\hat{i} + 8.0\hat{j} - 2.0\hat{k}$, all in meters. What is its average velocity during the 10 s?

7P. The position of an electron is given by $\vec{r} = 3.00t\hat{i} - 4.00t^2\hat{j} + 2.00\hat{k}$, with t in seconds and $\vec{r}$ in meters. (a) What is the electron's velocity $\vec{v}(t)$? At $t = 2.00$ s, what is $\vec{v}$ (b) in unit-vector notation and as (c) a magnitude and (d) an angle relative to the positive direction of the x axis?

8P. Oasis A is 90 km west of oasis B. A camel leaves oasis A and during a 50 h period walks 75 km in a direction 37° north of east. The camel then walks toward the south a distance of 65 km in a 35 h period, after which it rests for 5.0 h. (a) What is the camel's displacement with respect to oasis A after resting? (b) What is the camel's average velocity from the time it leaves oasis A until it finishes resting? (c) What is the camel's average speed from the time it leaves oasis A until it finishes resting? (d) If the camel is able to go without water for five days (120 h), what must its average velocity be after resting if it is to reach oasis B just in time?

SEC. 4-4   Average Acceleration and Instantaneous Acceleration

9E. A particle moves so that its position (in meters) as a function of time (in seconds) is $\vec{r} = \hat{i} + 4t^2\hat{j} + t\hat{k}$. Write expressions for (a) its velocity and (b) its acceleration as functions of time. **ssm**

10E. A proton initially has $\vec{v} = 4.0\hat{i} - 2.0\hat{j} + 3.0\hat{k}$ and then 4.0 s later has $\vec{v} = -2.0\hat{i} - 2.0\hat{j} + 5.0\hat{k}$ (in meters per second). For that 4.0 s, what is the proton's average acceleration $\vec{a}_{avg}$ (a) in unit-vector notation and (b) as a magnitude and a direction?

11E. The position $\vec{r}$ of a particle moving in an xy plane is given by $\vec{r} = (2.00t^3 - 5.00t)\hat{i} + (6.00 - 7.00t^4)\hat{j}$, with $\vec{r}$ in meters and t in seconds. Calculate (a) $\vec{r}$, (b) $\vec{v}$, and (c) $\vec{a}$ for $t = 2.00$ s. (d) What is the orientation of a line that is tangent to the particle's path at $t = 2.00$ s?

12E. An iceboat sails across the surface of a frozen lake with constant acceleration produced by the wind. At a certain instant the boat's velocity is $(6.30\hat{i} - 8.42\hat{j})$ m/s. Three seconds later, because of a wind shift, the boat is instantaneously at rest. What is its average acceleration for this 3 s interval?

13P. A particle leaves the origin with an initial velocity $\vec{v} = (3.00\hat{i})$ m/s and a constant acceleration $\vec{a} = (-1.00\hat{i} - 0.500\hat{j})$ m/s². When the particle reaches its maximum x coordinate, what are (a) its velocity and (b) its position vector? ssm ilw www

14P. The velocity $\vec{v}$ of a particle moving in the xy plane is given by $\vec{v} = (6.0t - 4.0t^2)\hat{i} + 8.0\hat{j}$, with $\vec{v}$ in meters per second and $t\ (> 0)$ in seconds. (a) What is the acceleration when $t = 3.0$ s? (b) When (if ever) is the acceleration zero? (c) When (if ever) is the velocity zero? (d) When (if ever) does the speed equal 10 m/s?

15P. A particle starts from the origin at $t = 0$ with a velocity of $8.0\hat{j}$ m/s and moves in the xy plane with a constant acceleration of $(4.0\hat{i} + 2.0\hat{j})$ m/s². At the instant the particle's x coordinate is 29 m, what are (a) its y coordinate and (b) its speed?

16P. Particle A moves along the line $y = 30$ m with a constant velocity $\vec{v}$ of magnitude 3.0 m/s and directed parallel to the positive x axis (Fig. 4-31). Particle B starts at the origin with zero speed and constant acceleration $\vec{a}$ (of magnitude 0.40 m/s²) at the same instant that particle A passes the y axis. What angle θ between $\vec{a}$ and the positive y axis would result in a collision between these two particles? (If your computation involves an equation with a term such as t^4, substitute $u = t^2$ and then consider solving the resulting quadratic equation to get u.)

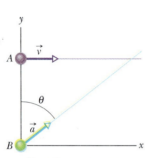

Fig. 4-31 Problem 16.

SEC. 4-6 Projectile Motion Analyzed

In some of these problems, exclusion of the effects of the air is unwarranted but helps simplify the calculations.

17E. A rifle is aimed horizontally at a target 30 m away. The bullet hits the target 1.9 cm below the aiming point. What are (a) the bullet's time of flight and (b) its speed as it emerges from the rifle? ssm

18E. A small ball rolls horizontally off the edge of a tabletop that is 1.20 m high. It strikes the floor at a point 1.52 m horizontally away from the edge of the table. (a) How long is the ball in the air? (b) What is its speed at the instant it leaves the table?

19E. A baseball leaves a pitcher's hand horizontally at a speed of 161 km/h. The distance to the batter is 18.3 m. (Ignore the effect of air resistance.) (a) How long does the ball take to travel the first half of that distance? (b) The second half? (c) How far does the ball fall freely during the first half? (d) During the second half? (e) Why aren't the quantities in (c) and (d) equal?

20E. A dart is thrown horizontally with an initial speed of 10 m/s toward point P, the bull's-eye on a dart board. It hits at point Q on the rim, vertically below P, 0.19 s later. (a) What is the distance PQ? (b) How far away from the dart board is the dart released?

21E. An electron, with an initial horizontal velocity of magnitude 1.00×10^9 cm/s, travels into the region between two horizontal metal plates that are electrically charged. In that region, it travels a horizontal distance of 2.00 cm and has a constant downward acceleration of magnitude 1.00×10^{17} cm/s² due to the charged plates. Find (a) the time required by the electron to travel the 2.00 cm and (b) the vertical distance it travels during that time. Also find the magnitudes of the (c) horizontal and (d) vertical velocity components of the electron as it emerges. ssm

22E. In the 1991 World Track and Field Championships in Tokyo, Mike Powell (Fig. 4-32) jumped 8.95 m, breaking the 23-year long-jump record set by Bob Beamon by a full 5 cm. Assume that Powell's speed on takeoff was 9.5 m/s (about equal to that of a sprinter) and that $g = 9.80$ m/s² in Tokyo. How much less was Powell's horizontal range than the maximum possible horizontal range (neglecting the effects of air) for a particle launched at the same speed of 9.5 m/s?

Fig. 4-32 Exercise 22.

23E. A stone is catapulted at time $t = 0$, with an initial velocity of magnitude 20.0 m/s and at an angle of 40.0° above the horizontal. What are the magnitudes of the (a) horizontal and (b) vertical components of its displacement from the catapult site at $t = 1.10$ s? Repeat for the (c) horizontal and (d) vertical components at $t = 1.80$ s, and for the (e) horizontal and (f) vertical components at $t = 5.00$ s. ssm

24P. A golf ball is struck at ground level. The speed of the golf ball as a function of the time is shown in Fig. 4-33, where $t = 0$ at the instant the ball is struck. (a) How far does the golf ball travel horizontally before returning to ground level? (b) What is the maximum height above ground level attained by the ball?

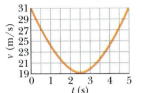

Fig. 4-33 Problem 24.

25P. A rifle that shoots bullets at 460 m/s is to be aimed at a target 45.7 m away and level with the rifle. How high above the target must the rifle barrel be pointed so that the bullet hits the target? ssm

26P. The pitcher in a slow-pitch softball game releases the ball at a point 3.0 ft above ground level. A stroboscopic plot of the position of the ball is shown in Fig. 4-34, where the readings are 0.25 s apart and the ball is released at $t = 0$. (a) What is the initial speed of the ball? (b) What is the speed of the ball at the instant it reaches its maximum height above ground level? (c) What is that maximum height?

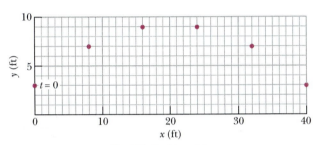

Fig. 4-34 Problem 26.

27P. Show that the maximum height reached by a projectile is $y_{max} = (v_0 \sin \theta_0)^2/2g$. ssm www

28P. You throw a ball toward a wall with a speed of 25.0 m/s and at an angle of 40.0° above the horizontal (Fig. 4-35). The wall is 22.0 m from the release point of the ball. (a) How far above the release point does the ball hit the wall? (b) What are the horizontal and vertical components of its velocity as it hits the wall? (c) When it hits, has it passed the highest point on its trajectory?

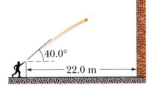

Fig. 4-35 Problem 28.

29P. A ball is shot from the ground into the air. At a height of 9.1 m, its velocity is observed to be $\vec{v} = 7.6\hat{i} + 6.1\hat{j}$ in meters per second ($\hat{i}$ horizontal, $\hat{j}$ upward). (a) To what maximum height does the ball rise? (b) What total horizontal distance does the ball travel? What are the (c) magnitude and (d) the direction of the ball's velocity just before it hits the ground? ilw

30P. Two seconds after being projected from ground level, a projectile is displaced 40 m horizontally and 53 m vertically above its point of projection. What are the (a) horizontal and (b) vertical components of the initial velocity of the projectile? (c) At the instant the projectile achieves its maximum height above ground level, how far is it displaced horizontally from its point of projection?

31P. A football player punts the football so that it will have a "hang time" (time of flight) of 4.5 s and land 46 m away. If the ball leaves the player's foot 150 cm above the ground, what must be (a) the magnitude and (b) the direction of the ball's initial velocity? ssm

32P. A lowly high diver pushes off horizontally with a speed of 2.00 m/s from the edge of a platform that is 10.0 m above the surface of the water. (a) At what horizontal distance from the edge of the platform is the diver 0.800 s after pushing off? (b) At what vertical distance above the surface of the water is the diver just then? (c) At what horizontal distance from the edge of the platform does the diver strike the water?

33P. A certain airplane has a speed of 290.0 km/h and is diving at an angle of 30.0° below the horizontal when the pilot releases a radar decoy (Fig. 4-36). The horizontal distance between the release point and the point where the decoy strikes the ground is 700 m. (a) How long is the decoy in the air? (b) How high was the released point? ilw

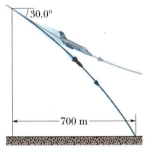

Fig. 4-36 Problem 33.

34P. The launching speed of a certain projectile is five times the speed it has at its maximum height. Calculate the elevation angle θ_0 at launching.

35P. A ball rolls horizontally off the top of a stairway with a speed of 1.52 m/s. The steps are 20.3 cm high and 20.3 cm wide. Which step does the ball hit first? ssm

36P. A soccer ball is kicked from the ground with an initial speed of 19.5 m/s at an upward angle of 45°. A player 55 m away in the direction of the kick starts running to meet the ball at that instant. What must be his average speed if he is to meet the ball just before it hits the ground? Neglect air resistance.

37P. An airplane, diving at an angle of 53.0° with the vertical, releases a projectile at an altitude of 730 m. The projectile hits the ground 5.00 s after being released. (a) What is the speed of the aircraft? (b) How far did the projectile travel horizontally during its flight? What were the (c) horizontal and (d) vertical components of its velocity just before striking the ground? ssm

38P. For women's volleyball the top of the net is 2.24 m above the floor and the court measures 9.0 m by 9.0 m on each side of the net. Using a jump serve, a player strikes the ball at a point that is 3.0 m above the floor and a horizontal distance of 8.0 m from the net. If the initial velocity of the ball is horizontal, (a) what minimum magnitude must it have if the ball is to clear the net and (b) what maximum magnitude can it have if the ball is to strike the floor inside the back line on the other side of the net?

39P. A batter hits a pitched ball when the center of the ball is 1.22 m above the ground. The ball leaves the bat at an angle of 45° with the ground. With that launch, the ball should have a horizontal range (returning to the *launch* level) of 107 m. (a) Does the ball clear a 7.32-m-high fence that is 97.5 m horizontally from the launch point? (b) Either way, find the distance between the top of the fence and the center of the ball when the ball reaches the fence. ssm www

40P. During a tennis match, a player serves the ball at 23.6 m/s, with the center of the ball leaving the racquet horizontally 2.37 m above the court surface. The net is 12 m away and 0.90 m high. When the ball reaches the net, (a) does the ball clear it and (b) what is the distance between the center of the ball and the top of the net? Suppose that, instead, the ball is served as before but now it leaves the racquet at 5.00° below the horizontal. When the ball reaches the net, (c) does the ball clear it and (d) what now is the distance between the center of the ball and the top of the net?

41P. A football kicker can give the ball an initial speed of 25 m/s. Within what two elevation angles must he kick the ball to score a field goal from a point 50 m in front of goalposts whose horizontal bar is 3.44 m above the ground? (If you want to work this out algebraically, use $\sin^2 \theta + \cos^2 \theta = 1$ to get a relation between $\tan^2 \theta$ and $1/\cos^2 \theta$, substitute, and then solve the resulting quadratic equation.) ssm

SEC. 4-7   Uniform Circular Motion

42E. What is the magnitude of the acceleration of a sprinter running at 10 m/s when rounding a turn with a radius of 25 m?

43E. An Earth satellite moves in a circular orbit 640 km above Earth's surface with a period of 98.0 min. What are (a) the speed and (b) the magnitude of the centripetal acceleration of the satellite? ssm

44E. A rotating fan completes 1200 revolutions every minute. Consider the tip of a blade, at a radius of 0.15 m. (a) Through what distance does the tip move in one revolution? What are (b) the tip's speed and (c) the magnitude of its acceleration? (d) What is the period of the motion?

45E. An astronaut is rotated in a horizontal centrifuge at a radius of 5.0 m. (a) What is the astronaut's speed if the centripetal acceleration has a magnitude of 7.0g? (b) How many revolutions per minute are required to produce this acceleration? (c) What is the period of the motion? ssm

46P. A carnival merry-go-round rotates about a vertical axis at a constant rate. A passenger standing on the edge of the merry-go-round has a constant speed of 3.66 m/s. For each of the following instantaneous situations, state how far the passenger is from the center of the merry-go-round, and in which direction. (a) The passenger has an acceleration of 1.83 m/s², east. (b) The passenger has an acceleration of 1.83 m/s², south.

47P. (a) What is the magnitude of the centripetal acceleration of an object on Earth's equator owing to the rotation of Earth? (b) What would the period of rotation of Earth have to be for objects on the equator to have a centripetal acceleration with a magnitude of 9.8 m/s²? ssm www

48P. The fast French train known as the TGV (Train à Grande Vitesse) has a scheduled average speed of 216 km/h. (a) If the train goes around a curve at that speed and the magnitude of the acceleration experienced by the passengers is to be limited to 0.050g, what is the smallest radius of curvature for the track that can be tolerated? (b) At what speed must the train go around a curve with a 1.00 km radius to be at the acceleration limit?

49P. A carnival Ferris wheel has a 15 m radius and completes five turns about its horizontal axis every minute. (a) What is the period of the motion? What is the centripetal acceleration of a passenger at (b) the highest point and (c) the lowest point, assuming the passenger is at a 15 m radius? ssm ilw

50P. When a large star becomes a *supernova*, its core may be compressed so tightly that it becomes a *neutron star*, with a radius of about 20 km (about the size of the San Francisco area). If a neutron star rotates once every second, (a) what is the speed of a particle on the star's equator and (b) what is the magnitude of the particle's centripetal acceleration? (c) If the neutron star rotates faster, do the answers to (a) and (b) increase, decrease, or remain the same?

51P. A boy whirls a stone in a horizontal circle of radius 1.5 m and at height 2.0 m above level ground. The string breaks, and the stone flies off horizontally and strikes the ground after traveling a horizontal distance of 10 m. What is the magnitude of the centripetal acceleration of the stone while in circular motion? ssm

52P. A particle P travels with constant speed on a circle of radius $r = 3.00$ m (Fig. 4-37) and completes one revolution in 20.0 s. The particle passes through O at time $t = 0$. State the following vectors in magnitude-angle notation (angle relative to the positive direction of x). With respect to O, find the particle's position vector at the times t of (a) 5.00 s, (b) 7.50 s, and (c) 10.0 s.

(d) For the 5.00 s interval from the end of the fifth second to the end of the tenth second, find the particle's displacement. (e) For the same interval, find its average velocity. Find its velocity at (f) the beginning and (g) the end of that 5.00 s interval. Next, find the acceleration at (h) the beginning and (i) the end of that interval.

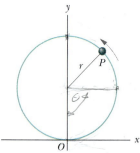

Fig. 4-37 Problem 52.

SEC. 4-8   Relative Motion in One Dimension

53E. A cameraman on a pickup truck is traveling westward at 20 km/h while he videotapes a cheetah that is moving westward 30 km/h faster than the truck. Suddenly, the cheetah stops, turns, and then runs at 45 km/h eastward, as measured by a suddenly nervous crew member who stands alongside the cheetah's path. The change in the animal's velocity takes 2.0 s. What is its acceleration from the perspective of (a) the cameraman and (b) the nervous crew member?

54E. A boat is traveling upstream at 14 km/h with respect to the water of a river. The water is flowing at 9 km/h with respect to the ground. (a) What is the velocity of the boat with respect to the ground? (b) A child on the boat walks from front to rear at 6 km/h with respect to the boat. What is the child's velocity with respect to the ground?

55P. A person walks up a stalled 15-m-long escalator in 90 s. When standing on the same escalator, now moving, the person is carried up in 60 s. How much time would it take that person to walk up the moving escalator? Does the answer depend on the length of the escalator? ssm

SEC. 4-9   Relative Motion in Two Dimensions

56E. In rugby a player can legally pass the ball to a teammate as long as the pass is not "forward" (it must not have a velocity component parallel to the length of the field and directed toward the other team's goal). Suppose a player runs parallel to the field's length and toward the other team's goal with a speed of 4.0 m/s while he passes the ball with a speed of 6.0 m/s relative to himself. What is the smallest angle from the forward direction that keeps the pass legal?

57E. Snow is falling vertically at a constant speed of 8.0 m/s. At what angle from the vertical do the snowflakes appear to be falling as viewed by the driver of a car traveling on a straight, level road with a speed of 50 km/h?   ssm

58E. Two highways intersect as shown in Fig. 4-38. At the instant shown, a police car P is 800 m from the intersection and moving at 80 km/h. Motorist M is 600 m from the intersection and moving at 60 km/h. (a) In unit-vector notation, what is the velocity of the motorist with respect to the police car? (b) For the instant shown in Fig. 4-38, how does the direction of the velocity found in (a) compare to the line of sight between the two cars? (c) If the cars maintain their velocities, do the answers to (a) and (b) change as the cars move nearer the intersection?

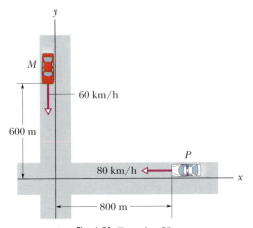

Fig. 4-38 Exercise 58.

59P. A train travels due south at 30 m/s (relative to the ground) in a rain that is blown toward the south by the wind. The path of each raindrop makes an angle of 70° with the vertical, as measured by an observer stationary on the ground. An observer on the train, however, sees the drops fall perfectly vertically. Determine the speed of the raindrops relative to the ground.   ssm

60P. Ship A is located 4.0 km north and 2.5 km east of ship B. Ship A has a velocity of 22 km/h toward the south and ship B has a velocity of 40 km/h in a direction 37° north of east. (a) What is the velocity of A relative to B? (Express your answer in terms of the unit vectors $\hat{i}$ and $\hat{j}$, where $\hat{i}$ is toward the east.) (b) Write an expression (in terms of $\hat{i}$ and $\hat{j}$) for the position of A relative to B as a function of t, where $t = 0$ when the ships are in the positions described above. (c) At what time is the separation between the ships least? (d) What is that least separation?

61P. Two ships, A and B, leave port at the same time. Ship A travels northwest at 24 knots and ship B travels at 28 knots in a direction

40° west of south. (1 knot = 1 nautical mile per hour; see Appendix D.) (a) What are the magnitude and direction of the velocity of ship A relative to B? (b) After what time will the ships be 160 nautical miles apart? (c) What will be the bearing of B (the direction of the position of B) relative to A at that time?   ssm   ilw

62P. A wooden boxcar is moving along a straight railroad track at speed v_1. A sniper fires a bullet (initial speed v_2) at it from a high-powered rifle. The bullet passes through both lengthwise walls of the car, its entrance and exit holes being exactly opposite each other as viewed from within the car. From what direction, relative to the track, is the bullet fired? Assume that the bullet is not deflected upon entering the car, but that its speed decreases by 20%. Take $v_1 = 85$ km/h and $v_2 = 650$ m/s. (Why don't you need to know the width of the boxcar?)

63P. A 200-m-wide river has a uniform flow speed of 1.1 m/s through a jungle and toward the east. An explorer wishes to leave a small clearing on the south bank and cross the river in a power-boat that moves at a constant speed of 4.0 m/s with respect to water. There is a clearing on the north bank 82 m upstream from a point directly opposite the clearing on the south bank. (a) In what direction must the boat be pointed in order to travel in a straight line and land in the clearing on the north bank? (b) How long will the boat take to cross the river and land in the clearing?

Additional Problem

64. *Curtain of death.* A large metallic asteroid strikes Earth and quickly digs a crater into the rocky material below ground level by launching rocks upward and outward. The following table gives five pairs of launch speeds and angles (from the horizontal) for such rocks, based on a model of crater formation. (Other rocks, with intermediate speeds and angles, are also launched.) Suppose that you are at $x = 20$ km when the asteroid strikes the ground at time $t = 0$ and position $x = 0$ (Fig. 4-39). (a) At $t = 20$ s, what are the x and y coordinates of the rocks headed in your direction from launches A through E? (b) Plot these coordinates and then sketch a curve through the points to include rocks with intermediate launch speeds and angles. The curve should give you an idea of what you would see as you look up into the approaching rocks and what dinosaurs must have seen during asteroid strikes long ago.

Launch	Speed (m/s)	Angle (degrees)
A	520	14.0
B	630	16.0
C	750	18.0
D	870	20.0
E	1000	22.0

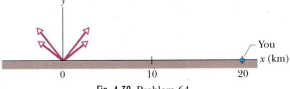

Fig. 4-39 Problem 64.

5 Force and Motion—I

On April 4, 1974, John Massis of Belgium managed to move two passenger cars belonging to New York's Long Island Railroad. He did so by clamping his teeth down on a bit that was attached to the cars with a rope and then leaning backward while pressing his feet against the railway ties. The cars together weighed about 80 tons.

Did Massis have to pull with superhuman force to accelerate them?

The answer is in this chapter.

5-1   What Causes an Acceleration?

If you see the velocity of a particle-like body change in either magnitude or direction, you know that something must have *caused* that change (that acceleration). Indeed, out of common experience, you know that the change in velocity must be due to an interaction between the body and something in its surroundings. For example, if you see a hockey puck that is sliding across an ice rink suddenly stop or suddenly change direction, you will suspect that the puck encountered a slight ridge on the ice surface.

An interaction that can cause an acceleration of a body is called a **force**, which is, loosely speaking, a push or pull on the body—the force is said to *act* on the body. For example, the bump on the hockey puck by the ridge is a push on the puck, causing an acceleration. The relationship between a force and the acceleration it causes was first understood by Isaac Newton (1642–1727) and is the subject of this chapter. The study of that relationship, as Newton presented it, is called *Newtonian mechanics*. We shall focus on its three primary laws of motion.

Newtonian mechanics does not apply to all situations. If the speeds of the interacting bodies are very large—an appreciable fraction of the speed of light—we must replace Newtonian mechanics with Einstein's special theory of relativity, which holds at any speed, including those near the speed of light. If the interacting bodies are on the scale of atomic structure (for example, they might be electrons within an atom), we must replace Newtonian mechanics with quantum mechanics. Physicists now view Newtonian mechanics as a special case of these two more comprehensive theories. Still, it is a very important special case because it applies to the motion of objects ranging in size from the very small (almost on the scale of atomic structure) to astronomical (objects such as galaxies and clusters of galaxies).

5-2   Newton's First Law

Before Newton formulated his mechanics, it was thought that some influence, a "force," was needed to keep a body moving at constant velocity. Similarly, a body was thought to be in its "natural state" when it was at rest. For it to move with constant velocity, it seemingly had to be propelled in some way, by a push or a pull. Otherwise, it would "naturally" stop moving.

These ideas were reasonable. If you send a puck sliding across a wooden floor, it does indeed slow and then stop. If you want to make it move across the floor with constant velocity, you have to continuously pull or push it.

Send a puck sliding over the ice of a skating rink, however, and it goes a lot farther. You can imagine longer and more slippery surfaces, over which the puck would slide farther and farther. In the limit you can think of a long, extremely slippery surface (said to be a **frictionless surface**), over which the puck would hardly slow. (We can in fact come close to this situation in the laboratory, by sending a puck sliding over a horizontal air table, across which it moves on a film of air.)

From these observations, we can conclude that a body will keep moving with constant velocity if no force acts on it. That leads us to the first of Newton's three laws of motion:

> **Newton's First Law:** If no force acts on a body, then the body's velocity cannot change; that is, the body cannot accelerate.

In other words, if the body is at rest, it stays at rest. If it is moving, it will continue to move with the same velocity (same magnitude *and* same direction).

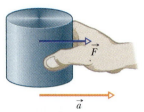

Fig. 5-1 A force $\vec{F}$ on the standard kilogram gives that body an acceleration $\vec{a}$.

5-3 Force

We now wish to define the unit of force. We know that a force can cause the acceleration of a body. Thus, we shall define the unit of force in terms of the acceleration that a force gives to a standard reference body. As the standard body, we shall use (or rather imagine that we use) the standard kilogram of Fig. 1-4. This body has been assigned, exactly and by definition, a mass of 1 kg.

We put the standard body on a horizontal frictionless table and pull the body to the right (Fig. 5-1) so that by trial and error, it eventually experiences a measured acceleration of 1 m/s². We then declare, as a matter of definition, that the force we are exerting on the standard body has a magnitude of 1 newton (abbreviated N).

We can exert a 2 N force on our standard body by pulling it so that its measured acceleration is 2 m/s², and so on. Thus in general, if our standard body of 1 kg mass has an acceleration of magnitude a, we know that force F must be acting on it and that the magnitude of the force (in newtons) is equal to the magnitude of the acceleration (in meters per second per second).

Thus, a force is measured by the acceleration it produces. However, acceleration is a vector quantity, with both magnitude and direction. Is force also a vector quantity? We can easily assign a direction to a force (just assign the direction of the acceleration), but that is not sufficient. We must prove by experiment that forces are vector quantities. Actually, that has been done: forces are indeed vector quantities; they have magnitudes and directions and they combine according to the vector rules of Chapter 3.

This means that when two or more forces act on a body, we can find their **net force** or **resultant force** by adding the individual forces vectorially. A single force with the magnitude and direction of the net force has the same effect on the body as all the individual forces together. This fact is called the **principle of superposition for forces.** The world would be quite strange if, for example, you and a friend were to pull on the standard body in the same direction, each with a force of 1 N, and yet somehow the net pull was 14 N.

In this book, forces are most often represented with a vector symbol such as $\vec{F}$, and a net force is represented with the vector symbol $\vec{F}_{net}$. As with other vectors, a force or a net force can have components along coordinate axes. When forces act only along a single axis, they are single-component forces. Then we can drop the overhead arrows on the force symbols and just use signs to indicate the directions of the forces along that axis.

Instead of what was previously given, the proper statement of Newton's First Law is in terms of a *net* force:

> **Newton's First Law:** If no *net* force acts on a body ($\vec{F}_{net} = 0$), then the body's velocity cannot change; that is, the body cannot accelerate.

There may be multiple forces acting on a body, but if their net (or resultant) force is zero, then the body cannot accelerate.

Inertial Reference Frames

Newton's first law is not true in all reference frames, but we can always find reference frames in which it (and the rest of Newtonian mechanics) is true. Such frames are called **inertial reference frames,** or simply **inertial frames.**

> An inertial reference frame is one in which Newton's laws hold.

For example, we can assume that the ground is an inertial frame provided that we can neglect Earth's actual astronomical motions (such as its rotation).

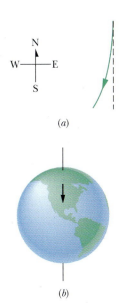

(a)

(b)

Fig. 5-2 (a) The path of a puck that is sent sliding due south along a long strip of frictionless ice, as seen by an observer on the ground. (b) The ground beneath the southward sliding puck rotates to the east as Earth rotates.

That assumption works well if we, say, send a puck sliding along a short strip of frictionless ice—an observer on the ground would find that the puck's motion obeys the laws of Newtonian mechanics. However, suppose we made the strip very long, extending, say, from north to south. Then an observer on the ground would find that the puck accelerates slightly toward the west as it moves south (Fig. 5-2a); yet, that observer would not be able to find a force that causes the westward acceleration. In this case, the ground is a **noninertial frame** because, for the puck's long path, Earth's rotation cannot be neglected. The surprising westward acceleration of the sliding puck relative to the ground is actually due to the eastward rotation of the ground beneath the puck (Fig. 5-2b).

In this book we usually assume that the ground is an inertial frame and that measurements of forces and accelerations are made from it. If measurements are made in, say, an elevator that is accelerating relative to the ground, then the measurements are being made in a noninertial frame and the results can be surprising. We see an example of this in Sample Problem 5-8.

✔**CHECKPOINT 1:** Which of the figure's six arrangements correctly show the vector addition of forces $\vec{F}_1$ and $\vec{F}_2$ to yield the third vector, which is meant to represent their net force $\vec{F}_{net}$?

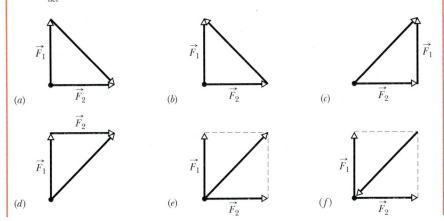

5-4 Mass

Everyday experience tells us that a given force produces different magnitudes of acceleration for different bodies. Put a baseball and a bowling ball on the floor and give both the same sharp kick. Even if you don't actually do this, you know the result: The baseball receives a noticeably larger acceleration than the bowling ball. The two accelerations differ because the mass of the baseball differs from the mass of the bowling ball—but what, exactly, is mass?

We can explain how to measure mass by imagining a series of experiments in an inertial frame. In the first experiment we exert a force on our standard body, whose mass m_0 is defined to be 1.0 kg. Suppose that the standard body accelerates at 1.0 m/s². We can then say the force on that body is 1.0 N.

We next apply that same force (we would need some way of being certain it is the same force) to a second body, body X, whose mass is not known. Suppose we find that this body X accelerates at 0.25 m/s². We know that a *less massive* baseball receives a *greater acceleration* than a more massive bowling ball when the same force (kick) is applied to both. Let us then make the following conjecture: The ratio of the masses of two bodies is equal to the inverse of the ratio of their accelerations when the same force is applied to both. For body X and the standard body, this tells us that

$$\frac{m_X}{m_0} = \frac{a_0}{a_X}.$$

Solving for m_X yields

$$m_X = m_0 \frac{a_0}{a_X} = (1.0 \text{ kg}) \frac{1.0 \text{ m/s}^2}{0.25 \text{ m/s}^2} = 4.0 \text{ kg}.$$

Our conjecture will be useful, of course, only if it continues to hold when we change the applied force to other values. For example, if we apply an 8.0 N force to the standard body, we obtain an acceleration of 8.0 m/s². When the 8.0 N force is applied to body X, we obtain an acceleration of 2.0 m/s². Our conjecture then gives us

$$m_X = m_0 \frac{a_0}{a_X} = (1.0 \text{ kg}) \frac{8.0 \text{ m/s}^2}{2.0 \text{ m/s}^2} = 4.0 \text{ kg},$$

consistent with our first experiment. Many experiments yielding similar results indicate that our conjecture provides a consistent and reliable means of assigning a mass to any given body.

Our measurement experiments indicate that mass is an *intrinsic* characteristic of a body—that is, a characteristic that automatically comes with the existence of the body. They also indicate that mass is a scalar quantity. However, the nagging question remains: What, exactly, is mass?

Since the word *mass* is used in everyday English, we should have some intuitive understanding of it, maybe something that we can physically sense. Is it a body's size, weight, or density? The answer is no, although those characteristics are sometimes confused with mass. We can say only that *the mass of a body is the characteristic that relates a force on the body to the resulting acceleration.* Mass has no more familiar definition; you can have a physical sensation of mass only when you attempt to accelerate a body, as in the kicking of a baseball or a bowling ball.

5-5 Newton's Second Law

All the definitions, experiments, and observations that we have discussed so far can be summarized in one neat statement:

> **Newton's Second Law:** The net force on a body is equal to the product of the body's mass and the acceleration of the body.

In equation form,

$$\vec{F}_{net} = m\vec{a} \qquad \text{(Newton's second law).} \qquad (5\text{-}1)$$

This equation is simple, but we must use it cautiously. First, we must be certain about which body we are applying it to. Then $\vec{F}_{net}$ must be the vector sum of *all* the forces that act on *that* body. Only forces that act on *that* body are to be included in the vector sum, not forces acting on other bodies that might be involved in the given situation. For example, if you are in a rugby scrum, the net force on *you* is the vector sum of all the pushes and pulls on *your* body. It does not include any push or pull on another player from you.

Like other vector equations, Eq. 5-1 is equivalent to three component equations, one written for each axis of an *xyz* coordinate system:

$$F_{net,x} = ma_x, \quad F_{net,y} = ma_y, \quad \text{and} \quad F_{net,z} = ma_z. \qquad (5\text{-}2)$$

TABLE 5-1   Units in Newton's Second Law (Eqs. 5-1 and 5-2)

System	Force	Mass	Acceleration
SI	newton (N)	kilogram (kg)	m/s^2
CGS^a	dyne	gram (g)	cm/s^2
British^b	pound (lb)	slug	ft/s^2

^{a}1 dyne = 1 g·cm/s^2.

^{b}1 lb = 1 slug·ft/s^2.

Each of these equations relates the net force component along an axis to the acceleration along that same axis. For example, the first equation tells us that the sum of all the force components along the x axis causes the x component a_x of the body's acceleration, but causes no acceleration in the y and z directions. Turned around, the acceleration component a_x is caused only by the sum of the force components along the x axis. In general,

> The acceleration component along a given axis is caused only by the sum of the force components along that *same* axis, and not by force components along any other axis.

Equation 5-1 tells us that if the net force on a body is zero, the body's acceleration $\vec{a} = 0$. If the body is at rest, it stays at rest; if it is moving, it continues to move at constant velocity. In such cases, any forces on the body *balance* one another, and both the forces and the body can be said to be in *equilibrium*. Commonly, the forces are also said to *cancel* one another, but the term "cancel" is tricky. It does *not* mean that the forces cease to exist (canceling forces is not like canceling dinner reservations). The forces still act on the body.

For SI units, Eq. 5-1 tells us that

$$1 \text{ N} = (1 \text{ kg})(1 \text{ m/s}^2) = 1 \text{ kg} \cdot \text{m/s}^2. \tag{5-3}$$

Some force units in other systems of units are given in Table 5-1 and Appendix D.

To solve problems with Newton's second law, we often draw a **free-body diagram** in which the only body shown is the one for which we are summing forces. A sketch of the body itself is preferred by some teachers but, to save space in these chapters, we shall usually represent the body with a dot. Also, each force on the body is drawn as a vector arrow with its tail on the body. A coordinate system is usually included, and the acceleration of the body is sometimes shown with a vector arrow (labeled as an acceleration).

A collection of two or more bodies is called a **system,** and any force on the bodies inside the system from bodies outside the system is called an **external force.** If the bodies are rigidly connected, then we can treat the system as one composite body, and the net force $\vec{F}_{net}$ on it is the vector sum of all external forces. (We do not include **internal forces**—that is, forces between two bodies inside the system.) For example, a connected railroad engine and car form a system. If, say, a tow line pulls on the front of the engine, then the force due to the tow line acts on the whole engine–car system. Just as for a single body, we can relate the net external force on a system to its acceleration with Newton's second law, $\vec{F}_{net} = m\vec{a}$, where m is the total mass of the system.

✓**CHECKPOINT 2:**   The figure here shows two horizontal forces acting on a block that is on a frictionless floor. Assume that a third horizontal force $\vec{F}_3$ also acts on the block. What are the magnitude and direction of $\vec{F}_3$ when the block is (a) stationary and (b) moving to the left with a constant speed of 5 m/s?

3 N ⟵   ⟶ 5 N

Sample Problem 5-1

In Figs. 5-3a to c, one or two forces act on a puck that moves over frictionless ice and along an x axis, in one-dimensional motion. The puck's mass is $m = 0.20$ kg. Forces $\vec{F}_1$ and $\vec{F}_2$ are directed along the axis and have magnitudes $F_1 = 4.0$ N and $F_2 = 2.0$ N. Force $\vec{F}_3$ is directed at angle $\theta = 30°$ and has magnitude $F_3 = 1.0$ N. In each situation, what is the acceleration of the puck?

SOLUTION: The Key Idea in each situation is that we can relate the acceleration $\vec{a}$ to the net force $\vec{F}_{net}$ acting on the puck with Newton's second law, $\vec{F}_{net} = m\vec{a}$. However, because the motion is along only the x axis, we can simplify each situation by writing the second law for x components only:

$$F_{net,x} = ma_x. \tag{5-4}$$

The free-body diagrams for the three situations are given in Figs. 5-3d to f, where the puck is represented by a dot.

For Fig. 5-3d, where only one horizontal force acts, Eq. 5-4 gives us

$$F_1 = ma_x,$$

which, with given data, yields

$$a_x = \frac{F_1}{m} = \frac{4.0 \text{ N}}{0.20 \text{ kg}} = 20 \text{ m/s}^2. \qquad \text{(Answer)}$$

The positive answer indicates that the acceleration is in the positive direction of the x axis.

In Fig. 5-3e, two horizontal forces act on the puck, $\vec{F}_1$ in the positive direction of x and $\vec{F}_2$ in the negative direction. Now Eq. 5-4 gives us

$$F_1 - F_2 = ma_x,$$

which, with given data, yields

$$a_x = \frac{F_1 - F_2}{m} = \frac{4.0 \text{ N} - 2.0 \text{ N}}{0.20 \text{ kg}} = 10 \text{ m/s}^2. \qquad \text{(Answer)}$$

Thus, the net force accelerates the puck in the positive direction of the x axis.

In Fig. 5-3f, force $\vec{F}_3$ is not directed along the direction of the puck's acceleration; only the x component $F_{3,x}$ is. (Force $\vec{F}_3$ is two-dimensional but the motion is only one-dimensional.) Thus, we

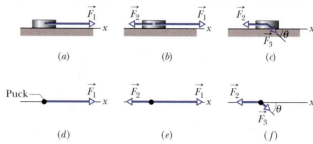

(a)   (b)   (c)

(d)   (e)   (f)

Fig. 5-3 Sample Problem 5-1. (a)–(c) In three situations, forces act on a puck that moves on frictionless ice along an x axis. (d)–(f) Free-body diagrams for the three situations.

write Eq. 5-4 as

$$F_{3,x} - F_2 = ma_x.$$

From the figure, we see that $F_{3,x} = F_3 \cos \theta$. Solving for the acceleration and substituting for $F_{3,x}$ yield

$$a_x = \frac{F_{3,x} - F_2}{m} = \frac{F_3 \cos \theta - F_2}{m}$$

$$= \frac{(1.0 \text{ N})(\cos 30°) - 2.0 \text{ N}}{0.20 \text{ kg}} = -5.7 \text{ m/s}^2. \qquad \text{(Answer)}$$

Thus, the net force accelerates the puck in the negative direction of the x axis.

✓**CHECKPOINT 3:**   The figure shows *overhead* views of four situations in which two forces accelerate the same block across a frictionless floor. Rank the situations according to the magnitudes of (a) the net force on the block and (b) the acceleration of the block, greatest first.

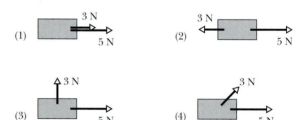

Sample Problem 5-2

In the overhead view of Fig. 5-4a, a 2.0 kg cookie tin is accelerated at 3.0 m/s² in the direction shown by $\vec{a}$, over a frictionless horizontal surface. The acceleration is caused by three horizontal forces, only two of which are shown: $\vec{F}_1$ of magnitude 10 N and $\vec{F}_2$ of magnitude 20 N. What is the third force $\vec{F}_3$ in unit-vector notation and as a magnitude and an angle?

SOLUTION: One Key Idea here is that the net force $\vec{F}_{net}$ on the tin is the sum of the three forces and is related to the acceleration $\vec{a}$ of the tin via Newton's second law ($\vec{F}_{net} = m\vec{a}$). Thus,

$$\vec{F}_1 + \vec{F}_2 + \vec{F}_3 = m\vec{a},$$

which gives us

$$\vec{F}_3 = m\vec{a} - \vec{F}_1 - \vec{F}_2. \tag{5-5}$$

A second Key Idea is that this is a two-dimensional problem; so we *cannot* find $\vec{F}_3$ merely by substituting the magnitudes for the vector quantities on the right side of Eq. 5-5. Instead, we must vectorially add $m\vec{a}$, $-\vec{F}_1$ (the reverse of $\vec{F}_1$), and $-\vec{F}_2$ (the reverse of $\vec{F}_2$), as is shown in Fig. 5-4b. This addition can be done directly on a vector-capable calculator because we know both magnitude and angle for each of these three vectors. However, here we shall evaluate the right side of Eq. 5-5 in terms of components, first along the x axis and then along the y axis.

Along the x axis we have

$$F_{3,x} = ma_x - F_{1,x} - F_{2,x}$$

$$= m(a \cos 50°) - F_1 \cos(-150°) - F_2 \cos 90°.$$

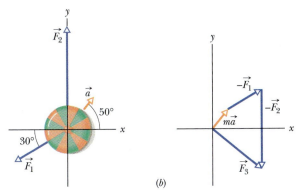

(a)   (b)

Fig. 5-4 Sample Problem 5-2. (a) An overhead view of two of three horizontal forces that act on a cookie tin, resulting in acceleration $\vec{a}$. $\vec{F_3}$ is not shown. (b) An arrangement of vectors $m\vec{a}$, $-\vec{F_1}$, and $-\vec{F_2}$ to find force $\vec{F_3}$.

Then, substituting known data, we find

$$F_{3,x} = (2.0 \text{ kg})(3.0 \text{ m/s}^2) \cos 50° - (10 \text{ N}) \cos(-150°)$$
$$- (20 \text{ N}) \cos 90°$$
$$= 12.5 \text{ N}.$$

Similarly, along the y axis we find

$$F_{3,y} = ma_y - F_{1,y} - F_{2,y}$$
$$= m(a \sin 50°) - F_1 \sin(-150°) - F_2 \sin 90°$$
$$= (2.0 \text{ kg})(3.0 \text{ m/s}^2) \sin 50° - (10 \text{ N}) \sin(-150°)$$
$$- (20 \text{ N}) \sin 90°$$
$$= -10.4 \text{ N}.$$

Thus, in unit-vector notation, we have

$$\vec{F_3} = F_{3,x}\hat{i} + F_{3,y}\hat{j} = (12.5 \text{ N})\hat{i} - (10.4 \text{ N})\hat{j}$$
$$\approx (13 \text{ N})\hat{i} - (10 \text{ N})\hat{j}. \qquad \text{(Answer)}$$

We can now use a vector-capable calculator to get the magnitude and the angle of $\vec{F_3}$. We can also use Eq. 3-6 to obtain the magnitude and the angle (from the positive direction of the x axis) as

$$F_3 = \sqrt{F_{3,x}^2 + F_{3,y}^2} = 16 \text{ N}$$

and

$$\theta = \tan^{-1} \frac{F_{3,y}}{F_{3,x}} = -40°. \qquad \text{(Answer)}$$

Sample Problem 5-3

In a two-dimensional tug-of-war, Alex, Betty, and Charles pull horizontally on an automobile tire at the angles shown in the overhead view of Fig. 5-5a. The tire remains stationary in spite of the three pulls. Alex pulls with force $\vec{F_A}$ of magnitude 220 N, and Charles pulls with force $\vec{F_C}$ of magnitude 170 N. The direction of $\vec{F_C}$ is not given. What is the magnitude of Betty's force $\vec{F_B}$?

SOLUTION: Because the three forces pulling on the tire do not accelerate the tire, the tire's acceleration is $\vec{a} = 0$ (that is, the forces are in equilibrium). The **Key Idea** here is that we can relate that acceleration to the net force $\vec{F}_{\text{net}}$ on the tire with Newton's second law ($\vec{F}_{\text{net}} = m\vec{a}$), which we can write as

$$\vec{F_A} + \vec{F_B} + \vec{F_C} = m(0) = 0,$$

or

$$\vec{F_B} = -\vec{F_A} - \vec{F_C}. \qquad (5\text{-}6)$$

The free-body diagram for the tire is shown in Fig. 5-5b, where we have conveniently centered a coordinate system on the tire and assigned ϕ to the angle of $\vec{F_C}$.

We want to solve for the magnitude of $\vec{F_B}$. Although we know both magnitude and direction for $\vec{F_A}$, we know only the magnitude of $\vec{F_C}$ and not its direction. Thus, with unknowns on both sides of Eq. 5-6, we cannot directly solve it on a vector-capable calculator. Instead we must rewrite Eq. 5-6 in terms of components for either the x or the y axis. Since $\vec{F_B}$ is directed along the y axis, we choose that axis and write

$$F_{By} = -F_{Ay} - F_{Cy}.$$

Evaluating these components with their angles and using the angle $133°$ ($= 180° - 47.0°$) for $\vec{F_A}$, we obtain

$$F_B \sin(-90°) = -F_A \sin 133° - F_C \sin \phi,$$

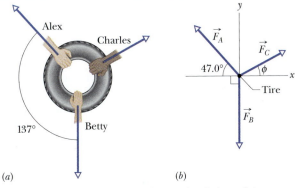

(a)   (b)

Fig. 5-5 Sample Problem 5-3. (a) An overhead view of three people pulling on a tire. (b) A free-body diagram for the tire.

which, with the given data for the magnitudes, yields

$$-F_B = -(220 \text{ N})(\sin 133°) - (170 \text{ N}) \sin \phi. \qquad (5\text{-}7)$$

However, we do not know ϕ.

We can find it by rewriting Eq. 5-6 for the x axis as

$$F_{Bx} = -F_{Ax} - F_{Cx}$$

and then as

$$F_B \cos(-90°) = -F_A \cos 133° - F_C \cos \phi,$$

which gives us

$$0 = -(220 \text{ N})(\cos 133°) - (170 \text{ N}) \cos \phi$$

and

$$\phi = \cos^{-1} -\frac{(220 \text{ N})(\cos 133°)}{170 \text{ N}} = 28.04°.$$

Inserting this into Eq. 5-7, we find

$$F_B = 241 \text{ N}. \qquad \text{(Answer)}$$

PROBLEM-SOLVING TACTICS

Tactic 1: *Dimensions and Vectors*
Many students fail to grasp the second Key Idea in Sample Problem 5-2, and that failure haunts them through the rest of this book. When you are dealing with forces, you cannot just add or subtract their magnitudes to find their net force unless they happen to be directed *along the same axis*. If they are not, you must use vector addition, either by means of a vector-capable calculator or by finding components along axes, as is done in Sample Problem 5-2.

Tactic 2: *Reading Force Problems*
Read the problem statement several times until you have a good mental picture of what the situation is, what data are given, and what is requested. If you know what the problem is about but don't know what to do next, put the problem aside and reread the text. If you are hazy about Newton's second law, reread that section. Study the sample problems. And remember, solving physics problems (like repairing cars and designing computer chips) takes training—you were not born with the ability.

Tactic 3: *Draw Two Types of Figures*
You may need two figures. One is a rough sketch of the actual situation. When you draw the forces, place the tail of each force vector either on the boundary of or within the body on which that force acts. The other figure is a free-body diagram: the forces on a *single* body are drawn, with the body represented by a dot or a sketch. Place the tail of each force vector on the dot or sketch.

Tactic 4: *What Is Your System?*
If you are using Newton's second law, you must know what body or system you are applying it to. In Sample Problem 5-1 it is the puck (not the ice). In Sample Problem 5-2, it is the cookie tin. In Sample Problem 5-3, it is the tire (not the people).

Tactic 5: *Choose Your Axes Wisely*
In Sample Problem 5-3, we saved a lot of work by choosing one of our coordinate axes to coincide with one of the forces (the y axis with $\vec{F}_B$).

5-6   Some Particular Forces

The Gravitational Force

A **gravitational force** $\vec{F}_g$ on a body is a pull that is directed toward a second body. In these early chapters, we do not discuss the nature of this force and we usually consider situations in which the second body is Earth. Thus, when we speak of *the* gravitational force $\vec{F}_g$ on a body, we usually mean a force that pulls on it directly toward the center of Earth—that is, directly down toward the ground. We shall assume that the ground is an inertial frame.

Suppose that the body, of mass m, is in free fall with the free-fall acceleration of magnitude g. Then, if we neglect the effects of the air, the only force acting on the body is the gravitational force $\vec{F}_g$. We can relate this downward force and downward acceleration with Newton's second law ($\vec{F} = m\vec{a}$). We place a vertical y axis along the body's path, with the positive direction upward. For this axis, Newton's second law can be written in the form $F_{\text{net},y} = ma_y$, which, in our situation, becomes

$$-F_g = m(-g)$$

or
$$F_g = mg. \qquad (5\text{-}8)$$

In other words, the magnitude of the gravitational force is equal to the product mg.

This same gravitational force, with the same magnitude, still acts on the body even when the body is not in free fall but, say, at rest on a pool table or moving across the table. (For the gravitational force to disappear, Earth would have to disappear.)

We can write Newton's second law for the gravitational force in these vector forms:

$$\vec{F}_g = -F_g\hat{j} = -mg\hat{j} = m\vec{g}, \qquad (5\text{-}9)$$

where $\hat{j}$ is the unit vector that points upward along a y axis, directly away from the ground, and $\vec{g}$ is the free-fall acceleration (written as a vector), directed downward.

Weight

The **weight** W of a body is the magnitude of the net force required to prevent the body from falling freely, as measured by someone on the ground. For example, to

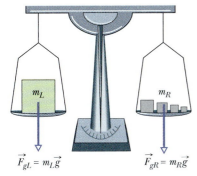

Fig. 5-6 An equal-arm balance. When the device is in balance, the gravitational force $\vec{F}_{gL}$ on the body being weighed (on the left pan) and the total gravitational force $\vec{F}_{gR}$ on the reference bodies (on the right pan) are equal. Thus, the mass of the body being weighed is equal to the total mass of the reference bodies.

keep a ball at rest in your hand while you stand on the ground, you must provide an upward force to balance the gravitational force on the ball from Earth. Suppose the magnitude of the gravitational force is 2.0 N. Then the magnitude of your upward force must be 2.0 N, and thus, the weight W of the ball is 2.0 N. We also say that the ball *weighs* 2.0 N, and speak about the ball *weighing* 2.0 N.

A ball with a weight of 3.0 N would require a greater force from you—namely, a 3.0 N force—to keep it at rest. The reason is that the gravitational force you must balance has a greater magnitude—namely, 3.0 N. We say that this second ball is *heavier* than the first ball.

Now let us generalize the situation. Consider a body that has an acceleration $\vec{a}$ of zero relative to the ground, which we again assume to be an inertial frame. Two forces act on the body: a downward gravitational force $\vec{F}_g$ and a balancing upward force of magnitude W. We can write Newton's second law for a vertical y axis, with the positive direction upward, as

$$F_{\text{net},y} = ma_y.$$

In our situation, this becomes

$$W - F_g = m(0) \tag{5-10}$$

or $\qquad W = F_g \qquad$ (weight, with ground as inertial frame). $\tag{5-11}$

This equation tells us (assuming the ground is an inertial frame) that

> The weight W of a body is equal to the magnitude F_g of the gravitational force on the body.

Substituting mg for F_g from Eq. 5-8, we find

$$W = mg \qquad \text{(weight)}, \tag{5-12}$$

which relates a body's weight to its mass.

To *weigh* a body means to measure its weight. One way to do this is to place the body on one of the pans of an equal-arm balance (Fig. 5-6) and then add reference bodies (whose masses are known) on the other pan until we strike a balance (so that the gravitational forces on the two sides match). The masses on the pans then match, and we know the mass m of the body. If we know the value of g for the location of the balance, we can find the weight of the body with Eq. 5-12.

We can also weigh a body with a spring scale (Fig. 5-7). The body stretches a spring, moving a pointer along a scale that has been calibrated and marked in either mass or weight units. (Most bathroom scales in the United States work this way and are marked in the force unit pounds.) If the scale is marked in mass units, it is accurate only where the value of g is the same as where the scale was calibrated.

The weight of a body must be measured when the body is not accelerating vertically relative to the ground. For example, you can measure your weight on a scale in your bathroom or on a fast train. However, if you repeat the measurement with the scale in an accelerating elevator, the reading on the scale differs from your weight because of the acceleration. Such a measurement is called an *apparent weight*.

Caution: The weight of a body is not the mass of the body. Weight is the magnitude of a force and is related to the mass by Eq. 5-12. If you move a body to a point where the value of g is different, the body's mass (an intrinsic property of the body) is not different but the weight is. For example, the weight of a bowling ball with a mass of 7.2 kg is 71 N on Earth but would be only 12 N on the Moon. The mass is the same on Earth and Moon, but the free-fall acceleration on the Moon is only 1.7 m/s².

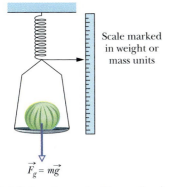

Scale marked in weight or mass units

Fig. 5-7 A spring scale. The reading is proportional to the *weight* of the object placed on the pan, and the scale gives that weight if marked in weight units. If, instead, it is marked in mass units, the reading is accurate only where the free-fall acceleration g is the same as where the scale was calibrated.

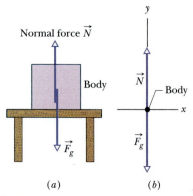

Fig. 5-8 (*a*) A body resting on a tabletop experiences a normal force $\vec{N}$ perpendicular to the tabletop. (*b*) The corresponding free-body diagram for the body.

The Normal Force

If you stand on a mattress, Earth pulls you downward, but you are stationary. The reason is that the mattress, because it deforms downward due to you, pushes up on you. Similarly, if you stand on a floor, it deforms (it is compressed, bent, or buckled ever so slightly), and it pushes up on you. Even a seemingly rigid concrete floor does this (if it is not sitting directly on the ground, enough people on the floor could actually break it).

The push on you from the mattress or floor is called a **normal force** and usually symbolized as $\vec{N}$. The name comes from the mathematical term *normal,* meaning perpendicular: The force on you from, say, the floor is perpendicular to the floor.

> When a body presses against a surface, the surface (even a seemingly rigid surface) deforms and pushes on the body with a normal force $\vec{N}$ that is perpendicular to the surface.

Figure 5-8*a* shows an example. A block of mass m lies on a table's horizontal surface and presses down on the table, deforming the table somewhat because of the gravitational force $\vec{F}_g$ on the block. The table pushes up on the block with normal force $\vec{N}$. The free-body diagram for the block is given in Fig. 5-8*b*. Forces $\vec{F}_g$ and $\vec{N}$ are the only two forces on the block and they are vertical. Thus, for the block we can write Newton's second law for a positive-upward y axis ($F_{\text{net},y} = ma_y$) as

$$N - F_g = ma_y.$$

From Eq. 5-8, we substitute mg for F_g, finding

$$N - mg = ma_y.$$

Then the magnitude of the normal force is

$$N = mg + ma_y = m(g + a_y) \tag{5-13}$$

for any vertical acceleration a_y of the table and block (they might be in an accelerating elevator). If the table and block are not accelerating relative to the ground, then $a_y = 0$ and Eq. 5-13 yields

$$N = mg. \tag{5-14}$$

✔**CHECKPOINT 4:**   In Fig. 5-8, is the magnitude of the normal force $\vec{N}$ greater than, less than, or equal to mg if the body and table are in an elevator that is moving upward (a) at constant speed and (b) at increasing speed?

Friction

If we slide or attempt to slide a body over a surface, the motion is resisted by a bonding between the body and the surface. (We discuss this bonding more in the next chapter.) The resistance is considered to be a single force $\vec{f}$, called the **frictional force,** or simply **friction.** This force is directed along the surface, opposite the direction of the intended motion (Fig. 5-9). Sometimes, to simplify a situation, friction is assumed to be negligible (the surface is *frictionless*).

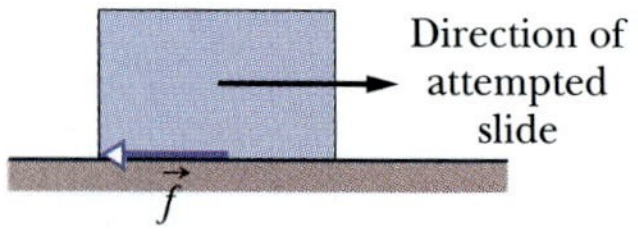

Tension

When a cord (or a rope, cable, or other such object) is attached to a body and pulled taut, the cord pulls on the body with a force $\vec{T}$ directed away from the body and along the cord (Fig. 5-10*a*). The force is often called a *tension force* because the

Fig. 5-9 A frictional force $\vec{f}$ opposes the attempted slide of a body over a surface.

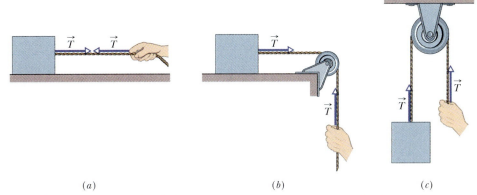

Fig. 5-10 (*a*) The cord, pulled taut, is under tension. If its mass is negligible, it pulls on the body and the hand with force $\vec{T}$, even if it runs around a massless, frictionless pulley as in (*b*) and (*c*).

cord is said to be in a state of *tension* (or to be *under tension*), which means that it is being pulled taut. The *tension in the cord* is the magnitude T of the force on the body. For example, if the force on the body has magnitude $T = 50$ N, then the tension in the cord is 50 N.

A cord is often said to be *massless* (meaning its mass is negligible compared to the body's mass) and unstretchable. The cord then exists only as a connection between two bodies. It pulls on both bodies with the same magnitude T, even if the bodies and the cord are accelerating and even if the cord runs around a *massless, frictionless pulley* (Figs. 5-10*b* and *c*). Such a pulley has negligible mass compared to the bodies and has negligible friction on its axle opposing its rotation. If the cord wraps halfway around a pulley, as in Fig. 5-10*c*, then the net force on the pulley from the cord has the magnitude $2T$.

✔**CHECKPOINT 5:**   The body that is suspended by a rope in Fig. 5-10*c* has a weight of 75 N. Is T equal to, greater than, or less than 75 N when the body is moving upward (a) at constant speed, (b) at increasing speed, and (c) at decreasing speed?

PROBLEM-SOLVING TACTICS

Tactic 6: *Normal Force*

Equation 5-14 for the normal force on a body holds only when $\vec{N}$ is directed upward and the body's vertical acceleration is zero, so we do *not* apply it for other orientations of $\vec{N}$ or when the vertical acceleration is not zero. Instead, we must derive a new expression for $\vec{N}$ from Newton's second law.

We are free to move $\vec{N}$ around in a figure as long as we main-

tain its orientation. For example, in Fig. 5-8*a* we can slide it downward so that its head is at the boundary of the body and the tabletop. However, $\vec{N}$ is least likely to be misinterpreted if its tail is at that boundary or somewhere within the body (as shown). An even better technique is to draw a free-body diagram as in Fig. 5-8*b*, with the tail of $\vec{N}$ directly on the dot or sketch representing the body.

Sample Problem 5-4

Let us return to John Massis and the railroad cars, and assume that Massis pulled (with his teeth) on his end of the rope with a constant force that was 2.5 times his body weight, at an angle θ of 30° from the horizontal. His mass m was 80 kg. The weight W of the cars was 700 kN, and he moved them 1.0 m along the rails. Assume that the rolling wheels encountered no retarding force from the rails. What was the speed of the cars at the end of the pull?

SOLUTION: A **Key Idea** here is that, from Newton's second law, the constant horizontal force on the cars from Massis causes a constant horizontal acceleration of the cars. Because the acceleration is constant and the motion is one-dimensional, we can use the equations of Table 2-1 to find the velocity v at the end of the pulling distance $d = 1.0$ m. We need an equation that contains v. Let us try Eq. 2-16,

$$v^2 = v_0^2 + 2a(x - x_0), \tag{5-15}$$

and place an x axis along the direction of motion, as shown in the free-body diagram of Fig. 5-11. We know that the initial velocity v_0 is 0 and the displacement $x - x_0$ is $d = 1.0$ m. However, we do not know the acceleration a along the x axis.

A second **Key Idea** is that we can relate a to the force on the cars from the rope by using Newton's second law. We can write that law for the x axis in Fig. 5-11 as $F_{net,x} = ma_x$ or, here,

$$F_{net,x} = Ma, \tag{5-16}$$

where M is the mass of the cars. The only force on the cars along the x axis is the horizontal component $T \cos \theta$ of the tension force $\vec{T}$ on the cars from the rope pulled by Massis. Thus, Eq. 5-16 becomes

$$T \cos \theta = Ma. \tag{5-17}$$

We know that T is 2.5 times the body weight of Massis. From Eq. 5-12, his weight is equal to mg, so we have

$$T = 2.5mg = (2.5)(80 \text{ kg})(9.8 \text{ m/s}^2) = 1960 \text{ N},$$

which is the force a good middle-weight weight lifter can produce—and far from a superhuman force.

We still need M in order to evaluate Eq. 5-17 for a. To find M, we again use Eq. 5-12, but now with the weight W of the cars:

$$M = \frac{W}{g} = \frac{7.0 \times 10^5 \text{ N}}{9.8 \text{ m/s}^2} = 7.143 \times 10^4 \text{ kg}.$$

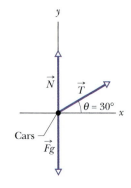

Fig. 5-11 Sample Problem 5-4. Free-body diagram for the passenger cars pulled by Massis. The vectors are not drawn to scale; the force $\vec{T}$ on the cars from the rope is much smaller than the normal force $\vec{N}$ on the cars from the rails and the gravitational force $\vec{F}_g$ on the cars.

By rearranging Eq. 5-17 and substituting for T, M, and θ, we get

$$a = \frac{T \cos \theta}{M} = \frac{(1960 \text{ N})(\cos 30°)}{7.143 \times 10^4 \text{ kg}} = 0.02376 \text{ m/s}^2.$$

Substituting this value and the other known values into Eq. 5-15 now gives us

$$v^2 = 0 + 2(0.02376 \text{ m/s}^2)(1.0 \text{ m})$$

and

$$v = 0.22 \text{ m/s.} \tag{Answer}$$

Massis would have done better if the rope had been attached higher on the car, so that it was horizontal. Can you see why?

5-7 Newton's Third Law

Two bodies are said to *interact* when they push or pull on each other—that is, when a force acts on each body due to the other body. For example, suppose that you position a book B so it leans against a crate C (Fig. 5-12*a*). Then the book and crate interact: There is a horizontal force $\vec{F}_{BC}$ on the book from the crate (or due to the crate) and a horizontal force $\vec{F}_{CB}$ on the crate from the book (or due to the book). This pair of forces is shown in Fig. 5-12*b*. Newton's third law states that

> **Newton's Third Law:** When two bodies interact, the forces on the bodies from each other are always equal in magnitude and opposite in direction.

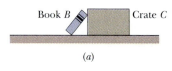

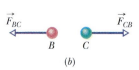

For the book and crate, we can write this law as the scalar relation

$$F_{BC} = F_{CB} \quad \text{(equal magnitudes)}$$

or as the vector relation

$$\vec{F}_{BC} = -\vec{F}_{CB} \quad \text{(equal magnitudes and opposite directions),}$$

Fig. 5-12 (*a*) Book B leans against crate C. (*b*) According to Newton's third law, the force $\vec{F}_{BC}$ on the book from the crate has the same magnitude but the opposite direction of the force $\vec{F}_{CB}$ on the crate from the book.

where the minus sign means that these two forces are in opposite directions. We can call the forces between two interacting bodies a **third-law force pair.** When any two bodies interact in any situation, a third-law force pair is present. The book and crate in Fig. 5-12*a* are stationary, but the third law would still hold if they were moving and even if they were accelerating.

As another example, let us find the third-law force pairs involving the cantaloupe in Fig. 5-13*a*, which lies on a table that stands on Earth. The cantaloupe interacts with the table and with Earth (this time, there are three bodies whose interactions we must sort out).

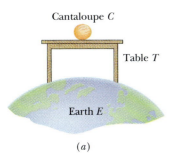

Cantaloupe C

Table T

Earth E

(a)

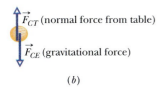

$\vec{F}_{CT}$ (normal force from table)

$\vec{F}_{CE}$ (gravitational force)

(b)

Cantaloupe

$\vec{F}_{CE}$

$\vec{F}_{EC}$

Earth

(c)

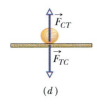

$\vec{F}_{CT}$

$\vec{F}_{TC}$

(d)

Fig. 5-13 (a) A cantaloupe lies on a table that stands on Earth. (b) The forces *on the cantaloupe* are $\vec{F}_{CT}$ and $\vec{F}_{CE}$. (c) The third-law force pair for the cantaloupe–Earth interaction. (d) The third-law force pair for the cantaloupe–table interaction.

Let's first focus only on the cantaloupe (Fig. 5-13b). Force $\vec{F}_{CT}$ is the normal force on the cantaloupe from the table, and force $\vec{F}_{CE}$ is the gravitational force on the cantaloupe due to Earth. Are they a third-law force pair? No, they are forces on a single body, the cantaloupe, and not on two interacting bodies.

To find a third-law pair, we should focus not on the cantaloupe but on the interaction between the cantaloupe and one of the other two bodies. First, in the cantaloupe–Earth interaction (Fig. 5-13c), Earth pulls on the cantaloupe with a gravitational force $\vec{F}_{CE}$ and the cantaloupe pulls on Earth with a gravitational force $\vec{F}_{EC}$. Are these forces a third-law force pair? Yes, they are forces on two interacting bodies, the force on each due to the other. Thus, by Newton's third law,

$$\vec{F}_{CE} = -\vec{F}_{EC} \qquad \text{(cantaloupe–Earth interaction)}.$$

Next, in the cantaloupe–table interaction, the force on the cantaloupe from the table is $\vec{F}_{CT}$ and the force on the table from the cantaloupe is $\vec{F}_{TC}$ (Fig. 5-13d). These forces are also a third-law force pair, and so

$$\vec{F}_{CT} = -\vec{F}_{TC} \qquad \text{(cantaloupe–table interaction)}.$$

✔**CHECKPOINT 6:** Suppose that the cantaloupe and table of Fig. 5-13 are in an elevator cab that begins to accelerate upward. (a) Do the magnitudes of forces $\vec{F}_{TC}$ and $\vec{F}_{CT}$ increase, decrease, or stay the same? (b) Are those two forces still equal in magnitude and opposite in direction? (c) Do the magnitudes of forces $\vec{F}_{CE}$ and $\vec{F}_{EC}$ increase, decrease, or stay the same? (d) Are those two forces still equal in magnitude and opposite in direction?

5-8 Applying Newton's Laws

The rest of this chapter consists of sample problems. You should pore over them, learning not just their particular answers but, instead, the procedures they show for attacking a problem. Especially important is knowing how to translate a sketch of a situation into a free-body diagram with appropriate axes, so that Newton's laws can be applied. We begin with a sample problem that is worked out in exhaustive detail, using a question-and-answer format.

Sample Problem 5-5

Figure 5-14 shows a block S (the *sliding block*) with mass $M = 3.3$ kg. The block is free to move along a horizontal frictionless surface such as an air table. This first block is connected by a cord that wraps over a frictionless pulley to a second block H (the *hanging block*), with mass $m = 2.1$ kg. The cord and pulley have negligible masses compared to the blocks (they are "massless"). The hanging block H falls as the sliding block S accelerates to the right. Find (a) the acceleration of the sliding block, (b) the acceleration of the hanging block, and (c) the tension in the cord.

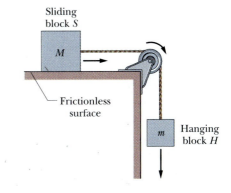

Sliding block S

M

Frictionless surface

m   Hanging block H

Fig. 5-14 Sample Problem 5-5. A block S of mass M is connected to a block H of mass m by a cord that wraps over a pulley.

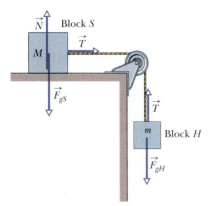

Fig. 5-15 The forces acting on the two blocks of Fig. 5-14.

Q *What is this problem all about?*

You are given two bodies, the sliding block and the hanging block, *and also Earth,* which pulls on both bodies. (Without Earth, nothing would happen here.) A total of five forces act on the blocks, as shown in Fig. 5-15:

1. The cord pulls to the right on sliding block *S* with a force of magnitude *T*.

2. The cord pulls upward on hanging block *H* with a force of the same magnitude *T*. This upward force keeps the hanging block from falling freely.

3. Earth pulls down on sliding block *S* with the gravitational force $\vec{F}_{gS}$, which has a magnitude equal to *Mg*.

4. Earth pulls down on hanging block *H* with the gravitational force $\vec{F}_{gH}$, which has a magnitude equal to *mg*.

5. The table pushes up on sliding block *S* with a normal force $\vec{N}$.

There is another thing that you should note. We assume that the cord does not stretch, so that if block *H* falls 1 mm in a certain time, block *S* moves 1 mm to the right in that same time. This means that the blocks move together and their accelerations have the same magnitude *a*.

Q *How do I classify this problem? Should it suggest a particular law of physics to me?*

Yes. Forces, masses, and accelerations are involved, and they should suggest Newton's second law of motion, $\vec{F}_{net} = m\vec{a}$. That is our starting Key Idea.

Q *If I apply Newton's second law to this problem, to what body should I apply it?*

We focus on two bodies in this problem, the sliding block and the hanging block. Although they are *extended objects* (they are not points), we can still treat each block as a particle because every small part of it (every atom, say) moves in exactly the same way. A second Key Idea is to apply Newton's second law separately to each block.

Q *What about the pulley?*

We cannot represent the pulley as a particle because different parts of it move in different ways. When we discuss rotation, we shall deal with pulleys in detail. Meanwhile, we eliminate the pul-

ley from consideration by assuming its mass is negligible compared with the masses of the two blocks. Then its function is just to change the cord's orientation.

Q *OK. Now how do I apply $\vec{F}_{net} = m\vec{a}$ to the sliding block?*

Represent block *S* as a particle of mass *M* and draw *all* the forces that act *on* it, as in Fig. 5-16*a*. This is the block's *free-body diagram.* There are three forces. Next, draw a set of axes. It makes sense to draw the *x* axis parallel to the table, in the direction in which the block moves.

Q *Thanks, but you still haven't told me how to apply $\vec{F}_{net} = m\vec{a}$ to the sliding block. All you have done is explain how to draw a free-body diagram.*

You are right, and here's the third Key Idea: The expression $\vec{F}_{net} = M\vec{a}$ is a vector equation, so we can write it as three component equations:

$$F_{net,x} = Ma_x \qquad F_{net,y} = Ma_y \qquad F_{net,z} = Ma_z \quad (5\text{-}18)$$

in which $F_{net,x}$, $F_{net,y}$, and $F_{net,z}$ are the components of the net force along the three axes. Now we apply each component equation to its corresponding direction.

Because block *S* does not accelerate vertically, $F_{net,y} = Ma_y$ becomes

$$N - F_{gS} = 0 \quad \text{or} \quad N = F_{gS}.$$

Thus in the *y* direction, the magnitude of the normal force on block *S* is equal to the magnitude of the gravitational force on that block. No force acts in the *z* direction, which is perpendicular to the page.

In the *x* direction, there is only one force component, which is *T*. Thus, $F_{net,x} = Ma_x$ becomes

$$T = Ma. \quad (5\text{-}19)$$

This equation contains two unknowns, *T* and *a*, so we cannot yet solve it. Recall, however, that we have not said anything about the hanging block.

Q *I agree. How do I apply $\vec{F}_{net} = m\vec{a}$ to the hanging block?*

We apply it just as we did for block *S*: Draw a free-body diagram for block *H*, as in Fig. 5-16*b*. Then apply $\vec{F}_{net} = m\vec{a}$ in component form. This time, because the acceleration is along the *y* axis, we use the second of Eqs. 5-18 ($F_{net,y} = ma_y$) to write

$$T - F_{gH} = ma_y.$$

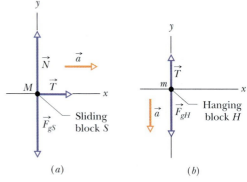

Fig. 5-16 (*a*) A free-body diagram for block *S* of Fig. 5-14. (*b*) A free-body diagram for the hanging block *H* of Fig. 5-14.

We can now substitute mg for F_{gH} and $-a$ for a_y (negative because block H accelerates downward, in the negative direction of the y axis). We find

$$T - mg = -ma. \qquad (5\text{-}20)$$

Now note that Eqs. 5-19 and 5-20 are simultaneous equations with the same two unknowns, T and a. Subtracting these equations eliminates T. Then solving for a yields

$$a = \frac{m}{M + m} g. \qquad (5\text{-}21)$$

Substituting this result into Eq. 5-19 yields

$$T = \frac{Mm}{M + m} g. \qquad (5\text{-}22)$$

Putting in the numbers gives, for these two quantities,

$$a = \frac{m}{M + m} g = \frac{2.1 \text{ kg}}{3.3 \text{ kg} + 2.1 \text{ kg}} (9.8 \text{ m/s}^2)$$
$$= 3.8 \text{ m/s}^2 \qquad \text{(Answer)}$$

and $\quad T = \dfrac{Mm}{M + m} g = \dfrac{(3.3 \text{ kg})(2.1 \text{ kg})}{3.3 \text{ kg} + 2.1 \text{ kg}} (9.8 \text{ m/s}^2)$
$$= 13 \text{ N}. \qquad \text{(Answer)}$$

Q *The problem is now solved, right?*

That's a fair question, but the problem is not really finished

until we have examined the results to see whether they make sense. (If you made these calculations on the job, wouldn't you want to see whether they made sense before you turned them in?)

Look first at Eq. 5-21. Note that it is dimensionally correct and that the acceleration a will always be less than g. This is as it must be, because the hanging block is not in free fall. The cord pulls upward on it.

Look now at Eq. 5-22, which we can rewrite in the form

$$T = \frac{M}{M + m} mg. \qquad (5\text{-}23)$$

In this form, it is easier to see that this equation is also dimensionally correct, because both T and mg have dimensions of forces. Equation 5-23 also lets us see that the tension in the cord is always less than mg, and thus always less than the gravitational force on the hanging block. That is a comforting thought because, if T were *greater* than mg, the hanging block would accelerate upward.

We can also check the results by studying special cases, in which we can guess what the answers must be. A simple example is to put $g = 0$, as if the experiment were carried out in interstellar space. We know that in that case, the blocks would not move from rest, there would be no forces on the ends of the cord, and so there would be no tension in the cord. Do the formulas predict this? Yes, they do. If you put $g = 0$ in Eqs. 5-21 and 5-22, you find $a = 0$ and $T = 0$. Two more special cases that you might try are $M = 0$ and $m \rightarrow \infty$.

Sample Problem 5-6

In Fig. 5-17a, a block B of mass $M = 15.0$ kg hangs by a cord from a knot K of mass m_K, which hangs from a ceiling by means of two other cords. The cords have negligible mass, and the magnitude of the gravitational force on the knot is negligible compared to the gravitational force on the block. What are the tensions in the three cords?

SOLUTION: Let's start with the block because it has only one attached cord. The free-body diagram in Fig. 5-17b shows the forces on the block: gravitational force $\vec{F}_g$ (with a magnitude of Mg) and force T_3 from the attached cord. A **Key Idea** is that we can relate these forces to the acceleration of the block via Newton's second law ($\vec{F}_{net} = m\vec{a}$). Because the forces are both vertical, we choose the vertical component version of the law ($F_{net,y} = ma_y$) and write

$$T_3 - F_g = Ma_y.$$

Substituting Mg for F_g and 0 for the block's acceleration a_y, we find

$$T_3 - Mg = M(0) = 0.$$

This means that the two forces on the block are in equilibrium. Substituting for M ($= 15.0$ kg) and g and solving for T_3 yield

$$T_3 = 147 \text{ N}. \qquad \text{(Answer)}$$

We next consider the knot in the free-body diagram of Fig. 5-17c, where the negligible gravitational force on the knot is not included. The **Key Idea** here is that we can relate the three other forces acting on the knot to the acceleration of the knot via New-

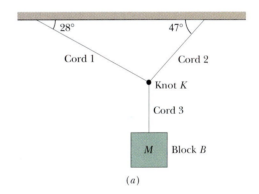

(a)

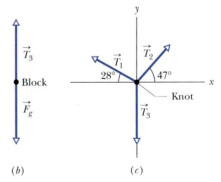

(b)　　　　　(c)

Fig. 5-17 Sample Problem 5-6. (a) A block of mass M hangs from three cords by means of a knot. (b) A free-body diagram for the block. (c) A free-body diagram for the knot.

ton's second law ($\vec{F}_{net} = m\vec{a}$) by writing

$$\vec{T}_1 + \vec{T}_2 + \vec{T}_3 = m_K \vec{a}_K.$$

Substituting 0 for the knot's acceleration $\vec{a}_K$ yields

$$\vec{T}_1 + \vec{T}_2 + \vec{T}_3 = 0, \qquad (5\text{-}24)$$

which means that the three forces on the knot are in equilibrium. Although we know both magnitude and angle for $\vec{T}_3$, we know only the angles and not the magnitudes for $\vec{T}_1$ and $\vec{T}_2$. With unknowns in two vectors, we cannot solve Eq. 5-24 for $\vec{T}_1$ or $\vec{T}_2$ directly on a vector-capable calculator.

Instead we rewrite Eq. 5-24 in terms of components along the x and y axes. For the x axis, we have

$$T_{1x} + T_{2x} + T_{3x} = 0,$$

which, using the given data, yields

$$-T_1 \cos 28° + T_2 \cos 47° + 0 = 0. \qquad (5\text{-}25)$$

(For the first term, we have two choices, either the one shown or

the equivalent $T_1 \cos 152°$, where $152°$ is the angle from the positive direction of the x axis.)

Similarly, for the y axis we rewrite Eq. 5-24 as

$$T_{1y} + T_{2y} + T_{3y} = 0$$

or

$$T_1 \sin 28° + T_2 \sin 47° - T_3 = 0.$$

Substituting our previous result for T_3 then gives us

$$T_1 \sin 28° + T_2 \sin 47° - 147 \text{ N} = 0. \qquad (5\text{-}26)$$

We cannot solve Eq. 5-25 or Eq. 5-26 separately because each contains two unknowns, but we can solve them simultaneously because they contain the same two unknowns. Doing so (either by substitution, by adding or subtracting the equations appropriately, or by using the equation-solving capability of a calculator), we discover

$$T_1 = 104 \text{ N} \quad \text{and} \quad T_2 = 134 \text{ N}. \qquad \text{(Answer)}$$

Thus, the tensions in the cords are 104 N in cord 1, 134 N in cord 2, and 147 N in cord 3.

Sample Problem 5-7

In Fig. 5-18a, a cord holds stationary a block of mass $m = 15$ kg, on a frictionless plane that is inclined at angle $\theta = 27°$.

(a) What are the magnitudes of the force $\vec{T}$ on the block from the cord and the normal force $\vec{N}$ on the block from the plane?

SOLUTION: Those two forces and the gravitational force $\vec{F}_g$ on the block are shown in the block's free-body diagram of Fig. 5-18b. Only these three forces act on the block. A **Key Idea** is that we can relate them to the block's acceleration via Newton's second law ($\vec{F}_{net} = m\vec{a}$), which we write as

$$\vec{T} + \vec{N} + \vec{F}_g = m\vec{a}.$$

Substituting 0 for the block's acceleration $\vec{a}$ yields

$$\vec{T} + \vec{N} + \vec{F}_g = 0, \qquad (5\text{-}27)$$

which says that the three forces are in equilibrium.

With two unknown vectors in Eq. 5-27, we cannot solve it for either vector directly on a vector-capable calculator. So, we must rewrite it in terms of components. We use a coordinate system with its x axis parallel to the plane, as shown in Fig. 5-18b; then two forces ($\vec{N}$ and $\vec{T}$) line up with the axes, making their components easy to find. To find the components of the gravitational force $\vec{F}_g$, we first note that the angle θ of the plane is also the angle between the y axis and the direction of $\vec{F}_g$ (see Fig. 5-18c). Component F_{gx} is then $-F_g \sin \theta$, which is equal to $-mg \sin \theta$; and component F_{gy} is then $-F_g \cos \theta$, which is equal to $-mg \cos \theta$.

Now, writing the x component version of Eq. 5-27 yields

$$T + 0 - mg \sin \theta = 0,$$

from which

$$T = mg \sin \theta$$
$$= (15 \text{ kg})(9.8 \text{ m/s}^2)(\sin 27°)$$
$$= 67 \text{ N}. \qquad \text{(Answer)}$$

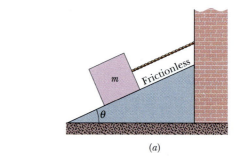

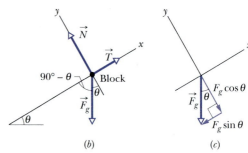

Fig. 5-18 Sample Problem 5-7 (a) A block of mass m held stationary by a cord. (b) A free-body diagram for the block. (c) The x and y components of $\vec{F}_g$.

Similarly, for the y axis, Eq. 5-27 yields

$$0 + N - mg \cos \theta = 0$$

or

$$N = mg \cos \theta$$
$$= (15 \text{ kg})(9.8 \text{ m/s}^2)(\cos 27°)$$
$$= 131 \text{ N} \approx 130 \text{ N}. \qquad \text{(Answer)}$$

(b) We now cut the cord. As the block then slides down the inclined plane, does it accelerate? If so, what is its acceleration?

SOLUTION: Cutting the cord removes force $\vec{T}$ from the block. Along the y axis, the normal force and component F_{gy} are still in equilibrium. However, along the x axis, only force component F_{gx} acts on the block; because it is directed down the plane (along the x axis), that component must cause the block to accelerate down the plane. Our Key Idea here is that we can relate F_{gx} to the acceleration a that it produces with Newton's second law written for x components ($F_{net,x} = ma_x$). We get

$$F_{gx} = ma$$

or

$$-mg \sin \theta = ma,$$

which gives us

$$a = -g \sin \theta. \qquad (5\text{-}28)$$

Substituting known data then yields

$$a = -(9.8 \text{ m/s}^2)(\sin 27°) = -4.4 \text{ m/s}^2. \qquad \text{(Answer)}$$

The magnitude of this acceleration a is less than the magnitude 9.8 m/s² of the free-fall acceleration because only a component of $\vec{F}_g$ (the component that is directed down the plane) is producing acceleration a.

✔CHECKPOINT 7: In the figure, horizontal force $\vec{F}$ is applied to a block on a ramp. (a) Is the component of $\vec{F}$ that is perpendicular to the ramp $F \cos \theta$ or $F \sin \theta$? (b) Does the presence of $\vec{F}$ increase or decrease the magnitude of the normal force on the block from the ramp?

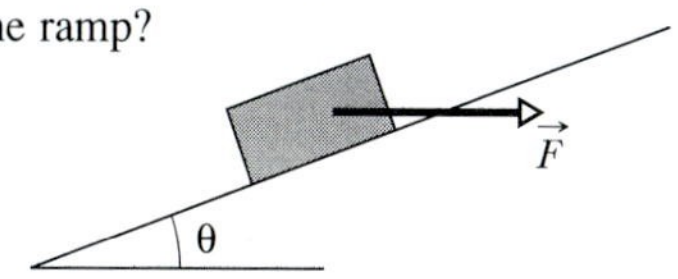

Sample Problem 5-8

In Fig. 5-19a, a passenger of mass m = 72.2 kg stands on a platform scale in an elevator cab. We are concerned with the scale readings when the cab is stationary, and when it is moving up or down.

(a) Find a general solution for the scale reading, whatever the vertical motion of the cab.

SOLUTION: One Key Idea here is that the scale reading is equal to the magnitude of the normal force $\vec{N}$ on the passenger from the scale. The only other force acting on the passenger is the gravitational force $\vec{F}_g$, as shown in the free-body diagram of the passenger in Fig. 5-19b.

A second Key Idea is that we can relate the forces on the passenger to the acceleration $\vec{a}$ of the passenger with Newton's second law ($\vec{F}_{net} = m\vec{a}$). However, recall that we can use this law only in an inertial frame. If the cab accelerates, then it is *not* an inertial frame. So, we choose the ground to be our inertial frame and make any measure of the passenger's acceleration relative to it.

Because the two forces on the passenger and the passenger's acceleration are all directed vertically, along the y axis shown in Fig. 5-19b, we can use Newton's second law written for y components ($F_{net,y} = ma_y$) to get

$$N - F_g = ma$$

or

$$N = F_g + ma. \qquad (5\text{-}29)$$

This tells us that the scale reading, which is equal to N, depends on the vertical acceleration a of the cab. Substituting mg for F_g gives us

$$N = m(g + a) \qquad \text{(Answer)} \quad (5\text{-}30)$$

for any choice of acceleration a.

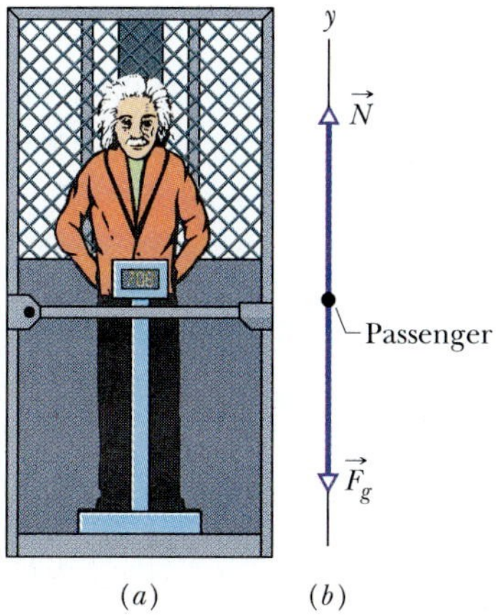

(b) What does the scale read if the cab is stationary or moving upward at a constant 0.50 m/s?

SOLUTION: The Key Idea here is that for any constant velocity (zero or otherwise), the acceleration a of the passenger is zero. Substituting this and other known values into Eq. 5-30, we find

$$N = (72.2 \text{ kg})(9.8 \text{ m/s}^2 + 0) = 708 \text{ N}. \qquad \text{(Answer)}$$

This is the weight of the passenger and is equal to the magnitude F_g of the gravitational force on him.

(c) What does the scale read if the cab accelerates upward at 3.20 m/s² and downward at 3.20 m/s²?

SOLUTION: For a = 3.20 m/s², Eq. 5-30 gives

$$N = (72.2 \text{ kg})(9.8 \text{ m/s}^2 + 3.20 \text{ m/s}^2)$$
$$= 939 \text{ N}, \qquad \text{(Answer)}$$

and for a = −3.20 m/s², it gives

$$N = (72.2 \text{ kg})(9.8 \text{ m/s}^2 - 3.20 \text{ m/s}^2)$$
$$= 477 \text{ N}. \qquad \text{(Answer)}$$

So, for an upward acceleration (either the cab's upward speed is increasing or its downward speed is decreasing), the scale reading

Fig. 5-19 Sample Problem 5-8. (a) A passenger stands on a platform scale that indicates his weight or apparent weight. (b) The free-body diagram for the passenger, showing the normal force $\vec{N}$ on him from the scale and the gravitational force $\vec{F}_g$.

is greater than the passenger's weight. That reading is a measurement of an apparent weight, because it is made in a noninertial frame. Similarly, for a downward acceleration (either the cab's upward speed is decreasing or its downward speed is increasing), the scale reading is less than the passenger's weight.

(d) During the upward acceleration in part (c), what is the magnitude F_{net} of the net force on the passenger, and what is the magnitude $a_{p,cab}$ of the passenger's acceleration as measured in the frame of the cab? Does $\vec{F}_{net} = m\vec{a}_{p,cab}$?

SOLUTION: One **Key Idea** here is that the magnitude F_g of the gravitational force on the passenger does not depend on the motion of the passenger or the cab, so, from part (b), F_g is 708 N. From part

(c), the magnitude N of the normal force on the passenger during the upward acceleration is the 939 N reading on the scale. Thus, the net force on the passenger is

$$F_{net} = N - F_g = 939 \text{ N} - 708 \text{ N} = 231 \text{ N}, \quad \text{(Answer)}$$

during the upward acceleration. However, the acceleration $a_{p,cab}$ of the passenger relative to the frame of the cab is zero. Thus, in the noninertial frame of the accelerating cab, F_{net} is not equal to $ma_{p,cab}$, and Newton's second law does not hold.

✔**CHECKPOINT 8:** In this sample problem what does the scale read if the elevator cable breaks, so that the cab falls freely; that is, what is the apparent weight of the passenger in free fall?

Sample Problem 5-9

In Fig. 5-20a, a constant horizontal force $\vec{F}_{ap}$ of magnitude 20 N is applied to block A of mass $m_A = 4.0$ kg, which pushes against block B of mass $m_B = 6.0$ kg. The blocks slide over a frictionless surface, along an x axis.

(a) What is the acceleration of the blocks?

SOLUTION: We shall first examine a solution with a serious error, then a dead-end solution, and then a successful solution.

    Serious Error: Because force $\vec{F}_{ap}$ is applied directly to block A, we use Newton's second law to relate that force to the acceleration $\vec{a}$ of block A. Because the motion is along the x axis, we use that law for x components ($F_{net,x} = ma_x$), writing it as

$$F_{ap} = m_A a.$$

However, this is seriously wrong because $\vec{F}_{ap}$ is not the only horizontal force acting on block A. There is also the force $\vec{F}_{AB}$ from block B (as shown in Fig. 5-20b).

    Dead-End Solution: Let us now include force $\vec{F}_{AB}$ by writing, again for the x axis,

$$F_{ap} - F_{AB} = m_A a.$$

(We use the minus sign to include the direction of $\vec{F}_{AB}$.) However, F_{AB} is a second unknown, so we cannot solve this equation for the desired acceleration a.

    Successful Solution: The **Key Idea** here is that, because of the direction in which force $\vec{F}_{ap}$ is applied, the two blocks form a rigidly connected system. We can relate the net force *on the system* to the acceleration *of the system* with Newton's second law. Here, once again for the x axis, we can write that law as

$$F_{ap} = (m_A + m_B)a,$$

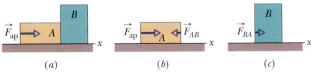

Fig. 5-20 Sample Problem 5-9. (a) A constant horizontal force $\vec{F}_{ap}$ is applied to block A, which pushes against block B. (b) Two horizontal forces act on block A: applied force $\vec{F}_{ap}$ and force $\vec{F}_{AB}$ from block B. (c) Only one horizontal force acts on block B: force $\vec{F}_{BA}$ from block A.

where now we properly apply $\vec{F}_{ap}$ to the system with total mass $m_A + m_B$. Solving for a and substituting known values, we find

$$a = \frac{F_{ap}}{m_A + m_B} = \frac{20 \text{ N}}{4.0 \text{ kg} + 6.0 \text{ kg}} = 2.0 \text{ m/s}^2. \quad \text{(Answer)}$$

Thus, the acceleration of the system and of each block is in the positive direction of the x axis and has the magnitude 2.0 m/s².

(b) What is the force $\vec{F}_{BA}$ on block B from block A (Fig. 5-20c)?

SOLUTION: The **Key Idea** here is that we can relate the net force on block B to the block's acceleration with Newton's second law. Here we can write that law, still for components along the x axis, as

$$F_{BA} = m_B a,$$

which, with known values, gives

$$F_{BA} = (6.0 \text{ kg})(2.0 \text{ m/s}^2) = 12 \text{ N}. \quad \text{(Answer)}$$

Thus, force $\vec{F}_{BA}$ is in the positive direction of the x axis and has a magnitude of 12 N.

REVIEW & SUMMARY

Newtonian Mechanics The velocity of a particle or a particle-like body can change (the particle can accelerate) when the particle is acted on by one or more **forces** (pushes or pulls) from other objects. *Newtonian mechanics* relates accelerations and forces.

Force Forces are vector quantities. Their magnitudes are defined in terms of the acceleration they would give the standard kilogram. A force that accelerates that standard body by exactly 1 m/s² is defined to have a magnitude of 1 N. The direction of a force is the direction of the acceleration it causes. Forces are combined according to the rules of vector algebra. The **net force** on a body is the vector sum of all the forces acting on it.

Mass   The **mass** of a body is the characteristic of that body that relates the body's acceleration to the force (or net force) causing the acceleration. Masses are scalar quantities.

Newton's First Law   If there is no net force on a body, the body must remain at rest if it is initially at rest, or move in a straight line at constant speed if it is in motion.

Inertial Reference Frames   Reference frames in which Newtonian mechanics holds are called *inertial reference frames* or simply *inertial frames*. We can approximate the ground as an inertial frame if Earth's motions can be neglected. Reference frames in which Newtonian mechanics does not hold are called *noninertial reference frames* or simply *noninertial frames*. An elevator accelerating relative to the ground is a noninertial frame.

Newton's Second Law   The net force $\vec{F}_{net}$ on a body with mass m is related to the body's acceleration $\vec{a}$ by

$$\vec{F}_{net} = m\vec{a}, \qquad (5\text{-}1)$$

which may be written in the component versions

$$F_{net,x} = ma_x \quad F_{net,y} = ma_y \quad \text{and} \quad F_{net,z} = ma_z. \qquad (5\text{-}2)$$

The second law indicates that in SI units

$$1 \text{ N} = 1 \text{ kg} \cdot \text{m/s}^2. \qquad (5\text{-}3)$$

A **free-body diagram** is helpful in solving problems with the second law: It is a stripped-down diagram in which only *one* body is considered. That body is represented by a sketch or simply a dot. The external forces on the body are drawn, and a coordinate system is superimposed, oriented so as to simplify the solution.

Some Particular Forces   A **gravitational force** $\vec{F}_g$ on a body is a pull by another body. In most situations in this book, the other body is Earth or some other astronomical body. For Earth, the force is directed down toward the ground, which is assumed to be an inertial frame. With that assumption, the magnitude of the force is

$$F_g = mg, \qquad (5\text{-}8)$$

where m is the body's mass and g is the magnitude of the free-fall acceleration.

The **weight** W of a body is the magnitude of the upward force needed to balance the gravitational force on the body due to Earth (or another astronomical body). It is related to the body's mass by

$$W = mg. \qquad (5\text{-}12)$$

A **normal force** $\vec{N}$ is the force on a body from a surface against which the body presses. The normal force is always perpendicular to the surface.

A **frictional force** $\vec{f}$ is the force on a body when the body slides or attempts to slide along a surface. The force is always parallel to the surface and directed so as to oppose the motion of the body. On a *frictionless surface,* the frictional force is negligible.

When a cord is under **tension,** it pulls on a body at each of its ends. The pull is directed along the cord, away from the point of attachment to each body. For a *massless* cord (a cord with negligible mass), the pulls at both ends of the cord have the same magnitude T, even if the cord runs around a *massless, frictionless pulley* (a pulley with negligible mass and negligible friction on its axle to oppose its rotation).

Newton's Third Law   If a force $\vec{F}_{BC}$ acts on body B due to body C, then there is a force $\vec{F}_{CB}$ on body C due to body B. The forces are equal in magnitude and opposite in direction:

$$\vec{F}_{BC} = -\vec{F}_{CB}.$$

QUESTIONS

1. Two horizontal forces,

$$\vec{F}_1 = (3 \text{ N})\hat{i} - (4 \text{ N})\hat{j} \quad \text{and} \quad \vec{F}_2 = -(1 \text{ N})\hat{i} - (2 \text{ N})\hat{j},$$

pull a banana split across a frictionless lunch counter. Without using a calculator, determine which of the vectors in the free-body diagram of Fig. 5-21 best represent (a) $\vec{F}_1$ and (b) $\vec{F}_2$. What is the net-force component along (c) the x axis and (d) the y axis? Into which quadrants do (e) the net-force vector and (f) the split's acceleration vector point?

2. At time $t = 0$, a single force $\vec{F}$ of constant magnitude begins to act on a rock that is moving along an x axis through deep space. The rock continues to move along that axis. (a) For time

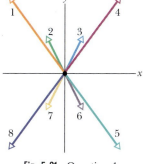

Fig. 5-21 Question 1.

$t > 0$, which of the following is a possible function $x(t)$ for the rock's position: (1) $x = 4t - 3$, (2) $x = -4t^2 + 6t - 3$, (3) $x = 4t^2 + 6t - 3$? (b) For which function is $\vec{F}$ directed opposite the rock's initial direction of motion?

3. Figure 5-22 shows overhead views of four situations in which forces act on a block that lies on a frictionless floor. If the force

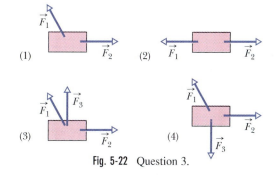

Fig. 5-22 Question 3.

magnitudes are chosen properly, in which situations is it possible that the block is (a) stationary and (b) moving with a constant velocity?

4. In Fig. 5-23, two forces $\vec{F}_1$ and $\vec{F}_2$ act on a "Rocky and Bullwinkle" lunch box as the lunch box slides at constant velocity over a frictionless lunchroom floor. We are to decrease the angle θ of $\vec{F}_1$ without changing the magnitude of $\vec{F}_1$. To keep the lunch box sliding at constant velocity, should we increase, decrease, or maintain the magnitude of $\vec{F}_2$?

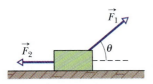

Fig. 5-23 Question 4.

5. Figure 5-24 gives the free-body diagram for four situations in which an object is pulled by several forces across a frictionless floor, as seen from overhead. In which situations does the object's acceleration $\vec{a}$ have (a) an x component and (b) a y component? (c) In each situation, give the direction of $\vec{a}$ by naming either a quadrant or a direction along an axis. (This can be done with a few mental calculations.)

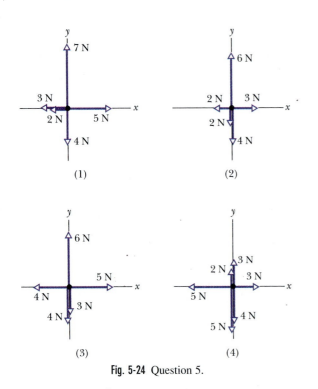

Fig. 5-24 Question 5.

6. Figure 5-25 gives three graphs of velocity component $v_x(t)$ and three graphs of velocity component $v_y(t)$. The graphs are not to scale. Which $v_x(t)$ graph and which $v_y(t)$ graph best correspond to each of the four situations in Question 5 and Fig. 5-24?

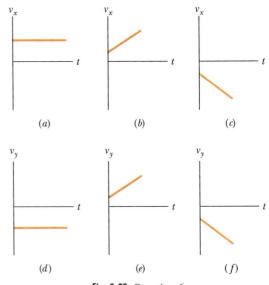

Fig. 5-25 Question 6.

7. The body that is suspended by a rope in Fig. 5-10c has a weight of 75 N. Is T equal to, greater than, or less than 75 N when the body is moving downward at (a) increasing speed and (b) decreasing speed?

8. A vertical force $\vec{F}$ is applied to a block of mass m that lies on a floor. What happens to the magnitude of the normal force $\vec{N}$ on the block from the floor as magnitude F is increased from zero if force $\vec{F}$ is (a) downward and (b) upward?

9. Figure 5-26 shows a train of four blocks being pulled across a frictionless floor by force $\vec{F}$. What total mass is accelerated to the right by (a) force $\vec{F}$, (b) cord 3, and (c) cord 1? (d) Rank the blocks according to their accelerations, greatest first. (e) Rank the cords according to their tension, greatest first. (Warm-up for Problems 34 and 36)

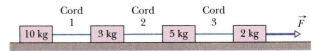

Fig. 5-26 Question 9.

10. Figure 5-27 shows a group of three blocks being pushed across a frictionless floor by horizontal force $\vec{F}$. What total mass is accelerated to the right by (a) force $\vec{F}$, (b) force $\vec{F}_{21}$ on block 2 from block 1, and (c) force $\vec{F}_{32}$ on block 3 from block 2? (d) Rank the blocks according to the magnitudes of their acceleration, greatest first. (e) Rank forces $\vec{F}$, $\vec{F}_{21}$, and $\vec{F}_{32}$ according to their magnitude, greatest first. (Warm-up for Problem 31)

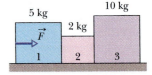

Fig. 5-27 Question 10.

11. In Fig. 5-28a, a toy box is on top of a (heavier) dog house, which sits on a wood floor. In Fig. 5-28b, these objects are represented by dots at the corresponding heights, and six vertical vectors (not to scale) are shown. Which of the vectors best represents

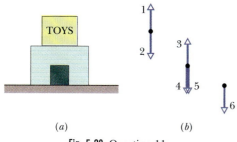

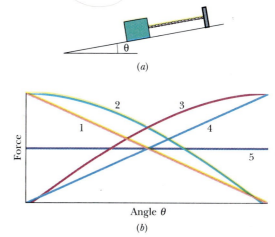

gravitational force $\vec{F}_g$ on the block that is along the ramp, (b) the tension in the cord, (c) the component of $\vec{F}_g$ that is perpendicular to the ramp, and (d) the normal force on the block from the ramp. (e) Which of the curves in Fig. 5-29*b* corresponds to each of the quantities in parts (a) through (d)?

(*a*)

(*b*)

Fig. 5-28 Question 11.

Fig. 5-29 Question 12.

(a) the gravitational force on the dog house, (b) the gravitational force on the toy box, (c) the force on the toy box from the dog house, (d) the force on the dog house from the toy box, (e) the force on the dog house from the floor, and (f) the force on the floor from the dog house? (g) Which of the forces are equal in magnitude? Which are (h) greatest and (i) least in magnitude?

12. In Fig. 5-29*a*, a block is attached by a rope to a bar that is itself rigidly attached to a ramp. Determine whether the magnitudes of the following increase, decrease, or remain the same as the angle θ of the ramp is increased from zero: (a) the component of the

---

EXERCISES & PROBLEMS

ssm Solution is in the Student Solutions Manual.
www Solution is available on the World Wide Web at:
          http://www.wiley.com/college/hrw
ilw Solution is available on the Interactive LearningWare.

SEC. 5-5   Newton's Second Law

1E. If the 1 kg standard body has an acceleration of 2.00 m/s² at 20° to the positive direction of the *x* axis, then what are (a) the *x* component and (b) the *y* component of the net force on it, and (c) what is the net force in unit-vector notation?

2E. Two horizontal forces act on a 2.0 kg chopping block that can slide over a frictionless kitchen counter, which lies in an *xy* plane. One force is $\vec{F}_1 = (3.0 \text{ N})\hat{\imath} + (4.0 \text{ N})\hat{\jmath}$. Find the acceleration of the chopping block in unit-vector notation when the other force is (a) $\vec{F}_2 = (-3.0 \text{ N})\hat{\imath} + (-4.0 \text{ N})\hat{\jmath}$, (b) $\vec{F}_2 = (-3.0 \text{ N})\hat{\imath} + (4.0 \text{ N})\hat{\jmath}$, and (c) $\vec{F}_2 = (3.0 \text{ N})\hat{\imath} + (-4.0 \text{ N})\hat{\jmath}$.

3E. Only two horizontal forces act on a 3.0 kg body. One force is 9.0 N, acting due east, and the other is 8.0 N, acting 62° north of west. What is the magnitude of the body's acceleration?

4E. While two forces act on it, a particle is to move at the constant velocity $\vec{v} = (3 \text{ m/s})\hat{\imath} - (4 \text{ m/s})\hat{\jmath}$. One of the forces is $\vec{F}_1 = (2 \text{ N})\hat{\imath} + (-6 \text{ N})\hat{\jmath}$. What is the other force?

5E. Three forces act on a particle that moves with unchanging velocity $\vec{v} = (2 \text{ m/s})\hat{\imath} - (7 \text{ m/s})\hat{\jmath}$. Two of the forces are $\vec{F}_1 = (2 \text{ N})\hat{\imath} + (3 \text{ N})\hat{\jmath} + (-2 \text{ N})\hat{k}$ and $\vec{F}_2 = (-5 \text{ N})\hat{\imath} + (8 \text{ N})\hat{\jmath} + (-2 \text{ N})\hat{k}$. What is the third force?

6P. Three astronauts, propelled by jet backpacks, push and guide a 120 kg asteroid toward a processing dock, exerting the forces shown in Fig. 5-30. What is the asteroid's acceleration (a) in unit-vector notation and as (b) a magnitude and (c) a direction?

7P. There are two forces on the 2.0 kg box in the overhead view of Fig. 5-31 but only one is shown. The figure also shows the acceleration of the box. Find the second force (a) in unit-vector notation and as (b) a magnitude and (c) a direction. ssm

8P. Figure 5-32 is an overhead view of a 12 kg tire that is to be pulled by three ropes. One force ($\vec{F}_1$, with magnitude 50 N) is indicated. Orient the other two forces $\vec{F}_2$ and $\vec{F}_3$ so that the magnitude of the resulting acceleration of the tire is least, and find that magnitude if (a) $F_2 = 30$ N, $F_3 = 20$ N; (b) $F_2 = 30$ N, $F_3 = 10$ N; and (c) $F_2 = F_3 = 30$ N.

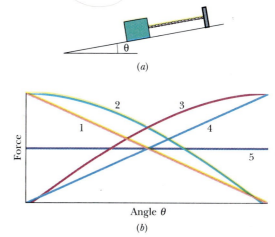

Fig. 5-30 Problem 6.

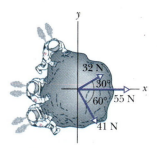

Fig. 5-31 Problem 7.

SEC. 5-6   Some Particular Forces

9E. (a) An 11.0 kg salami is supported by a cord that runs to a spring scale, which is supported

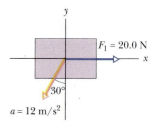

Fig. 5-32 Problem 8.

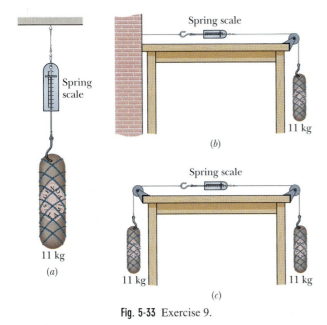

(a)

(b)

11 kg

11 kg

(c)

11 kg   11 kg

Fig. 5-33 Exercise 9.

by another cord from the ceiling (Fig. 5-33a). What is the reading on the scale, which is marked in weight units? (b) In Fig. 5-33b the salami is supported by a cord that runs around a pulley and to a scale. The opposite end of the scale is attached by a cord to a wall. What is the reading on the scale? (c) In Fig. 5-33c the wall has been replaced with a second 11.0 kg salami on the left, and the assembly is stationary. What is the reading on the scale now? ssm www

10E. A block with a weight of 3.0 N is at rest on a horizontal surface. A 1.0 N upward force is applied to the block by means of an attached vertical string. What are the magnitude and the direction of the force of the block on the horizontal surface?

11E. A certain particle has a weight of 22 N at a point where $g = 9.8$ m/s^2. What are its (a) weight and (b) mass at a point where $g = 4.9$ m/s^2? What are its (c) weight and (d) mass if it is moved to a point in space where $g = 0$? ssm

12E. Compute the weight of a 75 kg space ranger (a) on Earth, (b) on Mars, where $g = 3.8$ m/s^2, and (c) in interplanetary space, where $g = 0$. (d) What is the ranger's mass at each of these locations?

SEC. 5-8   Applying Newton's Laws

13E. When a nucleus captures a stray neutron, it must bring the neutron to a stop within the diameter of the nucleus by means of the *strong force*. That force, which "glues" the nucleus together, is approximately zero outside the nucleus. Suppose that a stray neutron with an initial speed of 1.4×10^7 m/s is just barely captured by a nucleus with diameter $d = 1.0 \times 10^{-14}$ m. Assuming that the strong force on the neutron is constant, find the magnitude of that force. The neutron's mass is 1.67×10^{-27} kg. ssm

14E. A 29.0 kg child, with a 4.50 kg backpack on his back, first stands on a sidewalk and then jumps up into the air. Find the magnitude and direction of the force on the sidewalk from the child

when the child is (a) standing still and (b) in the air. Now find the magnitude and direction of the *net* force on Earth due to the child when the child is (c) standing still and (d) in the air.

15E. Refer to Fig. 5-18. Let the mass of the block be 8.5 kg and the angle θ be 30°. Find (a) the tension in the cord and (b) the normal force acting on the block. (c) If the cord is cut, find the magnitude of the block's acceleration. ssm

16E. A 50 kg passenger rides in an elevator that starts from rest on the ground floor of a building at $t = 0$ and rises to the top floor during a 10 s interval. The acceleration of the elevator as a function of the time is shown in Fig. 5-34, where positive values of the acceleration mean that it is directed upward. Give the magnitude and direction of the following forces: (a) the maximum force on the passenger from the floor, (b) the minimum force on the passenger from the floor, and (c) the maximum force on the floor from the passenger.

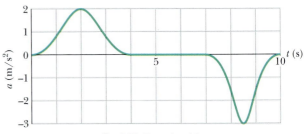

Fig. 5-34 Exercise 16.

17E. *Sunjamming.* A "sun yacht" is a spacecraft with a large sail that is pushed by sunlight. Although such a push is tiny in everyday circumstances, it can be large enough to send the spacecraft outward from the Sun on a cost-free but slow trip. Suppose that the spacecraft has a mass of 900 kg and receives a push of 20 N. (a) What is the magnitude of the resulting acceleration? If the craft starts from rest, (b) how far will it travel in 1 day and (c) how fast will it then be moving?

18E. The tension at which a fishing line snaps is commonly called the line's "strength." What minimum strength is needed for a line that is to stop a salmon of weight 85 N in 11 cm if the fish is initially drifting at 2.8 m/s? Assume a constant deceleration.

19E. An experimental rocket sled can be accelerated at a constant rate from rest to 1600 km/h in 1.8 s. What is the magnitude of the required net force if the sled has a mass of 500 kg? ssm

20E. A car that weighs 1.30×10^4 N is initially moving at a speed of 40 km/h when the brakes are applied and the car is brought to a stop in 15 m. Assuming that the force that stops the car is constant, find (a) the magnitude of that force and (b) the time required for the change in speed. If the initial speed is doubled, and the car experiences the same force during the braking, by what factors are (c) the stopping distance and (d) the stopping time multiplied? (There could be a lesson here about the danger of driving at high speeds.)

21E. An electron with a speed of 1.2×10^7 m/s moves horizontally into a region where a constant vertical force of 4.5×10^{-16} N acts on it. The mass of the electron is 9.11×10^{-31} kg. Determine the vertical distance the electron is deflected during the time it has moved 30 mm horizontally. ssm

22E. A car traveling at 53 km/h hits a bridge abutment. A passenger in the car moves forward a distance of 65 cm (with respect to the road) while being brought to rest by an inflated air bag. What magnitude of force (assumed constant) acts on the passenger's upper torso, which has a mass of 41 kg?

23E. Tarzan, who weighs 820 N, swings from a cliff at the end of a 20 m vine that hangs from a high tree limb and initially makes an angle of 22° with the vertical. Immediately after Tarzan steps off the cliff, the tension in the vine is 760 N. Choose a coordinate system for which the x axis points horizontally away from the edge of the cliff and the y axis points upward. (a) What is the force of the vine on Tarzan in unit-vector notation? (b) What is the net force acting on Tarzan in unit-vector notation? What are (c) the magnitude and (d) the direction of the net force acting on Tarzan? What are (e) the magnitude and (f) the direction of Tarzan's acceleration?

24P. A 50 kg skier is pulled up a frictionless ski slope that makes an angle of 8.0° with the horizontal by holding onto a tow rope that moves parallel to the slope. Determine the magnitude of the force of the rope on the skier at an instant when (a) the rope is moving with a constant speed of 2.0 m/s and (b) the rope is moving with a speed of 2.0 m/s but that speed is increasing at a rate of 0.10 m/s².

25P. A 40 kg girl and an 8.4 kg sled are on the frictionless ice of a frozen lake, 15 m apart but connected by a rope of negligible mass. The girl exerts a horizontal 5.2 N force on the rope. (a) What is the acceleration of the sled? (b) What is the acceleration of the girl? (c) How far from the girl's initial position do they meet? ssm

26P. You pull a short refrigerator with a constant force $\vec{F}$ across a greased (frictionless) floor, either with $\vec{F}$ horizontal (case 1) or with $\vec{F}$ tilted upward at an angle θ (case 2). (a) What is the ratio of the refrigerator's speed in case 2 to its speed in case 1 if you pull for a certain time t? (b) What is this ratio if you pull for a certain distance d?

27P. A firefighter with a weight of 712 N slides down a vertical pole with an acceleration of 3.00 m/s², directed downward. What are the magnitudes and directions of the vertical forces (a) on the firefighter from the pole and (b) on the pole from the firefighter? ilw

28P. For sport, a 12 kg armadillo runs onto a large pond of level, frictionless ice with an initial velocity of 5.0 m/s along the positive direction of an x axis. Take its initial position on the ice as being the origin. It slips over the ice while being pushed by a wind with a force of 17 N in the positive direction of the y axis. In unit-vector notation, what are the animal's (a) velocity and (b) position vector when it has slid for 3.0 s?

29P. A sphere of mass 3.0×10^{-4} kg is suspended from a cord. A steady horizontal breeze pushes the sphere so that the cord makes a constant angle of 37° with the vertical. Find (a) the magnitude of that push and (b) the tension in the cord. ssm ilw www

30P. A 40 kg skier comes directly down a frictionless ski slope that is inclined at an angle of 10° with the horizontal while a strong wind blows parallel to the slope. Determine the magnitude and direction of the force of the wind on the skier if (a) the magnitude of the skier's velocity is constant, (b) the magnitude of the skier's

velocity is increasing at a rate of 1.0 m/s², and (c) the magnitude of the skier's velocity is increasing at a rate of 2.0 m/s².

31P. Two blocks are in contact on a frictionless table. A horizontal force is applied to the larger block, as shown in Fig. 5-35. (a) If $m_1 = 2.3$ kg, $m_2 = 1.2$ kg, and $F = 3.2$ N, find the magnitude of the force between the two blocks. (b) Show that if a force of the same magnitude F is applied to the smaller block but in the opposite direction, the magnitude of the force between the blocks is 2.1 N, which is not the same value calculated in (a). (c) Explain the difference. ssm ilw

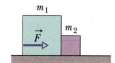
Fig. 5-35 Problem 31.

32P. A 1400 kg jet engine is fastened to the fuselage of a passenger jet by just three bolts (this is the usual practice). Assume that each bolt supports one-third of the load. (a) Calculate the force on each bolt as the plane waits in line for clearance to take off. (b) During flight, the plane encounters turbulence, which suddenly imparts an upward vertical acceleration of 2.6 m/s² to the plane. Calculate the force on each bolt now.

33P. An elevator and its load have a combined mass of 1600 kg. Find the tension in the supporting cable when the elevator, originally moving downward at 12 m/s, is brought to rest with constant acceleration in a distance of 42 m. ssm

34P. Figure 5-36 shows four penguins that are being playfully pulled along very slippery (frictionless) ice by a curator. The masses of three penguins and the tension in two of the cords are given. Find the penguin mass that is not given.

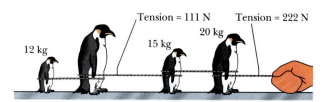
Fig. 5-36 Problem 34.

35P. An 80 kg person is parachuting and experiencing a downward acceleration of 2.5 m/s². The mass of the parachute is 5.0 kg. (a) What is the upward force on the open parachute from the air? (b) What is the downward force on the parachute from the person? ssm

36P. In Fig. 5-37, three blocks are connected and pulled to the right on a horizontal frictionless table by a force with a magnitude of $T_3 = 65.0$ N. If $m_1 = 12.0$ kg, $m_2 = 24.0$ kg, and $m_3 = 31.0$ kg, calculate (a) the acceleration of the system and the tensions (b) T_1 and (c) T_2 in the interconnecting cords.

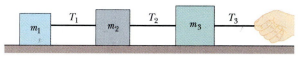
Fig. 5-37 Problem 36.

37P Imagine a landing craft approaching the surface of Callisto, one of Jupiter's moons. If the engine provides an upward force (thrust) of 3260 N, the craft descends at constant speed; if the engine provides only 2200 N, the craft accelerates downward at 0.39 m/s². (a) What is the weight of the landing craft in the vicinity of Callisto's surface? (b) What is the mass of the craft? (c) What is the magnitude of the free-fall acceleration near the surface of Callisto? ssm

38P. A worker drags a crate across a factory floor by pulling on a rope tied to the crate (Fig. 5-38). The worker exerts a force of 450 N on the rope, which is inclined at 38° to the horizontal, and the floor exerts a horizontal force of 125 N that opposes the motion. Calculate the magnitude of the acceleration of the crate if (a) its mass is 310 kg and (b) its weight is 310 N.

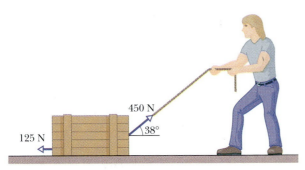

Fig. 5-38 Problem 38.

39P. A motorcycle and 60.0 kg rider accelerate at 3.0 m/s² up a ramp inclined 10° above the horizontal. (a) What is the magnitude of the net force acting on the rider? (b) What is the magnitude of the force on the rider from the motorcycle?

40P. An 85 kg man lowers himself to the ground from a height of 10.0 m by holding onto a rope that runs over a frictionless pulley to a 65 kg sandbag. With what speed does the man hit the ground if he started from rest?

41P. In Fig. 5-39, a chain consisting of five links, each of mass 0.100 kg, is lifted vertically with a constant acceleration of 2.50 m/s². Find the magnitudes of (a) the force on link 1 from link 2, (b) the force on link 2 from link 3, (c) the force on link 3 from link 4, and (d) the force on link 4 from link 5. Then find the magnitudes of (e) the force $\vec{F}$ on the top link from the person lifting the chain and (f) the *net* force accelerating each link. ssm

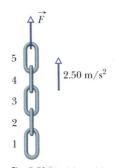

Fig. 5-39 Problem 41.

42P A Navy jet (Fig. 5-40) with a weight of 231 kN requires an airspeed of 85 m/s for liftoff. The engine develops a maximum force of 107 kN, but that is insufficient for reaching takeoff speed in the 90 m runway available on an aircraft carrier. What minimum force (assumed constant) is needed from the catapult that is used to help launch the jet? Assume that the catapult and the jet's engine each exert a constant force over the 90 m distance used for takeoff.

Fig. 5-40 Problem 42.

43P. A block of mass $m_1 = 3.70$ kg on a frictionless inclined plane of angle 30.0° is connected by a cord over a massless, frictionless pulley to a second block of mass $m_2 = 2.30$ kg hanging vertically (Fig. 5-41). What are (a) the magnitude of the acceleration of each block and (b) the direction of the acceleration of the hanging block? (c) What is the tension in the cord? ssm ilw www

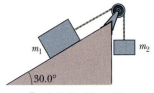

Fig. 5-41 Problem 43.

44P. In Fig. 5-42, a 1.0 kg pencil box on a 30° frictionless incline is connected to a 3.0 kg pen box on a horizontal frictionless surface. The pulley is frictionless and massless. (a) If the magnitude of $\vec{F}$ is 2.3 N, what is the tension in the connecting cord? (b) What is the largest value that the magnitude of $\vec{F}$ may have without the connecting cord becoming slack?

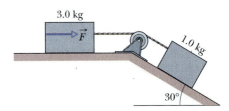

Fig. 5-42 Problem 44.

45P. A block is projected up a frictionless inclined plane with initial speed $v_0 = 3.50$ m/s. The angle of incline is $\theta = 32.0°$. (a) How far up the plane does it go? (b) How long does it take to get there? (c) What is its speed when it gets back to the bottom? ssm

46P. An interstellar ship has a mass of 1.20×10^6 kg and is initially at rest relative to a star system. (a) What constant acceleration is needed to bring the ship up to a speed of 0.10c (where c is the speed of light, 3.0×10^8 m/s) relative to the star system in 3.0 days? (b) What is that acceleration in g units? (c) What force is required for the acceleration? (d) If the engines are shut down when 0.10c is reached (the speed then remains constant), how long does the ship take (start to finish) to journey 5.0 light-months, the distance that light travels in 5.0 months?

47P. A 10 kg monkey climbs up a massless rope that runs over a frictionless tree limb and back down to a 15 kg package on the ground (Fig. 5-43). (a) What is the magnitude of the least acceleration the monkey must have if it is to lift the package off the ground? If, after the package has been lifted, the monkey stops its climb and holds onto the rope, what are (b) the magnitude and (c) the direction of the monkey's acceleration, and (d) what is the tension in the rope? **ssm**

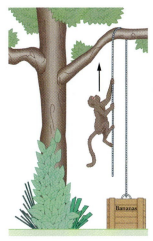

Fig. 5-43 Problem 47.

48P. In earlier days, horses pulled barges down canals in the manner shown in Fig. 5-44. Suppose that the horse pulls on the rope with a force of 7900 N at an angle of 18° to the direction of motion of the barge, which is headed straight along the canal. The mass of the barge is 9500 kg, and its acceleration is 0.12 m/s². What are (a) the magnitude and (b) the direction of the force on the barge from the water?

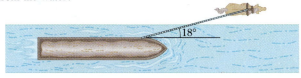

Fig. 5-44 Problem 48.

49P. In Fig. 5-45, a 5.00 kg block is pulled along a horizontal frictionless floor by a cord that exerts a force of magnitude $F = 12.0$ N at an angle $\theta = 25.0°$ above the horizontal. (a) What is the magnitude of the block's acceleration? (b) The force magnitude F is slowly increased. What is its value just before the block is lifted (completely) off the floor? (c) What is the magnitude of the block's acceleration just before it is lifted (completely) off the floor? **ssm**

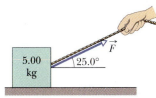

Fig. 5-45 Problem 49.

50P. Figure 5-46 shows a man sitting in a bosun's chair that dangles from a massless rope, which runs over a massless, frictionless pulley and back down to the man's hand. The combined mass of man and chair is 95.0 kg. With what force magnitude must the man pull on the rope if he is to rise (a) with a constant velocity and (b) with an upward acceleration of 1.30 m/s²? (*Hint:* A free-body diagram can really help.) Problem continues, next column.

Fig. 5-46 Problem 50.

Suppose, instead, that the rope on the right extends to the ground, where it is pulled by a co-worker. With what force magnitude must the co-worker pull for the man to rise (c) with a constant velocity and (d) with an upward acceleration of 1.30 m/s²? What is the magnitude of the force on the ceiling from the pulley system in (e) part a (f) part b, (g) part c, and (h) part d?

51P. A block of mass M is pulled along a horizontal frictionless surface by a rope of mass m, as shown in Fig. 5-47. A horizontal force $\vec{F}$ is applied to one end of the rope. (a) Show that the rope *must* sag, even if only by an imperceptible amount. Then, assuming that the sag is negligible, find (b) the acceleration of rope and block, (c) the force on the block from the rope, and (d) the tension in the rope at its midpoint. **ssm**

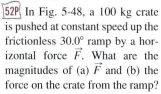

Fig. 5-47 Problem 51.

52P. In Fig. 5-48, a 100 kg crate is pushed at constant speed up the frictionless 30.0° ramp by a horizontal force $\vec{F}$. What are the magnitudes of (a) $\vec{F}$ and (b) the force on the crate from the ramp?

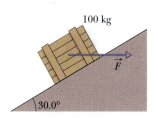

Fig. 5-48 Problem 52.

53P. A hot-air balloon of mass M is descending vertically with downward acceleration of magnitude a. How much mass (ballast) must be thrown out to give the balloon an upward acceleration of magnitude a (same magnitude but opposite direction)? Assume that the upward force from the air (the lift) does not change because of the decrease in mass. **ssm** **ilw** **www**

54P. Figure 5-49 shows a section of an alpine cable-car system. The maximum permissible mass of each car with occupants is 2800 kg. The cars, riding on a support cable, are pulled by a second cable attached to each pylon (support tower); assume the cables are straight. What is the difference in tension between adjacent sections of pull cable if the cars are at the maximum permissible mass and are being accelerated up the 35° incline at 0.81 m/s²?

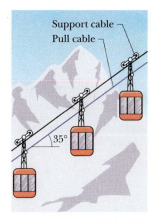

Fig. 5-49 Problem 54.

55P. An elevator with a weight of 27.8 kN is given an upward acceleration of 1.22 m/s² by a cable. (a) Calculate the tension in the cable. (b) What is the tension when the elevator is decelerating at the rate of 1.22 m/s² but is still moving upward?

56P. A lamp hangs vertically from a cord in a descending elevator that decelerates at 2.4 m/s². (a) If the tension in the cord is 89 N, what is the lamp's mass? (b) What is the cord's tension when the elevator ascends with an upward acceleration of 2.4 m/s²?

6 Force and Motion—II

Cats, who enjoy sleeping on window sills, are often kept in apartment buildings. When a cat accidentally falls out of a window and onto a sidewalk, the extent of injury (such as the number of fractured bones or the certainty of death) *decreases* with height if the fall is more than seven or eight floors. (There is even a record of a cat who fell 32 floors and suffered only slight damage to its thorax and one tooth.)

How can the danger possibly decrease with height?

The answer is in this chapter.

6-1 Friction

Frictional forces are unavoidable in our daily lives. If we were not able to counteract them, they would stop every moving object and bring to a halt every rotating shaft. About 20% of the gasoline used in an automobile is needed to counteract friction in the engine and in the drive train. On the other hand, if friction were totally absent, we could not get an automobile to go anywhere, and we could not walk or ride a bicycle. We could not hold a pencil and, if we could, it would not write. Nails and screws would be useless, woven cloth would fall apart, and knots would untie.

Here we deal with the frictional forces that exist between dry solid surfaces, either stationary relative to each other or moving across each other at slow speeds. Consider three simple thought experiments:

1. Send a book sliding across a long horizontal counter. As expected, the book slows and then stops. This means the book must have an acceleration parallel to the counter surface, in the direction opposite the book's velocity. From Newton's law, then, a force must act on the book parallel to the counter surface, in the direction opposite its velocity. That force is a frictional force.

2. Push horizontally on the book to make it travel at constant velocity along the counter. Can the force from you be the only horizontal force on the book? No, because then the book would accelerate. From Newton's second law, there must be a second force, directed opposite your force but with the same magnitude, so that the two forces balance. That second force is a frictional force, directed parallel to the counter.

3. Push horizontally on a heavy crate. The crate does not move. From Newton's second law, a second force must also act on the crate to counteract your force. Moreover, it must be directed opposite your force and have the same magnitude as your force, so that the two forces balance. That second force is a frictional force. Push even harder. The crate still does not move. Apparently the frictional force can change in magnitude so that the two forces still balance. Now push with all your strength. The crate begins to slide. Evidently, there is a maximum magnitude of the frictional force. When you exceed that maximum magnitude, the crate slides.

Figure 6-1 shows a similar situation in detail. In Fig. 6-1a, a block rests on a tabletop, with the gravitational force $\vec{F}_g$ balanced by a normal force $\vec{N}$. In Fig. 6-1b, you exert a force $\vec{F}$ on the block, attempting to pull it to the left. In response, a frictional force $\vec{f}_s$ is directed to the right, exactly balancing your force. The force $\vec{f}_s$ is called the **static frictional force.** The block does not move.

Figures 6-1c and 6-1d show that as you increase the magnitude of your applied force, the magnitude of the static frictional force $\vec{f}_s$ also increases and the block remains at rest. When the applied force reaches a certain magnitude, however, the block "breaks away" from its intimate contact with the tabletop and accelerates leftward (Fig. 6-1e). The frictional force that then opposes the motion is called the **kinetic frictional force** $\vec{f}_k$.

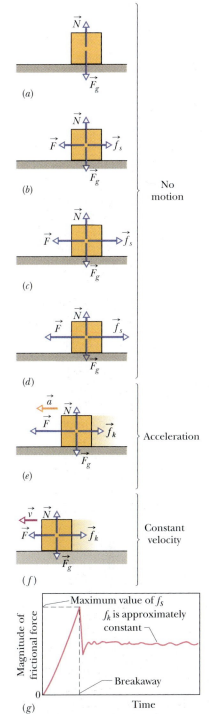

(a)

(b)

(c)

(d)

No motion

(e) Acceleration

(f) Constant velocity

(g)

Maximum value of f_s

f_k is approximately constant

Magnitude of frictional force

Breakaway

Time

Fig. 6-1 (a) The forces on a stationary block. (b–d) An external force $\vec{F}$, applied to the block, is balanced by a static frictional force $\vec{f}_s$. As F is increased, f_s also increases, until f_s reaches a certain maximum value. (e) The block then "breaks away," accelerating suddenly in the direction of $\vec{F}$. (f) If the block is now to move with constant velocity, F must be reduced from the maximum value it had just before the block broke away. (g) Some experimental results for the sequence (a) through (f).

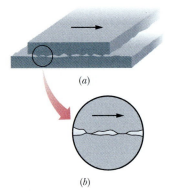

Fig. 6-2 The mechanism of sliding friction. (a) The upper surface is sliding to the right over the lower surface in this enlarged view. (b) A detail, showing two spots where cold-welding has occurred. Force is required to break the welds and maintain the motion.

Usually, the magnitude of the kinetic frictional force, which acts when there is motion, is less than the maximum magnitude of the static frictional force, which acts when there is no motion. Thus, if you wish the block to move across the surface with a constant speed, you must usually decrease the magnitude of the applied force once the block begins to move, as in Fig. 6-1f. As an example, Fig. 6-1g shows the results of an experiment in which the force on a block was slowly increased until breakaway occurred. Note the reduced force needed to keep the block moving at constant speed after breakaway.

A frictional force is, in essence, the vector sum of many forces acting between surface atoms of one body and those of another body. If two highly polished and carefully cleaned metal surfaces are brought together in a very good vacuum (to keep them clean), they cannot be made to slide over each other. Because the surfaces are so smooth, many atoms of one surface contact many atoms of the other surface, and the surfaces *cold-weld* together instantly, forming a single piece of metal. If a machinist's specially polished gage blocks are brought together in air, there is less atom-to-atom contact but the blocks stick firmly to each other and can be separated only by means of a wrenching motion. Usually, however, this much atom-to-atom contact is not possible. Even a highly polished metal surface is far from being flat on the atomic scale. Moreover, the surfaces of everyday objects have layers of oxides and other contaminants that reduce cold-welding.

When two ordinary surfaces are placed together, only the high points touch each other. (It is like having the Alps of Switzerland turned over and placed down on the Alps of Austria.) The actual *micro*scopic area of contact is much less than the apparent *macro*scopic contact area, perhaps by a factor of 10^4. Nonetheless, many contact points do cold-weld together. These welds produce static friction when an applied force attempts to slide the surfaces relative to each other.

If the applied force is great enough to pull one surface across the other, there is first a tearing of welds (at breakaway) and then a continuous re-forming and tearing apart of welds as movement occurs and chance contacts are made (Fig. 6-2). The kinetic frictional force $\vec{f}_k$ that opposes the motion is the vector sum of the forces at those many chance contacts.

If the two surfaces are pressed together harder, many more points cold-weld. Then, getting the surfaces to slide relative to each other requires a greater applied force: The static frictional force $\vec{f}_s$ has a greater maximum value. Once the surfaces are sliding, there are many more points of momentary cold-welding, so the kinetic frictional force $\vec{f}_k$ also has a greater magnitude.

Often, the sliding motion of one surface on another is "jerky," because the two surfaces alternately stick together and then slip. Such repetitive *stick-and-slip* can produce squeaking or squealing, as when tires skid on dry pavement, fingernails scratch along a chalkboard, or a rusty hinge is opened. It can also produce beautiful sounds, as when a bow is drawn properly across a violin string.

6-2 Properties of Friction

Experiment shows that when a dry and unlubricated body presses against a surface in the same condition, and a force $\vec{F}$ attempts to slide the body along the surface, the resulting frictional force has three properties:

Property 1. If the body does not move, then the static frictional force $\vec{f}_s$ and the component of $\vec{F}$ that is parallel to the surface balance each other. They are equal in magnitude, and $\vec{f}_s$ is directed opposite that component of $\vec{F}$.

Property 2. The magnitude of $\vec{f}_s$ has a maximum value $f_{s,\max}$ that is given by

$$f_{s,\max} = \mu_s N, \qquad (6\text{-}1)$$

where μ_s is the **coefficient of static friction** and N is the magnitude of the normal force on the body from the surface. If the magnitude of the component of $\vec{F}$ that is parallel to the surface exceeds $f_{s,max}$, then the body begins to slide along the surface.

Property 3. If the body begins to slide along the surface, the magnitude of the frictional force rapidly decreases to a value f_k given by

$$f_k = \mu_k N, \qquad (6\text{-}2)$$

where μ_k is the **coefficient of kinetic friction.** Thereafter, during the sliding, a kinetic frictional force $\vec{f}_k$ with magnitude given by Eq. 6-2 opposes the motion.

The magnitude N of the normal force appears in properties 2 and 3 as a measure of how firmly the body presses against the surface. If the body presses harder, then, by Newton's third law, N is greater. Properties 1 and 2 are worded in terms of a single applied force $\vec{F}$, but they also hold for the net force of several applied forces acting on the body. Equations 6-1 and 6-2 are *not* vector equations; the direction of $\vec{f}_s$ or $\vec{f}_k$ is always parallel to the surface and opposed to the attempted sliding, and the normal force $\vec{N}$ is perpendicular to the surface.

The coefficients μ_s and μ_k are dimensionless and must be determined experimentally. Their values depend on certain properties of both the body and the surface; hence, they are usually referred to with the preposition "between," as in "the value of μ_s *between* an egg and a Teflon-coated skillet is 0.04, but that *between* rock-climbing shoes and rock is as much as 1.2." We assume that the value of μ_k does not depend on the speed at which the body slides along the surface.

✔**CHECKPOINT 1:** A block lies on a floor. (a) What is the magnitude of the frictional force on it from the floor? (b) If a horizontal force of 5 N is now applied to the block, but the block does not move, what is the magnitude of the frictional force on it? (c) If the maximum value $f_{s,max}$ of the static frictional force on the block is 10 N, will the block move if the magnitude of the horizontally applied force is 8 N? (d) If the magnitude is 12 N? (e) What is the magnitude of the frictional force in part (c)?

Sample Problem 6-1

If a car's wheels are "locked" (kept from rolling) during emergency braking, the car slides along the road. Ripped-off bits of tire and small melted sections of road form the "skid marks" that reveal that cold-welding occurred during the slide. The record for the longest skid marks on a public road was reportedly set in 1960 by a Jaguar on the M1 highway in England (Fig. 6-3a)—the marks were 290 m long! Assuming that $\mu_k = 0.60$ and the car's acceleration was constant during the braking, how fast was the car going when the wheels became locked?

SOLUTION: One **Key Idea** here is that, because the acceleration is assumed constant, we can use the equations of Table 2-1 to find the car's initial speed v_0. Let us try Eq. 2-16,

$$v^2 = v_0^2 + 2a(x - x_0), \qquad (6\text{-}3)$$

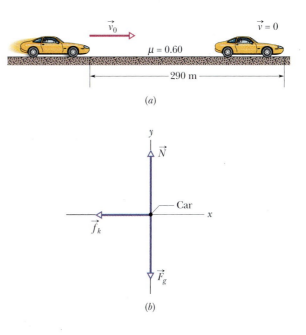

(a)

(b)

Fig. 6-3 Sample Problem 6-1. (a) A car, sliding to the right and finally stopping after a displacement of 290 m. (b) A free-body diagram for the car.

assuming that the car moved in the positive direction of an x axis. We know that the displacement $x - x_0$ was 290 m, we assume that the final speed v was 0, and we want v_0. However, we do not know the car's acceleration a.

To find a, we use another Key Idea: If we neglect the effects of the air on the car, the acceleration a was due only to a kinetic frictional force $\vec{f}_k$ on the car from the road, directed opposite the direction of the car's motion (Fig. 6-3b). We can relate this force to the acceleration by writing Newton's second law for x components ($F_{net,x} = ma_x$) as

$$-f_k = ma, \tag{6-4}$$

where m is the car's mass. The minus sign indicates the direction of the kinetic frictional force.

From Eq. 6-2, that frictional force has the magnitude $f_k = \mu_k N$, where N is the magnitude of the normal force on the car from the road. Because the car is not accelerating vertically, we know

from Fig. 6-3b and Newton's second law that the magnitude of $\vec{N}$ is equal to the magnitude of the gravitational force $\vec{F}_g$ on the car, which is mg. Thus, we have $N = mg$.

Now solving Eq. 6-4 for a and substituting $f_k = \mu_k N = \mu_k mg$ for f_k yield

$$a = -\frac{f_k}{m} = -\frac{\mu_k mg}{m} = -\mu_k g,$$

where the minus sign indicates that the acceleration is in the negative direction of the x axis, opposite the direction of the velocity. Next, substituting this for a in Eq. 6-3 with $v = 0$ and solving for v_0 give

$$v_0 = \sqrt{2\mu_k g(x - x_0)} = \sqrt{(2)(0.60)(9.8 \text{ m/s}^2)(290 \text{ m})}$$
$$= 58 \text{ m/s} = 210 \text{ km/h}. \tag{Answer}$$

We assumed that $v = 0$ at the far end of the skid marks. Actually, the marks ended only because the Jaguar left the road after 290 m. So v_0 was at least 210 km/h, and possibly much more.

Sample Problem 6-2

In Fig. 6-4a, a woman pulls a loaded sled of mass $m = 75$ kg along a horizontal surface at constant velocity. The coefficient of kinetic friction μ_k between the runners and the snow is 0.10, and the angle ϕ is 42°.

(a) What is the magnitude of the force $\vec{T}$ on the sled from the rope?

SOLUTION: We need three Key Ideas here:

1. Because the sled's velocity is constant, its acceleration is zero in spite of the woman's pull.

2. Acceleration is prevented by a kinetic frictional force $\vec{f}_k$ on the sled from the snow.

3. We can relate the (zero) acceleration of the sled to the forces on the sled, including the desired $\vec{T}$, with Newton's second law ($\vec{F}_{net} = m\vec{a}$).

Figure 6-4b shows the forces on the sled, including the gravitational force $\vec{F}_g$ and the normal force $\vec{N}$ from the snow surface. For these forces, Newton's second law, with $\vec{a} = 0$, gives us

$$\vec{T} + \vec{N} + \vec{F}_g + \vec{f}_k = 0. \tag{6-5}$$

We cannot solve Eq. 6-5 for $\vec{T}$ directly on a vector-capable calculator because we do not know the other vectors, so we rewrite it for components along the x and y axes of Fig. 6-4b. For the x axis, we get

$$T_x + 0 + 0 - f_k = 0$$

or

$$T \cos \phi - \mu_k N = 0, \tag{6-6}$$

where we have used Eq. 6-2 to substitute $\mu_k N$ for f_k. For the y axis, we have

$$T_y + N - F_g + 0 = 0$$

or

$$T \sin \phi + N - mg = 0, \tag{6-7}$$

where we have substituted mg for F_g.

Equations 6-6 and 6-7 are simultaneous equations with unknowns T and N. To solve them for T, we must first solve Eq. 6-6

for N and then substitute that expression into Eq. 6-7 to get

$$T = \frac{\mu_k mg}{\cos \phi + \mu_k \sin \phi}$$
$$= \frac{(0.10)(75 \text{ kg})(9.8 \text{ m/s}^2)}{\cos 42° + (0.10)(\sin 42°)}$$
$$= 91 \text{ N}. \tag{Answer}$$

(Instead, we could substitute known data into Eqs. 6-6 and 6-7 and then solve them by using a simultaneous-equations solver on a calculator.)

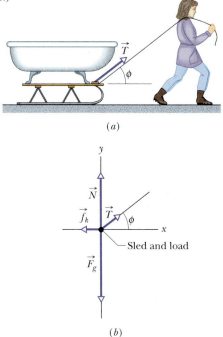

(a)

(b)

Fig. 6-4   Sample Problem 6-2. (a) A woman pulls a loaded sled at constant velocity via a force $\vec{T}$ on the sled from a rope. (b) A free-body diagram for the loaded sled.

(b) If the woman increases her pull on the rope, so that T is greater than 91 N, is the magnitude f_k of the frictional force greater than, less than, or the same as in (a)?

SOLUTION: The **Key Idea** here is that, by Eq. 6-2, the magnitude of f_k depends directly on the magnitude N of the normal force. Thus, we can answer if we find a relation between N and T. Equation 6-7 is such a relation. Rewriting it as

$$N = mg - T \sin \phi, \qquad (6\text{-}8)$$

we note that if T is increased, then N will decrease. (The physical reason is that the upward component of the rope's pull is greater, and thus the force on the sled from the snow is less.) Because $f_k = \mu_k N$, we see that f_k will be less than it was.

✓CHECKPOINT 2:   In the figure, horizontal force $\vec{F}_1$ of magnitude 10 N is applied to a box on a floor, but the box does not slide. Then, as the magnitude of vertically applied force $\vec{F}_2$ is increased from zero but before the box begins to slide, do the following quantities increase, decrease, or stay the same: (a) the magnitude of the frictional force on the box; (b) the magnitude of the normal force on the box from the floor; (c) the maximum value $f_{s,\text{max}}$ of the static frictional force on the box?

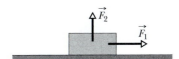

Sample Problem 6-3

Figure 6-5a shows a coin of mass m at rest on a book that has been tilted at an angle θ with the horizontal. By experimenting, you find that when θ is increased to 13°, the coin is on the *verge* of sliding down the book, which means that even a slight increase beyond 13° produces sliding. What is the coefficient of static friction μ_s between the coin and the book?

SOLUTION: If the book were frictionless, the coin would surely slide down it for any tilt of the book because of the gravitational force on the coin. Thus, one **Key Idea** here is that a frictional force $\vec{f}_s$ must be holding the coin in place. A second **Key Idea** is that, because the coin is *on the verge* of sliding *down* the book, that force is at its *maximum* magnitude $f_{s,\text{max}}$ and is directed *up* the book. Also, from Eq. 6-1, we know that $f_{s,\text{max}} = \mu_s N$, where N is the magnitude of the normal force $\vec{N}$ on the coin from the book. Thus,

$$f_s = f_{s,\text{max}} = \mu_s N,$$

from which
$$\mu_s = \frac{f_s}{N}. \qquad (6\text{-}9)$$

To evaluate this equation, we need to find the force magnitudes f_s and N. To do that, we use another **Key Idea**: When the coin is on the verge of sliding, it is stationary and thus its acceleration $\vec{a}$ is zero. We can relate this acceleration to the forces on the coin with Newton's second law ($\vec{F}_{\text{net}} = m\vec{a}$). As shown in the free-body diagram of the coin in Fig. 6-5b, those forces are (1) the frictional force $\vec{f}_s$, (2) the normal force $\vec{N}$, and (3) the gravitational force $\vec{F}_g$ on the coin, with magnitude equal to mg. Then, from Newton's second law with $\vec{a} = 0$, we have

$$\vec{f}_s + \vec{N} + \vec{F}_g = 0. \qquad (6\text{-}10)$$

To find f_s and N, we rewrite Eq. 6-10 for components along the x and y axes of the tilted coordinate system in Fig. 6-5b. For the x axis and with mg substituted for F_g, we have

$$f_s + 0 - mg \sin \theta = 0,$$

so
$$f_s = mg \sin \theta. \qquad (6\text{-}11)$$

Similarly, for the y axis we have

$$0 + N - mg \cos \theta = 0,$$

so
$$N = mg \cos \theta. \qquad (6\text{-}12)$$

Substituting Eqs. 6-11 and 6-12 into Eq. 6-9 produces

$$\mu_s = \frac{mg \sin \theta}{mg \cos \theta} = \tan \theta, \qquad (6\text{-}13)$$

which here means

$$\mu_s = \tan 13° = 0.23. \qquad (\text{Answer})$$

Actually you do not need to measure θ to get μ_s. Instead, measure the two lengths shown in Fig. 5-6a and then substitute h/d for $\tan \theta$ in Eq. 6-13.

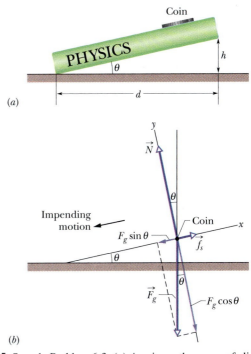

Fig. 6-5 Sample Problem 6-3. (a) A coin on the verge of sliding down a book. (b) A free-body diagram for the coin, showing the three forces (drawn to scale) that act on it. The gravitational force $\vec{F}_g$ is shown resolved into its components along the x and the y axes, whose orientations are chosen to simplify the problem. Component $F_g \sin \theta$ tends to slide the coin down the book. Component $F_g \cos \theta$ presses the coin onto the book.

Fig. 6-6 This skier crouches in an "egg position" so as to minimize her effective cross-sectional area and thus the air drag acting on her.

6-3 The Drag Force and Terminal Speed

A **fluid** is anything that can flow—generally either a gas or a liquid. When there is a relative velocity between a fluid and a body (either because the body moves through the fluid or because the fluid moves past the body), the body experiences a **drag force** $\vec{D}$ that opposes the relative motion and points in the direction in which the fluid flows relative to the body.

Here we examine only cases in which air is the fluid, the body is blunt (like a baseball) rather than slender (like a javelin), and the relative motion is fast enough so that the air becomes turbulent (breaks up into swirls) behind the body. In such cases, the magnitude of the drag force $\vec{D}$ is related to the relative speed v by an experimentally determined **drag coefficient** C according to

$$D = \tfrac{1}{2}C\rho A v^2, \tag{6-14}$$

where ρ is the air density (mass per volume) and A is the **effective cross-sectional area** of the body (the area of a cross section taken perpendicular to the velocity $\vec{v}$). The drag coefficient C (typical values range from 0.4 to 1.0) is not truly a constant for a given body, because if v varies significantly, the value of C can vary as well. Here, we ignore such complications.

Downhill speed skiers know well that drag depends on A and v^2. To reach high speeds a skier must reduce D as much as possible by, for example, riding the skis in the "egg position" (Fig. 6-6) to minimize A.

When a blunt body falls from rest through air, the drag force $\vec{D}$ is directed upward; its magnitude gradually increases from zero as the speed of the body increases. This upward force $\vec{D}$ opposes the downward gravitational force $\vec{F}_g$ on the body. We can relate these forces to the body's acceleration by writing Newton's second law for a vertical y axis ($F_{net,y} = ma_y$) as

$$D - F_g = ma, \tag{6-15}$$

where m is the mass of the body. As suggested in Fig. 6-7, if the body falls long enough, D eventually equals F_g. From Eq. 6-15, this means that $a = 0$ and so the body's speed no longer increases. The body then falls at a constant speed, called the **terminal speed** v_t.

To find v_t, we set $a = 0$ in Eq. 6-15 and substitute for D from Eq. 6-14, obtaining

$$\tfrac{1}{2}C\rho A v_t^2 - F_g = 0,$$

which gives

$$v_t = \sqrt{\frac{2F_g}{C\rho A}}. \tag{6-16}$$

Table 6-1 gives values of v_t for some common objects.

According to calculations* based on Eq. 6-14, a cat must fall about six floors to reach terminal speed. Until it does so, $F_g > D$ and the cat accelerates downward because of the net downward force. Recall from Chapter 2 that your body is an accelerometer, not a speedometer. Because the cat also senses the acceleration, it is frightened and keeps its feet underneath its body, its head tucked in, and its spine bent upward, making A small, v_t large, and injury on landing likely.

However, if the cat does reach v_t, the acceleration vanishes and the cat relaxes somewhat, stretching its legs and neck horizontally outward and straightening its spine (it then resembles a flying squirrel). These actions increase area A and thus

---

*W. O. Whitney and C. J. Mehlhaff, "High-Rise Syndrome in Cats." *The Journal of the American Veterinary Medical Association*, 1987, Vol. 191, pp. 1399–1403.

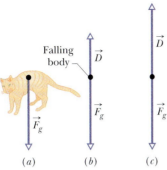

Fig. 6-7 The forces that act on a body falling through air: (a) the body when it has just begun to fall and (b) the free-body diagram a little later, after a drag force has developed. (c) The drag force has increased until it balances the gravitational force on the body. The body now falls at its constant terminal speed.

TABLE 6-1   Some Terminal Speeds in Air

Object	Terminal Speed (m/s)	95% Distance[a] (m)
Shot (from shot put)	145	2500
Sky diver (typical)	60	430
Baseball	42	210
Tennis ball	31	115
Basketball	20	47
Ping-Pong ball	9	10
Raindrop (radius = 1.5 mm)	7	6
Parachutist (typical)	5	3

[a]This is the distance through which the body must fall from rest to reach 95% of its terminal speed.

Source: Adapted from Peter J. Brancazio, Sport Science, 1984, Simon & Schuster, New York.

Fig. 6-8 A sky diver in a horizontal "spread eagle" maximizes the air drag.

also, by Eq. 6-14, the drag D. The cat begins to slow because now $D > F_g$ (the net force is upward), until a new, smaller v_t is reached. The decrease in v_t reduces the possibility of serious injury on landing. Just before the end of the fall, when it sees it is nearing the ground, the cat pulls its legs back beneath its body to prepare for the landing.

Humans often fall from great heights for the fun of skydiving. However, in April 1987, during a jump, sky diver Gregory Robertson noticed that fellow sky diver Debbie Williams had been knocked unconscious in a collision with a third sky diver and was unable to open her parachute. Robertson, who was well above Williams at the time and who had not yet opened his parachute for the 4 km plunge, reoriented his body head-down so as to minimize A and maximize his downward speed. Reaching an estimated v_t of 320 km/h, he caught up with Williams and then went into a horizontal "spread eagle" (as in Fig. 6-8) to increase D so that he could grab her. He opened her parachute and then, after releasing her, his own, a scant 10 s before impact. Williams received extensive internal injuries due to her lack of control on landing but survived.

Sample Problem 6-4

If a falling cat reaches a first terminal speed of 97 km/h while it is tucked in and then stretches out, doubling A, how fast is it falling when it reaches a new terminal speed?

SOLUTION: The Key Idea here is that the terminal speeds of the cat depend on (among other things) the effective cross-sectional areas A of the cat, according to Eq. 6-16. Thus, we can use that equation

to set up a ratio of speeds. We let v_{to} and v_{tn} represent the original and new terminal speeds, and A_o and A_n the original and new areas. Then by Eq. 6-16,

$$\frac{v_{tn}}{v_{to}} = \frac{\sqrt{2F_g/C\rho A_n}}{\sqrt{2F_g/C\rho A_o}} = \sqrt{\frac{A_o}{A_n}} = \sqrt{\frac{A_o}{2A_o}} = \sqrt{0.5} \approx 0.7,$$

which means that $v_{tn} \approx 0.7 v_{to}$, or about 68 km/h.

Sample Problem 6-5

A raindrop with radius $R = 1.5$ mm falls from a cloud that is at height $h = 1200$ m above the ground. The drag coefficient C for the drop is 0.60. Assume that the drop is spherical throughout its fall. The density of water ρ_w is 1000 kg/m³, and the density of air ρ_a is 1.2 kg/m³.

(a) What is the terminal speed of the drop?

SOLUTION: The Key Idea here is that the drop reaches a terminal speed v_t when the gravitational force on it is balanced by the air drag force on it, so its acceleration is zero. We could then apply Newton's second law and the drag force equation to find v_t, but Eq. 6-16 does all that for us.

To use Eq. 6-16, we need the drop's effective cross-sectional area A and the magnitude F_g of the gravitational force. Because the drop is spherical, A is the area of a circle (πR^2) with the same radius as the sphere. To find F_g, we use three facts: (1) $F_g = mg$, where m is the drop's mass; (2) the (spherical) drop's volume is $V = \frac{4}{3}\pi R^3$; and (3) the density of the water in the drop is the mass per volume, or $\rho_w = m/V$. Thus, we find

$$F_g = V\rho_w g = \tfrac{4}{3}\pi R^3 \rho_w g.$$

We next substitute this, the expression for A, and the given data into Eq. 6-16. Being careful to distinguish between the air density ρ_a and the water density ρ_w, we obtain

$$v_t = \sqrt{\frac{2F_g}{C\rho_a A}} = \sqrt{\frac{8\pi R^3 \rho_w g}{3C\rho_a \pi R^2}} = \sqrt{\frac{8R\rho_w g}{3C\rho_a}}$$

$$= \sqrt{\frac{(8)(1.5 \times 10^{-3}\text{ m})(1000\text{ kg/m}^3)(9.8\text{ m/s}^2)}{(3)(0.60)(1.2\text{ kg/m}^3)}}$$

$$= 7.4\text{ m/s} \approx 27\text{ km/h.}\qquad\text{(Answer)}$$

Note that the height of the cloud does not enter into the calculation. As Table 6-1 indicates, the raindrop reaches terminal speed after falling just a few meters.

(b) What would be the drop's speed just before impact if there were no drag force?

SOLUTION: The **Key Idea** here is that, with no drag force to reduce the drop's speed during the fall, the drop would fall with the constant free-fall acceleration g, so the constant-acceleration equations of Table 2-1 apply. Because we know the acceleration is g, the initial velocity v_0 is 0, and the displacement $x - x_0$ is h, we use Eq. 2-16 to find v:

$$v = \sqrt{2gh} = \sqrt{(2)(9.8\text{ m/s}^2)(1200\text{ m})}$$

$$= 153\text{ m/s} \approx 550\text{ km/h.}\qquad\text{(Answer)}$$

For that speed, Shakespeare would scarcely have written, "it droppeth as the gentle rain from heaven, upon the place beneath."

✔**CHECKPOINT 3:**   Near the ground, is the speed of large raindrops greater than, less than, or the same as the speed of small raindrops, assuming that all raindrops are spherical and have the same drag coefficient?

6-4 Uniform Circular Motion

From Section 4-7, recall that when a body moves in a circle (or a circular arc) at constant speed v, it is said to be in uniform circular motion. Also recall that the body has a centripetal acceleration (directed toward the center of the circle), of constant magnitude given by

$$a = \frac{v^2}{R}\qquad\text{(centripetal acceleration),}\qquad\text{(6-17)}$$

where R is the radius of the circle.

Let us examine two examples of uniform circular motion:

1. *Rounding a curve in a car.* You are sitting in the center of the rear seat of a car moving at a constant high speed along a flat road. When the driver suddenly turns left, rounding a corner in a circular arc, you slide across the seat toward the right and then jam against the car wall for the rest of the turn. What is going on?

   While the car moves in the circular arc, it is in uniform circular motion; that is, it has an acceleration that is directed toward the center of the circle. By Newton's second law, a force must cause this acceleration. Moreover, the force must also be directed toward the center of the circle. Thus, it is a **centripetal force,** where the adjective indicates the direction. In this example, the centripetal force is a frictional force on the tires from the road; it makes the turn possible.

   If you are to move in uniform circular motion along with the car, there must also be a centripetal force on you. However, apparently the frictional force on you from the seat was not great enough to make you go in a circle with the car. Thus, the seat slid beneath you, until the right wall of the car jammed into you. Then, its push on you provided the needed centripetal force on you, and you joined the car's uniform circular motion.

2. *Orbiting Earth.* This time you are a passenger in the space shuttle *Atlantis.* As it and you orbit Earth, you float through your cabin. What is going on?

Both you and the shuttle are in uniform circular motion and have accelerations directed toward the center of the circle. Again by Newton's second law, centripetal forces must cause these accelerations. This time the centripetal forces are gravitational pulls (the pull on you and the pull on the shuttle) by Earth, radially inward, toward the center of Earth.

In both car and shuttle you are in uniform circular motion, acted on by a centripetal force—yet your sensations in the two situations are quite different. In the car, jammed up against the wall, you are aware of being compressed by the wall. In the orbiting shuttle, however, you are floating around with no sensation of any force acting on you. Why this difference?

The difference is due to the nature of the two centripetal forces. In the car, the centripetal force is the push on the part of your body touching the car wall. You can sense the compression on that part of your body. In the shuttle, the centripetal force is Earth's gravitational pull on every atom of your body. Thus, there is no compression (or pull) on any one part of your body and no sensation of a force acting on you. (The sensation is said to be one of "weightlessness," but that description is tricky. The pull on you by Earth has certainly not disappeared and, in fact, is only a little less than it would be with you on the ground.)

Another example of a centripetal force is shown in Fig. 6-9. There a hockey puck moves around in a circle at constant speed v while tied to a string looped around a central peg. This time the centripetal force is the radially inward pull on the puck from the string. Without that force, the puck would slide off in a straight line instead of moving in a circle.

Note again that a centripetal force is not a new kind of force. The name merely indicates the direction of the force. It can, in fact, be a frictional force, a gravitational force, the force from a car wall or a string, or any other force. For any situation:

> A centripetal force accelerates a body by changing the direction of the body's velocity without changing the body's speed.

From Newton's second law and Eq. 6-17 ($a = v^2/R$), we can write the magnitude F of a centripetal force (or a net centripetal force) as

$$F = m\frac{v^2}{R} \qquad \text{(magnitude of centripetal force).} \qquad (6\text{-}18)$$

Because the speed v here is constant, so are also the magnitudes of the acceleration and the force.

However, the directions of the centripetal acceleration and force are not constant; they vary continuously so as to always point toward the center of the circle. For this reason, the force and acceleration vectors are sometimes drawn along a radial axis r that moves with the body and always extends from the center of the circle to the body, as in Fig. 6-9. The positive direction of the axis is radially outward, but the acceleration and force vectors point radially inward.

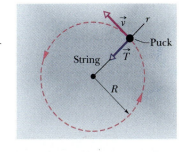

Fig. 6-9 An overhead view of a hockey puck of mass m moving with constant speed v in a circular path of radius R on a horizontal frictionless surface. The centripetal force on the puck is $\vec{T}$, the pull from the string, directed inward along the radial axis r extending through the puck.

Sample Problem 6-6

Igor is a cosmonaut-engineer on the *International Space Station*, in a circular orbit around Earth, at an altitude h of 520 km and with a constant speed v of 7.6 km/s. Igor's mass m is 79 kg.

(a) What is his acceleration?

SOLUTION: The Key Idea here is that Igor is in uniform circular motion and thus has a centripetal acceleration of magnitude given by Eq. 6-17 ($a = v^2/R$). The radius R of Igor's motion is $R_E + h$, where R_E is Earth's radius (6.37×10^6 m, from Appendix C). Thus,

$$a = \frac{v^2}{R} = \frac{v^2}{R_E + h}$$

$$= \frac{(7.6 \times 10^3 \text{ m/s})^2}{6.37 \times 10^6 \text{ m} + 0.52 \times 10^6 \text{ m}}$$

$$= 8.38 \text{ m/s}^2 \approx 8.4 \text{ m/s}^2. \qquad \text{(Answer)}$$

This is the value of the free-fall acceleration at Igor's altitude. If he were lifted to that altitude and released, instead of being put into orbit there, he would fall toward Earth's center, starting out with that value for his acceleration. The difference in the two situations is that when he orbits Earth, he always has a "sideways" motion as well: As he falls, he also moves to the side, so that he ends up moving along a curved path around Earth.

(b) What force does Earth exert on Igor?

SOLUTION: There are two Key Ideas here. First, there must be a centripetal force on Igor if he is to be in uniform circular motion. Second, that force is the gravitational force $\vec{F}_g$ on him from Earth, directed toward his center of rotation (at the center of Earth). From Newton's second law, written along the radial axis r, this force has the magnitude

$$F_g = ma = (79 \text{ kg})(8.38 \text{ m/s}^2)$$

$$= 662 \text{ N} \approx 660 \text{ N}. \qquad \text{(Answer)}$$

If Igor were to stand on a scale placed on the top of a tower with height $h = 520$ km, the scale would read 660 N. In orbit, the scale (if Igor could "stand" on it) would read zero because he and the scale are in free fall together, and therefore his feet do not actually press against it.

Sample Problem 6-7

In a 1901 circus performance, Allo "Dare Devil" Diavolo introduced the stunt of riding a bicycle in a loop-the-loop (Fig. 6-10a). Assuming that the loop is a circle with radius $R = 2.7$ m, what is the least speed v Diavolo could have at the top of the loop to remain in contact with it there?

SOLUTION: A Key Idea in analyzing Diavolo's stunt is to assume that he and his bicycle travel through the top of the loop as a single particle in uniform circular motion. Thus, at the top, the acceleration $\vec{a}$ of this particle must have the magnitude $a = v^2/R$ given by Eq. 6-17 and be directed downward, toward the center of the circular loop.

The forces on the particle when it is at the top of the loop are shown in the free-body diagram of Fig 6-10b. The gravitational force $\vec{F}_g$ is directed downward along a y axis. The normal force $\vec{N}$ on the particle from the loop is also directed downward. Thus, Newton's second law for y components ($F_{\text{net},y} = ma_y$) gives us

$$-N - F_g = m(-a).$$

This becomes

$$-N - mg = m\left(-\frac{v^2}{R}\right). \qquad (6\text{-}19)$$

Another Key Idea is that if the particle has the *least speed* v needed to remain in contact, then it is on the *verge of losing contact* with the loop (falling away from the loop), which means that $N = 0$. Substituting this value for N into Eq. 6-19, solving for v, and then substituting known values give us

$$v = \sqrt{gR} = \sqrt{(9.8 \text{ m/s}^2)(2.7 \text{ m})}$$

$$= 5.1 \text{ m/s}. \qquad \text{(Answer)}$$

Diavolo made certain that his speed at the top of the loop was greater than 5.1 m/s so that he did not lose contact with the loop and fall away from it. Note that this speed requirement is indepen-

(a)

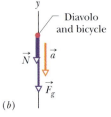

(b)

Fig. 6-10 Sample Problem 6-7. (*a*) Contemporary advertisement for Diavolo and (*b*) free-body diagram for the performer at the top of the loop.

dent of the mass of Diavolo and his bicycle. Had he feasted on, say, pierogies before his performance, he still would have had to exceed only 5.1 m/s.

✔**CHECKPOINT 4:** When you ride in a Ferris wheel at constant speed, what are the directions of your acceleration $\vec{a}$ and the normal force $\vec{N}$ on you (from the always upright seat) as you pass through (a) the highest point and (b) the lowest point of the ride?

Sample Problem 6-8

Even some seasoned roller-coaster riders blanch at the thought of riding the Rotor, which is essentially a large, hollow cylinder that is rotated rapidly around its central axis (Fig. 6-11). Before the ride begins, a rider enters the cylinder through a door on the side and stands on a floor, up against a canvas-covered wall. The door is closed, and as the cylinder begins to turn, the rider, wall, and floor move in unison. When the rider's speed reaches some predetermined value, the floor abruptly and alarmingly falls away. The rider does not fall with it but instead is pinned to the wall while the cylinder rotates, as if an unseen (and somewhat unfriendly) agent is pressing the body to the wall. Later, the floor is eased back to the rider's feet, the cylinder slows, and the rider sinks a few centimeters to regain footing on the floor. (Some riders consider all this to be fun.)

Suppose that the coefficient of static friction μ_s between the rider's clothing and the canvas is 0.40 and that the cylinder's radius R is 2.1 m.

(a) What minimum speed v must the cylinder and rider have if the rider is not to fall when the floor drops?

SOLUTION: We start with a question: What force can keep the rider from falling and how is it related to the speed v of her and the cylinder? To answer, we use three Key Ideas:

1. The gravitational force $\vec{F}_g$ on the rider tends to slide her down the wall, but she does not move because a frictional force from the wall acts upward on her (Fig 6-11).

2. If she is to be on the verge of sliding down, that upward force must be a *static* frictional force $\vec{f}_s$ at its maximum value $\mu_s N$, where N is the magnitude of the normal force $\vec{N}$ on her from the cylinder (Fig. 6-11).

3. This normal force is directed horizontally toward the central axis of the cylinder and is the centripetal force that causes the rider

to move in a circular path, with centripetal acceleration of magnitude $a = v^2/R$.

We want speed v in that last expression, for the condition that the rider is on the verge of sliding.

We first place a vertical y axis through the rider, with the positive direction upward. For Key Idea 1, we can then apply Newton's second law to the rider, writing it for y components ($F_{net,y} = ma_y$) as

$$f_s - mg = m(0),$$

where m is the rider's mass and mg is the magnitude of $\vec{F}_g$. For Key Idea 2, we substitute the maximum value $\mu_s N$ for f_s in this equation, getting

$$\mu_s N - mg = 0,$$

or

$$N = \frac{mg}{\mu_s}. \tag{6-20}$$

Next we place a radial r axis through the rider, with the positive direction outward. For Key Idea 3, we can then write Newton's second law for components along that axis as

$$-N = m\left(-\frac{v^2}{R}\right). \tag{6-21}$$

Substituting Eq. 6-20 for N and then solving for v, we find

$$v = \sqrt{\frac{gR}{\mu_s}} = \sqrt{\frac{(9.8 \text{ m/s}^2)(2.1 \text{ m})}{0.40}}$$

$$= 7.17 \text{ m/s} \approx 7.2 \text{ m/s}. \qquad \text{(Answer)}$$

Note that the result is independent of the rider's mass; it holds for anyone riding the Rotor, from a child to a sumo wrestler, which is why no one has to "weigh in" to ride the Rotor.

(b) If the rider's mass is 49 kg, what is the magnitude of the centripetal force on her?

SOLUTION: According to Eq. 6-21,

$$N = m\frac{v^2}{R} = (49 \text{ kg})\frac{(7.17 \text{ m/s})^2}{2.1 \text{ m}}$$

$$\approx 1200 \text{ N}. \qquad \text{(Answer)}$$

Although this force is directed toward the central axis, the rider has an overwhelming sensation that the force pinning her against the wall is directed radially outward. Her sensation stems from the fact that she is in a noninertial frame (she and it are accelerating). As measured from such frames, forces can be illusionary. The illusion is part of the Rotor's attraction.

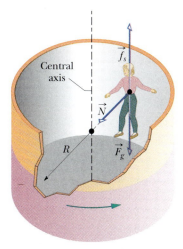

Fig. 6-11 Sample Problem 6-8. A Rotor in an amusement park, showing the forces on a rider. The centripetal force is the normal force with which the wall pushes inward on the rider.

✓CHECKPOINT 5:   If the Rotor initially moves at the minimum required speed for the rider not to fall and then its speed is increased in steps, do the following increase, decrease, or remain the same: (a) the magnitude of $\vec{f}_s$; (b) the magnitude of $\vec{N}$; (c) the value of $f_{s,max}$?

Sample Problem 6-9

Figure 6-12a represents a stock car of mass $m = 1600$ kg traveling at a constant speed $v = 20$ m/s around a flat, circular track of radius $R = 190$ m. For what value of μ_s between the track and the tires of the car will the car be on the verge of sliding off the track?

SOLUTION: We need to relate μ_s to the circular motion of the car. We start with four Key Ideas, all related to one force on the car:

1. A centripetal force must act on the car if the car is moving around a circular path; the force must be directed horizontally toward the center of the circle.

2. The only horizontal force acting on the car is a frictional force on the tires from the road. So, the required centripetal force is a frictional force.

3. Because the car is not sliding, the frictional force must be a *static* frictional force, the $\vec{f}_s$ shown in Fig 6-12a.

4. If the car is just on the verge of sliding, the magnitude f_s of the frictional force is just equal to the maximum value $f_{s,\max} = \mu_s N$, where N is the magnitude of the normal force on the car from the track.

Thus, to find μ_s, we start with the centripetal force on the car. Figure 6-12b is a free-body diagram for the car, shown on a radial axis r that always extends from the center of the circle through the

car as the car moves. The centripetal force $\vec{f}_s$ is directed inward along that axis, in the negative direction of the axis, and so is the car's centripetal acceleration $\vec{a}$ (of magnitude v^2/R). We can relate this force and acceleration by writing Newton's second law for components along the r axis ($F_{\text{net},r} = ma_r$) as

$$-f_s = m\left(-\frac{v^2}{R}\right). \qquad (6\text{-}22)$$

Substituting $f_{s,\max} = \mu_s N$ for f_s and solving for μ_s give us

$$\mu_s = \frac{mv^2}{NR}. \qquad (6\text{-}23)$$

Because the car does not accelerate vertically, the two vertical forces acting on it (see Fig. 6-12b) must balance; that is, the magnitude N of the normal force must equal the magnitude mg of the gravitational force. Substituting $N = mg$ into Eq. 6-23, we find

$$\mu_s = \frac{mv^2}{mgR} = \frac{v^2}{gR} \qquad (6\text{-}24)$$

$$= \frac{(20 \text{ m/s})^2}{(9.8 \text{ m/s}^2)(190 \text{ m})} = 0.21. \qquad \text{(Answer)}$$

This means that if $\mu_s = 0.21$, the car is on the verge of sliding off the track; if $\mu_s > 0.21$, the car is in no danger of sliding off; and if $\mu_s < 0.21$, the car will certainly slide off.

Equation 6-24 contains two important lessons for road engineers. First, the value of μ_s (required to prevent sliding) depends on the *square* of v. Much more friction is required when the turning speed is increased. You may have noted this effect if you have ever taken a flat turn too fast and suddenly felt the tires slip. Second, the mass m dropped out in our derivation of Eq. 6-24. Thus, Eq. 6-24 holds for a vehicle of any mass, from a kiddy car to a bicycle to a heavy truck.

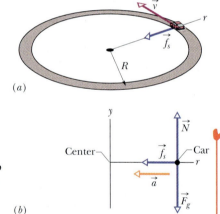

Fig. 6-12 Sample Problem 6-9. (a) A car moves around a flat curved road at constant speed v. The frictional force $\vec{f}_s$ provides the necessary centripetal force along a radial axis r. (b) A free-body diagram (not to scale) for the car, in the vertical plane containing r.

✓CHECKPOINT 6: In Fig. 6-12, suppose the car is on the verge of sliding when the radius of the circle is R_1. (a) If we double the car's speed, what is the least radius that would now keep the car from sliding? (b) If we also double the mass of the car (say by adding sandbags), what is the least radius that would now keep the car from sliding?

REVIEW & SUMMARY

Friction   When a force $\vec{F}$ tends to slide a body along a surface, a **frictional force** from the surface acts on the body. The frictional force is parallel to the surface and directed so as to oppose the sliding. It is due to bonding between the body and the surface.

If the body does not slide, the frictional force is a **static frictional force** $\vec{f}_s$. If there is sliding, the frictional force is a **kinetic frictional force** $\vec{f}_k$.

Three Properties of Friction

1. If the body does not move, then the static frictional force $\vec{f}_s$ and the component of $\vec{F}$ that is parallel to the surface are equal in

magnitude, and $\vec{f}_s$ is directed opposite that component. If that parallel component increases, magnitude f_s also increases.

2. The magnitude of $\vec{f}_s$ has a maximum value $f_{s,\max}$ that is given by

$$f_{s,\max} = \mu_s N, \qquad (6\text{-}1)$$

where μ_s is the **coefficient of static friction** and N is the magnitude of the normal force. If the component of $\vec{F}$ that is parallel to the surface exceeds $f_{s,\max}$, then the body slides on the surface.

3. If the body begins to slide on the surface, the magnitude of the frictional force rapidly decreases to a constant value f_k given by

$$f_k = \mu_k N, \qquad (6\text{-}2)$$

where μ_k is the **coefficient of kinetic friction.**

Drag Force   When there is a relative motion between air (or some other fluid) and a body, the body experiences a **drag force** $\vec{D}$ that opposes the relative motion and points in the direction in which the fluid flows relative to the body. The magnitude of $\vec{D}$ is related to the relative speed v by an experimentally determined **drag coefficient** C according to

$$D = \tfrac{1}{2}C\rho A v^2, \tag{6-14}$$

where ρ is the fluid density (mass per volume) and A is the **effective cross-sectional area** of the body (the area of a cross section taken perpendicular to the relative velocity $\vec{v}$).

Terminal Speed   When a blunt object has fallen far enough through air, the magnitudes of the drag force $\vec{D}$ and the gravitational force $\vec{F}_g$ on the body become equal. The body then falls at a con-

stant **terminal speed** v_t given by

$$v_t = \sqrt{\frac{2F_g}{C\rho A}}. \tag{6-16}$$

Uniform Circular Motion   If a particle moves in a circle or a circular arc with radius R at constant speed v, it is said to be in **uniform circular motion**. It then has a **centripetal acceleration** $\vec{a}$ with magnitude given by

$$a = \frac{v^2}{R}. \tag{6-17}$$

This acceleration is due to a net **centripetal force** on the particle, with magnitude given by

$$F = \frac{mv^2}{R}, \tag{6-18}$$

where m is the particle's mass. The vector quantities $\vec{a}$ and $\vec{F}$ are directed toward the center of curvature of the particle's path.

QUESTIONS

1. In three experiments, three different horizontal forces are applied to the same block lying on the same countertop. The force magnitudes are $F_1 = 12$ N, $F_2 = 8$ N, and $F_3 = 4$ N. In each experiment, the block remains stationary in spite of the applied force. Rank the forces according to (a) the magnitude f_s of the static frictional force on the block from the countertop and (b) the maximum value $f_{s,\mathrm{max}}$ of that force, greatest first.

2. In Fig. 6-13a, a "Batman" thermos is sent sliding leftward across a long plastic tray. What are the directions of the kinetic frictional forces on (a) the thermos and (b) the tray from each other? (c) Does the former increase or decrease the speed of the thermos relative to the floor? In Fig. 6-13b, the tray is now sent sliding leftward beneath the thermos. What now are the directions of the kinetic frictional forces on (d) the thermos and (e) the tray from each other? (f) Does the former increase or decrease the speed of the thermos relative to the floor? (g) *Do kinetic frictional forces always slow objects?*

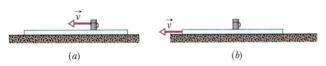

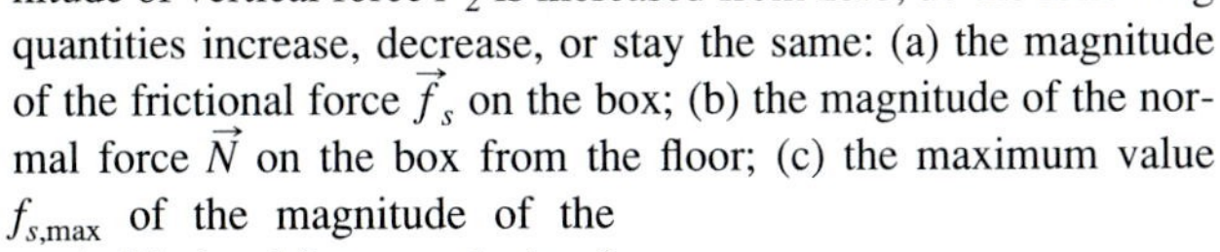

(a)   (b)

Fig. 6-13 Question 2.

3. In Fig. 6-14, horizontal force $\vec{F}_1$ of magnitude 10 N is applied to a box on a floor, but the box does not slide. Then, as the magnitude of vertical force $\vec{F}_2$ is increased from zero, do the following quantities increase, decrease, or stay the same: (a) the magnitude of the frictional force $\vec{f}_s$ on the box; (b) the magnitude of the normal force $\vec{N}$ on the box from the floor; (c) the maximum value $f_{s,\mathrm{max}}$ of the magnitude of the static frictional force on the box? (d) Does the box eventually slide?

4. If you press an apple crate against a wall so hard that the crate cannot slide down the wall, what is the direction of (a) the

static frictional force $\vec{f}_s$ on the crate from the wall and (b) the normal force $\vec{N}$ on the crate from the wall? If you increase your push, what happens to (c) f_s, (d) N, and (e) $f_{s,\mathrm{max}}$?

5. In Fig. 6-15, if the box is stationary and the angle θ of force $\vec{F}$ is increased, do the following quantities increase, decrease, or remain the same: (a) F_x; (b) f_s; (c) N; (d) $f_{s,\mathrm{max}}$? (e) If, instead, the box is sliding and θ is increased, does the magnitude of the frictional force on the box increase, decrease, or remain the same?

6. Repeat Question 5 for force $\vec{F}$ angled upward instead of downward as drawn.

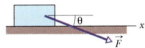

Fig. 6-15 Question 5.

7. Figure 6-16 shows a block of mass m on a slab of mass M and a horizontal force $\vec{F}$ applied to the block, causing it to slide over the slab. There is friction between the block and the slab (but not between the slab and the floor). (a) What mass determines the magnitude of the frictional force between the block and the slab? (b) At the block–slab interface, is the magnitude of the frictional force acting on the block greater than, less than, or equal to that of the frictional force acting on the slab? (c) What are the directions of those two frictional forces? (d) If we write Newton's second law for the slab, what mass should be multiplied by the acceleration of the slab? (Warm-up for Problem 27)

Fig. 6-16 Question 7.

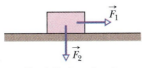

Fig. 6-14 Question 3.

8. *Follow-up to Problem 25.* Suppose the larger of the two blocks in Fig. 6-33 is, instead, fixed to the surface below it. Again, the smaller block is not to slip down the face of the larger block. (a) Is the magnitude of the frictional force between the blocks then greater than, less than, or the same as in the original problem? (b) Is the required minimum magnitude of the horizontal force $\vec{F}$ then greater than, less than, or the same as in the original problem?

9. Figure 6-17 shows the path of a park ride that travels at constant speed through five circular arcs of radii R_0, $2R_0$, and $3R_0$. Rank the arcs according to the magnitude of the centripetal force on a rider traveling in the arcs, greatest first.

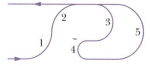

Fig. 6-17 Question 9.

10. A person riding a Ferris wheel moves through positions at (1) the top, (2) the bottom, and (3) midheight. If the wheel rotates at a constant rate, rank these three positions according to (a) the magnitude of the person's centripetal acceleration, (b) the magnitude of the net centripetal force on the person, and (c) the magnitude of the normal force on the person, greatest first.

EXERCISES & PROBLEMS

ssm Solution is in the Student Solutions Manual.
www Solution is available on the World Wide Web at:
http://www.wiley.com/college/hrw
ilw Solution is available on the Interactive Learning Ware.

SEC. 6-2 Properties of Friction

1E. A bedroom bureau with a mass of 45 kg, including drawers and clothing, rests on the floor. (a) If the coefficient of static friction between the bureau and the floor is 0.45, what is the magnitude of the minimum horizontal force that a person must apply to start the bureau moving? (b) If the drawers and clothing, with 17 kg mass, are removed before the bureau is pushed, what is the new minimum magnitude? ssm

2E. The coefficient of static friction between Teflon and scrambled eggs is about 0.04. What is the smallest angle from the horizontal that will cause the eggs to slide across the bottom of a Teflon-coated skillet?

3E. A baseball player with mass $m = 79$ kg, sliding into second base, is retarded by a frictional force of magnitude 470 N. What is the coefficient of kinetic friction μ_k between the player and the ground? ssm

4E. *The mysterious sliding stones.* Along the remote Racetrack Playa in Death Valley, California, stones sometimes gouge out prominent trails in the desert floor, as if they had been migrating (Fig. 6-18). For years curiosity mounted about why the stones moved. One explanation was that strong winds during the occasional rainstorms would drag the rough stones over ground softened by rain. When the desert dried out, the trails behind the stones were hard-baked in place. According to measurements, the coefficient of kinetic friction between the stones and the wet playa ground is about 0.80. What horizontal force is needed on a stone of typical mass 20 kg to maintain the stone's motion once a gust has started it moving? (Story continues with Exercise 32.)

5E. A person pushes horizontally with a force of 220 N on a 55 kg crate to move it across a level floor. The coefficient of kinetic friction is 0.35. (a) What is the magnitude of the frictional force? (b) What is the magnitude of the crate's acceleration? ssm ilw

6E. A house is built on the top of a hill with a nearby 45° slope (Fig. 6-19). An engineering study indicates that the slope angle should be reduced because the top layers of soil along the slope might slip past the lower layers. If the static coefficient of friction between two such layers is 0.5, what is the least angle ϕ through which the present slope should be reduced to prevent slippage?

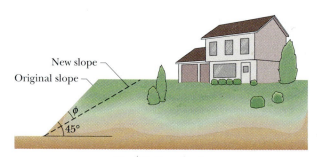

Fig. 6-19 Exercise 6.

7E. A 110 g hockey puck sent sliding over ice is stopped in 15 m by the frictional force on it from the ice. (a) If its initial speed is 6.0 m/s, what is the magnitude of the frictional force? (b) What is the coefficient of friction between the puck and the ice? ssm

8E. In Fig. 6-20, a 49 kg rock climber is climbing a "chimney" between two rock slabs. The static coefficient of friction between her shoes and the rock is 1.2; between her back and the rock it is 0.80. She has reduced her push against the rock until her back and her shoes are on the verge of slipping. (a) Draw a free-body diagram of the climber. (b) What is her push against the rock? (c) What fraction of her weight is supported by the frictional force on her shoes?

9P. A 12 N horizontal force $\vec{F}$ pushes a block weighing 5.0 N against a vertical wall (Fig. 6-21). The coefficient of static friction between the wall and the block is 0.60, and the coefficient of kinetic friction is 0.40. Assume that the block is not moving initially.

Fig. 6-18 Exercise 4.

Fig. 6-20 Exercise 8.

(a) Will the block move? (b) In unit-vector notation, what is the force on the block from the wall? ssm www

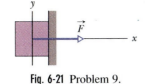

Fig. 6-21 Problem 9.

10P. A 2.5 kg block is initially at rest on a horizontal surface. A 6.0 N horizontal force and a vertical force $\vec{P}$ are applied to the block as shown in Fig. 6-22. The coefficients of friction for the block and surface are $\mu_s = 0.40$ and $\mu_k = 0.25$. Determine the magnitude and direction of the frictional force acting on the block if the magnitude of $\vec{P}$ is (a) 8.0 N, (b) 10 N, and (c) 12 N.

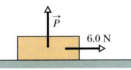

Fig. 6-22 Problem 10.

11P. A worker wishes to pile a cone of sand onto a circular area in his yard. The radius of the circle is R, and no sand is to spill onto the surrounding area (Fig. 6-23). If μ_s is the static coefficient of friction between each layer of sand along the slope and the sand beneath it (along which it might slip), show that the greatest volume of sand that can be stored in this manner is $\pi\mu_s R^3/3$. (The volume of a cone is $Ah/3$, where A is the base area and h is the cone's height.) ssm

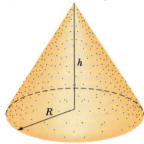

Fig. 6-23 Problem 11.

12P. A worker pushes horizontally on a 35 kg crate with a force of magnitude 110 N. The coefficient of static friction between the crate and the floor is 0.37. (a) What is the frictional force on the crate from the floor? (b) What is the maximum magnitude $f_{s,max}$ of the static frictional force under the circumstances? (c) Does the crate move? (d) Suppose, next, that a second worker pulls directly upward on the crate to help out. What is the least vertical pull that will allow the first worker's 110 N push to move the crate? (e) If, instead, the second worker pulls horizontally to help out, what is the least pull that will get the crate moving?

13P. A 68 kg crate is dragged across a floor by pulling on a rope attached to the crate and inclined 15° above the horizontal. (a) If the coefficient of static friction is 0.50, what minimum force magnitude is required from the rope to start the crate moving? (b) If $\mu_k = 0.35$, what is the magnitude of the initial acceleration of the crate? ssm

14P. A slide-loving pig slides down a certain 35° slide (Fig. 6-24) in twice the time it would take to slide down a frictionless 35° slide. What is the coefficient of kinetic friction between the pig and the slide?

15P. In Fig. 6-25, blocks A and B have weights of 44 N and 22 N,

Fig. 6-24 Problem 14.

respectively. (a) Determine the minimum weight of block C to keep A from sliding if μ_s between A and the table is 0.20. (b) Block C suddenly is lifted off A. What is the acceleration of block A if μ_k between A and the table is 0.15? ssm

16P. A 3.5 kg block is pushed along a horizontal floor by a force $\vec{F}$ of magnitude 15 N at an angle $\theta = 40°$ with the horizontal (Fig. 6-26). The coefficient of kinetic friction between the block and the floor is 0.25. Calculate the magnitudes of (a) the frictional force on the block from the floor and (b) the acceleration of the block.

17P. Figure 6-27 shows the cross section of a road cut into the side of a mountain. The solid line AA' represents a weak bedding plane along which sliding is possible. Block B directly above the highway is separated from uphill rock by a large crack (called a *joint*), so that only friction between the block and the bedding plane prevents sliding. The mass of the block is 1.8×10^7 kg, the *dip angle* θ of the bedding plane is 24°, and the coefficient of static friction between block and plane is 0.63. (a) Show that the block will not slide. (b) Water seeps into the joint and expands upon freezing, exerting on the block a force $\vec{F}$ parallel to AA'. What minimum value of F will trigger a slide?

18P. A loaded penguin sled weighing 80 N rests on a plane inclined at 20° to the horizontal (Fig. 6-28). Between the sled and the plane, the coefficient of static friction is 0.25, and the coefficient of kinetic friction is 0.15. (a) What is the minimum magnitude of the force $\vec{F}$, parallel to the plane, that will prevent the sled from slipping down the plane? (b) What is the minimum magnitude F that will start the sled moving up the plane? (c) What value of F is required to move the sled up the plane at constant velocity?

19P. Block B in Fig. 6-29 weighs 711 N. The coefficient of static friction between block and table is 0.25; assume that the cord between B and the knot is horizontal. Find the maximum weight of block A for which the system will be stationary. ssm

20P. A force $\vec{P}$, parallel to a sur-

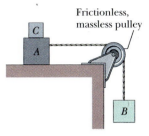

Fig. 6-25 Problem 15.

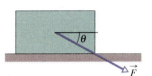

Fig. 6-26 Problem 16.

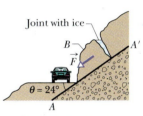

Fig. 6-27 Problem 17.

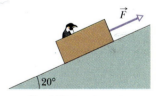

Fig. 6-28 Problem 18.

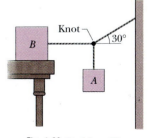

Fig. 6-29 Problem 19.

face inclined 15° above the horizontal, acts on a 45 N block, as shown in Fig. 6-30. The coefficients of friction for the block and surface are $\mu_s = 0.50$ and $\mu_k = 0.34$. If the block is initially at rest, determine the magnitude and

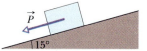

Fig. 6-30 Problem 20.

direction of the frictional force acting on the block for magnitudes of $\vec{P}$ of (a) 5.0 N, (b) 8.0 N, and (c) 15 N.

21P. Body A in Fig. 6-31 weighs 102 N, and body B weighs 32 N. The coefficients of friction between A and the incline are $\mu_s = 0.56$ and $\mu_k = 0.25$. Angle θ is 40°. Find the acceleration of A if (a) A is initially at rest, (b) A is initially moving up the incline, and (c) A is initially moving down the incline. ssm www

22P. In Fig. 6-31, two blocks are connected over a pulley. The mass of block A is 10 kg and the coefficient of kinetic friction between A and the incline is 0.20. Angle θ of the incline is 30°. Block A slides down the incline at constant speed. What is the mass of block B?

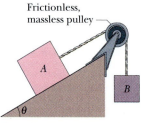

Frictionless, massless pulley

Fig. 6-31 Problems 21 and 22.

23P. Two blocks, of weights 3.6 N and 7.2 N, are connected by a massless string and slide down a 30° inclined plane. The coefficient of kinetic friction between the lighter block and the plane is 0.10; that between the heavier block and the plane is 0.20. Assuming that the lighter block leads, find (a) the magnitude of the acceleration of the blocks and (b) the tension in the string. (c) Describe the motion if, instead, the heavier block leads. ssm

24P. In Fig. 6-32, a box of Cheerios and a box of Wheaties are accelerated across a horizontal surface by a horizontal force $\vec{F}$ applied to the Cheerios box. The magnitude of the frictional force

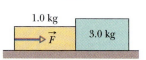

Fig. 6-32 Problem 24.

on the Cheerios box is 2.0 N, and the magnitude of the frictional force on the Wheaties box is 4.0 N. If the magnitude of $\vec{F}$ is 12 N, what is the magnitude of the force on the Wheaties box from the Cheerios box?

25P. The two blocks (with $m = 16$ kg and $M = 88$ kg) shown in Fig. 6-33 are not attached. The coefficient of static friction between the blocks is $\mu_s = 0.38$, but the surface beneath the larger block is frictionless. What is the minimum magnitude of the horizontal force $\vec{F}$ required to keep the smaller block from slipping down the larger block? (See Question 8 for a follow-up.) ilw

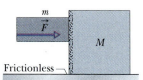

Fig. 6-33 Problem 25.

26P. In Fig. 6-34, a box of ant aunts (total mass $m_1 = 1.65$ kg) and a box of ant uncles (total mass $m_2 = 3.30$ kg) slide down

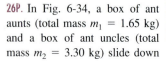

Fig. 6-34 Problem 26.

an inclined plane while attached by a massless rod parallel to the plane. The angle of incline is $\theta = 30°$. The coefficient of kinetic friction between the aunt box and the incline is $\mu_1 = 0.226$; that between the uncle box and the incline is $\mu_2 = 0.113$. Compute (a) the tension in the rod and (b) the common acceleration of the two boxes. (c) How would the answers to (a) and (b) change if the uncles trailed the aunts?

27P. A 40 kg slab rests on a frictionless floor. A 10 kg block rests on top of the slab (Fig. 6-35). The coefficient of static friction μ_s between the block and the slab is 0.60, whereas their kinetic friction coefficient μ_k is 0.40. The 10 kg block is pulled by a horizontal force with a magnitude of 100 N. What are the resulting accelerations of (a) the block and (b) the slab? ssm www

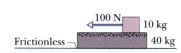

Fig. 6-35 Problem 27.

28P. A locomotive accelerates a 25-car train along a level track. Every car has a mass of 5.0×10^4 kg and is subject to a frictional force $f = 250v$, where the speed v is in meters per second and the force f is in newtons. At the instant when the speed of the train is 30 km/h, the magnitude of its acceleration is 0.20 m/s². (a) What is the tension in the coupling between the first car and the locomotive? (b) If this tension is equal to the maximum force the locomotive can exert on the train, what is the steepest grade up which the locomotive can pull the train at 30 km/h?

29P. In Fig. 6-36, a crate slides down an inclined right-angled trough. The coefficient of kinetic friction between the crate and the trough is μ_k. What is the acceleration of the crate in terms of μ_k, θ, and g? ssm

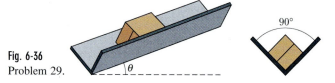

Fig. 6-36 Problem 29.

30P. An initially stationary box of sand is to be pulled across a floor by means of a cable in which the tension should not exceed 1100 N. The coefficient of static friction between the box and the floor is 0.35. (a) What should be the angle between the cable and the horizontal in order to pull the greatest possible amount of sand, and (b) what is the weight of the sand and box in that situation?

31P. A 1000 kg boat is traveling at 90 km/h when its engine is shut off. The magnitude of the frictional force $\vec{f}_k$ between boat and water is proportional to the speed v of the boat: $f_k = 70v$, where v is in meters per second and f_k is in newtons. Find the time required for the boat to slow to 45 km/h. ssm

SEC. 6-3   The Drag Force and Terminal Speed

32E. *Continuation of Exercise 4.* First reread the explanation of how the wind might drag desert stones across the playa. Now assume that Eq. 6-14 gives the magnitude of the air drag force on the typical 20 kg stone, which presents a vertical cross-sectional area to the wind of 0.040 m² and has a drag coefficient C of 0.80. Take the air density to be 1.21 kg/m³, and the coefficient of kinetic friction to be 0.80. (a) In kilometers per hour, what wind speed V along the ground is needed to maintain the stone's motion once it has started moving? Because winds along the ground are retarded by the ground, the wind speeds reported for storms are often measured at a height of 10 m. Assume wind speeds are 2.00 times those along the ground. (b) For your answer to (a), what wind speed would be reported for the storm and is that value reasonable for a high-speed wind in a storm? (Story continues with Problem 51.)

33E. Calculate the drag force on a missile 53 cm in diameter cruising with a speed of 250 m/s at low altitude, where the density of air is 1.2 kg/m³. Assume $C = 0.75$. ssm

34E. The terminal speed of a sky diver is 160 km/h in the spread-eagle position and 310 km/h in the nosedive position. Assuming that the diver's drag coefficient C does not change from one position to the other, find the ratio of the effective cross-sectional area A in the slower position to that in the faster position.

35P. Calculate the ratio of the drag force on a passenger jet flying with a speed of 1000 km/h at an altitude of 10 km to the drag force on a prop-driven transport flying at half the speed and half the altitude of the jet. At 10 km the density of air is 0.38 kg/m³, and at 5.0 km it is 0.67 kg/m³. Assume that the airplanes have the same effective cross-sectional area and the same drag coefficient C.

SEC. 6-4 Uniform Circular Motion

36E. During an Olympic bobsled run, the Jamaican team makes a turn of radius 7.6 m at a speed of 96.6 km/h. What is their acceleration in g-units?

37E. Suppose the coefficient of static friction between the road and the tires on a Formula One car is 0.6 during a Grand Prix auto race. What speed will put the car on the verge of sliding as it rounds a level curve of 30.5 m radius? ssm

38E. A roller-coaster car has a mass of 1200 kg when fully loaded with passengers. As the car passes over the top of a circular hill of radius 18 m, its speed is not changing. What are the magnitude and direction of the force of the track on the car at the top of the hill if the car's speed is (a) 11 m/s and (b) 14 m/s?

39E. What is the smallest radius of an unbanked (flat) track around which a bicyclist can travel if her speed is 29 km/h and the coefficient of static friction between tires and track is 0.32? ilw

40P. An amusement park ride consists of a car moving in a vertical circle on the end of a rigid boom of negligible mass. The combined weight of the car and riders is 5.0 kN, and the radius of the circle is 10 m. What are the magnitude and direction of the force of the boom on the car at the top of the circle if the car's speed there is (a) 5.0 m/s and (b) 12 m/s?

41P. A puck of mass m slides on a frictionless table while attached to a hanging cylinder of mass M by a cord through a hole in the table (Fig. 6-37). What speed keeps the cylinder at rest? ssm

42P. A bicyclist travels in a circle of radius 25.0 m at a constant speed of 9.00 m/s. The bicycle–rider mass is 85.0 kg. Calculate the magnitudes of (a) the force of friction on the bicycle from the road and (b) the *net* force on the bicycle from the road.

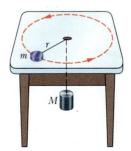

Fig. 6-37 Problem 41.

43P. A student of weight 667 N rides a steadily rotating Ferris wheel (the student sits upright). At the highest point, the magnitude of the normal force $\vec{N}$ on the student from the seat is 556 N. (a) Does the student feel "light" or "heavy" there? (b) What is the magnitude of $\vec{N}$ at the lowest point? (c) What is the magnitude N if the wheel's speed is doubled? ssm ilw www

44P. An old streetcar rounds a flat corner of radius 9.1 m, at 16 km/h. What angle with the vertical will be made by the loosely hanging hand straps?

45P. An airplane is flying in a horizontal circle at a speed of 480 km/h. If its wings are tilted 40° to the horizontal, what is the radius of the circle in which the plane is flying? (See Fig. 6-38.) Assume that the required force is provided entirely by an "aerodynamic lift" that is perpendicular to the wing surface. ssm

Fig. 6-38 Problem 45.

46P. A high-speed railway car goes around a flat, horizontal circle of radius 470 m at a constant speed. The magnitudes of the horizontal and vertical components of the force of the car on a 51.0 kg passenger are 210 N and 500 N, respectively. (a) What is the magnitude of the net force (of *all* the forces) on the passenger? (b) What is the speed of the car?

47P. As shown in Fig. 6-39, a 1.34 kg ball is connected by means of two massless strings to a vertical, rotating rod. The strings are tied to the rod and are taut. The tension in the upper string is 35 N. (a) Draw the free-body diagram for the ball. What are (b) the tension in the lower string, (c) the net force on the ball, and (d) the speed of the ball? ssm ilw

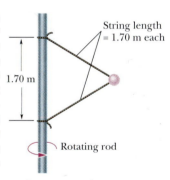

Fig. 6-39 Problem 47.

Additional Problem

48. *Continuation of Exercises 4 and 32.* Another explanation is that the stones move only when the water dumped on the playa during a storm freezes into a large, thin sheet. The stones are trapped in place in the sheet. Then, as air flows across the sheet during a wind, the air drag forces on the sheet and on the stones move them, with the stones gouging out the trails. The magnitude of the air drag force on this horizontal "ice sail" is given by $D_{ice} = 4C_{ice}\rho A_{ice}v^2$, where C_{ice} is the drag coefficient (about 2.0×10^{-3}), ρ is the air density (1.21 kg/m³), A_{ice} is the horizontal area of the ice, and v is the wind speed along the ice.

Assume the following: The sheet measures 400 m by 500 m by 4.0 mm and has a coefficient of kinetic friction of 0.10 with the ground and a density of 917 kg/m³. Also assume that 100 stones identical to the one in Exercise 4 are trapped in the ice. To maintain the motion of the sheet, what are the required wind speeds (a) near the sheet and (b) at a height of 10 m? (c) Are these reasonable values for high-speed winds in a storm?

7 Kinetic Energy and Work

In the weight-lifting competition of the 1996 Olympics, Andrey Chemerkin lifted a record-breaking 260.0 kg from the floor to over his head (about 2 m). In 1957 Paul Anderson stooped beneath a reinforced wood platform, placed his hands on a short stool to brace himself, and then pushed upward on the platform with his back, lifting the platform and its load about a centimeter. On the platform were auto parts and a safe filled with lead; the composite weight of the load was 27 900 N (6270 lb)!

Who did more work on the objects he lifted—Chemerkin or Anderson?

The answer is in this chapter.

7-1 Energy

Newton's laws of motion allow us to analyze many kinds of motion. However, the analysis is often complicated, requiring details about the motion that we simply do not know. Here is an example: A puck is sent sliding along an inclined frictionless track that includes several ups and downs (hills and valleys) of various shapes. The puck's initial speed is 4.0 m/s, and its initial height is 0.46 m. Using Newton's second law, can you calculate the puck's speed when it reaches the end of the track, at zero height? No, not without the details of how the incline varies all along the track, and then the calculations can be very complicated.

Long ago, scientists and engineers gradually began to realize that there is another, sometimes more powerful technique for analyzing motion. Moreover, this technique could be, and eventually was, extended to other situations, such as chemical reactions, geological processes, and biological functions, that do not involve motion. This other technique involves **energy**, which comes in a great many *forms* (or types). In fact, the term *energy* is so broad that a clear definition for it is difficult to write. Technically, energy is a scalar quantity that is associated with a state (or condition) of one or more objects. However, this definition is too vague to be of help to us now.

A looser definition might at least get us started. Energy is a number that we associate with a system of one or more objects. If a force changes one of the objects by, say, making it move, then the number changes. After countless experiments, scientists and engineers realized that if the scheme by which we assign these energy numbers is planned carefully, then the numbers can be used to predict the outcomes of experiments. (For example, they would allow us to easily find the puck's speed in the previous example.) However, learning how to use the numbering scheme is not easy and the scheme is not obvious. Thus, in this chapter we focus on only one form of energy—kinetic energy. Other forms of energy will show up throughout this book and your work in science or engineering.

Kinetic energy K is energy associated with the *state of motion* of an object. The faster the object moves, the greater is its kinetic energy. When the object is stationary, its kinetic energy is zero.

For an object of mass m whose speed v is well below the speed of light, we define kinetic energy as

$$K = \tfrac{1}{2}mv^2 \quad \text{(kinetic energy)}. \tag{7-1}$$

For example, a 3.0 kg duck flying past us at 2.0 m/s has a kinetic energy of 6.0 kg·m²/s²; that is, we associate that number with the duck's motion.

The SI unit of kinetic energy (and every other type of energy) is the **joule** (J), named for James Prescott Joule, an English scientist of the 1800s. It is defined directly from Eq. 7-1 in terms of the units for mass and velocity:

$$1 \text{ joule} = 1 \text{ J} = 1 \text{ kg} \cdot \text{m}^2/\text{s}^2. \tag{7-2}$$

Thus, the flying duck has a kinetic energy of 6.0 J.

Sample Problem 7-1

In 1896 in Waco, Texas, William Crush of the "Katy" railroad parked two locomotives at opposite ends of a 6.4-km-long track, fired them up, tied their throttles open, and then allowed them to crash head-on at full speed (Fig. 7-1) in front of 30,000 spectators. Hundreds of people were hurt by flying debris; several were killed. Assuming each locomotive weighed 1.2×10^6 N and its acceleration along the track was a constant 0.26 m/s², what was the total kinetic energy of the two locomotives just before the collision?

Fig. 7-1 Sample Problem 7-1. The aftermath of an 1896 crash of two locomotives.

SOLUTION: One Key Idea here is to find the kinetic energy of each locomotive with Eq. 7-1, but that means we need each locomotive's speed just before the collision and its mass. A second Key Idea is

that, because we can assume each locomotive had constant acceleration, we can use the equations in Table 2-1 to find its speed v just before the collision. We choose Eq. 2-16 because we know values for all the variables except v:

$$v^2 = v_0^2 + 2a(x - x_0).$$

With $v_0 = 0$ and $x - x_0 = 3.2 \times 10^3$ m (half the initial separation), this yields

$$v^2 = 0 + 2(0.26 \text{ m/s}^2)(3.2 \times 10^3 \text{ m}),$$

or $v = 40.8$ m/s

(about 150 km/h).

A third Key Idea is that we can find the mass of each locomotive by dividing its given weight by g:

$$m = \frac{1.2 \times 10^6 \text{ N}}{9.8 \text{ m/s}^2} = 1.22 \times 10^5 \text{ kg}.$$

Now, using Eq. 7-1, we find the total kinetic energy of the two locomotives just before the collision as

$$K = 2(\tfrac{1}{2}mv^2) = (1.22 \times 10^5 \text{ kg})(40.8 \text{ m/s})^2$$

$$= 2.0 \times 10^8 \text{ J}. \qquad \text{(Answer)}$$

Sitting near this collision was like sitting near an exploding bomb.

7-2 Work

If you accelerate an object to a greater speed by applying a force to the object, you increase the kinetic energy $K \ (= \tfrac{1}{2}mv^2)$ of the object. Similarly, if you decelerate the object to a lesser speed by applying a force, you decrease the kinetic energy of the object. We account for these changes in kinetic energy by saying that your force has transferred energy *to* the object from yourself or *from* the object to yourself.

In such a transfer of energy via a force, **work** W is said to be *done on the object by the force*. More formally, we define work as follows:

> ▶ Work W is energy transferred to or from an object by means of a force acting on the object. Energy transferred to the object is positive work, and energy transferred from the object is negative work.

"Work," then, is transferred energy; "doing work" is the act of transferring the energy. Work has the same units as energy and is a scalar quantity.

The term *transfer* can be misleading. It does not mean that anything material flows into or out of the object; that is, the transfer is not like a flow of water. Rather it is like the electronic transfer of money between two bank accounts: The number in one account goes up while the number in the other account goes down, with nothing material passing between the two accounts.

Note that we are not concerned here with the common meaning of the word "work," which implies that *any* physical or mental labor is work. For example, if you push hard against a wall, you tire because of the continuously repeated muscle contractions that are required, and you are, in the common sense, working. However, such effort does not cause an energy transfer to or from the wall and thus is not work done on the wall as defined here.

To avoid confusion in this chapter, we shall use the symbol W only for work and shall represent a weight with its equivalent mg.

7-3 Work and Kinetic Energy

Finding an Expression for Work

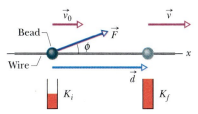

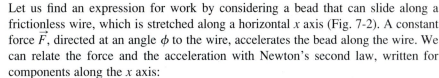

Let us find an expression for work by considering a bead that can slide along a frictionless wire, which is stretched along a horizontal x axis (Fig. 7-2). A constant force $\vec{F}$, directed at an angle ϕ to the wire, accelerates the bead along the wire. We can relate the force and the acceleration with Newton's second law, written for components along the x axis:

$$F_x = ma_x, \tag{7-3}$$

where m is the bead's mass. As the bead moves through a displacement $\vec{d}$, the force changes the bead's velocity from an initial value $\vec{v}_0$ to another value $\vec{v}$. Because the force is constant, we know that the acceleration is also constant. Thus, we can use Eq. 2-16 (one of the basic constant-acceleration equations of Chapter 2) to write, for components along the x axis,

$$v^2 = v_0^2 + 2a_x d. \tag{7-4}$$

Solving this equation for a_x, substituting into Eq. 7-3, and rearranging then give us

$$\tfrac{1}{2}mv^2 - \tfrac{1}{2}mv_0^2 = F_x d. \tag{7-5}$$

The first term on the left side of the equation is the kinetic energy K_f of the bead at the end of the displacement d, and the second term is the kinetic energy K_i of the bead at the start of the displacement. Thus, the left side of Eq. 7-5 tells us the kinetic energy has been changed by the force, and the right side tells us the change is equal to $F_x d$. Therefore, the work W done on the bead by the force (the energy transfer due to the force) is

$$W = F_x d. \tag{7-6}$$

If we know values for F_x and d, we can use this equation to calculate the work W done on the bead by the force.

Fig. 7-2 A constant force $\vec{F}$ directed at angle ϕ to the displacement $\vec{d}$ of a bead on a wire accelerates the bead along the wire, changing the velocity of the bead from $\vec{v}_0$ to $\vec{v}$. A "kinetic energy gauge" indicates the resulting change in the kinetic energy of the bead, from the value K_i to K_f.

> To calculate the work done on an object by a force during a displacement, we use only the force component along the object's displacement. The force component perpendicular to the displacement does zero work.

From Fig. 7-2, we see that we can write F_x as $F \cos \phi$, where ϕ is the angle between the directions of the displacement $\vec{d}$ and the force $\vec{F}$. We can rewrite Eq. 7-6 in a more general form as

$$W = Fd \cos \phi \qquad \text{(work done by a constant force).} \tag{7-7}$$

This equation is useful for calculating the work if we know values for F, d, and ϕ. Because the right side of this equation is equivalent to the scalar (or dot) product $\vec{F} \cdot \vec{d}$, we can also write

$$W = \vec{F} \cdot \vec{d} \qquad \text{(work done by a constant force).} \tag{7-8}$$

(You may wish to review the discussion of scalar products in Section 3-7.) Equation 7-8 is especially useful for calculating the work when $\vec{F}$ and $\vec{d}$ are given in unit-vector notation.

Cautions: There are two restrictions to using Eqs. 7-6 through 7-8 to calculate work done on an object by a force. First, the force must be a *constant force*; that is, it must not change in magnitude or direction as the object moves. (Later, we shall discuss what to do with a *variable force* that changes in magnitude.) Second, the object must be *particle-like*. This means that the object must be *rigid*; all parts of the object must move together, in the same direction. In this chapter we consider only particle-like objects, such as the bed and its rider being pushed in Fig. 7-3.

Fig. 7-3 A contestant in a bed race. We can approximate the bed and its rider as being a particle for the purpose of calculating the work done on them by the force applied by the student.

Signs for work. The work done on an object by a force can be either positive work or negative work. For example, if the angle ϕ in Eq. 7-7 is less than 90°, then cos ϕ is positive and thus so is the work. If ϕ is greater than 90° (up to 180°), then cos ϕ is negative and thus so is the work. (Can you see that the work is zero when $\phi = 90°$?) These results lead to a simple rule. To find the sign of the work done by a force, consider the vector component of the force along the displacement:

> ▶ A force does positive work when it has a vector component in the same direction as the displacement, and it does negative work when it has a vector component in the opposite direction. It does zero work when it has no such vector component.

Units for work. Work has the SI unit of the joule, the same as kinetic energy. However, from Eqs. 7-6 and 7-7 we can see that an equivalent unit is the newton-meter (N · m). The corresponding unit in the British system is the foot-pound (ft · lb). Extending Eq. 7-2, we have

$$1\text{ J} = \text{kg} \cdot \text{m}^2/\text{s}^2 = 1\text{ N} \cdot \text{m} = 0.738\text{ ft} \cdot \text{lb}. \tag{7-9}$$

Net work done by several forces. When two or more forces act on an object, the **net work** done on the object is the sum of the works done by the individual forces. We can calculate the net work in two ways. (1) We can find the work done by each force and then sum those works. (2) Alternatively, we can first find the net force $\vec{F}_{net}$ of those forces. Then we can use Eq. 7-7, substituting the magnitude F_{net} for F, and the angle between the directions of $\vec{F}_{net}$ and the displacement for ϕ; or we can use Eq. 7-8 with $\vec{F}_{net}$ substituted for $\vec{F}$.

Work–Kinetic Energy Theorem

Equation 7-5 relates the change in kinetic energy of the bead (from an initial $K_i = \frac{1}{2}mv_0^2$ to a later $K_f = \frac{1}{2}mv^2$) to the work W ($= F_x d$) done on the bead. For such particle-like objects, we can generalize that equation. Let ΔK be the change in the kinetic energy of the object, and W the net work done on it. Then we can write

$$\Delta K = K_f - K_i = W, \tag{7-10}$$

which says that

$$\begin{pmatrix} \text{change in the kinetic} \\ \text{energy of a particle} \end{pmatrix} = \begin{pmatrix} \text{net work done on} \\ \text{the particle} \end{pmatrix}.$$

We can also write

$$K_f = K_i + W, \tag{7-11}$$

which says that

$$\begin{pmatrix} \text{kinetic energy after} \\ \text{the net work is done} \end{pmatrix} = \begin{pmatrix} \text{kinetic energy} \\ \text{before the net work} \end{pmatrix} + \begin{pmatrix} \text{the net} \\ \text{work done} \end{pmatrix}.$$

These statements are known traditionally as the **work–kinetic energy theorem** for particles. They hold for both positive and negative work: If the net work done on a particle is positive, then the particle's kinetic energy increases by the amount of the work. If the net work done is negative, then the particle's kinetic energy decreases by the amount of the work.

For example, if the kinetic energy is initially 5 J and there is a net transfer of 2 J to the particle (positive net work), then the final kinetic energy is 7 J. If, instead, there is a net transfer of 2 J from the particle (negative net work), then the final kinetic energy is 3 J.

Sample Problem 7-2

Figure 7-4*a* shows two industrial spies sliding an initially stationary 225 kg floor safe a displacement $\vec{d}$ of magnitude 8.50 m, straight toward their truck. The push $\vec{F}_1$ of Spy 001 is 12.0 N, directed at an angle of 30° downward from the horizontal; the pull $\vec{F}_2$ of Spy 002 is 10.0 N, directed at 40° above the horizontal. The magnitudes and directions of these forces do not change as the safe moves, and the floor and safe make frictionless contact.

(a) What is the net work done on the safe by forces $\vec{F}_1$ and $\vec{F}_2$ during the displacement $\vec{d}$?

SOLUTION: We use two **Key Ideas** here. First, the net work *W* done on the safe by the two forces is the sum of the works they do individually. Second, because we can treat the safe as a particle and the forces are constant in both magnitude and direction, we can use either Eq. 7-7 ($W = Fd \cos \phi$) or Eq. 7-8 ($W = \vec{F} \cdot \vec{d}$) to calculate those works. Since we know the magnitudes and directions of the forces, we choose Eq. 7-7. From it and the free-body diagram for the safe in Fig. 7-4*b*, the work done by $\vec{F}_1$ is

$$W_1 = F_1 d \cos \phi_1 = (12.0 \text{ N})(8.50 \text{ m})(\cos 30°)$$
$$= 88.33 \text{ J},$$

and the work done by $\vec{F}_2$ is

$$W_2 = F_2 d \cos \phi_2 = (10.0 \text{ N})(8.50 \text{ m})(\cos 40°)$$
$$= 65.11 \text{ J}.$$

Thus, the net work *W* is

$$W = W_1 + W_2 = 88.33 \text{ J} + 65.11 \text{ J}$$
$$= 153.4 \text{ J} \approx 153 \text{ J}. \quad \text{(Answer)}$$

During the 8.50 m displacement, therefore, the spies transfer 153 J of energy to the kinetic energy of the safe.

(b) During the displacement, what is the work W_g done on the safe by the gravitational force $\vec{F}_g$ and what is the work W_N done on the safe by the normal force $\vec{N}$ from the floor?

SOLUTION: The **Key Idea** is that, because these forces are constant in both magnitude and direction, we can find the work they do with

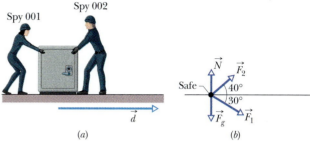

Spy 002
Spy 001

(a)

Safe

$\vec{N}$ $\vec{F}_2$
40°
30°
$\vec{F}_g$ $\vec{F}_1$

(b)

Fig. 7-4 Sample Problem 7-2. (*a*) Two spies move a floor safe through displacement $\vec{d}$. (*b*) A free-body diagram for the safe.

Eq. 7-7. Thus, with *mg* as the magnitude of the gravitational force, we write

$$W_g = mgd \cos 90° = mgd(0) = 0 \quad \text{(Answer)}$$

and

$$W_N = Nd \cos 90° = Nd(0) = 0. \quad \text{(Answer)}$$

We should have known this result. Because these forces are perpendicular to the displacement of the safe, they do zero work on the safe and do not transfer any energy to or from it.

(c) The safe is initially stationary. What is its speed v_f at the end of the 8.50 m displacement?

SOLUTION: Here the **Key Idea** is that the speed of the safe changes because its kinetic energy is changed when energy is transferred to it by $\vec{F}_1$ and $\vec{F}_2$. We relate the speed to the work done by combining Eqs. 7-10 and 7-1:

$$W = K_f - K_i = \tfrac{1}{2}mv_f^2 - \tfrac{1}{2}mv_i^2.$$

The initial speed v_i is zero, and we now know that the work done is 153.4 J. Solving for v_f and then substituting known data, we find that

$$v_f = \sqrt{\frac{2W}{m}} = \sqrt{\frac{2(153.4 \text{ J})}{225 \text{ kg}}}$$
$$= 1.17 \text{ m/s}. \quad \text{(Answer)}$$

Sample Problem 7-3

During a storm, a crate of crepe is sliding across a slick, oily parking lot through a displacement $\vec{d} = (-3.0 \text{ m})\hat{i}$ while a steady wind pushes against the crate with a force $\vec{F} = (2.0 \text{ N})\hat{i} + (-6.0 \text{ N})\hat{j}$. The situation and coordinate axes are shown in Fig. 7-5.

(a) How much work does this force from the wind do on the crate during the displacement?

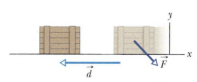

y

$\vec{F}$ *x*

$\vec{d}$

Fig. 7-5 Sample Problem 7-3. Constant force $\vec{F}$ slows a crate during displacement $\vec{d}$.

SOLUTION: The Key Idea here is that, because we can treat the crate as a particle and because the wind force is constant ("steady") in both magnitude and direction during the displacement, we can use either Eq. 7-7 ($W = Fd \cos \phi$) or Eq. 7-8 ($W = \vec{F} \cdot \vec{d}$) to calculate the work. Since we know $\vec{F}$ and $\vec{d}$ in unit-vector notation, we choose Eq. 7-8 and write

$$W = \vec{F} \cdot \vec{d} = [(2.0\ \text{N})\hat{i} + (-6.0\ \text{N})\hat{j}] \cdot [(-3.0\ \text{m})\hat{i}].$$

Of the possible unit-vector dot products, only $\hat{i} \cdot \hat{i}$, $\hat{j} \cdot \hat{j}$, and $\hat{k} \cdot \hat{k}$ are nonzero (see Appendix E). Here we obtain

$$W = (2.0\ \text{N})(-3.0\ \text{m})\hat{i} \cdot \hat{i} + (-6.0\ \text{N})(-3.0\ \text{m})\hat{j} \cdot \hat{i}$$
$$= (-6.0\ \text{J})(1) + 0 = -6.0\ \text{J}. \qquad \text{(Answer)}$$

Thus, the force does a negative 6.0 J of work on the crate, transfering 6.0 J of energy from the kinetic energy of the crate.

(b) If the crate has a kinetic energy of 10 J at the beginning of displacement $\vec{d}$, what is its kinetic energy at the end of $\vec{d}$?

SOLUTION: The Key Idea here is that, because the force does negative work on the crate, it reduces the crate's kinetic energy. Using the

work–kinetic energy theorem in the form of Eq. 7-11, we have

$$K_f = K_i + W = 10\ \text{J} + (-6.0\ \text{J}) = 4.0\ \text{J}. \quad \text{(Answer)}$$

Because the kinetic energy is decreased to 4.0 J, the crate has been slowed.

✔CHECKPOINT 2:   The figure shows four situations in which a force acts on a box while the box slides rightward a distance d across a frictionless floor. The magnitudes of the forces are identical; their orientations are as shown. Rank the situations according to the work done on the box during the displacement, from most positive to most negative.

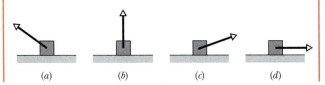

(a)   (b)   (c)   (d)

7-4   Work Done by a Gravitational Force

We next examine the work done on an object by a particular type of force—namely, the gravitational force acting on it. Figure 7-6 shows a particle-like tomato of mass m that is thrown upward with initial speed v_0 and thus with initial kinetic energy $K_i = \frac{1}{2}mv_0^2$. As the tomato rises, it is slowed by a gravitational force $\vec{F}_g$; that is, the tomato's kinetic energy decreases because $\vec{F}_g$ does work on the tomato as it rises.

Because we can treat the tomato as a particle, we can use Eq. 7-7 ($W = Fd \cos \phi$) to express the work done during a displacement $\vec{d}$. For the force magnitude F, we use mg as the magnitude of $\vec{F}_g$. Thus, the work W_g done by the gravitational force $\vec{F}_g$ is

$$W_g = mgd \cos \phi \qquad \text{(work done by gravitational force).} \qquad (7\text{-}12)$$

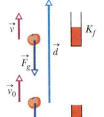

Fig. 7-6 A particle-like tomato of mass m thrown upward slows from velocity $\vec{v}_0$ to velocity $\vec{v}$ during displacement $\vec{d}$ because the gravitational force $\vec{F}_g$ acts on it. A kinetic energy gauge indicates the resulting change in the kinetic energy of the object, from K_i ($= \frac{1}{2}mv_0^2$) to K_f ($= \frac{1}{2}mv^2$).

For a rising object, force $\vec{F}_g$ is directed opposite the displacement $\vec{d}$, as indicated in Fig. 7-6. Thus, $\phi = 180°$ and

$$W_g = mgd \cos 180° = mgd(-1) = -mgd. \qquad (7\text{-}13)$$

The minus sign tells us that during the object's rise, the gravitational force on the object transfers energy in the amount mgd from the kinetic energy of the object. This is consistent with the slowing of the object as it rises.

After the object has reached its maximum height and is falling back down, the angle ϕ between force $\vec{F}_g$ and displacement $\vec{d}$ is zero. Thus,

$$W_g = mgd \cos 0° = mgd(+1) = +mgd. \qquad (7\text{-}14)$$

The plus sign tells us that the gravitational force now transfers energy in the amount mgd to the kinetic energy of the object. This is consistent with the speeding up of the object as it falls. (Actually, as we shall see in Chapter 8, energy transfers associated with lifting and lowering an object involve not just the object, but the full object–Earth system. Without Earth, of course, "lifting" would be meaningless.)

Work Done in Lifting and Lowering an Object

Now suppose we lift a particle-like object by applying a vertical force $\vec{F}$ to it. During the upward displacement, our applied force does positive work W_a on the object while the gravitational force does negative work W_g on it. Our force tends to transfer energy to the object while the gravitational force tends to transfer energy from it. By Eq. 7-10, the change ΔK in the kinetic energy of the object due to these two energy transfers is

$$\Delta K = K_f - K_i = W_a + W_g, \qquad (7\text{-}15)$$

in which K_f is the kinetic energy at the end of the displacement and K_i is that at the start of the displacement. This equation also applies if we lower the object, but then the gravitational force tends to transfer energy *to* the object while our force tends to transfer energy *from* it.

In one common situation the object is stationary before and after the lift—for example, when you lift a book from the floor to a shelf. Then K_f and K_i are both zero, and Eq. 7-15 reduces to

$$W_a + W_g = 0$$

or

$$W_a = -W_g. \qquad (7\text{-}16)$$

Note that we get the same result if K_f and K_i are not zero but are still equal. Either way, the result means that the work done by the applied force is the negative of the work done by the gravitational force; that is, the applied force transfers the same amount of energy to the object as the gravitational force transfers from the object. Using Eq. 7-12, we can rewrite Eq. 7-16 as

$$W_a = -mgd \cos \phi \qquad \text{(work in lifting and lowering; } K_f = K_i\text{)}, \qquad (7\text{-}17)$$

with ϕ being the angle between $\vec{F}_g$ and $\vec{d}$. If the displacement is vertically upward (Fig. 7-7a), then $\phi = 180°$ and the work done by our force equals mgd. If the displacement is vertically downward (Fig. 7-7b), then $\phi = 0°$ and the work done by the applied force equals $-mgd$.

Equations 7-16 and 7-17 apply to any situation in which an object is lifted or lowered, with the object stationary before and after the lift. They are independent of the magnitude of the force used. For example, when Chemerkin made his record-breaking lift, his force on the object he lifted varied considerably during the lift. Still, because the object was stationary before and after the lift, the work he did is given by Eqs. 7-16 and 7-17, where, in Eq. 7-17, mg is the weight of the object he lifted and d is the distance he lifted it.

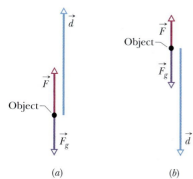

Fig. 7-7 (a) An applied force $\vec{F}$ lifts an object. The object's displacement $\vec{d}$ makes an angle $\phi = 180°$ with the gravitational force $\vec{F}_g$ on the object. The applied force does positive work on the object. (b) An applied force $\vec{F}$ lowers an object. The displacement $\vec{d}$ of the object makes an angle $\phi = 0°$ with the gravitational force $\vec{F}_g$. The applied force does negative work on the object.

Object

Object

Sample Problem 7-4

Let us return to the lifting feats of Andrey Chemerkin and Paul Anderson.

(a) Chemerkin made his record-breaking lift with rigidly connected objects (a barbell and disk weights) having a total mass $m = 260.0$ kg; he lifted them a distance of 2.0 m. During the lift, how much work was done on the objects by the gravitational force $\vec{F}_g$ acting on them?

SOLUTION: The **Key Idea** here is that we can treat the rigidly connected objects as a single particle and thus use Eq. 7-12 ($W_g = $

$mgd \cos \phi$) to find the work W_g done on them by $\vec{F}_g$. The total weight mg was 2548 N, the magnitude d of the displacement was 2.0 m, and the angle ϕ between the directions of the downward gravitational force and the upward displacement was 180°. Thus,

$$W_g = mgd \cos \phi = (2548 \text{ N})(2.0 \text{ m})(\cos 180°)$$
$$= -5100 \text{ J.} \qquad \text{(Answer)}$$

(b) How much work was done on the objects by Chemerkin's force during the lift?

SOLUTION: We do not have an expression for Chemerkin's force on the object, and even if we did, his force was certainly not constant, Thus, one Key Idea here is that we *cannot* just substitute his force into Eq. 7-7 to find his work. However, we know that the objects were stationary at the start and end of the lift. Therefore, as a second Key Idea, we know that the work W_{AC} done by Chemerkin's applied force was the negative of the work W_g done by the gravitational force $\vec{F}_g$. Equation 7-16 expresses this fact and gives us

$$W_{AC} = -W_g = +5100 \text{ J}. \qquad \text{(Answer)}$$

(c) While Chemerkin held the objects stationary above his head, how much work was done on them by his force?

SOLUTION: The Key Idea is that when he supported the objects, they were stationary. Thus, their displacement $d = 0$ and, by Eq. 7-7, the work done on them was zero (even though supporting them was a very tiring task).

(d) How much work was done by the force Paul Anderson applied to lift objects with a total weight of 27 900 N, a distance of 1.0 cm?

SOLUTION: Following the argument of parts (a) and (b), but now with $mg = 27\,900$ N and $d = 1.0$ cm, we find

$$W_{PA} = -W_g = -mgd \cos \phi = -mgd \cos 180°$$
$$= -(27\,900 \text{ N})(0.010 \text{ m})(-1) = 280 \text{ J}. \qquad \text{(Answer)}$$

Anderson's lift required a tremendous upward force but only a small energy transfer of 280 J, because of the short displacement involved. This photo shows another of his lifts.

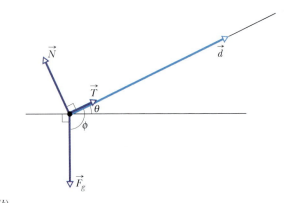

Sample Problem 7-5

An initially stationary 15.0 kg crate of cheese wheels is pulled, via a cable, a distance $L = 5.70$ m up a frictionless ramp, to a height h of 2.50 m, where it stops (Fig. 7-8a).

(a) How much work W_g is done on the crate by the gravitational force $\vec{F}_g$ during the lift?

SOLUTION: A Key Idea is that we can treat the crate as a particle and thus use Eq. 7-12 ($W_g = mgd \cos \phi$) to find the work W_g done by $\vec{F}_g$. However, we do not know the angle ϕ between the directions of $\vec{F}_g$ and displacement $\vec{d}$. From the crate's free-body diagram in Fig. 7-8b, we find that ϕ is $\theta + 90°$, where θ is the (unknown) angle of the ramp. Equation 7-12 then gives us

$$W_g = mgd \cos(\theta + 90°) = -mgd \sin \theta, \qquad (7\text{-}18)$$

where we have used a trigonometric identity to simplify the expression. The result seems to be useless because θ is unknown. But (continuing with physics courage) we see from Fig. 7-8a that $d \sin \theta = h$, where h is a known quantity (2.50 m). With this substitution, Eq. 7-18 then gives us

$$W_g = -mgh \qquad (7\text{-}19)$$
$$= -(15.0 \text{ kg})(9.8 \text{ m/s}^2)(2.50 \text{ m})$$
$$= -368 \text{ J}. \qquad \text{(Answer)}$$

Note that Eq. 7-19 tells us that the work W_g done by the gravitational force depends on the vertical displacement but (surprisingly) not on the horizontal displacement. (We return to this point in Chapter 8.)

(b) How much work W_T is done on the crate by the force $\vec{T}$ from the cable during the lift?

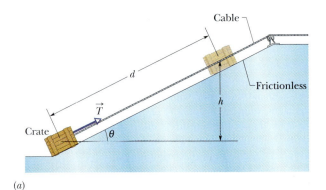

(a)

(b)

Fig. 7-8 Sample Problem 7-5. (a) A crate is pulled up a frictionless ramp by a force $\vec{T}$ parallel to the ramp. (b) A free-body diagram for the crate, showing also the displacement $\vec{d}$.

SOLUTION: We cannot just substitute the force magnitude T for F in Eq. 7-7 ($W = Fd \cos \phi$) because we do not know the value of T. However, a **Key Idea** to get us going is that we can treat the crate as a particle and then apply the work–kinetic energy theorem ($\Delta K = W$) to it. Because the crate is stationary before and after the lift, the change ΔK in its kinetic energy is zero. For the net work W done on the crate, we must sum the works done by all three forces acting on the crate. From (a), the work W_g done by the gravitational force $\vec{F}_g$ is -368 J. The work W_N done by the normal force $\vec{N}$ on the crate from the ramp is zero because $\vec{N}$ is perpendicular to the displacement. We want the work W_T done by $\vec{T}$. Thus,

the work–kinetic energy theorem gives us

$$\Delta K = W_T + W_g + W_N$$

or

$$0 = W_T - 368 \text{ J} + 0,$$

and so

$$W_T = 368 \text{ J}. \qquad \text{(Answer)}$$

✓**CHECKPOINT 3:** Suppose we raise the crate by the same height h but with a longer ramp. (a) Is the work done by force $\vec{T}$ now greater, smaller, or the same as before? (b) Is the magnitude of $\vec{T}$ needed to move the crate now greater, smaller, or the same as before?

Sample Problem 7-6

An elevator cab of mass $m = 500$ kg is descending with speed $v_i = 4.0$ m/s when its supporting cable begins to slip, allowing it to fall with constant acceleration $\vec{a} = \vec{g}/5$ (Fig. 7-9a).

(a) During the fall through a distance $d = 12$ m, what is the work W_g done on the cab by the gravitational force $\vec{F}_g$?

SOLUTION: The **Key Idea** here is that we can treat the cab as a particle and thus use Eq. 7-12 ($W_g = mgd \cos \phi$) to find the work W_g done on it by $\vec{F}_g$. From Fig. 7-9b (a free-body diagram of the cab with the cab's displacement $\vec{d}$ included), we see that the angle between the directions of $\vec{F}_g$ and the cab's displacement $\vec{d}$ is $0°$. Then, from Eq. 7-12, we find

$$W_g = mgd \cos 0° = (500 \text{ kg})(9.8 \text{ m/s}^2)(12 \text{ m})(1)$$
$$= 5.88 \times 10^4 \text{ J} \approx 59 \text{ kJ}. \qquad \text{(Answer)}$$

(b) During the 12 m fall, what is the work W_T done on the cab by the upward pull $\vec{T}$ of the elevator cable?

SOLUTION: A **Key Idea** here is that we can calculate the work W_T with Eq. 7-7 ($W = Fd \cos \phi$) if we first find an expression for the magnitude T of the cable's pull. A second **Key Idea** is that we can find that expression by writing Newton's second law for components along the y axis in Fig. 7-9b ($F_{net,y} = ma_y$). We get

$$T - F_g = ma.$$

Solving for T, substituting mg for F_g, and then substituting the result in Eq. 7-7, we obtain

$$W_T = Td \cos \phi = m(a + g)d \cos \phi.$$

Next, substituting $-g/5$ for the (downward) acceleration a and then $180°$ for the angle ϕ between the directions of forces $\vec{T}$ and $m\vec{g}$, we find

$$W_T = m\left(-\frac{g}{5} + g\right)d \cos \phi = \frac{4}{5}mgd \cos \phi$$
$$= \frac{4}{5}(500 \text{ kg})(9.8 \text{ m/s}^2)(12 \text{ m}) \cos 180°$$
$$= -4.70 \times 10^4 \text{ J} \approx -47 \text{ kJ}. \qquad \text{(Answer)}$$

Now note that W_T is not simply the negative of W_g, which we found in (a). The reason is that, because the cab accelerates during the fall, its speed changes during the fall, and thus its kinetic

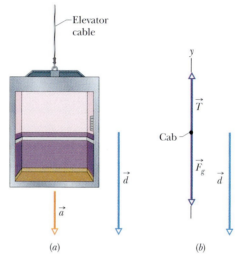

Fig. 7-9 Sample Problem 7-6. An elevator cab, descending with speed v_i, suddenly begins to accelerate downward. (a) It moves through a displacement $\vec{d}$ with constant acceleration $\vec{a} = \vec{g}/5$. (b) A free-body diagram for the cab, displacement included.

energy also changes. Therefore, Eq. 7-16 (which assumes that the initial and final kinetic energies are equal) does *not* apply here.

(c) What is the net work W done on the cab during the fall?

SOLUTION: The **Key Idea** here is that the net work is the sum of the works done by the forces acting on the cab:

$$W = W_g + W_T = 5.88 \times 10^4 \text{ J} - 4.70 \times 10^4 \text{ J}$$
$$= 1.18 \times 10^4 \text{ J} \approx 12 \text{ kJ}. \qquad \text{(Answer)}$$

(d) What is the cab's kinetic energy at the end of the 12 m fall?

SOLUTION: The **Key Idea** here is that the kinetic energy changes *because* of the net work done on the cab, according to Eq. 7-11 ($K_f = K_i + W$). From Eq. 7-1, we can write the kinetic energy at the start of the fall as $K_i = \frac{1}{2}mv_i^2$. We can then write Eq. 7-11 as

$$K_f = K_i + W = \frac{1}{2}mv_i^2 + W$$
$$= \frac{1}{2}(500 \text{ kg})(4.0 \text{ m/s})^2 + 1.18 \times 10^4 \text{ J}$$
$$= 1.58 \times 10^4 \text{ J} \approx 16 \text{ kJ}. \qquad \text{(Answer)}$$

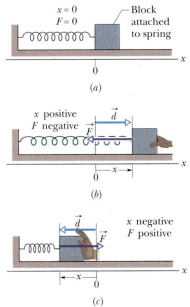

Fig. 7-10 (*a*) A spring in its relaxed state. The origin of an *x* axis has been placed at the end of the spring that is attached to a block. (*b*) The block is displaced by $\vec{d}$, and the spring is stretched by a positive amount *x*. Note the restoring force $\vec{F}$ exerted by the spring. (*c*) The spring is compressed by a negative amount *x*. Again, note the restoring force.

7-5   Work Done by a Spring Force

We next want to examine the work done on a particle-like object by a particular type of *variable force*—namely, a **spring force**, the force from a spring. Many forces in nature have the same mathematical form as the spring force. Thus, by examining this one force, you can gain an understanding of many others.

The Spring Force

Figure 7-10*a* shows a spring in its **relaxed state**—that is, neither compressed nor extended. One end is fixed, and a particle-like object, say, a block, is attached to the other, free end. If we stretch the spring by pulling the block to the right as in Fig. 7-10*b*, the spring pulls on the block toward the left. (Because a spring's force acts to restore the relaxed state, it is sometimes said to be a *restoring force*.) If we compress the spring by pushing the block to the left as in Fig. 7-10*c*, the spring now pushes on the block toward the right.

To a good approximation for many springs, the force $\vec{F}$ from a spring is proportional to the displacement $\vec{d}$ of the free end from its position when the spring is in the relaxed state. The *spring force* is given by

$$\vec{F} = -k\vec{d} \qquad \text{(Hooke's law)}, \qquad (7\text{-}20)$$

which is known as **Hooke's law** after Robert Hooke, an English scientist of the late 1600s. The minus sign in Eq. 7-20 indicates that the spring force is always opposite in direction from the displacement of the free end. The constant *k* is called the **spring constant** (or **force constant**) and is a measure of the stiffness of the spring. The larger *k* is, the stiffer the spring; that is, the stronger will be its pull or push for a given displacement. The SI unit for *k* is the newton per meter.

In Fig. 7-10 an *x* axis has been placed parallel to the length of a spring, with the origin ($x = 0$) at the position of the free end when the spring is in its relaxed state. For this common arrangement, we can write Eq. 7-20 as

$$F = -kx \qquad \text{(Hooke's law)}. \qquad (7\text{-}21)$$

If *x* is positive (the spring is stretched toward the right on the *x* axis), then *F* is negative (it is a pull toward the left). If *x* is negative (the spring is compressed toward the left), then *F* is positive (it is a push toward the right).

Note that a spring force is a *variable force* because its magnitude and direction depend on the position *x* of the free end; *F* can be symbolized as *F(x)*. Also note that Hooke's law is a *linear* relationship between *F* and *x*.

The Work Done by a Spring Force

To find an expression for the work done by the spring force as the block in Fig. 7-10*a* moves, let us make two simplifying assumptions about the spring. (1) It is *massless*; that is, its mass is negligible compared to the block's mass. (2) It is an *ideal spring*; that is, it obeys Hooke's law exactly. Let us also assume that the contact between the block and the floor is frictionless and that the block is particle-like.

We give the block a rightward jerk to get it moving, and then leave it alone. As the block moves rightward, the spring force $\vec{F}$ does work on the block, decreasing the kinetic energy and slowing the block. However, we *cannot* find this work by using Eq. 7-7 ($W = Fd \cos \phi$) because that equation assumes a constant force. The spring force is a variable force.

To find the work done by the spring, we use calculus. Let the block's initial position be x_i and its later position x_f. Then divide the distance between those two positions into many segments, each of tiny length Δx. Label these segments, starting from x_i, as segments 1, 2, and so on. As the block moves through a segment, the spring force hardly varies because the segment is so short that x hardly varies. Thus, we can approximate the force magnitude as being constant within the segment. Label these magnitudes as F_1 in segment 1, F_2 in segment 2, and so on.

With the force now constant in each segment, we *can* find the work done within each segment by using Eq. 7-7 ($W = Fd \cos \phi$). Here $\phi = 0$, so $\cos \phi = 1$. Then the work done is $F_1 \Delta x$ in segment 1, $F_2 \Delta x$ in segment 2, and so on. The net work W_s done by the spring, from x_i to x_f, is the sum of all these works:

$$W_s = \sum F_j \Delta x, \tag{7-22}$$

where j labels the segments. In the limit as Δx goes to zero, Eq. 7-22 becomes

$$W_s = \int_{x_i}^{x_f} F \, dx. \tag{7-23}$$

Substituting for F from Eq. 7-21, we find

$$W_s = \int_{x_i}^{x_f} (-kx) \, dx = -k \int_{x_i}^{x_f} x \, dx$$
$$= (-\tfrac{1}{2}k) \, [x^2]_{x_i}^{x_f} = (-\tfrac{1}{2}k)(x_f^2 - x_i^2). \tag{7-24}$$

Multiplied out, this yields

$$W_s = \tfrac{1}{2}kx_i^2 - \tfrac{1}{2}kx_f^2 \qquad \text{(work by a spring force).} \tag{7-25}$$

This work W_s done by the spring force can have a positive or negative value, depending on whether the *net* transfer of energy is to or from the block as the block moves from x_i to x_f. *Caution:* The final position x_f appears in the *second* term on the right side of Eq. 7-25. Therefore, Eq. 7-25 tells us:

▶ Work W_s is positive if the block ends up closer to the relaxed position ($x = 0$) than it was initially. It is negative if the block ends up farther away from $x = 0$. It is zero if the block ends up at the same distance from $x = 0$.

If $x_i = 0$ and if we call the final position x, then Eq. 7-25 becomes

$$W_s = -\tfrac{1}{2}kx^2 \qquad \text{(work by a spring force).} \tag{7-26}$$

The Work Done by an Applied Force

Now suppose that we displace the block along the x axis while continuing to apply a force $\vec{F}_a$ to it. During the displacement, our applied force does work W_a on the block while the spring force does work W_s. By Eq. 7-10, the change ΔK in the kinetic energy of the block due to these two energy transfers is

$$\Delta K = K_f - K_i = W_a + W_s, \tag{7-27}$$

in which K_f is the kinetic energy at the end of the displacement and K_i is that at the start of the displacement. If the block is stationary before and after the displacement, then K_f and K_i are both zero and Eq. 7-27 reduces to

$$W_a = -W_s. \tag{7-28}$$

▶ If a block that is attached to a spring is stationary before and after a displacement, then the work done on it by the applied force displacing it is the negative of the work done on it by the spring force.

Caution: If the block is not stationary before and after the displacement, then this statement is *not* true.

✓CHECKPOINT 4:   For three situations, the initial and final positions, respectively, along the x axis for the block in Fig. 7-10 are (a) -3 cm, 2 cm; (b) 2 cm, 3 cm; and (c) -2 cm, 2 cm. In each situation is the work done by the spring force on the block positive, negative, or zero?

Sample Problem 7-7

A package of spicy Cajun pralines lies on a frictionless floor, attached to the free end of a spring in the arrangement of Fig. 7-10a. An applied force of magnitude $F_a = 4.9$ N would be needed to hold the package stationary at $x_1 = 12$ mm.

(a) How much work does the spring force do on the package if the package is pulled rightward from $x_0 = 0$ to $x_2 = 17$ mm?

SOLUTION: A Key Idea here is that as the package moves from one position to another, the spring force does work on it as given by Eq. 7-25 or Eq. 7-26. We know that the initial position x_i is 0 and the final position x_f is 17 mm, but we do not know the spring constant k.

We can probably find k with Eq. 7-21 (Hooke's law), but we need a second Key Idea to use it: Were the package held stationary at $x_1 = 12$ mm, the spring force would have to balance the applied force (according to Newton's second law). Thus, the spring force F would have to be -4.9 N (toward the left in Fig. 7-10b), so Eq. 7-21 ($F = -kx$) gives us

$$k = -\frac{F}{x_1} = -\frac{-4.9 \text{ N}}{12 \times 10^{-3} \text{ m}} = 408 \text{ N/m}.$$

Now, with the package at $x_2 = 17$ mm, Eq. 7-26 yields

$$W_s = -\tfrac{1}{2}kx_2^2 = -\tfrac{1}{2}(408 \text{ N/m})(17 \times 10^{-3} \text{ m})^2$$
$$= -0.059 \text{ J}. \qquad \text{(Answer)}$$

(b) Next, the package is moved leftward to $x_3 = -12$ mm. How much work does the spring force do on the package during this displacement? Explain the sign of this work.

SOLUTION: The Key Idea here is the first one we noted in part (a). Now $x_i = +17$ mm and $x_f = -12$ mm, and Eq. 7-25 yields

$$W_s = \tfrac{1}{2}kx_i^2 - \tfrac{1}{2}kx_f^2 = \tfrac{1}{2}k(x_i^2 - x_f^2)$$
$$= \tfrac{1}{2}(408 \text{ N/m})[(17 \times 10^{-3} \text{ m})^2 - (-12 \times 10^{-3} \text{ m})^2]$$
$$= 0.030 \text{ J} = 30 \text{ mJ}. \qquad \text{(Answer)}$$

This work done on the block by the spring force is positive because the spring force does more positive work as the block moves from $x_i = +17$ mm to the spring's relaxed position than it does negative work as the block moves from the spring's relaxed position to $x_f = -12$ mm.

Sample Problem 7-8

In Fig. 7-11, a cumin canister of mass $m = 0.40$ kg slides across a horizontal frictionless counter with speed $v = 0.50$ m/s. It then runs into and compresses a spring of spring constant $k = 750$ N/m. When the canister is momentarily stopped by the spring, by what distance d is the spring compressed?

SOLUTION: There are three Key Ideas here:

1. The work W_s done on the canister by the spring force is related to the requested distance d by Eq. 7-26 ($W_s = -\tfrac{1}{2}kx^2$), with d replacing x.

2. The work W_s is also related to the kinetic energy of the canister by Eq. 7-10 ($K_f - K_i = W$).

3. The canister's kinetic energy has an initial value of $K = \tfrac{1}{2}mv^2$ and a value of zero when the canister is momentarily at rest.

Putting the first two of these ideas together, we write the work–kinetic energy theorem for the canister as

$$K_f - K_i = -\tfrac{1}{2}kd^2.$$

Substituting according to the third idea makes this

$$0 - \tfrac{1}{2}mv^2 = -\tfrac{1}{2}kd^2.$$

Simplifying, solving for d, and substituting known data then give us

$$d = v\sqrt{\frac{m}{k}} = (0.50 \text{ m/s})\sqrt{\frac{0.40 \text{ kg}}{750 \text{ N/m}}}$$
$$= 1.2 \times 10^{-2} \text{ m} = 1.2 \text{ cm}. \qquad \text{(Answer)}$$

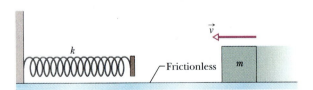

Fig. 7-11 Sample Problem 7-8. A canister of mass m moves at velocity $\vec{v}$ toward a spring with spring constant k.

7-6 Work Done by a General Variable Force

One-Dimensional Analysis

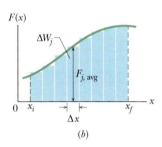

(a)

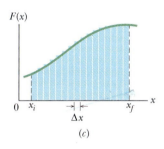

(b)

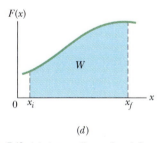

(c)

(d)

Fig. 7-12 (a) A one-dimensional force $\vec{F}$ plotted against the displacement x of a particle on which it acts. The particle moves from x_i to x_f. (b) Same as (a) but with the area under the curve divided into narrow strips. (c) Same as (b) but with the area divided into narrower strips. (d) The limiting case. The work done by the force is given by Eq. 7-32 and is represented by the shaded area between the curve and the x axis and between x_i and x_f.

Let us return to the situation of Fig. 7-2 but now consider the force to be directed along the x axis and the force magnitude to vary with position x. Thus, as the bead (particle) moves, the magnitude of the force doing work on it changes. Only the magnitude of this variable force changes, not its direction, and the magnitude at any position does not change with time.

Figure 7-12a shows a plot of such a *one-dimensional variable force*. We want an expression for the work done on the particle by this force as the particle moves from an initial point x_i to a final point x_f. However, we *cannot* use Eq. 7-7 because it applies only for a constant force $\vec{F}$. Here, again, we shall use calculus. We divide the area under the curve of Fig. 7-12a into a number of narrow strips of width Δx (Fig. 7-12b). We choose Δx small enough to permit us to take the force $F(x)$ as being reasonably constant over that interval. We let $F_{j,\text{avg}}$ be the average value of $F(x)$ within the jth interval. Then in Fig. 7-12b, $F_{j,\text{avg}}$ is the height of the jth strip.

With $F_{j,\text{avg}}$ considered constant, the increment (small amount) of work ΔW_j done by the force in the jth interval is now approximately given by Eq. 7-7 and is

$$\Delta W_j = F_{j,\text{avg}} \, \Delta x. \tag{7-29}$$

In Fig. 7-12b, ΔW_j is then equal to the area of the jth rectangular, shaded strip.

To approximate the total work W done by the force as the particle moves from x_i to x_f, we add the areas of all the strips between x_i and x_f in Fig. 7-12b:

$$W = \sum \Delta W_j = \sum F_{j,\text{avg}} \, \Delta x. \tag{7-30}$$

Equation 7-30 is an approximation because the broken "skyline" formed by the tops of the rectangular strips in Fig. 7-12b only approximates the actual curve of $F(x)$.

We can make the approximation better by reducing the strip width Δx and using more strips, as in Fig. 7-12c. In the limit, we let the strip width approach zero; the number of strips then becomes infinitely large and we have, as an exact result,

$$W = \lim_{\Delta x \to 0} \sum F_{j,\text{avg}} \, \Delta x. \tag{7-31}$$

This limit is exactly what we mean by the integral of the function $F(x)$ between the limits x_i and x_f. Thus, Eq. 7-31 becomes

$$W = \int_{x_i}^{x_f} F(x) \, dx \qquad \text{(work: variable force).} \tag{7-32}$$

If we know the function $F(x)$, we can substitute it into Eq. 7-32, introduce the proper limits of integration, carry out the integration, and thus find the work. (Appendix E contains a list of common integrals.) Geometrically, the work is equal to the area between the $F(x)$ curve and the x axis, between the limits x_i and x_f (shaded in Fig. 7-12d).

Three-Dimensional Analysis

Consider now a particle that is acted on by a three-dimensional force

$$\vec{F} = F_x \hat{i} + F_y \hat{j} + F_z \hat{k}, \tag{7-33}$$

in which the components F_x, F_y, and F_z can depend on the position of the particle; that is, they can be functions of that position. However, we make three simplifications: F_x may depend on x but not on y or z, F_y may depend on y but not on x or z, and F_z may depend on z but not on x or y. Now let the particle move through an incremental displacement

$$d\vec{r} = dx\hat{i} + dy\hat{j} + dz\hat{k}. \tag{7-34}$$

The increment of work dW done on the particle by $\vec{F}$ during the displacement $d\vec{r}$ is, by Eq. 7-8,

$$dW = \vec{F} \cdot d\vec{r} = F_x\,dx + F_y\,dy + F_z\,dz. \tag{7-35}$$

The work W done by $\vec{F}$ while the particle moves from an initial position r_i with coordinates (x_i, y_i, z_i) to a final position r_f with coordinates (x_f, y_f, z_f) is then

$$W = \int_{r_i}^{r_f} dW = \int_{x_i}^{x_f} F_x\,dx + \int_{y_i}^{y_f} F_y\,dy + \int_{z_i}^{z_f} F_z\,dz. \tag{7-36}$$

If $\vec{F}$ has only an x component, then the y and z terms in Eq. 7-36 are zero and the equation reduces to Eq. 7-32.

Work–Kinetic Energy Theorem with a Variable Force

Equation 7-32 gives the work done by a variable force on a particle in a one-dimensional situation. Let us now make certain that the work calculated with Eq. 7-32 is indeed equal to the change in kinetic energy of the particle, as the work–kinetic energy theorem states.

Consider a particle of mass m, moving along the x axis and acted on by a net force $F(x)$ that is directed along that axis. The work done on the particle by this force as the particle moves from an initial position x_i to a final position x_f is given by Eq. 7-32 as

$$W = \int_{x_i}^{x_f} F(x)\,dx = \int_{x_i}^{x_f} ma\,dx, \tag{7-37}$$

in which we use Newton's second law to replace $F(x)$ with ma. We can write the quantity $ma\,dx$ in Eq. 7-37 as

$$ma\,dx = m\frac{dv}{dt}\,dx. \tag{7-38}$$

From the "chain rule" of calculus, we have

$$\frac{dv}{dt} = \frac{dv}{dx}\frac{dx}{dt} = \frac{dv}{dx}v, \tag{7-39}$$

and Eq. 7-38 becomes

$$ma\,dx = m\frac{dv}{dx}v\,dx = mv\,dv. \tag{7-40}$$

Substituting Eq. 7-40 into Eq. 7-37 yields

$$W = \int_{v_i}^{v_f} mv\,dv = m\int_{v_i}^{v_f} v\,dv$$
$$= \tfrac{1}{2}mv_f^2 - \tfrac{1}{2}mv_i^2. \tag{7-41}$$

Note that when we change the variable from x to v we are required to express the limits on the integral in terms of the new variable. Note also that because the mass m is a constant, we are able to move it outside the integral.

Recognizing the terms on the right side of Eq. 7-41 as kinetic energies allows us to write this equation as

$$W = K_f - K_i = \Delta K,$$

which is the work–kinetic energy theorem.

Sample Problem 7-9

Force $\vec{F} = (3x^2 \text{ N})\hat{i} + (4 \text{ N})\hat{j}$, with x in meters, acts on a particle, changing only the kinetic energy of the particle. How much work is done on the particle as it moves from coordinates (2 m, 3 m) to (3 m, 0 m)? Does the speed of the particle increase, decrease, or remain the same?

SOLUTION: The Key Idea here is that the force is a variable force because its x component depends on the value of x. Thus, we cannot use Eqs. 7-7 and 7-8 to find the work done. Instead, we must use

Eq. 7-36 to integrate the force:

$$W = \int_2^3 3x^2 \, dx + \int_3^0 4 \, dy = 3 \int_2^3 x^2 \, dx + 4 \int_3^0 dy$$
$$= 3[\tfrac{1}{3}x^3]_2^3 + 4[y]_3^0 = [3^3 - 2^3] + 4[0 - 3]$$
$$= 7.0 \text{ J.} \qquad \text{(Answer)}$$

The positive result means that energy is transferred to the particle by force $\vec{F}$. Thus, the kinetic energy of the particle increases, and so must its speed.

7-7 Power

A contractor wishes to lift a load of bricks from the sidewalk to the top of a building by means of a winch. We can now calculate how much work the force applied by the winch must do on the load to make the lift. The contractor, however, is much more interested in the *rate* at which that work is done. Will the job take 5 minutes (acceptable) or a week (unacceptable)?

The time rate at which work is done by a force is said to be the **power** due to the force. If an amount of work W is done in an amount of time Δt by a force, the **average power** due to the force during that time interval is

$$P_{avg} = \frac{W}{\Delta t} \qquad \text{(average power).} \qquad (7\text{-}42)$$

The **instantaneous power** P is the instantaneous time rate of doing work, which we can write as

$$P = \frac{dW}{dt} \qquad \text{(instantaneous power).} \qquad (7\text{-}43)$$

Suppose we know the work $W(t)$ done by a force as a function of time. Then to get the instantaneous power P at, say, time $t = 3.0$ s during the work, we would first take the time derivative of $W(t)$, and then evaluate the result for $t = 3.0$ s.

The SI unit of power is the joule per second. This unit is used so often that it has a special name, the **watt** (W), after James Watt, who greatly improved the rate at which steam engines could do work. In the British system, the unit of power is the foot-pound per second. Often the horsepower is used. Some relations among these units are

$$1 \text{ watt} = 1 \text{ W} = 1 \text{ J/s} = 0.738 \text{ ft} \cdot \text{lb/s} \qquad (7\text{-}44)$$

and $\qquad 1 \text{ horsepower} = 1 \text{ hp} = 550 \text{ ft} \cdot \text{lb/s} = 746 \text{ W.} \qquad (7\text{-}45)$

Inspection of Eq. 7-42 shows that work can be expressed as power multiplied by time, as in the common unit, the kilowatt-hour. Thus,

$$1 \text{ kilowatt-hour} = 1 \text{ kW} \cdot \text{h} = (10^3 \text{ W})(3600 \text{ s}) \qquad (7\text{-}46)$$
$$= 3.60 \times 10^6 \text{ J} = 3.60 \text{ MJ.}$$

Perhaps because they appear on our utility bills, the watt and the kilowatt-hour have become identified as electrical units. They can be used equally well as units for other examples of power and work or energy. Thus, if you pick up this book from the floor and put it on a tabletop, you are free to report the work that you have done as $4 \times 10^{-6} \text{ kW} \cdot \text{h}$ (or more conveniently as 4 mW · h).

We can also express the rate at which a force does work on a particle (or particle-like object) in terms of that force and the particle's velocity. For a particle that is moving along a straight line (say, the x axis) and is acted on by a constant force $\vec{F}$ directed at some angle ϕ to that line, Eq. 7-43 becomes

$$P = \frac{dW}{dt} = \frac{F \cos \phi \, dx}{dt} = F \cos \phi \left(\frac{dx}{dt}\right),$$

or

$$P = Fv \cos \phi. \tag{7-47}$$

Reorganizing the right side of Eq. 7-47 as the dot product $\vec{F} \cdot \vec{v}$, we may also write Eq. 7-47 as

$$P = \vec{F} \cdot \vec{v} \qquad \text{(instantaneous power)}. \tag{7-48}$$

For example, the truck in Fig. 7-13 exerts a force $\vec{F}$ on the trailing load, which has velocity $\vec{v}$ at some instant. The instantaneous power due to $\vec{F}$ is the rate at which $\vec{F}$ does work on the load at that instant and is given by Eqs. 7-47 and 7-48. Saying that this power is "the power of the truck" is often acceptable, but we should keep in mind what is meant: Power is the rate at which the applied *force* does work.

✓CHECKPOINT 5:   A block moves with uniform circular motion because a cord tied to the block is anchored at the center of a circle. Is the power due to the force on the block from the cord positive, negative, or zero?

Fig. 7-13 The power due to the truck's applied force on the trailing load is the rate at which that force does work on the load.

Sample Problem 7-10

Figure 7-14 shows constant forces $\vec{F}_1$ and $\vec{F}_2$ acting on a box as the box slides rightward across a frictionless floor. Force $\vec{F}_1$ is horizontal, with magnitude 2.0 N; force $\vec{F}_2$ is angled upward by 60° to the floor and has magnitude 4.0 N. The speed v of the box at a certain instant is 3.0 m/s.

(a) What is the power due to each force acting on the box at that instant, and what is the net power? Is the net power changing at that instant?

SOLUTION: A **Key Idea** here is that we want an instantaneous power, not an average power over a time period. Also, we know the particle's velocity (rather than the work done on it). Therefore, we use Eq. 7-47 for each force. For force $\vec{F}_1$, at angle $\phi_1 = 180°$ to velocity

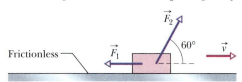

Fig. 7-14 Sample Problem 7-10. Two forces $\vec{F}_1$ and $\vec{F}_2$ act on a box that slides rightward across a frictionless floor. The velocity of the box is $\vec{v}$.

$\vec{v}$, we have

$$P_1 = F_1 v \cos \phi_1 = (2.0 \text{ N})(3.0 \text{ m/s}) \cos 180°$$

$$= -6.0 \text{ W}. \qquad \text{(Answer)}$$

This result tells us that force $\vec{F}_1$ is transferring energy *from* the box at the rate of 6.0 J/s.

For force $\vec{F}_2$, at angle $\phi_2 = 60°$ to velocity $\vec{v}$, we have

$$P_2 = F_2 v \cos \phi_2 = (4.0 \text{ N})(3.0 \text{ m/s}) \cos 60°$$

$$= 6.0 \text{ W}. \qquad \text{(Answer)}$$

This result tells us that force $\vec{F}_2$ is transferring energy *to* the box at the rate of 6.0 J/s.

A second Key Idea is that the net power is the sum of the individual powers:

$$P_{net} = P_1 + P_2$$

$$= -6.0 \text{ W} + 6.0 \text{ W} = 0, \qquad \text{(Answer)}$$

which tells us that the net rate of transfer of energy to or from the box is zero. Thus, the kinetic energy ($K = \frac{1}{2}mv^2$) of the box is not changing, and so the speed of the box will remain at 3.0 m/s. With

neither the forces $\vec{F}_1$ and $\vec{F}_2$ nor the velocity $\vec{v}$ changing, we see from Eq. 7-48 that P_1 and P_2 are constant and thus so is P_{net}.

(b) If the magnitude of $\vec{F}_2$ is, instead, 6.0 N, what now is the net power, and is it changing?

SOLUTION: The same Key Ideas as above give us, for the power now due to $\vec{F}_2$,

$$P_2 = F_2 v \cos \phi_2 = (6.0 \text{ N})(3.0 \text{ m/s}) \cos 60°$$

$$= 9.0 \text{ W}.$$

The power of force $\vec{F}_1$ is still $P_1 = -6.0$ W, so the net power is now

$$P_{net} = P_1 + P_2 = -6.0 \text{ W} + 9.0 \text{ W}$$

$$= 3.0 \text{ W}, \qquad \text{(Answer)}$$

which tells us that the net rate of transfer of energy to the box has a positive value. Thus, the kinetic energy of the box is increasing, and so also is the speed of the box. With the speed increasing, we see from Eq. 7-48 that the values of P_1 and P_2, and thus also of P_{net}, will be changing. Hence, this net power of 3.0 W is the net power only at the instant the speed is the given 3.0 m/s.

REVIEW & SUMMARY

Kinetic Energy   The **kinetic energy** K associated with the motion of a particle of mass m and speed v, where v is well below the speed of light, is

$$K = \tfrac{1}{2}mv^2 \qquad \text{(kinetic energy).} \qquad (7\text{-}1)$$

Work   Work W is energy transferred to or from an object via a force acting on the object. Energy transferred to the object is positive work, and energy transferred from the object is negative work.

Work Done by a Constant Force   The work done on a particle by a constant force $\vec{F}$ during displacement $\vec{d}$ of the particle is

$$W = Fd \cos \phi = \vec{F} \cdot \vec{d} \qquad \text{(work, constant force),} \qquad (7\text{-}7, 7\text{-}8)$$

in which ϕ is the constant angle between the directions of $\vec{F}$ and $\vec{d}$. Only the component of $\vec{F}$ that is along the displacement $\vec{d}$ can do work on the object. When two or more forces act on an object, their **net work** is the sum of the individual works by the forces, which is also equal to the work that would be done on the object by the net force $\vec{F}_{net}$ of those forces.

Work and Kinetic Energy   We can relate a change ΔK in kinetic energy of a particle to the net work W done on the particle with

$$\Delta K = K_f - K_i = W \qquad \text{(work–kinetic energy theorem),} \qquad (7\text{-}10)$$

in which K_i is the initial kinetic energy of the object and K_f is the kinetic energy after the work is done. Equation 7-10 rearranged gives us

$$K_f = K_i + W. \qquad (7\text{-}11)$$

Work Done by the Gravitational Force   The work W_g done by the gravitational force $\vec{F}_g$ on a particle-like object of mass m during a displacement $\vec{d}$ of the object is given by

$$W_g = mgd \cos \phi, \qquad (7\text{-}12)$$

in which ϕ is the angle between $\vec{F}_g$ and $\vec{d}$.

Work Done in Lifting and Lowering an Object   The work W_a done by an applied force during a lifting or lowering of a particle-like object is related to the work W_g done by the gravitational force and the change ΔK in the object's kinetic energy by

$$\Delta K = K_f - K_i = W_a + W_g. \qquad (7\text{-}15)$$

If the kinetic energy at the beginning of a lift equals that at the end of the lift, then Eq. 7-15 reduces to

$$W_a = -W_g, \qquad (7\text{-}16)$$

which tells us that the applied force transfers as much energy to the object as the gravitational force transfers from the object.

Spring Force   The force $\vec{F}$ from a spring is

$$\vec{F} = -k\vec{d} \qquad \text{(Hooke's law),} \qquad (7\text{-}20)$$

where $\vec{d}$ is the displacement of its free end from the position when the spring is in its **relaxed state** (neither compressed nor extended), and k is the **spring constant** (a measure of the spring's stiffness). If an x axis lies along the spring, with the origin at the location of the spring's free end when the spring is in its relaxed state, Eq. 7-20 can be written as

$$F = -kx \qquad \text{(Hooke's law).} \qquad (7\text{-}21)$$

A spring force is thus a variable force: It varies with the displacement of the spring's free end.

Work Done by a Spring Force

If an object is attached to the spring's free end, the work W_s done on the object by the spring force when the object is moved from an initial position x_i to a final position x_f is

$$W_s = \tfrac{1}{2}kx_i^2 - \tfrac{1}{2}kx_f^2. \qquad (7\text{-}25)$$

If $x_i = 0$ and $x_f = x$, then Eq. 7-25 becomes

$$W_s = -\tfrac{1}{2}kx^2. \qquad (7\text{-}26)$$

Work Done by a Variable Force

When the force $\vec{F}$ on a particle-like object depends on the position of the object, the work done by $\vec{F}$ on the object while the object moves from an initial position r_i with coordinates (x_i, y_i, z_i) to a final position r_f with coordinates (x_f, y_f, z_f) must be found by integrating the force. If we assume that component F_x may depend on x but not on y or z, component F_y may depend on y but not on x or z, and component F_z may depend on z but not on x or y, then the work is

$$W = \int_{x_i}^{x_f} F_x\,dx + \int_{y_i}^{y_f} F_y\,dy + \int_{z_i}^{z_f} F_z\,dz. \qquad (7\text{-}36)$$

If $\vec{F}$ has only an x component, then Eq. 7-36 reduces to

$$W = \int_{x_i}^{x_f} F(x)\,dx. \qquad (7\text{-}32)$$

Power

The **power** due to a force is the *rate* at which that force does work on an object. If the force does work W during a time interval Δt, the *average power* due to the force over that time interval is

$$P_{avg} = \frac{W}{\Delta t}. \qquad (7\text{-}42)$$

Instantaneous power is the instantaneous rate of doing work:

$$P = \frac{dW}{dt}. \qquad (7\text{-}43)$$

If the direction of a force $\vec{F}$ is at an angle ϕ to the direction of travel of the object, the instantaneous power is

$$P = Fv\cos\phi = \vec{F}\cdot\vec{v}, \qquad (7\text{-}47, 7\text{-}48)$$

in which $\vec{v}$ is the instantaneous velocity of the object.

QUESTIONS

1. Rank the following velocities according to the kinetic energy a particle will have with each velocity, greatest first: (a) $\vec{v} = 4\hat{i} + 3\hat{j}$, (b) $\vec{v} = -4\hat{i} + 3\hat{j}$, (c) $\vec{v} = -3\hat{i} + 4\hat{j}$, (d) $\vec{v} = 3\hat{i} - 4\hat{j}$, (e) $\vec{v} = 5\hat{i}$, and (f) $v = 5$ m/s at 30° to the horizontal.

2. Is positive or negative work done by a constant force $\vec{F}$ on a particle during a straight-line displacement $\vec{d}$ if (a) the angle between $\vec{F}$ and $\vec{d}$ is 30°; (b) the angle is 100°; (c) $\vec{F} = 2\hat{i} - 3\hat{j}$ and $\vec{d} = -4\hat{i}$?

3. Figure 7-15 shows six situations in which two forces act simultaneously on a box after the box has been sent sliding over a frictionless surface, either to the left or to the right. The forces are either 1 N or 2 N in magnitude, as indicated by the vector lengths.

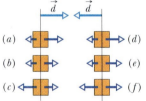

Fig. 7-15 Question 3.

For each situation, is the work done on the box by the net force during the indicated displacement $\vec{d}$ positive, negative, or zero?

4. Figure 7-16 shows the values of a force $\vec{F}$, directed along an x axis, that will act on a particle at the corresponding values of x. If the particle begins at rest at $x = 0$, what is the particle's coordinate when it has (a) its greatest kinetic energy, (b) its greatest speed, and (c) zero speed? (d) What is the particle's direction of travel after it reaches $x = 6$ m?

5. Figure 7-17 gives, for three situations, the position versus time of a box of contraband pulled by applied forces directed along an x axis on a frictionless surface. Line B is straight; the others are curved. Rank the situations according to the kinetic energy of the box at (a) time t_1 and (b) time t_2, greatest first. (c) Now rank them according to the net work done on the box by the applied forces during the time period t_1 to t_2, greatest first.

(d) For each situation, which of the following best describes the net work done by the applied forces during the period t_1 to t_2?

(1) Energy is transferred to the box.

(2) Energy is transferred from the box.

(3) The net work is zero.

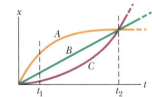

Fig. 7-17 Question 5.

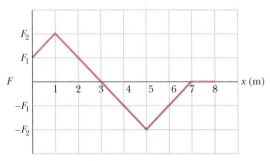

Fig. 7-16 Question 4.

6. Figure 7-18 shows four graphs (drawn to the same scale) of the x component of a variable force $\vec{F}$ (directed along an x axis) versus the position x of a particle on which the force acts. Rank the graphs according to the work done by $\vec{F}$ on the particle from $x = 0$ to x_1, from most positive work to most negative work.

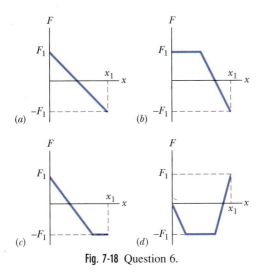

Fig. 7-18 Question 6.

7. In Fig. 7-19, a greased pig has a choice of three frictionless slides along which to slide to the ground. Rank the slides according to how much work the gravitational force does on the pig during the descent, greatest first.

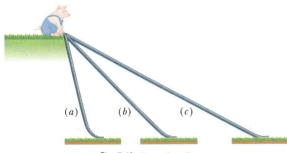

Fig. 7-19 Question 7.

8. You lift an armadillo to a shelf. Does the work done by your force on the armadillo depend on (a) the mass of the armadillo, (b) the weight of the armadillo, (c) the height of the shelf, (d) the time you take, or (e) whether you move the armadillo sideways or directly upward?

9. Figure 7-20 shows a bundle of magazines that is lifted by a cord through distance d. The table shows six pairs of values of the initial speed v_0 and final speed v (in meters per second) of the bundle at the beginning and end of distance d. Rank the pairs according to the work done by the cord's force on the bundle over distance d, most positive first, most negative last.

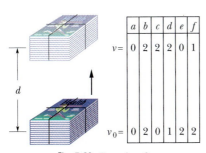

	a	b	c	d	e	f
$v=$	0	2	2	2	0	1
$v_0=$	0	2	0	1	2	2

Fig. 7-20 Question 9.

10. Spring A is stiffer than spring B; that is, $k_A > k_B$. The spring force of which spring does more work if the springs are compressed (a) the same distance and (b) by the same applied force?

11. A block is attached to a relaxed spring as in Fig. 7-21a. The spring constant k of the spring is such that for a certain rightward displacement $\vec{d}$ of the block, the spring force acting on the block has magnitude F_1 and has done work W_1 on the block. A second, identical spring is then attached on the opposite side of the block, as shown in Fig. 7-21b; both springs are in their relaxed state in the figure. If the block is again displaced by $\vec{d}$, (a) what is the magnitude of the net force on it from both springs and (b) how much work has been done on it by the spring forces?

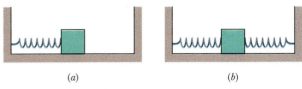

(a)   (b)

Fig. 7-21 Question 11.

12. Figure 7-22 gives the velocity versus time of a scooter car being moved along an axis by a varying applied force. The time axis shows six periods: $\Delta t_1 = \Delta t_2 = \Delta t_3 = \Delta t_6 = 2\Delta t_4 = \frac{2}{3}\Delta t_5$. (a) During which of the time periods is energy transferred *from* the scooter car by the applied force? (b) Rank the time periods according to the work done on the scooter car by the applied force during the period, most positive work first, most negative work last. (c) Rank the periods according to the rate at which the applied force transfers energy, greatest rate of transfer *to* the scooter car first, greatest rate of transfer *from* it last.

Fig. 7-22 Question 12.

13. In three situations, an initially stationary can of crayons is sent sliding over a frictionless floor by a different applied force. Plots of the resulting acceleration versus time for the three situations are given in Fig. 7-23. Rank the plots according to the work done by the applied force during the acceleration period, greatest first.

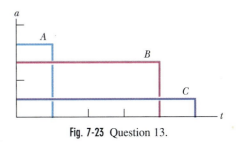

Fig. 7-23 Question 13.

EXERCISES & PROBLEMS

SEC. 7-1   Energy

1E. If an electron (mass $m = 9.11 \times 10^{-31}$ kg) in copper near the lowest possible temperature has a kinetic energy of 6.7×10^{-19} J, what is the speed of the electron? ssm

2E. On August 10, 1972, a large meteorite skipped across the atmosphere above western United States and Canada, much like a stone skipped across water. The accompanying fireball was so bright that it could be seen in the daytime sky (Fig. 7-24). The meteorite's mass was about 4×10^6 kg; its speed was about 15 km/s. Had it entered the atmosphere vertically, it would have hit Earth's surface with about the same speed. (a) Calculate the meteorite's loss of kinetic energy (in joules) that would have been associated with the vertical impact. (b) Express the energy as a multiple of the explosive energy of 1 megaton of TNT, which is 4.2×10^{15} J. (c) The energy associated with the atomic bomb explosion over Hiroshima was equivalent to 13 kilotons of TNT. To how many Hiroshima bombs would the meteorite impact have been equivalent?

Fig. 7-24 Exercise 2. A large meteorite skips across the atmosphere in the sky above the mountains (upper right).

3E. Calculate the kinetic energies of the following objects moving at the given speeds: (a) a 110 kg football linebacker running at 8.1 m/s; (b) a 4.2 g bullet at 950 m/s; (c) the aircraft carrier *Nimitz*, 91,400 tons at 32 knots.

4P. A father racing his son has half the kinetic energy of the son, who has half the mass of the father. The father speeds up by 1.0 m/s and then has the same kinetic energy as the son. What are the original speeds of (a) the father and (b) the son?

5P. A proton (mass $m = 1.67 \times 10^{-27}$ kg) is being accelerated along a straight line at 3.6×10^{15} m/s² in a machine. If the proton has an initial speed of 2.4×10^7 m/s and travels 3.5 cm, what then is (a) its speed and (b) the increase in its kinetic energy? ssm

SEC. 7-3   Work and Kinetic Energy

6E. A floating ice block is pushed through a displacement $\vec{d} = (15$ m$)\hat{i} - (12$ m$)\hat{j}$ along a straight embankment by rushing water, which exerts a force $\vec{F} = (210$ N$)\hat{i} - (150$ N$)\hat{j}$ on the block. How much work does the force do on the block during the displacement?

7E. To pull a 50 kg crate across a horizontal frictionless floor, a worker applies a force of 210 N, directed 20° above the horizontal. As the crate moves 3.0 m, what work is done on the crate by (a) the worker's force, (b) the gravitational force on the crate, and (c) the normal force on the crate from the floor? (d) What is the total work done on the crate? ssm

8E. A 1.0 kg standard body is at rest on a frictionless horizontal air track when a constant horizontal force $\vec{F}$ acting in the positive direction of an x axis along the track is applied to the body. A stroboscopic graph of the position of the body as it slides to the right is shown in Fig. 7-25. The force $\vec{F}$ is applied to the body at $t = 0$, and the graph records the position of the body at 0.50 s intervals. How much work is done on the body by the applied force $\vec{F}$ between $t = 0$ and $t = 2.0$ s?

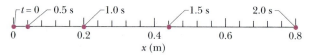

Fig. 7-25 Exercise 8.

9E. A luge and its rider, with a total mass of 85 kg, emerges from a downhill track onto a horizontal straight track with an initial speed of 37 m/s. If they stop at a constant deceleration of 2.0 m/s², (a) what magnitude F is required for the decelerating force, (b) what distance d do they travel while decelerating, and (c) what work W is done on them by the decelerating force? What are (d) F, (e) d, and (f) W for a deceleration of 4.0 m/s²? ilw

10P. A force acts on a 3.0 kg particle-like object in such a way that the position of the object as a function of time is given by $x = 3.0t - 4.0t^2 + 1.0t^3$, with x in meters and t in seconds. Find the work done on the object by the force from $t = 0$ to $t = 4.0$ s. (*Hint:* What are the speeds at those times?)

11P. Figure 7-26 shows three forces applied to a trunk that moves leftward by 3.00 m over a frictionless floor. The force magnitudes are $F_1 = 5.00$ N, $F_2 = 9.00$ N, and $F_3 = 3.00$ N. During the displacement, (a) what is the net work done on the trunk by the three forces and (b) does the kinetic energy of the trunk increase or decrease? ssm

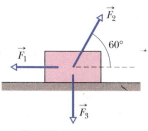

Fig. 7-26 Problem 11.

12P. The only force acting on a 2.0 kg canister that is moving in an xy plane has a magnitude of 5.0 N. The canister initially has a velocity of 4.0 m/s in the positive x direction, and some time later has a velocity of 6.0 m/s in the positive y direction. How much work is done on the canister by the 5.0 N force during this time?

13P. Figure 7-27 shows an overhead view of three horizontal forces acting on a cargo canister that was initially stationary but that now moves across a frictionless floor. The force magnitudes are $F_1 = 3.00$ N, $F_2 = 4.00$ N, and $F_3 = 10.0$ N. What is the net work done on the canister by the three forces during the first 4.00 m of displacement?   ssm

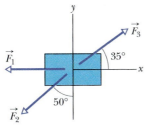

Fig. 7-27 Problem 13.

SEC. 7-4   Work Done by a Gravitational Force

14E. (a) In 1975 the roof of Montreal's Velodrome, with a weight of 360 kN, was lifted by 10 cm so that it could be centered. How much work was done on the roof by the forces making the lift? (b) In 1960 Mrs. Maxwell Rogers of Tampa, Florida, reportedly raised one end of a car that had fallen onto her son when a jack failed. If her panic lift effectively raised 4000 N (about $\frac{1}{4}$ of the car's weight) by 5.0 cm, how much work did her force do on the car?

15E. In Fig. 7-28, a cord runs around two massless, frictionless pulleys; a canister with mass $m = 20$ kg hangs from one pulley; and you exert a force $\vec{F}$ on the free end of the cord. (a) What must be the magnitude of $\vec{F}$ if you are to lift the canister at a constant speed? (b) To lift the canister by 2.0 cm, how far must you pull the free end of the cord? During that lift, what is the work done on the canister by (c) your force (via the cord) and (d) the gravitational force on the canister? (*Hint:* When a cord loops around a pulley as shown, it pulls on the pulley with a net force that is twice the tension in the cord.)

Fig. 7-28 Exercise 15.

16E. A 45 kg block of ice slides down a frictionless incline 1.5 m long and 0.91 m high. A worker pushes up against the ice, parallel to the incline, so that the block slides down at constant speed. (a) Find the magnitude of the worker's force. How much work is done on the block by (b) the worker's force, (c) the gravitational force on the block, (d) the normal force on the block from the surface of the incline, and (e) the net force on the block?

17P. A helicopter lifts a 72 kg astronaut 15 m vertically from the ocean by means of a cable. The acceleration of the astronaut is $g/10$. How much work is done on the astronaut by (a) the force from the helicopter and (b) the gravitational force on her? What are (c) the kinetic energy and (d) the speed of the astronaut just before she reaches the helicopter?   ssm   www

18P. A cave rescue team lifts an injured spelunker directly upward and out of a sinkhole by means of a motor-driven cable. The lift is performed in three stages, each requiring a vertical distance of 10.0 m: (a) the initially stationary spelunker is accelerated to a speed of 5.00 m/s; (b) he is then lifted at the constant speed of 5.00 m/s; (c) finally he is decelerated to zero speed. How much

work is done on the 80.0 kg rescuee by the force lifting him during each stage?

19P. A cord is used to vertically lower an initially stationary block of mass M at a constant downward acceleration of $g/4$. When the block has fallen a distance d, find (a) the work done by the cord's force on the block, (b) the work done by the gravitational force on the block, (c) the kinetic energy of the block, and (d) the speed of the block.   ssm

SEC. 7-5   Work Done by a Spring Force

20E. During spring semester at MIT, residents of the parallel buildings of the East Campus dorms battle one another with large catapults that are made with surgical hose mounted on a window frame. A balloon filled with dyed water is placed in a pouch attached to the hose, which is then stretched through the width of the room. Assume that the stretching of the hose obeys Hooke's law with a spring constant of 100 N/m. If the hose is stretched by 5.00 m and then released, how much work does the force from the hose do on the balloon in the pouch by the time the hose reaches its relaxed length?

21E. A spring with a spring constant of 15 N/cm has a cage attached to one end (Fig. 7-29). (a) How much work does the spring force do on the cage when the spring is stretched from its relaxed length by 7.6 mm? (b) How much additional work is done by the spring force when the spring is stretched by an additional 7.6 mm?   ssm   ilw   www

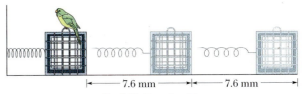

|← 7.6 mm →|← 7.6 mm →|

Fig. 7-29 Exercise 21.

22P. A 250 g block is dropped onto a relaxed vertical spring that has a spring constant of $k = 2.5$ N/cm (Fig. 7-30). The block becomes attached to the spring and compresses the spring 12 cm before momentarily stopping. While the spring is being compressed, what work is done on the block by (a) the gravitational force on it and (b) the spring force? (c) What is the speed of the block just before it hits the spring? (Assume that friction is negligible.) (d) If the speed at impact is doubled, what is the maximum compression of the spring?

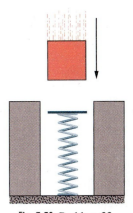

Fig. 7-30 Problem 22.

23P. The only force acting on a 2.0 kg body as it moves along the positive x axis has an x component $F_x = -6x$ N, where x is in meters. The velocity of the body at $x = 3.0$ m is 8.0 m/s. (a) What is the velocity of the body at $x = 4.0$ m? (b) At what positive value of x will the body have a velocity of 5.0 m/s?   ssm

SEC. 7-6 Work Done by a General Variable Force

24E. A 5.0 kg block moves in a straight line on a horizontal frictionless surface under the influence of a force that varies with position as shown in Fig. 7-31. How much work is done by the force as the block moves from the origin to $x = 8.0$ m?

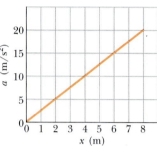

Fig. 7-31 Exercise 24.

25E. A 10 kg brick moves along an x axis. Its acceleration as a function of its position is shown in Fig. 7-32. What is the net work performed on the brick by the force causing the acceleration as the brick moves from $x = 0$ to $x = 8.0$ m? ssm ilw

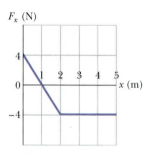

Fig. 7-32 Exercise 25.

26P. The only force acting on a 2.0 kg body as the body moves along the x axis varies as shown in Fig. 7-33. The velocity of the body at $x = 0$ is 4.0 m/s. (a) What is the kinetic energy of the body at $x = 3.0$ m? (b) At what value of x will the body have a kinetic energy of 8.0 J? (c) What is the maximum kinetic energy attained by the body between $x = 0$ and $x = 5.0$ m?

27P. The force on a particle is directed along an x axis and given by $F = F_0(x/x_0 - 1)$. Find the work done by the force in moving the particle from $x = 0$ to $x = 2x_0$ by (a) plotting $F(x)$ and measuring the work from the graph and (b) integrating $F(x)$. ssm

28P. A 1.5 kg block is initially at rest on a horizontal frictionless surface when a horizontal force in the positive direction of an x axis is applied to the block. The force is given by $\vec{F}(x) = (2.5 - x^2)\hat{i}$ N, where x is in meters and the initial position of the block is $x = 0$. (a) What is the kinetic energy of the block as it passes through $x = 2.0$ m? (b) What is the maximum kinetic energy of the block between $x = 0$ and $x = 2.0$ m?

29P. What work is done by a force $\vec{F} = (2x \text{ N})\hat{i} + (3 \text{ N})\hat{j}$, with x in meters, that moves a particle from a position $\vec{r}_i = (2 \text{ m})\hat{i} + (3 \text{ m})\hat{j}$ to a position $\vec{r}_f = -(4 \text{ m})\hat{i} - (3 \text{ m})\hat{j}$? ssm

SEC. 7-7 Power

30E. The loaded cab of an elevator has a mass of 3.0×10^3 kg and moves 210 m up the shaft in 23 s at constant speed. At what average rate does the force from the cable do work on the cab?

31E. A 100 kg block is pulled at a constant speed of 5.0 m/s across a horizontal floor by an applied force of 122 N directed 37° above the horizontal. What is the rate at which the force does work on the block? ssm ilw

32E. (a) At a certain instant, a particle-like object is acted on by a force $\vec{F} = (4.0 \text{ N})\hat{i} - (2.0 \text{ N})\hat{j} + (9.0 \text{ N})\hat{k}$ while having a velocity $\vec{v} = -(2.0 \text{ m/s})\hat{i} + (4.0 \text{ m/s})\hat{k}$. What is the instantaneous rate at which the force does work on the object? (b) At some other time, the velocity consists of only a y component. If the force is unchanged, and the instantaneous power is -12 W, what is the velocity of the object just then?

33P. A force of 5.0 N acts on a 15 kg body initially at rest. Compute the work done by the force in (a) the first, (b) the second, and (c) the third seconds and (d) the instantaneous power due to the force at the end of the third second. ssm

34P. A skier is pulled by a tow rope up a frictionless ski slope that makes an angle of 12° with the horizontal. The rope moves parallel to the slope with a constant speed of 1.0 m/s. The force of the rope does 900 J of work on the skier as the skier moves a distance of 8.0 m up the incline. (a) If the rope moved with a constant speed of 2.0 m/s, how much work would the force of the rope do on the skier as the skier moved a distance of 8.0 m up the incline? At what rate is the force of the rope doing work on the skier when the rope moves with a speed of (b) 1.0 m/s and (c) 2.0 m/s?

35P. A fully loaded, slow-moving freight elevator has a cab with a total mass of 1200 kg, which is required to travel upward 54 m in 3.0 min, starting and ending at rest. The elevator's counterweight has a mass of only 950 kg, so the elevator motor must help pull the cab upward. What average power is required of the force the motor exerts on the cab via the cable? ssm www

36P. A 0.30 kg ladle sliding on a horizontal frictionless surface is attached to one end of a horizontal spring (with $k = 500$ N/m) whose other end is fixed. The ladle has a kinetic energy of 10 J as it passes through its equilibrium position (the point at which the spring force is zero). (a) At what rate is the spring doing work on the ladle as the ladle passes through its equilibrium position? (b) At what rate is the spring doing work on the ladle when the spring is compressed 0.10 m and the ladle is moving away from the equilibrium position?

37P. The force (but not the power) required to tow a boat at constant velocity is proportional to the speed. If a speed of 4.0 km/h requires 7.5 kW, how much power does a speed of 12 km/h require? ssm

38P. Boxes are transported from one location to another in a warehouse by means of a conveyor belt that moves with a constant speed of 0.50 m/s. At a certain location the conveyor belt moves for 2.0 m up an incline that makes an angle of 10° with the horizontal, then for 2.0 m horizontally, and finally for 2.0 m down an incline that makes an angle of 10° with the horizontal. Assume that a 2.0 kg box rides on the belt without slipping. At what rate is the force of the conveyor belt doing work on the box (a) as the box moves up the 10° incline, (b) as the box moves horizontally, and (c) as the box moves down the 10° incline?

39P. A horse pulls a cart with a force of 40 lb at an angle of 30° above the horizontal and moves along at a speed of 6.0 mi/h. (a) How much work does the force do in 10 min? (b) What is the average power (in horsepower) of the force?

40P. An initially stationary 2.0 kg object accelerates horizontally and uniformly to a speed of 10 m/s in 3.0 s. (a) In that 3.0 s interval, how much work is done on the object by the force accelerating it? What is the instantaneous power due to that force (b) at the end of the interval and (c) at the end of the first half of the interval?

8 Potential Energy and Conservation of Energy

The prehistoric people of Easter Island carved hundreds of giant stone statues in their quarry and then moved them to sites all over the island. How they managed to move them by as much as 10 km without the use of sophisticated machines has been a hotly debated subject, with many fanciful theories about the source of the required energy.

How much energy *was* required to move one of the statues using only primitive means?

The answer is in this chapter.

8-1 Potential Energy

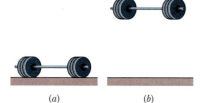

Fig. 8-1 When Chemerkin lifted the weights above his head, he increased the separation between the weights and Earth and thus changed the configuration of the weights–Earth system from that in (*a*) to that in (*b*).

In this chapter we continue the discussion of energy that we began in Chapter 7. To do so, we define a second form of energy: **potential energy** U is energy that can be associated with the configuration (or arrangement) of a system of objects that exert forces on one another. If the configuration of the system changes, then the potential energy of the system can also change.

One type of potential energy is the **gravitational potential energy** that is associated with the state of separation between objects, which attract one another via the gravitational force. For example, when Andrey Chemerkin lifted the record-breaking weights above his head in the 1996 Olympics, he increased the separation between the weights and Earth. The work his force did changed the gravitational potential energy of the weights–Earth system because it changed the configuration of the system—that is, the force shifted the relative locations of the weights and Earth (Fig. 8-1).

Another type of potential energy is **elastic potential energy,** which is associated with the state of compression or extension of an elastic (springlike) object. If you compress or extend a spring, you do work to change the relative locations of the coils within the spring. The result of the work done by your force is an increase in the elastic potential energy of the spring.

The idea of potential energy can be an enormously powerful tool in understanding situations involving the motion of an object. With it we can, in fact, easily solve problems that would require careful computer programming if we used only the ideas of earlier chapters.

Work and Potential Energy

In Chapter 7 we discussed the relation between work and a change in kinetic energy. Here we discuss the relation between work and a change in potential energy.

Let us throw a tomato upward (Fig. 8-2). We already know that as the tomato rises, the work W_g done on the tomato by the gravitational force is negative because the force transfers energy *from* the kinetic energy of the tomato. We can now finish the story by saying that this energy is transferred by the gravitational force *to* the gravitational potential energy of the tomato–Earth system.

The tomato slows, stops, and then begins to fall back down because of the gravitational force. During the fall, the transfer is reversed: The work W_g done on the tomato by the gravitational force is now positive—that force transfers energy *from* the gravitational potential energy of the tomato–Earth system *to* the kinetic energy of the tomato.

For either rise or fall, the change ΔU in gravitational potential energy is defined to equal the negative of the work done on the tomato by the gravitational force. Using the general symbol W for work, we write this as

$$\Delta U = -W. \tag{8-1}$$

This equation also applies to a block–spring system, as in Fig. 8-3. If we abruptly shove the block to send it moving rightward, the spring force acts leftward and thus does negative work on the block, transferring energy from the kinetic energy of the block to the elastic potential energy of the spring. The block slows and eventually stops, and then begins to move leftward because the spring force is still leftward. The transfer of energy is then reversed—it is from potential energy of the spring to kinetic energy of the block.

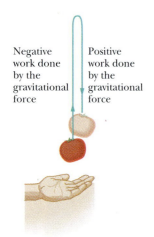

Negative work done by the gravitational force

Positive work done by the gravitational force

Fig. 8-2 A tomato is thrown upward. As it rises, the gravitational force does negative work on it, decreasing its kinetic energy. As the tomato descends, the gravitational force does positive work on it, increasing its kinetic energy.

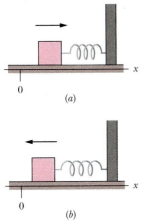

Fig. 8-3 A block, attached to a spring and initially at rest at $x = 0$, is set in motion toward the right. (*a*) As the block moves rightward (as indicated by the arrow), the spring force does negative work on it. (*b*) Then, as the block moves back toward $x = 0$, the spring force does positive work on it.

Conservative and Nonconservative Forces

Let us list the key elements of the two situations we just discussed:

1. The *system* consists of two or more objects.

2. A *force* acts between a particle-like object (tomato or block) in the system and the rest of the system.

3. When the system configuration changes, the force does *work* (call it W_1) on the particle-like object, transferring energy between the kinetic energy K of the object and some other form of energy of the system.

4. When the configuration change is reversed, the force reverses the energy transfer, doing work W_2 in the process.

In a situation in which $W_1 = -W_2$ is always true, the other form of energy is a potential energy, and the force is said to be a **conservative force.** As you might suspect, the gravitational force and the spring force are both conservative (since otherwise we could not have spoken of gravitational potential energy and elastic potential energy, as we did previously).

A force that is not conservative is called a **nonconservative force.** The kinetic frictional force and drag force are nonconservative. For an example, let us send a block sliding across a floor that is not frictionless. During the sliding, a kinetic frictional force from the floor does negative work on the block, slowing the block by transferring energy from its kinetic energy to a form of energy called *thermal energy* (which has to do with the random motions of atoms and molecules). We know from experiment that this energy transfer cannot be reversed (thermal energy cannot be transferred back to kinetic energy of the block by the kinetic frictional force). Thus, although we have a system (made up of the block and the floor), a force that acts between parts of the system, and a transfer of energy by the force, the force is not conservative. Therefore, thermal energy is not a potential energy.

When only conservative forces act on a particle-like object, we can greatly simplify otherwise difficult problems involving motion of the object. The next section, in which we develop a test for identifying conservative forces, provides one means for simplifying such problems.

8-2 Path Independence of Conservative Forces

The primary test for determining whether a force is conservative or nonconservative is this: Let the force act on a particle that moves along any *closed path,* beginning at some initial position and eventually returning to that position (so that the particle makes a *round trip* beginning and ending at the initial position). The force is conservative only if the total energy it transfers to and from the particle during the round trip along this and any other closed path is zero. In other words:

> The net work done by a conservative force on a particle moving around every closed path is zero.

We know from experiment that the gravitational force passes this *closed-path test.* An example is the tossed tomato of Fig. 8-2. The tomato leaves the launch point with speed v_0 and kinetic energy $\frac{1}{2}mv_0^2$. The gravitational force acting on the tomato slows it, stops it, and then causes it to fall back down. When the tomato returns to the launch point, it again has speed v_0 and kinetic energy $\frac{1}{2}mv_0^2$. Thus, the gravita-

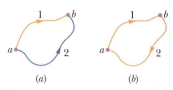

(a)   (b)

Fig. 8-4 (a) As a conservative force acts on it, a particle can move from point a to point b along either path 1 or path 2. (b) The particle moves in a round trip, from point a to point b along path 1 and then back to point a along path 2.

tional force transfers as much energy *from* the tomato during the ascent as it transfers *to* the tomato during the descent back to the launch point. The net work done on the tomato by the gravitational force during the round trip is zero.

An important result of the closed-path test is that

> The work done by a conservative force on a particle moving between two points does not depend on the path taken by the particle.

For example, suppose that a particle moves from point a to point b in Fig. 8-4a along either path 1 or path 2. If only a conservative force acts on the particle, then the work done on the particle is the same along the two paths. In symbols, we can write this result as

$$W_{ab,1} = W_{ab,2}, \tag{8-2}$$

where the subscript ab indicates the initial and final points, respectively, and the subscripts 1 and 2 indicate the path.

This result is powerful, because it allows us to simplify difficult problems when only a conservative force is involved. Suppose you need to calculate the work done by a conservative force along a given path between two points, and the calculation is difficult or even impossible without additional information. You can find the work by substituting some other path between those two points for which the calculation is easier and possible. Sample Problem 8-1 gives an example, but first we need to prove Eq. 8-2.

Proof of Equation 8-2

Figure 8-4b shows an arbitrary round trip for a particle that is acted upon by a single force. The particle moves from an initial point a to point b along path 1, and then back to point a along path 2. The force does work on the particle as the particle moves along each path. Without worrying about where positive work is done and where negative work is done, let us just represent the work done from a to b along path 1 as $W_{ab,1}$ and the work done from b back to a along path 2 as $W_{ba,2}$. If the force is conservative, then the net work done during the round trip must be zero:

$$W_{ab,1} + W_{ba,2} = 0,$$

and thus

$$W_{ab,1} = -W_{ba,2}. \tag{8-3}$$

In words, the work done along the outward path must be the negative of the work done along the path back.

Let us now consider the work $W_{ab,2}$ done on the particle by the force when the particle moves from a to b along path 2, as indicated in Fig. 8-4a. If the force is conservative, that work is the negative of $W_{ba,2}$:

$$W_{ab,2} = -W_{ba,2}. \tag{8-4}$$

Substituting $W_{ab,2}$ for $-W_{ba,2}$ in Eq. 8-3, we obtain

$$W_{ab,1} = W_{ab,2},$$

which is what we set out to prove.

✓CHECKPOINT 1: The figure shows three paths connecting points a and b. A single force $\vec{F}$ does the indicated work on a particle moving along each path in the indicated direction. On the basis of this information, is force $\vec{F}$ conservative?

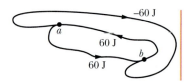

Sample Problem 8-1

Figure 8-5a shows a 2.0 kg block of slippery cheese that slides along a frictionless track from point a to point b. The cheese travels through a total distance of 2.0 m along the track, and a net vertical distance of 0.80 m. How much work is done on the cheese by the gravitational force during the slide?

SOLUTION: A Key Idea here is that we *cannot* use Eq. 7-12 ($W_g = mgd \cos \phi$) to calculate the work done by the gravitational force $\vec{F}_g$ as the cheese moves along the track. The reason is that the angle ϕ between the directions of $\vec{F}_g$ and the displacement $\vec{d}$ varies along the track in an unknown way. (Even if we did know the shape of the track and could calculate ϕ along it, the calculation could be very difficult.)

A second Key Idea is that because $\vec{F}_g$ is a conservative force, we can find the work by choosing some other path between a and b—one that makes the calculation easy. Let us choose the dashed path in Fig. 8-5b; it consists of two straight segments. Along the horizontal segment, the angle ϕ is a constant 90°. Even though we do not know the displacement along that horizontal segment, Eq. 7-12 tells us that the work W_h done there is

$$W_h = mgd \cos 90° = 0.$$

Along the vertical segment, the displacement d is 0.80 m and, with $\vec{F}_g$ and $\vec{d}$ both downward, the angle ϕ is a constant 0°. Thus, Eq. 7-12 gives us, for the work W_v done along the vertical part of

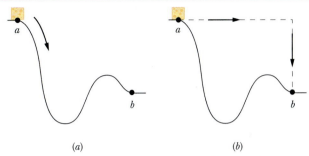

Fig. 8-5 Sample Problem 8-1. (*a*) A block of cheese slides along a frictionless track from point a to point b. (*b*) Finding the work done on the cheese by the gravitational force is easier along the dashed path than along the actual path taken by the cheese; the result is the same for both paths.

the dashed path,

$$W_v = mgd \cos 0°$$
$$= (2.0 \text{ kg})(9.8 \text{ m/s}^2)(0.80 \text{ m})(1) = 15.7 \text{ J}.$$

The total work done on the cheese by $\vec{F}_g$ as the cheese moves from point a to point b along the dashed path is then

$$W = W_h + W_v = 0 + 15.7 \text{ J} \approx 16 \text{ J}. \quad \text{(Answer)}$$

This is also the work done as the cheese moves along the track from a to b.

8-3 Determining Potential Energy Values

Here we find equations that give the value of the two types of potential energy discussed in this chapter: gravitational potential energy and elastic potential energy. However, first we must find a general relation between a conservative force and the associated potential energy.

Consider a particle-like object that is part of a system in which a conservative force $\vec{F}$ acts. When that force does work W on the object, the change ΔU in the potential energy associated with the system is the negative of the work done. We wrote this fact as Eq. 8-1 ($\Delta U = -W$). For the most general case, in which the force may vary with position, we may write the work W as in Eq. 7-32:

$$W = \int_{x_i}^{x_f} F(x) \, dx. \quad (8\text{-}5)$$

This equation gives the work done by the force when the object moves from point x_i to point x_f, changing the configuration of the system. (Because the force is conservative, the work is the same for all paths between those two points.)

Substituting Eq. 8-5 into Eq. 8-1, we find that the change in potential energy due to the change in configuration is

$$\Delta U = -\int_{x_i}^{x_f} F(x) \, dx. \quad (8\text{-}6)$$

This is the general relation we sought. Let's put it to use.

Gravitational Potential Energy

We first consider a particle with mass m moving vertically along a y axis (the positive direction is upward). As the particle moves from point y_i to point y_f, the gravitational force $\vec{F}_g$ does work on it. To find the corresponding change in the gravitational potential energy of the particle–Earth system, we use Eq. 8-6 with two changes: (1) We integrate along the y axis instead of the x axis, because the gravitational force acts vertically. (2) We substitute $-mg$ for the force symbol F, because $\vec{F}_g$ has the magnitude mg and is directed down the y axis. We then have

$$\Delta U = -\int_{y_i}^{y_f} (-mg)\, dy = mg \int_{y_i}^{y_f} dy = mg \left[\, y \,\right]_{y_i}^{y_f},$$

which yields

$$\Delta U = mg(y_f - y_i) = mg\, \Delta y. \tag{8-7}$$

Only *changes* ΔU in gravitational potential energy (or any other type of potential energy) are physically meaningful. However, to simplify a calculation or a discussion, we sometimes would like to say that a certain gravitational potential value U is associated with a certain particle–Earth system when the particle is at a certain height y. To do so, we rewrite Eq. 8-7 as

$$U - U_i = mg(y - y_i). \tag{8-8}$$

Then we take U_i to be the gravitational potential energy of the system when it is in a **reference configuration** in which the particle is at a **reference point** y_i. Usually we take $U_i = 0$ and $y_i = 0$. Doing this changes Eq. 8-8 to

$$U(y) = mgy \qquad \text{(gravitational potential energy).} \tag{8-9}$$

This equation tells us:

> ►The gravitational potential energy associated with a particle–Earth system depends only on the vertical position y (or height) of the particle relative to the reference position $y = 0$, not on the horizontal position.

Elastic Potential Energy

We next consider the block–spring system shown in Fig. 8-3, with the block moving on the end of a spring of spring constant k. As the block moves from point x_i to point x_f, the spring force $F = -kx$ does work on the block. To find the corresponding change in the elastic potential energy of the block–spring system, we substitute $-kx$ for $F(x)$ in Eq. 8-6. We then have

$$\Delta U = -\int_{x_i}^{x_f} (-kx)\, dx = k \int_{x_i}^{x_f} x\, dx = \tfrac{1}{2} k \left[\, x^2 \,\right]_{x_i}^{x_f},$$

or

$$\Delta U = \tfrac{1}{2} k x_f^2 - \tfrac{1}{2} k x_i^2. \tag{8-10}$$

To associate a potential energy value U with the block at position x, we choose the reference configuration to be when the spring is at its relaxed length and the block is at $x_i = 0$. Then the elastic potential energy U_i is 0, and Eq. 8-10 becomes

$$U - 0 = \tfrac{1}{2} k x^2 - 0,$$

which gives us

$$\cdot \; U(x) = \tfrac{1}{2} k x^2 \qquad \text{(elastic potential energy).} \tag{8-11}$$

✔CHECKPOINT 2: A particle is to move along the x axis from $x = 0$ to x_1 while a conservative force, directed along the x axis, acts on the particle. The figure shows three situations in which the x component of that force varies with x. The force has the same maximum magnitude F_1 in all three situations. Rank the situations according to the change in the associated potential energy during the particle's motion, most positive first.

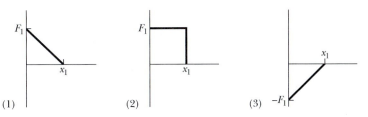

(1)   (2)   (3)

PROBLEM-SOLVING TACTICS

Tactic 1: *Using the Term "Potential Energy"*

A potential energy is associated with a system as a whole. However, you might see statements that associate it with only part of the system. For example, you might read, "An apple hanging in a tree has a gravitational potential energy of 30 J." Such statements are often acceptable, but you should always keep in mind that the potential energy is actually associated with a system—here the apple–Earth system. Also keep in mind that assigning a particular potential energy value, such as 30 J here, to an object or even a system makes sense *only* if the reference potential energy value is known, as explored in Sample Problem 8-2.

Sample Problem 8-2

A 2.0 kg sloth hangs 5.0 m above the ground (Fig. 8-6).

(a) What is the gravitational potential energy U of the sloth–Earth system if we take the reference point $y = 0$ to be (1) at the ground, (2) at a balcony floor that is 3.0 m above the ground, (3) at the limb, and (4) 1.0 m above the limb? Take the gravitational potential energy to be zero at $y = 0$.

SOLUTION: The **Key Idea** here is that once we have chosen the reference point for $y = 0$, we can calculate the gravitational potential energy U of the system *relative to that reference point* with Eq. 8-9. For example, for choice (1) the sloth is at $y = 5.0$ m, and

$$U = mgy = (2.0 \text{ kg})(9.8 \text{ m/s}^2)(5.0 \text{ m})$$
$$= 98 \text{ J}. \qquad \text{(Answer)}$$

For the other choices, the values of U are

(2) $U = mgy = mg(2.0 \text{ m}) = 39 \text{ J}$,
(3) $U = mgy = mg(0) = 0 \text{ J}$,
(4) $U = mgy = mg(-1.0 \text{ m}) = -19.6 \text{ J} \approx -20 \text{ J}$.   (Answer)

(b) The sloth drops to the ground. For each choice of reference point, what is the change ΔU in the potential energy of the sloth–Earth system due to the fall?

SOLUTION: The **Key Idea** here is that the *change* in potential energy does not depend on the choice of the reference point for $y = 0$; instead, it depends on the change in height Δy. For all four situations, we have the same $\Delta y = -5.0$ m. Thus, for (1) to (4), Eq. 8-7 tells us that

$$\Delta U = mg\,\Delta y = (2.0 \text{ kg})(9.8 \text{ m/s}^2)(-5.0 \text{ m})$$
$$= -98 \text{ J}. \qquad \text{(Answer)}$$

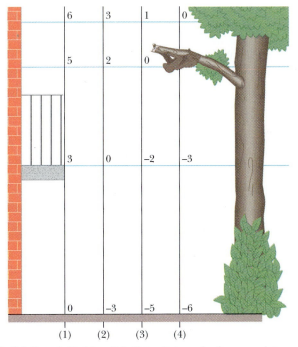

Fig. 8-6 Sample Problem 8-2. Four choices of reference point $y = 0$. Each y axis is marked in units of meters. The choice affects the value of the potential energy U of the sloth–Earth system. However, it does not affect the change ΔU in potential energy of the system if the sloth moves by, say, falling.

In olden days, a native Alaskan would be tossed via a blanket to be able to see farther over the flat terrain. Nowadays, it is done just for fun. During the ascent of the child in the photograph, energy is transferred from kinetic energy to gravitational potential energy. The maximum height is reached when that transfer is complete. Then the transfer is reversed during the fall.

8-4 Conservation of Mechanical Energy

The **mechanical energy** E_{mec} of a system is the sum of its potential energy U and the kinetic energy K of the objects within it:

$$E_{mec} = K + U \qquad \text{(mechanical energy).} \qquad (8\text{-}12)$$

In this section, we examine what happens to this mechanical energy when only conservative forces cause energy transfers within the system—that is, when frictional and drag forces do not act on the objects in the system. Also, we shall assume that the system is *isolated* from its environment; that is, no *external force* from an object outside the system causes energy changes inside the system.

When a conservative force does work W on an object within the system, it transfers energy between kinetic energy K of the object and potential energy U of the system. From Eq. 7-10, the change ΔK in kinetic energy is

$$\Delta K = W \qquad (8\text{-}13)$$

and from Eq. 8-1, the change ΔU in potential energy is

$$\Delta U = -W. \qquad (8\text{-}14)$$

Combining Eqs. 8-13 and 8-14, we find that

$$\Delta K = -\Delta U. \qquad (8\text{-}15)$$

In words, one of these energies increases exactly as much as the other decreases.

We can rewrite Eq. 8-15 as

$$K_2 - K_1 = -(U_2 - U_1), \qquad (8\text{-}16)$$

where the subscripts refer to two different instants and thus to two different arrangements of the objects in the system. Rearranging Eq. 8-16 yields

$$K_2 + U_2 = K_1 + U_1 \qquad \text{(conservation of mechanical energy).} \qquad (8\text{-}17)$$

In words, this equation says:

$$\left(\begin{array}{c} \text{the sum of } K \text{ and } U \text{ for} \\ \text{any state of a system} \end{array} \right) = \left(\begin{array}{c} \text{the sum of } K \text{ and } U \text{ for} \\ \text{any other state of the system} \end{array} \right),$$

when the system is isolated and only conservative forces act on the objects in the system. In other words:

> In an isolated system where only conservative forces cause energy changes, the kinetic energy and potential energy can change, but their sum, the mechanical energy E_{mec} of the system, cannot change.

This result is called the **principle of conservation of mechanical energy.** (Now you can see where *conservative* forces got their name.) With the aid of Eq. 8-15, we can write this principle in one more form, as

$$\Delta E_{mec} = \Delta K + \Delta U = 0. \qquad (8\text{-}18)$$

The principle of conservation of mechanical energy allows us to solve problems that would be quite difficult to solve using only Newton's laws:

> When the mechanical energy of a system is conserved, we can relate the sum of kinetic energy and potential energy at one instant to that at another instant *without considering the intermediate motion* and *without finding the work done by the forces involved.*

Fig. 8-7 A pendulum, with its mass concentrated in a bob at the lower end, swings back and forth. One full cycle of the motion is shown. During the cycle the values of the potential and kinetic energies of the pendulum–Earth system vary as the bob rises and falls, but the mechanical energy E_{mec} of the system remains constant. The energy E_{mec} can be described as continuously shifting between the kinetic and potential forms. In stages (a) and (e), all the energy is kinetic energy. The bob then has its greatest speed and is at its lowest point. In stages (c) and (g), all the energy is potential energy. The bob then has zero speed and is at its highest point. In stages (b), (d), (f), and (h), half the energy is kinetic energy and half is potential energy. If the swinging involved a frictional force at the point where the pendulum is attached to the ceiling, or a drag force due to the air, then E_{mec} would not be conserved, and eventually the pendulum would stop.

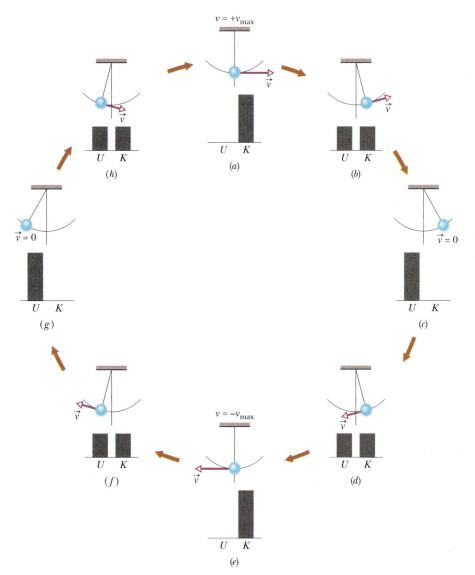

Figure 8-7 shows an example in which the principle of conservation of mechanical energy can be applied: As a pendulum swings, the energy of the pendulum–Earth system is transferred back and forth between kinetic energy K and gravitational potential energy U, with the sum $K + U$ being constant. If we know the gravitational potential energy when the pendulum bob is at its highest point (Fig. 8-7c), Eq. 8-17 gives us the kinetic energy of the bob at the lowest point (Fig. 8-7e).

For example, let us choose the lowest point as the reference point, with the gravitational potential energy $U_2 = 0$. Suppose then that the potential energy at the highest point is $U_1 = 20$ J relative to the reference point. Because the bob momentarily stops at its highest point, the kinetic energy there is $K_1 = 0$. Substituting these values into Eq. 8-17 gives us the kinetic energy K_2 at the lowest point:

$$K_2 + 0 = 0 + 20 \text{ J} \quad \text{or} \quad K_2 = 20 \text{ J.}$$

Note that we get this result without considering the motion between the highest and lowest points (such as in Fig. 8-7d) and without finding the work done by any forces involved in the motion.

CHECKPOINT 3:   The figure shows four situations—one in which an initially stationary block is dropped and three in which the block is allowed to slide down frictionless ramps. (a) Rank the situations according to the kinetic energy of the block at point B, greatest first. (b) Rank them according to the speed of the block at point B, greatest first.

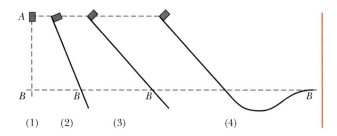

(1)   (2)   (3)   (4)

Sample Problem 8-3

In Fig. 8-8, a child of mass m is released from rest at the top of a water slide, at height $h = 8.5$ m above the bottom of the slide. Assuming that the slide is frictionless because of the water on it, find the child's speed at the bottom of the slide.

SOLUTION: A **Key Idea** here is that we cannot find her speed at the bottom by using her acceleration along the slide as we might have in earlier chapters because we do not know the slope (angle) of the slide. However, because that speed is related to her kinetic energy, perhaps we can use the principle of conservation of mechanical energy to get the speed. Then we would not need to know the slope or anything about the slide's shape. A second **Key Idea** is that mechanical energy is conserved in an isolated system when only conservative forces cause energy transfers. Let's check.

Forces: Two forces act on the child. The *gravitational force,* a conservative force, does work on her. The *normal force* on her from the slide does no work, because its direction at any point during the descent is always perpendicular to the direction in which the child moves.

System: Because the only force doing work on the child is the gravitational force, we choose the child–Earth system as our system, which we take to be isolated.

Thus, we have only a conservative force doing work in an isolated system, so we *can* use the principle of conservation of mechanical energy. Let the mechanical energy be $E_{mec,t}$ when the child is at the top of the slide and $E_{mec,b}$ when she is at the bottom. Then the conservation principle tells us

$$E_{mec,b} = E_{mec,t}.$$

Expanding this to show both kinds of mechanical energy, we have

$$K_b + U_b = K_t + U_t,$$

or $$\tfrac{1}{2}mv_b^2 + mgy_b = \tfrac{1}{2}mv_t^2 + mgy_t.$$

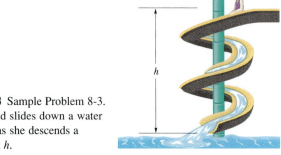

Fig. 8-8 Sample Problem 8-3. A child slides down a water slide as she descends a height h.

Dividing by m and rearranging yield
$$v_b^2 = v_t^2 + 2g(y_t - y_b).$$
Putting $v_t = 0$ and $y_t - y_b = h$ leads to
$$v_b = \sqrt{2gh} = \sqrt{(2)(9.8 \text{ m/s}^2)(8.5 \text{ m})}$$
$$= 13 \text{ m/s.} \qquad \text{(Answer)}$$

This is the same speed that the child would reach if she fell 8.5 m. On an actual slide, some frictional forces would act and the child would not be moving quite so fast.

Although this problem is hard to solve directly with Newton's laws, using conservation of mechanical energy makes the solution much easier. However, if we were asked to find the time taken for the child to reach the bottom of the slide, energy methods would be of no use; we would need to know the shape of the slide, and we would have a difficult problem.

Now that we have worked this sample problem, return to the puck example in the first paragraph of Chapter 7. See if you can show that the puck's speed at the end of the track is 5.0 m/s.

Sample Problem 8-4

A 61.0 kg bungee-cord jumper is on a bridge 45.0 m above a river. The elastic bungee cord has a relaxed length of $L = 25.0$ m. Assume that the cord obeys Hooke's law, with a spring constant of 160 N/m. If the jumper stops before reaching the water, what is the height h of her feet above the water at her lowest point?

SOLUTION: Figure 8-9 shows the jumper at the lowest point, with her feet at height h and with the cord stretched by distance d from its relaxed length. If we knew d, we could find h. One **Key Idea** is that perhaps we can solve for d by applying the principle of conserva-

tion of mechanical energy, between her initial point (on the bridge) and her lowest point. In that case, a second **Key Idea** is that mechanical energy is conserved in an isolated system when only conservative forces cause energy transfers. Let's check.

Forces: The gravitational force does work on the jumper throughout her fall. Once the bungee cord becomes taut, the spring-like force from it does work on her, transferring energy to elastic potential energy of the cord. The force from the cord also pulls on the bridge, which is attached to Earth. The gravitational force and the spring-like force are conservative.

Fig. 8-9 Sample Problem 8-4. A bungee-cord jumper at the lowest point of the jump.

change in the elastic potential energy of the bungee cord, and ΔU_g is the change in the jumper's gravitational potential energy. All these changes must be computed between her initial point and her lowest point. Because she is stationary (at least momentarily) both initially and at her lowest point, $\Delta K = 0$. From Fig. 8-9, we see that the change Δy in her height is $-(L + d)$, so we have

$$\Delta U_g = mg\,\Delta y = -mg(L + d),$$

where m is her mass. Also from Fig. 8-9, we see that the bungee cord is stretched by distance d. Thus, we also have

$$\Delta U_e = \tfrac{1}{2}kd^2.$$

Inserting these expressions and the given data into Eq. 8-19, we obtain

$$0 + \tfrac{1}{2}kd^2 - mg(L + d) = 0$$

or

$$\tfrac{1}{2}kd^2 - mgL - mgd = 0$$

and then

$$\tfrac{1}{2}(160\ \text{N/m})d^2 - (61.0\ \text{kg})(9.8\ \text{m/s}^2)(25.0\ \text{m})$$
$$- (61.0\ \text{kg})(9.8\ \text{m/s}^2)d = 0.$$

Solving this quadratic equation yields

$$d = 17.9\ \text{m}.$$

The jumper's feet are then a distance of $(L + d) = 42.9$ m below their initial height. Thus,

$$h = 45.0\ \text{m} - 42.9\ \text{m} = 2.1\ \text{m}. \qquad \text{(Answer)}$$

System: The jumper–Earth–cord system includes all these forces and energy transfers, and we can take it to be isolated. Thus, we *can* apply the principle of conservation of mechanical energy to the system. From Eq. 8-18, we can write the principle as

$$\Delta K + \Delta U_e + \Delta U_g = 0, \qquad (8\text{-}19)$$

where ΔK is the change in the jumper's kinetic energy, ΔU_e is the

---

PROBLEM-SOLVING TACTICS

Tactic 2: *Conservation of Mechanical Energy*
Asking the following questions will help you to solve problems involving the conservation of mechanical energy.

For what system is mechanical energy conserved? You should be able to separate your system from its environment. Imagine drawing a closed surface such that whatever is inside is your system and whatever is outside is the environment of that system. In Sample Problem 8-3 the system is the *child + Earth* and in Sample Problem 8-4, it is the *jumper + Earth + cord*.

Is friction or drag present? If friction or drag is present, mechanical energy is not conserved.

Is your system isolated? Conservation of mechanical energy applies only to isolated systems. That means that no *external forces* (forces exerted by objects outside the system) should do work on the objects in the system.

What are the initial and final states of your system? The system changes from some initial state (or configuration) to some final state. You apply the principle of conservation of mechanical energy by saying that E_{mec} has the same value in both these states. Be very clear about what these two states are.

8-5 Reading a Potential Energy Curve

Once again we consider a particle that is part of a system in which a conservative force acts. This time suppose that the particle is constrained to move along an x axis while the conservative force does work on it. We can learn a lot about the motion of the particle from a plot of the system's potential energy $U(x)$. However, before we discuss such plots, we need one more relationship.

Finding the Force Analytically

Equation 8-6 tells us how to find the change ΔU in potential energy between two points in a one-dimensional situation if we know the force $F(x)$. Now we want to go the other way; that is, we know the potential energy function $U(x)$ and want to find the force.

For one-dimensional motion, the work W done by a force that acts on a particle as the particle moves through a distance Δx is $F(x)\,\Delta x$. We can then write Eq. 8-1 as

$$\Delta U(x) = -W = -F(x)\,\Delta x.$$

Solving for $F(x)$ and passing to the differential limit yield

$$F(x) = -\frac{dU(x)}{dx} \qquad \text{(one-dimensional motion),} \qquad (8\text{-}20)$$

which is the relation we sought.

We can check this result by putting $U(x) = \frac{1}{2}kx^2$, which is the elastic potential energy function for a spring force. Equation 8-20 then yields, as expected, $F(x) = -kx$, which is Hooke's law. Similarly, we can substitute $U(x) = mgx$, which is the gravitational potential energy function for a particle–Earth system, with a particle of mass m at height x above Earth's surface. Equation 8-20 then yields $F = -mg$, which is the gravitational force on the particle.

The Potential Energy Curve

Figure 8-10a is a plot of a potential energy function $U(x)$ for a system in which a particle is in one-dimensional motion while a conservative force $F(x)$ does work on it. We can easily find $F(x)$ by (graphically) taking the slope of the $U(x)$ curve at various points. (Equation 8-20 tells us that $F(x)$ is negative the slope of the $U(x)$ curve.) Figure 8-10b is a plot of $F(x)$ found in this way.

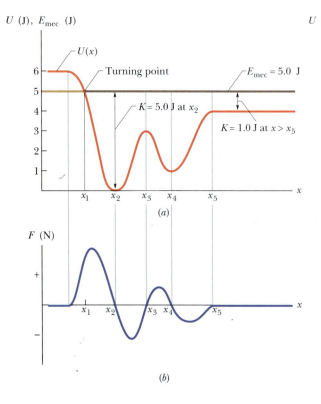

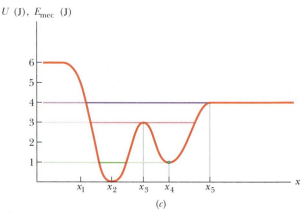

Fig. 8-10 (a) A plot of $U(x)$, the potential energy function of a system containing a particle confined to move along the x axis. There is no friction, so mechanical energy is conserved. (b) A plot of the force $F(x)$ acting on the particle, derived from the potential energy plot by taking its slope at various points. (c) The $U(x)$ plot of (a) with three different possible values of E_{mec} shown.

Turning Points

In the absence of a nonconservative force, the mechanical energy E of the system has a constant value given by

$$U(x) + K(x) = E_{mec}. \qquad (8\text{-}21)$$

Here $K(x)$ is the *kinetic energy function* of the particle (this $K(x)$ gives the kinetic energy as a function of the particle's location x). We may rewrite Eq. 8-21 as

$$K(x) = E_{mec} - U(x). \qquad (8\text{-}22)$$

Suppose that E_{mec} (which has a constant value, remember) happens to be 5.0 J. It would be represented in Fig. 8-10a by a horizontal line that runs through the value 5.0 J on the energy axis. (It is, in fact, shown there.)

Equation 8-22 tells us how to determine the kinetic energy K for any location x of the particle: On the $U(x)$ curve, find U for that location x and then subtract U from E_{mec}. For example, if the particle is at any point to the right of x_5, then $K = 1.0$ J. The value of K is greatest (5.0 J) when the particle is at x_2, and least (0 J) when the particle is at x_1.

Since K can never be negative (because v^2 is always positive), the particle can never move to the left of x_1, where $E_{mec} - U$ is negative. Instead, as the particle moves toward x_1 from x_2, K decreases (the particle slows) until $K = 0$ at x_1 (the particle stops there).

Note that when the particle reaches x_1, the force on the particle, given by Eq. 8-20, is positive (because the slope dU/dx is negative). This means that the particle does not remain at x_1 but instead begins to move to the right, opposite its earlier motion. Hence x_1 is a **turning point,** a place where $K = 0$ (because $U = E$) and the particle changes direction. There is no turning point (where $K = 0$) on the right side of the graph. When the particle heads to the right, it will continue indefinitely.

Equilibrium Points

Figure 8-10c shows three different values for E_{mec} superposed on the plot of the same potential energy function $U(x)$. Let us see how they would change the situation. If $E_{mec} = 4.0$ J (purple line), the turning point shifts from x_1 to a point between x_1 and x_2. Also, at any point to the right of x_5, the system's mechanical energy is equal to its potential energy; thus, the particle has no kinetic energy and (by Eq. 8-20) no force acts on it, and so it must be stationary. A particle at such a position is said to be in **neutral equilibrium.** (A marble placed on a horizontal tabletop is in that state.)

If $E_{mec} = 3.0$ J (pink line), there are two turning points: One is between x_1 and x_2, and the other is between x_4 and x_5. In addition, x_3 is a point at which $K = 0$. If the particle is located exactly there, the force on it is also zero, and the particle remains stationary. However, if it is displaced even slightly in either direction, a nonzero force pushes it farther in the same direction, and the particle continues to move. A particle at such a position is said to be in **unstable equilibrium.** (A marble balanced on top of a bowling ball is an example.)

Next consider the particle's behavior if $E_{mec} = 1.0$ J (green line). If we place it at x_4, it is stuck there. It cannot move left or right on its own because to do so would require a negative kinetic energy. If we push it slightly left or right, a restoring force appears that moves it back to x_4. A particle at such a position is said to be in **stable equilibrium.** (A marble placed at the bottom of a hemispherical bowl is an example.) If we place the particle in the cuplike *potential well* centered at x_2, it is between two turning points. It can still move somewhat, but only partway to x_1 or x_3.

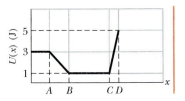

✔CHECKPOINT 4:   The figure gives the potential energy function $U(x)$ for a system in which a particle is in one-dimensional motion. (a) Rank regions AB, BC, and CD according to the magnitude of the force on the particle, greatest first. (b) What is the direction of the force when the particle is in region AB?

8-6   Work Done on a System by an External Force

In Chapter 7, we defined work as being energy transferred to or from an object by means of a force acting on the object. We can now extend that definition to an external force acting on a system of objects.

> Work is energy transferred to or from a system by means of an external force acting on that system.

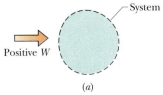

Figure 8-11a represents positive work (a transfer of energy *to* a system), and Fig. 8-11b represents negative work (a transfer of energy *from* a system). When more than one force acts on a system, their *net work* is the energy transferred to or from the system.

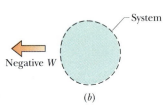

These transfers are like transfers of money to and from a bank account. If a system consists of a single particle or particle-like object, as in Chapter 7, the work done on the system by a force can change only the kinetic energy of the system. The energy statement for such transfers is the work–kinetic energy theorem of Eq. 7-10 ($\Delta K = W$); that is, a single particle has only one energy account, called kinetic energy. External forces can transfer energy into or out of that account. If a system is more complicated, however, an external force can change other forms of energy (such as potential energy); that is, a more complicated system can have multiple energy accounts.

Fig. 8-11 (a) Positive work W on an arbitrary system means a transfer of energy to the system. (b) Negative work W means a transfer of energy from the system.

Let us find energy statements for such systems by examining two basic situations, one that does not involve friction and one that does.

No Friction Involved

To compete in a bowling-ball hurling contest, you first squat and cup your hands under the ball on the floor. Then you rapidly straighten up while also pulling your hands up sharply, launching the ball upward at about face level. During your upward motion, your applied force on the ball obviously does work; that is, it is an external force that transfers energy, but to what system?

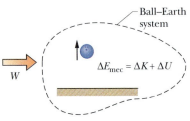

To answer, we check to see which energies change. There is a change ΔK in the ball's kinetic energy and, because the ball and Earth become more separated, there is a change ΔU in the gravitational potential energy of the ball–Earth system. To include both changes, we need to consider the ball–Earth system. Then your force is an external force doing work on that system, and the work is

$$W = \Delta K + \Delta U, \tag{8-23}$$

Fig. 8-12 Positive work W is done on a system of a bowling ball and Earth, causing a change ΔE_{mec} in the mechanical energy of the system, a change ΔK in the ball's kinetic energy, and a change ΔU in the system's gravitational potential energy.

or $$W = \Delta E_{mec} \qquad \text{(work done on system, no friction involved),} \tag{8-24}$$

where ΔE_{mec} is the change in the mechanical energy of the system. These two equations, which are represented in Fig. 8-12, are equivalent energy statements for work done on a system by an external force when friction is not involved.

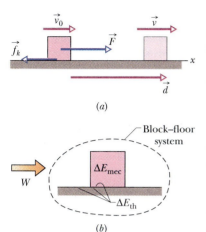

(a)

(b)

Fig. 8-13 (a) A block is pulled across a floor by force $\vec{F}$ while a kinetic frictional force opposes the motion. The block has velocity $\vec{v}_0$ at the start of a displacement $\vec{d}$ and a velocity $\vec{v}$ at the end of the displacement. (b) Positive work W is done on the block–floor system by force $\vec{F}$, resulting in a change ΔE_{mec} in the block's mechanical energy and a change ΔE_{th} in the thermal energy of the block and floor.

Friction Involved

We next consider the example in Fig. 8-13a. A constant horizontal force $\vec{F}$ pulls a block along an x axis and through a displacement of magnitude d, increasing the block's velocity from $\vec{v}_0$ to $\vec{v}$. During the motion, a constant kinetic frictional force $\vec{f}_k$ from the floor acts on the block. Let us first choose the block as our system and apply Newton's second law to it. We can write that law for components along the x axis ($F_{\text{net},x} = ma_x$) as

$$F - f_k = ma. \tag{8-25}$$

Because the forces are constant, the acceleration $\vec{a}$ is also constant. Thus, we can use Eq. 2-16 to write

$$v^2 = v_0^2 + 2ad.$$

Solving this equation for a, substituting the result into Eq. 8-25, and rearranging then give us

$$Fd = \tfrac{1}{2}mv^2 - \tfrac{1}{2}mv_0^2 + f_k d \tag{8-26}$$

or, because $\tfrac{1}{2}mv^2 - \tfrac{1}{2}mv_0^2 = \Delta K$ for the block,

$$Fd = \Delta K + f_k d. \tag{8-27}$$

In a more general situation (say, one in which the block is moving up a ramp), there can be a change in potential energy. To include such a possible change, we generalize Eq. 8-27 by writing

$$Fd = \Delta E_{\text{mec}} + f_k d. \tag{8-28}$$

By experiment we find that the block and the portion of the floor along which it slides become warmer as the block slides. As we shall discuss in Chapter 19, the temperature of an object is related to the object's thermal energy E_{th} (the energy associated with the random motion of the atoms and molecules in the object). Here, the thermal energy of the block and floor increase because (1) there is friction between them and (2) there is sliding. Recall that friction is due to the cold-welding between two surfaces. As the block slides over the floor, the sliding causes repeated tearing and reforming of the welds between the block and the floor, which makes the block and floor warmer. Thus, the sliding increases their thermal energy E_{th}.

Through experiment, we find that the increase ΔE_{th} in thermal energy is equal to the product of the magnitudes f_k and d:

$$\Delta E_{\text{th}} = f_k d \qquad \text{(increase in thermal energy by sliding).} \tag{8-29}$$

Thus, we can rewrite Eq. 8-28 as

$$Fd = \Delta E_{\text{mec}} + \Delta E_{\text{th}}. \tag{8-30}$$

Fd is the work W done by the external force $\vec{F}$ (the energy transferred by the force), but on which system is the work done (where are the energy transfers made)? To answer, we check to see which energies change. The block's mechanical energy changes, and the thermal energies of the block and floor also change. Therefore, the work done by force $\vec{F}$ is done on the block–floor system. That work is

$$W = \Delta E_{\text{mec}} + \Delta E_{\text{th}} \qquad \text{(work done on system, friction involved).} \tag{8-31}$$

This equation, which is represented in Fig. 8-13b, is the energy statement for the work done on a system by an external force when friction is involved.

✔CHECKPOINT 5:   In three trials, a block is pushed by a horizontal applied force across a floor that is not frictionless, as in Fig. 8-13a. The magnitudes F of the applied force and the result of the pushing on the block's speed are given in the table. In all three trials, the block is pushed through the same distance d. Rank the three trials according to the change in the thermal energy of the block and floor that occurs in that distance d, greatest first.

Trial	F	Result on Block's Speed
a	5.0 N	decreases
b	7.0 N	remains constant
c	8.0 N	increases

Sample Problem 8-5

The giant stone statues of Easter Island were most likely moved by the prehistoric islanders by cradling each statue in a wooden sled and then pulling the sled over a "runway" consisting of almost identical logs acting as rollers. In a modern reenactment of this technique, 25 men were able to move a 9000 kg Easter Island-type statue 45 m over level ground in 2 min.

(a) Estimate the work the net force $\vec{F}$ from the men did during the 45 m displacement of the statue, and determine the system on which that force did the work.

SOLUTION: One Key Idea is that we can calculate the work done with Eq. 7-7 ($W = Fd \cos \phi$). Here d is the distance 45 m, F is the magnitude of the net force on the statue from the 25 men, and $\phi = 0°$. Let us estimate that each man pulled with a force magnitude equal to twice his weight, which we take to be the same value mg for all the men. Thus, the magnitude of the net force was $F = (25)(2mg) = 50mg$. Estimating a man's mass as 80 kg, we can then write Eq. 7-7 as

$$W = Fd \cos \phi = 50mgd \cos \phi$$
$$= (50)(80 \text{ kg})(9.8 \text{ m/s}^2)(45 \text{ m}) \cos 0°$$
$$= 1.8 \times 10^6 \text{ J} \approx 2 \text{ MJ}. \qquad \text{(Answer)}$$

The Key Idea in determining the system on which the work is done is to see which energies change. Because the statue moved, there was certainly a change ΔK in its kinetic energy during the motion. We can easily guess that there must have been considerable kinetic friction between the sled, logs, and ground, resulting in a change ΔE_{th} in their thermal energies. Thus, the system on which the work was done consisted of the statue, sled, logs, and ground.

(b) What was the increase ΔE_{th} in the thermal energy of the system during the 45 m displacement?

SOLUTION: The Key Idea here is that we can relate ΔE_{th} to the work W done by $\vec{F}$ with the energy statement of Eq. 8-31 for a system that involves friction:

$$W = \Delta E_{mec} + \Delta E_{th}.$$

We know the value of W from (a). The change ΔE_{mec} in the crate's mechanical energy was zero because the statue was stationary at the beginning and the end of the move and did not change in elevation. Thus, we find

$$\Delta E_{th} = W = 1.8 \times 10^6 \text{ J} \approx 2 \text{ MJ}. \qquad \text{(Answer)}$$

(c) Estimate the work that would have been done by the 25 men if they had moved the statue 10 km across level ground on Easter Island. Also estimate the total change ΔE_{th} that would have occurred in the statue–sled–logs–ground system.

SOLUTION: The Key Ideas here are the same as in (a) and (b). Thus we calculate W as in (a), but with 1×10^4 m now substituted for d. Also, we again equate ΔE_{th} to W. We get

$$W = \Delta E_{th} = 3.9 \times 10^8 \text{ J} \approx 400 \text{ MJ}. \qquad \text{(Answer)}$$

This would have been a staggering amount of energy for the men to have transferred during the movement of a statue. Still, the 25 men *could* have moved the statue 10 km, and the required energy does not suggest some mysterious source.

Sample Problem 8-6

A food shipper pushes a wood crate of cabbage heads (total mass $m = 14$ kg) across a concrete floor with a constant horizontal force $\vec{F}$ of magnitude 40 N. In a straight-line displacement of magnitude $d = 0.50$ m, the speed of the crate decreases from $v_0 = 0.60$ m/s to $v = 0.20$ m/s.

(a) How much work is done by force $\vec{F}$, and on what system does it do the work?

SOLUTION: One Key Idea is that Eq. 7-7 holds here: The work W done by $\vec{F}$ can be calculated as

$$W = Fd \cos \phi = (40 \text{ N})(0.50 \text{ m}) \cos 0°$$
$$= 20 \text{ J}. \qquad \text{(Answer)}$$

The Key Idea in determining the system on which the work is done is to see which energies change. Because the crate's speed changes, there is certainly a change ΔK in the crate's kinetic energy. Is there friction between the floor and the crate, and thus a change in thermal energy? Note that $\vec{F}$ and the crate's velocity have the same direction. Thus, a Key Idea here is that if there is no friction, then $\vec{F}$ should be accelerating the crate to a *greater* speed. However, the crate is *slowing*, so there must be friction and a change ΔE_{th} in thermal energy of the crate and the floor. Therefore, the system on which the work is done is the crate–floor system, because both energy changes occur in that system.

(b) What is the increase ΔE_{th} in the thermal energy of the crate and floor?

SOLUTION: The Key Idea here is that we can relate ΔE_{th} to the work W done by $\vec{F}$ with the energy statement of Eq. 8-31 for a system that involves friction:

$$W = \Delta E_{mec} + \Delta E_{th}. \qquad (8\text{-}32)$$

We know the value of W from (a). The change ΔE_{mec} in the crate's mechanical energy is just the change in its kinetic energy because no potential energy changes occur, so we have

$$\Delta E_{mec} = \Delta K = \tfrac{1}{2}mv^2 - \tfrac{1}{2}mv_0^2.$$

Substituting this into Eq. 8-32 and solving for ΔE_{th}, we find

$$\Delta E_{th} = W - (\tfrac{1}{2}mv^2 - \tfrac{1}{2}mv_0^2) = W - \tfrac{1}{2}m(v^2 - v_0^2)$$

$$= 20 \text{ J} - \tfrac{1}{2}(14 \text{ kg})[(0.20 \text{ m/s})^2 - (0.60 \text{ m/s})^2]$$

$$= 22.2 \text{ J} \approx 22 \text{ J}. \qquad \text{(Answer)}$$

8-7 Conservation of Energy

We now have discussed several situations in which energy is transferred to or from objects and systems, much like money is transferred between accounts. In each situation we assume that the energy that was involved could always be accounted for; that is, energy could not magically appear or disappear. In more formal language, we assumed (correctly) that energy obeys a law called the **law of conservation of energy,** which is concerned with the **total energy** E of a system. That total is the sum of the system's mechanical energy, thermal energy, and any form of *internal energy* in addition to thermal energy. (We have not yet discussed other forms of internal energy.) The law states that

> ▶ The total energy E of a system can change only by amounts of energy that are transferred to or from the system.

The only type of energy transfer that we have considered is work W done on a system. Thus, for us at this point, this law states that

$$W = \Delta E = \Delta E_{mec} + \Delta E_{th} + \Delta E_{int}, \qquad (8\text{-}33)$$

where ΔE_{mec} is any change in the mechanical energy of the system, ΔE_{th} is any change in the thermal energy of the system, and ΔE_{int} is any change in any other form of internal energy of the system. Included in ΔE_{mec} are changes ΔK in kinetic energy and changes ΔU in potential energy (elastic, gravitational, or any other form we might find).

This law of conservation of energy is *not* something we have derived from basic physics principles. Rather, it is a law based on countless experiments. Scientists and engineers have never found an exception to it.

Isolated System

If a system is isolated from its environment, then there can be no energy transfers to or from it. For that case, the law of conservation of energy states:

> ▶ The total energy E of an isolated system cannot change.

Many energy transfers may be going on *within* an isolated system, between, say, kinetic energy and a potential energy or kinetic energy and thermal energy. However, the total of all the forms of energy in the system cannot change.

We can use the rock climber in Fig. 8-14 as an example, approximating her, her gear, and Earth as an isolated system. As she rappels down the rock face, changing the configuration of the system, she needs to control the transfer of energy from the gravitational potential energy of the system. (That energy cannot just disappear.) Some of it is transferred to her kinetic energy. However, she obviously does not want very much transferred to that form or she will be moving too quickly, so she has wrapped the rope around metal rings to produce friction between the rope and the rings as she moves down. The sliding of the rings on the rope then transfers the gravitational potential energy of the system to thermal energy of the rings and rope in a way that she can control. The total energy of the climber–gear–Earth system (the total of its gravitational potential energy, kinetic energy, and thermal energy) does not change during her descent.

For an isolated system, the law of conservation of energy can be written in two ways. First, by setting $W = 0$ in Eq. 8-33, we get

Fig. 8-14 To descend, the rock climber must transfer energy from the gravitational potential energy of a system consisting of her, her gear, and Earth. She has wrapped the rope around metal rings so that the rope rubs against the rings. This allows most of the transferred energy to go to the thermal energy of the rope and rings rather than to her kinetic energy.

$$\Delta E_{mec} + \Delta E_{th} + \Delta E_{int} = 0 \qquad \text{(isolated system).} \qquad (8\text{-}34)$$

We can also let $\Delta E_{mec} = E_{mec,2} - E_{mec,1}$, where the subscripts 1 and 2 refer to two different instants, say before and after a certain process has occurred. Then Eq. 8-34 becomes

$$E_{mec,2} = E_{mec,1} - \Delta E_{th} - \Delta E_{int}. \qquad (8\text{-}35)$$

Equation 8-35 tells us:

> In an isolated system, we can relate the total energy at one instant to the total energy at another instant *without considering the energies at intermediate times.*

This fact can be a very powerful tool in solving problems about isolated systems when you need to relate energies of a system before and after a certain process occurs in the system.

In Section 8-4, we discussed a special situation for isolated systems—namely, the situation in which nonconservative forces (such as a kinetic frictional force) do not act within them. In that special situation, ΔE_{th} and ΔE_{int} are both zero, and so Eq. 8-35 reduces to Eq. 8-18. In other words, the mechanical energy of an isolated system is conserved when nonconservative forces do not act within it.

Power

Now that you have seen how energy can be transferred from one form to another, we can expand the definition of power given in Section 7-7. There it is the rate at which work is done by a force. In a more general sense, power P is the rate at which energy is transferred by a force from one form to another. If an amount of energy ΔE is transferred in an amount of time Δt, the **average power** due to the force is

$$P_{avg} = \frac{\Delta E}{\Delta t}. \qquad (8\text{-}36)$$

Similarly, the **instantaneous power** due to the force is

$$P = \frac{dE}{dt}. \qquad (8\text{-}37)$$

Sample Problem 8-7

In Fig. 8-15, a 2.0 kg package of tamale slides along a floor with speed $v_1 = 4.0$ m/s. It then runs into and compresses a spring, until the package momentarily stops. Its path to the initially relaxed spring is frictionless, but as it compresses the spring, a kinetic frictional force from the floor, of magnitude 15 N, acts on it. The spring constant is 10,000 N/m. By what distance d is the spring compressed when the package stops?

SOLUTION: A starting Key Idea is to examine all the forces acting on the package, and then to determine whether we have an isolated system or a system on which an external force is doing work.

Forces: The normal force on the package from the floor does no work on the package, because its direction is always perpendicular to that of the package's displacement. For the same reason, the gravitational force on the package does no work. As the spring is compressed, however, a spring force does work on the package, transferring energy to elastic potential energy of the spring. The spring force also pushes against a rigid wall. Because there is friction between the package and the floor, the sliding of the package across the floor increases their thermal energies.

System: The package–spring–floor–wall system includes all these forces and energy transfers in one isolated system. Therefore, a second Key Idea is that, because the system is isolated, its total energy cannot change. We can then apply the law of conservation of energy in the form of Eq. 8-35 to the system:

$$E_{mec,2} = E_{mec,1} - \Delta E_{th}. \qquad (8\text{-}38)$$

Let subscript 1 correspond to the initial state of the sliding package, and subscript 2 correspond to the state in which the package is momentarily stopped and the spring is compressed by dis-

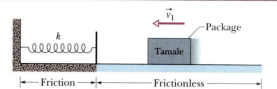

Fig. 8-15 Sample Problem 8-7. A package slides across a frictionless floor with velocity $\vec{v}_1$ toward a spring of spring constant k. When the package reaches the spring, a frictional force from the floor acts on it.

tance d. For both states the mechanical energy of the system is the sum of the package's kinetic energy ($K = \frac{1}{2}mv^2$) and the spring's potential energy ($U = \frac{1}{2}kx^2$). For state 1, $U = 0$ (because the spring is not compressed), and the package's speed is v_1. Thus, we have

$$E_{mec,1} = K_1 + U_1 = \tfrac{1}{2}mv_1^2 + 0.$$

For state 2, $K = 0$ (because the package is stopped), and the compression distance is d. Therefore, we have

$$E_{mec,2} = K_2 + U_2 = 0 + \tfrac{1}{2}kd^2.$$

Finally, by Eq. 8-29, we can substitute $f_k d$ for the change ΔE_{th} in the thermal energy of the package and the floor. We can now rewrite Eq. 8-38 as

$$\tfrac{1}{2}kd^2 = \tfrac{1}{2}mv_1^2 - f_k d.$$

Rearranging and substituting known data give us

$$5000d^2 + 15d - 16 = 0.$$

Solving this quadratic equation yields

$$d = 0.055 \text{ m} = 5.5 \text{ cm}. \qquad \text{(Answer)}$$

Sample Problem 8-8

In Fig. 8-16, a circus beagle of mass $m = 6.0$ kg runs onto the left end of a curved ramp with speed $v_0 = 7.8$ m/s at height $y_0 = 8.5$ m above the floor. It then slides to the right and comes to a momentary stop when it reaches a height $y = 11.1$ m from the floor. The ramp is not frictionless. What is the increase ΔE_{th} in the thermal energy of the beagle and ramp because of the sliding?

SOLUTION: A Key Idea to get us started is to examine all the forces on the beagle, and then see if we have an isolated system or a system on which an external force is doing work.

Forces: The normal force on the beagle from the ramp does no work on the beagle, because its direction is always perpendicular to that of the beagle's displacement. The gravitational force on the beagle does do work as the beagle's elevation changes. Because there is friction between the beagle and the ramp, the sliding increases their thermal energy.

System: The beagle–ramp–Earth system includes all these forces and energy transfers in one isolated system. Then a second Key Idea is that, because the system is isolated, its total energy cannot change. We can apply the law of conservation of energy in the form of Eq. 8-34 to this system:

$$\Delta E_{mec} + \Delta E_{th} = 0, \qquad (8\text{-}39)$$

Fig. 8-16 Sample Problem 8-8. A beagle slides along a curved ramp, starting with speed v_0 at height y_0, and reaching a height y at which it momentarily stops.

where the energy changes occur between the initial state and the state when the beagle stops momentarily. Also, the change ΔE_{mec} is the sum of the change ΔK in the kinetic energy of the beagle and the change ΔU in the gravitational potential energy of the system, where

$$\Delta K = 0 - \tfrac{1}{2}mv_0^2$$

and

$$\Delta U = mgy - mgy_0.$$

Substituting these expressions into Eq. 8-39 and solving for ΔE_{th} yield

$$\begin{aligned}\Delta E_{th} &= \tfrac{1}{2}mv_0^2 - mg(y - y_0) \\ &= \tfrac{1}{2}(6.0 \text{ kg})(7.8 \text{ m/s})^2 \\ &\quad - (6.0 \text{ kg})(9.8 \text{ m/s}^2)(11.1 \text{ m} - 8.5 \text{ m}) \\ &\approx 30 \text{ J.} \quad\quad\quad\quad\quad\quad \text{(Answer)}\end{aligned}$$

REVIEW & SUMMARY

Conservative Forces   A force is a **conservative force** if the net work it does on a particle moving around every closed path, from an initial point and then back to that point, is zero. Equivalently, it is conservative if the net work it does on a particle moving between two points does not depend on the path taken by the particle. The gravitational force and the spring force are conservative forces; the kinetic frictional force is a **nonconservative force.**

Potential Energy   A **potential energy** is energy that is associated with the configuration of a system in which a conservative force acts. When the conservative force does work W on a particle within the system, the change ΔU in the potential energy of the system is

$$\Delta U = -W. \quad\quad\quad\quad (8\text{-}1)$$

If the particle moves from point x_i to point x_f, the change in the potential energy of the system is

$$\Delta U = -\int_{x_i}^{x_f} F(x) \, dx. \quad\quad\quad\quad (8\text{-}6)$$

Gravitational Potential Energy   The potential energy associated with a system consisting of Earth and a nearby particle is **gravitational potential energy.** If the particle moves from height y_i to height y_f, the change in the gravitational potential energy of the particle–Earth system is

$$\Delta U = mg(y_f - y_i) = mg \, \Delta y. \quad\quad\quad\quad (8\text{-}7)$$

If the **reference position** of the particle is set as $y_i = 0$ and the corresponding gravitational potential energy of the system is set as $U_i = 0$, then the gravitational potential energy U when the particle is at any height y is

$$U(y) = mgy. \quad\quad\quad\quad (8\text{-}9)$$

Elastic Potential Energy   **Elastic potential energy** is the energy associated with the state of compression or extension of an elastic object. For a spring that exerts a spring force $F = -kx$ when its free end has displacement x, the elastic potential energy is

$$U(x) = \tfrac{1}{2}kx^2. \quad\quad\quad\quad (8\text{-}11)$$

The reference configuration has the spring at its relaxed length, at which $x = 0$ and $U = 0$.

Mechanical Energy   The **mechanical energy** E_{mec} of a system is the sum of its kinetic energy K and its potential energy U:

$$E_{mec} = K + U. \quad\quad\quad\quad (8\text{-}12)$$

An *isolated system* is one in which no *external force* causes energy changes. If only conservative forces do work within an isolated system, then the mechanical energy E_{mec} of the system cannot change. This **principle of conservation of mechanical energy** is written as

$$K_2 + U_2 = K_1 + U_1, \quad\quad\quad\quad (8\text{-}17)$$

in which the subscripts refer to different instants during an energy transfer process. This conservation principle can also be written as

$$\Delta E_{mec} = \Delta K + \Delta U = 0. \quad\quad\quad\quad (8\text{-}18)$$

Potential Energy Curves   If we know the **potential energy function** $U(x)$ for a system in which a one-dimensional force F acts on a particle, we can find the force as

$$F(x) = -\frac{dU(x)}{dx}. \quad\quad\quad\quad (8\text{-}20)$$

If $U(x)$ is given on a graph, then at any value of x, the force F is the negative of the slope of the curve there and the kinetic energy of the particle is given by

$$K(x) = E_{mec} - U(x), \quad\quad\quad\quad (8\text{-}22)$$

where E_{mec} is the mechanical energy of the system. A **turning point** is a point x at which the particle reverses its motion (there, $K = 0$). The particle is in **equilibrium** at points where the slope of the $U(x)$ curve is zero (there, $F(x) = 0$).

Work Done on a System by an External Force   Work W is energy transferred to or from a system by means of an external force acting on the system. When more than one force acts on a system, their *net work* is the transferred energy. When friction is not involved, the work done on the system and the change ΔE_{mec} in the mechanical energy of the system are equal:

$$W = \Delta E_{mec} = \Delta K + \Delta U. \quad\quad\quad\quad (8\text{-}24, 8\text{-}23)$$

When a kinetic frictional force acts within the system, then the thermal energy E_{th} of the system changes. (This energy is associated with the random motion of atoms and molecules in the system.) The work done on the system is then

$$W = \Delta E_{mec} + \Delta E_{th}. \quad\quad\quad\quad (8\text{-}31)$$

The change ΔE_{th} is related to the magnitude f_k of the frictional force and the magnitude d of the displacement caused by the external force by

$$\Delta E_{th} = f_k d. \quad\quad\quad\quad (8\text{-}29)$$

Conservation of Energy The total energy E of a system (the sum of its mechanical energy and its internal energies, including thermal energy) can change only by amounts of energy that are transferred to or from the system. This experimental fact is known as the **law of conservation of energy.** If work W is done on the system, then

$$W = \Delta E = \Delta E_{mec} + \Delta E_{th} + \Delta E_{int}. \qquad (8\text{-}33)$$

If the system is isolated ($W = 0$), this gives

$$\Delta E_{mec} + \Delta E_{th} + \Delta E_{int} = 0 \qquad (8\text{-}34)$$

and

$$E_{mec,2} = E_{mec,1} - \Delta E_{th} - \Delta E_{int}, \qquad (8\text{-}35)$$

where the subscripts 1 and 2 refer to two different instants.

Power The **power** due to a force is the *rate* at which that force transfers energy. If an amount of energy ΔE is transferred by a force in an amount of time Δt, the **average power** of the force is

$$P_{avg} = \frac{\Delta E}{\Delta t}. \qquad (8\text{-}36)$$

The **instantaneous power** due to a force is

$$P = \frac{dE}{dt}. \qquad (8\text{-}37)$$

QUESTIONS

1. Figure 8-17 shows one direct path and four indirect paths from point i to point f. Along the direct path and three of the indirect paths, only a conservative force F_c acts on a certain object. Along the fourth indirect path, both F_c and a nonconservative force F_{nc} act on the object. The change ΔE_{mec} in the object's mechanical energy (in joules) in going from i to f is indicated along each straight line segment of the indirect paths. (a) What is ΔE_{mec} in moving from i to f along the direct path? (b) What is ΔE_{mec} due to F_{nc} along the one path where it acts?

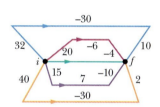

Fig. 8-17 Question 1.

2. A spring is initially stretched by 3.0 cm from its relaxed length. Here are four choices for changing the initial stretch: (a) to a stretch of 2.0 cm, (b) to a compression of 2.0 cm, (c) to a compression of 4.0 cm, and (d) to a stretch of 4.0 cm. Rank the choices according to the changes they make in the elastic potential energy of the spring, most positive first, most negative last.

3. A coconut is thrown from a cliff edge toward a wide, flat valley, with initial speed $v_0 = 8$ m/s. Rank the following choices for the launch direction according to (a) the initial kinetic energy of the coconut and (b) its kinetic energy just before hitting the valley bottom, greatest first: (1) $\vec{v}_0$ almost vertically upward, (2) $\vec{v}_0$ angled upward by 45°, (3) $\vec{v}_0$ horizontal, (4) $\vec{v}_0$ angled downward by 45°, and (5) $\vec{v}_0$ almost vertically downward.

4. In Fig. 8-18, a brave skater slides down three slopes of frictionless ice whose vertical heights d are identical. Rank the slopes according to (a) the work done on the skater by the gravitational force during the descent on each slope and (b) the change in her kinetic energy produced along the slope, greatest first.

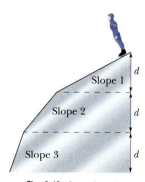

Fig. 8-18 Question 4.

5. In Fig. 8-19, a small, initially stationary block is released on a frictionless ramp at a height of 3.0 m. Hill heights along the ramp are as shown. The hills have identical circular tops (assume that the block does not fly off any hill). (a) Which hill is the first the block cannot cross? (b) What does it do after failing to cross that hill? On which hilltop is (c) the centripetal acceleration of the block greatest and (d) the normal force on the block least?

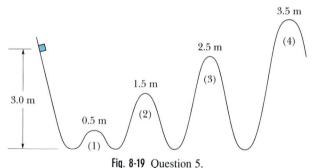

Fig. 8-19 Question 5.

6. Figure 8-20 gives the potential energy function of a particle. (a) Rank regions AB, BC, CD, and DE according to the magnitude of the force on the particle, greatest first. What value must the mechanical energy E_{mec} of the particle not exceed if the particle is to be (b) trapped in the potential well at the left, (c) trapped in the

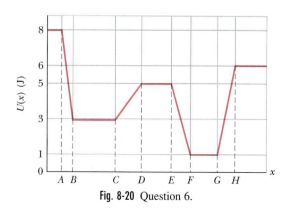

Fig. 8-20 Question 6.

potential well at the right, and (d) able to move between the two potential wells but not to the right of point *H*? For the situation of (d), in which of regions *BC*, *DE*, and *FG* will the particle have (e) the greatest kinetic energy and (f) the least speed?

7. In Fig. 8-21, a block is released from rest on a track with an initial gravitational potential energy U_i. The curved portions on the track are frictionless, but the horizontal portion, of length *L*, produces a frictional force *f* on the block. (a) How much energy is transferred to thermal energy if the block passes once through length *L*? How many times does the block pass through that length if the initial potential energy U_i is equal to (b) 0.50*fL*, (c) 1.25*fL*, and (d) 2.25*fL*? (e) For each of those three situations, does the block come to a stop at the center of the horizontal portion, to the left of center, or to the right of center? (Warm-up for Problem 63)

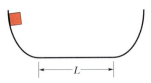

Fig. 8-21 Question 7.

8. In Fig. 8-22, a block slides along a track that descends through distance *h*. The track is frictionless except for the lower section. There the block slides to a stop in a certain distance *D* because of friction. (a) If we decrease *h*, will the block now slide to a stop in a distance that is greater than, less than, or equal to *D*? (b) If, instead, we increase the mass of the block, will the stopping distance now be greater than, less than, or equal to *D*?

Fig. 8-22 Question 8.

9. In Fig. 8-23, a block slides from *A* to *C* along a frictionless ramp, and then it passes through horizontal region *CD*, where a frictional force acts on it. Is the block's kinetic energy increasing, decreasing, or constant in (a) region *AB*, (b) region *BC*, and (c) region *CD*? (d) Is the block's mechanical energy increasing, decreasing, or constant in those regions?

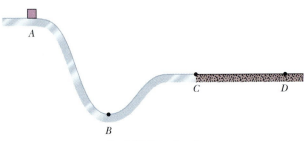

Fig. 8-23 Question 9.

EXERCISES & PROBLEMS

> **ssm** Solution is in the Student Solutions Manual.
> **www** Solution is available on the World Wide Web at:
>   http://www.wiley.com/college/hrw
> **ilw** Solution is available on the Interactive LearningWare.

SEC. 8-3   Determining Potential Energy Values

1E. What is the spring constant of a spring that stores 25 J of elastic potential energy when compressed by 7.5 cm from its relaxed length? **ssm**

2E. You drop a 2.00 kg textbook to a friend who stands on the ground 10.0 m below the textbook with outstretched hands 1.50 m above the ground (Fig. 8-24). (a) How much work W_g is done on the textbook by the gravitational force as it drops to your friend's hands? (b) What is the change ΔU in the gravitational potential energy of the textbook–Earth system during the drop? If the gravitational potential energy *U* of that system is taken to be zero at ground level, what is *U*

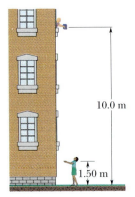

Fig. 8-24 Exercises 2 and 10.

when the textbook (c) is released and (d) reaches the hands? Now take *U* to be 100 J at ground level and again find (e) W_g, (f) ΔU, (g) *U* at the release point, and (h) *U* at the hands.

3E. In Fig. 8-25, a 2.00 g ice flake is released from the edge of a hemispherical bowl whose radius *r* is 22.0 cm. The flake–bowl contact is frictionless. (a) How much work is done on the flake by the gravitational force during the flake's descent to the bottom of the bowl? (b) What is the change in the potential energy of the flake–Earth system during that descent? (c) If that potential energy is taken to be zero at the bottom of the bowl, what is its value when the flake is released? (d) If, instead, the potential energy is taken to be zero at the release point, what is its value when the flake reaches the bottom of the bowl? (e) If the mass of the flake were doubled, would the magnitudes of the answers to (a) through (d) increase, decrease, or remain the same? **ssm**

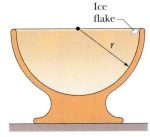

Fig. 8-25 Exercises 3 and 9.

4E. In Fig. 8-26, a frictionless roller coaster of mass *m* tops the first hill with speed v_0. How much work does the gravitational force do on it from that point to (a) point *A*, (b) point *B*, and (c) point *C*? If the gravitational potential energy of the coaster–Earth system is taken to be zero at point C, what is its value when the coaster is at (d) point B and (e) point A? (f) If mass m were doubled, would the change in the gravitational potential energy of the system between points A and B increase, decrease, or remain the same?

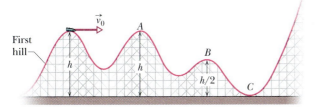

Fig. 8-26 Exercises 4 and 12.

5E. Figure 8-27 shows a ball with mass m attached to the end of a thin rod with length L and negligible mass. The other end of the rod is pivoted so that the ball can move in a vertical circle. The rod is held in the horizontal position as shown and then given enough of a downward push to cause the ball to swing down and around and just reach the vertically upward position, with zero speed there. How much work is done on the ball by the gravitational force from the initial point to (a) the lowest point, (b) the highest point, and (c) the point on the right at which the ball is level with the initial point? If the gravitational potential energy of the ball–Earth system is taken to be zero at the initial point, what is its value when the ball reaches (d) the low-est point, (e) the highest point, and (f) the point on the right that is level with the initial point? (g) Suppose the rod were pushed harder so that the ball passed through the highest point with a nonzero speed. Would the change in the gravitational potential en-ergy from the lowest point to the highest point then be greater, less, or the same? **ssm**

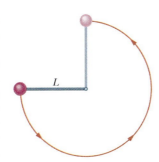

Fig. 8-27 Exercises 5 and 11.

6P. In Fig. 8-28, a small block of mass m can slide along the fric-tionless loop-the-loop. The block is released from rest at point P, at height $h = 5R$ above the bottom of the loop. How much work does the gravitational force do on the block as the block travels from point P to (a) point Q and (b) the top of the loop? If the gravita-tional potential energy of the block–Earth system is taken to be zero at the bottom of the loop,

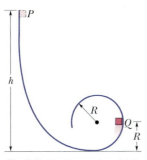

Fig. 8-28 Problems 6 and 20.

what is that potential energy when the block is (c) at point P, (d) at point Q, and (e) at the top of the loop? (f) If, instead of being released, the block is given some initial speed downward along the track, do the answers to (a) through (e) increase, decrease, or re-main the same?

7P. A 1.50 kg snowball is fired from a cliff 12.5 m high with an initial velocity of 14.0 m/s, directed 41.0° above the horizontal.

(a) How much work is done on the snowball by the gravitational force during its flight to the flat ground below the cliff? (b) What is the change in the gravitational potential energy of the snowball–Earth system during the flight? (c) If that gravitational potential energy is taken to be zero at the height of the cliff, what is its value when the snowball reaches the ground? **ssm**

8P. Figure 8-29 shows a thin rod, of length L and negligible mass, that can pivot about one end to rotate in a vertical circle. A heavy ball of mass m is attached to the other end. The rod is pulled aside through an angle θ and released. As the ball descends to its lowest point, (a) how much work does the gravitational force do on it and (b) what is the change in the gravitational potential energy of the ball–Earth system? (c) If the gravitational potential energy is taken to be zero at the lowest point, what is its value just as the ball is released? (d) Do the mag-nitudes of the answers to (a) through (c) increase, decrease, or remain the same if angle θ is in-creased?

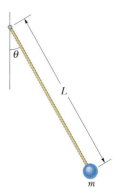

Fig. 8-29 Problems 8 and 14.

SEC. 8-4 Conservation of Mechanical Energy

9E. (a) In Exercise 3, what is the speed of the flake when it reaches the bottom of the bowl? (b) If we substituted a second flake with twice the mass, what would its speed be? (c) If, instead, we gave the flake an initial downward speed along the bowl, would the answer to (a) increase, decrease, or remain the same? **ssm www**

10E. (a) In Exercise 2, what is the speed of the textbook when it reaches the hands? (b) If we substituted a second textbook with twice the mass, what would its speed be? (c) If, instead, the text-book were thrown down, would the answer to (a) increase, de-crease, or remain the same?

11E. (a) In Exercise 5, what initial speed must be given the ball so that it reaches the vertically upward position with zero speed? What then is its speed at (b) the lowest point and (c) the point on the right at which the ball is level with the initial point? (d) If the ball's mass were doubled, would the answers to (a) through (c) increase, decrease, or remain the same? **ssm**

12E. In Exercise 4, what is the speed of the coaster at (a) point A, (b) point B, and (c) point C? (d) How high will it go on the last hill, which is too high for it to cross? (e) If we substitute a second coaster with twice the mass, what then are the answers to (a) through (d)?

13E. In Fig. 8-30, a runaway truck with failed brakes is moving downgrade at 130 km/h just before the driver steers the truck up a

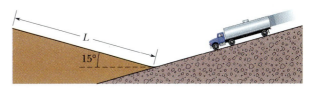

Fig. 8-30 Exercise 13.

frictionless emergency escape ramp with an inclination of 15°. The truck's mass is 5000 kg. (a) What minimum length L must the ramp have if the truck is to stop (momentarily) along it? (Assume the truck is a particle, and justify that assumption.) Does the minimum length L increase, decrease, or remain the same if (b) the truck's mass is decreased and (c) its speed is decreased? ssm

14P. (a) In Problem 8, what is the speed of the ball at the lowest point if $L = 2.00$ m, $\theta = 30.0°$, and $m = 5.00$ kg? (b) Does the speed increase, decrease, or remain the same if the mass is increased?

15P. (a) In Problem 7, using energy techniques rather than the techniques of Chapter 4, find the speed of the snowball as it reaches the ground below the cliff. What is that speed (b) if the launch angle is changed to 41.0° *below* the horizontal and (c) if the mass is changed to 2.50 kg? ssm

16P. Figure 8-31 shows an 8.00 kg stone at rest on a spring. The spring is compressed 10.0 cm by the stone. (a) What is the spring constant? (b) The stone is pushed down an additional 30.0 cm and released. What is the elastic potential energy of the compressed spring just before that release?
(c) What is the change in the gravitational potential energy of the stone–Earth system when the stone moves from the release point to its maximum height?
(d) What is that maximum height, measured from the release point?

Fig. 8-31 Problem 16.

17P. A 5.0 g marble is fired vertically upward using a spring gun. The spring must be compressed 8.0 cm if the marble is to just reach a target 20 m above the marble's position on the compressed spring. (a) What is the change ΔU_g in the gravitational potential energy of the marble–Earth system during the 20 m ascent? (b) What is the change ΔU_s in the elastic potential energy of the spring during its launch of the marble? (c) What is the spring constant of the spring? ssm www

18P. Figure 8-32 shows a pendulum of length L. Its bob (which effectively has all the mass) has speed v_0 when the cord makes an angle θ_0 with the vertical. (a) Derive an expression for the speed of the bob when it is in its lowest position. What is the least value that v_0 can have if the pendulum is to swing down and then up (b) to a horizontal position, and (c) to a vertical position with the cord remaining straight? (d) Do the answers to (b) and (c) increase, decrease, or remain the same if θ_0 is increased by a few degrees?

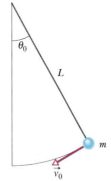

Fig. 8-32 Problem 18.

19P. A 2.00 kg block is placed against a spring on a frictionless 30.0° incline (Fig. 8-33). (The block is not attached to the spring.) The spring, whose spring constant is 19.6 N/cm, is compressed 20.0 cm and then released. (a) What is the elastic potential energy of the compressed spring? (b) What is the change in the gravitational potential energy of the block–Earth system as the block

moves from the release point to its highest point on the incline? (c) How far along the incline is the highest point from the release point? ilw

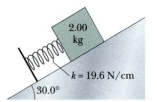

Fig. 8-33 Problem 19.

20P. In Problem 6, what are (a) the horizontal component and (b) the vertical component of the *net* force acting on the block at point Q? (c) At what height h should the block be released from rest so that it is on the verge of losing contact with the track at the top of the loop? (*On the verge of losing contact* means that the normal force on the block from the track has just then become zero.) (d) Graph the magnitude of the normal force on the block at the top of the loop versus initial height h, for the range $h = 0$ to $h = 6R$.

21P. In Fig. 8-34, a 12 kg block is released from rest on a 30° frictionless incline. Below the block is a spring that can be compressed 2.0 cm by a force of 270 N. The block momentarily stops when it compresses the spring by 5.5 cm. (a) How far does the block move down the incline from its rest position to this stopping point? (b) What is the speed of the block just as it touches the spring? ssm

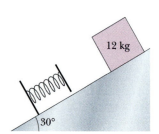

Fig. 8-34 Problem 21.

22P. At $t = 0$ a 1.0 kg ball is thrown from the top of a tall tower with velocity $\vec{v} = (18$ m/s$)\hat{i} + (24$ m/s$)\hat{j}$. What is the change in the potential energy of the ball–Earth system between $t = 0$ and $t = 6.0$ s?

23P. The string in Fig. 8-35 is $L = 120$ cm long, has a ball attached to one end, and is fixed at its other end. The distance d to the fixed peg at point P is 75.0 cm. When the initially stationary ball is released with the string horizontal as shown, it will swing along the dashed arc. What is its speed when it reaches (a) its lowest point and (b) its highest point after the string catches on the peg? ilw

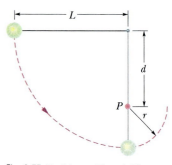

Fig. 8-35 Problems 23 and 29.

24P. A 60 kg skier starts from rest at a height of 20 m above the end of a ski-jump ramp as shown in Fig. 8-36. As the skier leaves the ramp, his velocity makes an angle of 28° with the horizontal.

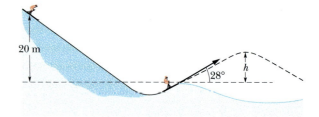

Fig. 8-36 Problem 24.

Neglect the effects of air resistance and assume the ramp is frictionless. (a) What is the maximum height h of his jump above the end of the ramp? (b) If he increased his weight by putting on a backpack, would h then be greater, less, or the same?

25P. A 2.0 kg block is dropped from a height of 40 cm onto a spring of spring constant $k =$ 1960 N/m (Fig. 8-37). Find the maximum distance the spring is compressed. ssm

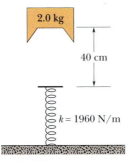

Fig. 8-37 Problem 25.

26P. Tarzan, who weighs 688 N, swings from a cliff at the end of a convenient vine that is 18 m long (Fig. 8-38). From the top of the cliff to the bottom of the swing, he descends by 3.2 m. The vine will break if the force on it exceeds 950 N. (a) Does the vine break? (b) If no, what is the greatest force on it during the swing? If yes, at what angle with the vertical does it break?

27P. Two children are playing a game in which they try to hit a small box on the floor with a marble fired from a spring-loaded gun that is mounted on a table. The target box is 2.20 m horizontally from the edge of the table; see Fig. 8-39. Bobby compresses the spring 1.10 cm, but the center of the marble falls 27.0 cm short of the center of the box. How far should Rhoda compress the spring to score a direct hit? Assume that neither the spring nor the ball encounters friction in the gun. ssm

Fig. 8-38 Problem 26.

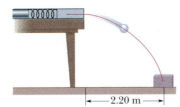

Fig. 8-39 Problem 27

28P. A 700 g block is released from rest at height h_0 above a vertical spring with spring constant $k = 400$ N/m and negligible mass. The block sticks to the spring and momentarily stops after compressing the spring 19.0 cm. How much work is done (a) by the block on the spring and (b) by the spring on the block? (c) What is the value of h_0? (d) If the block were released from height $2h_0$ above the spring, what would be the maximum compression of the spring?

29P. In Fig. 8-35 show that, if the ball is to swing completely around the fixed peg, then $d > 3L/5$. (*Hint:* The ball must still be moving at the top of its swing. Do you see why?) ssm www

30P. To make a pendulum, a 300 g ball is attached to one end of a string that has a length of 1.4 m and negligible mass. (The other end of the string is fixed.) The ball is pulled to one side until the string makes an angle of 30.0° with the vertical; then (with the string taut) the ball is released from rest. Find (a) the speed of the ball when the string makes an angle of 20.0° with the vertical and (b) the maximum speed of the ball. (c) What is the angle between the string and the vertical when the speed of the ball is one-third its maximum value?

31P. A rigid rod of length L and negligible mass has a ball with mass m attached to one end and its other end fixed, to form a pendulum. The pendulum is inverted, with the rod straight up, and then released. At the lowest point, what are (a) the ball's speed and (b) the tension in the rod? (c) The pendulum is next released at rest from a horizontal position. At what angle from the vertical does the tension in the rod equal the weight of the ball? ssm

32P. In Fig. 8-40, a spring with spring constant $k = 170$ N/m is at the top of a 37.0° frictionless incline. The lower end of the incline is 1.00 m from the end of the spring, which is at its relaxed length. A 2.00 kg canister is pushed against the spring until the spring is compressed 0.200 m and released from rest. (a) What

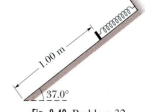

Fig. 8-40 Problem 32.

is the speed of the canister at the instant the spring returns to its relaxed length (which is when the canister loses contact with the spring)? (b) What is the speed of the canister when it reaches the lower end of the incline?

33P*. In Fig. 8-41, a chain is held on a frictionless table with one-fourth of its length hanging over the edge. If the chain has length L and mass m, how much work is required to pull the hanging part back onto the table? ssm

Fig. 8-41 Problem 33.

34P. A spring with spring constant $k = 400$ N/m is placed in a vertical orientation with its lower end supported by a horizontal surface. The upper end is depressed 25.0 cm, and a block with a weight of 40.0 N is placed (unattached) on the depressed spring. The system is then released from rest. Assume the gravitational potential energy U_g of the block is zero at the release point ($y = 0$) and calculate the gravitational potential energy, the elastic potential energy U_e, and the kinetic energy K of the block for y equal to (a) 0, (b) 5.00 cm, (c) 10.0 cm, (d) 15.0 cm, (e) 20.0 cm, (f) 25.0 cm, and (g) 30.0 cm. Also, (h) how far above its point of release does the block rise?

35P*. A boy is seated on the top of a hemispherical mound of ice (Fig. 8-42). He is given a very small push and starts sliding down the ice. Show that he leaves the ice at a point whose height is $2R/3$ if the ice is frictionless. (*Hint:* The normal force vanishes as he leaves the ice.) ssm

Fig. 8-42 Problem 35.

SEC. 8-5   Reading a Potential Energy Curve

36E. A conservative force $F(x)$ acts on a 2.0 kg particle that moves along the x axis. The potential energy $U(x)$ associated with $F(x)$ is graphed in Fig. 8-43. When the particle is at $x = 2.0$ m, its velocity is -1.5 m/s. (a) What are the magnitude and direction of $F(x)$ at this position? (b) Between what limits of x does the particle move? (c) What is its speed at $x = 7.0$ m?

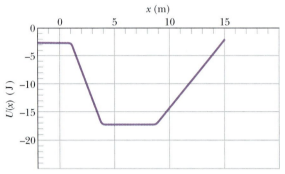

Fig. 8-43 Exercise 36.

37P. The potential energy of a diatomic molecule (a two-atom system like H_2 or O_2) is given by

$$U = \frac{A}{r^{12}} - \frac{B}{r^6},$$

where r is the separation of the two atoms of the molecule and A and B are positive constants. This potential energy is associated with the force that binds the two atoms together. (a) Find the *equilibrium separation*, that is, the distance between the atoms at which the force on each atom is zero. Is the force repulsive (the atoms are pushed apart) or attractive (they are pulled together) if their separation is (b) smaller and (c) larger than the equilibrium separation?   ssm

38P. A single conservative force $F(x)$ acts on a 1.0 kg particle that moves along an x axis. The potential energy $U(x)$ associated with $F(x)$ is given by

$$U(x) = -4x\, e^{-x/4} \text{ J},$$

where x is in meters. At $x = 5.0$ m the particle has a kinetic energy of 2.0 J. (a) What is the mechanical energy of the system? (b) Make a plot of $U(x)$ as a function of x for $0 \le x \le 10$ m, and on the same graph draw the line that represents the mechanical energy of the system. Use part (b) to determine (c) the least value of x and (d) the greatest value of x between which the particle can move. Use part (b) to determine (e) the maximum kinetic energy of the particle and (f) the value of x at which it occurs. (g) Determine the equation for $F(x)$ as a function of x. (h) For what (finite) value of x does $F(x) = 0$?

SEC. 8-6   Work Done on a System by an External Force

39E. A collie drags its bed box across a floor by applying a horizontal force of 8.0 N. The kinetic frictional force acting on the box has magnitude 5.0 N. As the box is dragged through 0.70 m along the way, what are (a) the work done by the collie's applied force and (b) the increase in thermal energy of the bed and floor?

40E. The temperature of a plastic cube is monitored while the cube is pushed 3.0 m across a floor at constant speed by a horizontal force of 15 N. The monitoring reveals that the thermal energy of the cube increases by 20 J. What is the increase in the thermal energy of the floor along which the cube slides?

41P. A 3.57 kg block is drawn at constant speed 4.06 m along a horizontal floor by a rope. The force on the block from the rope has a magnitude of 7.68 N and is directed 15.0° above the horizontal. What are (a) the work done by the rope's force, (b) the increase in thermal energy of the block–floor system, and (c) the coefficient of kinetic friction between the block and floor?   ssm

42P. A worker pushed a 27 kg block 9.2 m along a level floor at constant speed with a force directed 32° below the horizontal. If the coefficient of kinetic friction between block and floor was 0.20, what were (a) the work done by the worker's force and (b) the increase in thermal energy of the block–floor system?

SEC. 8-7   Conservation of Energy

43E. A 25 kg bear slides, from rest, 12 m down a lodgepole pine tree, moving with a speed of 5.6 m/s just before hitting the ground. (a) What change occurs in the gravitational potential energy of the bear–Earth system during the slide? (b) What is the kinetic energy of the bear just before hitting the ground? (c) What is the average frictional force that acts on the sliding bear?   ssm   ilw

44E. A 30 g bullet, with a horizontal velocity of 500 m/s, comes to a stop 12 cm within a solid wall. (a) What is the change in its mechanical energy? (b) What is the magnitude of the average force from the wall stopping it?

45E. A 60 kg skier leaves the end of a ski-jump ramp with a velocity of 24 m/s directed 25° above the horizontal. Suppose that as a result of air drag the skier returns to the ground with a speed of 22 m/s, landing 14 m vertically below the end of the ramp. From the launch to the return to the ground, by how much is the mechanical energy of the skier–Earth system reduced because of air drag?

46E. A 75 g Frisbee is thrown from a point 1.1 m above the ground with a speed of 12 m/s. When it has reached a height of 2.1 m, its speed is 10.5 m/s. What was the reduction in the mechanical energy of the Frisbee–Earth system because of air drag?

47E. An outfielder throws a baseball with an initial speed of 81.8 mi/h. Just before an infielder catches the ball at the same level, the ball's speed is 110 ft/s. In foot-pounds, by how much is the mechanical energy of the ball–Earth system reduced because of air drag? (The weight of a baseball is 9.0 oz.)

48E. Approximately 5.5×10^6 kg of water fall 50 m over Niagara Falls each second. (a) What is the decrease in the gravitational potential energy of the water–Earth system each second? (b) If all this energy could be converted to electrical energy (it cannot be), at what rate would electrical energy be supplied? (The mass of 1 m³ of water is 1000 kg.) (c) If the electrical energy were sold at 1 cent/kW · h, what would be the yearly cost?

49E. During a rockslide, a 520 kg rock slides from rest down a hillside that is 500 m long and 300 m high. The coefficient of kinetic friction between the rock and the hill surface is 0.25. (a) If the gravitational potential energy U of the rock–Earth system is zero at the bottom of the hill, what is the value of U just before the slide? (b) How much energy is transferred to thermal energy during the slide? (c) What is the kinetic energy of the rock as it reaches the bottom of the hill? (d) What is its speed then?

50P. You push a 2.0 kg block against a horizontal spring, compressing the spring by 15 cm. Then you release the block, and the spring sends it sliding across a tabletop. It stops 75 cm from where you released it. The spring constant is 200 N/m. What is the coefficient of kinetic friction between the block and the table?

51P. As Fig. 8-44 shows, a 3.5 kg block is accelerated by a compressed spring whose spring constant is 640 N/m. After leaving the spring at the spring's relaxed length, the block travels over a horizontal surface, with a coefficient of kinetic friction of 0.25, for a distance of 7.8 m before stopping. (a) What is the increase in the thermal energy of the block–floor system? (b) What is the maximum kinetic energy of the block? (c) Through what distance is the spring compressed before the block begins to move? ssm www

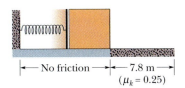

Fig. 8-44 Problem 51.

52P. In Fig. 8-45, a block is moved down an incline a distance of 5.0 m from point A to point B by a force $\vec{F}$ that is parallel to the incline and has magnitude 2.0 N. The magnitude of the frictional force acting on the block is 10 N. If the kinetic energy of the block increases by 35 J between A and B, how much work is done on the block by the gravitational force as the block moves from A to B?

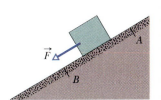

Fig. 8-45 Problem 52.

53P. A certain spring is found *not* to conform to Hooke's law. The force (in newtons) it exerts when stretched a distance x (in meters) is found to have magnitude $52.8x + 38.4x^2$ in the direction opposing the stretch. (a) Compute the work required to stretch the spring from $x = 0.500$ m to $x = 1.00$ m. (b) With one end of the spring fixed, a particle of mass 2.17 kg is attached to the other end of the spring when it is extended by an amount $x = 1.00$ m. If the particle is then released from rest, what is its speed at the instant the spring has returned to the configuration in which the extension is $x = 0.500$ m? (c) Is the force exerted by the spring conservative or nonconservative? Explain. ssm

54P. A 4.0 kg bundle starts up a 30° incline with 128 J of kinetic energy. How far will it slide up the incline if the coefficient of kinetic friction between bundle and incline is 0.30?

55P. Two snowy peaks are 850 m and 750 m above the valley between them. A ski run extends down from the top of the higher peak and then back up to the top of the lower one, with a total length of 3.2 km and an average slope of 30° (Fig. 8-46). (a) A

skier starts from rest at the top of the higher peak. At what speed will he arrive at the top of the lower peak if he coasts without using ski poles? Ignore friction. (b) Approximately what coefficient of kinetic friction between snow and skis would make him stop just at the top of the lower peak? ssm

Fig. 8-46 Problem 55.

56P. A girl whose weight is 267 N slides down a 6.1 m playground slide that makes an angle of 20° with the horizontal. The coefficient of kinetic friction between slide and child is 0.10. (a) How much energy is transferred to thermal energy? (b) If the girl starts at the top with a speed of 0.457 m/s, what is her speed at the bottom?

57P. In Fig. 8-47, a 2.5 kg block slides head on into a spring with a spring constant of 320 N/m. When the block stops, it has compressed the spring by 7.5 cm. The coefficient of kinetic friction between the block and the horizontal surface is 0.25. While the block is in contact with the spring and being brought to rest, what are (a) the work done by the spring force and (b) the increase in thermal energy of the block–floor system? (c) What is the block's speed just as the block reaches the spring? ilw

Fig. 8-47 Problem 57.

58P. A factory worker accidentally releases a 180 kg crate that was being held at rest at the top of a 3.7 m-long-ramp inclined at 39° to the horizontal. The coefficient of kinetic friction between the crate and the ramp, and between the crate and the horizontal factory floor, is 0.28. (a) How fast is the crate moving as it reaches the bottom of the ramp? (b) How far will it subsequently slide across the factory floor? (Assume that the crate's kinetic energy does not change as it moves from the ramp onto the floor.) (c) Do the answers to (a) and (b) increase, decrease, or remain the same if we halve the mass of the crate?

59P. In Fig. 8-48, a block slides along a track from one level to a higher level, by moving through an intermediate valley. The track is frictionless until the block reaches the higher level. There a frictional force stops the block in a distance d. The block's initial speed v_0 is 6.0 m/s; the height difference h is 1.1 m; and the coefficient of kinetic friction μ is 0.60. Find d.

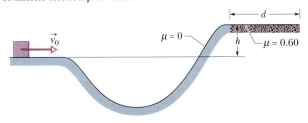

Fig. 8-48 Problem 59.

60P. A cookie jar is moving up a 40° incline. At a point 55 cm from the bottom of the incline (measured along the incline), it has a speed of 1.4 m/s. The coefficient of kinetic friction between jar and incline is 0.15. (a) How much farther up the incline will the jar move? (b) How fast will it be going when it has slid back to the bottom of the incline? (c) Do the answers to (a) and (b) increase, decrease, or remain the same if we decrease the coefficient of kinetic friction (but do not change the given speed or location)?

61P. A stone with weight w is thrown vertically upward into the air from ground level with initial speed v_0. If a constant force f due to air drag acts on the stone throughout its flight, (a) show that the maximum height reached by the stone is

$$h = \frac{v_0^2}{2g(1 + f/w)}.$$

(b) Show that the stone's speed is

$$v = v_0 \left(\frac{w - f}{w + f}\right)^{1/2}$$

just before impact with the ground. ssm

62P. A playground slide is in the form of an arc of a circle with a maximum height of 4.0 m, with a radius of 12 m, and with the ground tangent to the circle (Fig. 8-49). A 25 kg child starts from rest at the top of the slide and has a speed of 6.2 m/s at the bottom. (a) What is the length of the slide? (b) What average frictional force acts on the child over this distance? If, instead of the ground, a vertical line through the *top of the slide* is tangent to the circle, what are (c) the length of the slide and (d) the average frictional force on the child?

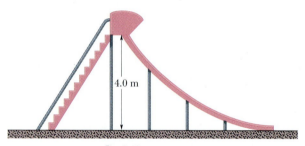

4.0 m

Fig. 8-49 Problem 62.

63P. A particle can slide along a track with elevated ends and a flat central part, as shown in Fig. 8-50. The flat part has length L. The curved portions of the track are frictionless, but for the flat part the coefficient of kinetic friction is $\mu_k = 0.20$. The particle is released from rest at point A, which is a height $h = L/2$ above the flat part of the track. Where does the particle finally stop?

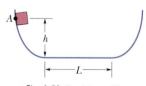

Fig. 8-50 Problem 63.

64P. The cable of the 1800 kg elevator cab in Fig. 8-51 snaps when the cab is at rest at the first floor, where the cab bottom is a distance $d = 3.7$ m above a cushioning spring whose spring constant is $k = 0.15$ MN/m. A safety device clamps the cab against guide rails so that a constant frictional force of 4.4 kN opposes the cab's mo-

tion. (a) Find the speed of the cab just before it hits the spring. (b) Find the maximum distance x that the spring is compressed (the frictional force still acts during this compression). (c) Find the distance that the cab will bounce back up the shaft. (d) Using conservation of energy, find the approximate total distance that the cab will move before coming to rest. (Assume that the frictional force on the cab is negligible when the cab is stationary.)

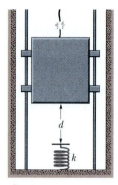

Fig. 8-51 Problem 64.

65P. At a certain factory, 300 kg crates are dropped vertically from a packing machine onto a conveyor belt moving at 1.20 m/s (Fig. 8-52). (A motor maintains the belt's constant speed.) The coefficient of kinetic friction between the belt and each crate is 0.400. After a short time, slipping between the belt and the crate ceases, and the crate then moves along with the belt. For the period of time during which the crate is being brought to rest relative to the belt, calculate, for a coordinate system at rest in the factory, (a) the kinetic energy supplied to the crate, (b) the magnitude of the kinetic frictional force acting on the crate, and (c) the energy supplied by the motor. (d) Explain why the answers to (a) and (c) are different.

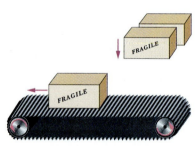

Fig. 8-52 Problem 65.

Additional Problems

66. The maximum force that you can exert on an object with one of your back teeth is about 750 N. Suppose that as you gradually bite on a clump of elastic licorice, the licorice resists its compression by one of the teeth with a spring-like force of spring constant 2.5×10^5 N/m. Find (a) the distance the licorice is compressed by your tooth and (b) the work your tooth does on the licorice during the compression. (c) Plot the magnitude of your force versus the compression distance. (d) If there is a potential energy associated with this compression, plot it versus compression distance.

In the 1990s the pelvis of a particular *Triceratops* dinosaur was found to have deep bite marks. The shape of the marks suggested that they were made by a *Tyrannosaurus rex* dinosaur. To test the idea, researchers made a replica of a *T. rex* tooth from bronze and aluminum and then used a hydraulic press to gradually drive the replica into cow bone to the depth seen in the *Triceratops* bone. A graph of the force required for the penetration versus the depth of penetration is given in Fig. 8-53 for one trial; the required force increased with depth because, as the nearly conical tooth penetrated the bone, more of the tooth came in contact with the bone. (e) How much work was done by the hydraulic press—and thus, presumably, by the *T. rex*—in such a penetration? (f) Is there a potential energy associated with this penetration? The large biting force and energy expenditure attributed to the *T. rex* by this research (and unsurpassed by modern animals) suggest that the animal was a predator and not a scavenger (as had been argued by some researchers).

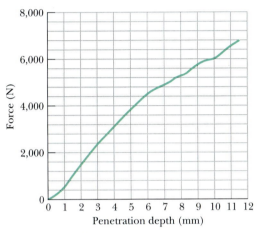

Fig. 8-53 Problem 66.

67. *Fly-Fishing and Speed Amplification.* If you throw a loose fishing fly, it will travel horizontally only about 1 m. However, if you throw that fly attached to fishing line by casting the line with a rod, the fly will easily travel horizontally to the full length of the line, say, 20 m.

The cast is depicted in Fig. 8-54: Initially (Fig. 8-54a) the line of length L is extended horizontally leftward and moving rightward with a speed v_0. As the fly at the end of the line moves forward, the line doubles over, with the upper section still moving and the lower section stationary (Fig. 8-54b). The upper section decreases in length as the lower section increases in length (Fig. 8-54c), until the line is extended horizontally rightward and there is only a lower section (Fig. 8-54d). If air drag is neglected, the initial kinetic energy of the line in Fig. 8-54a becomes progressively concentrated in the fly and the decreasing portion of the line that is still moving, resulting in an amplification (increase) in the speed of the fly and that portion.

(a) Using the x axis indicated, show that when the fly position is x, the length of the still-moving (upper) section of line is $(L - x)/2$. (b) Assuming that the line is uniform with a linear density ρ (mass per unit length), what is the mass of the still-moving section? Next, let m_f represent the mass of the fly, and assume that the kinetic energy of the moving section does not change from its initial value (when the moving section had length L and speed v_0) even though the length of the moving section is decreasing during the cast. (c) Find an expression for the speed of the still-moving section and the fly.

Assume that initial speed $v_0 = 6.0$ m/s, line length $L = 20$ m, fly mass $m_f = 0.80$ g, and linear density $\rho = 1.3$ g/m. (d) Plot the fly's speed v versus its position x. (e) What is the fly's speed just as the line approaches its final horizontal orientation and the fly is about to flip over and stop? (In more realistic calculations, air drag reduces this final speed.)

Speed amplification can also be produced with a bullwhip and even a rolled-up, wet towel that is popped against a victim in a common locker-room prank. (Adapted from "The Mechanics of Flycasting: The Flyline," by Graig A. Spolek, *American Journal of Physics*, Sept. 1986, Vol. 54, No. 9, pp. 832–836.)

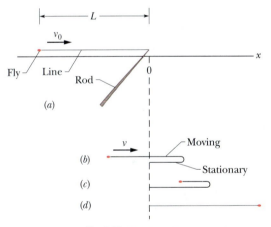

Fig. 8-54 Problem 67.

9 Systems of Particles

If you leap forward, chances are that your head and torso will follow a parabolic path, like a baseball thrown in from the outfield. However, when a skilled ballet dancer leaps across the stage in a *grand jeté*, the path taken by her head and torso is nearly horizontal during much of the jump. She seems to be floating across the stage. The audience may not know much about projectile motion, but they still sense that something unusual has happened.

How does the ballerina seemingly "turn off" the gravitational force?

The answer is in this chapter.

9-1 A Special Point

Physicists love to look at something complicated and find in it something simple and familiar. Here is an example. If you flip a baseball bat into the air, its motion as it turns is clearly more complicated than that of, say, a nonspinning tossed ball (Fig. 9-1a), which moves like a particle. Every part of the bat moves in a different way from every other part, so you cannot represent the bat as a tossed particle; instead, it is a system of particles.

However, if you look closely, you will find that one special point of the bat moves in a simple parabolic path, just as a particle would if tossed into the air (Fig. 9-1b). In fact, that special point moves as though (1) the bat's total mass were concentrated there and (2) the gravitational force on the bat acted only there. That special point is said to be the **center of mass** of the bat. In general:

> The center of mass of a body or a system of bodies is the point that moves as though all of the mass were concentrated there and all external forces were applied there.

The center of mass of a baseball bat lies along the bat's central axis. You can locate it by balancing the bat horizontally on an outstretched finger: The center of mass is on the bat's axis just above your finger.

9-2 The Center of Mass

We shall now spend some time determining how to find the center of mass in various systems. We start with a system of a few particles, and then we consider a system of a great many particles (as in a baseball bat).

Systems of Particles

Figure 9-2a shows two particles of masses m_1 and m_2 separated by a distance d. We have arbitrarily chosen the origin of the x axis to coincide with the particle of mass m_1. We *define* the position of the center of mass (com) of this two-particle system to be

$$x_{\text{com}} = \frac{m_2}{m_1 + m_2} d. \qquad (9-1)$$

Suppose, as an example, that $m_2 = 0$. Then there is only one particle, of mass m_1, and the center of mass must lie at the position of that particle; Eq. 9-1 dutifully

(a)

(b)

Fig. 9-1 (a) A ball tossed into the air follows a parabolic path. (b) The center of mass (the black dot) of a baseball bat that is flipped into the air does also, but all other points of the bat follow more complicated curved paths.

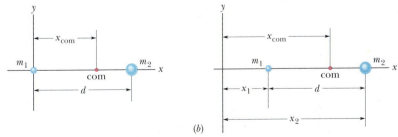
(a)   (b)

Fig. 9-2 (a) Two particles of masses m_1 and m_2 are separated by a distance d. The dot labeled com shows the position of the center of mass, calculated from Eq. 9-1. (b) The same as (a) except that the origin is located farther from the particles. The position of the center of mass is calculated from Eq. 9-2. The location of the center of mass (with respect to the particles) is the same in both cases.

reduces to $x_{com} = 0$. If $m_1 = 0$, there is again only one particle (of mass m_2), and we have, as we expect, $x_{com} = d$. If $m_1 = m_2$, the masses of the particles are equal and the center of mass should be halfway between them; Eq. 9-1 reduces to $x_{com} = \frac{1}{2}d$, again as we expect. Finally, Eq. 9-1 tells us that if neither m_1 nor m_2 is zero, x_{com} can have only values that lie between zero and d; that is, the center of mass must lie somewhere between the two particles.

Figure 9-2b shows a more generalized situation, in which the coordinate system has been shifted leftward. The position of the center of mass is now defined as

$$x_{com} = \frac{m_1 x_1 + m_2 x_2}{m_1 + m_2}. \tag{9-2}$$

Note that if we put $x_1 = 0$, then x_2 becomes d and Eq. 9-2 reduces to Eq. 9-1, as it must. Note also that in spite of the shift of the coordinate system, the center of mass is still the same distance from each particle.

We can rewrite Eq. 9-2 as

$$x_{com} = \frac{m_1 x_1 + m_2 x_2}{M}, \tag{9-3}$$

in which M is the total mass of the system. (Here, $M = m_1 + m_2$.) We can extend this equation to a more general situation in which n particles are strung out along the x axis. Then the total mass is $M = m_1 + m_2 + \cdots + m_n$, and the location of the center of mass is

$$x_{com} = \frac{m_1 x_1 + m_2 x_2 + m_3 x_3 + \cdots + m_n x_n}{M}$$

$$= \frac{1}{M} \sum_{i=1}^{n} m_i x_i. \tag{9-4}$$

Here the subscript i is a running number, or index, that takes on all integer values from 1 to n. It identifies the various particles, their masses, and their x coordinates.

If the particles are distributed in three dimensions, the center of mass must be identified by three coordinates. By extension of Eq. 9-4, they are

$$x_{com} = \frac{1}{M} \sum_{i=1}^{n} m_i x_i, \qquad y_{com} = \frac{1}{M} \sum_{i=1}^{n} m_i y_i, \qquad z_{com} = \frac{1}{M} \sum_{i=1}^{n} m_i z_i. \tag{9-5}$$

We can also define the center of mass with the language of vectors. First recall that the position of a particle at coordinates x_i, y_i, and z_i is given by a position vector:

$$\vec{r}_i = x_i \hat{i} + y_i \hat{j} + z_i \hat{k}. \tag{9-6}$$

Here the index identifies the particle, and $\hat{i}$, $\hat{j}$, and $\hat{k}$ are unit vectors pointing, respectively, in the positive direction of the x, y, and z axes. Similarly, the position of the center of mass of a system of particles is given by a position vector:

$$\vec{r}_{com} = x_{com}\hat{i} + y_{com}\hat{j} + z_{com}\hat{k}. \tag{9-7}$$

The three scalar equations of Eq. 9-5 can now be replaced by a single vector equation,

$$\vec{r}_{com} = \frac{1}{M} \sum_{i=1}^{n} m_i \vec{r}_i, \tag{9-8}$$

where again M is the total mass of the system. You can check that this equation is correct by substituting Eqs. 9-6 and 9-7 into it, and then separating out the x, y, and z components. The scalar relations of Eq. 9-5 result.

Solid Bodies

An ordinary object, such as a baseball bat, contains so many particles (atoms) that we can best treat it as a continuous distribution of matter. The "particles" then become differential mass elements dm, the sums of Eq. 9-5 become integrals, and the coordinates of the center of mass are defined as

$$x_{\text{com}} = \frac{1}{M} \int x \, dm, \qquad y_{\text{com}} = \frac{1}{M} \int y \, dm, \qquad z_{\text{com}} = \frac{1}{M} \int z \, dm, \quad (9\text{-}9)$$

where M is now the mass of the object.

Evaluating these integrals for most common objects (like a television set or a moose) would be difficult, so here we shall consider only *uniform* objects. Such an object has *uniform density*, or mass per unit volume; that is, the density ρ (Greek letter rho) is the same for any given element of the object as for the whole object:

$$\rho = \frac{dm}{dV} = \frac{M}{V}, \tag{9-10}$$

where dV is the volume occupied by a mass element dm, and V is the total volume of the object. If we substitute $dm = M/V \, dV$ from Eq. 9-10 into Eq. 9-9, we find that

$$x_{\text{com}} = \frac{1}{V} \int x \, dV, \qquad y_{\text{com}} = \frac{1}{V} \int y \, dV, \qquad z_{\text{com}} = \frac{1}{V} \int z \, dV. \quad (9\text{-}11)$$

You can bypass one or more of these integrals if an object has a point, a line, or a plane of symmetry. The center of mass of such an object then lies at that point, on that line, or in that plane. For example, the center of mass of a uniform sphere (which has a point of symmetry) is at the center of the sphere (which is the point of symmetry). The center of mass of a uniform cone (whose axis is a line of symmetry) lies on the axis of the cone. The center of mass of a banana (which has a plane of symmetry that splits it into two equal parts) lies somewhere in that plane.

The center of mass of an object need not lie within the object. There is no dough at the center of mass of a doughnut, and no iron at the center of mass of a horseshoe.

✔**CHECKPOINT 1:**   The figure shows a uniform square plate from which four identical squares at the corners will be removed. (a) Where is the center of mass of the plate originally? Where is it after the removal of (b) square 1; (c) squares 1 and 2; (d) squares 1 and 3; (e) squares 1, 2, and 3; (f) all four squares? Answer in terms of quadrants, axes, or points (without calculation, of course).

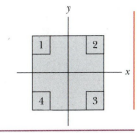

Sample Problem 9-1

Three particles of masses $m_1 = 1.2$ kg, $m_2 = 2.5$ kg, and $m_3 = 3.4$ kg form an equilateral triangle of edge length $a = 140$ cm. Where is the center of mass of this three-particle system?

SOLUTION: A Key Idea to get us started is that we are dealing with particles instead of an extended solid body, so we can use Eq. 9-5 to locate their center of mass. The particles are in the plane of the equilateral triangle, so we need only the first two equations. A second Key Idea is that we can simplify the calculations by choosing the x and y axes so that one of the particles is located at the origin

and the x axis coincides with one of the triangle's sides (Fig. 9-3). The three particles then have the following coordinates:

Particle	Mass (kg)	x (cm)	y (cm)
1	1.2	0	0
2	2.5	140	0
3	3.4	70	121

The total mass M of the system is 7.1 kg.

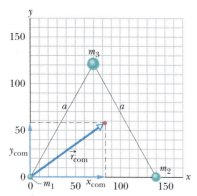

Fig. 9-3 Sample Problem 9-1. Three particles form an equilateral triangle of edge length a. The center of mass is located by the position vector $\vec{r}_{com}$.

From Eq. 9-5, the coordinates of the center of mass are

$$x_{com} = \frac{1}{M}\sum_{i=1}^{3} m_i x_i = \frac{m_1 x_1 + m_2 x_2 + m_3 x_3}{M}$$

$$= \frac{(1.2\ \text{kg})(0) + (2.5\ \text{kg})(140\ \text{cm}) + (3.4\ \text{kg})(70\ \text{cm})}{7.1\ \text{kg}}$$

$$= 83\ \text{cm} \qquad\qquad\qquad \text{(Answer)}$$

and

$$y_{com} = \frac{1}{M}\sum_{i=1}^{3} m_i y_i = \frac{m_1 y_1 + m_2 y_2 + m_3 y_3}{M}$$

$$= \frac{(1.2\ \text{kg})(0) + (2.5\ \text{kg})(0) + (3.4\ \text{kg})(121\ \text{cm})}{7.1\ \text{kg}}$$

$$= 58\ \text{cm}. \qquad\qquad\qquad \text{(Answer)}$$

In Fig. 9-3, the center of mass is located by the position vector $\vec{r}_{com}$, which has components x_{com} and y_{com}.

Sample Problem 9-2

Figure 9-4a shows a uniform metal plate P of radius $2R$ from which a disk of radius R has been stamped out (removed) in an assembly line. Using the xy coordinate system shown, locate the center of mass com$_P$ of the plate.

SOLUTION: First, let us roughly locate the center of plate P by using the **Key Idea** of symmetry. We note that the plate is symmetric about the x axis (we get the portion below that axis by rotating the upper portion about the axis). Thus, com$_P$ must be on the x axis. The plate (with the disk removed) is not symmetric about the y axis. However, because there is somewhat more mass on the right of the y axis, com$_P$ must be somewhat to the right of that axis. Thus, the location of com$_P$ should be roughly as indicated in Fig. 9-4a.

Another **Key Idea** here is that plate P is an extended solid body, so we can use Eqs. 9-11 to find the actual coordinates of com$_P$. However, that procedure is difficult. A much easier way is to use this **Key Idea**: In working with centers of mass, we can assume that the mass of a *uniform* object is concentrated in a particle at the object's center of mass. Here is how we do so:

First, put the stamped-out disk (call it disk S) back into place (Fig. 9-4b) to form the original composite plate (call it plate C). Because of its circular symmetry, the center of mass com$_S$ for disk S is at the center of S, at $x = -R$ (as shown). Similarly, the center of mass com$_C$ for composite plate C is at the center of C, at the origin (as shown). We then have the following:

Plate	Center of Mass	Location of com	Mass
P	com$_P$	$x_P = ?$	m_P
S	com$_S$	$x_S = -R$	m_S
C	com$_C$	$x_C = 0$	$m_C = m_S + m_P$

Now we use the **Key Idea** of concentrated mass: Assume that mass m_S of disk S is concentrated in a particle at $x_S = -R$, and mass m_P is concentrated in a particle at x_P (Fig. 9-4c). Next treat these two particles as a two-particle system, using Eq. 9-2 to find their

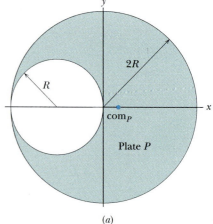

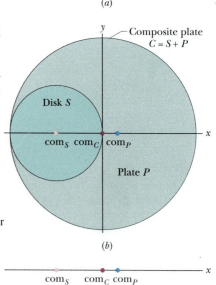

Fig. 9-4 Sample Problem 9-2.
(a) Plate P is a metal plate of radius $2R$, with a circular hole of radius R. The center of mass of P is at point com$_P$. (b) Disk S has been put back into place to form a composite plate C. The center of mass com$_S$ of disk S and the center of mass com$_C$ of plate C are shown. (c) The center of mass com$_{S+P}$ of the combination of S and P coincides with com$_C$, which is at $x = 0$.

center of mass x_{S+P}. We get

$$x_{S+P} = \frac{m_S x_S + m_P x_P}{m_S + m_P}. \tag{9-12}$$

Next note that the combination of disk S and plate P is composite plate C. Thus, the position x_{S+P} of com_{S+P} must coincide with the position x_C of com_C, which is at the origin; so $x_{S+P} = x_C = 0$. Substituting this into Eq. 9-12 and solving for x_P, we get

$$x_P = -x_S \frac{m_S}{m_P}. \tag{9-13}$$

Now we seem to have a problem, because we do not know the masses in Eq. 9-13. However, we can relate the masses to the

face areas of S and P by noting that

$$\text{mass} = \text{density} \times \text{volume}$$
$$= \text{density} \times \text{thickness} \times \text{area}.$$

Then $\dfrac{m_S}{m_P} = \dfrac{\text{density}_S}{\text{density}_P} \times \dfrac{\text{thickness}_S}{\text{thickness}_P} \times \dfrac{\text{area}_S}{\text{area}_P}.$

Because the plate is uniform, the densities and thicknesses are equal; we are left with

$$\frac{m_S}{m_P} = \frac{\text{area}_S}{\text{area}_P} = \frac{\text{area}_S}{\text{area}_C - \text{area}_S} = \frac{\pi R^2}{\pi (2R)^2 - \pi R^2} = \frac{1}{3}.$$

Substituting this and $x_S = -R$ into Eq. 9-13, we have

$$x_P = \tfrac{1}{3} R. \tag{Answer}$$

PROBLEM-SOLVING TACTICS

Tactic 1: *Center-of-Mass Problems*

Sample Problems 9-1 and 9-2 provide three strategies for simplifying center-of-mass problems. (1) Make full use of the symmetry of the object, be it about a point, a line, or a plane. (2) If the object can be divided into several parts, treat each of these parts as a particle, located at its own center of mass. (3) Choose your axes

wisely: If your system is a group of particles, choose one of the particles as your origin. If your system is a body with a line of symmetry, let that be your x or y axis. The choice of origin is completely arbitrary; the location of the center of mass is the same regardless of the origin from which it is measured.

9-3 Newton's Second Law for a System of Particles

If you roll a cue ball at a second billiard ball that is at rest, you expect that the two-ball system will continue to have some forward motion after impact. You would be surprised, for example, if both balls came back toward you or if both moved to the right or to the left.

What continues to move forward, its steady motion completely unaffected by the collision, is the center of mass of the two-ball system. If you focus on this point—which is always halfway between these bodies because they have identical masses—you can easily convince yourself by trial at a billiard table that this is so. No matter whether the collision is glancing, head on, or somewhere in between, the center of mass continues to move forward, as if the collision had never occurred. Let us look into this center-of-mass motion in more detail.

To do so, we replace the pair of billiard balls with an assemblage of n particles of (possibly) different masses. We are interested not in the individual motions of these particles but *only* in the motion of their center of mass. Although the center of mass is just a point, it moves like a particle whose mass is equal to the total mass of the system; we can assign a position, a velocity, and an acceleration to it. We state (and shall prove next) that the (vector) equation that governs the motion of the center of mass of such a system of particles is

$$\vec{F}_{\text{net}} = M\vec{a}_{\text{com}} \qquad \text{(system of particles).} \tag{9-14}$$

This equation is Newton's second law for the motion of the center of mass of a system of particles. Note that it has the same form ($\vec{F}_{\text{net}} = m\vec{a}$) that holds for the motion of a single particle. However, the three quantities that appear in Eq. 9-14

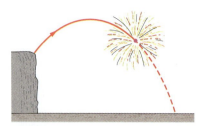

Fig. 9-5 A fireworks rocket explodes in flight. In the absence of air drag, the center of mass of the fragments would continue to follow the original parabolic path, until fragments began to hit the ground.

must be evaluated with some care:

1. $\vec{F}_{net}$ is the net force of *all external forces* that act on the system. Forces on one part of the system from another (*internal forces*) are not included in Eq. 9-14.

2. M is the *total mass* of the system. We assume that no mass enters or leaves the system as it moves, so that M remains constant. The system is said to be **closed.**

3. $\vec{a}_{com}$ is the acceleration of the *center of mass* of the system. Equation 9-14 gives no information about the acceleration of any other point of the system.

Equation 9-14 is equivalent to three equations involving the components of $\vec{F}_{net}$ and $\vec{a}_{com}$ along the three coordinate axes. These equations are

$$F_{net,x} = Ma_{com,x} \qquad F_{net,y} = Ma_{com,y} \qquad F_{net,z} = Ma_{com,z}. \qquad (9\text{-}15)$$

Now we can go back and examine the behavior of the billiard balls. Once the cue ball has begun to roll, no net external force acts on the (two-ball) system. Thus, because $\vec{F}_{net} = 0$, Eq. 9-14 tells us that $\vec{a}_{com} = 0$ also. Because acceleration is the rate of change of velocity, we conclude that the velocity of the center of mass of the system of two balls does not change. When the two balls collide, the forces that come into play are *internal* forces, on one ball from the other. Such forces do not contribute to the net force $\vec{F}_{net}$, which remains zero. Thus, the center of mass of the system, which was moving forward before the collision, must continue to move forward after the collision, with the same speed and in the same direction.

Equation 9-14 applies not only to a system of particles but also to a solid body, such as the bat of Fig. 9-1*b*. In that case, M in Eq. 9-14 is the mass of the bat and $\vec{F}_{net}$ is the gravitational force on the bat. Equation 9-14 then tells us that $\vec{a}_{com} = \vec{g}$. In other words, the center of mass of the bat moves as if the bat were a single particle of mass M, with force $\vec{F}_g$ acting on it.

Figure 9-5 shows another interesting case. Suppose that at a fireworks display, a rocket is launched on a parabolic path. At a certain point, it explodes into fragments. If the explosion had not occurred, the rocket would have continued along the trajectory shown in the figure. The forces of the explosion are *internal* to the system (first the rocket and then its fragments); that is, they are forces on parts of the system from other parts. If we ignore air drag, the net *external* force $\vec{F}_{net}$ acting on the system is the gravitational force on the system, regardless of whether the rocket explodes. Thus, from Eq. 9-14, the acceleration $\vec{a}_{com}$ of the center of mass of the fragments (while they are in flight) remains equal to $\vec{g}$. This means that the center of mass of the fragments follows the same parabolic trajectory that the rocket would have followed had it not exploded.

When a ballet dancer leaps across the stage in a grand jeté, she raises her arms and stretches her legs out horizontally as soon as her feet leave the stage (Fig. 9-6). These actions shift her center of mass upward through her body. Although the shifting center of mass faithfully follows a parabolic path across the stage, its movement relative to the body decreases the height that is attained by her head and torso, relative to that of a normal jump. The result is that the head and torso follow a nearly horizontal path, giving an illusion that the dancer is floating.

Proof of Equation 9-14

Now let us prove this important equation. From Eq. 9-8 we have, for a system of n particles,

$$M\vec{r}_{com} = m_1\vec{r}_1 + m_2\vec{r}_2 + m_3\vec{r}_3 + \cdots + m_n\vec{r}_n, \qquad (9\text{-}16)$$

Path of head

Path of center of mass

Fig. 9-6 A grand jeté. (Adapted from *The Physics of Dance,* by Kenneth Laws, Schirmer Books, 1984.)

in which M is the system's total mass and $\vec{r}_{com}$ is the vector locating the position of the system's center of mass.

Differentiating Eq. 9-16 with respect to time gives

$$M\vec{v}_{com} = m_1\vec{v}_1 + m_2\vec{v}_2 + m_3\vec{v}_3 + \cdots + m_n\vec{v}_n. \qquad (9\text{-}17)$$

Here $\vec{v}_i\ (= d\vec{r}_i/dt)$ is the velocity of the ith particle, and $\vec{v}_{com}\ (= d\vec{r}_{com}/dt)$ is the velocity of the center of mass.

Differentiating Eq. 9-17 with respect to time leads to

$$M\vec{a}_{com} = m_1\vec{a}_1 + m_2\vec{a}_2 + m_3\vec{a}_3 + \cdots + m_n\vec{a}_n. \qquad (9\text{-}18)$$

Here $\vec{a}_i\ (= d\vec{v}_i/dt)$ is the acceleration of the ith particle, and $\vec{a}_{com}\ (= d\vec{v}_{com}/dt)$ is the acceleration of the center of mass. Although the center of mass is just a geometrical point, it has a position, a velocity, and an acceleration, as if it were a particle.

From Newton's second law, $m_i\vec{a}_i$ is equal to the resultant force $\vec{F}_i$ that acts on the ith particle. Thus, we can rewrite Eq. 9-18 as

$$M\vec{a}_{com} = \vec{F}_1 + \vec{F}_2 + \vec{F}_3 + \cdots + \vec{F}_n. \qquad (9\text{-}19)$$

Among the forces that contribute to the right side of Eq. 9-19 will be forces that the particles of the system exert on each other (internal forces) and forces exerted on the particles from outside the system (external forces). By Newton's third law, the internal forces form third-law force pairs and cancel out in the sum that appears on the right side of Eq. 9-19. What remains is the vector sum of all the *external* forces that act on the system. Equation 9-19 then reduces to Eq. 9-14, the relation that we set out to prove.

✔**CHECKPOINT 2:**   Two skaters on frictionless ice hold opposite ends of a pole of negligible mass. An axis runs along the pole, and the origin of the axis is at the center of mass of the two-skater system. One skater, Fred, weighs twice as much as the other skater, Ethel. Where do the skaters meet if (a) Fred pulls hand over hand along the pole so as to draw himself to Ethel, (b) Ethel pulls hand over hand to draw herself to Fred, and (c) both skaters pull hand over hand?

Sample Problem 9-3

The three particles in Fig. 9-7a are initially at rest. Each experiences an *external* force due to bodies outside the three-particle system. The directions are indicated, and the magnitudes are $F_1 = 6.0$ N, $F_2 = 12$ N, and $F_3 = 14$ N. What is the acceleration of the center of mass of the system, and in what direction does it move?

SOLUTION: The position of the center of mass, calculated by the method of Sample Problem 9-1, is marked by a dot in the figure. One **Key Idea** here is that we can treat the center of mass as if it

were a real particle, with a mass equal to the system's total mass $M = 16$ kg. We can also treat the three external forces as if they act at the center of mass (Fig. 9-7b).

A second **Key Idea** is that we can now apply Newton's second law ($\vec{F}_{net} = m\vec{a}$) to the center of mass, writing

$$\vec{F}_{net} = M\vec{a}_{com} \tag{9-20}$$

or

$$\vec{F}_1 + \vec{F}_2 + \vec{F}_3 = M\vec{a}_{com}$$

so

$$\vec{a}_{com} = \frac{\vec{F}_1 + \vec{F}_2 + \vec{F}_3}{M}. \tag{9-21}$$

Equation 9-20 tells us that the acceleration $\vec{a}_{com}$ of the center of mass is in the same direction as the net external force $\vec{F}_{net}$ on the system (Fig. 9-7b). Because the particles are initially at rest, the center of mass must also be at rest. As the center of mass then begins to accelerate, it must move off in the common direction of $\vec{a}_{com}$ and $\vec{F}_{net}$.

We can evaluate the right side of Eq. 9-21 directly on a vector-capable calculator, or we can rewrite Eq. 9-21 in component form, find the components of $\vec{a}_{com}$, and then find $\vec{a}_{com}$. Along the x axis, we have

$$a_{com,x} = \frac{F_{1x} + F_{2x} + F_{3x}}{M}$$

$$= \frac{-6.0 \text{ N} + (12 \text{ N}) \cos 45° + 14 \text{ N}}{16 \text{ kg}} = 1.03 \text{ m/s}^2.$$

Along the y axis, we have

$$a_{com,y} = \frac{F_{1y} + F_{2y} + F_{3y}}{M}$$

$$= \frac{0 + (12 \text{ N}) \sin 45° + 0}{16 \text{ kg}} = 0.530 \text{ m/s}^2.$$

From these components, we find that $\vec{a}_{com}$ has the magnitude

$$a_{com} = \sqrt{(a_{com,x})^2 + (a_{com,y})^2}$$

$$= 1.16 \text{ m/s}^2 \approx 1.2 \text{ m/s}^2 \tag{Answer}$$

and the angle (from the positive direction of the x axis)

$$\theta = \tan^{-1} \frac{a_{com,y}}{a_{com,x}} = 27°. \tag{Answer}$$

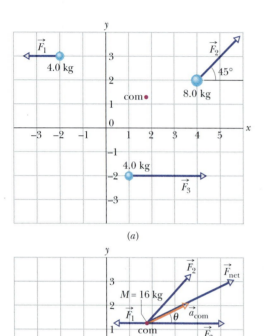

(a)

(b)

Fig. 9-7 Sample Problem 9-3. (*a*) Three particles, initially at rest in the positions shown, are acted on by the external forces shown. The center of mass (com) of the system is marked. (*b*) The forces are now transferred to the center of mass of the system, which behaves like a particle with a mass M equal to the total mass of the system. The net external force $\vec{F}_{net}$ and the acceleration $\vec{a}_{com}$ of the center of mass are shown.

9-4   Linear Momentum

Momentum is a word that has several meanings in everyday language but only a single precise meaning in physics. The **linear momentum** of a particle is a vector $\vec{p}$, defined as

$$\vec{p} = m\vec{v} \qquad \text{(linear momentum of a particle)}, \tag{9-22}$$

in which m is the mass of the particle and $\vec{v}$ is its velocity. (The adjective *linear* is often dropped, but it serves to distinguish $\vec{p}$ from *angular* momentum, which is introduced in Chapter 12 and which is associated with rotation.) Since m is always a positive scalar quantity, Eq. 9-22 tells us that $\vec{p}$ and $\vec{v}$ have the same direction. From Eq. 9-22, the SI unit for momentum is the kilogram-meter per second.

Newton actually expressed his second law of motion in terms of momentum:

▶ The time rate of change of the momentum of a particle is equal to the net force acting on the particle and is in the direction of that force.

In equation form this becomes

$$\vec{F}_{net} = \frac{d\vec{p}}{dt}. \tag{9-23}$$

Substituting for $\vec{p}$ from Eq. 9-22 gives

$$\vec{F}_{net} = \frac{d\vec{p}}{dt} = \frac{d}{dt}(m\vec{v}) = m\frac{d\vec{v}}{dt} = m\vec{a}.$$

Thus, the relations $\vec{F}_{net} = d\vec{p}/dt$ and $\vec{F}_{net} = m\vec{a}$ are equivalent expressions of Newton's second law of motion for a particle.

✔CHECKPOINT 3:   The figure gives the linear momentum versus time for a particle moving along an axis. A force directed along the axis acts on the particle. (a) Rank the four regions indicated according to the magnitude of the force, greatest first. (b) In which region is the particle slowing?

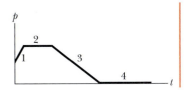

9-5   The Linear Momentum of a System of Particles

Now consider a system of n particles, each with its own mass, velocity, and linear momentum. The particles may interact with each other, and external forces may act on them as well. The system as a whole has a total linear momentum $\vec{P}$, which is defined to be the vector sum of the individual particles' linear momenta. Thus,

$$\vec{P} = \vec{p}_1 + \vec{p}_2 + \vec{p}_3 + \cdots + \vec{p}_n$$
$$= m_1\vec{v}_1 + m_2\vec{v}_2 + m_3\vec{v}_3 + \cdots + m_n\vec{v}_n. \tag{9-24}$$

If we compare this equation with Eq. 9-17, we see that

$$\vec{P} = M\vec{v}_{com} \qquad \text{(linear momentum, system of particles)}, \tag{9-25}$$

which gives us another way to define the linear momentum of a system of particles:

▶ The linear momentum of a system of particles is equal to the product of the total mass M of the system and the velocity of the center of mass.

If we take the time derivative of Eq. 9-25, we find

$$\frac{d\vec{P}}{dt} = M\frac{d\vec{v}_{com}}{dt} = M\vec{a}_{com}. \tag{9-26}$$

Comparing Eqs. 9-14 and 9-26 allows us to write Newton's second law for a system of particles in the equivalent form

$$\vec{F}_{net} = \frac{d\vec{P}}{dt} \qquad \text{(system of particles),} \qquad (9\text{-}27)$$

where $\vec{F}_{net}$ is the net external force acting on the system. This equation is the generalization of the single-particle equation $\vec{F}_{net} = d\vec{p}/dt$ to a system of many particles.

Sample Problem 9-4

Figure 9-8a shows a 2.0 kg toy race car before and after taking a turn on a track. Its speed is 0.50 m/s before the turn and 0.40 m/s after the turn. What is the change $\Delta\vec{P}$ in the linear momentum of the car due to the turn?

SOLUTION: We treat the car as a system of particles. Then a Key Idea is that to get $\Delta\vec{P}$, we need the car's linear momenta before and after the turn. However, that means we first need its velocity $\vec{v}_i$ before the turn and its velocity $\vec{v}_f$ after the turn. Using the coordinate system of Fig. 9-8a, we write $\vec{v}_i$ and $\vec{v}_f$ as

$$\vec{v}_i = -(0.50 \text{ m/s})\hat{j} \quad \text{and} \quad \vec{v}_f = (0.40 \text{ m/s})\hat{i}.$$

Equation 9-25 then gives us the linear momentum $\vec{P}_i$ before the turn and the linear momentum $\vec{P}_f$ after the turn:

$$\vec{P}_i = M\vec{v}_i = (2.0 \text{ kg})(-0.50 \text{ m/s})\hat{j} = (-1.0 \text{ kg} \cdot \text{m/s})\hat{j}$$

and

$$\vec{P}_f = M\vec{v}_f = (2.0 \text{ kg})(0.40 \text{ m/s})\hat{i} = (0.80 \text{ kg} \cdot \text{m/s})\hat{i}.$$

A second Key Idea is that, because these linear momenta are not directed along the same axis, we *cannot* find the change in linear momentum $\Delta\vec{P}$ by merely subtracting the magnitude of $\vec{P}_i$ from the magnitude of $\vec{P}_f$. Instead, we must write the change in linear momentum as the vector equation

$$\Delta\vec{P} = \vec{P}_f - \vec{P}_i \qquad (9\text{-}28)$$

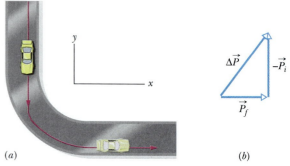

(a)   (b)

Fig. 9-8 Sample Problem 9-4. (*a*) A toy car takes a turn. (*b*) The change $\Delta\vec{P}$ in the car's linear momentum is the vector sum of its final linear momentum $\vec{P}_f$ and the negative of its initial linear momentum $\vec{P}_i$.

and then as

$$\Delta\vec{P} = (0.80 \text{ kg} \cdot \text{m/s})\hat{i} - (-1.0 \text{ kg} \cdot \text{m/s})\hat{j}$$
$$= (0.8\hat{i} + 1.0\hat{j}) \text{ kg} \cdot \text{m/s}. \qquad \text{(Answer)}$$

Figure 9-8b shows $\Delta\vec{P}$, $\vec{P}_f$, and $-\vec{P}_i$. Note that we subtract $\vec{P}_i$ from $\vec{P}_f$ in Eq. 9-28 by adding $-\vec{P}_i$ to $\vec{P}_f$.

9-6   Conservation of Linear Momentum

Suppose that the net external force acting on a system of particles is zero (the system is isolated) and that no particles leave or enter the system (the system is closed). Putting $\vec{F}_{net} = 0$ in Eq. 9-27 then yields $d\vec{P}/dt = 0$, or

$$\vec{P} = \text{constant} \qquad \text{(closed, isolated system).} \qquad (9\text{-}29)$$

In words,

> If no net external force acts on a system of particles, the total linear momentum $\vec{P}$ of the system cannot change.

This result is called the **law of conservation of linear momentum**. It can also be written as

$$\vec{P}_i = \vec{P}_f \qquad \text{(closed, isolated system).} \qquad (9\text{-}30)$$

In words, this equation says that, for a closed, isolated system,

$$\begin{pmatrix} \text{total linear momentum} \\ \text{at some initial time } t_i \end{pmatrix} = \begin{pmatrix} \text{total linear momentum} \\ \text{at some later time } t_f \end{pmatrix}.$$

Equations 9-29 and 9-30 are vector equations and, as such, each is equivalent to three equations corresponding to the conservation of linear momentum in three mutually perpendicular directions as in, say, the *xyz* coordinate system. Depending on the forces acting on a system, linear momentum might be conserved in one or two directions but not in all directions. However,

▶ If the component of the net *external* force on a closed system is zero along an axis, then the component of the linear momentum of the system along that axis cannot change.

As an example, suppose that you toss a grapefruit across a room. During its flight, the only external force acting on the grapefruit (which we take as the system) is the gravitational force $\vec{F}_g$, which is directed vertically downward. Thus, the vertical component of the linear momentum of the grapefruit changes, but since no horizontal external force acts on the grapefruit, the horizontal component of the linear momentum cannot change.

Note that we focus on the external forces acting on a closed system. Although internal forces can change the linear momentum of portions of the system, they cannot change the total linear momentum of the entire system.

✔**CHECKPOINT 4:**   An initially stationary device lying on a frictionless floor explodes into two pieces, which then slide across the floor. One piece slides in the positive direction of an *x* axis. (a) What is the sum of the momenta of the two pieces after the explosion? (b) Can the second piece move at an angle to the *x* axis? (c) What is the direction of the momentum of the second piece?

Sample Problem 9-5

A ballot box with mass $m = 6.0$ kg slides with speed $v = 4.0$ m/s across a frictionless floor in the positive direction of an *x* axis. It suddenly explodes into two pieces. One piece, with mass $m_1 = 2.0$ kg, moves in the positive direction of the *x* axis with speed $v_1 = 8.0$ m/s. What is the velocity of the second piece, with mass m_2?

SOLUTION: There are two Key Ideas here. First, we could get the velocity of the second piece if we knew its momentum, because we already know its mass is $m_2 = m - m_1 = 4.0$ kg. Second, we can relate the momenta of the two pieces to the original momentum of the box if momentum is conserved. Let's check.

Our reference frame will be that of the floor. Our system, which consists initially of the box and then of the two pieces, is closed but is not isolated, because the box and pieces each experience a normal force from the floor and a gravitational force. However, those forces are both vertical and thus cannot change the horizontal component of the momentum of the system. Neither can the forces produced by the explosion, because those forces are internal to the system. Thus, the horizontal component of the momentum of the system is conserved, and we can apply Eq. 9-30 along the *x* axis.

The initial momentum of the system is that of the box:

$$\vec{P}_i = m\vec{v}.$$

Similarly, we can write the final momenta of the two pieces as

$$\vec{P}_{f1} = m_1\vec{v}_1 \quad \text{and} \quad \vec{P}_{f2} = m_2\vec{v}_2.$$

The final total momentum $\vec{P}_f$ of the system is the vector sum of the momenta of the two pieces:

$$\vec{P}_f = \vec{P}_{f1} + \vec{P}_{f2} = m_1\vec{v}_1 + m_2\vec{v}_2.$$

Since all the velocities and momenta in this problem are vectors along the *x* axis, we can write them in terms of their *x* components. Doing so while applying Eq. 9-30, we now obtain

$$P_i = P_f$$

or

$$mv = m_1v_1 + m_2v_2.$$

Inserting known data, we find

$$(6.0 \text{ kg})(4.0 \text{ m/s}) = (2.0 \text{ kg})(8.0 \text{ m/s}) + (4.0 \text{ kg})v_2$$

and thus

$$v_2 = 2.0 \text{ m/s.} \qquad \text{(Answer)}$$

Since the result is positive, the second piece moves in the positive direction of the *x* axis.

Sample Problem 9-6

Figure 9-9a shows a space hauler and cargo module, of total mass M traveling along an x axis in deep space. They have an initial velocity $\vec{v}_i$ of magnitude 2100 km/h relative to the Sun. With a small explosion, the hauler ejects the cargo module, of mass $0.20M$ (Fig. 9-9b). The hauler then travels 500 km/h faster than the module along the x axis; that is, the relative speed v_{rel} between the hauler and the module is 500 km/h. What then is the velocity $\vec{v}_{HS}$ of the hauler relative to the Sun?

SOLUTION: The Key Idea here is that, because the hauler–module system is closed and isolated, its total linear momentum is conserved; that is,

$$\vec{P}_i = \vec{P}_f, \tag{9-31}$$

where the subscripts i and f refer to values before and after the ejection, respectively. Because the motion is along a single axis, we can write momenta and velocities in terms of their x components. Before the ejection, we have

$$P_i = Mv_i. \tag{9-32}$$

Let v_{MS} be the velocity of the ejected module relative to the Sun. The total linear momentum of the system after the ejection is then

$$P_f = (0.20M)v_{MS} + (0.80M)v_{HS}, \tag{9-33}$$

where the first term on the right is the linear momentum of the module and the second term is that of the hauler.

We do not know the velocity v_{MS} of the module relative to the Sun, but we can relate it to the known velocities with

$$\begin{pmatrix} \text{velocity of} \\ \text{hauler relative} \\ \text{to Sun} \end{pmatrix} = \begin{pmatrix} \text{velocity of} \\ \text{hauler relative} \\ \text{to module} \end{pmatrix} + \begin{pmatrix} \text{velocity of} \\ \text{module relative} \\ \text{to Sun} \end{pmatrix}.$$

In symbols, this gives us

$$v_{HS} = v_{rel} + v_{MS} \tag{9-34}$$

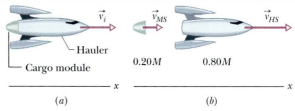

Fig. 9-9 Sample Problem 9-6. (a) A space hauler, with a cargo module, moving at initial velocity $\vec{v}_i$. (b) The hauler has ejected the cargo module. Now the module has velocity $\vec{v}_{MS}$ and the hauler has velocity $\vec{v}_{HS}$.

or

$$v_{MS} = v_{HS} - v_{rel}.$$

Substituting this expression for v_{MS} into Eq. 9-33, and then substituting Eqs. 9-32 and 9-33 into Eq. 9-31, we find

$$Mv_i = 0.20M(v_{HS} - v_{rel}) + 0.80Mv_{HS},$$

which gives us

$$v_{HS} = v_i + 0.20v_{rel},$$

or

$$\begin{aligned} v_{HS} &= 2100 \text{ km/h} + (0.20)(500 \text{ km/h}) \\ &= 2200 \text{ km/h}. \end{aligned} \tag{Answer}$$

✔CHECKPOINT 5:   The table gives velocities of the hauler and module (after ejection and relative to the Sun), and the relative speed between the hauler and the module for three situations. What are the missing values?

	Velocities (km/h)		Relative Speed
	Module	Hauler	(km/h)
(a)	1500	2000	
(b)		3000	400
(c)	1000		600

Sample Problem 9-7

A firecracker placed inside a coconut of mass M, initially at rest on a frictionless floor, blows the coconut into three pieces that slide across the floor. An overhead view is shown in Fig. 9-10a. Piece C, with mass $0.30M$, has final speed $v_{fC} = 5.0$ m/s.

(a) What is the speed of piece B, with mass $0.20M$?

SOLUTION: A Key Idea here is to see whether linear momentum is conserved. We note that (1) the coconut and its pieces form a closed system, (2) the explosion forces are internal to that system, and (3) no net external force acts on the system. Therefore, the linear momentum of the system is conserved.

To get started, we superimpose an xy coordinate system as shown in Fig. 9-10b, with the negative direction of the x axis coinciding with the direction of $\vec{v}_{fA}$. The x axis is at 80° with the direction of $\vec{v}_{fC}$ and 50° with the direction of $\vec{v}_{fB}$.

A second Key Idea is that linear momentum is conserved separately along both x and y axes. Let's use the y axis and write

$$P_{iy} = P_{fy}, \tag{9-35}$$

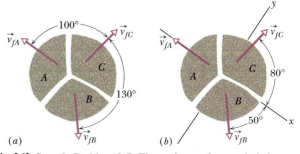

Fig. 9-10 Sample Problem 9-7. Three pieces of an exploded coconut move off in three directions along a frictionless floor. (a) An overhead view of the event. (b) The same with a two-dimensional axis system imposed.

where subscript i refers to the initial value (before the explosion), and subscript y refers to the y component of $\vec{P}_i$ or $\vec{P}_f$.

The component P_{iy} of the initial linear momentum is zero, because the coconut is initially at rest. To get an expression for P_{fy}, we find the y component of the final linear momentum of each piece, using the y-component version of Eq. 9-22 ($p_y = mv_y$):

$$p_{fA,y} = 0,$$

$$p_{fB,y} = -0.20Mv_{fB,y} = -0.20Mv_{fB}\sin 50°,$$

$$p_{fC,y} = 0.30Mv_{fC,y} = 0.30Mv_{fC}\sin 80°.$$

(Note that $p_{fA,y} = 0$ because of our choice of axes.) Equation 9-35 can now be written as

$$P_{iy} = P_{fy} = p_{fA,y} + p_{fB,y} + p_{fC,y}.$$

Then, with $v_{fC} = 5.0$ m/s, we have

$$0 = 0 - 0.20Mv_{fB}\sin 50° + (0.30M)(5.0 \text{ m/s})\sin 80°,$$

from which we find

$$v_{fB} = 9.64 \text{ m/s} \approx 9.6 \text{ m/s.} \qquad \text{(Answer)}$$

(b) What is the speed of piece A?

SOLUTION: Because linear momentum is also conserved along the x axis, we have

$$P_{ix} = P_{fx}, \qquad (9-36)$$

where $P_{ix} = 0$ because the coconut is initially at rest. To get P_{fx},

we find the x components of the final momenta, using the fact that piece A must have a mass of $0.50M$ ($= M - 0.20M - 0.30M$):

$$p_{fA,x} = -0.50Mv_{fA},$$

$$p_{fB,x} = 0.20Mv_{fB,x} = 0.20Mv_{fB}\cos 50°,$$

$$p_{fC,x} = 0.30Mv_{fC,x} = 0.30Mv_{fC}\cos 80°.$$

Equation 9-36 can now be written as

$$P_{ix} = P_{fx} = p_{fA,x} + p_{fB,x} + p_{fC,x}.$$

Then, with $v_{fC} = 5.0$ m/s and $v_{fB} = 9.64$ m/s, we have

$$0 = -0.50Mv_{fA} + 0.20M(9.64 \text{ m/s})\cos 50°$$
$$+ 0.30M(5.0 \text{ m/s})\cos 80°,$$

from which we find

$$v_{fA} = 3.0 \text{ m/s.} \qquad \text{(Answer)}$$

✔**CHECKPOINT 6:** Suppose that the exploding coconut is accelerating in the negative direction of y in Fig. 9-10 (say it is on a ramp that slants downward in that direction). Is linear momentum conserved along (a) the x axis (as stated in Eq. 9-36) and (b) the y axis (as stated in Eq. 9-35)?

PROBLEM-SOLVING TACTICS

Tactic 2: *Conservation of Linear Momentum*
For problems involving the conservation of linear momentum, first make sure that you have chosen a closed, isolated system. *Closed* means that no matter (no particles) passes through the system boundary in any direction. *Isolated* means that the net external force acting on the system is zero. If it is not isolated, then remember that each component of linear momentum is conserved separately if the corresponding component of the net external force is zero. So, you might conserve one component and not another.

Next, select two appropriate states of the system (which you may choose to call the initial state and the final state) and write expressions for the linear momentum of the system in each of these two states. In writing these expressions, make sure that you know what inertial reference frame you are using, and make sure also that you include the entire system, not missing any part of it and not including objects that do not belong to your system.

Finally, set your expressions for $\vec{P}_i$ and $\vec{P}_f$ equal to each other and solve for what is requested.

9-7 Systems with Varying Mass: A Rocket

In the systems we have dealt with so far, we have assumed that the total mass of the system remains constant. Sometimes, as in a rocket (Fig. 9-11), it does not. Most of the mass of a rocket on its launching pad is fuel, all of which will eventually be burned and ejected from the nozzle of the rocket engine.

We handle the variation of the mass of the rocket as the rocket accelerates by applying Newton's second law, not to the rocket alone but to the rocket and its ejected combustion products taken together. The mass of *this* system does *not* change as the rocket accelerates.

Finding the Acceleration

Assume that we are at rest relative to an inertial reference frame, watching a rocket accelerate through deep space with no gravitational or atmospheric drag forces acting

Fig. 9-11 Liftoff of Project Mercury spacecraft.

Fig. 9-12 (*a*) An accelerating rocket of mass *M* at time *t*, as seen from an inertial reference frame. (*b*) The same but at time *t* + *dt*. The exhaust products released during interval *dt* are shown.

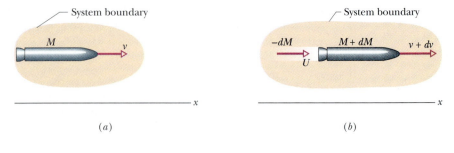

(*a*)   (*b*)

on it. For this one-dimensional motion, let *M* be the mass of the rocket and *v* its velocity at an arbitrary time *t* (see Fig. 9-12*a*).

Figure 9-12*b* shows how things stand a time interval *dt* later. The rocket now has velocity *v* + *dv* and mass *M* + *dM*, where the change in mass *dM* is a *negative quantity*. The exhaust products released by the rocket during interval *dt* have mass −*dM* and velocity *U* relative to our inertial reference frame.

Our system consists of the rocket and the exhaust products released during interval *dt*. The system is closed and isolated, so the linear momentum of the system must be conserved during *dt*; that is,

$$P_i = P_f, \tag{9-37}$$

where the subscripts *i* and *f* indicate the values at the beginning and end of time interval *dt*. We can rewrite Eq. 9-37 as

$$Mv = -dM\,U + (M + dM)(v + dv), \tag{9-38}$$

where the first term on the right is the linear momentum of the exhaust products released during interval *dt* and the second term is the linear momentum of the rocket at the end of interval *dt*.

We can simplify Eq. 9-38 by using the relative speed v_{rel} between the rocket and the exhaust products, which is related to the velocities relative to the frame with

$$\left(\begin{array}{c}\text{velocity of rocket} \\ \text{relative to frame}\end{array}\right) = \left(\begin{array}{c}\text{velocity of rocket} \\ \text{relative to products}\end{array}\right) + \left(\begin{array}{c}\text{velocity of products} \\ \text{relative to frame}\end{array}\right).$$

In symbols, this means

$$(v + dv) = v_{\text{rel}} + U,$$

or

$$U = v + dv - v_{\text{rel}}. \tag{9-39}$$

Substituting this result for *U* into Eq. 9-38 yields, with a little algebra,

$$-dM\,v_{\text{rel}} = M\,dv. \tag{9-40}$$

Dividing each side by *dt* gives us

$$-\frac{dM}{dt}v_{\text{rel}} = M\frac{dv}{dt}. \tag{9-41}$$

We replace *dM/dt* (the rate at which the rocket loses mass) by −*R*, where *R* is the (positive) mass rate of fuel consumption, and we recognize that *dv/dt* is the acceleration of the rocket. With these changes, Eq. 9-41 becomes

$$Rv_{\text{rel}} = Ma \qquad \text{(first rocket equation).} \tag{9-42}$$

Equation 9-42 holds at any instant, with the mass *M*, the fuel consumption rate *R*, and the acceleration *a* evaluated at that instant.

The left side of Eq. 9-42 has the dimensions of a force (kg · m/s² = N) and depends only on design characteristics of the rocket engine, namely, the rate *R* at which it consumes fuel mass and the speed v_{rel} with which that mass is ejected relative to the rocket. We call this term Rv_{rel} the **thrust** of the rocket engine and represent it with T. Newton's second law emerges clearly if we write Eq. 9-42 as $T = Ma$, in which a is the acceleration of the rocket at the time that its mass is M.

Finding the Velocity

How will the velocity of a rocket change as it consumes its fuel? From Eq. 9-40 we have

$$dv = -v_{rel}\frac{dM}{M}.$$

Integrating leads to

$$\int_{v_i}^{v_f} dv = -v_{rel}\int_{M_i}^{M_f}\frac{dM}{M},$$

in which M_i is the initial mass of the rocket and M_f its final mass. Evaluating the integrals then gives

$$v_f - v_i = v_{rel}\ln\frac{M_i}{M_f} \qquad \text{(second rocket equation)} \qquad (9\text{-}43)$$

for the increase in the speed of the rocket during the change in mass from M_i to M_f. (The symbol "ln" in Eq. 9-43 means the *natural logarithm*.) We see here the advantage of multistage rockets, in which M_f is reduced by discarding successive stages when their fuel is depleted. An ideal rocket would reach its destination with only its payload remaining.

Sample Problem 9-8

A rocket whose initial mass M_i is 850 kg consumes fuel at the rate $R = 2.3$ kg/s. The speed v_{rel} of the exhaust gases relative to the rocket engine is 2800 m/s.

(a) What thrust does the rocket engine provide?

SOLUTION: The Key Idea here is that the thrust T is equal to the product of the fuel consumption rate R and the relative speed v_{rel} at which exhaust gases are expelled:

$$T = Rv_{rel} = (2.3 \text{ kg/s})(2800 \text{ m/s})$$
$$= 6440 \text{ N} \approx 6400 \text{ N}. \qquad \text{(Answer)}$$

(b) What is the initial acceleration of the rocket?

SOLUTION: We can relate the thrust T of a rocket to the magnitude a of the resulting acceleration with $T = Ma$, where M is the rocket's mass. The Key Idea, however, is that M decreases and a increases as fuel is consumed. Because we want the initial value of a here, we must use the initial value M_i of the mass, finding that

$$a = \frac{T}{M_i} = \frac{6440 \text{ N}}{850 \text{ kg}} = 7.6 \text{ m/s}^2. \qquad \text{(Answer)}$$

To be launched from Earth's surface, a rocket must have an initial acceleration greater than $g = 9.8$ m/s². Put another way, the thrust T of the rocket engine must exceed the initial gravitational force

on the rocket, which here has the magnitude $M_i g$, which gives us (850 kg)(9.8 m/s²), or 8330 N. Because the acceleration or thrust requirement is not met (here $T = 6400$ N), our rocket could not be launched from Earth's surface by itself; it would require another, more powerful, rocket.

(c) Suppose, instead, that the rocket is launched from a spacecraft already in deep space, where we can neglect any gravitational force acting on it. The mass M_f of the rocket when its fuel is exhausted is 180 kg. What is its speed relative to the spacecraft at that time? Assume that the spacecraft is so massive that the launch does not alter its speed.

SOLUTION: The Key Idea here is that the rocket's final speed v_f (when the fuel is exhausted) depends on the ratio M_i/M_f of its initial mass to its final mass, as given by Eq. 9-43. With the initial speed $v_i = 0$, we have

$$v_f = v_{rel}\ln\frac{M_i}{M_f}$$
$$= (2800 \text{ m/s})\ln\frac{850 \text{ kg}}{180 \text{ kg}}$$
$$= (2800 \text{ m/s})\ln 4.72 \approx 4300 \text{ m/s}. \qquad \text{(Answer)}$$

Note that the ultimate speed of the rocket can exceed the exhaust speed v_{rel}.

9-8   External Forces and Internal Energy Changes

As the ice skater in Fig. 9-13a pushes herself away from a railing, there is a force $\vec{F}$ on her from the railing at angle ϕ to the horizontal. This force accelerates her, increasing her speed until she leaves the railing (Fig. 9-13b). Thus, her kinetic energy is increased via the force. This example differs in two ways from earlier examples in which an object's kinetic energy is changed via a force:

1. Previously, each part of an object moved rigidly in the same direction. Here the skater's arm does not move like the rest of her body.

2. Previously, energy was transferred between the object (or system) and its environment via an external force; that is, the force did work. Here the energy is transferred internally (from one part of the system to another) via the external force $\vec{F}$. In particular, the energy is transferred from internal biochemical energy of the skater's muscles to kinetic energy of her body as a whole.

We want to relate the external force $\vec{F}$ to that internal energy transfer.

In spite of the two differences listed here, we can relate $\vec{F}$ to the change in kinetic energy much as we did for a particle in Section 7-3. To do so, we first mentally concentrate the mass M of the skater at her center of mass so that we can treat her as a particle at the center of mass. Then we treat the external force $\vec{F}$ as acting on the particle (Fig. 9-13c). The horizontal force component $F \cos \phi$ accelerates that particle, resulting in a change ΔK in its kinetic energy during a displacement of magnitude d. As we prove later, we can relate these quantities with

$$\Delta K = Fd \cos \phi. \qquad (9\text{-}44)$$

We can imagine cases in which the external force would also change the height of the skater's center of mass and thus cause a change ΔU in the gravitational potential energy of the skater–Earth system. To include such a change ΔU, we can rewrite Eq. 9-44 as

$$\Delta K + \Delta U = Fd \cos \phi. \qquad (9\text{-}45)$$

The left side here is ΔE_{mec}, the change in the mechanical energy of the system. Thus, in the general case, we have

$$\Delta E_{mec} = Fd \cos \phi \qquad \text{(external force, change in } E_{mec}\text{)}. \qquad (9\text{-}46)$$

Next let us consider the energy transfers in the ice skater, which we now take as the system. Although an external force acts on the system, the force does not transfer energy to or from the system. Thus, the total energy E of the system cannot change: $\Delta E = 0$. We know that when the ice skater pushes off from the rail, there is a change not only in the mechanical energy of her center of mass, but also in the energy of her muscles. Without going into the details of the change in muscular energy, let us just represent it with ΔE_{int} (for a change in internal energy). Then we can write $\Delta E = 0$ as

$$\Delta E_{int} + \Delta E_{mec} = 0, \qquad (9\text{-}47)$$

or

$$\Delta E_{int} = -\Delta E_{mec}. \qquad (9\text{-}48)$$

This equation means that as E_{mec} increases for the ice skater, E_{int} decreases by just as much. Substituting for ΔE_{mec} in Eq. 9-48 from Eq. 9-46, we have

$$\Delta E_{int} = -Fd \cos \phi \qquad \text{(external force, internal energy change)}. \qquad (9\text{-}49)$$

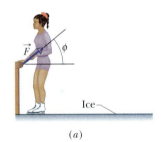

(a)

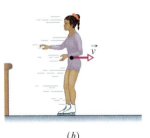

(b)

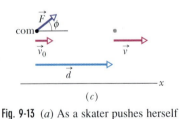

(c)

Fig. 9-13 (a) As a skater pushes herself away from a railing, the force on her from the railing is $\vec{F}$. (b) After the skater leaves the railing, her center of mass has velocity $\vec{v}$. (c) External force $\vec{F}$ is taken to act at the skater's center of mass, at angle ϕ with a horizontal x axis. When the center of mass goes through displacement $\vec{d}$, its velocity is changed from $\vec{v}_0$ to $\vec{v}$ by the horizontal component of $\vec{F}$.

Fig. 9-14 A vehicle accelerates to the right using four-wheel drive. The road exerts four frictional forces (two of them shown) on the bottom surfaces of the tires. Taken together, these four forces make up the net external force $\vec{F}$ acting on the car.

This equation relates the change ΔE_{int} in internal energy that is made via the external force $\vec{F}$. If $\vec{F}$ is not constant in magnitude, we can replace the symbol F in Eqs. 9-46 and 9-49 with the symbol F_{avg} for the average magnitude of $\vec{F}$.

Although we have used an ice skater to derive Eqs. 9-46 and 9-49, the equations hold for other objects in which a change ΔE_{int} in the internal energy of a system is made via an external force. For example, consider a vehicle with four-wheel drive (all four wheels are made to turn by the engine) as its speed is increased. During the acceleration, the engine causes the tires to push backward on the road's surface. This push produces frictional forces that act on each tire in the forward direction (Fig. 9-14). The net external force $\vec{F}$, which is the sum of these frictional forces, gives the car's center of mass an acceleration $\vec{a}$. Thus, there is a transfer of energy from the internal energy stored in the fuel to the kinetic energy of the car. If $\vec{F}$ is constant, then for a given displacement $\vec{d}$ of the car's center of mass along level road, we can relate the change ΔK in the car's kinetic energy to the external force $\vec{F}$ with Eq. 9-45, with $\Delta U = 0$ and $\phi = 0$.

If the driver applies the brakes, Eq. 9-45 still holds. Now $\vec{F}$ due to the frictional forces is toward the rear, and $\phi = 180°$. Energy is now transferred from kinetic energy of the car's center of mass to thermal energy of the brakes.

Proof of Equation 9-44

Let us return to the ice skater and Fig. 9-13. Suppose that during displacement $\vec{d}$ of her center of mass, the velocity of her center of mass changes from $\vec{v}_0$ to $\vec{v}$. Then the magnitude of $\vec{v}$ is, by Eq. 2-16,

$$v^2 = v_0^2 + 2a_x d, \tag{9-50}$$

where a_x is her acceleration. Multiplying both sides of Eq. 9-50 by the skater's mass M and rearranging yield

$$\tfrac{1}{2}Mv^2 - \tfrac{1}{2}Mv_0^2 = Ma_x d. \tag{9-51}$$

The left side of Eq. 9-51 is the difference between the final kinetic energy K_f of the center of mass and the initial kinetic energy K_i. This difference is the change ΔK in the kinetic energy of the center of mass due to the force $\vec{F}$. Making that substitution, and substituting the product $F \cos \phi$ for the product Ma_x according to Newton's second law, we get

$$\Delta K = Fd \cos \phi,$$

as we intended to prove.

Sample Problem 9-9

When a click beetle is upside down on its back, it jumps upward by suddenly arching its back, transferring energy stored in a muscle to mechanical energy. This launching mechanism produces an audible click, giving the beetle its name. Videotape of a certain click-beetle jump shows that the center of mass of a click beetle of mass $m = 4.0 \times 10^{-6}$ kg moved directly upward by 0.77 mm during the launch and then to a maximum height of $h = 0.30$ m. What was the average magnitude of the external force $\vec{F}$ on the beetle's back from the floor during the launch?

SOLUTION: The Key Idea here is that during the launch, energy is transferred from an internal energy of the beetle to the mechanical energy of the beetle–Earth system, in the amount ΔE_{mec}. The transfer is made via the external force $\vec{F}$. To find the magnitude of the force with Eq. 9-46, we first need an expression for ΔE_{mec}.

Let the system's mechanical energy be $E_{mec,0}$ just before the launch and $E_{mec,1}$ at the end of the launch. Then the change ΔE_{mec} is

$$\Delta E_{mec} = E_{mec,1} - E_{mec,0}. \tag{9-52}$$

We must now find expressions for $E_{mec,0}$ and $E_{mec,1}$. Let the gravitational potential energy of the beetle–Earth system be $U_0 = 0$ when the beetle is on the floor. Also, just before the launch, the kinetic energy of the beetle's center of mass is $K_0 = 0$, so the beginning mechanical energy $E_{mec,0}$ is 0. Unfortunately, when we try to find $E_{mec,1}$, we get stuck because we do not know the beetle's kinetic energy K_1 or speed v_1 at the end of the launch.

We get unstuck with a second Key Idea: The mechanical energy of the system does not change from the time the beetle is launched until it reaches its maximum height. We can find this mechanical energy $E_{mec,1}$ at the maximum height because we know the beetle's speed ($v = 0$) and height ($y = h$) there. Thus, we have

$$E_{mec,1} = K + U = \tfrac{1}{2}mv^2 + mgy = 0 + mgh = mgh.$$

Substituting this and $E_{mec,0} = 0$ into Eq. 9-52 now gives us

$$\Delta E_{mec} = mgh - 0 = mgh. \tag{9-53}$$

We can now use Eq. 9-46 to relate this change in mechanical energy to the external force. We write

$$\Delta E_{mec} = F_{avg}d \cos \phi. \tag{9-54}$$

F_{avg} is the average magnitude of the external force on the beetle, d is the magnitude (0.77 mm) of the displacement of the beetle's center of mass during the launch (when the external force acts on it), and ϕ ($= 0°$) is the angle between the directions of the external force and the displacement.

Solving Eq. 9-54 for F_{avg} and using Eq. 9-53 yield

$$\begin{aligned} F_{avg} &= \frac{\Delta E_{mec}}{d \cos \phi} = \frac{mgh}{d \cos \phi} \\ &= \frac{(4.0 \times 10^{-6} \text{ kg})(9.8 \text{ m/s}^2)(0.30 \text{ m})}{(7.7 \times 10^{-4} \text{ m})(\cos 0°)} \\ &= 1.5 \times 10^{-2} \text{ N}. \end{aligned} \qquad \text{(Answer)}$$

This force magnitude may seem small, but to the click beetle it is enormous because, as you can show, it gives the beetle an acceleration of over $380g$ during the launch.

REVIEW & SUMMARY

Center of Mass   The **center of mass** of a system of n particles is defined to be the point whose coordinates are given by

$$x_{com} = \frac{1}{M} \sum_{i=1}^{n} m_i x_i, \quad y_{com} = \frac{1}{M} \sum_{i=1}^{n} m_i y_i, \quad z_{com} = \frac{1}{M} \sum_{i=1}^{n} m_i z_i, \tag{9-5}$$

or

$$\vec{r}_{com} = \frac{1}{M} \sum_{i=1}^{n} m_i \vec{r}_i, \tag{9-8}$$

where M is the total mass of the system. If the mass is continuously distributed, the center of mass is given by

$$x_{com} = \frac{1}{M} \int x \, dm, \quad y_{com} = \frac{1}{M} \int y \, dm, \quad z_{com} = \frac{1}{M} \int z \, dm. \tag{9-9}$$

If the density (mass per unit volume) is uniform, then Eq. 9-9 can be rewritten as

$$x_{com} = \frac{1}{V} \int x \, dV, \quad y_{com} = \frac{1}{V} \int y \, dV, \quad z_{com} = \frac{1}{V} \int z \, dV, \tag{9-11}$$

where V is the volume occupied by M.

Newton's Second Law for a System of Particles   The motion of the center of mass of any system of particles is governed by **Newton's second law for a system of particles**, which is

$$\vec{F}_{net} = M\vec{a}_{com}. \tag{9-14}$$

Here $\vec{F}_{net}$ is the net force of all the *external* forces acting on the system, M is the total mass of the system, and $\vec{a}_{com}$ is the acceleration of the system's center of mass.

Linear Momentum and Newton's Second Law   For a single particle, we define a quantity $\vec{p}$ called its **linear momentum** as

$$\vec{p} = m\vec{v}, \tag{9-22}$$

and can write Newton's second law in terms of this momentum:

$$\vec{F}_{net} = \frac{d\vec{p}}{dt}. \tag{9-23}$$

For a system of particles these relations become

$$\vec{P} = M\vec{v}_{com} \quad \text{and} \quad \vec{F}_{net} = \frac{d\vec{P}}{dt}. \tag{9-25, 9-27}$$

Conservation of Linear Momentum   If a system is isolated so that no net *external* force acts on the system, the linear momentum $\vec{P}$ of the system remains constant:

$$\vec{P} = \text{constant} \qquad \text{(closed, isolated system).} \tag{9-29}$$

This can also be written as

$$\vec{P}_i = \vec{P}_f \qquad \text{(closed, isolated system),} \tag{9-30}$$

where the subscripts refer to the values of $\vec{P}$ at some initial time and at a later time. Equations 9-29 and 9-30 are equivalent statements of the **law of conservation of linear momentum.**

Variable-Mass Systems   If a system has varying mass, we redefine the system, enlarging its boundaries until they encompass a larger system whose mass *does* remain constant; then we apply the law of conservation of linear momentum. For a rocket, this means that the system includes both the rocket and its exhaust gases. Analysis of such a system shows that in the absence of external forces a rocket accelerates at an instantaneous rate given by

$$Rv_{rel} = Ma \qquad \text{(first rocket equation),} \tag{9-42}$$

in which M is the rocket's instantaneous mass (including unexpended fuel), R is the fuel consumption rate, and v_{rel} is the fuel's exhaust speed relative to the rocket. The term Rv_{rel} is the **thrust** of the rocket engine. For a rocket with constant R and v_{rel}, whose speed changes from v_i to v_f when its mass changes from M_i to M_f,

$$v_f - v_i = v_{\text{rel}} \ln \frac{M_i}{M_f} \qquad \text{(second rocket equation).} \quad \text{(9-43)}$$

External Forces and Internal Energy Changes Energy can be transferred inside a system between an internal energy and mechanical energy via an external force $\vec{F}$. The change ΔE_{int} in the

internal energy is related by

$$\Delta E_{\text{int}} = -Fd \cos \phi \quad \text{(9-49)}$$

to the external force, the displacement $\vec{d}$ of the system's center of mass, and the angle ϕ between the directions of $\vec{F}$ and $\vec{d}$. The change in the mechanical energy is

$$\Delta E_{\text{mec}} = \Delta K + \Delta U = Fd \cos \phi \quad \text{(9-46, 9-45)}$$

QUESTIONS

1. Figure 9-15 shows four uniform square metal plates that have had a section removed. The origin of the x and y axes is at the center of the original plate, and in each case the center of mass of the removed section was at the origin. In each case, where is the center of mass of the remaining section of the plate? Answer in terms of quadrants, lines, or points.

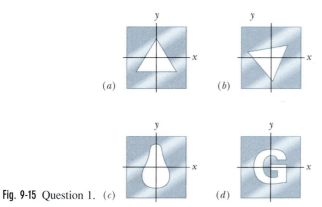

Fig. 9-15 Question 1.

2. Some skilled basketball players seem to hang in midair during a jump at the basket, allowing them more time to shift the ball from hand to hand and then into the basket. If a player raises arms or legs during a jump, does the player's time in the air increase, decrease, or stay the same?

3. In Fig. 9-16, a penguin stands at the left edge of a uniform sled of length L, which lies on frictionless ice. The sled and penguin have equal masses. (a) Where is the center of mass of the sled? (b) How far and in what direction is the center of the sled from the center of mass of the sled–penguin system?

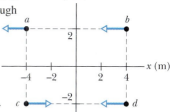

Fig. 9-16 Question 3.

The penguin then waddles to the right edge of the sled, and the sled slides on the ice. (c) Does the center of mass of the sled–penguin system move leftward, rightward, or not at all? (d) Now how far and in what direction is the center of the sled from the center of mass of the sled–penguin system? (e) How far does the penguin move relative to the sled? Relative to the center of mass of the sled–penguin system, how far does (f) the center of the sled move and (g) the penguin move? (Warm-up for Problem 19)

4. In Question 3 and Fig. 9-16, suppose that the sled and penguin are initially moving rightward at speed v_0. (a) As the penguin waddles to the right edge of the sled, is the speed v of the sled less than, greater than, or equal to v_0? (b) If the penguin then waddles back to the left edge, during that motion is the speed v of the sled less than, greater than, or equal to v_0?

5. Figure 9-17 shows an overhead view of four particles of equal mass sliding over a frictionless surface at constant velocity. The directions of the velocities are indicated; their magnitudes are equal. Consider pairing the particles. Which pairs form a system with a center of mass that (a) is stationary, (b) is stationary and at the origin, and (c) passes through the origin?

Fig. 9-17 Question 5.

6. Figure 9-18 shows an overhead view of three particles on which external forces act. The magnitudes and directions of the forces on two of the particles are indicated. What are the magnitude and direction of the force acting on the third particle if the center of mass of the three-particle system is (a) stationary, (b) moving at a constant velocity rightward, and (c) accelerating rightward?

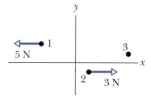

Fig. 9-18 Question 6.

7. A container sliding along an x axis on a frictionless surface explodes into three pieces. The pieces then move along the x axis in the directions indicated in Fig. 9-19. The following table gives four sets of magnitudes (in kg · m/s) for the momenta $\vec{p}_1$, $\vec{p}_2$, and $\vec{p}_3$ of the pieces. Rank the sets according to the initial speed of the container, greatest first.

	p_1	p_2	p_3		p_1	p_2	p_3
(a)	10	2	6	(b)	10	6	2
(c)	2	10	6	(d)	6	2	10

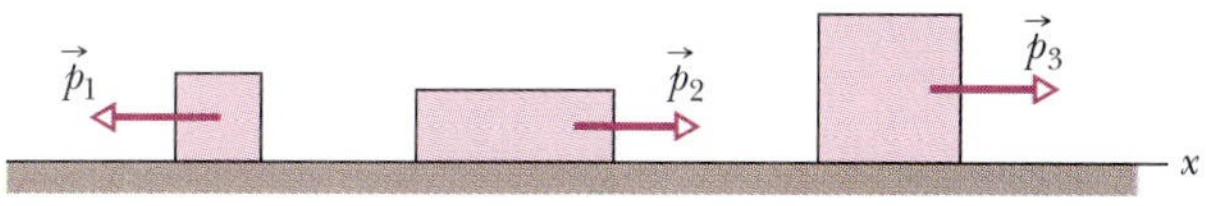

Fig. 9-19 Question 7.

8. A spaceship that is moving along an x axis separates into two parts, like the hauler in Fig. 9-9. (a) Which of the graphs in

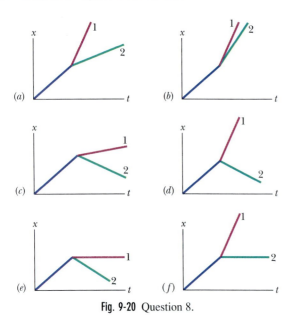

Fig. 9-20 could possibly give the position versus time for the ship and the two parts? (b) Which of the numbered lines pertains to the trailing part? (c) Rank the possible graphs according to the relative speed between the parts, greatest first.

9. Consider a box, like that in Sample Problem 9-5, which explodes into two pieces while moving with a constant positive velocity along an x axis. If one piece, with mass m_1, ends up with positive velocity $\vec{v}_1$, then the second piece, with mass m_2, could end up with (a) a positive velocity $\vec{v}_2$ (Fig. 9-21a), (b) a negative velocity $\vec{v}_2$ (Fig. 9-21b), or (c) zero velocity (Fig. 9-21c). Rank those three possible results for the second piece according to the corresponding magnitude of $\vec{v}_1$, greatest first.

Fig. 9-20 Question 8.

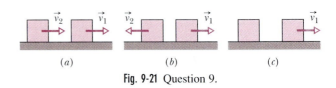

Fig. 9-21 Question 9.

EXERCISES & PROBLEMS

SEC. 9-2   The Center of Mass

1E. (a) How far is the center of mass of the Earth–Moon system from the center of Earth? (Appendix C gives the masses of Earth and the Moon and the distance between the two.) (b) Express the answer to (a) as a fraction of Earth's radius R_e. **ssm**

2E. A distance of 1.131×10^{-10} m lies between the centers of the carbon and oxygen atoms in a carbon monoxide (CO) gas molecule. Locate the center of mass of a CO molecule relative to the carbon atom. (Find the masses of C and O in Appendix F.)

3E. What are (a) the x coordinate and (b) the y coordinate of the center of mass of the three-particle system shown in Fig. 9-22? (c) What happens to the center of mass as the mass of the top-most particle is gradually increased? **ssm**

4E. Three thin rods, each of length L, are arranged in an inverted **U**, as shown in Fig. 9-23. The two rods on the arms of the **U** each have mass M; the third rod has mass $3M$. Where is the center of mass of the assembly?

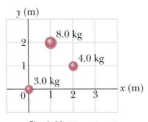

Fig. 9-22 Exercise 3.

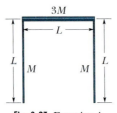

Fig. 9-23 Exercise 4.

5E. A uniform square plate 6 m on a side has had a square piece 2 m on a side cut out of it (Fig. 9-24). The center of that piece is at $x = 2$ m, $y = 0$. The center of the square plate is at $x = y = 0$. Find (a) the x coordinate and (b) the y coordinate of the center of mass of the remaining piece.

6P. Figure 9-25 shows the dimensions of a composite slab; half the slab is made of aluminum (density $= 2.70$ g/cm^3) and half is made of iron (density $= 7.85$ g/cm^3). Where is the center of mass of the slab?

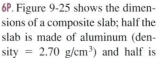

Fig. 9-24 Exercise 5.

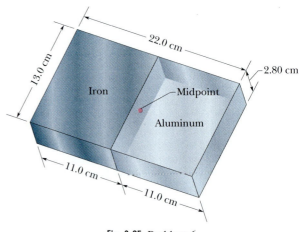

Fig. 9-25 Problem 6.

7P. In the ammonia (NH_3) molecule (see Fig. 9-26), the three hydrogen (H) atoms form an equilateral triangle; the center of the triangle is 9.40×10^{-11} m from each hydrogen atom. The nitrogen (N) atom is at the apex of a pyramid, with the three hydrogen atoms forming the base. The nitrogen-to-hydrogen atomic mass ratio is 13.9, and the nitrogen-to-hydrogen distance is 10.14×10^{-11} m. Locate the center of mass of the molecule relative to the nitrogen atom. ilw

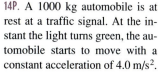

Fig. 9-26 Problem 7.

8P. Figure 9-27 shows a cubical box that has been constructed from metal plate of uniform density and negligible thickness. The box is open at the top and has edge length 40 cm. Find (a) the x coordinate, (b) the y coordinate, and (c) the z coordinate of the center of mass of the box.

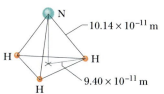

Fig. 9-27 Problem 8.

9P*. A right cylindrical can with mass M, height H, and uniform density is initially filled with soda of mass m (Fig. 9-28). We punch small holes in the top and bottom to drain the soda; we then consider the height h of the center of mass of the can and any soda within it. What is h (a) initially and (b) when all the soda has drained? (c) What happens to h during the draining of the soda? (d) If x is the height of the remaining soda at any given instant, find x (in terms of M, H, and m) when the center of mass reaches its lowest point. ssm

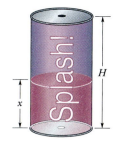

Fig. 9-28 Problem 9.

SEC. 9-3 Newton's Second Law for a System of Particles

10E. Two skaters, one with mass 65 kg and the other with mass 40 kg, stand on an ice rink holding a pole of length 10 m and negligible mass. Starting from the ends of the pole, the skaters pull themselves along the pole until they meet. How far does the 40 kg skater move?

11E. An old Chrysler with mass 2400 kg is moving along a straight stretch of road at 80 km/h. It is followed by a Ford with mass 1600 kg moving at 60 km/h. How fast is the center of mass of the two cars moving? ssm

12E. A man of mass m clings to a rope ladder suspended below a balloon of mass M; see Fig. 9-29. The balloon is stationary with respect to the ground. (a) If the man begins to climb the ladder at speed v (with respect to the ladder), in what direction and with what speed (with respect to the ground) will the balloon move?

(b) What is the state of the motion after the man stops climbing?

13P. A stone is dropped at $t = 0$. A second stone, with twice the mass of the first, is dropped from the same point at $t = 100$ ms. (a) How far below the release point is the center of mass of the two stones at $t = 300$ ms? (Neither stone has yet reached the ground.) (b) How fast is the center of mass of the two-stone system moving at that time? ilw

14P. A 1000 kg automobile is at rest at a traffic signal. At the instant the light turns green, the automobile starts to move with a constant acceleration of 4.0 m/s².

Fig. 9-29 Exercise 12.

At the same instant a 2000 kg truck, traveling at a constant speed of 8.0 m/s, overtakes and passes the automobile. (a) How far is the center of mass of the automobile–truck system from the traffic light at $t = 3.0$ s? (b) What is the speed of the center of mass of the automobile–truck system then?

15P. A shell is shot with an initial velocity $\vec{v}_0$ of 20 m/s, at an angle of 60° with the horizontal. At the top of the trajectory, the shell explodes into two fragments of equal mass (Fig. 9-30). One fragment, whose speed immediately after the explosion is zero, falls vertically. How far from the gun does the other fragment land, assuming that the terrain is level and that air drag is negligible? ssm

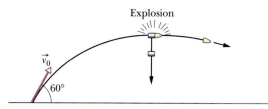

Fig. 9-30 Problem 15.

16P. A big olive ($m = 0.50$ kg) lies at the origin and a big Brazil nut ($M = 1.5$ kg) lies at the point (1.0, 2.0) m in an xy plane. At $t = 0$, a force $\vec{F}_o = (2\hat{i} + 3\hat{j})$ N begins to act on the olive, and a force $\vec{F}_n = (-3\hat{i} - 2\hat{j})$ N begins to act on the nut. In unit-vector notation, what is the displacement of the center of mass of the olive–nut system at $t = 4.0$ s, with respect to its position at $t = 0$?

17P. Two identical containers of sugar are connected by a massless cord that passes over a massless, frictionless pulley with a diameter of 50 mm (Fig. 9-31). The two containers are at the same level. Each originally has a mass of 500 g. (a) What is the horizontal position of their center of mass? (b) Now 20 g of sugar is transferred from one container to the other, but the containers are prevented from moving. What is the

Fig. 9-31 Problem 17.

new horizontal position of their center of mass, relative to the central axis through the lighter container? (c) The two containers are now released. In what direction does the center of mass move? (d) What is its acceleration? ssm

18P. Ricardo, of mass 80 kg, and Carmelita, who is lighter, are enjoying Lake Merced at dusk in a 30 kg canoe. When the canoe is at rest in the placid water, they exchange seats, which are 3.0 m apart and symmetrically located with respect to the canoe's center. Ricardo notices that the canoe moves 40 cm relative to a submerged log during the exchange and calculates Carmelita's mass, which she has not told him. What is it?

19P. In Fig. 9-32*a*, a 4.5 kg dog stands on an 18 kg flatboat and is 6.1 m from the shore. He walks 2.4 m along the boat toward shore and then stops. Assuming there is no friction between the boat and the water, find how far the dog is then from the shore. (*Hint:* See Fig. 9-32*b*. The dog moves leftward and the boat moves rightward, but does the center of mass of the *boat* + *dog* system move?) ssm www

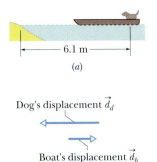

Fig. 9-32 Problem 19.

SEC. 9-5   The Linear Momentum of a System of Particles

20E. How fast must an 816 kg VW Beetle travel (a) to have the same linear momentum as a 2650 kg Cadillac going 16 km/h and (b) to have the same kinetic energy?

21E. Suppose that your mass is 80 kg. How fast would you have to run to have the same linear momentum as a 1600 kg car moving at 1.2 km/h?

22E. A 0.70 kg ball is moving horizontally with a speed of 5.0 m/s when it strikes a vertical wall. The ball rebounds with a speed of 2.0 m/s. What is the magnitude of the change in linear momentum of the ball?

23P. A 2100 kg truck traveling north at 41 km/h turns east and accelerates to 51 km/h. (a) What is the change in the kinetic energy of the truck? What are (b) the magnitude and (c) the direction of the change in the linear momentum of the truck? ilw

24P. A 0.165 kg cue ball with an initial speed of 2.00 m/s bounces off the rail in a game of pool, as shown from an overhead view in Fig. 9-33. For *x* and *y* axes located as shown, the bounce reverses the *y* component of the ball's velocity but does not alter the *x* component. (a) What is θ in Fig. 9-33? (b) What is the change in the ball's linear momentum in unit-vector notation? (The fact that the ball rolls is not relevant to either question.)

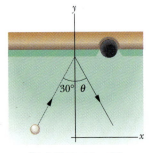

Fig. 9-33 Problem 24.

25P. An object is tracked by a radar station and found to have a position vector given by $\vec{r} = (3500 - 160t)\hat{i} + 2700\hat{j} + 300\hat{k}$,

with $\vec{r}$ in meters and *t* in seconds. The radar station's *x* axis points east, its *y* axis north, and its *z* axis vertically up. If the object is a 250 kg meteorological missile, what are (a) its linear momentum, (b) its direction of motion, and (c) the net force on it?

26P. A 0.30 kg softball has a velocity of 15 m/s at an angle of 35° below the horizontal just before making contact with the bat. What is the magnitude of the change in momentum of the ball while it is in contact with the bat if the ball leaves the bat with a velocity of (a) 20 m/s, vertically downward and (b) 20 m/s, horizontally away from the batter and back toward the pitcher?

SEC. 9-6   Conservation of Linear Momentum

27E. A 91 kg man lying on a surface of negligible friction shoves a 68 g stone away from him, giving it a speed of 4.0 m/s. What velocity does the man acquire as a result? ssm

28E. Two blocks of masses 1.0 kg and 3.0 kg are connected by a spring and rest on a frictionless surface. They are given velocities toward each other such that the 1.0 kg block travels initially at 1.7 m/s toward the center of mass, which remains at rest. What is the initial velocity of the other block?

29E. A 75 kg man is riding on a 39 kg cart traveling at a speed of 2.3 m/s. He jumps off with zero horizontal speed relative to the ground. What is the resulting change in the speed of the cart? ssm

30E. A mechanical toy slides along an *x* axis on a frictionless surface with a velocity of $(-0.40 \text{ m/s})\hat{i}$ when two internal springs separate the toy into three parts, as given in the table. What is the velocity of part *A*?

Part	Mass (kg)	Velocity (m/s)
A	0.50	?
B	0.60	$0.20\hat{i}$
C	0.20	$0.30\hat{i}$

31E. A space vehicle is traveling at 4300 km/h relative to Earth when the exhausted rocket motor is disengaged and sent backward with a speed of 82 km/h relative to the command module. The mass of the motor is four times the mass of the module. What is the speed of the command module relative to Earth just after the separation?

32E. A railroad flatcar of weight *W* can roll without friction along a straight horizontal track. Initially, a man of weight *w* is standing on the car, which is moving to the right with speed v_0 (see Fig. 9-34). What is the change in velocity of the car if the man runs to the left (in the figure) so that his speed relative to the car is v_{rel}?

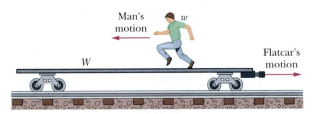

Fig. 9-34 Exercise 32.

33P. The last stage of a rocket, which is traveling at a speed of 7600 m/s, consists of two parts that are clamped together: a rocket case with a mass of 290.0 kg and a payload capsule with a mass of 150.0 kg. When the clamp is released, a compressed spring causes the two parts to separate with a relative speed of 910.0 m/s. What are the speeds of (a) the rocket case and (b) the payload after they have separated? Assume that all velocities are along the same line. Find the total kinetic energy of the two parts (c) before and (d) after they separate; account for any difference. ssm

34P. A 4.0 kg mess kit sliding on a frictionless surface explodes into two 2.0 kg parts, one moving at 3.0 m/s, due north, and the other at 5.0 m/s, 30° north of east. What is the original speed of the mess kit?

35P. A certain radioactive nucleus can transform to another nucleus by emitting an electron and a neutrino. (The *neutrino* is one of the fundamental particles of physics.) Suppose that in such a transformation, the initial nucleus is stationary, the electron and neutrino are emitted along perpendicular paths, and the magnitudes of the linear momenta are 1.2×10^{-22} kg·m/s for the electron and 6.4×10^{-23} kg·m/s for the neutrino. As a result of the emissions, the new nucleus moves (recoils). (a) What is the magnitude of its linear momentum? What is the angle between its path and the path of (b) the electron and (c) the neutrino? (d) What is its kinetic energy if its mass is 5.8×10^{-26} kg? ilw

36P. Particle A and particle B are held together with a compressed spring between them. When they are released, the spring pushes them apart and they then fly off in opposite directions, free of the spring. The mass of A is 2.00 times the mass of B, and the energy stored in the spring was 60 J. Assume that the spring has negligible mass and that all its stored energy is transferred to the particles. Once that transfer is complete, what are the kinetic energies of (a) particle A and (b) particle B?

37P. A 20.0 kg body is moving in the positive x direction with a speed of 200 m/s when, owing to an internal explosion, it breaks into three parts. One part, with a mass of 10.0 kg, moves away from the point of explosion with a speed of 100 m/s in the positive y direction. A second fragment, with a mass of 4.00 kg, moves in the negative x direction with a speed of 500 m/s. (a) What is the velocity of the third (6.00 kg) fragment? (b) How much energy is released in the explosion? Ignore effects due to the gravitational force. ssm ilw www

38P. An object, with mass m and speed v relative to an observer, explodes into two pieces, one three times as massive as the other; the explosion takes place in deep space. The less massive piece stops relative to the observer. How much kinetic energy is added to the system in the explosion, as measured in the observer's reference frame?

39P. A vessel at rest explodes, breaking into three pieces. Two pieces, having equal mass, fly off perpendicular to one another with the same speed of 30 m/s. The third piece has three times the mass of each other piece. What are the magnitude and direction of its velocity immediately after the explosion? ssm

40P. An 8.0 kg body is traveling at 2.0 m/s with no external force acting on it. At a certain instant an internal explosion occurs, splitting the body into two chunks of 4.0 kg mass each. The explosion gives the chunks an additional 16 J of kinetic energy. Neither chunk

leaves the line of original motion. Determine the speed and direction of motion of each of the chunks after the explosion.

SEC. 9-7 Systems with Varying Mass: A Rocket

41E. A 6090 kg space probe, moving nose-first toward Jupiter at 105 m/s relative to the Sun, fires its rocket engine, ejecting 80.0 kg of exhaust at a speed of 253 m/s relative to the space probe. What is the final velocity of the probe? ssm

42E. A rocket is moving away from the solar system at a speed of 6.0×10^3 m/s. It fires its engine, which ejects exhaust with a speed of 3.0×10^3 m/s relative to the rocket. The mass of the rocket at this time is 4.0×10^4 kg, and its acceleration is 2.0 m/s². (a) What is the thrust of the engine? (b) At what rate, in kilograms per second, is exhaust ejected during the firing?

43E. A rocket, which is in deep space and initially at rest relative to an inertial reference frame, has a mass of 2.55×10^5 kg, of which 1.81×10^5 kg is fuel. The rocket engine is then fired for 250 s, during which fuel is consumed at the rate of 480 kg/s. The speed of the exhaust products relative to the rocket is 3.27 km/s. (a) What is the rocket's thrust? After the 250 s firing, what are (b) the mass and (c) the speed of the rocket? ssm ilw

44E. Consider a rocket that is in deep space and at rest relative to an inertial reference frame. The rocket's engine is to be fired for a certain interval. What must be the rocket's *mass ratio* (ratio of initial to final mass) over that interval if the rocket's original speed relative to the inertial frame is to be equal to (a) the exhaust speed (speed of the exhaust products relative to the rocket) and (b) 2.0 times the exhaust speed?

45E. During a lunar mission, it is necessary to increase the speed of a spacecraft by 2.2 m/s when it is moving at 400 m/s relative to the Moon. The speed of the exhaust products from the rocket engine is 1000 m/s relative to the spacecraft. What fraction of the initial mass of the spacecraft must be burned and ejected to accomplish the speed increase?

46E. A railroad car moves at a constant speed of 3.20 m/s under a grain elevator. Grain drops into it at the rate of 540 kg/min. What is the magnitude of the force needed to keep the car moving at constant speed if friction is negligible?

47P. In Fig. 9-35, two long barges are moving in the same direction in still water, one with a speed of 10 km/h and the other with a speed of 20 km/h. While they are passing each other, coal is shov-

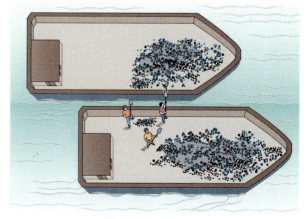

Fig. 9-35 Problem 47.

eled from the slower to the faster one at a rate of 1000 kg/min. How much additional force must be provided by the driving engines of (a) the fast barge and (b) the slow barge if neither is to change speed? Assume that the shoveling is always perfectly sideways and that the frictional forces between the barges and the water do not depend on the mass of the barges. ssm

48P. A 6100 kg rocket is set for vertical firing from the ground. If the exhaust speed is 1200 m/s, how much gas must be ejected each second if the thrust (a) is to equal the magnitude of the gravitational force on the rocket and (b) is to give the rocket an initial upward acceleration of 21 m/s^2?

SEC. 9-8   External Forces and Internal Energy Changes

49E. In 1981, Daniel Goodwin climbed 443 m up the *exterior* of the Sears Building in Chicago using suction cups and metal clips. (a) Approximate his mass and then compute how much energy he had to transfer from biomechanical (internal) energy to the gravitational potential energy of the Earth–Goodwin system to lift his center of mass to that height. (b) How much energy would he have had to transfer if he had, instead, taken the stairs inside the building (to the same height)?

50E. The summit of Mount Everest is 8850 m above sea level. (a) How much energy would a 90 kg climber expend against the gravitational force on him in climbing to the summit from sea level? (b) How many candy bars, at 1.25 MJ per bar, would supply an energy equivalent to this? Your answer should suggest that work done against the gravitational force is a very small part of the energy expended in climbing a mountain.

51E. A sprinter who weighs 670 N runs the first 7.0 m of a race in 1.6 s, starting from rest and accelerating uniformly. What are the sprinter's (a) speed and (b) kinetic energy at the end of the 1.6 s? (c) What average power does the sprinter generate during the 1.6 s interval?

52E. The luxury liner *Queen Elizabeth 2* has a diesel-electric powerplant with a maximum power of 92 MW at a cruising speed of 32.5 knots. What forward force is exerted on the ship at this speed? (1 knot = 1.852 km/h.)

53E. A swimmer moves through the water at an average speed of 0.22 m/s. The average drag force opposing this motion is 110 N. What average power is required of the swimmer?

54E. An automobile with passengers has weight 16,400 N and is moving at 113 km/h when the driver brakes to a stop. The frictional force on the wheels from the road has a magnitude of 8230 N. Find the stopping distance.

55E. A 55 kg woman leaps vertically from a crouching position in which her center of mass is 40 cm above the ground. As her feet leave the floor her center of mass is 90 cm above the ground; it rises to 120 cm at the top of her leap. (a) As she is pressing down on the ground during the leap, what is the average magnitude of the force on her from the ground? (b) What maximum speed does she attain? ssm   www

56P. A 1500 kg automobile starts from rest on a horizontal road and gains a speed of 72 km/h in 30 s. (a) What is the kinetic energy of the auto at the end of the 30 s? (b) What is the average power required of the car during the 30 s interval? (c) What is the instantaneous power at the end of the 30 s interval, assuming that the acceleration is constant?

57P. A locomotive with a power capability of 1.5 MW can accelerate a train from a speed of 10 m/s to 25 m/s in 6.0 min. (a) Calculate the mass of the train. Find (b) the speed of the train and (c) the force accelerating the train as functions of time (in seconds) during the 6.0 min interval. (d) Find the distance moved by the train during the interval.

58P. Resistance to the motion of an automobile consists of road friction, which is almost independent of speed, and air drag, which is proportional to speed-squared. For a certain car with a weight of 12,000 N, the total resistant force F is given by $F = 300 + 1.8v^2$, where F is in newtons and v is in meters per second. Calculate the power (in horsepower) required to accelerate the car at 0.92 m/s^2 when the speed is 80 km/h.

Additional Problem

59. Some solidified lava contains a pattern of horizontal bubble layers separated vertically with few intermediate bubbles. (Researchers must slice open solidified lava to see these bubbles.) Apparently, as the lava was cooling, bubbles rising from the bottom of the lava separated into these layers and then were locked into place when the lava solidified. Similar layering of bubbles has been studied in certain creamy stouts poured fresh from a tap into a clear glass. The rising bubbles quickly become sorted into layers (Fig. 9-36). The bubbles trapped within a layer rise at speed v_t; the free bubbles between the layers rise at a greater speed v_f. Bubbles breaking free from the top of one layer rise to join the bottom of the next layer. Assume that the rate at which a layer loses height at its top is $dy/dt = v_f$ and the rate at which it gains height at its bottom is $dy/dt = v_f$. Also assume that $v_f = 2.0v_t = 1.0$ cm/s. What are the speed and direction of motion of the layer's center of mass?

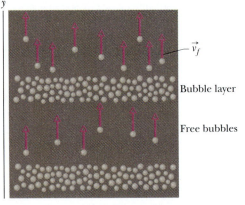

Fig. 9-36 Problem 59.

10 Collisions

Ronald McNair, a physicist and one of the astronauts killed in the explosion of the *Challenger* space shuttle, held a black belt in karate. Here he breaks several concrete slabs with one blow. In such karate demonstrations, a pine board or a concrete "patio block" is typically used. When struck, the board or block bends, storing energy like a stretched spring does, until a critical energy is reached. Then the object breaks. The energy necessary to break the board is about three times that for the block.

Why then is the board considerably easier to break?

The answer is in this chapter.

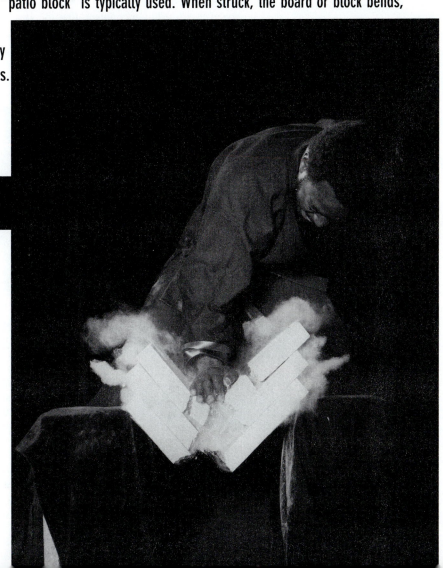

10-1 What Is a Collision?

In everyday language, a *collision* occurs when objects crash into each other. Although we will refine that definition, it conveys the meaning well enough and covers common collisions, such as those between billiard balls, a hammer and a nail, and— too commonly—automobiles. Figure 10-1*a* shows the lasting result of a collision (an impressive crash) that occurred about 20,000 years ago. Collisions range from the microscopic scale of subatomic particles (Fig. 10-1*b*) to the astronomic scale of colliding stars and colliding galaxies. Even when they occur on a human scale, they are often too brief to be visible, although they involve significant distortion of the *colliding bodies* (Fig. 10-1*c*).

We shall use the following more formal definition of collision:

> ▶ A **collision** is an isolated event in which two or more bodies (the colliding bodies) exert relatively strong forces on each other for a relatively short time.

We must be able to distinguish times that are *before, during,* and *after* a collision, as suggested in Fig. 10-2. That figure shows a system of two colliding bodies and indicates that the forces the bodies exert on each other are internal to the system.

Note that our formal definition of collision does not require the "crash" of our informal definition. When a space probe swings around a large planet to pick up speed (a *slingshot* encounter), that too is a collision. The probe and planet do not actually "touch," but a collision does not require contact, and a collision force does not have to be a force involving contact; it can just as easily be a gravitational force, as in this case.

(*a*)

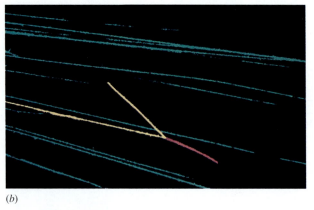

(*b*)

(*c*)

Fig. 10-1 Collisions range widely in scale. (*a*) Meteor Crater in Arizona is about 1200 m wide and 200 m deep. (*b*) An alpha particle coming in from the left (whose trail is colored yellow in this false-color photograph) bounces off a nitrogen nucleus that had been stationary and that now moves toward the bottom right (red trail). (*c*) In a tennis match, the ball is in contact with the racquet for about 4 ms in each collision (for a cumulative time of only 1 s per set).

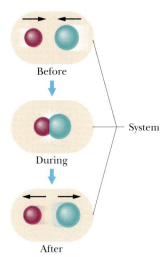

Before

During

After

Fig. 10-2 A flowchart showing the system in which a collision occurs.

Many physicists today spend their time playing what we can call "the collision game." A principal goal of this game is to find out as much as possible about the forces that act during a collision, from knowledge of the state of the particles before and after the collision. Virtually all our understanding of the subatomic world—electrons, protons, neutrons, muons, quarks, and the like—comes from experiments involving collisions. The rules of the game are the laws of conservation of momentum and of energy.

10-2 Impulse and Linear Momentum

Single Collision

Figure 10-3 shows the third-law force pair, $\vec{F}(t)$ and $-\vec{F}(t)$, that acts during a simple head-on collision between two particle-like bodies of different masses. These forces will change the linear momentum of both bodies; the amount of the change will depend not only on the average values of the forces, but also on the time Δt during which they act. To see this quantitatively, let us apply Newton's second law in the form $\vec{F} = d\vec{p}/dt$ to, say, body R on the right in Fig. 10-3. We have

$$d\vec{p} = \vec{F}(t)\, dt, \qquad (10\text{-}1)$$

in which $\vec{F}(t)$ is a time-varying force with magnitude given by the curve in Fig. 10-4a. Let us integrate Eq. 10-1 over the interval Δt—that is, from an initial time t_i (just before the collision) to a final time t_f (just after the collision). We obtain

$$\int_{\vec{p}_i}^{\vec{p}_f} d\vec{p} = \int_{t_i}^{t_f} \vec{F}(t)\, dt. \qquad (10\text{-}2)$$

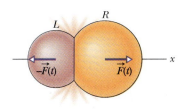

Fig. 10-3 Two particle-like bodies L and R collide with each other. During the collision, body L exerts force $\vec{F}(t)$ on body R, and body R exerts force $-\vec{F}(t)$ on body L. Forces $\vec{F}(t)$ and $-\vec{F}(t)$ are a third-law force pair. Their magnitudes vary with time during the collision, but at any given instant those magnitudes are equal.

The left side of this equation is $\vec{p}_f - \vec{p}_i$, the change in linear momentum of body R. The right side, which is a measure of both the strength and the duration of the collision force, is called the **impulse** $\vec{J}$ of the collision. Thus,

$$\vec{J} = \int_{t_i}^{t_f} \vec{F}(t)\, dt \qquad \text{(impulse defined).} \qquad (10\text{-}3)$$

Equation 10-3 tells us that the magnitude of the impulse is equal to the area under the $F(t)$ curve of Fig. 10-4a. Because $F(t)$ on body R and $-F(t)$ on body L are third-law force pairs, their impulses have the same magnitudes but opposite directions.

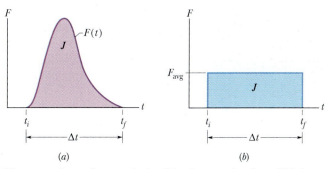

(a)   (b)

Fig. 10-4 (a) The curve shows the magnitude of the time-varying force $F(t)$ that acts on body R during the collision of Fig. 10-3. The area under the curve is equal to the magnitude of the impulse $\vec{J}$ on body R in the collision. (b) The height of the rectangle represents the average force F_{avg} acting on body R over the time interval Δt. The area within the rectangle is equal to the area under the curve in (a) and thus is also equal to the magnitude of the impulse $\vec{J}$ in the collision.

From Eqs. 10-2 and 10-3 we see that the change in the linear momentum of each body in a collision is equal to the impulse that acts on that body:

$$\vec{p}_f - \vec{p}_i = \Delta \vec{p} = \vec{J} \qquad \text{(impulse–linear momentum theorem).} \qquad (10\text{-}4)$$

Equation 10-4 is called the **impulse–linear momentum theorem**; it tells us that impulse and linear momentum are both vectors and have the same units and dimensions. Equation 10-4 can also be written in component form as

$$p_{fx} - p_{ix} = \Delta p_x = J_x, \qquad (10\text{-}5)$$

$$p_{fy} - p_{iy} = \Delta p_y = J_y, \qquad (10\text{-}6)$$

and

$$p_{fz} - p_{iz} = \Delta p_z = J_z. \qquad (10\text{-}7)$$

If F_{avg} is the average magnitude of the force in Fig. 10-4a, we can write the magnitude of the impulse as

$$J = F_{avg} \, \Delta t, \qquad (10\text{-}8)$$

where Δt is the duration of the collision. The value of F_{avg} must be such that the area within the rectangle of Fig. 10-4b is equal to the area under the actual $F(t)$ curve of Fig. 10-4a.

✔ **CHECKPOINT 1:**   A paratrooper whose chute fails to open lands in snow; he is hurt slightly. Had he landed on bare ground, the stopping time would have been 10 times shorter and the collision lethal. Does the presence of the snow increase, decrease, or leave unchanged the values of (a) the paratrooper's change in momentum, (b) the impulse stopping the paratrooper, and (c) the force stopping the paratrooper?

Series of Collisions

Now let's consider the force on a body when it undergoes a series of identical, repeated collisions. For example, as a prank, we might adjust one of those machines that fires tennis balls to fire them at a rapid rate directly at a wall. Each collision would produce a force on the wall, but that is not the force we are seeking. We want the average force F_{avg} on the wall during the bombardment—that is, the average force during a large number of collisions.

In Fig. 10-5, a steady stream of projectile bodies, with identical mass m and linear momenta $m\vec{v}$, moves along an x axis and collides with a target body that is fixed in place. Let n be the number of projectiles that collide in a time interval Δt. Because the motion is along only the x axis, we can use the components of the momenta along that axis. Thus, each projectile has initial momentum mv and undergoes a change Δp in linear momentum because of the collision. The total change in linear momentum for n projectiles during interval Δt is $n \, \Delta p$. The resulting impulse J on the target during Δt is along the x axis and has the same magnitude of $n \, \Delta p$ but is in the opposite direction. We can write this relation in component form as

$$J = -n \, \Delta p, \qquad (10\text{-}9)$$

where the minus sign indicates that J and Δp have opposite directions.

By rearranging Eq. 10-8 and substituting Eq. 10-9, we find the average force F_{avg} acting on the target during the collisions:

$$F_{avg} = \frac{J}{\Delta t} = -\frac{n}{\Delta t} \Delta p = -\frac{n}{\Delta t} m \, \Delta v. \qquad (10\text{-}10)$$

This equation gives us F_{avg} in terms of $n/\Delta t$, the rate at which the projectiles collide with the target, and Δv, the change in the velocity of those projectiles.

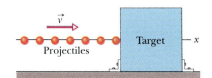

Fig. 10-5 A steady stream of projectiles, with identical linear momenta, collides with a target, which is fixed in place. The average force F_{avg} on the target is to the right and has a magnitude that depends on the rate at which the projectiles collide or, equivalently, the rate at which mass collides.

If the projectiles stop upon impact, then in Eq. 10-10 we can substitute, for Δv,

$$\Delta v = v_f - v_i = 0 - v = -v, \tag{10-11}$$

where v_i (= v) and v_f (= 0) are the velocities before and after the collision, respectively. If, instead, the projectiles bounce (rebound) directly backward from the target with no change in speed, then $v_f = -v$ and we can substitute

$$\Delta v = v_f - v_i = -v - v = -2v. \tag{10-12}$$

In time interval Δt, an amount of mass $\Delta m = nm$ collides with the target. With this result, we can rewrite Eq. 10-10 as

$$F_{avg} = -\frac{\Delta m}{\Delta t}\Delta v. \tag{10-13}$$

This equation gives the average force F_{avg} in terms of $\Delta m/\Delta t$, the rate at which mass collides with the target. Here again we can substitute for Δv from Eq. 10-11 or 10-12 depending on what the projectiles do.

✓CHECKPOINT 2:   The figure shows an overhead view of a ball bouncing from a vertical wall without any change in its speed. Consider the change $\Delta\vec{p}$ in the ball's linear momentum. (a) Is Δp_x positive, negative, or zero? (b) Is Δp_y positive, negative, or zero? (c) What is the direction of $\Delta\vec{p}$?

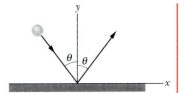

Sample Problem 10-1

A pitched 140 g baseball, in horizontal flight with a speed v_i of 39.0 m/s, is struck by a bat. After leaving the bat, the ball travels in the opposite direction with speed v_f, also 39.0 m/s.

(a) What impulse J acts on the ball while it is in contact with the bat during the collision?

SOLUTION: The **Key Idea** here is that we can calculate the impulse from the change it produces in the ball's linear momentum, using Eq. 10-4 for one-dimensional motion. Let us choose the direction in which the ball is initially moving to be the negative direction. From Eq. 10-4 we have

$$J = p_f - p_i = mv_f - mv_i$$
$$= (0.140 \text{ kg})(39.0 \text{ m/s}) - (0.140 \text{ kg})(-39.0 \text{ m/s})$$
$$= 10.9 \text{ kg} \cdot \text{m/s}. \qquad \text{(Answer)}$$

With our sign convention, the initial velocity of the ball is negative and the final velocity is positive. The impulse turns out to be positive, which tells us that the direction of the impulse vector acting on the ball is the direction in which the bat is swinging.

(b) The impact time Δt for the baseball–bat collision is 1.20 ms. What average force acts on the baseball?

SOLUTION: The **Key Idea** here is that the average force of the collision is the ratio of the impulse J to the duration Δt of the collision (see Eq. 10-8). Thus,

$$F_{avg} = \frac{J}{\Delta t} = \frac{10.9 \text{ kg} \cdot \text{m/s}}{0.00120 \text{ s}}$$
$$= 9080 \text{ N}. \qquad \text{(Answer)}$$

Note that this is the *average* force; the *maximum* force is larger. The sign of the average force on the ball from the bat is positive, which means that the direction of the force vector is the same as that of the impulse vector.

In defining a collision, we assumed that no net external force acts on the colliding bodies. That is not true in this case, because the gravitational force always acts on the ball, whether the ball is in flight or in contact with the bat. However, this force, with a magnitude of $mg = 1.37$ N, is negligible compared to the average force exerted by the bat, which has a magnitude of 9080 N. We are quite safe in treating the collision as "isolated."

(c) Now suppose the collision is not head-on, and the ball leaves the bat with a speed v_f of 45.0 m/s at an upward angle of 30.0° (Fig. 10-6). What now is the impulse on the ball?

SOLUTION: The **Key Idea** here is that now the collision is two-dimensional because the ball's outward path is not along the same axis as its incoming path. Thus, we must use vectors to find the impulse

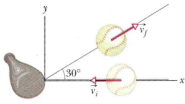

Fig. 10-6 Sample Problem 10-1. A bat collides with a pitched baseball, sending the ball off at an angle of 30° up from the horizontal.

$\vec{J}$. From Eq. 10-4, we can write

$$\vec{J} = \Delta\vec{p} = \vec{p}_f - \vec{p}_i = m\vec{v}_f - m\vec{v}_i.$$

Thus,
$$\vec{J} = m(\vec{v}_f - \vec{v}_i). \qquad (10\text{-}14)$$

We can evaluate the right side of this equation directly on a vector-capable calculator, since we know that the mass m is 0.140 kg, the final velocity $\vec{v}_f$ is 45.0 m/s at 30.0°, and the initial velocity $\vec{v}_i$ is 39.0 m/s at 180°.

Instead, we can evaluate Eq. 10-14 in component form. To do so, we first place an xy coordinate system as shown in Fig. 10-6. Then along the x axis we have

$$J_x = p_{fx} - p_{ix} = m(v_{fx} - v_{ix})$$
$$= (0.140 \text{ kg})[(45.0 \text{ m/s})(\cos 30.0°) - (-39.0 \text{ m/s})]$$
$$= 10.92 \text{ kg} \cdot \text{m/s}.$$

Along the y axis,

$$J_y = p_{fy} - p_{iy} = m(v_{fy} - v_{iy})$$
$$= (0.140 \text{ kg})[(45.0 \text{ m/s})(\sin 30.0°) - 0]$$
$$= 3.150 \text{ kg} \cdot \text{m/s}.$$

The impulse is then

$$\vec{J} = (10.9\hat{i} + 3.15\hat{j}) \text{ kg} \cdot \text{m/s}, \qquad \text{(Answer)}$$

and the magnitude and direction of $\vec{J}$ are

$$J = \sqrt{J_x^2 + J_y^2} = 11.4 \text{ kg} \cdot \text{m/s}$$

and
$$\theta = \tan^{-1}\frac{J_y}{J_x} = 16°. \qquad \text{(Answer)}$$

10-3   Momentum and Kinetic Energy in Collisions

Consider a system of two colliding bodies. If there is to be a collision, then at least one of the bodies must be moving, so the system has a certain kinetic energy and a certain linear momentum before the collision. During the collision, the kinetic energy and linear momentum of each body are changed by the impulse from the other body. For the rest of this chapter we shall discuss these changes—and also the changes in the kinetic energy and linear momentum of the system as a whole—without knowing the details of the impulses that determine the changes. The discussion will be limited to collisions in systems that are **closed** (no mass enters or leaves them) and **isolated** (no net external forces act on the bodies within the system).

Kinetic Energy

If the total kinetic energy of the system of two colliding bodies is unchanged by the collision, then the kinetic energy of the system is *conserved* (it is the same before and after the collision). Such a collision is called an **elastic collision.** In everyday collisions of common bodies, such as two cars or a ball and a bat, some energy is always transferred from kinetic energy to other forms of energy, such as thermal energy or energy of sound. Thus, the kinetic energy of the system is *not* conserved. Such a collision is called an **inelastic collision.**

However, in some situations, we can *approximate* a collision of common bodies as elastic. Suppose that you drop a Superball onto a hard floor. If the collision between the ball and floor (or Earth) were elastic, the ball would lose no kinetic energy because of the collision and would rebound to its original height. However, the actual rebound height is somewhat short, showing that at least some kinetic energy is lost in the collision and thus that the collision is somewhat inelastic. Still, we might choose to neglect that small loss of kinetic energy to approximate the collision as elastic.

A dropped golf ball will lose more of its kinetic energy and will rebound to only 60% of its original height. This collision is noticeably inelastic and cannot be approximated as elastic. If you drop a ball of wet putty onto the floor, it sticks to the floor and does not rebound at all. Because the putty sticks, this collision is called a **completely inelastic collision.** Figure 10-7 shows a more dramatic example of a completely inelastic collision. In such collisions, the bodies always stick together and lose kinetic energy.

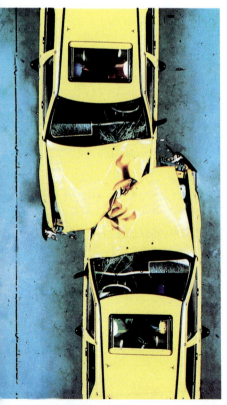

Fig. 10-7 Two cars after an almost head-on, almost completely inelastic collision.

Linear Momentum

Regardless of the details of the impulses in a collision, and regardless of what happens to the total kinetic energy of the system, the total linear momentum $\vec{P}$ of a closed, isolated system *cannot* change. The reason is that $\vec{P}$ can be changed only by external forces (from outside the system), but the forces in the collision are internal forces (inside the system). Thus, we have this important rule:

> In a closed, isolated system containing a collision, the linear momentum of each colliding body may change but the total linear momentum $\vec{P}$ of the system cannot change, whether the collision is elastic or inelastic.

This is actually another statement of the **law of conservation of linear momentum** that we first discussed in Section 9-6. In the next two sections we apply this law to some specific collisions, first inelastic and then elastic.

10-4 Inelastic Collisions in One Dimension

One-Dimensional Collision

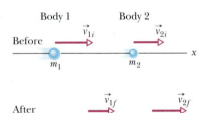

Fig. 10-8 Bodies 1 and 2 move along an x axis, before and after they have an inelastic collision.

Figure 10-8 shows two bodies just before and just after they have a *one-dimensional collision* (meaning that the motions before and after the collision are along a single axis). The velocities before the collision (subscript i) and after the collision (subscript f) are indicated. The two bodies form our system, which is closed and isolated. We can write the law of conservation of linear momentum for this two-body system as

$$\begin{pmatrix} \text{total momentum } \vec{P}_i \\ \text{before the collision} \end{pmatrix} = \begin{pmatrix} \text{total momentum } \vec{P}_f \\ \text{after the collision} \end{pmatrix},$$

which we can symbolize as

$$\vec{p}_{1i} + \vec{p}_{2i} = \vec{p}_{1f} + \vec{p}_{2f} \qquad \text{(conservation of linear momentum).} \qquad (10\text{-}15)$$

Because the motion is one-dimensional, we can drop the overhead arrows for vectors and use only components along the axis. Thus, from $p = mv$, we can rewrite Eq. 10-15 as

$$m_1 v_{1i} + m_2 v_{2i} = m_1 v_{1f} + m_2 v_{2f}. \qquad (10\text{-}16)$$

If we know values for, say, the masses, the initial velocities, and one of the final velocities, we can find the other final velocity with Eq. 10-16.

Completely Inelastic Collision

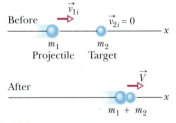

Fig. 10-9 A completely inelastic collision between two bodies. Before the collision, the body with mass m_2 is at rest and the body with mass m_1 moves directly toward it. After the collision, the stuck-together bodies move with the same velocity $\vec{V}$.

Figure 10-9 shows two bodies before and after they have a completely inelastic collision (meaning they stick together). The body with mass m_2 happens to be initially at rest ($v_{2i} = 0$). We can refer to that body as the *target*, and the incoming body as the *projectile*. After the collision, the stuck-together bodies move with velocity V. For this situation, we can rewrite Eq. 10-16 as

$$m_1 v_{1i} = (m_1 + m_2)V \qquad (10\text{-}17)$$

or

$$V = \frac{m_1}{m_1 + m_2} v_{1i}. \qquad (10\text{-}18)$$

If we know values for, say, the masses and the initial velocity v_{1i} of the projectile, we can find the final velocity V with Eq. 10-18. Note that V must be less than v_{1i}, because the mass ratio $m_1/(m_1 + m_2)$ must be less than unity.

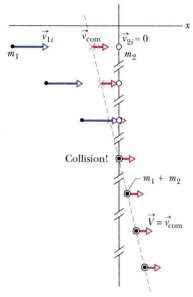

Fig. 10-10 Some freeze-frames of the two-body system in Fig. 10-9, which undergoes a completely inelastic collision. The system's center of mass is shown in each freeze-frame. The velocity $\vec{v}_{com}$ of the center of mass is unaffected by the collision. Because the bodies stick together after the collision, their common velocity $\vec{V}$ must be equal to $\vec{v}_{com}$.

Velocity of Center of Mass

In a closed, isolated system, the velocity $\vec{v}_{com}$ of the center of mass of the system cannot be changed by a collision because, with the system isolated, there is no net external force to change it. To get an expression for $\vec{v}_{com}$, let us return to the two-body system and one-dimensional collision of Fig. 10-8. From Eq. 9-25 ($\vec{P} = M\vec{v}_{com}$), we can relate $\vec{v}_{com}$ to the total linear momentum $\vec{P}$ of that two-body system by writing

$$\vec{P} = M\vec{v}_{com} = (m_1 + m_2)\vec{v}_{com}. \tag{10-19}$$

The total linear momentum $\vec{P}$ is conserved during the collision; so it is given by either side of Eq. 10-15. Let us use the left side to write

$$\vec{P} = \vec{p}_{1i} + \vec{p}_{2i}. \tag{10-20}$$

Substituting this for $\vec{P}$ in Eq. 10-19 and solving for $\vec{v}_{com}$ give us

$$\vec{v}_{com} = \frac{\vec{P}}{m_1 + m_2} = \frac{\vec{p}_{1i} + \vec{p}_{2i}}{m_1 + m_2}. \tag{10-21}$$

The right side of this equation is a constant, and $\vec{v}_{com}$ has that same constant value before and after the collision.

For example, Fig. 10-10 shows, in a series of freeze-frames, the motion of the center of mass for the completely inelastic collision of Fig. 10-9. Body 2 is the target, and its initial linear momentum in Eq. 10-21 is $\vec{p}_{2i} = m_2\vec{v}_{2i} = 0$. Body 1 is the projectile, and its initial linear momentum in Eq. 10-21 is $\vec{p}_{1i} = m_1\vec{v}_{1i}$. Note that as the series of freeze-frames progresses to and then beyond the collision, the center of mass moves at a constant velocity to the right. After the collision, the common final speed V of the bodies is equal to $\vec{v}_{com}$ because then the center of mass travels with the stuck-together bodies.

✔CHECKPOINT 3: Body 1 and body 2 are in a completely inelastic one-dimensional collision. What is their final momentum if their initial momenta are, respectively, (a) 10 kg · m/s and 0; (b) 10 kg · m/s and 4 kg · m/s; (c) 10 kg · m/s and −4 kg · m/s?

Sample Problem 10-2

The *ballistic pendulum* was used to measure the speeds of bullets before electronic timing devices were developed. The version shown in Fig. 10-11 consists of a large block of wood of mass $M = 5.4$ kg, hanging from two long cords. A bullet of mass $m = 9.5$ g is fired into the block, coming quickly to rest. The *block + bullet* then swing upward, their center of mass rising a vertical distance $h = 6.3$ cm before the pendulum comes momentarily to rest at the end of its arc. What is the speed of the bullet just prior to the collision?

SOLUTION: We can see that the bullet's speed v must determine the rise height h. However, a **Key Idea** is that we cannot use the conservation of mechanical energy to relate these two quantities because surely energy is transferred from mechanical energy to other forms (such as thermal energy and energy to break apart the wood) as the bullet penetrates the block. Another **Key Idea** helps—we can split this complicated motion into two steps that we can separately analyze: (1) the bullet–block collision and (2) the bullet–block rise, during which mechanical energy *is* conserved.

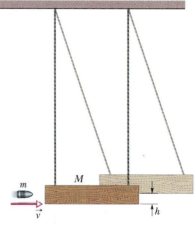

Fig. 10-11 Sample Problem 10-2. A ballistic pendulum, used to measure the speeds of bullets.

Step 1. Because the collision within the bullet–block system is so brief, we can make two important assumptions: (1) During the collision, the gravitational force on the block and the force on the block from the cords are still balanced. Thus, during the collision, the net external impulse on the bullet–block system is zero. Therefore, the system is isolated and its total linear momentum is conserved. (2) The collision is one-dimensional in the sense that the direction of the bullet and block *just after the collision* is in the bullet's original direction of motion.

Because the collision is one-dimensional, the block is initially at rest, and the bullet sticks in the block, we use Eq. 10-18 to express the conservation of linear momentum. Replacing the symbols there with the corresponding symbols here, we have

$$V = \frac{m}{m + M} v. \qquad (10\text{-}22)$$

Step 2. As the bullet and block now swing up together, the mechanical energy of the bullet–block–Earth system is conserved. (This mechanical energy is not changed by the force of the cords on the block, because that force is always directed perpendicular to the block's direction of travel.) Let's take the block's initial level as our reference level of zero gravitational potential energy. Then conservation of mechanical energy means that the system's kinetic energy at the start of the swing must equal its gravitational potential energy at the highest point of the swing. Because the speed of the bullet and block at the start of the swing is the speed V immediately after the collision, we may write this conservation as

$$\tfrac{1}{2}(m + M)V^2 = (m + M)gh.$$

Substituting for V from Eq. 10-22 leads to

$$
\begin{aligned}
v &= \frac{m + M}{m} \sqrt{2gh} \\
&= \left(\frac{0.0095 \text{ kg} + 5.4 \text{ kg}}{0.0095 \text{ kg}}\right) \sqrt{(2)(9.8 \text{ m/s}^2)(0.063 \text{ m})} \\
&= 630 \text{ m/s.} \qquad \text{(Answer)}
\end{aligned}
$$

The ballistic pendulum is a kind of "transformer," exchanging the high speed of a light object (the bullet) for the low—and thus more easily measurable—speed of a massive object (the block).

Sample Problem 10-3

A karate expert strikes downward with his fist (of mass $m_1 = 0.70$ kg), breaking a 0.14 kg board (Fig. 10-12a). He then does the same to a 3.2 kg concrete block. The spring constants k for bending are 4.1×10^4 N/m for the board and 2.6×10^6 N/m for the block.

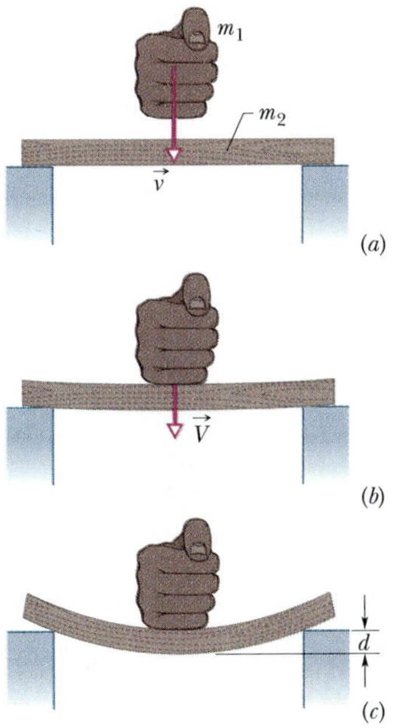

Fig. 10-12 Sample Problem 10-3. (a) A karate expert strikes at a flat object with speed v. (b) Fist and object undergo a completely inelastic collision, and bending begins. The *fist + object* then have speed V. (c) The object breaks when its center has been deflected by an amount d.

Breaking occurs at a deflection d of 16 mm for the board and 1.1 mm for the block (Fig. 10-12c). (The data are taken from "The Physics of Karate," by S. R. Wilk, R. E. McNair, and M. S. Feld, *American Journal of Physics,* September 1983.)

(a) Just before the object (board or block) breaks, what is the energy stored in it?

SOLUTION: The **Key Idea** here is that we can treat the bending as the compression of a spring for which Hooke's law applies. The stored potential energy is then, from Eq. 8-11, $U = \tfrac{1}{2}kd^2$. For the board,

$$
\begin{aligned}
U &= \tfrac{1}{2}(4.1 \times 10^4 \text{ N/m})(0.016 \text{ m})^2 \\
&= 5.248 \text{ J} \approx 5.2 \text{ J.} \qquad \text{(Answer)}
\end{aligned}
$$

For the block,

$$
\begin{aligned}
U &= \tfrac{1}{2}(2.6 \times 10^6 \text{ N/m})(0.0011 \text{ m})^2 \\
&= 1.573 \text{ J} \approx 1.6 \text{ J.} \qquad \text{(Answer)}
\end{aligned}
$$

(b) What is the lowest fist speed v_{fist} required to break the object (board or block)? Assume the following: The collisions are completely inelastic collisions of only the fist and the object. Bending begins just after the collision. Mechanical energy is conserved from the beginning of the bending until just before the object breaks. The speed of the fist and object is negligible at that point.

SOLUTION: The **Key Idea** here is that we can split up this complicated motion into three steps that we can separately analyze:

1. The completely inelastic one-dimensional collision of the fist and the object transfers energy to the kinetic energy of the fist–object system.

2. That energy is then transferred to the potential energy U stored by the bending.

3. The object breaks when U reaches the value calculated in part (a).

In step 1, we can use Eq. 10-18 to relate the fist speed v_{fist} just before the collision to the fist–object speed V_{fo} just after the collision and as the bending begins. With the notation we are using here, Eq. 10-18 becomes

$$V_{fo} = \frac{m_1}{m_1 + m_2} v_{fist}. \qquad (10\text{-}23)$$

In step 2, the mechanical energy of the fist–object system is conserved during the bending (up to the break). (Because the downward deflection of the object is small, the change in the gravitational potential energies of fist and object during the deflection are small enough to be neglected.) We can write the conservation of mechanical energy during the bending as

$$\begin{pmatrix}\text{kinetic energy at}\\\text{start of bending}\end{pmatrix} = \begin{pmatrix}\text{potential energy of bending}\\\text{just before the break}\end{pmatrix},$$

or

$$\tfrac{1}{2}(m_1 + m_2)V_{fo}^2 = U. \qquad (10\text{-}24)$$

Substituting for V_{fo} from Eq. 10-23 and solving for v_{fist}, we find

$$v_{fist} = \frac{1}{m_1} \sqrt{2U(m_1 + m_2)}. \qquad (10\text{-}25)$$

For step 3, we substitute the proper mass value and the breaking value U as obtained in part (a). We find that for the board,

$$v_{fist} = 4.2 \text{ m/s} \qquad \text{(Answer)}$$

and for the block,

$$v_{fist} = 5.0 \text{ m/s}. \qquad \text{(Answer)}$$

Thus, from the answer to (a) we see that breaking a board requires the greater energy. However, from the answers to (b) we see why a board is easier to break: The required fist speed is less. The reason lies in Eq. 10-23. If we decrease the target mass in the collision, we increase the speed V_{fo} given to the object. Thus, we also increase the fraction of the fist's energy that is transferred to the object. (One reason why breaking a pencil in the arrangement of Fig. 10-12 would be easy is because of the pencil's small mass.)

10-5 Elastic Collisions in One Dimension

Stationary Target

As we discussed in Section 10-3, everyday collisions are inelastic but we can approximate some of them as being elastic; that is, we can approximate that the total kinetic energy of the colliding bodies is conserved and is not transferred to other forms of energy:

$$\begin{pmatrix}\text{total kinetic energy}\\\text{before the collision}\end{pmatrix} = \begin{pmatrix}\text{total kinetic energy}\\\text{after the collision}\end{pmatrix}.$$

This does not mean that the kinetic energy of each colliding body cannot change. Rather, it means this:

> In an elastic collision, the kinetic energy of each colliding body may change, but the total kinetic energy of the system does not change.

For example, the collision of a cue ball with an object ball in a game of pool can be approximated as being an elastic collision. If the collision is head-on (the cue ball heads directly toward the object ball), the kinetic energy of the cue ball can be transferred almost entirely to the object ball. (Still, the fact that the collision makes a sound means that at least a little of the kinetic energy is transferred to the energy of the sound.)

Figure 10-13 shows two bodies before and after they have a one-dimensional collision, like a head-on collision between pool balls. A projectile body of mass m_1 and initial velocity v_{1i} moves toward a target body of mass m_2 that is initially at rest ($v_{2i} = 0$). Let's assume that this two-body system is closed and isolated. Then the net linear momentum of the system is conserved, and from Eq. 10-15 we can write that conservation as

$$m_1 v_{1i} = m_1 v_{1f} + m_2 v_{2f} \qquad \text{(linear momentum).} \qquad (10\text{-}26)$$

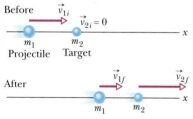

Fig. 10-13 Body 1 moves along an x axis before having an elastic collision with body 2, which is initially at rest. Both bodies move along that axis after the collision.

If the collision is also elastic, then the total kinetic energy is conserved and we can write that conservation as

$$\tfrac{1}{2}m_1 v_{1i}^2 = \tfrac{1}{2}m_1 v_{1f}^2 + \tfrac{1}{2}m_2 v_{2f}^2 \qquad \text{(kinetic energy).} \qquad (10\text{-}27)$$

In each of these equations, the subscript i identifies the initial velocities and the subscript f the final velocities of the bodies. If we know the masses of the bodies and if we also know v_{1i}, the initial velocity of body 1, the only unknown quantities are v_{1f} and v_{2f}, the final velocities of the two bodies. With two equations at our disposal, we should be able to find these two unknowns.

To do so, we rewrite Eq. 10-26 as

$$m_1(v_{1i} - v_{1f}) = m_2 v_{2f} \qquad (10\text{-}28)$$

and Eq. 10-27 as*

$$m_1(v_{1i} - v_{1f})(v_{1i} + v_{1f}) = m_2 v_{2f}^2. \qquad (10\text{-}29)$$

After dividing Eq. 10-29 by Eq. 10-28 and doing some more algebra, we obtain

$$v_{1f} = \frac{m_1 - m_2}{m_1 + m_2} v_{1i} \qquad (10\text{-}30)$$

and

$$v_{2f} = \frac{2m_1}{m_1 + m_2} v_{1i}. \qquad (10\text{-}31)$$

We note from Eq. 10-31 that v_{2f} is always positive (the target body with mass m_2 always moves forward). From Eq. 10-30 we see that v_{1f} may be of either sign (the projectile body with mass m_1 moves forward if $m_1 > m_2$ but rebounds if $m_1 < m_2$).

Let us look at a few special situations.

1. **Equal masses**   If $m_1 = m_2$, Eqs. 10-30 and 10-31 reduce to

$$v_{1f} = 0 \quad \text{and} \quad v_{2f} = v_{1i},$$

which we might call a pool player's result. It predicts that after a head-on collision of bodies with equal masses, body 1 (initially moving) stops dead in its tracks and body 2 (initially at rest) takes off with the initial speed of body 1. In head-on collisions, bodies of equal mass simply exchange velocities. This is true even if the target particle (body 2) is not initially at rest.

2. **A massive target**   In Fig. 10-13, a massive target means that $m_2 \gg m_1$. For example, we might fire a golf ball at a cannonball. Equations 10-30 and 10-31 then reduce to

$$v_{1f} \approx -v_{1i} \quad \text{and} \quad v_{2f} \approx \left(\frac{2m_1}{m_2}\right) v_{1i}. \qquad (10\text{-}32)$$

This tells us that body 1 (the golf ball) simply bounces back along its incoming path, its speed essentially unchanged. Body 2 (the cannonball) moves forward at a low speed, because the quantity in parentheses in Eq. 10-32 is much less than unity. All this is what we should expect.

3. **A massive projectile**   This is the opposite case; that is, $m_1 \gg m_2$. This time, we fire a cannonball at a golf ball. Equations 10-30 and 10-31 reduce to

$$v_{1f} \approx v_{1i} \quad \text{and} \quad v_{2f} \approx 2v_{1i}. \qquad (10\text{-}33)$$

---

*In this step, we use the identity $a^2 - b^2 = (a - b)(a + b)$. It reduces the amount of algebra needed to solve the simultaneous equations, Eqs. 10-28 and 10-29.

Equation 10-33 tells us that body 1 (the cannonball) simply keeps on going, scarcely slowed by the collision. Body 2 (the golf ball) charges ahead at twice the speed of the cannonball.

You may wonder: "Why twice the speed?" As a starting point in thinking about the matter, recall the collision described by Eq. 10-32, in which the velocity of the incident light body (the golf ball) changed from $+v$ to $-v$, a velocity *change* of $2v$. The same *change* in velocity (but now from zero to $2v$) occurs in this example also.

Moving Target

Fig. 10-14 Two bodies headed for a one-dimensional elastic collision.

Now that we have examined the elastic collision of a projectile and a stationary target, let us examine the situation in which both bodies are moving before they undergo an elastic collision.

For the situation of Fig. 10-14, the conservation of linear momentum is written as

$$m_1 v_{1i} + m_2 v_{2i} = m_1 v_{1f} + m_2 v_{2f}, \tag{10-34}$$

and the conservation of kinetic energy is written as

$$\tfrac{1}{2} m_1 v_{1i}^2 + \tfrac{1}{2} m_2 v_{2i}^2 = \tfrac{1}{2} m_1 v_{1f}^2 + \tfrac{1}{2} m_2 v_{2f}^2. \tag{10-35}$$

To solve these simultaneous equations for v_{1f} and v_{2f}, we first rewrite Eq. 10-34 as

$$m_1(v_{1i} - v_{1f}) = -m_2(v_{2i} - v_{2f}), \tag{10-36}$$

and Eq. 10-35 as

$$m_1(v_{1i} - v_{1f})(v_{1i} + v_{1f}) = -m_2(v_{2i} - v_{2f})(v_{2i} + v_{2f}). \tag{10-37}$$

After dividing Eq. 10-37 by Eq. 10-36 and doing some more algebra, we obtain

$$v_{1f} = \frac{m_1 - m_2}{m_1 + m_2} v_{1i} + \frac{2m_2}{m_1 + m_2} v_{2i} \tag{10-38}$$

and

$$v_{2f} = \frac{2m_1}{m_1 + m_2} v_{1i} + \frac{m_2 - m_1}{m_1 + m_2} v_{2i}. \tag{10-39}$$

Note that the assignment of subscripts 1 and 2 to the bodies is arbitrary. If we exchange those subscripts in Fig. 10-14 and in Eqs. 10-38 and 10-39, we end up with the same set of equations. Note also that if we set $v_{2i} = 0$, body 2 becomes a stationary target as in Fig. 10-13, and Eqs. 10-38 and 10-39 reduce to Eqs. 10-30 and 10-31, respectively.

✔**CHECKPOINT 4:**   What is the final linear momentum of the target in Fig. 10-13 if the initial linear momentum of the projectile is 6 kg · m/s and the final linear momentum of the projectile is (a) 2 kg · m/s and (b) −2 kg · m/s? (c) What is the final kinetic energy of the target if the initial and final kinetic energies of the projectile are, respectively, 5 J and 2 J?

Sample Problem 10-4

Two metal spheres, suspended by vertical cords, initially just touch, as shown in Fig. 10-15. Sphere 1, with mass $m_1 = 30$ g, is pulled to the left to height $h_1 = 8.0$ cm, and then released from rest. After swinging down, it undergoes an elastic collision with sphere 2, whose mass $m_2 = 75$ g. What is the velocity v_{1f} of sphere 1 just after the collision?

SOLUTION: A first **Key Idea** is that we can split this complicated motion into two steps that we can separately analyze: (1) the descent of sphere 1 and (2) the two-sphere collision.

Step 1.   The **Key Idea** here is that as sphere 1 swings down, the mechanical energy of the sphere–Earth system is conserved.

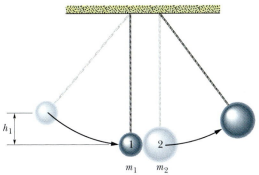

Fig. 10-15 Sample Problem 10-4. Two metal spheres suspended by cords just touch when they are at rest. Sphere 1, with mass m_1, is pulled to the left to height h_1 and then released.

(The mechanical energy is not changed by the force of the cord on sphere 1 because that force is always directed perpendicular to the sphere's direction of travel.) Let's take the lowest level as our reference level of zero gravitational potential energy. Then the kinetic energy of sphere 1 at the lowest level must equal the gravitational potential energy of the sys-

tem when sphere 1 is at the initial height. Thus,

$$\tfrac{1}{2}m_1 v_{1i}^2 = m_1 g h_1,$$

which we solve for the speed v_{1i} of sphere 1 just before the collision:

$$v_{1i} = \sqrt{2gh_1} = \sqrt{(2)(9.8 \text{ m/s}^2)(0.080 \text{ m})} = 1.252 \text{ m/s}.$$

Step 2. Here we can make two assumptions in addition to the assumption that the collision is elastic. First, we can assume that the collision is one-dimensional because the motions of the spheres are approximately horizontal from just before the collision to just after it. Second, because the collision is so brief, we can assume that the two-sphere system is closed and isolated. This gives the **Key Idea** that the total linear momentum of the system is conserved. Thus, we can use Eq. 10-30 to find the velocity of sphere 1 just after the collision:

$$v_{1f} = \frac{m_1 - m_2}{m_1 + m_2} v_{1i} = \frac{0.030 \text{ kg} - 0.075 \text{ kg}}{0.030 \text{ kg} + 0.075 \text{ kg}} (1.252 \text{ m/s})$$
$$= -0.537 \text{ m/s} \approx -0.54 \text{ m/s}. \qquad \text{(Answer)}$$

The minus sign tells us that sphere 1 moves to the left just after the collision.

10-6 Collisions in Two Dimensions

When two bodies collide, the impulses of one on the other determine the directions in which they then travel. In particular, when the collision is not head-on, the bodies do not end up traveling along their initial axis. For such two-dimensional collisions in a closed, isolated system, the total linear momentum must still be conserved:

$$\vec{P}_{1i} + \vec{P}_{2i} = \vec{P}_{1f} + \vec{P}_{2f}. \qquad (10\text{-}40)$$

If the collision is also elastic (a special case), then the total kinetic energy is also conserved:

$$K_{1i} + K_{2i} = K_{1f} + K_{2f}. \qquad (10\text{-}41)$$

Equation 10-40 is often more useful for analyzing a two-dimensional collision if we write it in terms of components on an xy coordinate system. For example, Fig. 10-16 shows a *glancing collision* (it is not head-on) between a projectile body and a target body initially at rest. The impulses between the bodies have sent the bodies off at angles θ_1 and θ_2 to the x axis, along which the projectile initially traveled. In this situation we would rewrite Eq. 10-40 for components along the x axis as

$$m_1 v_{1i} = m_1 v_{1f} \cos \theta_1 + m_2 v_{2f} \cos \theta_2, \qquad (10\text{-}42)$$

and along the y axis as

$$0 = -m_1 v_{1f} \sin \theta_1 + m_2 v_{2f} \sin \theta_2. \qquad (10\text{-}43)$$

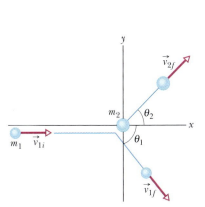

Fig. 10-16 An elastic collision between two bodies in which the collision is not head-on. The body with mass m_2 (the target) is initially at rest.

We can also write Eq. 10-41 (for the special case of an elastic collision) in terms of speeds:

$$\tfrac{1}{2}m_1 v_{1i}^2 = \tfrac{1}{2}m_1 v_{1f}^2 + \tfrac{1}{2}m_2 v_{2f}^2 \qquad \text{(kinetic energy).} \qquad (10\text{-}44)$$

Equations 10-42 to 10-44 contain seven variables: two masses, m_1 and m_2; three speeds, v_{1i}, v_{1f}, and v_{2f}; and two angles, θ_1 and θ_2. If we know any four of these quantities, we can solve the three equations for the remaining three quantities.

✔**CHECKPOINT 5:** In Fig. 10-16, suppose that the projectile has an initial momentum of 6 kg · m/s, a final x component of momentum of 4 kg · m/s, and a final y component of momentum of -3 kg · m/s. For the target, what then are (a) the final x component of momentum and (b) the final y component of momentum?

Sample Problem 10-5

Two skaters collide and embrace, in a completely inelastic collision. Thus, they stick together after impact, as suggested by Fig. 10-17, where the origin is placed at the point of collision. Alfred, whose mass m_A is 83 kg, is originally moving east with speed $v_A = 6.2$ km/h. Barbara, whose mass m_B is 55 kg, is originally moving north with speed $v_B = 7.8$ km/h.

(a) What is the velocity $\vec{V}$ of the couple after they collide?

SOLUTION: One **Key Idea** here is the assumption that the two skaters form a closed, isolated system; that is, we assume no *net* external force acts on them. In particular, we neglect any frictional force on their skates from the ice. With that assumption, we can apply the conservation of the total linear momentum $\vec{P}$ to the system by writing $\vec{P}_i = \vec{P}_f$ as

$$m_A\vec{v}_A + m_B\vec{v}_B = (m_A + m_B)\vec{V}. \tag{10-45}$$

Solving for $\vec{V}$ gives us

$$\vec{V} = \frac{m_A\vec{v}_A + m_B\vec{v}_B}{m_A + m_B}.$$

We can solve this directly on a vector-capable calculator by substituting given data for the symbols on the right side. We can also solve it by applying a second **Key Idea** (one we have used before) and then some algebra: The idea is that the total linear momentum of the system is conserved separately for components along the x axis and y axis shown in Fig. 10-17. Writing Eq. 10-45 in component form for the x axis yields

$$m_Av_A + m_B(0) = (m_A + m_B)V \cos \theta, \tag{10-46}$$

and for the y axis

$$m_A(0) + m_Bv_B = (m_A + m_B)V \sin \theta. \tag{10-47}$$

We cannot solve either of these equations separately because they both contain two unknowns (V and θ), but we can solve them simultaneously by dividing Eq. 10-47 by Eq. 10-46. We get

$$\tan \theta = \frac{m_Bv_B}{m_Av_A} = \frac{(55 \text{ kg})(7.8 \text{ km/h})}{(83 \text{ kg})(6.2 \text{ km/h})} = 0.834.$$

Thus,

$$\theta = \tan^{-1} 0.834 = 39.8° \approx 40°. \quad \text{(Answer)}$$

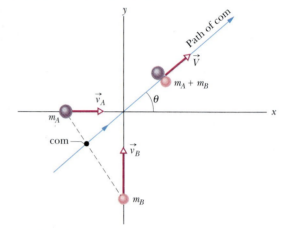

Fig. 10-17 Sample Problem 10-5. Two skaters, Alfred (A) and Barbara (B), represented by spheres in this simplified overhead view, have a completely inelastic collision. Afterward, they move off together at angle θ, with speed V. The path of their center of mass is shown. The position of the center of mass for the indicated positions of the skaters before the collision is also shown.

From Eq. 10-47, with $m_A + m_B = 138$ kg, we then have

$$V = \frac{m_Bv_B}{(m_A + m_B) \sin \theta} = \frac{(55 \text{ kg})(7.8 \text{ km/h})}{(138 \text{ kg})(\sin 39.8°)}$$
$$= 4.86 \text{ km/h} \approx 4.9 \text{ km/h}. \quad \text{(Answer)}$$

(b) What is the velocity $\vec{v}_{\text{com}}$ of the center of mass of the two skaters before the collision and after the collision?

SOLUTION: For the situation after the collision, we use the **Key Idea** that because the skaters are stuck together, their center of mass must travel with them, as shown in Fig. 10-17. Thus, the velocity $\vec{v}_{\text{com}}$ of their center of mass is equal to $\vec{V}$, as calculated in (a).

To find $\vec{v}_{\text{com}}$ before the collision, we use another **Key Idea**: The $\vec{v}_{\text{com}}$ of a system can be changed only by a net external force, not an internal force. However, here we assumed that the skaters form an isolated system (no net external force acts on the system). Therefore, $\vec{v}_{\text{com}}$ cannot change because of the collision (which produces only internal forces). Thus, before and after the collision, we have

$$\vec{v}_{\text{com}} = \vec{V}. \quad \text{(Answer)}$$

REVIEW & SUMMARY

Collisions In a **collision,** two bodies exert strong forces on each other for a relatively short time. These forces are internal to the two-body system and are significantly larger than any external force during the collision.

Impulse and Linear Momentum Applying Newton's second law in momentum form to a particle-like body involved in a collision leads to the **impulse–linear momentum theorem:**

$$\vec{p}_f - \vec{p}_i = \Delta\vec{p} = \vec{J}, \tag{10-4}$$

where $\vec{p}_f - \vec{p}_i = \Delta\vec{p}$ is the change in the body's linear momentum, and $\vec{J}$ is the **impulse** due to the force $\vec{F}(t)$ exerted on the body by the other body in the collision:

$$\vec{J} = \int_{t_i}^{t_f} \vec{F}(t)\, dt. \tag{10-3}$$

If F_{avg} is the average magnitude of $\vec{F}(t)$ during the collision and Δt is the duration of the collision, then for one-dimensional motion

$$J = F_{avg}\,\Delta t. \tag{10-8}$$

When a steady stream of bodies, each with mass m and speed v, collides with a body whose position is fixed, the average force on the fixed body is

$$F_{avg} = -\frac{n}{\Delta t}\,\Delta p = -\frac{n}{\Delta t}\,m\,\Delta v, \tag{10-10}$$

where $n/\Delta t$ is the rate at which the bodies collide with the fixed body, and Δv is the change in velocity of each colliding body. This average force can also be written as

$$F_{avg} = -\frac{\Delta m}{\Delta t}\,\Delta v, \tag{10-13}$$

where $\Delta m/\Delta t$ is the rate at which mass collides with the fixed body. In Eqs. 10-10 and 10-13, $\Delta v = -v$ if the bodies stop upon impact, or $\Delta v = -2v$ if they bounce directly backward with no change in their speed.

Inelastic Collision—One Dimension In an *inelastic collision* of two bodies, the kinetic energy of the two-body system is not conserved. If the system is closed and isolated, then the total linear momentum of the system *must* be conserved, which we can write in vector form as

$$\vec{p}_{1i} + \vec{p}_{2i} = \vec{p}_{1f} + \vec{p}_{2f}, \tag{10-15}$$

where subscripts i and f refer to values just before and just after the collision, respectively.

If the motion of the bodies is along a single axis, the collision is one-dimensional and we can write Eq. 10-15 in terms of velocity components along that axis:

$$m_1 v_{1i} + m_2 v_{2i} = m_1 v_{1f} + m_2 v_{2f}. \tag{10-16}$$

If the bodies stick together, the collision is a *completely inelastic collision* and the bodies have the same final velocity V (because they *are* stuck together).

Motion of the Center of Mass The center of mass of a closed, isolated system of two colliding bodies is not affected by the collision. In particular, the velocity $\vec{v}_{com}$ of the center of mass cannot be changed by the collision and is related to the constant total momentum $\vec{P}$ of the system by

$$\vec{v}_{com} = \frac{\vec{P}}{m_1 + m_2} = \frac{\vec{p}_{1i} + \vec{p}_{2i}}{m_1 + m_2}. \tag{10-21}$$

Elastic Collisions—One Dimension An *elastic collision* is a special type of collision in which the kinetic energy of the system of colliding bodies is conserved. Some collisions in the everyday world can be approximated as being elastic collisions. If the system is closed and isolated, its linear momentum is also conserved. For a one-dimensional collision in which body 2 is a target and body 1 is an incoming projectile, conservation of kinetic energy and linear momentum yield the following expressions for the velocities immediately after the collision:

$$v_{1f} = \frac{m_1 - m_2}{m_1 + m_2}\,v_{1i} \tag{10-30}$$

and

$$v_{2f} = \frac{2m_1}{m_1 + m_2}\,v_{1i}. \tag{10-31}$$

If both bodies are moving prior to the collision, their velocities immediately after the collision are given by

$$v_{1f} = \frac{m_1 - m_2}{m_1 + m_2}\,v_{1i} + \frac{2m_2}{m_1 + m_2}\,v_{2i} \tag{10-38}$$

and

$$v_{2f} = \frac{2m_1}{m_1 + m_2}\,v_{1i} + \frac{m_2 - m_1}{m_1 + m_2}\,v_{2i}. \tag{10-39}$$

Note the symmetry of subscripts 1 and 2 in Eqs. 10-38 and 10-39.

Collisions in Two Dimensions If two bodies collide and their motion is not along a single axis (the collision is not head-on), then the collision is two-dimensional. If the two-body system is closed and isolated, then the law of conservation of momentum applies to the collision and can be written as

$$\vec{P}_{1i} + \vec{P}_{2i} = \vec{P}_{1f} + \vec{P}_{2f}, \tag{10-40}$$

In component form, the law gives two equations that describe the collision (one equation for each of the two dimensions). If the collision is also elastic (a special case), then the conservation of kinetic energy during the collision gives a third equation:

$$K_{1i} + K_{2i} = K_{1f} + K_{2f}. \tag{10-41}$$

QUESTIONS

1. Figure 10-18 shows three graphs of force magnitude versus time for a body involved in a collision. Rank the graphs according to the magnitude of the impulse on the body, greatest first.

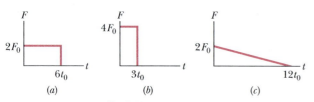

Fig. 10-18 Question 1.

2. Two objects that are moving along an xy plane on a frictionless floor collide. Assume that they form a closed, isolated system. The following table gives some of the momentum components (in kilogram-meters per second) before and after the collision. What are the missing values?

Situation	Object	Before p_x	Before p_y	After p_x	After p_y
1	A	3	4	7	2
	B	2	2		
2	C	−4	5	3	
	D		−2	4	2
3	E	−6			3
	F	6	2	−4	−3

3. The following table gives, for three situations, the masses (in kilograms) and velocities (in meters per second) of the two bodies in Fig. 10-14. For which situations is the center of mass of the two-body system stationary?

Situation	m_1	v_1	m_2	v_2
a	2	3	4	−3
b	6	2	3	−4
c	4	3	4	−3

4. Figure 10-19 shows four graphs of position versus time for two bodies and their center of mass. The two bodies form a closed, isolated system and undergo a completely inelastic, one-dimensional collision along an x axis. In graph 1, are (a) the two bodies and (b) the center of mass moving in the positive or negative direction of the x axis? (c) Which graphs correspond to a physically impossible situation? Explain.

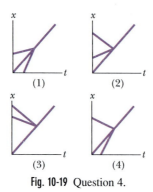

Fig. 10-19 Question 4.

5. In Fig. 10-20, blocks A and B have linear momenta with directions as shown and with magnitudes of 9 kg · m/s and 4 kg · m/s, respectively. (a) What is the direction of motion of the center of mass of the two-block system over the frictionless floor? (b) If the blocks stick together during their collision, in what direction do they move? (c) If, instead, block A ends up moving to the left, is the magnitude of its momentum then smaller than, more than, or the same as that of block B?

Fig. 10-20 Question 5.

6. A projectile body moving in the positive direction of an x axis on a frictionless floor runs into an initially stationary target body (as in Fig. 10-13) in a one-dimensional collision. Assume the particles form a closed, isolated system. Nine choices for a graph of the momenta of the bodies versus time (before and after the collision) are given in Fig. 10-21. Determine which choices represent physically impossible situations and explain why.

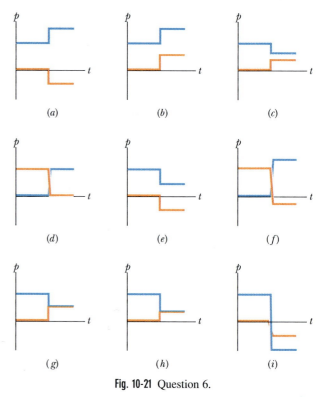

Fig. 10-21 Question 6.

7. Two bodies have undergone an elastic one-dimensional collision along an x axis. Figure 10-22 is a graph of position versus time for those bodies and for their center of mass. (a) Were both bodies initially moving, or was one initially stationary? Which line segment corresponds to the motion of the center of mass (b) before the collision and (c) after the col-

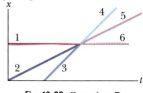

Fig. 10-22 Question 7.

lision? (d) Is the mass of the body that was moving faster before the collision greater than, less than, or equal to that of the other body?

8. Drop, in succession, a baseball and a basketball from about shoulder height above a hard floor, and note how high each re-bounds. Then align the baseball above the basketball (with a small separation as in Fig. 10-23*a*) and drop them simultaneously. (Be prepared to duck, and guard your face.) (a) Is the rebound height of the basketball now higher or lower than before (Fig. 10-23*b*)? (b) Is the rebound height of the baseball less than or greater than the sum of the individual baseball and basketball rebound heights? (See also Problem 45.)

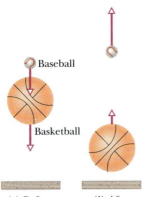

Fig. 10-23 Question 8 and Problem 45.

(a) Before   (b) After

Baseball

Basketball

9. A projectile hockey puck *A*, with initial momentum 5 kg · m/s along an *x* axis, collides with an initially stationary hockey puck *B*. The pucks slide over frictionless ice and are shown in an over-head view in Fig. 10-24. Also shown are three general choices for the path taken by puck *A* after the collision. Which choice is ap-propriate if the momentum of puck *B* after the collision has an *x* component of (a) 5 kg · m/s, (b) more than 5 kg · m/s, and (c) less than 5 kg · m/s? Can that *x* component be (d) 1 kg · m/s or (e) −1 kg · m/s?

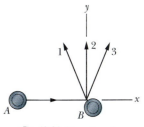

Fig. 10-24 Question 9.

10. Two bodies that form a closed, isolated system undergo a col-lision along a frictionless floor. Which of the three choices in Fig. 10-25 best represents the paths of those bodies and the path of their center of mass as seen from overhead?

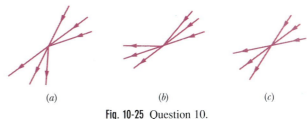

(a)   (b)   (c)

Fig. 10-25 Question 10.

EXERCISES & PROBLEMS

SEC. 10-2   Impulse and Linear Momentum

1E. A cue stick strikes a stationary pool ball, with an average force of 50 N over a time of 10 ms. If the ball has mass 0.20 kg, what speed does it have just after impact?   **ssm**

2E. The National Transportation Safety Board is testing the crash-worthiness of a new car. The 2300 kg vehicle, moving at 15 m/s, is allowed to collide with a bridge abutment, which stops it in 0.56 s. What is the magnitude of the average force that acts on the car during the impact?

3E. A 150 g baseball pitched at a speed of 40 m/s is hit straight back to the pitcher at a speed of 60 m/s. What is the magnitude of the average force on the ball from the bat if the bat is in contact with the ball for 5.0 ms?

4E. Until he was in his seventies, Henri LaMothe excited audiences by belly-flopping from a height of 12 m into 30 cm of water (Fig. 10-26). Assuming that he stops just as he reaches the bottom of the water and estimating his mass, find the magnitudes of (a) the av-erage force and (b) the average impulse on him from the water.

5E. A force that averages 1200 N is applied to a 0.40 kg steel ball moving at 14 m/s in a collision lasting 27 ms. If the force is in a direction opposite the initial velocity of the ball, find the final speed and direction of the ball.   **ssm**

Fig. 10-26 Exercise 4.

6E. In February 1955, a paratrooper fell 370 m from an airplane without being able to open his chute but happened to land in snow, suffering only minor injuries. Assume that his speed at impact was 56 m/s (terminal speed), that his mass (including gear) was 85 kg, and that the magnitude of the force on him from the snow was at the survivable limit of 1.2×10^5 N. What are (a) the minimum depth of snow that would have stopped him safely and (b) the magnitude of the impulse on him from the snow?

7E. A 1.2 kg ball drops vertically onto a floor, hitting with a speed of 25 m/s. It rebounds with an initial speed of 10 m/s. (a) What impulse acts on the ball during the contact? (b) If the ball is in contact with the floor for 0.020 s, what is the magnitude of the average force on the floor from the ball?

8P. It is well known that bullets and other missiles fired at Superman simply bounce off his chest (Fig. 10-27). Suppose that a gangster sprays Superman's chest with 3 g bullets at the rate of 100 bullets/min, and the speed of each bullet is 500 m/s. Suppose too that the bullets rebound straight back with no change in speed. What is the magnitude of the average force on Superman's chest from the stream of bullets?

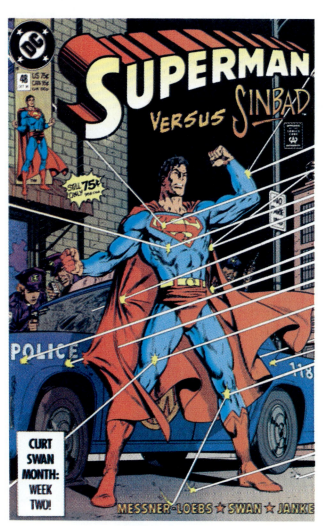

Fig. 10-27 Problem 8.

9P. A 1400 kg car moving at 5.3 m/s is initially traveling north in the positive y direction. After completing a 90° right-hand turn to the positive x direction in 4.6 s, the inattentive operator drives into a tree, which stops the car in 350 ms. In unit-vector notation, what is the impulse on the car (a) due to the turn and (b) due to the collision? What is the magnitude of the average force that acts on the car (c) during the turn and (d) during the collision? (e) What is the angle between the average force in (c) and the positive x direction? ssm www

10P. A 0.30 kg softball has a velocity of 12 m/s at an angle of 35° below the horizontal just before making contact with a bat. The ball leaves the bat 2.0 ms later with a vertical velocity of magnitude 10 m/s as shown in Fig. 10-28. What is the magnitude of the average force of the bat on the ball during the ball–bat contact?

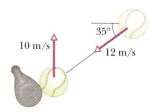

Fig. 10-28 Problem 10.

11P. The magnitude of an unbalanced force on a 10 kg object increases at a constant rate from zero to 50 N in 4.0 s, causing the initially stationary object to move. What is the object's speed at the end of the 4.0 s? ssm

12P. During a violent thunderstorm, hail of diameter 1.0 cm falls directly downward at a speed of 25 m/s. There are estimated to be 120 hailstones per cubic meter of air. (a) What is the mass of each hailstone (density = 0.92 g/cm³)? (b) Assuming that the hail does not bounce, find the magnitude of the average force on a flat roof measuring 10 m × 20 m due to the impact of the hail. (*Hint:* During impact, the force on a hailstone from the roof is approximately equal to the net force on the hailstone, because the gravitational force on it is small.)

13P. A pellet gun fires ten 2.0 g pellets per second with a speed of 500 m/s. The pellets are stopped by a rigid wall. What are (a) the momentum of each pellet, (b) the kinetic energy of each pellet, and (c) the magnitude of the average force on the wall from the stream of pellets? (d) If each pellet is in contact with the wall for 0.6 ms, what is the magnitude of the average force on the wall from each pellet during contact? (e) Why is this average force so different from the average force calculated in (c)? ssm

14P. Figure 10-29 shows an approximate plot of force magnitude versus time during the collision of a 58 g Superball with a wall. The initial velocity of the ball is 34 m/s perpendicular to the wall; it rebounds directly back with approximately the same speed, also perpendicular to the wall. What is F_{max}, the maximum magnitude of the force on the ball from the wall during the collision?

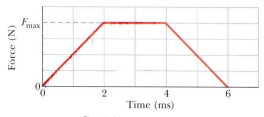

Fig. 10-29 Problem 14.

15P. A spacecraft is separated into two parts by detonating the explosive bolts that hold them together. The masses of the parts are 1200 kg and 1800 kg; the magnitude of the impulse on each part from the bolts is 300 N · s. With what relative speed do the two parts separate because of the detonation? ssm

16P. A ball having a mass of 150 g strikes a wall with a speed of 5.2 m/s and rebounds with only 50% of its initial kinetic energy. (a) What is the speed of the ball immediately after rebounding? (b) What is the magnitude of the impulse on the wall from the ball? (c) If the ball was in contact with the wall for 7.6 ms, what was the magnitude of the average force on the ball from the wall during this time interval?

17P. In the overhead view of Fig. 10-30, a 300 g ball with a speed v of 6.0 m/s strikes a wall at an angle θ of 30° and then rebounds with the same speed and angle. It is in contact with the wall for 10 ms. (a) What is the impulse on the ball from the wall? (b) What is the average force on the wall from the ball?

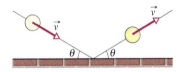

Fig. 10-30 Problem 17.

18P. A 2500 kg unmanned space probe is moving in a straight line at a constant speed of 300 m/s. Control rockets on the space probe execute a burn in which a thrust of 3000 N acts for 65.0 s. (a) What is the change in the magnitude of the probe's linear momentum if the thrust is backward, forward, or directly sideways? (b) What is the change in kinetic energy under the same three conditions? Assume that the mass of the ejected burn products is negligible compared to the mass of the space probe.

19P. A soccer player kicks a soccer ball of mass 0.45 kg that is initially at rest. The player's foot is in contact with the ball for 3.0×10^{-3} s, and the force of the kick is given by

$$F(t) = [(6.0 \times 10^6)t - (2.0 \times 10^9)t^2] \text{ N},$$

for $0 \le t \le 3.0 \times 10^{-3}$ s, where t is in seconds. Find the magnitudes of the following: (a) the impulse on the ball due to the kick, (b) the average force on the ball from the player's foot during the period of contact, (c) the maximum force on the ball from the player's foot during the period of contact, and (d) the ball's speed immediately after it loses contact with the player's foot. ssm

SEC. 10-4 Inelastic Collisions in One Dimension

20E. A 5.20 g bullet moving at 672 m/s strikes a 700 g wooden block at rest on a frictionless surface. The bullet emerges, traveling in the same direction with its speed reduced to 428 m/s. (a) What is the resulting speed of the block? (b) What is the speed of the bullet–block center of mass?

21E. A 6.0 kg box sled is coasting across frictionless ice at a speed of 9.0 m/s when a 12 kg package is dropped into it from above. What is the new speed of the sled? ssm

22E. A bullet of mass 10 g strikes a ballistic pendulum of mass 2.0 kg. The center of mass of the pendulum rises a vertical distance

of 12 cm. Assuming that the bullet remains embedded in the pendulum, calculate the bullet's initial speed.

23E. Meteor Crater in Arizona (Fig. 10-1a) is thought to have been formed by the impact of a meteor with Earth some 20,000 years ago. The mass of the meteor is estimated at 5×10^{10} kg, and its speed at 7200 m/s. What speed would such a meteor give Earth in a head-on collision? ssm

24E. A bullet of mass 4.5 g is fired horizontally into a 2.4 kg wooden block at rest on a horizontal surface. The coefficient of kinetic friction between block and surface is 0.20. The bullet stops in the block, which slides straight ahead for 1.8 m (without rotation). (a) What is the speed of the block immediately after the bullet stops relative to it? (b) At what speed is the bullet fired?

25P. Two cars A and B slide on an icy road as they attempt to stop at a traffic light. The mass of A is 1100 kg, and the mass of B is 1400 kg. The coefficient of kinetic friction between the locked wheels of either car and the road is 0.13. Car A succeeds in stopping at the light, but car B cannot stop and rear-ends car A. After the collision, A stops 8.2 m ahead of its position at impact, and B 6.1 m ahead; see Fig. 10-31. Both drivers had their brakes locked throughout the incident. From the distance each car moved after the collision, find the speed of (a) car A and (b) car B immediately after impact. (c) Use conservation of linear momentum to find the speed at which car B struck car A. On what grounds can the use of linear momentum conservation be criticized here?

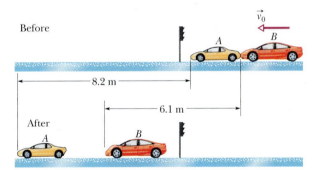

Fig. 10-31 Problem 25.

26P. In Fig. 10-32a, a 3.50 g bullet is fired horizontally at two blocks at rest on a frictionless tabletop. The bullet passes through the first block, with mass 1.20 kg, and embeds itself in the second, with mass 1.80 kg. Speeds of 0.630 m/s and 1.40 m/s, respectively, are thereby given to the blocks (Fig. 10-32b). Neglecting the mass

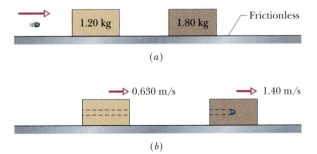

Fig. 10-32 Problem 26.

removed from the first block by the bullet, find (a) the speed of the bullet immediately after it emerges from the first block and (b) the bullet's original speed.

27P. A box is put on a scale that is marked in units of mass and adjusted to read zero when the box is empty. A stream of marbles is then poured into the box from a height h above its bottom at a rate of R (marbles per second). Each marble has mass m. (a) If the collisions between the marbles and the box are completely inelastic, find the scale reading at time t after the marbles begin to fill the box. (b) Determine a numerical answer when $R = 100$ s^{-1}, $h = 7.60$ m, $m = 4.50$ g, and $t = 10.0$ s. **ssm**

28P. A 5.0 kg block with a speed of 3.0 m/s collides with a 10 kg block that has a speed of 2.0 m/s in the same direction. After the collision, the 10 kg block is observed to be traveling in the original direction with a speed of 2.5 m/s. (a) What is the velocity of the 5.0 kg block immediately after the collision? (b) By how much does the total kinetic energy of the system of two blocks change because of the collision? (c) Suppose, instead, that the 10 kg block ends up with a speed of 4.0 m/s. What then is the change in the total kinetic energy? (d) Account for the result you obtained in (c). **ilw**

29P. A railroad freight car of mass 3.18×10^4 kg collides with a stationary caboose car. They couple together, and 27.0% of the initial kinetic energy is transferred to thermal energy, sound, vibrations, and so on. Find the mass of the caboose. **ssm www**

30P. A 10 g bullet moving directly upward at 1000 m/s strikes and passes through the center of mass of a 5.0 kg block initially at rest (Fig. 10-33). The bullet emerges from the block moving directly upward at 400 m/s. To what maximum height does the block then rise above its initial position?

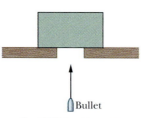

Fig. 10-33 Problem 30.

31P. In Fig. 10-34, a ball of mass m is shot with speed v_i into the barrel of a spring gun of mass M initially at rest on a frictionless surface. The ball sticks in the barrel at the point of maximum compression of the spring. Assume that the increase in thermal energy due to friction between the ball and the barrel is negligible. (a) What is the speed of the spring gun after the ball stops in the barrel? (b) What fraction of the initial kinetic energy of the ball is stored in the spring? **ssm**

Fig. 10-34 Problem 31.

32P. A 4.0 kg physics book and a 6.0 kg calculus book, connected by a spring, are stationary on a horizontal frictionless surface. The spring constant is 8000 N/m. The books are pushed together, compressing the spring, and then they are released from rest. When the spring has returned to its unstretched length, the speed of the calculus book is 4.0 m/s. How much energy is stored in the spring at the instant the books are released?

33P. A block of mass $m_1 = 2.0$ kg slides along a frictionless table with a speed of 10 m/s. Directly in front of it, and moving in the same direction, is a block of mass $m_2 = 5.0$ kg moving at 3.0 m/s.

A massless spring with spring constant $k = 1120$ N/m is attached to the near side of m_2, as shown in Fig. 10-35. When the blocks collide, what is the maximum compression of the spring? (*Hint:* At the moment of maximum compression of the spring, the two blocks move as one. Find the velocity by noting that the collision is completely inelastic at this point.) **ilw**

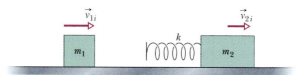

Fig. 10-35 Problem 33.

34P. A 1.0 kg block at rest on a horizontal frictionless surface is connected to an unstretched spring ($k = 200$ N/m) whose other end is fixed (Fig. 10-36). A 2.0 kg block moving at 4.0 m/s collides with the 1.0 kg block. If the two blocks stick together after the one-dimensional collision, what maximum compression of the spring occurs when the blocks momentarily stop?

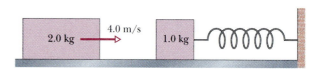

Fig. 10-36 Problem 34.

SEC. 10-5 Elastic Collisions in One Dimension

35E. The blocks in Fig. 10-37 slide without friction. (a) What is the velocity $\vec{v}$ of the 1.6 kg block after the collision? (b) Is the collision elastic? (c) Suppose the initial velocity of the 2.4 kg block is the reverse of what is shown. Can the velocity $\vec{v}$ of the 1.6 kg block after the collision be in the direction shown? **ssm**

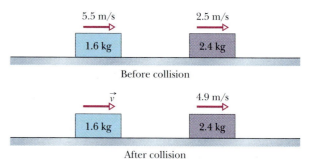

Fig. 10-37 Exercise 35.

36E. An electron undergoes a one-dimensional elastic collision with an initially stationary hydrogen atom. What percentage of the electron's initial kinetic energy is transferred to kinetic energy of the hydrogen atom? (The mass of the hydrogen atom is 1840 times the mass of the electron.)

37E. A cart with mass 340 g moving on a frictionless linear air track at an initial speed of 1.2 m/s undergoes an elastic collision with an initially stationary cart of unknown mass. After the collision, the first cart continues in its original direction at 0.66 m/s. (a) What is the mass of the second cart? (b) What is its speed after impact? (c) What is the speed of the two-cart center of mass?   **ssm**

38E. Spacecraft *Voyager 2* (of mass m and speed v relative to the Sun) approaches the planet Jupiter (of mass M and speed V_J relative to the Sun) as shown in Fig. 10-38. The spacecraft rounds the planet and departs in the opposite direction. What is its speed, relative to the Sun, after this slingshot encounter, which can be analyzed as a collision? Assume $v = 12$ km/s and $V_J = 13$ km/s (the orbital speed of Jupiter). The mass of Jupiter is very much greater than the mass of the spacecraft ($M \gg m$).

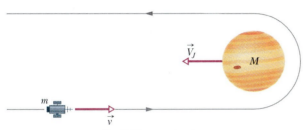

Fig. 10-38 Exercise 38.

39E. An alpha particle (mass 4 u) experiences an elastic head-on collision with a gold nucleus (mass 197 u) that is originally at rest. (The symbol u represents the atomic mass unit.) What percentage of its original kinetic energy does the alpha particle lose?   **ilw**

40P. A steel ball of mass 0.500 kg is fastened to a cord that is 70.0 cm long and fixed at the far end. The ball is then released when the cord is horizontal (Fig. 10-39). At the bottom of its path, the ball strikes a 2.50 kg steel block initially at rest on a frictionless surface. The collision is elastic. Find (a) the speed of the ball and (b) the speed of the block, both just after the collision.

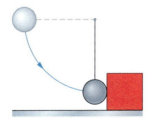

Fig. 10-39 Problem 40.

41P. A body of mass 2.0 kg makes an elastic collision with another body at rest and continues to move in the original direction but with one-fourth of its original speed. (a) What is the mass of the other body? (b) What is the speed of the two-body center of mass if the initial speed of the 2.0 kg body was 4.0 m/s?   **ssm**   **www**

42P. In the two-sphere arrangement of Sample Problem 10-4, assume that sphere 1 has a mass of 50 g and an initial height of 9.0 cm and that sphere 2 has a mass of 85 g. After the collision, what height is reached by (a) sphere 1 and (b) sphere 2? After the next (elastic) collision, what height is reached by (c) sphere 1 and (d) sphere 2? (*Hint:* Do not use rounded-off values.)

43P. Two titanium spheres approach each other head-on with the same speed and collide elastically. After the collision, one of the spheres, whose mass is 300 g, remains at rest. (a) What is the mass of the other sphere? (b) What is the speed of the two-sphere center of mass if the initial speed of each sphere is 2.0 m/s?   **ssm**

44P. In Fig. 10-40, block 1 of mass m_1 is at rest on a long frictionless table that is up against a wall. Block 2 of mass m_2 is placed between block 1 and the wall and sent sliding to the left, toward block 1, with constant speed v_{2i}. Assuming that all collisions are elastic, find the value of m_2 (in terms of m_1) for which both blocks move with the same velocity after block 2 has collided once with block 1 and once with the wall. Assume the wall to have infinite mass.

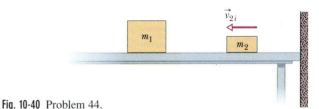

Fig. 10-40 Problem 44.

45P. A small ball of mass m is aligned above a larger ball of mass M (with a slight separation, as in Fig. 10-23a), and the two are dropped simultaneously from height h. (Assume the radius of each ball is negligible compared to h.) (a) If the larger ball rebounds elastically from the floor and then the small ball rebounds elastically from the larger ball, what ratio m/M results in the larger ball stopping upon its collision with the small ball? (The answer is approximately the mass ratio of a baseball to a basketball, as in Question 8.) (b) What height does the small ball then reach?   **ssm**

SEC. 10-6   Collisions in Two Dimensions

46E. Two 2.0 kg bodies, A and B, collide. The velocities before the collision are $\vec{v}_A = 15\hat{i} + 30\hat{j}$ and $\vec{v}_B = -10\hat{i} + 5.0\hat{j}$. After the collision, $\vec{v}'_A = -5.0\hat{i} + 20\hat{j}$. All speeds are given in meters per second. (a) What is the final velocity of B? (b) How much kinetic energy is gained or lost in the collision?

47E. An alpha particle collides with an oxygen nucleus that is initially at rest. The alpha particle is scattered at an angle of $64.0°$ from its initial direction of motion, and the oxygen nucleus recoils at an angle of $51.0°$ on the opposite side of that initial direction. The final speed of the nucleus is 1.20×10^5 m/s. Find (a) the final speed and (b) the initial speed of the alpha particle. (In atomic mass units, the mass of an alpha particle is 4.0 u, and the mass of an oxygen nucleus is 16 u.)   **ilw**

48E. A proton with a speed of 500 m/s collides elastically with another proton initially at rest. The projectile and target protons then move along perpendicular paths, with the projectile path at $60°$ from the original direction. After the collision, what are the speeds of (a) the target proton and (b) the projectile proton?

49E. In a game of pool, the cue ball strikes another ball of the same mass and initially at rest. After the collision, the cue ball moves at 3.50 m/s along a line making an angle of $22.0°$ with its original direction of motion, and the second ball has a speed of 2.00 m/s. Find (a) the angle between the direction of motion of the second ball and the original direction of motion of the cue ball and (b) the original speed of the cue ball. (c) Is kinetic energy (of the centers of mass, don't consider the rotation) conserved?   **ssm**

50P. Two balls A and B, having different but unknown masses, collide. Initially, A is at rest and B has speed v. After the collision, B has speed $v/2$ and moves perpendicularly to its original motion. (a) Find the direction in which ball A moves after the collision. (b) Show that you cannot determine the speed of A from the information given.

51P. After a completely inelastic collision, two objects of the same mass and same initial speed are found to move away together at $\frac{1}{2}$ their initial speed. Find the angle between the initial velocities of the objects. **ssm**

52P. A billiard ball moving at a speed of 2.2 m/s strikes an identical stationary ball a glancing blow. After the collision, one ball is found to be moving at a speed of 1.1 m/s in a direction making a 60° angle with the original line of motion. (a) Find the velocity of the other ball. (b) Can the collision be inelastic, given these data?

53P. In Fig. 10-41, ball 1 with an initial speed of 10 m/s collides elastically with stationary balls 2 and 3, whose centers are on a line perpendicular to the initial velocity of ball 1 and that are initially in contact with each other. The three balls are identical. Ball 1 is aimed directly at the contact point, and all motion is frictionless. After the collision, what are the velocities of (a) ball 2, (b) ball 3, and (c) ball 1? (*Hint:* With friction absent, each impulse is directed along the line connecting the centers of the colliding balls, normal to the colliding surfaces.) **ssm**

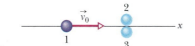

Fig. 10-41 Problem 53.

54P. Two 30 kg children, each with a speed of 4.0 m/s, are sliding on a frictionless frozen pond when they collide and stick together because they have Velcro straps on their jackets. The two children then collide and stick to a 75 kg man who was sliding at 2.0 m/s. After this collision, the three-person composite is stationary. What is the angle between the initial velocity vectors of the two children?

55P. Show that if a neutron is scattered through 90° in an elastic collision with an initially stationary deuteron, the neutron loses $\frac{2}{3}$ of its initial kinetic energy to the deuteron. (In atomic mass units, the mass of a neutron is 1.0 u and the mass of a deuteron is 2.0 u.) **ssm**

Additional Problems

56. Basilisk lizards can run across the top of a water surface (Fig. 10-42). With each step, a lizard first slaps its foot against the water and then pushes it down into the water rapidly enough to form an air cavity around the top of the foot. To avoid having to pull the foot back up against water drag in order to complete the step, the lizard withdraws the foot before water can flow into the air cavity. During this full action of slap, downward push, and withdrawal, the average upward impulse on the lizard must match the downward impulse due to the gravitational force if the lizard is not to sink. Suppose that the mass of a basilisk lizard is 90.0 g, the mass of each foot is 3.00 g, the speed of a foot as it slaps the water is 1.50 m/s, and the time for a single step is 0.600 s. (a) What is the magnitude of the impulse on the lizard during the slap? (Assume this impulse is directly upward.) (b) During the 0.600 s duration of a step, what is the downward impulse on the lizard due to the gravitational force? (c) Which action, the slap or the push, provides the primary support for the lizard, or are they approximately equal in their support?

Fig. 10-42 Problem 56. Basilisk lizard running across water.

57. *Tyrannosaurus rex* may have known from experience not to run particularly fast because of the danger of tripping, in which case their short forearms would have been no help in cushioning the fall. Suppose a *T. Rex* of mass m trips while walking, toppling over, with its center of mass falling freely a distance of 1.5 m. Then its center of mass descends an additional 0.30 m owing to compression of its body and the ground. (a) In multiples of the dinosaur's weight, what is the approximate magnitude of the average vertical force on the dinosaur during its collision with the ground (during the descent of 0.30 m)? Now assume that the dinosaur is running at a speed of 19 m/s (fast) when it trips, falls to the ground, and then slides to a stop with a coefficient of kinetic friction of 0.6. Assume also that the average vertical force during the collision and sliding is that in (a). What, approximately, are (b) the magnitude of the average total force on it from the ground (again in multiples of its weight) and (c) the sliding distance? The force magnitudes of (a) and (b) strongly suggest that the collision would injure the torso of the dinosaur. The head, which would fall farther, would suffer even greater injury.

11 Rotation

In judo, a weaker and smaller fighter who understands physics can defeat a stronger and larger fighter who does not. This fact is demonstrated by the basic "hip throw," in which a fighter rotates the fighter's opponent around his hip and—if the throw is successful—onto the mat. Without the proper use of physics, the throw requires considerable strength and can easily fail.

What is the advantage offered by physics?

The answer is in this chapter.

(a)

(b)

Fig. 11-1 Figure skater Michelle Kwan in motion of (a) pure translation in a fixed direction and (b) pure rotation about a vertical axis.

11-1 Translation and Rotation

The graceful movement of figure skaters can be used to illustrate two kinds of pure, or unmixed, motion. Figure 11-1a shows a skater gliding across the ice in a straight line with constant speed. Her motion is one of pure **translation.** Figure 11-1b shows her spinning at a constant rate about a vertical axis, in a motion of pure **rotation.**

Translation is motion along a straight line, which has been our focus up to now. Rotation is the motion of wheels, gears, motors, planets, the hands of clocks, the rotors of jet engines, and the blades of helicopters. It is our focus in this chapter.

11-2 The Rotational Variables

We wish to examine the rotation of a rigid body about a fixed axis. A **rigid body** is a body that can rotate with all its parts locked together and without any change in its shape. A **fixed axis** means that the rotation occurs about an axis that does not move. Thus, we shall not examine an object like the Sun, because the parts of the Sun (a ball of gas) are not locked together. We also shall not examine an object like a bowling ball rolling along a bowling alley, because the ball rotates about an axis that moves (the ball's motion is a mixture of rotation and translation).

Figure 11-2 shows a rigid body of arbitrary shape in rotation about a fixed axis, called the **axis of rotation** or the **rotation axis.** Every point of the body moves in a circle whose center lies on the axis of rotation, and every point moves through the same angle during a particular time interval. In pure translation, every point of the body moves in a straight line, and every point moves through the same *linear distance* during a particular time interval. (Comparisons between angular and linear motion will appear throughout this chapter.)

We deal now—one at a time—with the angular equivalents of the linear quantities position, displacement, velocity, and acceleration.

Angular Position

Figure 11-2 shows a *reference line*, fixed in the body, perpendicular to the rotation axis, and rotating with the body. The **angular position** of this line is the angle of the line relative to a fixed direction, which we take as the **zero angular position.** In Fig. 11-3, the angular position θ is measured relative to the positive direction of the x axis. From geometry, we know that θ is given by

$$\theta = \frac{s}{r} \qquad \text{(radian measure).} \tag{11-1}$$

Here s is the length of arc (or the arc distance) along a circle and between the x axis (the zero angular position) and the reference line; r is the radius of that circle.

An angle defined in this way is measured in **radians** (rad) rather than in revolutions (rev) or degrees. The radian, being the ratio of two lengths, is a pure number and thus has no dimension. Because the circumference of a circle of radius r is $2\pi r$, there are 2π radians in a complete circle:

$$1 \text{ rev} = 360° = \frac{2\pi r}{r} = 2\pi \text{ rad,} \tag{11-2}$$

and thus

$$1 \text{ rad} = 57.3° = 0.159 \text{ rev.} \tag{11-3}$$

We do *not* reset θ to zero with each complete rotation of the reference line about the rotation axis. If the reference line completes two revolutions from the zero angular position, then the angular position θ of the line is $\theta = 4\pi$ rad.

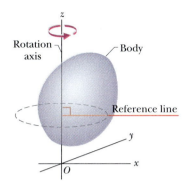

Fig. 11-2 A rigid body of arbitrary shape in pure rotation about the z axis of a coordinate system. The position of the *reference line* with respect to the rigid body is arbitrary, but it is perpendicular to the rotation axis. It is fixed in the body and rotates with the body.

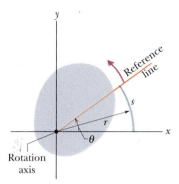

Fig. 11-3 The rotating rigid body of Fig. 11-2 in cross section, viewed from above. The plane of the cross section is perpendicular to the rotation axis, which now extends out of the page, toward you. In this position of the body, the reference line makes an angle θ with the x axis.

For pure translational motion along the x direction, we can know all there is to know about a moving body if we are given $x(t)$, its position as a function of time. Similarly, for pure rotation, we can know all there is to know about a rotating body if we are given $\theta(t)$, the angular position of the body's reference line as a function of time.

Angular Displacement

If the body of Fig. 11-3 rotates about the rotation axis as in Fig. 11-4, changing the angular position of the reference line from θ_1 to θ_2, the body undergoes an **angular displacement** $\Delta\theta$ given by

$$\Delta\theta = \theta_2 - \theta_1. \tag{11-4}$$

This definition of angular displacement holds not only for the rigid body as a whole but also for *every particle within that body* because the particles are all locked together.

If a body is in translational motion along an x axis, its displacement Δx is either positive or negative, depending on whether the body is moving in the positive or negative direction of the axis. Similarly, the angular displacement $\Delta\theta$ of a rotating body is either positive or negative, according to the following rule:

> ⬤ An angular displacement in the counterclockwise direction is positive, and one in the clockwise direction is negative.

The phrase "*clocks are negative*" can help you remember this rule.

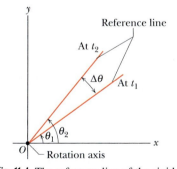

Fig. 11-4 The reference line of the rigid body of Figs. 11-2 and 11-3 is at angular position θ_1 at time t_1 and at angular position θ_2 at a later time t_2. The quantity $\Delta\theta\ (= \theta_2 - \theta_1)$ is the angular displacement that occurs during the interval $\Delta t\ (= t_2 - t_1)$. The body itself is not shown.

Angular Velocity

Suppose (see Fig. 11-4) that our rotating body is at angular position θ_1 at time t_1 and at angular position θ_2 at time t_2. We define the **average angular velocity** of the body in the time interval Δt from t_1 to t_2 to be

$$\omega_{avg} = \frac{\theta_2 - \theta_1}{t_2 - t_1} = \frac{\Delta\theta}{\Delta t}, \tag{11-5}$$

in which $\Delta\theta$ is the angular displacement that occurs during Δt (ω is the lowercase Greek letter omega).

The (**instantaneous**) **angular velocity** ω, with which we shall be most concerned, is the limit of the ratio in Eq. 11-5 as Δt approaches zero. Thus,

$$\omega = \lim_{\Delta t \to 0} \frac{\Delta\theta}{\Delta t} = \frac{d\theta}{dt}. \qquad (11\text{-}6)$$

If we know $\theta(t)$, we can find the angular velocity ω by differentiation.

Equations 11-5 and 11-6 hold not only for the rotating rigid body as a whole but also for *every particle of that body* because the particles are all locked together. The unit of angular velocity is commonly the radian per second (rad/s) or the revolution per second (rev/s). Another measure of angular velocity was used during at least the first three decades of rock: Music was produced by vinyl (phonograph) records that were played on turntables at "$33\frac{1}{3}$ rpm" or "45 rpm," meaning at $33\frac{1}{3}$ rev/min or 45 rev/min.

If a particle moves in translation along an x axis, its linear velocity v is either positive or negative, depending on whether the particle is moving in the positive or negative direction of the axis. Similarly, the angular velocity ω of a rotating rigid body is either positive or negative, depending on whether the body is rotating counterclockwise (positive) or clockwise (negative). ("Clocks are negative" still works.) The magnitude of an angular velocity is called the **angular speed,** which is also represented with ω.

Angular Acceleration

If the angular velocity of a rotating body is not constant, then the body has an angular acceleration. Let ω_2 and ω_1 be its angular velocities at times t_2 and t_1, respectively. The **average angular acceleration** of the rotating body in the interval from t_1 to t_2 is defined as

$$\alpha_{\text{avg}} = \frac{\omega_2 - \omega_1}{t_2 - t_1} = \frac{\Delta\omega}{\Delta t}, \qquad (11\text{-}7)$$

in which $\Delta\omega$ is the change in the angular velocity that occurs during the time interval Δt. The (**instantaneous**) **angular acceleration** α, with which we shall be most concerned, is the limit of this quantity as Δt approaches zero. Thus,

$$\alpha = \lim_{\Delta t \to 0} \frac{\Delta\omega}{\Delta t} = \frac{d\omega}{dt}. \qquad (11\text{-}8)$$

Equations 11-7 and 11-8 hold not only for the rotating rigid body as a whole but also for *every particle of that body*. The unit of angular acceleration is commonly the radian per second-squared (rad/s^2) or the revolution per second-squared (rev/s^2).

Sample Problem 11-1

The disk in Fig. 11-5a is rotating about its central axis like a merry-go-round. The angular position $\theta(t)$ of a reference line on the disk is given by

$$\theta = -1.00 - 0.600t + 0.250t^2, \qquad (11\text{-}9)$$

with t in seconds, θ in radians, and the zero angular position as indicated in the figure.

(a) Graph the angular position of the disk versus time from $t = -3.0$ s to $t = 6.0$ s. Sketch the disk and its angular position reference line at $t = -2.0$ s, 0 s, and 4.0 s, and when the curve crosses the t axis.

SOLUTION: The Key Idea here is that the angular position of the disk is the angular position $\theta(t)$ of its reference line, which is given by

Eq. 11-9 as a function of time. So we graph Eq. 11-9; the result is shown in Fig. 11-5*b*.

To sketch the disk and its reference line at a particular time, we need to determine θ for that time. To do so, we substitute the time into Eq. 11-9. For $t = -2.0$ s, we get

$$\theta = -1.00 - (0.600)(-2.0) + (0.250)(-2.0)^2$$

$$= 1.2 \text{ rad} = 1.2 \text{ rad} \frac{360°}{2\pi \text{ rad}} = 69°.$$

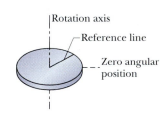

Rotation axis
Reference line
Zero angular position

(a)

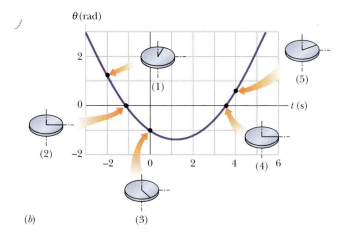

θ (rad)

(1)
(5)
t (s)
(2)
(4) 6
(3)

(b)

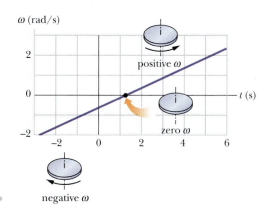

ω (rad/s)

positive ω
t (s)
zero ω
negative ω

(c)

Fig. 11-5 Sample Problem 11-1. (*a*) A rotating disk. (*b*) A plot of the disk's angular position $\theta(t)$. Five sketches indicate the angular position of the reference line on the disk for five points on the curve. (*c*) A plot of the disk's angular velocity $\omega(t)$. Positive values of ω correspond to counterclockwise rotation, and negative values to clockwise rotation.

This means that at $t = -2.0$ s the reference line on the disk is rotated counterclockwise from the zero angular position by 1.2 rad or 69° (counterclockwise because θ is positive). Sketch 1 in Fig. 11-5*b* shows this angular position of the reference line.

Similarly, for $t = 0$, we find $\theta = -1.00$ rad $= -57°$, which means that the reference line is rotated clockwise from the zero angular position by 1.0 rad or 57°, as shown in sketch 3. For $t = 4.0$ s, we find $\theta = 0.60$ rad $= 34°$ (sketch 5). Drawing sketches for when the curve crosses the t axis is easy, because then $\theta = 0$ and the reference line is momentarily aligned with the zero angular position (sketches 2 and 4).

(b) At what time t_{min} does $\theta(t)$ reach the minimum value shown in Fig. 11-5*b*? What is that minimum value?

SOLUTION: The **Key Idea** here is that to find the extreme value (here the minimum) of a function, we take the first derivative of the function and set the result to zero. The first derivative of $\theta(t)$ is

$$\frac{d\theta}{dt} = -0.600 + 0.500t. \quad (11\text{-}10)$$

Setting this to zero and solving for t give us the time at which $\theta(t)$ is minimum:

$$t_{min} = 1.20 \text{ s.} \quad \text{(Answer)}$$

To get the minimum value of θ, we next substitute t_{min} into Eq. 11-9, finding

$$\theta = -1.36 \text{ rad} \approx -77.9°. \quad \text{(Answer)}$$

This *minimum* of $\theta(t)$ (the bottom of the curve in Fig. 11-5*b*) corresponds to the *maximum clockwise* rotation of the disk from the zero angular position, somewhat more than is shown in sketch 3.

(c) Graph the angular velocity ω of the disk versus time from $t = -3.0$ s to $t = 6.0$ s. Sketch the disk and indicate the direction of turning and the sign of ω at $t = -2.0$ s and 4.0 s, and also at t_{min}.

SOLUTION: The **Key Idea** here is that, from Eq. 11-6, the angular velocity ω is equal to $d\theta/dt$ as given in Eq. 11-10. So, we have

$$\omega = -0.600 + 0.500t. \quad (11\text{-}11)$$

The graph of this function $\omega(t)$ is shown in Fig. 11-5*c*.

To sketch the disk at $t = -2.0$ s, we substitute that value into Eq. 11-11, obtaining

$$\omega = -1.6 \text{ rad/s.} \quad \text{(Answer)}$$

The minus sign tells us that at $t = -2.0$ s, the disk is turning clockwise, as suggested by the lowest sketch in Fig. 11-5*c*.

Substituting $t = 4.0$ s into Eq. 11-11 gives us

$$\omega = 1.4 \text{ rad/s.} \quad \text{(Answer)}$$

The implied plus sign tells us that at $t = 4.0$ s, the disk is turning counterclockwise (the highest sketch in Fig. 11-5*c*).

For t_{min}, we already know that $d\theta/dt = 0$. So, we must also have $\omega = 0$. That is, the disk momentarily stops when the reference line reaches the minimum value of θ in Fig. 11-5*b*, as suggested by the center sketch in Fig. 11-5*c*.

(d) Use the results in parts (a) through (c) to describe the motion of the disk from $t = -3.0$ s to $t = 6.0$ s.

SOLUTION: When we first observe the disk at $t = -3.0$ s, it has a positive angular position and is turning clockwise but slowing. It stops at angular position $\theta = -1.36$ rad and then begins to turn counterclockwise, with its angular position eventually becoming positive again.

✔CHECKPOINT 1:   A disk can rotate about its central axis like the one in Fig. 11-5a. Which of the following pairs of values for its initial and final angular positions, respectively, give a negative angular displacement: (a) -3 rad, $+5$ rad, (b) -3 rad, -7 rad, (c) 7 rad, -3 rad?

11-3   Are Angular Quantities Vectors?

We can describe the position, velocity, and acceleration of a single particle by means of vectors. If the particle is confined to a straight line, however, we do not really need vector notation. Such a particle has only two directions available to it, and we can indicate these directions with plus and minus signs.

In the same way, a rigid body rotating about a fixed axis can rotate only clockwise or counterclockwise as seen along the axis, and again we can select between the two directions by means of plus and minus signs. The question arises: "Can we treat the angular displacement, velocity, and acceleration of a rotating body as vectors?" The answer is a qualified "yes" (see the caution below, in connection with angular displacements).

Consider the angular velocity. Figure 11-6a shows a vinyl record rotating on a turntable. The record has a constant angular speed ω ($= 33\frac{1}{3}$ rev/min) in the clockwise direction. We can represent its angular velocity as a vector $\vec{\omega}$ pointing along the axis of rotation, as in Fig. 11-6b. Here's how: We choose the length of this vector according to some convenient scale, for example, with 1 cm corresponding to 10 rev/min. Then we establish a direction for the vector $\vec{\omega}$ by using a **right-hand rule,** as Fig. 11-6c shows: Curl your right hand about the rotating record, your fingers pointing *in the direction of rotation.* Your extended thumb will then point in the direction of the angular velocity vector. If the record were to rotate in the opposite sense, the right-hand rule would tell you that the angular velocity vector then points in the opposite direction.

It is not easy to get used to representing angular quantities as vectors. We instinctively expect that something should be moving *along* the direction of a vector. That is not the case here. Instead, something (the rigid body) is rotating *around* the direction of the vector. In the world of pure rotation, a vector defines an axis of rotation, not a direction in which something moves. Nonetheless, the vector also defines the motion. Furthermore, it obeys all the rules for vector manipulation discussed in Chapter 3. The angular acceleration $\vec{\alpha}$ is another vector, and it too obeys those rules.

Fig. 11-6 (a) A record rotating about a vertical axis that coincides with the axis of the spindle. (b) The angular velocity of the rotating record can be represented by the vector $\vec{\omega}$, lying along the axis and pointing down, as shown. (c) We establish the direction of the angular velocity vector as downward by using the right-hand rule. When the fingers of the right hand curl around the record and point the way it is moving, the extended thumb points in the direction of $\vec{\omega}$.

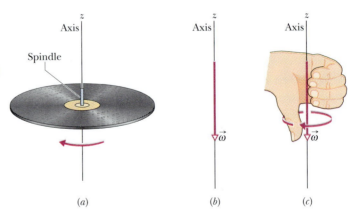

(a)   (b)   (c)

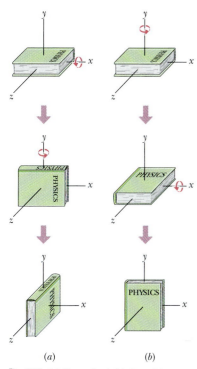

Fig. 11-7 (*a*) From its initial position, at the top, the book is given two successive 90° rotations, first about the (horizontal) *x* axis and then about the (vertical) *y* axis. (*b*) The book is given the same rotations, but in the reverse order.

In this chapter we consider only rotations that are about a fixed axis. For such situations, we need not consider vectors—we can represent angular velocity with ω and angular acceleration with α, and we can indicate direction with an implied plus sign for counterclockwise or an explicit minus sign for clockwise.

Now for the caution: Angular *displacements* (unless they are very small) *cannot* be treated as vectors. Why not? We can certainly give them both magnitude and direction, as we did for the angular velocity vector in Fig. 11-6. However, to be represented as a vector, a quantity must *also* obey the rules of vector addition, one of which says that if you add two vectors, the order in which you add them does not matter. Angular displacements fail this test.

Figure 11-7 gives an example. An initially horizontal book is given two 90° angular displacements, first in the order of Fig. 11-7a and then in the order of Fig. 11-7b. Although the two angular displacements are identical, their order is not, and the book ends up with different orientations. Thus, the addition of the two angular displacements depends on their order and they cannot be vectors.

11-4 Rotation with Constant Angular Acceleration

In pure translation, motion with a *constant linear acceleration* (for example, that of a falling body) is an important special case. In Table 2-1, we displayed a series of equations that hold for such motion.

In pure rotation, the case of *constant angular acceleration* is also important, and a parallel set of equations holds for this case also. We shall not derive them here, but simply write them from the corresponding linear equations, substituting equivalent angular quantities for the linear ones. This is done in Table 11-1, which lists both sets of equations (Eqs. 2-11 and 2-15 to 2-18; 11-12 to 11-16).

Recall that Eqs. 2-11 and 2-15 are basic equations for constant linear acceleration—the other equations in the Linear list can be derived from them. Similarly, Eqs. 11-12 and 11-13 are the basic equations for constant angular acceleration, and the other equations in the Angular list can be derived from them. To solve a simple problem involving constant angular acceleration, you can usually use an equation from the Angular list (*if* you have the list). Choose an equation for which the only unknown variable will be the variable requested in the problem. A better plan is to remember only Eqs. 11-12 and 11-13, and then solve them as simultaneous equations whenever needed. An example is given in Sample Problem 11-3.

> ✔**CHECKPOINT 2:** In four situations, a rotating body has angular position $\theta(t)$ given by (a) $\theta = 3t - 4$, (b) $\theta = -5t^3 + 4t^2 + 6$, (c) $\theta = 2/t^2 - 4/t$, and (d) $\theta = 5t^2 - 3$. To which situations do the angular equations of Table 11-1 apply?

TABLE 11-1 Equations of Motion for Constant Linear Acceleration and for Constant Angular Acceleration

Equation Number	Linear Equation	Missing Variable		Angular Equation	Equation Number
(2-11)	$v = v_0 + at$	$x - x_0$	$\theta - \theta_0$	$\omega = \omega_0 + \alpha t$	(11-12)
(2-15)	$x - x_0 = v_0 t + \frac{1}{2}at^2$	v	ω	$\theta - \theta_0 = \omega_0 t + \frac{1}{2}\alpha t^2$	(11-13)
(2-16)	$v^2 = v_0^2 + 2a(x - x_0)$	t	t	$\omega^2 = \omega_0^2 + 2\alpha(\theta - \theta_0)$	(11-14)
(2-17)	$x - x_0 = \frac{1}{2}(v_0 + v)t$	a	α	$\theta - \theta_0 = \frac{1}{2}(\omega_0 + \omega)t$	(11-15)
(2-18)	$x - x_0 = vt - \frac{1}{2}at^2$	v_0	ω_0	$\theta - \theta_0 = \omega t - \frac{1}{2}\alpha t^2$	(11-16)

Sample Problem 11-2

A grindstone (Fig. 11-8) rotates at constant angular acceleration $\alpha = 0.35$ rad/s^2. At time $t = 0$, it has an angular velocity of $\omega_0 = -4.6$ rad/s and a reference line on it is horizontal, at the angular position $\theta_0 = 0$.

(a) At what time after $t = 0$ is the reference line at the angular position $\theta = 5.0$ rev?

SOLUTION: The **Key Idea** here is that the angular acceleration is constant, so we can use the rotation equations of Table 11-1. We choose Eq. 11-13,

$$\theta - \theta_0 = \omega_0 t + \tfrac{1}{2}\alpha t^2,$$

because the only unknown variable it contains is the desired time t. Substituting known values and setting $\theta_0 = 0$ and $\theta = 5.0$ rev $= 10\pi$ rad give us

$$10\pi \text{ rad} = (-4.6 \text{ rad/s})t + \tfrac{1}{2}(0.35 \text{ rad/s}^2)t^2.$$

(We converted 5.0 rev to 10π rad to keep the units consistent.) Solving this quadratic equation for t, we find

$$t = 32 \text{ s.} \qquad \text{(Answer)}$$

(b) Describe the grindstone's rotation between $t = 0$ and $t = 32$ s.

SOLUTION: The wheel is initially rotating in the negative (clockwise) direction with angular velocity $\omega_0 = -4.6$ rad/s, but its angular acceleration α is positive. This initial opposition of the signs of angular velocity and angular acceleration means that the wheel

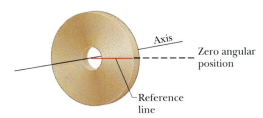

Fig. 11-8 Sample Problem 11-2. A grindstone. At $t = 0$ the reference line (which we imagine to be marked on the stone) is horizontal.

slows in its rotation in the negative direction, stops, and then reverses to rotate in the positive direction. After the reference line comes back through its initial orientation of $\theta = 0$, the wheel turns an additional 5.0 rev by time $t = 32$ s.

(c) At what time t does the grindstone momentarily stop?

SOLUTION: We again go to the table of equations for constant angular acceleration, and again we need an equation that contains only the desired unknown variable t. However, now we use another **Key Idea**. The equation must also contain the variable ω, so that we can set it to 0 and then solve for the corresponding time t. We choose Eq. 11-12, which yields

$$t = \frac{\omega - \omega_0}{\alpha} = \frac{0 - (-4.6 \text{ rad/s})}{0.35 \text{ rad/s}^2} = 13 \text{ s.} \quad \text{(Answer)}$$

Sample Problem 11-3

While you are operating a Rotor (the rotating cylindrical ride discussed in Sample Problem 6-8), you spot a passenger in acute distress and decrease the angular speed of the cylinder from 3.40 rad/s to 2.00 rad/s in 20.0 rev, at constant angular acceleration. (The passenger is obviously more of a "translation person" than a "rotation person.")

(a) What is the constant angular acceleration during this decrease in angular speed?

SOLUTION: Assume that the rotation is in the counterclockwise direction, and let the acceleration begin at time $t = 0$, at angular position θ_0. The **Key Idea** here is that, because the angular acceleration is constant, we can relate the cylinder's angular acceleration to its angular velocity and angular displacement via the basic equations for constant angular acceleration (Eqs. 11-12 and 11-13). The initial angular velocity is $\omega_0 = 3.40$ rad/s, the angular displacement is $\theta - \theta_0 = 20.0$ rev, and the angular velocity at the end of that displacement is $\omega = 2.00$ rad/s. But we do not know the angular acceleration α and time t, which are in both basic equations.

To eliminate the unknown t, we use Eq. 11-12 to write

$$t = \frac{\omega - \omega_0}{\alpha},$$

which we then substitute into Eq. 11-13 to write

$$\theta - \theta_0 = \omega_0 \left(\frac{\omega - \omega_0}{\alpha} \right) + \tfrac{1}{2}\alpha \left(\frac{\omega - \omega_0}{\alpha} \right)^2.$$

Solving for α, substituting known data, and converting 20 rev to 125.7 rad, we find

$$\alpha = \frac{\omega^2 - \omega_0^2}{2(\theta - \theta_0)} = \frac{(2.00 \text{ rad/s})^2 - (3.40 \text{ rad/s})^2}{2(125.7 \text{ rad})}$$

$$= -0.0301 \text{ rad/s}^2. \qquad \text{(Answer)}$$

(b) How much time did the speed decrease take?

SOLUTION: Now that we know α, we can use Eq. 11-12 to solve for t:

$$t = \frac{\omega - \omega_0}{\alpha} = \frac{2.00 \text{ rad/s} - 3.40 \text{ rad/s}}{-0.0301 \text{ rad/s}^2}$$

$$= 46.5 \text{ s.} \qquad \text{(Answer)}$$

This sample problem is actually an angular version of Sample Problem 2-5. The data and symbols differ, but the techniques of solution are identical.

11-5 Relating the Linear and Angular Variables

In Section 4-7, we discussed uniform circular motion, in which a particle travels at constant linear speed v along a circle and around an axis of rotation. When a rigid body, such as a merry-go-round, rotates around an axis, each particle in the body moves in its own circle around that axis. Since the body is rigid, all the particles make one revolution in the same amount of time; that is, they all have the same angular speed ω.

However, the farther a particle is from the axis, the greater the circumference of its circle is, and so the faster its linear speed v must be. You can notice this on a merry-go-round. You turn with the same angular speed ω regardless of your distance from the center, but your linear speed v increases noticeably if you move to the outside edge of the merry-go-round.

We often need to relate the linear variables s, v, and a for a particular point in a rotating body to the angular variables θ, ω, and α for that body. The two sets of variables are related by r, the *perpendicular distance* of the point from the rotation axis. This perpendicular distance is the distance between the point and the rotation axis, measured along a perpendicular to the axis. It is also the radius r of the circle traveled by the point around the axis of rotation.

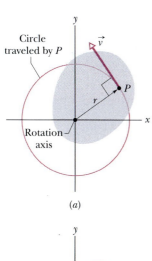

(a)

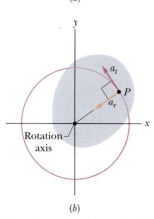

(b)

Fig. 11-9 The rotating rigid body of Fig. 11-2, shown in cross section viewed from above. Every point of the body (such as P) moves in a circle around the rotation axis. (a) The linear velocity $\vec{v}$ of every point is tangent to the circle in which the point moves. (b) The linear acceleration $\vec{a}$ of the point has (in general) two components: a tangential component a_t and a radial component a_r.

The Position

If a reference line on a rigid body rotates through an angle θ, a point within the body at a position r from the rotation axis moves a distance s along a circular arc, where s is given by Eq. 11-1:

$$s = \theta r \qquad \text{(radian measure).} \tag{11-17}$$

This is the first of our linear–angular relations. *Caution:* The angle θ here must be measured in radians because Eq. 11-17 is itself the definition of angular measure in radians.

The Speed

Differentiating Eq. 11-17 with respect to time—with r held constant—leads to

$$\frac{ds}{dt} = \frac{d\theta}{dt} r.$$

However, ds/dt is the linear speed (the magnitude of the linear velocity) of the point in question, and $d\theta/dt$ is the angular speed ω of the rotating body. So

$$v = \omega r \qquad \text{(radian measure).} \tag{11-18}$$

Caution: The angular speed ω must be expressed in radian measure.

Equation 11-18 tells us that since all points within the rigid body have the same angular speed ω, points with greater radius r have greater linear speed v. Figure 11-9a reminds us that the linear velocity is always tangent to the circular path of the point in question.

If the angular speed ω of the rigid body is constant, then Eq. 11-18 tells us that the linear speed v of any point within it is also constant. Thus, each point within the body undergoes uniform circular motion. The period of revolution T for the motion of each point and for the rigid body itself is given by Eq. 4-33:

$$T = \frac{2\pi r}{v}. \tag{11-19}$$

This equation tells us that the time for one revolution is the distance $2\pi r$ traveled in one revolution divided by the speed at which that distance is traveled. Substituting for v from Eq. 11-18 and canceling r, we find also that

$$T = \frac{2\pi}{\omega} \qquad \text{(radian measure)}. \tag{11-20}$$

This equivalent equation says that the time for one revolution is the angular distance 2π rad traveled in one revolution divided by the angular speed (or rate) at which that angle is traveled.

The Acceleration

Differentiating Eq. 11-18 with respect to time—again with r held constant—leads to

$$\frac{dv}{dt} = \frac{d\omega}{dt} r. \tag{11-21}$$

Here we run up against a complication. In Eq. 11-21, dv/dt represents only the part of the linear acceleration that is responsible for changes in the *magnitude* v of the linear velocity $\vec{v}$. Like $\vec{v}$, that part of the linear acceleration is tangent to the path of the point in question. We call it the *tangential component* a_t of the linear acceleration of the point, and we write

$$a_t = \alpha r \qquad \text{(radian measure)}, \tag{11-22}$$

where $\alpha = d\omega/dt$. *Caution:* The angular acceleration α in Eq. 11-22 must be expressed in radian measure.

In addition, as Eq. 4-32 tells us, a particle (or point) moving in a circular path has a *radial component* of linear acceleration, $a_r = v^2/r$ (directed radially inward), that is responsible for changes in the *direction* of the linear velocity $\vec{v}$. By substituting for v from Eq. 11-18, we can write this component as

$$a_r = \frac{v^2}{r} = \omega^2 r \qquad \text{(radian measure)}. \tag{11-23}$$

Thus, as Fig. 11-9b shows, the linear acceleration of a point on a rotating rigid body has, in general, two components. The radially inward component a_r (given by Eq. 11-23) is present whenever the angular velocity of the body is not zero. The tangential component a_t (given by Eq. 11-22) is present whenever the angular acceleration is not zero.

✔**CHECKPOINT 3:** A cockroach rides the rim of a rotating merry-go-round. If the angular speed of this system (*merry-go-round + cockroach*) is constant, does the cockroach have (a) radial acceleration and (b) tangential acceleration? If the angular speed is decreasing, does the cockroach have (c) radial acceleration and (d) tangential acceleration?

Sample Problem 11-4

Figure 11-10 shows a centrifuge used to accustom astronaut trainees to high accelerations. The radius r of the circle traveled by an astronaut is 15 m.

(a) At what constant angular speed must the centrifuge rotate if the astronaut is to have a linear acceleration of magnitude $11g$?

Fig. 11-10 Sample Problem 11-4. A centrifuge is used to accustom astronauts to the large acceleration experienced during a liftoff.

we have

$$\omega = \sqrt{\frac{a_r}{r}} = \sqrt{\frac{(11)(9.8 \text{ m/s}^2)}{15 \text{ m}}}$$

$$= 2.68 \text{ rad/s} \approx 26 \text{ rev/min.} \qquad \text{(Answer)}$$

(b) What is the tangential acceleration of the astronaut if the centrifuge accelerates at a constant rate from rest to the angular speed of (a) in 120 s?

SOLUTION: The Key Idea here is that the tangential acceleration a_t, which is the linear acceleration along the circular path, is related to the angular acceleration α by Eq. 11-22 ($a_t = \alpha r$). Also, because the angular acceleration is constant, we can use Eq. 11-12 ($\omega = \omega_0 + \alpha t$) from Table 11-1 to find α from the given angular speeds. Putting these two equations together, we find

$$a_t = \alpha r = \frac{\omega - \omega_0}{t} r$$

$$= \frac{2.68 \text{ rad/s} - 0}{120 \text{ s}} (15 \text{ m}) = 0.34 \text{ m/s}^2$$

$$= 0.034g. \qquad \text{(Answer)}$$

SOLUTION: The Key Idea is this: Because the angular speed is constant, the angular acceleration α ($= d\omega/dt$) is zero and so is the tangential component of the linear acceleration ($a_t = \alpha r$). This leaves only the radial component. From Eq. 11-23 ($a_r = \omega^2 r$), with $a_r = 11g$,

Although the final radial acceleration $a_r = 11g$ is large (and alarming), the astronaut's tangential acceleration a_t during the speed-up is not.

PROBLEM-SOLVING TACTICS

Tactic 1: *Units for Angular Variables*
In Eq. 11-1 ($\theta = s/r$), we began the use of radian measure for all angular variables whenever we are using equations that contain both angular and linear variables. Thus, we must express angular displacements in radians, angular velocities in rad/s and rad/min, and angular accelerations in rad/s^2 and rad/min^2. Equations 11-17, 11-18, 11-20, 11-22, and 11-23 are marked to emphasize this. The only exceptions to this rule are equations that involve *only* angular variables, such as the angular equations listed in Table 11-1. Here

you are free to use any unit you wish for the angular variables; that is, you may use radians, degrees, or revolutions, as long as you use them consistently.

In equations where radian measure must be used, you need not keep track of the unit "radian" (rad) algebraically, as you must do for other units. You can add or delete it at will, to suit the context. In Sample Problem 11-4a the unit was added to the answer; in Sample Problem 11-4b it was omitted from the answer.

11-6 Kinetic Energy of Rotation

The rapidly rotating blade of a table saw certainly has kinetic energy due to that rotation. How can we express the energy? We cannot apply the familiar formula $K = \frac{1}{2}mv^2$ to the saw as a whole because that would only give us the kinetic energy of the saw's center of mass, which is zero.

Instead, we shall treat the table saw (and any other rotating rigid body) as a collection of particles with different speeds. We can then add up the kinetic energies of all the particles to find the kinetic energy of the body as a whole. In this way we obtain, for the kinetic energy of a rotating body,

$$K = \frac{1}{2}m_1 v_1^2 + \frac{1}{2}m_2 v_2^2 + \frac{1}{2}m_3 v_3^2 + \cdots$$
$$= \sum \frac{1}{2}m_i v_i^2, \qquad (11\text{-}24)$$

in which m_i is the mass of the ith particle and v_i is its speed. The sum is taken over all the particles in the body.

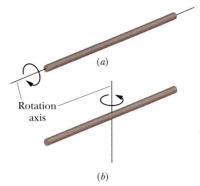

Fig. 11-11 A long rod is much easier to rotate about (*a*) its central (longitudinal) axis than about (*b*) an axis through its center and perpendicular to its length because the mass is distributed closer to the rotation axis in (*a*) than in (*b*).

The problem with Eq. 11-24 is that v_i is not the same for all particles. We solve this problem by substituting for v from Eq. 11-18 ($v = \omega r$), so that we have

$$K = \sum \tfrac{1}{2} m_i (\omega r_i)^2 = \tfrac{1}{2} \left(\sum m_i r_i^2 \right) \omega^2, \qquad (11\text{-}25)$$

in which ω *is* the same for all particles.

The quantity in parentheses on the right side of Eq. 11-25 tells us how the mass of the rotating body is distributed about its axis of rotation. We call that quantity the **rotational inertia** (or **moment of inertia**) I of the body with respect to the axis of rotation. It is a constant for a particular rigid body and a particular rotation axis. (That axis must always be specified if the value of I is to be meaningful.)

We may now write

$$I = \sum m_i r_i^2 \qquad \text{(rotational inertia)} \qquad (11\text{-}26)$$

and substitute into Eq. 11-25, obtaining

$$K = \tfrac{1}{2} I \omega^2 \qquad \text{(radian measure)} \qquad (11\text{-}27)$$

as the expression we seek. Because we have used the relation $v = \omega r$ in deriving Eq. 11-27, ω must be expressed in radian measure. The SI unit for I is the kilogram–square meter (kg · m^2).

Equation 11-27, which gives the kinetic energy of a rigid body in pure rotation, is the angular equivalent of the formula $K = \tfrac{1}{2} M v_{\text{com}}^2$, which gives the kinetic energy of a rigid body in pure translation. In both formulas there is a factor of $\tfrac{1}{2}$. Where mass M appears in one equation, I (which involves both mass and its distribution) appears in the other. Finally, each equation contains as a factor the square of a speed — translational or rotational as appropriate. The kinetic energies of translation and of rotation are not different kinds of energy. They are both kinetic energy, expressed in ways that are appropriate to the motion at hand.

We noted previously that the rotational inertia of a rotating body involves not only its mass but also how that mass is distributed. Here is an example that you can literally feel. Rotate a long, fairly heavy rod (a pole, a length of lumber, or something similar), first around its central (longitudinal) axis (Fig. 11-11*a*) and then around an axis perpendicular to the rod and through the center (Fig. 11-11*b*). Both rotations involve the very same mass, but the first rotation is much easier than the second. The reason is that the mass is distributed much closer to the rotation axis in the first rotation. As a result, the rotational inertia of the rod is much smaller in Fig. 11-11*a* than in Fig. 11-11*b*. In general, smaller rotational inertia means easier rotation.

✔**CHECKPOINT 4:**   The figure shows three small spheres that rotate about a vertical axis. The perpendicular distance between the axis and the center of each sphere is given. Rank the three spheres according to their rotational inertia about that axis, greatest first.

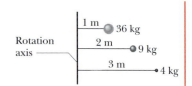

11-7 Calculating the Rotational Inertia

If a rigid body consists of a few particles, we can calculate its rotational inertia about a given rotation axis with Eq. 11-26 ($I = \sum m_i r_i^2$), that is, we can find the product mr^2 for each particle and then sum the products. (Recall that r is the perpendicular distance a particle is from the given rotation axis.)

TABLE 11-2   Some Rotational Inertias

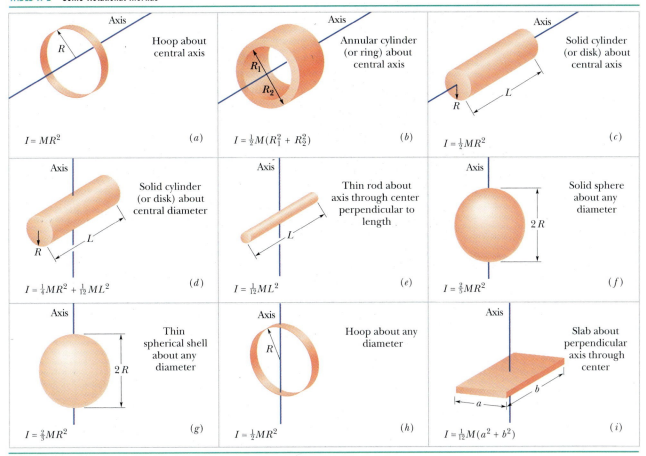

If a rigid body consists of a great many adjacent particles (it is *continuous,* like a Frisbee), using Eq. 11-26 would require a computer. Thus, instead, we replace the sum in Eq. 11-26 with an integral and define the rotational inertia of the body as

$$I = \int r^2 \, dm \qquad \text{(rotational inertia, continuous body).} \qquad (11\text{-}28)$$

Table 11-2 gives the results of such integration for nine common body shapes and the indicated axes of rotation.

Parallel-Axis Theorem

Suppose we want to find the rotational inertia I of a body of mass M about a given axis. In principle, we can always find I with the integration of Eq. 11-28. However, there is a shortcut if we happen to already know the rotational inertia I_{com} of the body about a *parallel* axis that extends through the body's center of mass. Let h be the perpendicular distance between the given axis and the axis through the center of mass (remember these two axes must be parallel). Then the rotational inertia I about the given axis is

$$I = I_{\text{com}} + Mh^2 \qquad \text{(parallel-axis theorem).} \qquad (11\text{-}29)$$

This equation is known as the **parallel-axis theorem.** We shall now prove it and then put it to use in Checkpoint 5 and Sample Problem 11-5.

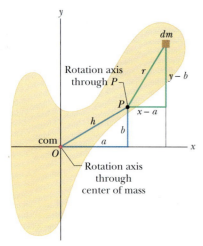

Fig. 11-12 A rigid body in cross section, with its center of mass at O. The parallel-axis theorem (Eq. 11-29) relates the rotational inertia of the body about an axis through O to that about a parallel axis through a point such as P, a distance h from the body's center of mass. Both axes are perpendicular to the plane of the figure.

Proof of the Parallel-Axis Theorem

Let O be the center of mass of the arbitrarily shaped body shown in cross section in Fig. 11-12. Place the origin of the coordinates at O. Consider an axis through O perpendicular to the plane of the figure, and another axis through point P parallel to the first axis. Let the x and y coordinates of P be a and b.

Let dm be a mass element with the general coordinates x and y. The rotational inertia of the body about the axis through P is then, from Eq. 11-28,

$$I = \int r^2 \, dm = \int [(x - a)^2 + (y - b)^2] \, dm,$$

which we can rearrange as

$$I = \int (x^2 + y^2) \, dm - 2a \int x \, dm - 2b \int y \, dm + \int (a^2 + b^2) \, dm. \quad (11\text{-}30)$$

From the definition of the center of mass (Eq. 9-9), the middle two integrals of Eq. 11-30 give the coordinates of the center of mass (multiplied by a constant) and thus must each be zero. Because $x^2 + y^2$ is equal to R^2, where R is the distance from O to dm, the first integral is simply I_{com}, the rotational inertia of the body about an axis through its center of mass. Inspection of Fig. 11-12 shows that the last term in Eq. 11-30 is Mh^2, where M is the body's total mass. Thus, Eq. 11-30 reduces to Eq. 11-29, which is the relation that we set out to prove.

✔**CHECKPOINT 5:** The figure shows a booklike object (one side is longer than the other) and four choices of rotation axes, all perpendicular to the face of the object. Rank the choices according to the rotational inertia of the object about the axis, greatest first.

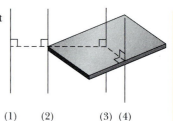

(1)   (2)   (3) (4)

Sample Problem 11-5

Figure 11-13a shows a rigid body consisting of two particles of mass m connected by a rod of length L and negligible mass.

(a) What is the rotational inertia I_{com} of this body about an axis through its center of mass, perpendicular to the rod as shown?

SOLUTION: The **Key Idea** is that because we have only two particles with mass, we can find the body's rotational inertia I_{com} by using Eq. 11-26 rather than by integration. For the two particles, each at perpendicular distance $\frac{1}{2}L$ from the rotation axis, we have

$$I = \sum m_i r_i^2 = (m)(\tfrac{1}{2}L)^2 + (m)(\tfrac{1}{2}L)^2$$

$$= \tfrac{1}{2}mL^2. \quad \text{(Answer)}$$

(b) What is the rotational inertia I of the body about an axis through the left end of the rod and parallel to the first axis (Fig. 11-13b)?

SOLUTION: This situation is simple enough that we can find I using either of two **Key Ideas**. The first is identical to the one we used in (a). The only difference here is that the perpendicular distance r_i is zero for the particle on the left, and L for the particle on the right. Now Eq. 11-26 gives us

$$I = m(0)^2 + mL^2 = mL^2. \quad \text{(Answer)}$$

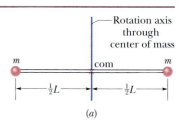

(a)

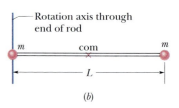

(b)

Fig. 11-13 Sample Problem 11-5. A rigid body of two particles of mass m, which are joined by a rod of negligible mass.

The second **Key Idea** provides a more powerful technique: Because we already know I_{com} about an axis through the center of mass and because the axis here is parallel to that "com axis," we can apply the parallel-axis theorem (Eq. 11-29). We find

$$I = I_{\text{com}} + Mh^2 = \tfrac{1}{2}mL^2 + (2m)(\tfrac{1}{2}L)^2$$

$$= mL^2. \quad \text{(Answer)}$$

Sample Problem 11-6

Large machine components that undergo prolonged, high-speed rotation are first examined for the possibility of failure in a *spin test system*. In this system, a component is *spun up* (brought up to high speed) while inside a cylindrical arrangement of lead bricks and containment liner, all within a steel shell that is closed by a lid clamped into place. If the rotation causes the component to shatter, the soft lead bricks are supposed to catch the pieces so that the failure can then be analyzed.

In early 1985, Test Devices, Inc. (www.testdevices.com) was spin testing a sample of a solid steel rotor (a disk) of mass $M = 272$ kg and radius $R = 38.0$ cm. When the sample reached an angular speed ω of 14 000 rev/min, the test engineers heard a dull thump from the test system, which was located one floor down and one room over from them. Investigating, they found that lead bricks had been thrown out in the hallway leading to the test room, a door to the room had been hurled into the adjacent parking lot, one lead brick had shot from the test site through the wall of a neighbor's kitchen, the structural beams of the test building had been damaged, the concrete floor beneath the spin chamber had been shoved downward by about 0.5 cm, and the 900 kg lid had been blown upward through the ceiling and had then crashed back onto the test equipment (Fig. 11-14). The exploding pieces had not penetrated the room of the test engineers only by luck.

How much energy was released in the explosion of the rotor?

SOLUTION: The **Key Idea** here is that this released energy was equal to the rotational kinetic energy K of the rotor just as it reached the angular speed of 14 000 rev/min. We can find K with Eq. 11-27 ($K = \frac{1}{2}I\omega^2$), but first we need an expression for the rotational inertia I. Because the rotor was a disk that rotated like a merry-go-round, I is given by the expression in Table 11-2c ($I = \frac{1}{2}MR^2$).

Fig. 11-14 Sample Problem 11-6. Some of the destruction caused by the explosion of a rapidly rotating steel disk.

Thus, we have
$$I = \tfrac{1}{2}MR^2 = \tfrac{1}{2}(272 \text{ kg})(0.38 \text{ m})^2 = 19.64 \text{ kg} \cdot \text{m}^2.$$

The angular speed of the rotor was
$$\omega = (14\,000 \text{ rev/min})(2\pi \text{ rad/rev})\left(\frac{1 \text{ min}}{60 \text{ s}}\right)$$
$$= 1.466 \times 10^3 \text{ rad/s}.$$

Now we can use Eq. 11-27 to write
$$K = \tfrac{1}{2}I\omega^2 = \tfrac{1}{2}(19.64 \text{ kg} \cdot \text{m}^2)(1.466 \times 10^3 \text{ rad/s})^2$$
$$= 2.1 \times 10^7 \text{ J}. \qquad \text{(Answer)}$$

Being near this explosion was like being near an exploding bomb.

11-8 Torque

A doorknob is located as far as possible from the door's hinge line for a good reason. If you want to open a heavy door, you must certainly apply a force; that alone, however, is not enough. Where you apply that force and in what direction you push are also important. If you apply your force nearer to the hinge line than the knob, or at any angle other than 90° to the plane of the door, you must use a greater force to move the door than if you apply the force at the knob and perpendicular to the door's plane.

Figure 11-15a shows a cross section of a body that is free to rotate about an axis passing through O and perpendicular to the cross section. A force $\vec{F}$ is applied at point P, whose position relative to O is defined by a position vector $\vec{r}$. The directions of vectors $\vec{F}$ and $\vec{r}$ make an angle ϕ with each other. (For simplicity, we consider only forces that have no component parallel to the rotation axis; thus, $\vec{F}$ is in the plane of the page.)

To determine how $\vec{F}$ results in a rotation of the body around the rotation axis, we resolve $\vec{F}$ into two components (Fig. 11-15b). One component, called the *radial component F_r*, points along $\vec{r}$. This component does not cause rotation, because it acts along a line that extends through O. (If you pull on a door parallel to the plane of the door, you do not rotate the door.) The other component of $\vec{F}$, called the

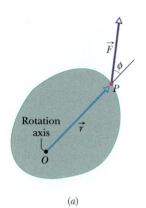

(a)

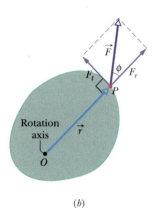

(b)

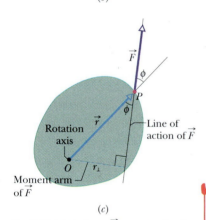

(c)

Fig. 11-15 (a) A force $\vec{F}$ acts at point P on a rigid body that is free to rotate about an axis through O; the axis is perpendicular to the plane of the cross section shown here. (b) The torque due to this force is $(r)(F \sin \phi)$. We can also write it as rF_t, where F_t is the tangential component of $\vec{F}$. (c) The torque can also be written as $r_\perp F$, where $r_\perp$ is the moment arm of $\vec{F}$.

tangential component F_t, is perpendicular to $\vec{r}$ and has magnitude $F_t = F \sin \phi$. This component *does* cause rotation. (If you pull on a door perpendicular to its plane, you can rotate the door.)

The ability of $\vec{F}$ to rotate the body depends not only on the magnitude of its tangential component F_t, but also on just how far from O the force is applied. To include both these factors, we define a quantity called **torque** τ as the product of the two factors and write it as

$$\tau = (r)(F \sin \phi). \tag{11-31}$$

Two equivalent ways of computing the torque are

$$\tau = (r)(F \sin \phi) = rF_t \tag{11-32}$$

and

$$\tau = (r \sin \phi)(F) = r_\perp F, \tag{11-33}$$

where $r_\perp$ is the perpendicular distance between the rotation axis at O and an extended line running through the vector $\vec{F}$ (Fig. 11-15c). This extended line is called the **line of action** of $\vec{F}$, and $r_\perp$ is called the **moment arm** of $\vec{F}$. Figure 11-15b shows that we can describe r, the magnitude of $\vec{r}$, as being the moment arm of the force component F_t.

Torque, which comes from the Latin word meaning "to twist," may be loosely identified as the turning or twisting action of the force $\vec{F}$. When you apply a force to an object—such as a screwdriver or torque wrench—with the purpose of turning that object, you are applying a torque. The SI unit of torque is the newton-meter (N · m). *Caution*: The newton-meter is also the unit of work. Torque and work, however, are quite different quantities and must not be confused. Work is often expressed in joules (1 J = 1 N · m), but torque never is.

In the next chapter we shall discuss torque in a general way as being a vector quantity. Here, however, because we consider only rotation around a single axis, we do not need vector notation. Instead, a torque has either a positive or negative value depending on the direction of rotation it would give a body initially at rest: If the body would rotate counterclockwise, the torque is positive. If the object would rotate clockwise, the torque is negative. (The phrase "clocks are negative" from Section 11-2 still works.)

Torques obey the superposition principle that we discussed in Chapter 5 for forces: When several torques act on a body, the **net torque** (or **resultant torque**) is the sum of the individual torques. The symbol for net torque is τ_{net}.

✓CHECKPOINT 6: The figure shows an overhead view of a meter stick that can pivot about the dot at the position marked 20 (for 20 cm). All five horizontal forces on the stick have the same magnitude. Rank those forces according to the magnitude of the torque that they produce, greatest first.

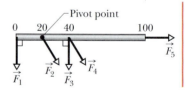

11-9 Newton's Second Law for Rotation

A torque can cause rotation of a rigid body, as when you use a torque to rotate a door. Here we want to relate the net torque τ_{net} on a rigid body to the angular acceleration α it causes about a rotation axis. We do so by analogy with Newton's second law ($F_{\text{net}} = ma$) for the acceleration a of a body of mass m due to a net force F_{net} along a coordinate axis. We replace F_{net} with τ_{net}, m with I, and a with α, writing

$$\tau_{\text{net}} = I\alpha \qquad \text{(Newton's second law for rotation)}, \tag{11-34}$$

where α must be in radian measure.

Proof of Equation 11-34

We prove Eq. 11-34 by first considering the simple situation shown in Fig. 11-16. The rigid body there consists of a particle of mass m on one end of a massless rod of length r. The rod can move only by rotating about its other end, around a rotation axis (an axle) that is perpendicular to the plane of the page. Thus, the particle can move only in a circular path that has the rotation axis at its center.

A force $\vec{F}$ acts on the particle. However, because the particle can move only along the circular path, only the tangential component F_t of the force (the component that is tangent to the circular path) can accelerate the particle along the path. We can relate F_t to the particle's tangential acceleration a_t along the path with Newton's second law, writing

$$F_t = ma_t.$$

The torque acting on the particle is, from Eq. 11-32,

$$\tau = F_t r = ma_t r.$$

From Eq. 11-22 ($a_t = \alpha r$) we can write this as

$$\tau = m(\alpha r)r = (mr^2)\alpha. \qquad (11\text{-}35)$$

The quantity in parentheses on the right side of Eq. 11-35 is the rotational inertia of the particle about the rotation axis (see Eq. 11-26). Thus, Eq. 11-35 reduces to

$$\tau = I\alpha \qquad \text{(radian measure).} \qquad (11\text{-}36)$$

For the situation in which more than one force is applied to the particle, we can generalize Eq. 11-36 as

$$\tau_{\text{net}} = I\alpha \qquad \text{(radian measure),} \qquad (11\text{-}37)$$

which we set out to prove. We can extend this equation to any rigid body rotating about a fixed axis, because any such body can always be analyzed as an assembly of single particles.

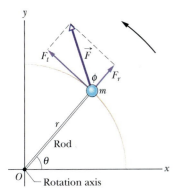

Fig. 11-16 A simple rigid body, free to rotate about an axis through O, consists of a particle of mass m fastened to the end of a rod of length r and negligible mass. An applied force $\vec{F}$ causes the body to rotate.

✔**CHECKPOINT 7:**   The figure shows an overhead view of a meter stick that can pivot about the point indicated, which is to the left of the stick's midpoint. Two horizontal forces, $\vec{F}_1$ and $\vec{F}_2$, are applied to the stick. Only $\vec{F}_1$ is shown. Force $\vec{F}_2$ is perpendicular to the stick and is applied at the right end. If the stick is not to turn, (a) what should be the direction of $\vec{F}_2$, and (b) should F_2 be greater than, less than, or equal to F_1?

Sample Problem 11-7

Figure 11-17a shows a uniform disk, with mass $M = 2.5$ kg and radius $R = 20$ cm, mounted on a fixed horizontal axle. A block with mass $m = 1.2$ kg hangs from a massless cord that is wrapped around the rim of the disk. Find the acceleration of the falling block, the angular acceleration of the disk, and the tension in the cord. The cord does not slip, and there is no friction at the axle.

SOLUTION: One **Key Idea** here is that, taking the block as a system, we can relate its acceleration a to the forces acting on it with Newton's second law ($\vec{F}_{\text{net}} = m\vec{a}$). Those forces are shown in the block's free-body diagram in Fig. 11-17b: The force from the cord is $\vec{T}$ and the gravitational force is $\vec{F}_g$, of magnitude mg. We can

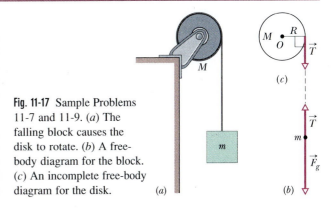

Fig. 11-17 Sample Problems 11-7 and 11-9. (a) The falling block causes the disk to rotate. (b) A free-body diagram for the block. (c) An incomplete free-body diagram for the disk.

now write Newton's second law for components along a vertical y axis ($F_{net,y} = ma_y$) as

$$T - mg = ma. \tag{11-38}$$

However, we cannot solve this equation for a because it also contains the unknown T.

Previously, when we got stuck on the y axis, we would switch to the x axis. Here, we switch to the rotation of the disk and use this **Key Idea**: Taking the disk as a system, we can relate its angular acceleration α to the torque acting on it with Newton's second law for rotation ($\tau_{net} = I\alpha$). To calculate the torques and the rotational inertia I, we take the rotation axis to be perpendicular to the disk and through its center, at point O in Fig. 11-17c.

The torques are then given by Eq. 11-32 ($\tau = rF_t$). The gravitational force on the disk and the force on the disk from the axle both act at the center of the disk and thus at distance $r = 0$, so their torques are zero. The force $\vec{T}$ on the disk due to the cord acts at distance $r = R$ and is tangent to the rim of the disk. Therefore, its torque is $-RT$, negative because the torque rotates the disk clockwise from rest. From Table 11-2c, the rotational inertia I of the disk is $\frac{1}{2}MR^2$. Thus we can write $\tau_{net} = I\alpha$ as

$$-RT = \tfrac{1}{2}MR^2\alpha. \tag{11-39}$$

This equation seems useless because it has two unknowns, α and T, neither of which is the desired a. However, mustering physics courage, we can make it useful with a third **Key Idea**: Because the cord does not slip, the linear acceleration a of the block and

the (tangential) linear acceleration a_t of the rim of the disk are equal. Then, by Eq. 11-22 ($a_t = \alpha r$) we see that here $\alpha = a/R$. Substituting this in Eq. 11-39 yields

$$T = -\tfrac{1}{2}Ma. \tag{11-40}$$

Now combining Eqs. 11-38 and 11-40 leads to

$$a = -g\,\frac{2m}{M + 2m} = -(9.8 \text{ m/s}^2)\frac{(2)(1.2 \text{ kg})}{2.5 \text{ kg} + (2)(1.2 \text{ kg})}$$
$$= -4.8 \text{ m/s}^2. \tag{Answer}$$

We then use Eq. 11-40 to find T:

$$T = -\tfrac{1}{2}Ma = -\tfrac{1}{2}(2.5 \text{ kg})(-4.8 \text{ m/s}^2)$$
$$= 6.0 \text{ N}. \tag{Answer}$$

As we should expect, the acceleration of the falling block is less than g, and the tension in the cord ($= 6.0$ N) is less than the gravitational force on the hanging block ($= mg = 11.8$ N). We see also that the acceleration of the block and the tension depend on the mass of the disk but not on its radius. As a check, we note that the formulas derived above predict $a = -g$ and $T = 0$ for the case of a massless disk ($M = 0$). This is what we would expect; the block simply falls as a free body, trailing the string behind it.

From Eq. 11-22, the angular acceleration of the disk is

$$\alpha = \frac{a}{R} = \frac{-4.8 \text{ m/s}^2}{0.20 \text{ m}} = -24 \text{ rad/s}^2. \tag{Answer}$$

Sample Problem 11-8

To throw an 80 kg opponent with a basic judo hip throw, you intend to pull his uniform with a force $\vec{F}$ and a moment arm $d_1 = 0.30$ m from a pivot point (rotation axis) on your right hip (Fig. 11-18). You wish to rotate him about the pivot point with an angular acceleration α of -6.0 rad/s^2—that is, with an angular acceleration that is *clockwise* in the figure. Assume that his rotational inertia I relative to the pivot point is 15 kg · m^2.

(a) What must the magnitude of $\vec{F}$ be if, before you throw him, you bend your opponent forward to bring his center of mass to your hip (Fig. 11-18a)?

SOLUTION: One **Key Idea** here is that we can relate your pull $\vec{F}$ on him to the given angular acceleration α via Newton's second law for rotation ($\tau_{net} = I\alpha$). As his feet leave the floor, we can assume that only three forces act on him: your pull $\vec{F}$, a force $\vec{N}$ on him from you at the pivot point (this force is not indicated in Fig. 11-18), and the gravitational force $\vec{F}_g$. To use $\tau_{net} = I\alpha$, we need the corresponding three torques, each about the pivot point.

From Eq. 11-33 ($\tau = r_\perp F$), the torque due to your pull $\vec{F}$ is equal to $-d_1F$, where d_1 is the moment arm $r_\perp$ and the sign indicates the clockwise rotation this torque tends to cause. The torque due to $\vec{N}$ is zero, because $\vec{N}$ acts at the pivot point and thus has moment arm $r_\perp = 0$.

To evaluate the torque due to $\vec{F}_g$, we need a **Key Idea** from Chapter 9: We can assume that $\vec{F}_g$ acts at your opponent's center of mass. With the center of mass at the pivot point, $\vec{F}_g$ has moment

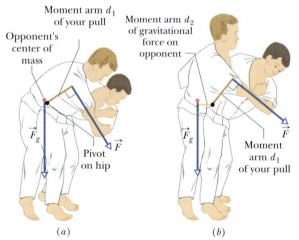

Fig. 11-18 Sample Problem 11-8. A judo hip throw (a) correctly executed and (b) incorrectly executed.

arm $r_\perp = 0$ and thus the torque due to $\vec{F}_g$ is zero. Thus, the only torque on your opponent is due to your pull $\vec{F}$, and we can write $\tau_{net} = I\alpha$ as

$$-d_1F = I\alpha.$$

We then find

$$F = \frac{-I\alpha}{d_1} = \frac{-(15 \text{ kg} \cdot \text{m}^2)(-6.0 \text{ rad/s}^2)}{0.30 \text{ m}}$$
$$= 300 \text{ N}. \tag{Answer}$$

(b) What must the magnitude of $\vec{F}$ be if your opponent remains upright before you throw him, so that $\vec{F}_g$ has a moment arm $d_2 = 0.12$ m from the pivot point (Fig. 11-18b)?

SOLUTION: The Key Ideas we need here are similar to those in (a) with one exception: Because the moment arm for $\vec{F}_g$ is no longer zero, the torque due to $\vec{F}_g$ is now equal to d_2mg, and is positive because the torque attempts counterclockwise rotation. Now we write $\tau_{\text{net}} = I\alpha$ as

$$-d_1F + d_2mg = I\alpha,$$

which gives

$$F = -\frac{I\alpha}{d_1} + \frac{d_2mg}{d_1}.$$

From (a), we know that the first term on the right is equal to 300 N. Substituting this and the given data, we have

$$F = 300 \text{ N} + \frac{(0.12 \text{ m})(80 \text{ kg})(9.8 \text{ m/s}^2)}{0.30 \text{ m}}$$

$$= 613.6 \text{ N} \approx 610 \text{ N}. \qquad \text{(Answer)}$$

The results indicate that you will have to pull much harder if you do not initially bend your opponent to bring his center of mass to your hip. A good judo fighter knows this lesson from physics. (An analysis of the physics of judo and aikido is given in "The Amateur Scientist" by J. Walker, *Scientific American,* July 1980, Vol. 243, pp. 150–161.)

11-10 Work and Rotational Kinetic Energy

As we discussed in Chapter 7, when a force F causes a rigid body of mass m to accelerate along a coordinate axis, it does work W on the body. Thus, the body's kinetic energy ($K = \frac{1}{2}mv^2$) can change. Suppose it is the only energy of the body that changes. Then we relate the change ΔK in kinetic energy to the work W with the work–kinetic energy theorem (Eq. 7-10), writing

$$\Delta K = K_f - K_i = \tfrac{1}{2}mv_f^2 - \tfrac{1}{2}mv_i^2 = W \qquad \text{(work–kinetic energy theorem).} \quad (11\text{-}41)$$

For motion confined to an x axis, we can calculate the work with Eq. 7-32,

$$W = \int_{x_i}^{x_f} F \, dx \qquad \text{(work, one-dimensional motion).} \quad (11\text{-}42)$$

This reduces to $W = Fd$ when F is constant and the body's displacement is d. The rate at which the work is done is the power, which we can find with Eqs. 7-43 and 7-48,

$$P = \frac{dW}{dt} = Fv \qquad \text{(power, one-dimensional motion).} \quad (11\text{-}43)$$

Now let us consider a rotational situation that is similar. When a torque accelerates a rigid body in rotation about a fixed axis, it does work W on the body. Therefore, the body's rotational kinetic energy ($K = \frac{1}{2}I\omega^2$) can change. Suppose that it is the only energy of the body that changes. Then we can still relate the change ΔK in kinetic energy to the work W with the work–kinetic energy theorem, except now the kinetic energy is a rotational kinetic energy:

$$\Delta K = K_f - K_i = \tfrac{1}{2}I\omega_f^2 - \tfrac{1}{2}I\omega_i^2 = W \qquad \text{(work–kinetic energy theorem).} \quad (11\text{-}44)$$

Here, I is the rotational inertia of the body about the fixed axis and ω_i and ω_f are the angular speeds of the body before and after the work is done, respectively.

Also, we can calculate the work with a rotational equivalent of Eq. 11-42,

$$W = \int_{\theta_i}^{\theta_f} \tau \, d\theta \qquad \text{(work, rotation about fixed axis),} \quad (11\text{-}45)$$

where τ is the torque doing the work W, and θ_i and θ_f are the body's angular positions before and after the work is done, respectively. When τ is constant, Eq. 11-45 reduces to

$$W = \tau(\theta_f - \theta_i) \qquad \text{(work, constant torque).} \quad (11\text{-}46)$$

The rate at which the work is done is the power, which we can find with the rotational equivalent of Eq. 11-43,

$$P = \frac{dW}{dt} = \tau\omega \qquad \text{(power, rotation about fixed axis).} \qquad (11\text{-}47)$$

Table 11-3 summarizes the equations that apply to the rotation of a rigid body about a fixed axis and the corresponding equations for translational motion.

Proof of Eqs. 11-44 through 11-47

Let us again consider the situation of Fig. 11-16, in which force $\vec{F}$ rotates a rigid body consisting of a single particle of mass m fastened to the end of a massless rod. During the rotation, force $\vec{F}$ does work on the body. Let us assume that the only energy of the body that is changed by $\vec{F}$ is the kinetic energy. Then we can apply the work–kinetic energy theorem of Eq. 11-41:

$$\Delta K = K_f - K_i = W. \qquad (11\text{-}48)$$

Using $K = \frac{1}{2}mv^2$ and Eq. 11-18 ($v = \omega r$), we can rewrite Eq. 11-48 as

$$\Delta K = \frac{1}{2}mr^2\omega_f^2 - \frac{1}{2}mr^2\omega_i^2 = W. \qquad (11\text{-}49)$$

From Eq. 11-26, the rotational inertia for this one-particle body is $I = mr^2$. Substituting this into Eq. 11-49 yields

$$\Delta K = \frac{1}{2}I\omega_f^2 - \frac{1}{2}I\omega_i^2 = W,$$

which is Eq. 11-44. We derived it for a rigid body with one particle, but it holds for any rigid body rotated about a fixed axis.

We next relate the work W done on the body in Fig. 11-16 to the torque τ on the body due to force $\vec{F}$. When the particle moves a distance ds along its circular path, only the tangential component F_t of the force accelerates the particle along the path. Therefore, only F_t does work on the particle. We write that work dW as $F_t\,ds$. However, we can replace ds with $r\,d\theta$, where $d\theta$ is the angle through which the particle moves. Thus we have

$$dW = F_t r\,d\theta. \qquad (11\text{-}50)$$

From Eq. 11-32, we see that the product $F_t r$ is equal to the torque τ, so we can rewrite Eq. 11-50 as

$$dW = \tau\,d\theta. \qquad (11\text{-}51)$$

TABLE 11-3   Some Corresponding Relations for Translational and Rotational Motion

Pure Translation (Fixed Direction)		Pure Rotation (Fixed Axis)	
Position	x	Angular position	θ
Velocity	$v = dx/dy$	Angular velocity	$\omega = d\theta/dt$
Acceleration	$a = dv/dt$	Angular acceleration	$\alpha = d\omega/dt$
Mass	m	Rotational inertia	I
Newton's second law	$F_{\text{net}} = ma$	Newton's second law	$\tau_{\text{net}} = I\alpha$
Work	$W = \int F\,dx$	Work	$W = \int \tau\,d\theta$
Kinetic energy	$K = \frac{1}{2}mv^2$	Kinetic energy	$K = \frac{1}{2}I\omega^2$
Power (constant force)	$P = Fv$	Power (constant torque)	$P = \tau\omega$
Work–kinetic energy theorem	$W = \Delta K$	Work–kinetic energy theorem	$W = \Delta K$

The work done during a finite angular displacement from θ_i to θ_f is then

$$W = \int_{\theta_i}^{\theta_f} \tau \, d\theta,$$

which is Eq. 11-45. It holds for any rigid body rotating about a fixed axis. Equation 11-46 comes directly from Eq. 11-45.

We can find the power P for rotational motion from Eq. 11-51:

$$P = \frac{dW}{dt} = \tau \frac{d\theta}{dt} = \tau\omega,$$

which is Eq. 11-47.

Sample Problem 11-9

Let the disk in Sample Problem 11-7 and Fig. 11-17 start from rest at time $t = 0$. What is its rotational kinetic energy K at $t = 2.5$ s?

SOLUTION: We can find K with Eq. 11-27 ($K = \frac{1}{2}I\omega^2$). We already know that I is equal to $\frac{1}{2}MR^2$, but we do not yet know ω at $t = 2.5$ s. A **Key Idea**, though, is that the angular acceleration α has the constant value of -24 rad/s^2, so we can apply the equations for constant angular acceleration in Table 11-1. Because we want ω and know α and ω_0 ($= 0$), we use Eq. 11-12:

$$\omega = \omega_0 + \alpha t = 0 + \alpha t = \alpha t.$$

Substituting $\omega = \alpha t$ and $I = \frac{1}{2}MR^2$ into Eq. 11-27, we find

$$K = \tfrac{1}{2}I\omega^2 = \tfrac{1}{2}(\tfrac{1}{2}MR^2)(\alpha t)^2 = \tfrac{1}{4}M(R\alpha t)^2$$
$$= \tfrac{1}{4}(2.5 \text{ kg})[(0.20 \text{ m})(-24 \text{ rad/s}^2)(2.5 \text{ s})]^2$$
$$= 90 \text{ J}. \qquad \text{(Answer)}$$

We can also get this answer with a different **Key Idea**: We can find the disk's kinetic energy from the work done on the disk. First, we relate the *change* in the kinetic energy of the disk to the net work W done on the disk, using the work–kinetic energy theorem of Eq. 11-44 ($K_f - K_i = W$). With K substituted for K_f and 0 for

K_i, we get

$$K = K_i + W = 0 + W = W. \qquad (11\text{-}52)$$

Next we want to find the work W. We can relate W to the torques acting on the disk with Eq. 11-45 or 11-46. The only torque causing angular acceleration and doing work is the torque due to force $\vec{T}$ on the disk from the cord. From Sample Problem 11-7, this torque is equal to $-TR$. Another **Key Idea** is that because α is constant, this torque also must be constant. Thus, we can use Eq. 11-46 to write

$$W = \tau(\theta_f - \theta_i) = -TR(\theta_f - \theta_i). \qquad (11\text{-}53)$$

We need one more **Key Idea**: Because α is constant, we can use Eq. 11-13 to find $\theta_f - \theta_i$. With $\omega_i = 0$, we have

$$\theta_f - \theta_i = \omega_i t + \tfrac{1}{2}\alpha t^2 = 0 + \tfrac{1}{2}\alpha t^2 = \tfrac{1}{2}\alpha t^2.$$

Now we substitute this into Eq. 11-53 and then substitute the result into Eq. 11-52. With $T = 6.0$ N and $\alpha = -24$ rad/s^2 (from Sample Problem 11-7), we have

$$K = W = -TR(\theta_f - \theta_i) = -TR(\tfrac{1}{2}\alpha t^2) = -\tfrac{1}{2}TR\alpha t^2$$
$$= -\tfrac{1}{2}(6.0 \text{ N})(0.20 \text{ m})(-24 \text{ rad/s}^2)(2.5 \text{ s})^2$$
$$= 90 \text{ J}. \qquad \text{(Answer)}$$

Sample Problem 11-10

A rigid sculpture consists of a thin hoop (of mass m and radius $R = 0.15$ m) and a thin radial rod (of mass m and length $L = 2.0R$), arranged as shown in Fig. 11-19. The sculpture can pivot around a horizontal axis in the plane of the hoop, passing through its center.

(a) In terms of m and R, what is the sculpture's rotational inertia I about the rotation axis?

SOLUTION: A **Key Idea** here is that we can separately find the rotational inertias of the hoop and the rod and then add the results to get the sculpture's total rotational inertia I. From Table 11-2h, the hoop has rotational inertia $I_{\text{hoop}} = \frac{1}{2}mR^2$ about its diameter. From Table 11-2e, the rod has rotational inertia $I_{\text{com}} = mL^2/12$ about an axis through its center of mass and parallel to the sculpture's rotation axis. To find its rotational inertia I_{rod} about that rotation axis, we

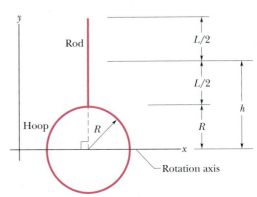

Fig. 11-19 Sample Problem 11-10. A rigid sculpture consisting of a hoop and a rod can rotate around a horizontal axis.

use Eq. 11-29, the parallel-axis theorem:

$$I_{rod} = I_{com} + mh_{com}^2 = \frac{mL^2}{12} + m\left(R + \frac{L}{2}\right)^2$$

$$= 4.33mR^2,$$

where we have used the fact that $L = 2.0R$ and where the perpendicular distance between the rod's center of mass and the rotation axis is $h = R + L/2$. Thus, the rotational inertia I of the sculpture about the rotation axis is

$$I = I_{hoop} + I_{rod} = \tfrac{1}{2}mR^2 + 4.33mR^2$$

$$= 4.83mR^2 \approx 4.8mR^2. \qquad \text{(Answer)}$$

(b) Starting from rest, the sculpture rotates around the rotation axis from the initial upright orientation of Fig. 11-19. What is its angular speed ω about the axis when it is inverted?

SOLUTION: Three Key Ideas are required here:

1. We can relate the sculpture's speed ω to its rotational kinetic energy K with Eq. 11-27 ($K = \tfrac{1}{2}I\omega^2$).

2. We can relate K to the gravitational potential energy U of the sculpture via the conservation of the sculpture's mechanical energy E during the rotation. Thus, during the rotation, E does not change ($\Delta E = 0$) as energy is transferred from U to K.

3. For the gravitational potential energy we can treat the rigid sculpture as a particle located at the center of mass, with the total mass $2m$ concentrated there.

We can write the conservation of mechanical energy ($\Delta E = 0$) as

$$\Delta K + \Delta U = 0. \qquad (11\text{-}54)$$

As the sculpture rotates from its initial position at rest to its inverted

position, when the angular speed is ω, the change ΔK in its kinetic energy is

$$\Delta K = K_f - K_i = \tfrac{1}{2}I\omega^2 - 0 = \tfrac{1}{2}I\omega^2. \qquad (11\text{-}55)$$

From Eq. 8-7 ($\Delta U = mg\,\Delta y$), the corresponding change ΔU in the gravitational potential energy is

$$\Delta U = (2m)g\,\Delta y_{com}, \qquad (11\text{-}56)$$

where $2m$ is the sculpture's total mass, and Δy_{com} is the vertical displacement of its center of mass during the rotation.

To find Δy_{com}, we first find the initial location y_{com} of the center of mass in Fig. 11-19. The hoop (with mass m) is centered at $y = 0$. The rod (with mass m) is centered at $y = R + L/2$. Thus, from Eq. 9-5, the sculpture's center of mass is at

$$y_{com} = \frac{m(0) + m(R + L/2)}{2m} = \frac{0 + m(R + 2R/2)}{2m} = R.$$

When the sculpture is inverted, the center of mass is this same distance R from the rotation axis but *below* it. Therefore, the vertical displacement of the center of mass from the initial position to the inverted position is $\Delta y_{com} = -2R$.

Now let's pull these results together. Substituting Eqs. 11-55 and 11-56 into 11-54 gives us

$$\tfrac{1}{2}I\omega^2 + (2m)g\,\Delta y_{com} = 0.$$

Substituting $I = 4.83mR^2$ from (a) and $\Delta y_{com} = -2R$ from above and solving for ω, we find

$$\omega = \sqrt{\frac{8g}{4.83R}} = \sqrt{\frac{(8)(9.8 \text{ m/s}^2)}{(4.83)(0.15 \text{ m})}}$$

$$= 10 \text{ rad/s.} \qquad \text{(Answer)}$$

REVIEW & SUMMARY

Angular Position   To describe the rotation of a rigid body about a fixed axis, called the **rotation axis,** we assume a **reference line** is fixed in the body, perpendicular to that axis and rotating with the body. We measure the **angular position** θ of this line relative to a fixed direction. When θ is measured in **radians,**

$$\theta = \frac{s}{r} \quad \text{(radian measure)}, \qquad (11\text{-}1)$$

where s is the arc length of a circular path of radius r and angle θ. Radian measure is related to angle measure in revolutions and degrees by

$$1 \text{ rev} = 360° = 2\pi \text{ rad.} \qquad (11\text{-}2)$$

Angular Displacement   A body that rotates about a rotation axis, changing its angular position from θ_1 to θ_2, undergoes an **angular displacement**

$$\Delta\theta = \theta_2 - \theta_1, \qquad (11\text{-}4)$$

where $\Delta\theta$ is positive for counterclockwise rotation and negative for clockwise rotation.

Angular Velocity and Speed   If a body rotates through an angular displacement $\Delta\theta$ in a time interval Δt, its **average angular velocity** ω_{avg} is

$$\omega_{avg} = \frac{\Delta\theta}{\Delta t}. \qquad (11\text{-}5)$$

The **(instantaneous) angular velocity** ω of the body is

$$\omega = \frac{d\theta}{dt}. \qquad (11\text{-}6)$$

Both ω_{avg} and ω are vectors, with directions given by the **right-hand rule** of Fig. 11-6. They are positive for counterclockwise rotation and negative for clockwise rotation. The magnitude of the body's angular velocity is the **angular speed.**

Angular Acceleration   If the angular velocity of a body changes from ω_1 to ω_2 in a time interval $\Delta t = t_2 - t_1$, the **average angular acceleration** α_{avg} of the body is

$$\alpha_{avg} = \frac{\omega_2 - \omega_1}{t_2 - t_1} = \frac{\Delta\omega}{\Delta t}. \tag{11-7}$$

The **(instantaneous) angular acceleration** α of a body is

$$\alpha = \frac{d\omega}{dt}. \tag{11-8}$$

Both α_{avg} and α are vectors.

The Kinematic Equations for Constant Angular Acceleration

Constant angular acceleration (α = constant) is an important special case of rotational motion. The appropriate kinematic equations, given in Table 11-1, are

$$\omega = \omega_0 + \alpha t, \tag{11-12}$$

$$\theta - \theta_0 = \omega_0 t + \tfrac{1}{2}\alpha t^2, \tag{11-13}$$

$$\omega^2 = \omega_0^2 + 2\alpha(\theta - \theta_0), \tag{11-14}$$

$$\theta - \theta_0 = \tfrac{1}{2}(\omega_0 + \omega)t, \tag{11-15}$$

$$\theta - \theta_0 = \omega t - \tfrac{1}{2}\alpha t^2. \tag{11-16}$$

Linear and Angular Variables Related

A point in a rigid rotating body, at a *perpendicular distance r* from the rotation axis, moves in a circle with radius r. If the body rotates through an angle θ, the point moves along an arc with length s given by

$$s = \theta r \qquad \text{(radian measure)}, \tag{11-17}$$

where θ is in radians.

The linear velocity $\vec{v}$ of the point is tangent to the circle; the point's linear speed v is given by

$$v = \omega r \qquad \text{(radian measure)}, \tag{11-18}$$

where ω is the angular speed (in radians per second) of the body.

The linear acceleration $\vec{a}$ of the point has both *tangential* and *radial* components. The tangential component is

$$a_t = \alpha r \qquad \text{(radian measure)}, \tag{11-22}$$

where α is the magnitude of the angular acceleration (in radians per second-squared) of the body. The radial component of $\vec{a}$ is

$$a_r = \frac{v^2}{r} = \omega^2 r \qquad \text{(radian measure)}. \tag{11-23}$$

If the point moves in uniform circular motion, the period T of the motion for the point and the body is

$$T = \frac{2\pi r}{v} = \frac{2\pi}{\omega} \qquad \text{(radian measure)}. \tag{11-19, 11-20}$$

Rotational Kinetic Energy and Rotational Inertia

The kinetic energy K of a rigid body rotating about a fixed axis is given by

$$K = \tfrac{1}{2}I\omega^2 \qquad \text{(radian measure)}, \tag{11-27}$$

in which I is the **rotational inertia** of the body, defined as

$$I = \sum m_i r_i^2 \tag{11-26}$$

for a system of discrete particles and as

$$I = \int r^2 \, dm \tag{11-28}$$

for a body with continuously distributed mass. The r and r_i in these expressions represent the perpendicular distance from the axis of rotation to each mass element in the body.

The Parallel-Axis Theorem

The *parallel-axis theorem* relates the rotational inertia I of a body about any axis to that of the same body about a parallel axis through the center of mass:

$$I = I_{com} + Mh^2. \tag{11-29}$$

Here h is the perpendicular distance between the two axes.

Torque

Torque is a turning or twisting action on a body about a rotation axis due to a force $\vec{F}$. If $\vec{F}$ is exerted at a point given by the position vector $\vec{r}$ relative to the axis, then the magnitude of the torque is

$$\tau = rF_t = r_\perp F = rF \sin\phi, \tag{11-32, 11-33, 11-31}$$

where F_t is the component of $\vec{F}$ perpendicular to $\vec{r}$, and ϕ is the angle between $\vec{r}$ and $\vec{F}$. The quantity $r_\perp$ is the perpendicular distance between the rotation axis and an extended line running through the $\vec{F}$ vector. This line is called the **line of action** of $\vec{F}$, and $r_\perp$ is called the **moment arm** of $\vec{F}$. Similarly, r is the moment arm of F_t.

The SI unit of torque is the newton-meter ($\text{N} \cdot \text{m}$). A torque τ is positive if it tends to rotate a body at rest counterclockwise and negative if it tends to rotate the body in the clockwise direction.

Newton's Second Law in Angular Form

The rotational analog of Newton's second law is

$$\tau_{net} = I\alpha, \tag{11-37}$$

where τ_{net} is the net torque acting on a particle or rigid body, I is the rotational inertia of the particle or body about the rotation axis, and α is the resulting angular acceleration about that axis.

Work and Rotational Kinetic Energy

The equations used for calculating work and power in rotational motion correspond to equations used for translational motion and are

$$W = \int_{\theta_i}^{\theta_f} \tau \, d\theta \tag{11-45}$$

and

$$P = \frac{dW}{dt} = \tau\omega. \tag{11-47}$$

When τ is constant, Eq. 11-45 reduces to

$$W = \tau(\theta_f - \theta_i). \tag{11-46}$$

The form of the work–kinetic energy theorem used for rotating bodies is

$$\Delta K = K_f - K_i = \tfrac{1}{2}I\omega_f^2 - \tfrac{1}{2}I\omega_i^2 = W. \tag{11-44}$$

QUESTIONS

1. Figure 11-20b is a graph of the angular position of the rotating disk of Fig. 11-20a. Is the angular velocity of the disk positive, negative, or zero at (a) $t = 1$ s, (b) $t = 2$ s, and (c) $t = 3$ s? (d) Is the angular acceleration positive or negative?

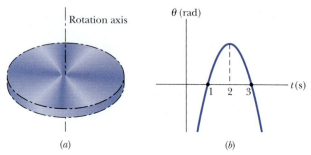

Fig. 11-20 Question 1.

2. Figure 11-21 is a graph of the angular velocity of the rotating disk of Fig. 11-20a. What are the (a) initial and (b) final directions of rotation? (c) Does the disk momentarily stop? (d) Is the angular acceleration positive or negative? (e) Is the angular acceleration constant or varying?

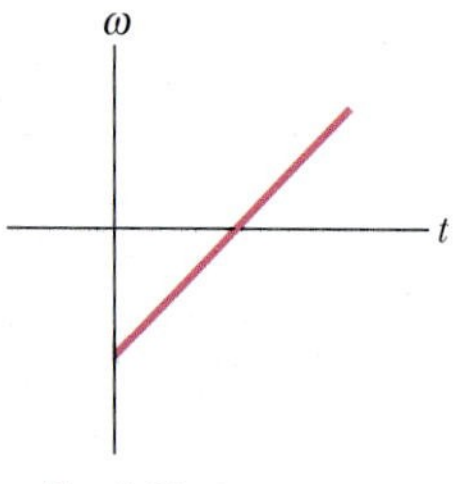

3. Hold your right arm downward, palm toward your thigh. Keeping your wrist rigid, (1) lift the arm until it is horizontal and forward, (2) move it horizontally until it is pointed toward the right, and (3) then bring it down to your side. Your palm faces forward. If you start over, but reverse the steps, why does your palm *not* face forward?

Fig. 11-21 Question 2.

4. For which of the following expressions for $\omega(t)$ of a rotating object do the angular equations of Table 11-1 apply? (a) $\omega = 3$; (b) $\omega = 4t^2 + 2t - 6$; (c) $\omega = 3t - 4$; (d) $\omega = 5t^2 - 3$, all with ω in radians per second and t in seconds.

5. Figure 11-22 is a graph of the angular velocity versus time for the rotating disk of Fig. 11-20a. For a point on the rim of the disk, rank the four instants a, b, c, and d according to the magnitude of (a) the tangential acceleration and (b) the radial acceleration, greatest first.

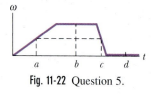

Fig. 11-22 Question 5.

6. The overhead view of Fig. 11-23 is a snapshot of a disk turning counterclockwise like a merry-go-round. The angular speed ω of the disk is decreasing (the disk is turning counterclockwise slower and slower). The figure shows the

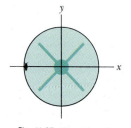

Fig. 11-23 Question 6.

position of a cockroach that rides the rim of the disk. At the instant of the snapshot, what are the directions of (a) the cockroach's radial acceleration and (b) its tangential acceleration?

7. Figure 11-24 shows an assembly of three small spheres of the same mass that are attached to a massless rod with the indicated spacings. Consider the rotational inertia I of the assembly about each sphere, in turn. Then rank the spheres according to the rotational inertia about them, greatest first.

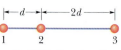

Fig. 11-24 Question 7.

8. Figure 11-25a shows an overhead view of a horizontal bar that can pivot about the point indicated. Two horizontal forces act on the bar, but the bar is stationary. If the angle between the bar and force $\vec{F}_2$ is now decreased from the initial 90° and the bar is still not to turn, should the magnitude of $\vec{F}_2$ be made larger, made smaller, or left the same?

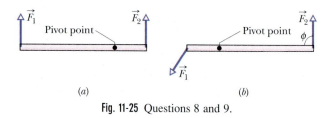

Fig. 11-25 Questions 8 and 9.

9. Figure 11-25b shows an overhead view of a horizontal bar that is rotated about the pivot point by two horizontal forces, $\vec{F}_1$ and $\vec{F}_2$, at opposite ends of the bar. The direction of $\vec{F}_2$ is at angle ϕ to the bar. Rank the following values of ϕ according to the magnitude of the angular acceleration of the bar, greatest first: 90°, 70°, and 110°.

10. In the overhead view of Fig. 11-26, five forces of the same magnitude act on a merry-go-round for the strange; it is a square that can rotate about point P at midlength along one of the edges. Rank the forces acting on it according to the magnitude of the torque they create about point P, greatest first.

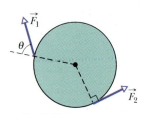

Fig. 11-26 Question 10.

11. In Fig. 11-27, two forces $\vec{F}_1$ and $\vec{F}_2$ act on a disk that turns about its center like a merry-go-round. The forces maintain the indicated angles during the rotation, which is counterclockwise and at a constant rate. However, we are to decrease the angle θ of $\vec{F}_1$ without changing the magnitude of $\vec{F}_1$. (a) To keep the angular speed constant, should we increase, decrease, or maintain

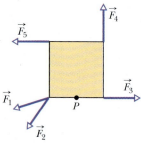

Fig. 11-27 Question 11.

the magnitude of $\vec{F}_2$? Do forces (b) $\vec{F}_1$ and (c) $\vec{F}_2$ tend to rotate the disk clockwise or counterclockwise?

12. Figure 11-28 gives the angular velocity ω versus time t for a merry-go-round being turned by a varying applied force. The magnitudes of the curve's slopes in time intervals 1, 3, 4, and 6 are equal. (a) During which of the time intervals is energy transferred *from* the merry-go-round by the applied force? (b) Rank the time intervals according to the work done on the merry-go-round by the applied force during the interval, most positive work first, most negative work last. (c) Rank the intervals according to the rate at which the applied force transfers energy, greatest rate of transfer *to* the merry-go-round first, greatest rate of transfer *from* it last.

Fig. 11-28 Question 12.

EXERCISES & PROBLEMS

SEC. 11-2 The Rotational Variables

1E. During a time interval t the flywheel of a generator turns through the angle $\theta = at + bt^3 - ct^4$, where a, b, and c are constants. Write expressions for the wheel's (a) angular velocity and (b) angular acceleration.

2E. What is the angular speed of (a) the second hand, (b) the minute hand, and (c) the hour hand of a smoothly running analog watch? Answer in radians per second.

3E. Our Sun is 2.3×10^4 ly (light-years) from the center of our Milky Way galaxy and is moving in a circle around that center at a speed of 250 km/s. (a) How long does it take the Sun to make one revolution about the galactic center? (b) How many revolutions has the Sun completed since it was formed about 4.5×10^9 years ago? ssm

4E. The angular position of a point on the rim of a rotating wheel is given by $\theta = 4.0t - 3.0t^2 + t^3$, where θ is in radians and t is in seconds. What are the angular velocities at (a) $t = 2.0$ s and (b) $t = 4.0$ s? (c) What is the average angular acceleration for the time interval that begins at $t = 2.0$ s and ends at $t = 4.0$ s? What are the instantaneous angular accelerations at (d) the beginning and (e) the end of this time interval?

5E. The angular position of a point on a rotating wheel is given by $\theta = 2 + 4.0t^2 + 2t^3$, where θ is in radians and t is in seconds. At $t = 0$, what are (a) the point's angular position and (b) its angular velocity? (c) What is its angular velocity at $t = 4.0$ s? (d) Calculate its angular acceleration at $t = 2.0$ s. (e) Is its angular acceleration constant? ssm

6P. The wheel in Fig. 11-29 has eight equally spaced spokes and a radius of 30 cm. It is mounted on a fixed axle and is spinning at

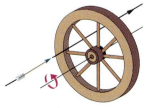

Fig. 11-29 Problem 6.

2.5 rev/s. You want to shoot a 20-cm-long arrow parallel to this axle and through the wheel without hitting any of the spokes. Assume that the arrow and the spokes are very thin. (a) What minimum speed must the arrow have? (b) Does it matter where between the axle and rim of the wheel you aim? If so, what is the best location?

7P. A diver makes 2.5 revolutions on the way from a 10-m-high platform to the water. Assuming zero initial vertical velocity, find the diver's average angular velocity during a dive. ilw

SEC. 11-4 Rotation with Constant Angular Acceleration

8E. The angular speed of an automobile engine is increased at a constant rate from 1200 rev/min to 3000 rev/min in 12 s. (a) What is its angular acceleration in revolutions per minute-squared? (b) How many revolutions does the engine make during this 12 s interval?

9E. A record turntable rotating at $33\frac{1}{3}$ rev/min slows down and stops in 30 s after the motor is turned off. (a) Find its (constant) angular acceleration in revolutions per minute-squared. (b) How many revolutions does it make in this time? ssm

10E. A disk, initially rotating at 120 rad/s, is slowed down with a constant angular acceleration of magnitude 4.0 rad/s². (a) How much time does the disk take to stop? (b) Through what angle does the disk rotate during that time?

11E. A heavy flywheel rotating on its central axis is slowing down because of friction in its bearings. At the end of the first minute of slowing, its angular speed is 0.90 of its initial angular speed of 250 rev/min. Assuming a constant angular acceleration, find its angular speed at the end of the second minute.

12E. Starting from rest, a disk rotates about its central axis with constant angular acceleration. In 5.0 s, it rotates 25 rad. During that time, what are the magnitudes of (a) the angular acceleration and (b) the average angular velocity? (c) What is the instantaneous angular velocity of the disk at the end of the 5.0 s? (d) With the angular acceleration unchanged, through what additional angle will the disk turn during the next 5.0 s?

13P. A wheel has a constant angular acceleration of 3.0 rad/s². During a certain 4.0 s interval, it turns through an angle of 120 rad.

Assuming that the wheel starts from rest, how long is it in motion at the start of this 4.0 s interval? ssm www

14P. A wheel, starting from rest, rotates with a constant angular acceleration of 2.00 rad/s^2. During a certain 3.00 s interval, it turns through 90.0 rad. (a) How long is the wheel turning before the start of the 3.00 s interval? (b) What is the angular velocity of the wheel at the start of the 3.00 s interval?

15P. At $t = 0$, a flywheel has an angular velocity of 4.7 rad/s, an angular acceleration of -0.25 rad/s^2, and a reference line at $\theta_0 = 0$. (a) Through what maximum angle θ_{max} will the reference line turn in the positive direction? At what times t will the reference line be at (b) $\theta = \frac{1}{2}\theta_{max}$ and (c) $\theta = -10.5$ rad (consider both positive and negative values of t)? (d) Graph θ versus t, and indicate the answers to (a), (b), and (c) on the graph.

16P. A disk rotates about its central axis starting from rest and accelerates with constant angular acceleration. At one time it is rotating at 10 rev/s; 60 revolutions later, its angular speed is 15 rev/s. Calculate (a) the angular acceleration, (b) the time required to complete the 60 revolutions, (c) the time required to reach the 10 rev/s angular speed, and (d) the number of revolutions from rest until the time the disk reaches the 10 rev/s angular speed.

17P. A flywheel turns through 40 rev as it slows from an angular speed of 1.5 rad/s to a stop. (a) Assuming a constant angular acceleration, find the time for it to come to rest. (b) What is its angular acceleration? (c) How much time is required for it to complete the first 20 of the 40 revolutions? ilw

18P. A wheel rotating about a fixed axis through its center has a constant angular acceleration of 4.0 rad/s^2. In a certain 4.0 s interval the wheel turns through an angle of 80 rad. (a) What is the angular velocity of the wheel at the start of the 4.0 s interval? (b) Assuming that the wheel starts from rest, how long is it in motion at the start of the 4.0 s interval?

SEC. 11-5 Relating the Linear and Angular Variables

19E. What is the linear acceleration of a point on the rim of a 30-cm-diameter record rotating at a constant angular speed of $33\frac{1}{3}$ rev/min? ssm

20E. A vinyl record on a turntable rotates at $33\frac{1}{3}$ rev/min. (a) What is its angular speed in radians per second? What is the linear speed of a point on the record at the needle when the needle is (b) 15 cm and (c) 7.4 cm from the turntable axis?

21E. What is the angular speed of a car traveling at 50 km/h and rounding a circular turn of radius 110 m?

22E. A flywheel with a diameter of 1.20 m is rotating at an angular speed of 200 rev/min. (a) What is the angular speed of the flywheel in radians per second? (b) What is the linear speed of a point on the rim of the flywheel? (c) What constant angular acceleration (in revolutions per minute-squared) will increase the wheel's angular speed to 1000 rev/min in 60 s? (d) How many revolutions does the wheel make during that 60 s?

23E. An astronaut is being tested in a centrifuge. The centrifuge has a radius of 10 m and, in starting, rotates according to $\theta = 0.30t^2$, where t is in seconds and θ is in radians. When $t = 5.0$ s, what are the magnitudes of the astronaut's (a) angular velocity,

(b) linear velocity, (c) tangential acceleration, and (d) radial acceleration?

24E. What are the magnitudes of (a) the angular velocity, (b) the radial acceleration, and (c) the tangential acceleration of a spaceship taking a circular turn of radius 3220 km at a speed of 29 000 km/h?

25P. An early method of measuring the speed of light makes use of a rotating slotted wheel. A beam of light passes through a slot at the outside edge of the wheel, as in Fig. 11-30, travels to a distant mirror, and returns to the wheel just in time to pass through the next slot in the wheel. One such slotted wheel has a radius of 5.0 cm and 500 slots at its edge. Measurements taken when the mirror is $L = 500$ m from the wheel indicate a speed of light of 3.0×10^5 km/s. (a) What is the (constant) angular speed of the wheel? (b) What is the linear speed of a point on the edge of the wheel? ssm

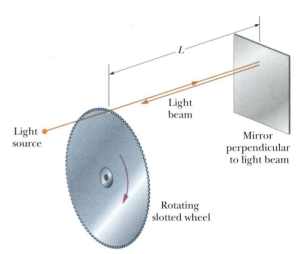

Fig. 11-30 Problem 25.

26P. The flywheel of a steam engine runs with a constant angular velocity of 150 rev/min. When steam is shut off, the friction of the bearings and of the air stops the wheel in 2.2 h. (a) What is the constant angular acceleration, in revolutions per minute-squared, of the wheel during the slowdown? (b) How many rotations does the wheel make before stopping? (c) At the instant the flywheel is turning at 75 rev/min, what is the tangential component of the linear acceleration of a flywheel particle that is 50 cm from the axis of rotation? (d) What is the magnitude of the net linear acceleration of the particle in (c)?

27P. (a) What is the angular speed ω about the polar axis of a point on Earth's surface at a latitude of 40° N? (Earth rotates about that axis.) (b) What is the linear speed v of the point? What are (c) ω and (d) v for a point at the equator? ssm

28P. A gyroscope flywheel of radius 2.83 cm is accelerated from rest at 14.2 rad/s^2 until its angular speed is 2760 rev/min. (a) What is the tangential acceleration of a point on the rim of the flywheel during this spin-up process? (b) What is the radial acceleration of this point when the flywheel is spinning at full speed? (c) Through what distance does a point on the rim move during the spin-up?

29P. In Fig. 11-31, wheel A of radius $r_A = 10$ cm is coupled by belt B to wheel C of radius $r_C = 25$ cm. The angular speed of wheel A is increased from rest at a constant rate of 1.6 rad/s^2. Find the time for wheel C to reach a rotational speed of 100 rev/min, assuming the belt does not slip. (*Hint:* If the belt does not slip, the linear speeds at the rims of the two wheels must be equal.) ssm www

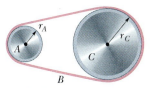

Fig. 11-31 Problem 29.

30P. An object rotates about a fixed axis, and the angular position of a reference line on the object is given by $\theta = 0.40e^{2t}$, where θ is in radians and t is in seconds. Consider a point on the object that is 4.0 cm from the axis of rotation. At $t = 0$, what are the magnitudes of the point's (a) tangential component of acceleration and (b) radial component of acceleration?

31P. A pulsar is a rapidly rotating neutron star that emits a radio beam like a lighthouse emits a light beam. We receive a radio pulse for each rotation of the star. The period T of rotation is found by measuring the time between pulses. The pulsar in the Crab nebula (Fig. 11-32) has a period of rotation of $T = 0.033$ s that is increas-

Fig. 11-32 Problem 31. The Crab nebula resulted from a star whose explosion was seen in 1054. In addition to the gaseous debris seen here, the explosion left a spinning neutron star at its center. The star has a diameter of only 30 km.

ing at the rate of 1.26×10^{-5} s/y. (a) What is the pulsar's angular acceleration? (b) If its angular acceleration is constant, how many years from now will the pulsar stop rotating? (c) The pulsar originated in a supernova explosion seen in the year 1054. What was the initial T for the pulsar? (Assume constant angular acceleration since the pulsar originated.) ssm

32P. A record turntable is rotating at $33\frac{1}{3}$ rev/min. A watermelon seed is on the turntable 6.0 cm from the axis of rotation. (a) Calculate the acceleration of the seed, assuming that it does not slip. (b) What is the minimum value of the coefficient of static friction between the seed and the turntable if the seed is not to slip? (c) Suppose that the turntable achieves its angular speed by starting from rest and undergoing a constant angular acceleration for 0.25 s. Calculate the minimum coefficient of static friction required for the seed not to slip during the acceleration period.

SEC. 11-6 Kinetic Energy of Rotation

33E. Calculate the rotational inertia of a wheel that has a kinetic energy of 24 400 J when rotating at 602 rev/min. ssm

34P. The oxygen molecule O_2 has a mass of 5.30×10^{-26} kg and a rotational inertia of 1.94×10^{-46} kg·m^2 about an axis through the center of the line joining the atoms and perpendicular to that line. Suppose the center of mass of an O_2 molecule in a gas has a translational speed of 500 m/s and the molecule has a rotational kinetic energy that is $\frac{2}{3}$ of the translational kinetic energy of its center of mass. What then is the molecule's angular speed about the center of mass?

SEC. 11-7 Calculating the Rotational Inertia

35E. Two uniform solid cylinders, each rotating about its central (longitudinal) axis, have the same mass of 1.25 kg and rotate with the same angular speed of 235 rad/s, but they differ in radius. What is the rotational kinetic energy of (a) the smaller cylinder, of radius 0.25 m, and (b) the larger cylinder, of radius 0.75 m? ssm

36E. A communications satellite is a solid cylinder with mass 1210 kg, diameter 1.21 m, and length 1.75 m. Prior to launching from the shuttle cargo bay, it is set spinning at 1.52 rev/s about the cylinder axis (Fig. 11-33). Calculate the satellite's (a) rotational inertia about the rotation axis and (b) rotational kinetic energy.

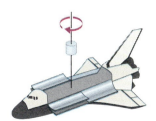

Fig. 11-33 Exercise 36.

37E. In Fig. 11-34, two particles, each with mass m, are fastened to each other, and to a rotation axis at O, by two thin rods, each with length d and mass M. The combination rotates around the rotation axis with angular velocity ω. In terms of these symbols, and measured about O, what are the combination's (a) rotational inertia and (b) kinetic energy?

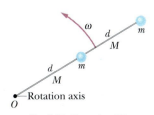

Fig. 11-34 Exercise 37.

38E. Each of the three helicopter rotor blades shown in Fig. 11-35 is 5.20 m long and has a mass of 240 kg. The rotor is rotating at 350 rev/min. (a) What is the rotational inertia of the rotor assembly about the axis of rotation? (Each blade can be considered to be a thin rod rotated about one end.) (b) What is the total kinetic energy of rotation?

Fig. 11-35 Exercise 38.

39E. Calculate the rotational inertia of a meter stick, with mass 0.56 kg, about an axis perpendicular to the stick and located at the 20 cm mark. (Treat the stick as a thin rod.) ssm

40P. Four identical particles of mass 0.50 kg each are placed at the vertices of a 2.0 m × 2.0 m square and held there by four massless rods, which form the sides of the square. What is the rotational inertia of this rigid body about an axis that (a) passes through the midpoints of opposite sides and lies in the plane of the square, (b) passes through the midpoint of one of the sides and is perpendicular to the plane of the square, and (c) lies in the plane of the square and passes through two diagonally opposite particles?

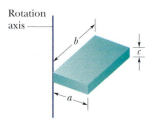

Fig. 11-36 Problem 41.

41P. The uniform solid block in Fig. 11-36 has mass M and edge lengths a, b, and c. Calculate its rotational inertia about an axis through one corner and perpendicular to the large faces. ssm

42P. The masses and coordinates of four particles are as follows: 50 g, $x = 2.0$ cm, $y = 2.0$ cm; 25 g, $x = 0$, $y = 4.0$ cm; 25 g, $x = -3.0$ cm, $y = -3.0$ cm; 30 g, $x = -2.0$ cm, $y = 4.0$ cm. What are the rotational inertias of this collection about the (a) x, (b) y, and (c) z axes? (d) Suppose the answers to (a) and (b) are A and B, respectively. Then what is the answer to (c) in terms of A and B?

43P. (a) Show that the rotational inertia of a solid cylinder of mass M and radius R about its central axis is equal to the rotational inertia of a thin hoop of mass M and radius $R/\sqrt{2}$ about its central axis. (b) Show that the rotational inertia I of any given body of mass M about any given axis is equal to the rotational inertia of an *equivalent hoop* about that axis, if the hoop has the same mass M and a radius k given by

$$k = \sqrt{\frac{I}{M}}.$$

The radius k of the equivalent hoop is called the *radius of gyration* of the given body. ssm

44P. Delivery trucks that operate by making use of energy stored in a rotating flywheel have been used in Europe. The trucks are charged by using an electric motor to get the flywheel up to its top speed of 200π rad/s. One such flywheel is a solid, uniform cylinder with a mass of 500 kg and a radius of 1.0 m. (a) What is the kinetic energy of the flywheel after charging? (b) If the truck operates with an average power requirement of 8.0 kW, for how many minutes can it operate between chargings?

SEC. 11-8 Torque

45E. A small ball of mass 0.75 kg is attached to one end of a 1.25-m-long massless rod, and the other end of the rod is hung from a pivot. When the resulting pendulum is 30° from the vertical, what is the magnitude of the torque about the pivot? ssm

46E. The length of a bicycle pedal arm is 0.152 m, and a downward force of 111 N is applied to the pedal by the rider's foot. What is the magnitude of the torque about the pedal arm's pivot point when the arm makes an angle of (a) 30°, (b) 90°, and (c) 180° with the vertical?

47P. The body in Fig. 11-37 is pivoted at O, and two forces act on it as shown. (a) Find an expression for the net torque on the body about the pivot. (b) If $r_1 =$ 1.30 m, $r_2 = 2.15$ m, $F_1 = 4.20$ N, $F_2 = 4.90$ N, $\theta_1 = 75.0°$, and $\theta_2 = 60.0°$, what is the net torque about the pivot? ssm ilw

Fig. 11-37 Problem 47.

48P. The body in Fig. 11-38 is pivoted at O. Three forces act on it in the directions shown: $F_A = 10$ N at point A, 8.0 m from O; $F_B = 16$ N at point B, 4.0 m from O; and $F_C = 19$ N at point C, 3.0 m from O. What is the net torque about O?

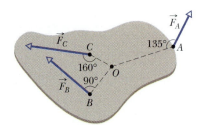

Fig. 11-38 Problem 48.

SEC. 11-9 Newton's Second Law for Rotation

49E. During the launch from a board, a diver's angular speed about her center of mass changes from zero to 6.20 rad/s in 220 ms. Her rotational inertia about her center of mass is 12.0 kg · m². During the launch, what are the magnitudes of (a) her average angular acceleration and (b) the average external torque on her from the board? ssm ilw

50E. A torque of 32.0 N · m on a certain wheel causes an angular acceleration of 25.0 rad/s². What is the wheel's rotational inertia?

51E. A thin spherical shell has a radius of 1.90 m. An applied torque of 960 N · m gives the shell an angular acceleration of 6.20 rad/s² about an axis through the center of the shell. What are (a) the rotational inertia of the shell about that axis and (b) the mass of the shell? ssm

52E. In Fig. 11-39, a cylinder having a mass of 2.0 kg can rotate about its central axis through point O. Forces are applied as shown: $F_1 = 6.0$ N, $F_2 = 4.0$ N, $F_3 = 2.0$ N, and $F_4 = 5.0$ N. Also, $R_1 = $ 5.0 cm and $R_2 = 12$ cm. Find the magnitude and direction of the angular acceleration of the cylinder. (During the rotation, the forces maintain their same angles relative to the cylinder.)

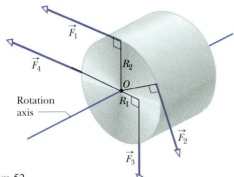

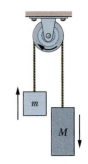

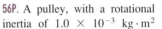

rest, the heavier block falls 75.0 cm in 5.00 s (without the cord slipping on the pulley). (a) What is the magnitude of the blocks' acceleration? What is the tension in the part of the cord that supports (b) the heavier block and (c) the lighter block? (d) What is the magnitude of the pulley's angular acceleration? (e) What is its rotational inertia? **ssm** **www**

Fig. 11-42 Problem 55.

56P. A pulley, with a rotational inertia of 1.0×10^{-3} kg·m² about its axle and a radius of 10 cm, is acted on by a force applied tangentially at its rim. The force magnitude varies in time as $F = 0.50t + 0.30t^2$, with F in newtons and t in seconds. The pulley is initially at rest. At $t = 3.0$ s what are (a) its angular acceleration and (b) its angular speed?

57P. Figure 11-43 shows two blocks, each of mass m, suspended from the ends of a rigid massless rod of length $L_1 + L_2$, with $L_1 = 20$ cm and $L_2 = 80$ cm. The rod is held horizontally on the fulcrum and then released. What are the magnitudes of the initial accelerations of (a) the block closer to the fulcrum and (b) the other block?

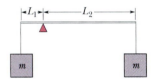

Fig. 11-43 Problem 57.

Fig. 11-39 Problem 52.

53P. Figure 11-40 shows the massive shield door at a neutron test facility at Lawrence Livermore Laboratory; this is the world's heaviest hinged door. The door has a mass of 44,000 kg, a rotational inertia about a vertical axis through its huge hinges of 8.7×10^4 kg·m², and a (front) face width of 2.4 m. Neglecting friction, what steady force, applied at its outer edge and perpendicular to the plane of the door, can move it from rest through an angle of 90° in 30 s?

SEC. 11-10   Work and Rotational Kinetic Energy

58E. (a) If $R = 12$ cm, $M = 400$ g, and $m = 50$ g in Fig. 11-17, find the speed of the block after it has descended 50 cm starting from rest. Solve the problem using energy conservation principles. (b) Repeat (a) with $R = 5.0$ cm.

59E. An automobile crankshaft transfers energy from the engine to the axle at the rate of 100 hp (= 74.6 kW) when rotating at a speed of 1800 rev/min. What torque (in newton-meters) does the crankshaft deliver?

60E. A 32.0 kg wheel, essentially a thin hoop with radius 1.20 m, is rotating at 280 rev/min. It must be brought to a stop in 15.0 s. (a) How much work must be done to stop it? (b) What is the required average power?

61P. A thin rod of length L and mass m is suspended freely from one end. It is pulled to one side and then allowed to swing like a pendulum, passing through its lowest position with angular speed ω. In terms of these symbols and g, and neglecting friction and air resistance, find (a) the rod's kinetic energy at its lowest position and (b) how far above that position the center of mass rises.

62P. Calculate (a) the torque, (b) the energy, and (c) the average power required to accelerate Earth in 1 day from rest to its present angular speed about its axis.

63P. A meter stick is held vertically with one end on the floor and is then allowed to fall. Find the speed of the other end when it hits the floor, assuming that the end on the floor does not slip. (*Hint:* Consider the stick to be a thin rod and use the conservation of energy principle.) **ssm** **ilw**

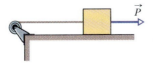

Fig. 11-40 Problem 53.

54P. A wheel of radius 0.20 m is mounted on a frictionless horizontal axis. The rotational inertia of the wheel about the axis is 0.050 kg·m². A massless cord wrapped around the wheel is attached to a 2.0 kg block that slides on a horizontal frictionless surface. If a horizontal force of magnitude $P = 3.0$ N is applied to the block as shown in Fig. 11-41, what is the magnitude of the angular acceleration of the wheel? Assume the string does not slip on the wheel.

55P. In Fig. 11-42, one block has mass $M = 500$ g, the other has mass $m = 460$ g, and the pulley, which is mounted in horizontal frictionless bearings, has a radius of 5.00 cm. When released from

Fig. 11-41 Problem 54.

64P. A uniform cylinder of radius 10 cm and mass 20 kg is mounted so as to rotate freely about a horizontal axis that is parallel to and 5.0 cm from the central longitudinal axis of the cylinder. (a) What is the rotational inertia of the cylinder about the axis of rotation? (b) If the cylinder is released from rest with its central longitudinal axis at the same height as the axis about which the cylinder rotates, what is the angular speed of the cylinder as it passes through its lowest position?

65P. A rigid body is made of three identical thin rods, each with length L, fastened together in the form of a letter **H** (Fig. 11-44). The body is free to rotate about a horizontal axis that runs along the length of one of the legs of the **H**.

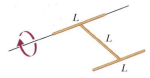

Fig. 11-44 Problem 65.

The body is allowed to fall from rest from a position in which the plane of the **H** is horizontal. What is the angular speed of the body when the plane of the **H** is vertical? ssm www

66P. A uniform spherical shell of mass M and radius R rotates about a vertical axis on frictionless bearings (Fig. 11-45). A massless cord passes around the equator of the shell, over a pulley of rotational inertia I and radius r, and is attached to a small object of mass m. There is no friction on the pulley's axle; the cord does not slip on the pulley. What is the speed of the object after it falls a distance h from rest? Use energy considerations.

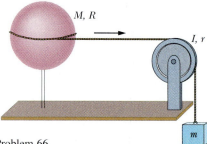

Fig. 11-45 Problem 66.

67P. A tall, cylinder-shaped chimney falls over when its base is ruptured. Treat the chimney as a thin rod of length H, and let θ be the angle the chimney makes with the vertical. In terms of these symbols and g, express the following: (a) the angular speed of the chimney, (b) the radial acceleration of the chimney's top, and (c) the tangential acceleration of the top. (*Hint:* Use energy considerations, not a torque.) In part (c), recall that $\alpha = d\omega/dt$.) (d) At what angle θ does the tangential acceleration equal g? ssm

Additional Problems

68. At 7:14 A.M. on June 30, 1908, a huge explosion occurred above remote central Siberia, at latitude 61° N and longitude 102° E; the fireball thus created was the brightest flash seen by anyone before nuclear weapons. The *Tunguska Event*, which according to one chance witness "covered an enormous part of the sky," was probably the explosion of a *stony asteroid* about 140 m wide. (a) Considering only Earth's rotation, determine how much later the asteroid would have had to arrive to put the explosion above Helsinki at longitude 25° E. This would have obliterated the city. (b) If the asteroid had, instead, been a *metallic asteroid,* it could have reached Earth's surface. How much later would such an asteroid have had to arrive to put the impact in the Atlantic Ocean at longitude 20° W? (The resulting tsunamis would have wiped out coastal civilization on both sides of the Atlantic.)

69. The method in which the massive lintels (top stones) were lifted to the top of the upright stones at Stonehenge has long been debated. One possible method was tested in a small Czech town. A concrete block of mass 5124 kg was pulled up along two oak beams whose top surfaces had been debarked and then lubricated with fat (Fig. 11-46). The beams were 10 m long, and each extended from the ground to the top of one of the two upright pillars onto which the lintel was to be raised. The pillars were 3.9 m high; the coefficient of static friction between the block and the beams was 0.22. The pull on the block was via ropes that wrapped around the block and also around the top ends of two spruce logs of length 4.5 m. A riding platform was strung at the opposite end of each log. When enough workers sat or stood on a riding platform, the corresponding spruce log would pivot about the top of its upright pillar and pull one side of the block a short distance up a beam. For each log, the ropes were approximately perpendicular to the log; the distance between the pivot point and the point where the ropes wrapped around the log was 0.70 m. Assuming that each worker had a mass of 85 kg, find the least number of workers needed on the two platforms so that the block begins to move up along the beams. (About half this number could actually move the block by moving first one side of the block and then the other side.)

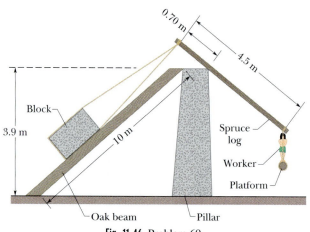

Fig. 11-46 Problem 69.

70. Cheetahs in full run have been reported to have an astounding speed of 114 km/h (about 71 mi/h) by observers moving along with the cheetahs in Jeep-like vehicles. Imagine trying to measure a cheetah's speed by keeping your Jeep abreast of the animal while also glancing at your speedometer, which is registering 114 km/h. You keep the Jeep at a constant 8.0 m from the cheetah, but the noise of the Jeep causes the cheetah to continuously veer away from the Jeep, along a circular path of radius 92 m. Thus, you travel along a circular path of radius 100 m. (a) What is the angular speed of you and the cheetah around the circular paths? (b) What is the linear speed of the cheetah along its path? (If you did not account for the circular motion, you would conclude erroneously that the cheetah's speed is 114 km/h, and that type of error was apparently made in the published reports.)

12 Rolling, Torque, and Angular Momentum

In 1897, a European "aerialist" made the first triple somersault during the flight from a swinging trapeze to the hands of a partner. For the next 85 years aerialists attempted to complete a *quadruple* somersault, but not until 1982 was it done before an audience: Miguel Vazquez of the Ringling Bros. and Barnum & Bailey Circus rotated his body in four complete circles in midair before his brother Juan caught him. Both were stunned by their success.

Why was the feat so difficult, and what feature of physics made it (finally) possible?

The answer is in this chapter.

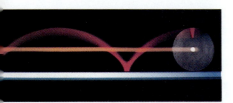

12-1 Rolling

When a bicycle moves along a straight track, the center of each wheel moves forward in pure translation. A point on the rim of the wheel, however, traces out a more complex path, as Fig. 12-1 shows. In what follows, we analyze the motion of a rolling wheel first by viewing it as a combination of pure translation and pure rotation, and then by viewing it as rotation alone.

Fig. 12-1 A time exposure photograph of a rolling disk. Small lights have been attached to the disk, one at its center and one at its edge. The latter traces out a curve called a *cycloid*.

Rolling as Rotation and Translation Combined

Imagine that you are watching the wheel of a bicycle, which passes you at constant speed while *rolling smoothly* (that is, without sliding) along a street. As shown in Fig. 12-2, the center of mass O of the wheel moves forward at constant speed v_{com}. The point P on the street where the wheel makes contact also moves forward at speed v_{com}, so that it always remains directly below O.

During a time interval t, you see both O and P move forward by a distance s. The bicycle rider sees the wheel rotate through an angle θ about the center of the wheel, with the point of the wheel that was touching the street at the beginning of t moving through arc length s. Equation 11-17 relates the arc length s to the rotation angle θ:

$$s = \theta R, \tag{12-1}$$

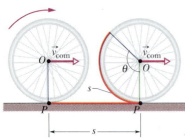

Fig. 12-2 The center of mass O of a rolling wheel moves a distance s at velocity $\vec{v}_{com}$, while the wheel rotates through angle θ. The point P at which the wheel makes contact with the surface over which the wheel rolls also moves a distance s.

where R is the radius of the wheel. The linear speed v_{com} of the center of the wheel (the center of mass of this uniform wheel) is ds/dt. The angular speed ω of the wheel about its center is $d\theta/dt$. Thus, differentiating Eq. 12-1 with respect to time (with R held constant) gives us

$$v_{com} = \omega R \qquad \text{(smooth rolling motion).} \tag{12-2}$$

Figure 12-3 shows that the rolling motion of a wheel is a combination of purely translational and purely rotational motions. Figure 12-3a shows the purely rotational motion (as if the rotation axis through the center were stationary): Every point on the wheel rotates about the center with angular speed ω. (This is the type of motion we considered in Chapter 11.) Every point on the outside edge of the wheel has linear speed v_{com} given by Eq. 12-2. Figure 12-3b shows the purely translational

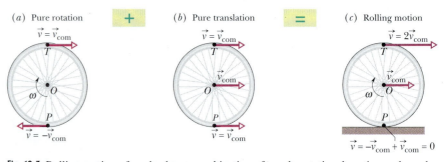

Fig. 12-3 Rolling motion of a wheel as a combination of purely rotational motion and purely translational motion. (a) The purely rotational motion: All points on the wheel move with the same angular speed ω. Points on the outside edge of the wheel all move with the same linear speed $v = v_{com}$. The linear velocities $\vec{v}$ of two such points, at top (T) and bottom (P) of the wheel, are shown. (b) The purely translational motion: All points on the wheel move to the right with the same linear velocity $\vec{v}_{com}$ as the center of the wheel. (c) The rolling motion of the wheel is the combination of (a) and (b).

Fig. 12-4 A photograph of a rolling bicycle wheel. The spokes near the top of the wheel are more blurred than those near the bottom of the wheel because they are moving faster, as Fig. 12-3c shows.

motion (as if the wheel did not rotate at all): Every point on the wheel moves to the right with speed v_{com}.

The combination of Figs. 12-3a and 12-3b yields the actual rolling motion of the wheel, Fig. 12-3c. Note that in this combination of motions, the portion of the wheel at the bottom (at point P) is stationary and the portion at the top (at point T) is moving at speed $2v_{com}$, faster than any other portion of the wheel. These results are demonstrated in Fig. 12-4, which is a time exposure of a rolling bicycle wheel. You can tell that the wheel is moving faster near its top than near its bottom because the spokes are more blurred at the top than at the bottom.

The motion of any round body rolling smoothly over a surface can be separated into purely rotational and purely translational motions, as in Figs. 12-3a and 12-3b.

Rolling as Pure Rotation

Figure 12-5 suggests another way to look at the rolling motion of a wheel—namely, as pure rotation about an axis that always extends through the point where the wheel contacts the street as the wheel moves. We consider the rolling motion to be pure rotation about an axis passing through point P in Fig. 12-3c and perpendicular to the plane of the figure. The vectors in Fig. 12-5 then represent the instantaneous velocities of points on the rolling wheel.

Question: What angular speed about this new axis will a stationary observer assign to a rolling bicycle wheel?

Answer: The same angular speed ω that the rider assigns to the wheel as she or he observes it in pure rotation about an axis through its center of mass.

To verify this answer, let us use it to calculate the linear speed of the top of the rolling wheel from the point of view of a stationary observer. If we call the wheel's radius R, the top is a distance 2R from the axis through P in Fig. 12-5, so the linear speed at the top should be (using Eq. 12-2)

$$v_{top} = (\omega)(2R) = 2(\omega R) = 2v_{com},$$

in exact agreement with Fig. 12-3c. You can similarly verify the linear speeds shown for the portion of the wheel at points O and P in Fig. 12-3c.

✓CHECKPOINT 1:   The rear wheel on a clown's bicycle has twice the radius of the front wheel. (a) When the bicycle is moving, is the linear speed at the very top of the rear wheel greater than, less than, or the same as that of the front wheel? (b) Is the angular speed of the rear wheel greater than, less than, or the same as that of the front wheel?

12-2 The Kinetic Energy of Rolling

Let us now calculate the kinetic energy of the rolling wheel as measured by the stationary observer. If we view the rolling as pure rotation about an axis through P in Fig. 12-5, then from Eq. 11-27 we have

$$K = \tfrac{1}{2}I_P\omega^2, \tag{12-3}$$

in which ω is the angular speed of the wheel and I_P is the rotational inertia of the wheel about the axis through P. From the parallel-axis theorem of Eq. 11-29 ($I = I_{com} + Mh^2$), we have

$$I_P = I_{com} + MR^2, \tag{12-4}$$

in which M is the mass of the wheel, I_{com} is its rotational inertia about an axis through

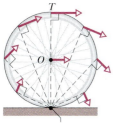

Rotation axis at P

Fig. 12-5 Rolling can be viewed as pure rotation, with angular speed ω, about an axis that always extends through P. The vectors show the instantaneous linear velocities of selected points on the rolling wheel. You can obtain the vectors by combining the translational and rotational motions as in Fig. 12-3.

its center of mass, and R (the wheel's radius) is the perpendicular distance h. Substituting Eq. 12-4 into Eq. 12-3, we obtain

$$K = \tfrac{1}{2}I_{com}\omega^2 + \tfrac{1}{2}MR^2\omega^2,$$

and using the relation $v_{com} = \omega R$ (Eq. 12-2) yields

$$K = \tfrac{1}{2}I_{com}\omega^2 + \tfrac{1}{2}Mv_{com}^2. \tag{12-5}$$

We can interpret the term $\tfrac{1}{2}I_{com}\omega^2$ as the kinetic energy associated with the rotation of the wheel about an axis through its center of mass (Fig. 12-3a), and the term $\tfrac{1}{2}Mv_{com}^2$ as the kinetic energy associated with the translational motion of the wheel's center of mass (Fig. 12-3b). Thus, we have the following rule:

> ▶ A rolling object has two types of kinetic energy: a rotational kinetic energy ($\tfrac{1}{2}I_{com}\omega^2$) due to its rotation about its center of mass and a translational kinetic energy ($\tfrac{1}{2}Mv_{com}^2$) due to translation of its center of mass.

Sample Problem 12-1

A uniform solid cylindrical disk, of mass $M = 1.4$ kg and radius $R = 8.5$ cm, rolls smoothly across a horizontal table at a speed of 15 cm/s. What is its kinetic energy K?

SOLUTION: Equation 12-5 gives the kinetic energy of a rolling object, but we need three Key Ideas to use it.

1. When we speak of the speed of a rolling object, we always mean the speed of the center of mass, so here $v_{com} = 15$ cm/s.

2. Equation 12-5 requires the angular speed ω of the rolling object, which we can relate to v_{com} with Eq. 12-2, writing $\omega = v_{com}/R$.

3. Equation 12-5 also requires the rotational inertia I_{com} of the object about its center of mass. From Table 11-2c, we find that for a solid disk $I_{com} = \tfrac{1}{2}MR^2$.

Now Eq. 12-5 gives us

$$\begin{aligned}
K &= \tfrac{1}{2}I_{com}\omega^2 + \tfrac{1}{2}Mv_{com}^2 \\
&= (\tfrac{1}{2})(\tfrac{1}{2}MR^2)(v_{com}/R)^2 + \tfrac{1}{2}Mv_{com}^2 = \tfrac{3}{4}Mv_{com}^2 \\
&= \tfrac{3}{4}(1.4 \text{ kg})(0.15 \text{ m/s})^2 \\
&= 0.024 \text{ J} = 24 \text{ mJ}. \qquad\qquad \text{(Answer)}
\end{aligned}$$

12-3 The Forces of Rolling

Friction and Rolling

If a wheel rolls at constant speed, as in Fig. 12-2, it has no tendency to slide at the point of contact P, and thus no frictional force acts there. However, if a net force acts on the rolling wheel to speed it up or to slow it, then that net force causes acceleration $\vec{a}_{com}$ of the center of mass along the direction of travel. It also causes the wheel to rotate faster or slower, which means it causes an angular acceleration α about the center of mass. These accelerations tend to make the wheel slide at P. Thus, a frictional force must act on the wheel at P to oppose that tendency.

If the wheel does not slide, the force is a static frictional force $\vec{f}_s$ and the motion is smooth rolling. We can then relate the magnitudes of the linear acceleration $\vec{a}_{com}$ and the angular acceleration α by differentiating Eq. 12-2 with respect to time (with R held constant). On the left side, dv_{com}/dt is a_{com}; and on the right side $d\omega/dt$ is α. So, for smooth rolling we have

$$a_{com} = \alpha R \qquad \text{(smooth rolling motion).} \tag{12-6}$$

If the wheel does slide when the net force acts on it, the frictional force that acts at P in Fig. 12-2 is a kinetic frictional force $\vec{f}_k$. The motion then is not smooth rolling, and Eq. 12-6 does not apply to the motion. In this chapter we discuss only smooth rolling motion.

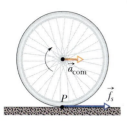

Fig. 12-6 A wheel rolls horizontally without sliding while accelerating with linear acceleration $\vec{a}_{com}$. A static frictional force $\vec{f}_s$ acts on the wheel at P, opposing its tendency to slide.

Figure 12-6 shows an example in which a wheel is being made to rotate faster while rolling along a flat surface, as on a bicycle wheel at the start of a race. The faster rotation tends to make the bottom of the wheel slide to the left at point P. A frictional force at P, directed to the right, opposes this tendency to slide. If the wheel does not slide, that frictional force is a static frictional force $\vec{f}_s$ (as shown), the motion is smooth rolling, and Eq. 12-6 applies to the motion. (Without friction, bicycle races would be stationary and very boring.)

If the wheel in Fig. 12-6 were made to rotate slower, as on a slowing bicycle, we would change the figure in two ways: The directions of the center-of-mass acceleration $\vec{a}_{com}$ and the frictional force $\vec{f}_s$ at point P would now be to the left.

Rolling down a Ramp

Figure 12-7 shows a round uniform body of mass M and radius R rolling smoothly down a ramp at angle θ, along an x axis. We want to find expressions for the body's acceleration $a_{com,x}$ down the ramp. We do this by using Newton's second law in both its linear version ($F_{net} = Ma$) and its angular version ($\tau_{net} = I\alpha$).

We start by drawing the forces on the body as shown in Fig. 12-7:

1. The gravitational force $\vec{F}_g$ on the body is directed downward. The tail of the vector is placed at the center of mass of the body. The component along the ramp is $F_g \sin \theta$, which is equal to $Mg \sin \theta$.

2. A normal force $\vec{N}$ is perpendicular to the ramp. It acts at the point of contact P, but in Fig. 12-7 the vector has been shifted along its direction until its tail is at the body's center of mass.

3. A static frictional force $\vec{f}_s$ acts at the point of contact P and is directed up the ramp. (Do you see why? If the body were to slide at P, it would slide *down* the ramp. Thus, the frictional force opposing the sliding must be *up* the ramp.)

We can write Newton's second law for components along the x axis in Fig. 12-7 ($F_{net,x} = ma_x$) as

$$f_s - Mg \sin \theta = Ma_{com,x}. \tag{12-7}$$

This equation contains two unknowns, f_s and $a_{com,x}$. (We should *not* assume that f_s is at its maximum value $f_{s,max}$. All we know is that the value of f_s is just right for the body to roll smoothly down the ramp, without sliding.)

We now wish to apply Newton's second law in angular form to the body's rotation about its center of mass. First, we shall use Eq. 11-33 ($\tau = r_\perp F$) to write the torques on the body about that point. The frictional force $\vec{f}_s$ has moment arm R and thus produces a torque Rf_s, which is positive because it tends to rotate the body counterclockwise in Fig. 12-7. Forces $\vec{F}_g$ and $\vec{N}$ have zero moment arms about the center of mass and thus produce zero torques. So we can write the angular form of Newton's second law ($\tau_{net} = I\alpha$) about an axis through the body's center of mass as

$$Rf_s = I_{com}\alpha. \tag{12-8}$$

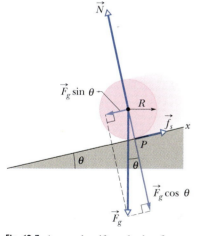

Fig. 12-7 A round uniform body of radius R rolls down a ramp. The forces that act on the body are the gravitational force $\vec{F}_g$, a normal force $\vec{N}$, and a frictional force $\vec{f}_s$ pointing up the ramp. (For clarity, vector $\vec{N}$ has been shifted in the direction it points until its tail is at the center of the body.)

This equation contains two unknowns, f_s and α.

Because the body is rolling smoothly, we can use Eq. 12-6 ($a_{com} = \alpha R$) to relate the unknowns $a_{com,x}$ and α. But we must be cautious because here $a_{com,x}$ is negative (in the negative direction of the x axis) and α is positive (counterclockwise). Thus we substitute $-a_{com,x}/R$ for α in Eq. 12-8. Then, solving for f_s, we obtain

$$f_s = -I_{com}\frac{a_{com,x}}{R^2}. \tag{12-9}$$

Substituting the right side of Eq. 12-9 for f_s in Eq. 12-7, we then find

$$a_{\text{com},x} = -\frac{g \sin \theta}{1 + I_{\text{com}}/MR^2}. \qquad (12\text{-}10)$$

We can use this equation to find the linear acceleration $a_{\text{com},x}$ of any body rolling along an incline of angle θ with the horizontal.

✔**CHECKPOINT 2:**   Disks A and B are identical and roll across a floor with equal speeds. Then disk A rolls up an incline, reaching a maximum height h, and disk B moves up an incline that is identical except that it is frictionless. Is the maximum height reached by disk B greater than, less than, or equal to h?

Sample Problem 12-2

A uniform ball, of mass $M = 6.00$ kg and radius R, rolls smoothly from rest down a ramp at angle $\theta = 30.0°$ (Fig. 12-7).

(a) The ball descends a vertical height $h = 1.20$ m to reach the bottom of the ramp. What is its speed at the bottom?

SOLUTION: One **Key Idea** here is that we can relate the ball's speed at the bottom to its kinetic energy K_f there. A second **Key Idea** is that the mechanical energy E of the ball–Earth system is conserved as the ball rolls down the ramp. The reason is that the only force doing work on the ball is the gravitational force, a conservative force. The normal force on the ball from the ramp does zero work because it is perpendicular to the ball's path. The frictional force on the ball from the ramp does not transfer any energy to thermal energy because the ball does not slide (it *rolls smoothly*).

Therefore, we can write the conservation of mechanical energy ($E^f = E^i$) as

$$K_f + U_f = K_i + U_i, \qquad (12\text{-}11)$$

where subscripts f and i refer to the final values (at the bottom) and initial values (at rest), respectively. The gravitational potential energy is initially $U_i = Mgh$ (where M is the ball's mass) and finally $U_f = 0$. The kinetic energy is initially $K_i = 0$. For the final kinetic energy K_f, we need a third **Key Idea**: Because the ball rolls, the kinetic energy involves both translation *and* rotation, so we include them both with the right side of Eq. 12-5. Substituting all these expressions into Eq. 12-11 gives us

$$(\tfrac{1}{2}I_{\text{com}}\omega^2 + \tfrac{1}{2}Mv_{\text{com}}^2) + 0 = 0 + Mgh, \qquad (12\text{-}12)$$

where I_{com} is the ball's rotational inertia about an axis through its center of mass, v_{com} is the requested speed at the bottom, and ω is the angular speed there.

Because the ball rolls smoothly, we can use Eq. 12-2 to sub-

stitute v_{com}/R for ω to reduce the unknowns in Eq. 12-12. Doing so, substituting $\tfrac{2}{5}MR^2$ for I_{com} (from Table 11-2f), and then solving for v_{com} give us

$$v_{\text{com}} = \sqrt{(\tfrac{10}{7})gh} = \sqrt{(\tfrac{10}{7})(9.8 \text{ m/s}^2)(1.2 \text{ m})}$$

$$= 4.1 \text{ m/s}. \qquad \text{(Answer)}$$

Note that the answer does not depend on the mass M or radius R of the ball.

(b) What are the magnitude and direction of the friction force on the ball as it rolls down the ramp?

SOLUTION: The **Key Idea** here is that, because the ball rolls smoothly, Eq. 12-9 gives the frictional force on the ball. However, first we need the ball's acceleration $a_{\text{com},x}$ from Eq. 12-10:

$$a_{\text{com},x} = -\frac{g \sin \theta}{1 + I_{\text{com}}/MR^2} = -\frac{g \sin \theta}{1 + \tfrac{2}{5}MR^2/MR^2}$$

$$= -\frac{(9.8 \text{ m/s}^2) \sin 30.0°}{1 + \tfrac{2}{5}} = -3.50 \text{ m/s}^2.$$

Note that we needed neither mass M nor radius R to find $a_{\text{com},x}$. Thus, any size ball with any uniform mass would have this acceleration down a 30.0° ramp, provided the ball rolls smoothly.

We can now solve Eq. 12-9 as

$$f_s = -I_{\text{com}}\frac{a_{\text{com},x}}{R^2} = -\tfrac{2}{5}MR^2\frac{a_{\text{com},x}}{R^2} = -\tfrac{2}{5}Ma_{\text{com},x}$$

$$= -\tfrac{2}{5}(6.00 \text{ kg})(-3.50 \text{ m/s}^2) = 8.40 \text{ N}. \qquad \text{(Answer)}$$

Note that here we needed mass M but not radius R. Thus, a 30.0° ramp would produce a frictional force of magnitude 8.40 N on any size ball with a uniform mass of 6.00 kg, for smooth rolling.

12-4 The Yo-Yo

A yo-yo is a physics lab that you can fit in your pocket. If a yo-yo rolls down its string for a distance h, it loses potential energy in amount mgh but gains kinetic energy in both translational ($\tfrac{1}{2}mv_{\text{com}}^2$) and rotational ($\tfrac{1}{2}I_{\text{com}}\omega^2$) forms. As it climbs back up, it loses kinetic energy and regains potential energy.

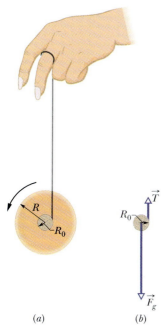

Fig. 12-8 (*a*) A yo-yo, shown in cross section. The string, of assumed negligible thickness, is wound around an axle of radius R_0. (*b*) A free-body diagram for the falling yo-yo. Only the axle is shown.

In a modern yo-yo, the string is not tied to the axle but is looped around it. When the yo-yo "hits" the bottom of its string, an upward force on the axle from the string stops the descent. The yo-yo then spins, axle inside loop, with only rotational kinetic energy. The yo-yo keeps spinning ("sleeping") until you "wake it" by jerking on the string, causing the string to catch on the axle and the yo-yo to climb back up. The rotational kinetic energy of the yo-yo at the bottom of its string (and thus the sleeping time) can be considerably increased by throwing the yo-yo downward so that it starts down the string with initial speeds v_{com} and ω instead of rolling down from rest.

To find an expression for the linear acceleration a_{com} of a yo-yo rolling down a string, we could use Newton's second law just as we did for the body rolling down a ramp in Fig. 12-7. The analysis is the same except for the following:

1. Instead of rolling down a ramp at angle θ with the horizontal, the yo-yo rolls down a string at angle $\theta = 90°$ with the horizontal.

2. Instead of rolling on its outer surface at radius R, the yo-yo rolls on an axle of radius R_0 (Fig. 12-8*a*).

3. Instead of being slowed by frictional force $\vec{f}_s$, the yo-yo is slowed by the force $\vec{T}$ on it from the string (Fig. 12-8*b*).

The analysis would again lead us to Eq. 12-10. Therefore, let us just change the notation in Eq. 12-10 and set $\theta = 90°$ to write the linear acceleration as

$$a_{com} = -\frac{g}{1 + I_{com}/MR_0^2}, \tag{12-13}$$

where I_{com} is the yo-yo's rotational inertia about its center, and M is its mass. A yo-yo has the same downward acceleration when it is climbing back up the string, because the forces on it are still those shown in Fig. 12-8*b*.

12-5 Torque Revisited

In Chapter 11 we defined torque τ for a rigid body that can rotate around a fixed axis, with each particle in the body forced to move in a path that is a circle about that axis. We now expand the definition of torque to apply it to an individual particle that moves along any path relative to a fixed *point* (rather than a fixed axis). The path need no longer be a circle, and we must write the torque as a vector $\vec{\tau}$ that may have any direction.

Figure 12-9*a* shows such a particle at point A in the xy plane. A single force $\vec{F}$ in that plane acts on the particle, and the particle's position relative to the origin O is given by position vector $\vec{r}$. The torque $\vec{\tau}$ acting on the particle relative to the fixed point O is a vector quantity defined as

$$\vec{\tau} = \vec{r} \times \vec{F} \quad \text{(torque defined).} \tag{12-14}$$

We can evaluate the vector (or cross) product in this definition of $\vec{\tau}$ by using the rules for such products given in Section 3-7. To find the direction of $\vec{\tau}$, we slide the vector $\vec{F}$ (without changing its direction) until its tail is at the origin O, so that the two vectors in the vector product are tail to tail as in Fig. 12-9*b*. We then use the right-hand rule for vector products in Fig. 3-20*a*, sweeping the fingers of the right hand from $\vec{r}$ (the first vector in the product) into $\vec{F}$ (the second vector). The outstretched right thumb then gives the direction of $\vec{\tau}$. In Fig. 12-9*b*, the direction of $\vec{\tau}$ is in the positive direction of the z axis.

To determine the magnitude of $\vec{\tau}$, we apply the general result of Eq. 3-27

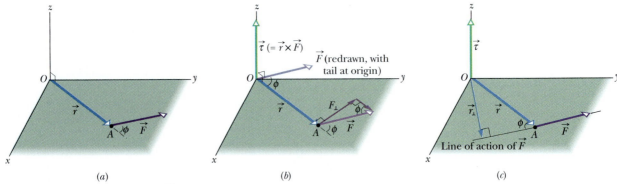

Fig. 12-9 Defining torque. (a) A force $\vec{F}$, lying in the xy plane, acts on a particle at point A. (b) This force produces a torque $\vec{\tau} (= \vec{r} \times \vec{F})$ on the particle with respect to the origin O. By the right-hand rule for vector (cross) products, the torque vector points in the positive direction of z. Its magnitude is given by $rF_\perp$ in (b) and by $r_\perp F$ in (c).

$(c = ab \sin \phi)$, finding

$$\tau = rF \sin \phi, \tag{12-15}$$

where ϕ is the angle between the directions of $\vec{r}$ and $\vec{F}$ when the vectors are tail to tail. From Fig. 12-9b, we see that Eq. 12-15 can be rewritten as

$$\tau = rF_\perp, \tag{12-16}$$

where $F_\perp$ $(= F \sin \phi)$ is the component of $\vec{F}$ perpendicular to $\vec{r}$. From Fig. 12-9c, we see that Eq. 12-15 can also be rewritten as

$$\tau = r_\perp F, \tag{12-17}$$

where $r_\perp$ $(= r \sin \phi)$ is the moment arm of $\vec{F}$ (the perpendicular distance between O and the line of action of $\vec{F}$).

Sample Problem 12-3

In Fig. 12-10a, three forces, each of magnitude 2.0 N, act on a particle. The particle is in the xz plane at point A given by position vector $\vec{r}$, where $r = 3.0$ m and $\theta = 30°$. Force $\vec{F}_1$ is parallel to the x axis, force $\vec{F}_2$ is parallel to the z axis, and force $\vec{F}_3$ is parallel to the y axis. What is the torque, about the origin O, due to each force?

SOLUTION: The **Key Idea** here is that, because the three force vectors do not lie in a plane, we cannot evaluate their torques as in Chapter 11. Instead, we must use vector (or cross) products, with magnitudes given by Eq. 12-15 ($\tau = rF \sin \phi$) and directions given by the right-hand rule for vector products.

Because we want the torques with respect to the origin O, the vector $\vec{r}$ required for each cross product is the given position vector. To determine the angle ϕ between the direction of $\vec{r}$ and the direction of each force, we shift the force vectors of Fig. 12-10a, each in turn, so that their tails are at the origin. Figures 12-10b, c,

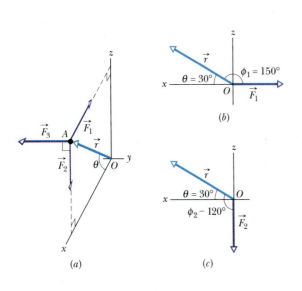

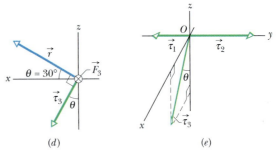

Fig. 12-10 Sample Problem 12-3. (a) A particle at point A is acted on by three forces, each parallel to a coordinate axis. The angle ϕ (used in finding torque) is shown (b) for $\vec{F}_1$ and (c) for $\vec{F}_2$. (d) Torque $\vec{\tau}_3$ is perpendicular to both $\vec{r}$ and $\vec{F}_3$ (force $\vec{F}_3$ is directed into the plane of the figure). (e) The torques (relative to the origin O) acting on the particle.

and *d*, which are direct views of the *xz* plane, show the shifted force vectors $\vec{F}_1$, $\vec{F}_2$, and $\vec{F}_3$, respectively. (Note how much easier the angles are to see.) In Fig. 12-10*d*, the angle between the directions of $\vec{r}$ and $\vec{F}_3$ is 90° and the symbol $\otimes$ means $\vec{F}_3$ is directed into the page. If it were directed out of the page, it would be represented with the symbol $\odot$.

Now, applying Eq. 12-15 for each force, we find the magnitudes of the torques to be

$$\tau_1 = rF_1 \sin \phi_1 = (3.0 \text{ m})(2.0 \text{ N})(\sin 150°) = 3.0 \text{ N} \cdot \text{m},$$

$$\tau_2 = rF_2 \sin \phi_2 = (3.0 \text{ m})(2.0 \text{ N})(\sin 120°) = 5.2 \text{ N} \cdot \text{m},$$

and

$$\tau_3 = rF_3 \sin \phi_3 = (3.0 \text{ m})(2.0 \text{ N})(\sin 90°)$$
$$= 6.0 \text{ N} \cdot \text{m}. \qquad \text{(Answer)}$$

To find the directions of these torques, we use the right-hand rule, placing the fingers of the right hand so as to rotate $\vec{r}$ into $\vec{F}$ through the *smaller* of the two angles between their directions. The thumb points in the direction of the torque. Thus $\vec{\tau}_1$ is directed into the page in Fig. 12-10*b*; $\vec{\tau}_2$ is directed out of the page in Fig. 12-10*c*; and $\vec{\tau}_3$ is directed as shown in Fig. 12-10*d*. All three torque vectors are shown in Fig. 12-10*e*.

✓**CHECKPOINT 3:** The position vector $\vec{r}$ of a particle points along the positive direction of the *z* axis. If the torque on the particle is (a) zero, (b) in the negative direction of *x*, and (c) in the negative direction of *y*, in what direction is the force producing the torque?

PROBLEM-SOLVING TACTICS

Tactic 1: *Vector Products and Torques*

Equation 12-15 for torques is our first application of the vector (or cross) product. You might want to review Section 3-7, where the rules for the vector product are given. In that section, Problem-Solving Tactic 5 lists many common errors in finding the direction of a vector product.

Keep in mind that a torque is calculated *with respect to* (or

about) a point, which must be known if the value of the torque is to be meaningful. Changing the point can change the torque in both magnitude and direction. For example, in Sample Problem 12-3, the torques due to the three forces are calculated about the origin *O*. You can show that the torques due to the same three forces are all zero if they are calculated about point *A* (at the position of the particle), because then *r* = 0 for each force.

12-6 Angular Momentum

Recall that the concept of linear momentum $\vec{p}$ and the principle of conservation of linear momentum are extremely powerful tools. They allow us to predict the outcome of, say, a collision of two cars without knowing the details of the collision. Here we begin a discussion of the angular counterpart of $\vec{p}$. We shall end the chapter by discussing the angular counterpart of the conservation principle.

Figure 12-11 shows a particle of mass *m* with linear momentum $\vec{p}$ $(= m\vec{v})$ as it passes through point *A* in the *xy* plane. The **angular momentum** $\vec{\ell}$ of this particle with respect to the origin *O* is a vector quantity defined as

$$\vec{\ell} = \vec{r} \times \vec{p} = m(\vec{r} \times \vec{v}) \qquad \text{(angular momentum defined),} \qquad (12\text{-}18)$$

where $\vec{r}$ is the position vector of the particle with respect to *O*. As the particle moves relative to *O* in the direction of its momentum $\vec{p}$ $(= m\vec{v})$, position vector $\vec{r}$ rotates around *O*. Note carefully that to have angular momentum about *O*, the particle does *not* itself have to rotate around *O*. Comparison of Eqs. 12-14 and 12-18 shows that angular momentum bears the same relation to linear momentum that torque does to force. The SI unit of angular momentum is the kilogram-meter-squared per second (kg · m²/s), equivalent to the joule-second (J · s).

To find the direction of the angular momentum vector $\vec{\ell}$ in Fig. 12-11, we slide the vector $\vec{p}$ until its tail is at the origin *O*. Then we use the right-hand rule for

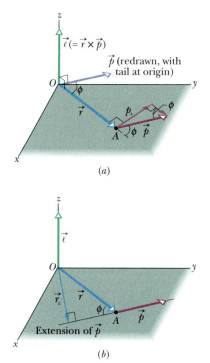

Fig. 12-11 Defining angular momentum. A particle passing through point *A* has linear momentum $\vec{p}$ $(= m\vec{v})$, with the vector $\vec{p}$ lying in the *xy* plane. The particle has angular momentum $\vec{\ell}$ $(= \vec{r} \times \vec{p})$ with respect to the origin *O*. By the right-hand rule, the angular momentum vector points in the positive direction of *z*. (*a*) The magnitude of $\vec{\ell}$ is given by $\ell = rp_\perp = rmv_\perp$. (*b*) The magnitude of $\vec{\ell}$ is also given by $\ell = r_\perp p = r_\perp mv$.

vector products, sweeping the fingers from $\vec{r}$ into $\vec{p}$. The outstretched thumb then shows that the direction of $\vec{\ell}$ is in the positive direction of the z axis in Fig. 12-11. This positive direction is consistent with the counterclockwise rotation of the particle's position vector $\vec{r}$ about the z axis, as the particle continues to move. (A negative direction of $\vec{\ell}$ would be consistent with a clockwise rotation of $\vec{r}$ about the z axis.)

To find the magnitude of $\vec{\ell}$, we use the general result of Eq. 3-27, to write

$$\ell = rmv \sin \phi, \qquad (12\text{-}19)$$

where ϕ is the angle between $\vec{r}$ and $\vec{p}$ when these two vectors are tail to tail. From Fig. 12-11a, we see that Eq. 12-19 can be rewritten as

$$\ell = rp_\perp = rmv_\perp, \qquad (12\text{-}20)$$

where $p_\perp$ is the component of $\vec{p}$ perpendicular to $\vec{r}$ and $v_\perp$ is the component of $\vec{v}$ perpendicular to $\vec{r}$. From Fig. 12-11b, we see that Eq. 12-19 can also be rewritten as

$$\ell = r_\perp p = r_\perp mv, \qquad (12\text{-}21)$$

where $r_\perp$ is the perpendicular distance between O and the extension of $\vec{p}$.

Just as is true for torque, angular momentum has meaning only with respect to a specified origin. Moreover, if the particle in Fig. 12-11 did not lie in the xy plane, or if the linear momentum $\vec{p}$ of the particle did not also lie in that plane, the angular momentum $\vec{\ell}$ would not be parallel to the z axis. The direction of the angular momentum vector is always perpendicular to the plane formed by the position and linear momentum vectors $\vec{r}$ and $\vec{p}$.

✔**CHECKPOINT 4:** In part a of the figure, particles 1 and 2 move around point O in opposite directions, in circles with radii 2 m and 4 m. In part b, particles 3 and 4 travel in the same direction, along straight lines at perpendicular distances of 4 m and 2 m from point O. Particle 5 moves directly away from O. All five particles have the same mass and the same constant speed.
(a) Rank the particles according to the magnitudes of their angular momentum about point O, greatest first.
(b) Which particles have negative angular momentum about point O?

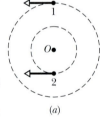

(a)

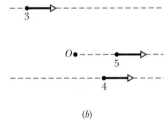

(b)

Sample Problem 12-4

Figure 12-12 shows an overhead view of two particles moving at constant momentum along horizontal paths. Particle 1, with momentum magnitude $p_1 = 5.0$ kg·m/s, has position vector $\vec{r}_1$ and will pass 2.0 m from point O. Particle 2, with momentum magnitude $p_2 = 2.0$ kg·m/s, has position vector $\vec{r}_2$ and will pass 4.0 m from point O. What is the net angular momentum $\vec{L}$ about point O of the two-particle system?

SOLUTION: The **Key Idea** here is that to find $\vec{L}$, we can first find the individual angular momenta $\vec{\ell}_1$ and $\vec{\ell}_2$ and then add them. To evaluate their magnitudes, we can use any one of Eqs. 12-18 through 12-21. However, Eq. 12-21 is easiest, because we are given the perpendicular distances $r_{1\perp}$ (= 2.0 m) and $r_{2\perp}$ (= 4.0 m) and the

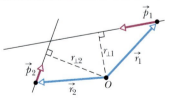

Fig. 12-12 Sample Problem 12-4. Two particles pass near point O.

momentum magnitudes p_1 and p_2. We are not given all the values required for the other equations.

For particle 1, Eq. 12-21 yields

$$\ell_1 = r_{1\perp}p_1 = (2.0 \text{ m})(5.0 \text{ kg} \cdot \text{m/s})$$
$$= 10 \text{ kg} \cdot \text{m}^2/\text{s}.$$

To find the direction of vector $\vec{\ell}_1$, we use Eq. 12-18 and the right-hand rule for vector products. For $\vec{r}_1 \times \vec{p}_1$, the vector product is out of the page, perpendicular to the plane of Fig. 12-12. This is the positive direction, consistent with the counterclockwise rotation of the particle's position vector $\vec{r}_1$ around O as particle 1 moves. Thus, the angular momentum vector for particle 1 is

$$\ell_1 = +10 \text{ kg} \cdot \text{m}^2/\text{s}.$$

Similarly, the magnitude of $\vec{\ell}_2$ is

$$\ell_2 = r_{\perp 2}p_2 = (4.0 \text{ m})(2.0 \text{ kg} \cdot \text{m/s})$$
$$= 8.0 \text{ kg} \cdot \text{m}^2/\text{s},$$

and the vector product $\vec{r}_2 \times \vec{p}_2$ is into the page, which is the negative direction, consistent with the clockwise rotation of $\vec{r}_2$ around O as particle 2 moves. Thus, the angular momentum vector for particle 2 is

$$\ell_2 = -8.0 \text{ kg} \cdot \text{m}^2/\text{s}.$$

The net angular momentum for the two-particle system is then

$$L = \ell_1 + \ell_2 = +10 \text{ kg} \cdot \text{m}^2/\text{s} + (-8.0 \text{ kg} \cdot \text{m}^2/\text{s})$$
$$= +2.0 \text{ kg} \cdot \text{m}^2/\text{s}. \qquad \text{(Answer)}$$

The plus sign means that the system's net angular momentum about point O is out of the page.

12-7 Newton's Second Law In Angular Form

Newton's second law written in the form

$$\vec{F}_{\text{net}} = \frac{d\vec{p}}{dt} \qquad \text{(single particle)} \qquad (12\text{-}22)$$

expresses the close relation between force and linear momentum for a single particle. We have seen enough of the parallelism between linear and angular quantities to be pretty sure that there is also a close relation between torque and angular momentum. Guided by Eq. 12-22, we can even guess that it must be

$$\vec{\tau}_{\text{net}} = \frac{d\vec{\ell}}{dt} \qquad \text{(single particle)}. \qquad (12\text{-}23)$$

Equation 12-23 is indeed an angular form of Newton's second law for a single particle:

> The (vector) sum of all the torques acting on a particle is equal to the time rate of change of the angular momentum of that particle.

Equation 12-23 has no meaning unless the torques $\vec{\tau}$ and the angular momentum $\vec{\ell}$ are defined with respect to the same origin.

Proof of Equation 12-23

We start with Eq. 12-18, the definition of the angular momentum of a particle:

$$\vec{\ell} = m(\vec{r} \times \vec{v}),$$

where $\vec{r}$ is the position vector of the particle and $\vec{v}$ is the velocity of the particle. Differentiating* each side with respect to time t yields

$$\frac{d\vec{\ell}}{dt} = m\left(\vec{r} \times \frac{d\vec{v}}{dt} + \frac{d\vec{r}}{dt} \times \vec{v} \right). \qquad (12\text{-}24)$$

However, $d\vec{v}/dt$ is the acceleration $\vec{a}$ of the particle, and $d\vec{r}/dt$ is its velocity $\vec{v}$. Thus, we can rewrite Eq. 12-24 as

$$\frac{d\vec{\ell}}{dt} = m(\vec{r} \times \vec{a} + \vec{v} \times \vec{v}).$$

---

*In differentiating a vector product, be sure not to change the order of the two quantities (here $\vec{r}$ and $\vec{v}$) that form that product. (See Eq. 3-28.)

Now $\vec{v} \times \vec{v} = 0$ (the vector product of any vector with itself is zero because the angle between the two vectors is necessarily zero). This leads to

$$\frac{d\vec{\ell}}{dt} = m(\vec{r} \times \vec{a}) = \vec{r} \times m\vec{a}.$$

We now use Newton's second law ($\vec{F}_{net} = m\vec{a}$) to replace $m\vec{a}$ with its equal, the vector sum of the forces that act on the particle, obtaining

$$\frac{d\vec{\ell}}{dt} = \vec{r} \times \vec{F}_{net} = \sum(\vec{r} \times \vec{F}). \qquad (12\text{-}25)$$

Here the symbol Σ indicates that we must sum the vector products $\vec{r} \times \vec{F}$ for all the forces. However, from Eq. 12-14, we know that each one of those vector products is the torque associated with one of the forces. Therefore, Eq. 12-25 tells us that

$$\vec{\tau}_{net} = \frac{d\vec{\ell}}{dt}.$$

This is Eq. 12-23, the relation that we set out to prove.

✔**CHECKPOINT 5:**   The figure shows the position vector $\vec{r}$ of a particle at a certain instant, and four choices for the direction of a force that is to accelerate the particle. All four choices lie in the xy plane. (a) Rank the choices according to the magnitude of the time rate of change ($d\vec{\ell}/dt$) they produce in the angular momentum of the particle about point O, greatest first. (b) Which choice results in a negative rate of change about O?

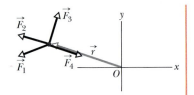

Sample Problem 12-5

In Fig. 12-13, a penguin of mass m falls from rest at point A, a horizontal distance D from the origin O of an xyz coordinate system. (The positive direction of the z axis is directly outward from the plane of the figure.)

(a) What is the angular momentum $\vec{\ell}$ of the falling penguin about O?

SOLUTION: One **Key Idea** here is that we can treat the penguin as a particle, and thus its angular momentum $\vec{\ell}$ is given by Eq. 12-18 ($\vec{\ell} = \vec{r} \times \vec{p}$), where $\vec{r}$ is the penguin's position vector (extending from O to the penguin) and $\vec{p}$ is the penguin's linear momentum. The second **Key Idea** is that the penguin has *angular* momentum about O even though it moves in a straight line, because $\vec{r}$ rotates about O as the penguin falls.

To find the magnitude of $\vec{\ell}$, we can use any one of the scalar equations derived from Eq. 12-18—namely, Eqs. 12-19 through 12-21. However, Eq. 12-21 ($\ell = r_\perp mv$) is easiest because the perpendicular distance $r_\perp$ between O and an extension of vector $\vec{p}$ is the given distance D. A third **Key Idea** is an old one: The speed of an object that has fallen from rest for a time t is $v = gt$. We can now write Eq. 12-21 in terms of given quantities as

$$\ell = r_\perp mv = Dmgt. \qquad \text{(Answer)}$$

To find the direction of $\vec{\ell}$, we use the right-hand rule for the vector product $\vec{r} \times \vec{p}$ in Eq. 12-18. Mentally shift $\vec{p}$ until its tail

Fig. 12-13 Sample Problem 12-5. A penguin falls vertically from point A. The torque $\vec{\tau}$ and the angular momentum $\vec{\ell}$ of the falling penguin with respect to the origin O are directed into the plane of the figure at O.

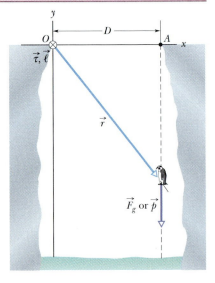

is at the origin, and then use the fingers of your right hand to rotate $\vec{r}$ into $\vec{p}$ through the smaller angle between the two vectors. Your outstretched thumb then points into the plane of the figure, indicating that the product $\vec{r} \times \vec{p}$ and thus also $\vec{\ell}$ are directed into that plane, in the negative direction of the z axis. We represent $\vec{\ell}$ with an encircled cross $\otimes$ at O. The vector $\vec{\ell}$ changes with time in magnitude only; its direction remains unchanged.

(b) About the origin O, what is the torque $\vec{\tau}$ on the penguin due to the gravitational force $\vec{F}_g$?

SOLUTION: One **Key Idea** here is that the torque is given by Eq. 12-14 ($\vec{\tau} = \vec{r} \times \vec{F}$), where now the force is $\vec{F}_g$. An associated **Key Idea** is that $\vec{F}_g$ causes a torque on the penguin, even though the penguin moves in a straight line, because $\vec{r}$ rotates about O as the penguin moves.

To find the magnitude of $\vec{\tau}$, we can use any one of the scalar equations derived from Eq. 12-14—namely, Eqs. 12-15 through 12-17. However, Eq. 12-17 ($\tau = r_\perp F$) is easiest because the perpendicular distance $r_\perp$ between O and the line of action of $\vec{F}_g$ is the given distance D. So, substituting D and using mg for the magnitude of $\vec{F}_g$, we can write Eq. 12-17 as

$$\tau = DF_g = Dmg. \qquad \text{(Answer)}$$

Using the right-hand rule for the vector product $\vec{r} \times \vec{F}$ in Eq. 12-14, we find that the direction of $\vec{\tau}$ is the negative direction of the z axis, the same as $\vec{\ell}$.

The results we obtained in parts (a) and (b) must be consistent with Newton's second law in the angular form of Eq. 12-23 ($\vec{\tau}_{\text{net}} = d\vec{\ell}/dt$). To check the magnitudes we got, we write Eq. 12-23 in component form for the z axis and then substitute our result $\ell = Dmgt$. We find

$$\tau = \frac{d\ell}{dt} = \frac{d(Dmgt)}{dt} = Dmg,$$

which is the magnitude we found for $\vec{\tau}$. To check the directions, we note that Eq. 12-23 tells us that $\vec{\tau}$ and $d\vec{\ell}/dt$ must have the same direction. So $\vec{\tau}$ and $\vec{\ell}$ must also have the same direction, which is what we found.

12-8 The Angular Momentum of a System of Particles

Now we turn our attention to the angular momentum of a system of particles with respect to an origin. The total angular momentum $\vec{L}$ of the system is the (vector) sum of the angular momenta $\vec{\ell}$ of the individual particles:

$$\vec{L} = \vec{\ell}_1 + \vec{\ell}_2 + \vec{\ell}_3 + \cdots + \vec{\ell}_n = \sum_{i=1}^{n} \vec{\ell}_i, \qquad (12\text{-}26)$$

in which i ($= 1, 2, 3, \ldots$) labels the particles.

With time, the angular momenta of individual particles may change, either because of interactions within the system (between the individual particles) or because of influences that may act on the system from the outside. We can find the change in $\vec{L}$ as these changes take place by taking the time derivative of Eq. 12-26. Thus,

$$\frac{d\vec{L}}{dt} = \sum_{i=1}^{n} \frac{d\vec{\ell}_i}{dt}. \qquad (12\text{-}27)$$

From Eq. 12-23, we see that $d\vec{\ell}_i/dt$ is equal to the net torque $\vec{\tau}_{\text{net},i}$ on the ith particle. We can rewrite Eq. 12-27 as

$$\frac{d\vec{L}}{dt} = \sum_{i=1}^{n} \vec{\tau}_{\text{net},i}. \qquad (12\text{-}28)$$

That is, the rate of change of the system's angular momentum $\vec{L}$ is equal to the vector sum of the torques on its individual particles. Those torques include *internal torques* (due to forces between the particles) and *external torques* (due to forces on the particles from bodies external to the system). However, the forces between the particles always come in third-law force pairs so their torques sum to zero. Thus, the only torques that can change the total angular momentum $\vec{L}$ of the system are the external torques acting on the system.

Let $\vec{\tau}_{\text{net}}$ represent the net external torque, the vector sum of all external torques on all particles in the system. Then we can write Eq. 12-28 as

$$\vec{\tau}_{\text{net}} = \frac{d\vec{L}}{dt} \qquad \text{(system of particles)}. \qquad (12\text{-}29)$$

This equation is Newton's second law for rotation in angular form, for a system of particles. It says:

> ►The net external torque $\vec{\tau}_{net}$ acting on a system of particles is equal to the time rate of change of the system's total angular momentum $\vec{L}$.

Equation 12-29 is analogous to $\vec{F}_{net} = d\vec{P}/dt$ (Eq. 9-23) but requires extra caution: Torques and the system's angular momentum must be measured relative to the same origin. If the center of mass of the system is not accelerating relative to an inertial frame, that origin can be any point. However, if the center of mass of the system *is* accelerating, the origin can be only at that center of mass. As an example, consider a wheel as the system of particles. If the wheel is rotating about an axis that is fixed relative to the ground, then the origin for applying Eq. 12-29 can be any point that is stationary relative to the ground. However, if the wheel is rotating about an axis that is accelerating (such as when the wheel rolls down a ramp), then the origin can be only at the wheel's center of mass.

12-9 The Angular Momentum of a Rigid Body Rotating About a Fixed Axis

We next evaluate the angular momentum of a system of particles that form a rigid body that rotates about a fixed axis. Figure 12-14a shows such a body. The fixed axis of rotation is the z axis, and the body rotates about it with constant angular speed ω. We wish to find the angular momentum of the body about that axis.

We can find the angular momentum by summing the z components of the angular momenta of the mass elements in the body. In Fig. 12-14a, a typical mass element, of mass Δm_i, moves around the z axis in a circular path. The position of the mass element is located relative to the origin O by position vector $\vec{r}_i$. The radius of the mass element's circular path is $r_{\perp i}$, the perpendicular distance between the element and the z axis.

The magnitude of the angular momentum $\vec{\ell}_i$ of this mass element, with respect to O, is given by Eq. 12-19:

$$\ell_i = (r_i)(p_i)(\sin 90°) = (r_i)(\Delta m_i\, v_i),$$

where p_i and v_i are the linear momentum and linear speed of the mass element, and 90° is the angle between $\vec{r}_i$ and $\vec{p}_i$. The angular momentum vector $\vec{\ell}_i$ for the mass element in Fig. 12-14a is shown in Fig. 12-14b; its direction must be perpendicular to those of $\vec{r}_i$ and $\vec{p}_i$.

We are interested in the component of $\vec{\ell}_i$ that is parallel to the rotation axis, here the z axis. That z component is

$$\ell_{iz} = \ell_i \sin \theta = (r_i \sin \theta)(\Delta m_i\, v_i) = r_{\perp i}\, \Delta m_i\, v_i.$$

The z component of the angular momentum for the rotating rigid body as a whole is found by adding up the contributions of all the mass elements that make up the body. Thus, because $v = \omega r_\perp$, we may write

$$L_z = \sum_{i=1}^{n} \ell_{iz} = \sum_{i=1}^{n} \Delta m_i\, v_i r_{\perp i} = \sum_{i=1}^{n} \Delta m_i(\omega r_{\perp i})r_{\perp i}$$

$$= \omega \left(\sum_{i=1}^{n} \Delta m_i\, r_{\perp i}^2 \right). \tag{12-30}$$

We can remove ω from the summation here because it has the same value for all points of the rotating rigid body.

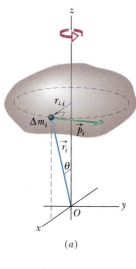

(a)

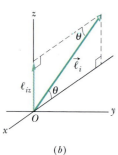

(b)

Fig. 12-14 (a) A rigid body rotates about the z axis with angular speed ω. A mass element of mass Δm_i within the body moves about the z axis in a circle with radius $r_{\perp i}$. The mass element has linear momentum $\vec{p}_i$, and it is located relative to the origin O by position vector $\vec{r}_i$. Here the mass element is shown when $r_{\perp i}$ is parallel to the x axis. (b) The angular momentum $\vec{\ell}_i$, with respect to O, of the mass element in (a). The z component ℓ_{iz} is also shown.

TABLE 12-1   More Corresponding Variables and Relations for Translational and Rotational Motion[a]

	Translational		Rotational	
Force	$\vec{F}$	Torque	$\vec{\tau} (= \vec{r} \times \vec{F})$	
Linear momentum	$\vec{p}$	Angular momentum	$\vec{\ell} (= \vec{r} \times \vec{p})$	
Linear momentum[b]	$\vec{P} (= \Sigma \vec{p}_i)$	Angular momentum[b]	$\vec{L} (= \Sigma \vec{\ell}_i)$	
Linear momentum[b]	$\vec{P} = M\vec{v}_{com}$	Angular momentum[c]	$L = I\omega$	
Newton's second law[b]	$\vec{F}_{net} = \dfrac{d\vec{P}}{dt}$	Newton's second law[b]	$\vec{\tau}_{net} = \dfrac{d\vec{L}}{dt}$	
Conservation law[d]	$\vec{P} = $ a constant	Conservation law[d]	$\vec{L} = $ a constant	

[a]See also Table 11-3.   [b]For systems of particles, including rigid bodies.
[c]For a rigid body about a fixed axis, with L being the component along that axis.
[d]For a closed, isolated system.

The quantity $\Sigma \, \Delta m_i \, r_{\perp i}^2$ in Eq. 12-30 is the rotational inertia I of the body about the fixed axis (see Eq. 11-26). Thus Eq. 12-30 reduces to

$$L = I\omega \qquad \text{(rigid body, fixed axis).} \qquad (12\text{-}31)$$

We have dropped the subscript z, but you must remember that the angular momentum defined by Eq. 12-31 is the angular momentum about the rotation axis. Also, I in that equation is the rotational inertia about that same axis.

Table 12-1, which supplements Table 11-3, extends our list of corresponding linear and angular relations.

✔**CHECKPOINT 6:**   In the figure, a disk, a hoop, and a solid sphere are made to spin about fixed central axes (like a top) by means of strings wrapped around them, with the strings producing the same constant tangential force $\vec{F}$ on all three objects. The three objects have the same mass and radius, and they are initially stationary. Rank the objects according to (a) their angular momentum about their central axes and (b) their angular speed, greatest first, when the strings have been pulled for a certain time t.

 Disk
 Hoop
 Sphere

Sample Problem 12-6

George Washington Gale Ferris, Jr., a civil engineering graduate from Rensselaer Polytechnic Institute, built the original Ferris wheel (Fig. 12-15) for the 1893 World's Columbian Exposition in Chicago. The wheel, an astounding engineering construction at the time, carried 36 wooden cars, each holding as many as 60 passengers, around a circle of radius $R = 38$ m. The mass of each car was about 1.1×10^4 kg. The mass of the wheel's structure was about 6.0×10^5 kg, which was mostly in the circular grid from which the cars were suspended. The cars were loaded 6 at a time, and once all 36 cars were full, the wheel made a complete rotation at an angular speed ω_F in about 2 min.

Fig. 12-15 Sample Problem 12-6. The original Ferris wheel, built in 1893 near the University of Chicago, towered over the surrounding buildings.

(a) Estimate the magnitude L of the angular momentum of the wheel and its passengers while the wheel rotated at ω_F.

SOLUTION: The **Key Idea** here is that we can treat the wheel, cars, and passengers as a rigid object rotating about a fixed axis, at the wheel's axle. Then Eq. 12-31 ($L = I\omega$) gives the magnitude of the angular momentum of that object. We need to find the rotational inertia I of this object and the angular speed ω_F.

To find I, let us start with the loaded cars. Because we can treat them as particles, at distance R from the axis of rotation, we know from Eq. 11-26 that their rotational inertia is $I_{pc} = M_{pc}R^2$, where M_{pc} is their total mass. Let us assume that the 36 cars are each filled with 60 passengers, each of mass 70 kg. Then their total mass is

$$M_{pc} = 36[1.1 \times 10^4 \text{ kg} + 60(70 \text{ kg})] = 5.47 \times 10^5 \text{ kg}$$

and their rotational inertia is

$$I_{pc} = M_{pc}R^2 = (5.47 \times 10^5 \text{ kg})(38 \text{ m})^2 = 7.90 \times 10^8 \text{ kg} \cdot \text{m}^2.$$

Next we consider the structure of the wheel. Let us assume that the rotational inertia of the structure is due mainly to the circular grid suspending the cars. Further, let us assume that the grid forms a hoop of radius R, with a mass M_{hoop} of 3.0×10^5 kg (half the wheel's mass). From Table 11-2a, the rotational inertia of the hoop is

$$I_{\text{hoop}} = M_{\text{hoop}}R^2 = (3.0 \times 10^5 \text{ kg})(38 \text{ m})^2$$
$$= 4.33 \times 10^8 \text{ kg} \cdot \text{m}^2.$$

The combined rotational inertia I of the cars, passengers, and hoop is then

$$I = I_{pc} + I_{\text{hoop}} = 7.90 \times 10^8 \text{ kg} \cdot \text{m}^2 + 4.33 \times 10^8 \text{ kg} \cdot \text{m}^2$$
$$= 1.22 \times 10^9 \text{ kg} \cdot \text{m}^2.$$

To find the rotational speed ω_F, we use Eq. 11-5 ($\omega_{\text{avg}} = \Delta\theta/\Delta t$). Here the wheel goes through an angular displacement of $\Delta\theta = 2\pi$ rad in a time period $\Delta t = 2$ min. Thus, we have

$$\omega_F = \frac{2\pi \text{ rad}}{(2 \text{ min})(60 \text{ s/min})} = 0.0524 \text{ rad/s}.$$

Now we can find the magnitude L of the angular momentum with Eq. 12-31:

$$L = I\omega_F = (1.22 \times 10^9 \text{ kg} \cdot \text{m}^2)(0.0524 \text{ rad/s})$$
$$= 6.39 \times 10^7 \text{ kg} \cdot \text{m}^2/\text{s} \approx 6.4 \times 10^7 \text{ kg} \cdot \text{m}^2/\text{s}. \quad \text{(Answer)}$$

(b) Assume that the fully loaded wheel is rotated from rest to ω_F in a time period $\Delta t_1 = 5.0$ s. What is the magnitude τ_{avg} of the average net external torque acting on it during Δt_1?

SOLUTION: The **Key Idea** here is that the average net external torque is related to the change ΔL in the angular momentum of the loaded wheel by Eq. 12-29 ($\vec{\tau}_{\text{net}} = d\vec{L}/dt$). Because the wheel rotates about a fixed axis to reach angular speed ω_F in time period Δt_1, we can rewrite Eq. 12-29 as $\tau_{\text{avg}} = \Delta L/\Delta t_1$. The change ΔL is from zero to the answer for part (a). Thus, we have

$$\tau_{\text{avg}} = \frac{\Delta L}{\Delta t_1} = \frac{6.39 \times 10^7 \text{ kg} \cdot \text{m}^2/\text{s} - 0}{5.0 \text{ s}}$$
$$\approx 1.3 \times 10^7 \text{ N} \cdot \text{m}. \quad \text{(Answer)}$$

12-10 Conservation of Angular Momentum

So far we have discussed two powerful conservation laws, the conservation of energy and the conservation of linear momentum. Now we meet a third law of this type, involving the conservation of angular momentum. We start from Eq. 12-29 ($\vec{\tau}_{\text{net}} = d\vec{L}/dt$), which is Newton's second law in angular form. If no net external torque acts on the system, this equation becomes $d\vec{L}/dt = 0$, or

$$\vec{L} = \text{a constant} \quad \text{(isolated system)}. \quad (12\text{-}32)$$

This result, called the **law of conservation of angular momentum,** can also be written as

$$\left(\begin{array}{c} \text{net angular momentum} \\ \text{at some initial time } t_i \end{array} \right) = \left(\begin{array}{c} \text{net angular momentum} \\ \text{at some later time } t_f \end{array} \right),$$

or

$$\vec{L}_i = \vec{L}_f \quad \text{(isolated system)}. \quad (12\text{-}33)$$

Equations 12-32 and 12-33 tell us:

> If the net external torque acting on a system is zero, the angular momentum $\vec{L}$ of the system remains constant, no matter what changes take place within the system.

Equations 12-32 and 12-33 are vector equations; as such, they are equivalent to three component equations corresponding to the conservation of angular momentum in three mutually perpendicular directions. Depending on the torques acting on a system, the angular momentum of the system might be conserved in only one or two directions but not in all directions:

> ► If the component of the net *external* torque on a system along a certain axis is zero, then the component of the angular momentum of the system along that axis cannot change, no matter what changes take place within the system.

We can apply this law to the isolated body in Fig. 12-14, which rotates around the z axis. Suppose that the initially rigid body somehow redistributes its mass relative to that rotation axis, changing its rotational inertia about that axis. Equations 12-32 and 12-33 state that the angular momentum of the body cannot change. Substituting Eq. 12-31 (for the angular momentum along the rotational axis) into Eq. 12-33, we write this conservation law as

$$I_i \omega_i = I_f \omega_f. \tag{12-34}$$

Here the subscripts refer to the values of the rotational inertia I and angular speed ω before and after the redistribution of mass.

Like the other two conservation laws that we have discussed, Eqs. 12-32 and 12-33 hold beyond the limitations of Newtonian mechanics. They hold for particles whose speeds approach that of light (where the theory of special relativity reigns), and they remain true in the world of subatomic particles (where quantum physics reigns). No exceptions to the law of conservation of angular momentum have ever been found.

We now discuss four examples involving this law.

1. **The spinning volunteer**   Figure 12-16 shows a student seated on a stool that can rotate freely about a vertical axis. The student, who has been set into rotation at a modest initial angular speed ω_i, holds two dumbbells in his outstretched hands. His angular momentum vector $\vec{L}$ lies along the vertical rotation axis, pointing upward.

   The instructor now asks the student to pull in his arms; this action reduces his rotational inertia from its initial value I_i to a smaller value I_f because he moves mass closer to the rotation axis. His rate of rotation increases markedly, from ω_i to ω_f. The student can then slow down by extending his arms once more.

   No net external torque acts on the system consisting of the student, stool, and dumbbells. Thus, the angular momentum of that system about the rotation

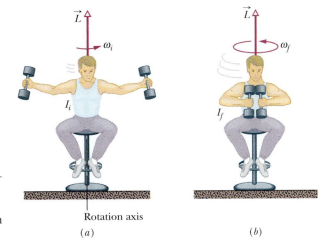

Fig. 12-16 (*a*) The student has a relatively large rotational inertia about the rotation axis and a relatively small angular speed. (*b*) By decreasing his rotational inertia, the student automatically increases his angular speed. The angular momentum $\vec{L}$ of the rotating system remains unchanged.

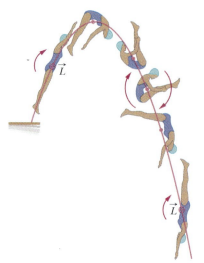

Fig. 12-17 The diver's angular momentum $\vec{L}$ is constant throughout the dive, being represented by the tail $\otimes$ of an arrow that is perpendicular to the plane of the figure. Note also that her center of mass (see the dots) follows a parabolic path.

axis must remain constant, no matter how the student maneuvers the dumbbells. In Fig. 12-16a, the student's angular speed ω_i is relatively low and his rotational inertia I_i is relatively high. According to Eq. 12-34, his angular speed in Fig. 12-16b must be greater to compensate for the decreased rotational inertia.

2. **The springboard diver**   Figure 12-17 shows a diver doing a forward one-and-a-half-somersault dive. As you should expect, her center of mass follows a parabolic path. She leaves the springboard with a definite angular momentum $\vec{L}$ about an axis through her center of mass, represented by a vector pointing into the plane of Fig. 12-17, perpendicular to the page. When she is in the air, no net external torque acts on her about her center of mass, so her angular momentum about her center of mass cannot change. By pulling her arms and legs into the closed *tuck position,* she can considerably reduce her rotational inertia about the same axis and thus, according to Eq. 12-34, considerably increase her angular speed. Pulling out of the tuck position (into the *open layout position*) at the end of the dive increases her rotational inertia and thus slows her rotation rate so she can enter the water with little splash. Even in a more complicated dive involving both twisting and somersaulting, the angular momentum of the diver must be conserved, in both magnitude *and* direction, throughout the dive.

3. **Spacecraft orientation**   Figure 12-18, which represents a spacecraft with a rigidly mounted flywheel, suggests a scheme (albeit crude) for orientation control. The *spacecraft + flywheel* form an isolated system. Therefore, if the system's total angular momentum $\vec{L}$ is zero because neither spacecraft nor flywheel is turning, it must remain zero (as long as the system remains isolated).

To change the orientation of the spacecraft, the flywheel is made to rotate (Fig. 12-18a). The spacecraft will start to rotate in the opposite sense to maintain the system's angular momentum at zero. When the flywheel is then brought to rest, the spacecraft will also stop rotating but will have changed its orientation (Fig. 12-18b). Throughout, the angular momentum of the system *spacecraft + flywheel* never differs from zero.

Interestingly, the spacecraft *Voyager 2,* on its 1986 flyby of the planet Uranus, was set into unwanted rotation by this flywheel effect every time its tape recorder was turned on at high speed. The ground staff at the Jet Propulsion Laboratory had to program the on-board computer to turn on counteracting thruster jets every time the tape recorder was turned on or off.

4. **The incredible shrinking star**   When the nuclear fire in the core of a star burns low, the star may eventually begin to collapse, building up pressure in its interior. The collapse may go so far as to reduce the radius of the star from something like that of the Sun to the incredibly small value of a few kilometers. The star then becomes a *neutron star*—its material has been compressed to an incredibly dense gas of neutrons.

During this shrinking process, the star is an isolated system and its angular momentum $\vec{L}$ cannot change. Because its rotational inertia is greatly reduced, its angular speed is correspondingly greatly increased, to as much as 600 to 800 revolutions per *second.* For comparison, the Sun, a typical star, rotates at about one revolution per month.

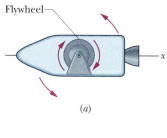

(a)

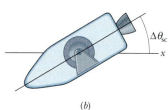

(b)

Fig. 12-18 (a) An idealized spacecraft containing a flywheel. If the flywheel is made to rotate clockwise as shown, the spacecraft itself will rotate counterclockwise. (b) When the flywheel is braked to a stop, the spacecraft will also stop rotating but will be reoriented by the angle $\Delta\theta_{sc}$.

✓**CHECKPOINT 7:**   A rhinoceros beetle rides the rim of a small disk that rotates like a merry-go-round. If the beetle crawls toward the center of the disk, do the following (each relative to the central axis) increase, decrease, or remain the same: (a) the rotational inertia of the beetle–disk system, (b) the angular momentum of the system, and (c) the angular speed of the beetle and disk?

Sample Problem 12-7

Figure 12-19a shows a student, again sitting on a stool that can rotate freely about a vertical axis. The student, initially at rest, is holding a bicycle wheel whose rim is loaded with lead and whose rotational inertia I_{wh} about its central axis is 1.2 kg · m². The wheel is rotating at an angular speed ω_{wh} of 3.9 rev/s; as seen from overhead, the rotation is counterclockwise. The axis of the wheel is vertical, and the angular momentum $\vec{L}_{wh}$ of the wheel points vertically upward. The student now inverts the wheel (Fig. 12-19b) so that, as seen from overhead, it is rotating clockwise. Its angular momentum is then $-\vec{L}_{wh}$. The inversion results in the student, the stool, and the wheel's center rotating together as a composite rigid body about the stool's rotation axis, with rotational inertia $I_b =$ 6.8 kg · m². (The fact that the wheel is also rotating about its center does not affect the mass distribution of this composite body; thus, I_b has the same value whether or not the wheel rotates.) With what angular speed ω_b and in what direction does the composite body rotate after the inversion of the wheel?

SOLUTION: The Key Ideas here are these:

1. The angular speed ω_b we seek is related to the final angular momentum $\vec{L}_b$ of the composite body about the stool's rotation axis by Eq. 12-31 ($L = I\omega$).

2. The initial angular speed ω_{wh} of the wheel is related to the angular momentum $\vec{L}_{wh}$ of the wheel's rotation about its center by the same equation.

3. The vector addition of $\vec{L}_b$ and $\vec{L}_{wh}$ gives the total angular momentum $\vec{L}_{tot}$ of the system of student, stool, and wheel.

4. As the wheel is inverted, no net *external* torque acts on that system to change $\vec{L}_{tot}$ about any vertical axis. (Torques due to forces between the student and the wheel as the student inverts the wheel are *internal* to the system.) So, the system's total angular momentum is conserved about any vertical axis.

The conservation of $\vec{L}_{tot}$ is represented with vectors in Fig. 12-19c. We can also write it in terms of components along a vertical axis as

$$L_{b,f} + L_{wh,f} = L_{b,i} + L_{wh,i}, \qquad (12\text{-}35)$$

where i and f refer to the initial state (before inversion of the wheel) and the final state (after inversion). Because inversion of the wheel inverted the angular momentum vector of the wheel's rotation, we substitute $-L_{wh,i}$ for $L_{wh,f}$. Then, if we set $L_{b,i} = 0$ (because the student, the stool, and the wheel's center were initially at rest),

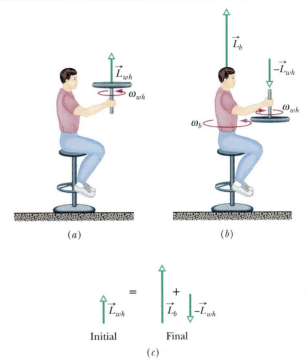

Fig. 12-19 Sample Problem 12-7. (a) A student holds a bicycle wheel rotating around the vertical. (b) The student inverts the wheel, setting himself into rotation. (c) The net angular momentum of the system must remain the same in spite of the inversion.

Eq. 12-35 yields

$$L_{b,f} = 2L_{wh,i}.$$

Using Eq. 12-31, we next substitute $I_b\omega_b$ for $L_{b,f}$ and $I_{wh}\omega_{wh}$ for $L_{wh,i}$ and solve for ω_b, finding

$$\omega_b = \frac{2I_{wh}}{I_b}\omega_{wh}$$

$$= \frac{(2)(1.2 \text{ kg} \cdot \text{m}^2)(3.9 \text{ rev/s})}{6.8 \text{ kg} \cdot \text{m}^2} = 1.4 \text{ rev/s}. \quad \text{(Answer)}$$

This positive result tells us that the student rotates counterclockwise about the stool axis as seen from overhead. If the student wishes to stop rotating, he has only to invert the wheel once more.

Sample Problem 12-8

During a jump to his partner, an aerialist is to make a quadruple somersault lasting a time $t = 1.87$ s. For the first and last quarter-revolution, he is in the extended orientation shown in Fig. 12-20, with rotational inertia $I_1 = 19.9$ kg · m² around his center of mass (the dot). During the rest of the flight he is in a tight tuck, with rotational inertia $I_2 = 3.93$ kg · m². What must be his angular speed ω_2 around his center of mass during the tuck?

SOLUTION: Obviously he must turn fast enough to complete the 4.0 rev required for a quadruple somersault in the given 1.87 s. To do so, he increases his angular speed to ω_2 by tucking. We can relate ω_2 to his initial angular speed ω_1 with this Key Idea: His angular momentum about his center of mass is conserved throughout the free flight because there is no net external torque about his center of mass to change it. From Eq. 12-34, we can write the

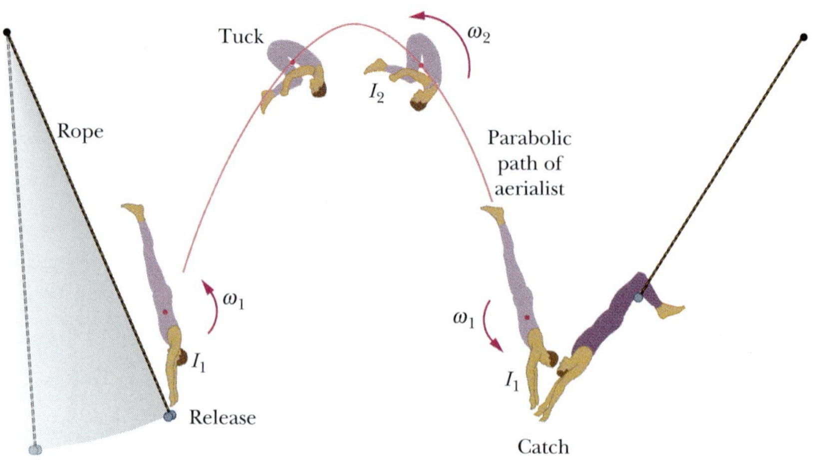

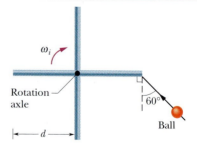

Fig. 12-20 Sample Problem 12-8. An aerialist performing a multiple somersault to a partner.

conservation of angular momentum ($L_1 = L_2$) as

$$I_1 \omega_1 = I_2 \omega_2$$

or

$$\omega_1 = \frac{I_2}{I_1} \omega_2. \qquad (12\text{-}36)$$

A second **Key Idea** is that these angular speeds are related to the angles through which he must rotate and the time available to do so. At the start and at the end, he must rotate in the extended orientation for a total angle of $\theta_1 = 0.500$ rev (two quarter-turns) in a time we shall call t_1. In the tuck, he must rotate through an angle of $\theta_2 = 3.50$ rev in a time t_2. From Eq. 11-5 ($\omega_{avg} = \Delta\theta/\Delta t$), we can write

$$t_1 = \frac{\theta_1}{\omega_1} \quad \text{and} \quad t_2 = \frac{\theta_2}{\omega_2}.$$

Thus, his total flight time is

$$t = t_1 + t_2 = \frac{\theta_1}{\omega_1} + \frac{\theta_2}{\omega_2}, \qquad (12\text{-}37)$$

which we know to be 1.87 s. Now substituting from Eq. 12-36 for ω_1 yields

$$t = \frac{\theta_1 I_1}{\omega_2 I_2} + \frac{\theta_2}{\omega_2} = \frac{1}{\omega_2}\left(\theta_1 \frac{I_1}{I_2} + \theta_2\right).$$

Inserting the known data, we obtain

$$1.87 \text{ s} = \frac{1}{\omega_2}\left((0.500 \text{ rev})\frac{19.9 \text{ kg}\cdot\text{m}^2}{3.93 \text{ kg}\cdot\text{m}^2} + 3.50 \text{ rev}\right),$$

which gives us

$$\omega_2 = 3.23 \text{ rev/s.} \qquad \text{(Answer)}$$

This angular speed is so fast that the aerialist cannot clearly see his surroundings or fine-tune his rotation by adjusting his tuck. The possibility of an aerialist making a four-and-a-half somersault flight, which would require a greater value of ω_2 and thus a smaller I_2 via a tighter tuck, seems very small.

Sample Problem 12-9

(This final sample problem of the chapter is long and challenging, but it is helpful because it pulls together many ideas of Chapters 11 and 12.) In the overhead view of Fig. 12-21, four thin, uniform rods, each of mass M and length $d = 0.50$ m, are rigidly connected to a vertical axle to form a turnstile. The turnstile rotates clockwise about the axle, which is attached to a floor, with initial angular velocity $\omega_i = -2.0$ rad/s. A mud ball of mass $m = \frac{1}{3}M$ and initial speed $v_i = 12$ m/s is thrown along the path shown and sticks to the end of one rod. What is the final angular velocity ω_f of the ball–turnstile system?

SOLUTION: A **Key Idea** here can be stated in a question-and-answer format. The question is this: Does the system have a quantity that is conserved during the collision and that involves angular velocity, so that we can solve for ω_f? To answer, let us check the conservation possibilities:

1. The total kinetic energy K is *not* conserved, because the collision between ball and rod is completely inelastic (the ball

Fig. 12-21 Sample Problem 12-9. An overhead view of four rigidly connected rods rotating freely around a central axle, and the path a mud ball takes to stick onto one of the rods.

sticks). So, some energy must be transferred from kinetic energy to other types of energy (such as thermal energy). For the same reason, total mechanical energy is not conserved.

2. The total linear momentum $\vec{P}$ is also *not* conserved, because during the collision an external force acts on the turnstile at the attachment of the axle to the floor. (This is the force that keeps the turnstile from moving across the floor when it is hit by the mud ball.)

3. The total angular momentum $\vec{L}$ of the system about the axle *is* conserved because there is no net external torque to change $\vec{L}$. (The forces in the collision produce only internal torques; the external force on the turnstile acts at the axle, has zero moment arm, and thus does not produce an external torque.)

We can write the conservation of the system's total angular momentum ($L_f = L_i$) about the axle as

$$L_{ts,f} + L_{ball,f} = L_{ts,i} + L_{ball,i}, \qquad (12\text{-}38)$$

where ts stands for "turnstile." The final angular velocity ω_f is contained in the terms $L_{ts,f}$ and $L_{ball,f}$ because those final angular momenta depend on how fast the turnstile and ball are rotating. To find ω_f, we consider first the turnstile and then the ball, and then we return to Eq. 12-38.

Turnstile: The **Key Idea** here is that, because the turnstile is a rotating rigid object, Eq. 12-31 ($L = I\omega$) gives its angular momentum. Thus, we can write its final and initial angular momenta about the axle as

$$L_{ts,f} = I_{ts}\omega_f \quad \text{and} \quad L_{ts,i} = I_{ts}\omega_i. \qquad (12\text{-}39)$$

Because the turnstile consists of four rods, each rotating around an end, the rotational inertia I_{ts} of the turnstile is four times the rotational inertia I_{rod} of each rod about its end. From Table 11-2e, we know that the rotational inertia I_{com} of a rod about its center is $\frac{1}{12}Md^2$, where M is its mass and d is its length. To get I_{rod} we use the parallel-axis theorem of Eq. 11-29 ($I = I_{com} + Mh^2$). Here perpendicular distance h is $d/2$. Thus, we find

$$I_{rod} = \tfrac{1}{12}Md^2 + M\left(\frac{d}{2}\right)^2 = \tfrac{1}{3}Md^2.$$

With four rods in the turnstile, we then have

$$I_{ts} = \tfrac{4}{3}Md^2. \qquad (12\text{-}40)$$

Ball: Before the collision, the ball is like a particle moving along a straight line, as in Fig. 12-11. So, to find the ball's initial angular momentum $L_{ball,i}$ about the axle, we can use any of Eqs. 12-18 through 12-21, but Eq. 12-20 ($\ell = rmv_\perp$) is easiest. Here ℓ is $L_{ball,i}$; just before the ball hits, its radial distance r from the axle is d and the component $v_\perp$ of the ball's velocity perpendicular to r is $v_i \cos 60°$.

To give a sign to this angular momentum, we mentally draw a position vector from the turnstile's axle to the ball. As the ball approaches the turnstile, this position vector rotates counterclockwise about the axle, so the ball's angular momentum is a positive quantity. We can now rewrite $\ell = rmv_\perp$ as

$$L_{ball,i} = mdv_i \cos 60°. \qquad (12\text{-}41)$$

After the collision, the ball is like a particle rotating in a circle of radius d. So, from Eq. 11-26 ($I = \Sigma m_i r_i^2$), we have $I_{ball} = md^2$ about the axle. Then from Eq. 12-31 ($L = I\omega$), we can write the final angular momentum of the ball about the axle as

$$L_{ball,f} = I_{ball}\omega_f = md^2\omega_f. \qquad (12\text{-}42)$$

Return to Eq. 12-38: Substituting from Eqs. 12-39 through 12-42 into Eq.12-38, we have

$$\tfrac{4}{3}Md^2\omega_f + md^2\omega_f = \tfrac{4}{3}Md^2\omega_i + mdv_i \cos 60°.$$

Substituting $M = 3m$ and solving for ω_f, we find

$$\omega_f = \frac{1}{5d}(4d\omega_i + v_i \cos 60°)$$

$$= \frac{1}{5(0.50\ \text{m})}[4(0.50\ \text{m})(-2.0\ \text{rad/s}) + (12\ \text{m/s})(\cos 60°)]$$

$$= 0.80\ \text{rad/s}. \qquad \text{(Answer)}$$

Thus, the turnstile is now turning counterclockwise.

REVIEW & SUMMARY

Rolling Bodies For a wheel of radius R that is rolling smoothly (no sliding),

$$v_{com} = \omega R, \qquad (12\text{-}2)$$

where v_{com} is the linear speed of the wheel's center and ω is the angular speed of the wheel about its center. The wheel may also be viewed as rotating instantaneously about the point P of the "road" that is in contact with the wheel. The angular speed of the wheel about this point is the same as the angular speed of the wheel about its center. The rolling wheel has kinetic energy

$$K = \tfrac{1}{2}I_{com}\omega^2 + \tfrac{1}{2}Mv_{com}^2, \qquad (12\text{-}5)$$

where I_{com} is the rotational moment of the wheel about its center and M is the mass of the wheel. If the wheel is being accelerated

but is still rolling smoothly, the acceleration of the center of mass $\vec{a}_{com}$ is related to the angular acceleration α about the center with

$$a_{com} = \alpha R. \qquad (12\text{-}6)$$

If the wheel rolls smoothly down a ramp of angle θ, its acceleration along an x axis extending up the ramp is

$$a_{com,x} = -\frac{g \sin \theta}{1 + I_{com}/MR^2}. \qquad (12\text{-}10)$$

Torque as a Vector In three dimensions, *torque* $\vec{\tau}$ is a vector quantity defined relative to a fixed point (usually an origin); it is

$$\vec{\tau} = \vec{r} \times \vec{F}, \qquad (12\text{-}14)$$

where $\vec{F}$ is a force applied to a particle and $\vec{r}$ is a position vector locating the particle relative to the fixed point (or origin). The magnitude of $\vec{\tau}$ is given by

$$\tau = rF\sin\phi = rF_{\perp} = r_{\perp}F, \quad \text{(12-15, 12-16, 12-17)}$$

where ϕ is the angle between $\vec{F}$ and $\vec{r}$, $F_{\perp}$ is the component of $\vec{F}$ perpendicular to $\vec{r}$, and $r_{\perp}$ is the moment arm of $\vec{F}$. The direction of $\vec{\tau}$ is given by the right-hand rule for cross products.

Angular Momentum of a Particle   The *angular momentum* $\vec{\ell}$ of a particle with linear momentum $\vec{p}$, mass m, and linear velocity $\vec{v}$ is a vector quantity defined relative to a fixed point (usually an origin); it is

$$\vec{\ell} = \vec{r} \times \vec{p} = m(\vec{r} \times \vec{v}). \quad \text{(12-18)}$$

The magnitude of $\vec{\ell}$ is given by

$$\ell = rmv\sin\phi \quad \text{(12-19)}$$
$$= rp_{\perp} = rmv_{\perp} \quad \text{(12-20)}$$
$$= r_{\perp}p = r_{\perp}mv, \quad \text{(12-21)}$$

where ϕ is the angle between $\vec{r}$ and $\vec{p}$, $p_{\perp}$ and $v_{\perp}$ are the components of $\vec{p}$ and $\vec{v}$ perpendicular to $\vec{r}$, and $r_{\perp}$ is the perpendicular distance between the fixed point and the extension of $\vec{p}$. The direction of $\vec{\ell}$ is given by the right-hand rule for cross products.

Newton's Second Law in Angular Form   Newton's second law for a particle can be written in angular form as

$$\vec{\tau}_{net} = \frac{d\vec{\ell}}{dt}, \quad \text{(12-23)}$$

where $\vec{\tau}_{net}$ is the net torque acting on the particle, and $\vec{\ell}$ is the angular momentum of the particle.

Angular Momentum of a System of Particles   The angular momentum $\vec{L}$ of a system of particles is the vector sum of the angular momenta of the individual particles:

$$\vec{L} = \vec{\ell}_1 + \vec{\ell}_2 + \cdots + \vec{\ell}_n = \sum_{i=1}^{n} \vec{\ell}_i. \quad \text{(12-26)}$$

The time rate of change of this angular momentum is equal to the net external torque on the system (the vector sum of the torques due to interactions of the particles of the system with particles external to the system):

$$\vec{\tau}_{net} = \frac{d\vec{L}}{dt} \quad \text{(system of particles).} \quad \text{(12-29)}$$

Angular Momentum of a Rigid Body   For a rigid body rotating about a fixed axis, the component of its angular momentum parallel to the rotation axis is

$$L = I\omega \quad \text{(rigid body, fixed axis).} \quad \text{(12-31)}$$

Conservation of Angular Momentum   The angular momentum $\vec{L}$ of a system remains constant if the net external torque acting on the system is zero:

$$\vec{L} = \text{a constant} \quad \text{(isolated system)} \quad \text{(12-32)}$$

or $\qquad \vec{L}_i = \vec{L}_f \quad \text{(isolated system).} \quad \text{(12-33)}$

This is the **law of conservation of angular momentum.** It is one of the fundamental conservation laws of nature, having been verified even in situations (involving high-speed particles or subatomic dimensions) in which Newton's laws are not applicable.

QUESTIONS

1. In Fig. 12-22, a block slides down a frictionless ramp and a sphere rolls without sliding down a ramp of the same angle θ. The block and sphere have the same mass, start from rest at point A, and descend through point B. (a) In that descent, is the work done by the gravitational force on the block greater than, less than, or the same as the work done by the gravitational force on the sphere? At B, which object has more (b) translational kinetic energy and (c) speed down the ramp?

2. A cannonball rolls down an incline without sliding. If the roll is now repeated with an incline that is less steep but of the same

height as the first incline, are (a) the ball's time to reach the bottom and (b) its translational kinetic energy at the bottom greater than, less than, or the same as previously?

3. In Fig. 12-23, a woman rolls a cylindrical drum by pushing a board across its top. The drum moves through the distance $L/2$, which is half the board's length. The drum rolls smoothly without sliding or bouncing, and the board does not slide over the drum. (a) What length of board rolls over the top of the drum? (b) How far does the woman walk?

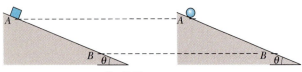

Fig. 12-22 Question 1.

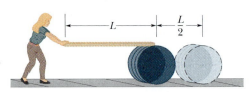

Fig. 12-23 Question 3.

4. The position vector $\vec{r}$ of a particle relative to a certain point has a magnitude of 3 m, and the force $\vec{F}$ on the particle has a magnitude of 4 N. What is the angle between the directions of $\vec{r}$ and $\vec{F}$ if the magnitude of the associated torque equals (a) zero and (b) 12 N·m?

5. Figure 12-24 shows a particle moving at constant velocity $\vec{v}$ and five points with their xy coordinates. Rank the points according to the magnitude of the angular momentum of the particle measured about them, greatest first.

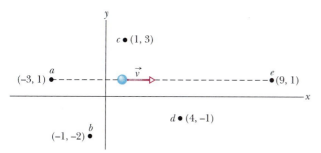

Fig. 12-24 Question 5.

6. (a) In Checkpoint 4, what is the torque on particles 1 and 2 about point O due to the centripetal forces that cause those particles to circle at constant speed? (b) As particles 3, 4, and 5 move from the left to the right of point O, do their individual angular momenta increase, decrease, or stay the same?

7. Figure 12-25 shows three particles of the same mass and the same constant speed moving as indicated by the velocity vectors. Points a, b, c, and d form a square, with point e at the center. Rank the points according to the magnitude of the net angular momentum of the three-particle system when measured about the points, greatest first.

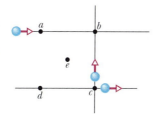

Fig. 12-25 Question 7.

8. A bola, which consists of three heavy balls connected to a common point by identical lengths of sturdy string, is readied for launch by holding one of the balls overhead and rotating the wrist, causing the other two balls to rotate in a horizontal circle about the hand. The bola is then released, and its configuration rapidly changes from that in the overhead view of Fig. 12-26a to that of Fig. 12-26b. Thus, the rotation is initially around axis 1 through the ball that was held. Then it is around axis 2 through the center of mass. Are (a) the angular momentum and (b) the angular speed around axis 2 greater than, less than, or the same as that around axis 1?

9. A rhinoceros beetle rides the rim of a horizontal disk rotating counterclockwise like a merry-go-round. If it then walks along the rim in the direction of the rotation, will the magnitudes of the following increase, decrease, or remain the same: (a) the angular momentum of the beetle–disk system, (b) the angular momentum and angular velocity of the beetle, and (c) the angular momentum and angular velocity of the disk? (d) What are your answers if the beetle walks in the direction opposite the rotation?

10. Figure 12-27 shows an overhead view of a rectangular slab that can spin like a merry-go-round about its center at O. Also shown are seven paths along which wads of bubble gum can be thrown (all with the same speed and mass) to stick onto the stationary slab. (a) Rank the paths according to the angular speed that the slab (and gum) will have after the gum sticks, greatest first. (b) For which paths will the angular momentum of the slab (and gum) about O be negative from the view of Fig. 12-27?

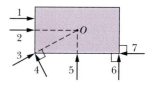

Fig. 12-27 Question 10.

11. In Fig. 12-28, three forces of the same magnitude are applied to a particle at the origin ($\vec{F}_1$ acts directly into the plane of the figure). Rank the forces according to the magnitudes of the torques they create about (a) point P_1, (b) point P_2, and (c) point P_3, greatest first.

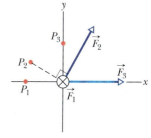

Fig. 12-28 Question 11.

12. Figure 12-29 shows two particles A and B at xyz coordinates (1 m, 1 m, 0) and (1 m, 0, 1 m). Acting on each particle are three numbered forces, all of the same magnitude and each directed parallel to an axis. (a) Which of the forces produce a torque about the origin that is directed parallel to y? (b) Rank the forces according to the magnitudes of the torques they produce on the particles about the origin, greatest first.

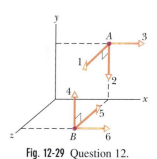

Fig. 12-29 Question 12.

13. Figure 11-24 in Chapter 11 shows an assembly of three small spheres of the same mass that are attached to a massless rod with the indicated spacings. The assembly is to be rotated at 3.0 rad/s about an axis through one of the spheres and perpendicular to the plane of the page. There are, of course, three such choices of the axis. Rank those choices according to (a) the magnitude of the angular momentum the assembly will have about the chosen rotation axis and (b) the rotational kinetic energy the assembly will have, greatest first.

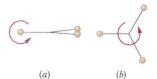

Fig. 12-26 Question 8.   (a)   (b)

EXERCISES & PROBLEMS

SEC. 12-1   Rolling

1E. An automobile traveling 80.0 km/h has tires of 75.0 cm diameter. (a) What is the angular speed of the tires about their axles? (b) If the car is brought to a stop uniformly in 30.0 complete turns of the tires (without skidding), what is the magnitude of the angular acceleration of the wheels? (c) How far does the car move during the braking?

2P. Consider a 66-cm-diameter tire on a car traveling at 80 km/h on a level road in the positive direction of an x axis. Relative to a woman in the car, what are (a) the linear velocity $\vec{v}$ and (b) the magnitude a of the linear acceleration of the center of the wheel? What are (c) $\vec{v}$ and (d) a for a point at the top of the tire? What are (e) $\vec{v}$ and (f) a for a point at the bottom of the tire?

    Now repeat the questions relative to a hitchhiker sitting near the road: What are (g) $\vec{v}$ at the wheel's center, (h) a at the wheel's center, (i) $\vec{v}$ at the tire top, (j) a at the tire top, (k) $\vec{v}$ at the tire bottom, and (l) a at the tire bottom?

SEC. 12-2   The Kinetic Energy of Rolling

3E. A 140 kg hoop rolls along a horizontal floor so that its center of mass has a speed of 0.150 m/s. How much work must be done on the hoop to stop it? ssm

4E. A thin-walled pipe rolls along the floor. What is the ratio of its translational kinetic energy to its rotational kinetic energy about an axis parallel to its length and through its center of mass?

5E. A 1000 kg car has four 10 kg wheels. When the car is moving, what fraction of the total kinetic energy of the car is due to rotation of the wheels about their axles? Assume that the wheels have the same rotational inertia as uniform disks of the same mass and size. Why do you not need the radius of the wheels? ssm ilw www

6P. A body of radius R and mass m is rolling smoothly with speed v on a horizontal surface. It then rolls up a hill to a maximum height h. (a) If $h = 3v^2/4g$, what is the body's rotational inertia about the rotational axis through its center of mass? (b) What might the body be?

SEC. 12-3   The Forces of Rolling

7E. A uniform solid sphere rolls down an incline. (a) What must be the incline angle if the linear acceleration of the center of the sphere is to have a magnitude of $0.10g$? (b) If a frictionless block were to slide down the incline at that angle, would its acceleration magnitude be more than, less than, or equal to $0.10g$? Why?

8P. A constant horizontal force of magnitude 10 N is applied to a wheel of mass 10 kg and radius 0.30 m as shown in Fig. 12-30. The wheel rolls smoothly on the horizontal surface, and the acceleration of its center of mass has magnitude 0.60 m/s². (a) What are the magnitude and direction of the frictional force on the wheel?

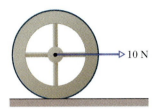

Fig. 12-30 Problem 8.

(b) What is the rotational inertia of the wheel about the rotation axis through its center of mass?

9P. A solid ball starts from rest at the upper end of the track shown in Fig. 12-31 and rolls without slipping until it rolls off the right-hand end. If $H = 6.0$ m and $h = 2.0$ m and the track is horizontal at the right-hand end, how far horizontally from point A does the ball land on the floor? ssm

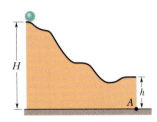

Fig. 12-31 Problem 9.

10P. A small sphere, with radius r and mass m, rolls without slipping on the inside of a large fixed hemisphere with radius R and a vertical axis of symmetry. It starts at the top from rest. (a) What is its kinetic energy at the bottom? (b) What fraction of its kinetic energy at the bottom is associated with rotation about an axis through its center of mass? (c) Assuming $r \ll R$, find the magnitude of the normal force on the hemisphere from the ball when the ball reaches the bottom.

11P. A solid cylinder of radius 10 cm and mass 12 kg starts from rest and rolls without slipping a distance of 6.0 m down a house roof that is inclined at 30°. (See Fig. 12-32.) (a) What is the angular speed of the cylinder about its center as it leaves the house roof? (b) The roof's edge is 5.0 m high. How far horizontally from the roof's edge does the cylinder hit the level ground? ilw

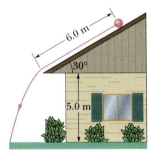

Fig. 12-32 Problem 11.

12P. A small solid marble of mass m and radius r will roll without slipping along the loop-the-loop track shown in Fig. 12-33 if it is released from rest somewhere on the straight section of track. (a) From what initial height h above the bottom of the track must the marble be released if it is to be on the verge of leaving the track at the top of the loop? (The radius of the loop-the-loop is R; assume $R \gg r$.) (b) If the marble is released from height $6R$ above the bottom of the track, what is the horizontal component of the force acting on it at point Q?

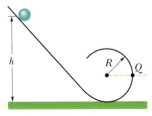

Fig. 12-33 Problem 12.

13P. A hollow sphere of radius 0.15 m, with rotational inertia $I = 0.040$ kg·m^2 about a line through its center of mass, rolls without slipping up a surface inclined at 30° to the horizontal. At a certain initial position, the sphere's total kinetic energy is 20 J. (a) How much of this initial kinetic energy is rotational? (b) What is the speed of the center of mass of the sphere at the initial position? What are (c) the total kinetic energy of the sphere and (d) the speed of its center of mass after it has moved 1.0 m up along the incline from its initial position?

14P. A bowler throws a bowling ball of radius $R = 11$ cm along a lane. The ball slides on the lane, with initial speed $v_{com,0} = 8.5$ m/s and initial angular speed $\omega_0 = 0$. The coefficient of kinetic

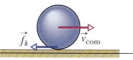

Fig. 12-34 Problem 14.

friction between the ball and the lane is 0.21. The kinetic frictional force $\vec{f}_k$ acting on the ball (Fig. 12-34) causes a linear acceleration of the ball while producing a torque that causes an angular acceleration of the ball. When speed v_{com} has decreased enough and angular speed ω has increased enough, the ball stops sliding and then rolls smoothly. (a) What then is v_{com} in terms of ω? During the sliding, what are the ball's (b) linear acceleration and (c) angular acceleration? (d) How long does the ball slide? (e) How far does the ball slide? (f) What is the speed of the ball when smooth rolling begins?

SEC. 12-4 The Yo-Yo

15E. A yo-yo has a rotational inertia of 950 g·cm^2 and a mass of 120 g. Its axle radius is 3.2 mm, and its string is 120 cm long. The yo-yo rolls from rest down to the end of the string. (a) What is the magnitude of its linear acceleration? (b) How long does it take to reach the end of the string? As it reaches the end of the string, what are its (c) linear speed, (d) translational kinetic energy, (e) rotational kinetic energy, and (f) angular speed? ssm

16P. Suppose that the yo-yo in Exercise 15, instead of rolling from rest, is thrown so that its initial speed down the string is 1.3 m/s. (a) How long does the yo-yo take to reach the end of the string? As it reaches the end of the string, what are its (b) total kinetic energy, (c) linear speed, (d) translational kinetic energy, (e) angular speed, and (f) rotational kinetic energy?

SEC. 12-5 Torque Revisited

17E. Show that, if $\vec{r}$ and $\vec{F}$ lie in a given plane, the torque $\vec{\tau} = \vec{r} \times \vec{F}$ has no component in that plane.

18E. What are the magnitude and direction of the torque about the origin on a plum located at coordinates $(-2.0$ m, 0, 4.0 m) due to force $\vec{F}$ whose only component is (a) $F_x = 6.0$ N, (b) $F_x = -6.0$ N, (c) $F_z = 6.0$ N, and (d) $F_z = -6.0$ N?

19E. What are the magnitude and direction of the torque about the origin on a particle located at coordinates $(0, -4.0$ m, 3.0 m) due to (a) force $\vec{F}_1$ with components $F_{1x} = 2.0$ N and $F_{1y} = F_{1z} = 0$, and (b) force $\vec{F}_2$ with components $F_{2x} = 0$, $F_{2y} = 2.0$ N, and $F_{2z} = 4.0$ N?

20P. Force $\vec{F} = (2.0$ N$)\hat{i} - (3.0$ N$)\hat{k}$ acts on a pebble with position vector $\vec{r} = (0.50$ m$)\hat{j} - (2.0$ m$)\hat{k}$, relative to the origin. What is

the resulting torque acting on the pebble about (a) the origin and (b) a point with coordinates $(2.0$ m, 0, -3.0 m)?

21P. Force $\vec{F} = (-8.0$ N$)\hat{i} + (6.0$ N$)\hat{j}$ acts on a particle with position vector $\vec{r} = (3.0$ m$)\hat{i} + (4.0$ m$)\hat{j}$. What are (a) the torque on the particle about the origin and (b) the angle between the directions of $\vec{r}$ and $\vec{F}$? ssm

22P. What is the torque about the origin on a jar of jalapeño peppers located at coordinates $(3.0$ m, -2.0 m, 4.0 m) due to (a) force $\vec{F}_1 = (3.0$ N$)\hat{i} - (4.0$ N$)\hat{j} + (5.0$ N$)\hat{k}$, (b) force $\vec{F}_2 = (-3.0$ N$)\hat{i} - (4.0$ N$)\hat{j} - (5.0$ N$)\hat{k}$, and (c) the vector sum of $\vec{F}_1$ and $\vec{F}_2$? (d) Repeat part (c) about a point with coordinates $(3.0$ m, 2.0 m, 4.0 m) instead of about the origin.

SEC. 12-6 Angular Momentum

23E. Two objects are moving as shown in Fig. 12-35. What is their total angular momentum about point O? ilw

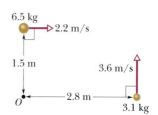

Fig. 12-35 Exercise 23.

24E. In Fig. 12-36, a particle P with mass 2.0 kg has position vector $\vec{r}$ of magnitude 3.0 m and velocity $\vec{v}$ of magnitude 4.0 m/s. A force $\vec{F}$ of magnitude 2.0 N acts on the particle. All three vectors lie in the xy plane oriented as shown. About the origin, what are (a) the angular momentum of the particle and (b) the torque acting on the particle?

25E. At a certain time, a 0.25 kg object has a position vector $\vec{r} = 2.0\hat{i} - 2.0\hat{k}$, in meters. At that instant, its velocity in meters per second is $\vec{v} = -5.0\hat{i} + 5.0\hat{k}$, and the force in newtons acting on it is $\vec{F} = 4.0\hat{j}$. (a) What is the angular momentum of the object about the origin? (b) What torque acts on it? ssm

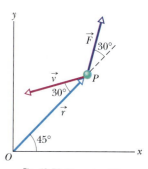

Fig. 12-36 Exercise 24.

26P. A 2.0 kg particle-like object moves in a plane with velocity components $v_x = 30$ m/s and $v_y = 60$ m/s as it passes through the point with (x, y) coordinates of $(3.0, -4.0)$ m. Just then, what is its angular momentum relative to (a) the origin and (b) the point $(-2.0, -2.0)$ m?

27P. Two particles, each of mass m and speed v, travel in opposite directions along parallel lines separated by a distance d. (a) In terms of m, v, and d, find an expression for the magnitude L of the angular momentum of the two-particle system around a point midway between the two lines. (b) Does the expression change if the point about which L is calculated is not midway between the lines? (c) Now reverse the direction of travel for one of the particles and repeat (a) and (b). ssm

28P. A 4.0 kg particle moves in an xy plane. At the instant when the particle's position and velocity are $\vec{r} = (2.0\hat{i} + 4.0\hat{j})$ m and $\vec{v} = -4.0\hat{j}$ m/s, the force on the particle is $\vec{F} = -3.0\hat{i}$ N. At this instant, determine (a) the particle's angular momentum about the origin, (b) the particle's angular momentum about the point $x = 0$,

$y = 4.0$ m, (c) the torque acting on the particle about the origin, and (d) the torque acting on the particle about the point $x = 0$, $y = 4.0$ m.

SEC. 12-7   Newton's Second Law in Angular Form

29E. A 3.0 kg particle with velocity $\vec{v} = (5.0$ m/s$)\hat{i} - (6.0$ m/s$)\hat{j}$ is at $x = 3.0$ m, $y = 8.0$ m. It is pulled by a 7.0 N force in the negative x direction. (a) What is the angular momentum of the particle about the origin? (b) What torque about the origin acts on the particle? (c) At what rate is the angular momentum of the particle changing with time?   ssm   ilw

30E. A particle is acted on by two torques about the origin: $\vec{\tau}_1$ has a magnitude of 2.0 N·m and is directed in the positive direction of the x axis, and $\vec{\tau}_2$ has a magnitude of 4.0 N·m and is directed in the negative direction of the y axis. What are the magnitude and direction of $d\vec{\ell}/dt$, where $\vec{\ell}$ is the angular momentum of the particle about the origin?

31E. What torque about the origin acts on a particle moving in the xy plane, clockwise about the origin, if the particle has the following magnitudes of angular momentum about the origin:
(a) 4.0 kg·m^2/s,
(b) $4.0t^2$ kg·m^2/s,
(c) $4.0\sqrt{t}$ kg·m^2/s,
(d) $4.0/t^2$ kg·m^2/s?

32P. At time $t = 0$, a 2.0 kg particle has position vector $\vec{r} = (4.0$ m$)\hat{i} - (2.0$ m$)\hat{j}$ relative to the origin. Its velocity just then is given by $\vec{v} = (-6.0t^2$ m/s$)\hat{i}$. About the origin and for $t > 0$, what are (a) the particle's angular momentum and (b) the torque acting on the particle? (c) Repeat (a) and (b) about a point with coordinates $(-2.0$ m, -3.0 m, 0) instead of about the origin.

SEC. 12-9   The Angular Momentum of a Rigid Body Rotating About a Fixed Axis

33E. The angular momentum of a flywheel having a rotational inertia of 0.140 kg·m^2 about its central axis decreases from 3.00 to 0.800 kg·m^2/s in 1.50 s. (a) What is the magnitude of the average torque acting on the flywheel about its central axis during this period? (b) Assuming a constant angular acceleration, through what angle does the flywheel turn? (c) How much work is done on the wheel? (d) What is the average power of the flywheel?   ssm

34E. A sanding disk with rotational inertia 1.2×10^{-3} kg·m^2 is attached to an electric drill whose motor delivers a torque of 16 N·m. Find (a) the angular momentum of the disk about its central axis and (b) the angular speed of the disk 33 ms after the motor is turned on.

35E. Three particles, each of mass m, are fastened to each other and to a rotation axis at O by three massless strings, each with length d as shown in Fig. 12-37. The combination rotates around the rotational axis with angular velocity ω in such a way that the particles remain in a straight line. In terms of m, d, and ω, and rel-

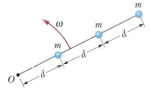

Fig. 12-37 Exercise 35.

ative to point O, what are (a) the rotational inertia of the combination, (b) the angular momentum of the middle particle, and (c) the total angular momentum of the three particles?   ssm

36P. An impulsive force $F(t)$ acts for a short time Δt on a rotating rigid body with rotational inertia I. Show that

$$\int \tau \, dt = F_{avg}R \, \Delta t = I(\omega_f - \omega_i),$$

where τ is the torque due to the force, R is the moment arm of the force, F_{avg} is the average value of the force during the time it acts on the body, and ω_i and ω_f are the angular velocities of the body just before and just after the force acts. (The quantity $\int \tau \, dt = F_{avg}R \, \Delta t$ is called the *angular impulse,* in analogy with $F_{avg} \, \Delta t$, the linear impulse.)

37P*. Two cylinders having radii R_1 and R_2 and rotational inertias I_1 and I_2 about their central axes are supported by axles perpendicular to the plane of Fig. 12-38. The large cylinder is initially rotating clockwise with angular velocity ω_0. The small cylinder is moved to the right until it touches the large cylinder and is caused to rotate by the frictional force between the two. Eventually, slipping ceases, and the two cylinders rotate at constant rates in opposite directions. Find the final angular velocity ω_2 of the small cylinder in terms of I_1, I_2, R_1, R_2, and ω_0. (*Hint:* Neither angular momentum nor kinetic energy is conserved. Apply the angular impulse equation of Problem 36.)   ssm   www

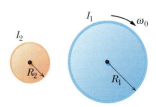

Fig. 12-38 Problem 37.

38P. Figure 12-39 shows a rigid structure consisting of a circular hoop of radius R and mass m, and a square made of four thin bars, each of length R and mass m. The rigid structure rotates at a constant speed about a vertical axis, with a period of rotation of 2.5 s. Assuming $R = 0.50$ m and $m = 2.0$ kg, calculate (a) the structure's rotational inertia about the axis of rotation and (b) its angular momentum about that axis.

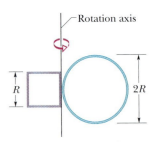

Fig. 12-39 Problem 38.

SEC. 12-10   Conservation of Angular Momentum

39E. A man stands on a platform that is rotating (without friction) with an angular speed of 1.2 rev/s; his arms are outstretched and he holds a brick in each hand. The rotational inertia of the system consisting of the man, bricks, and platform about the central axis is 6.0 kg·m^2. If by moving the bricks the man decreases the rotational inertia of the system to 2.0 kg·m^2, (a) what is the resulting angular speed of the platform and (b) what is the ratio of the new kinetic energy of the system to the original kinetic energy? (c) What provided the added kinetic energy?   ssm

40E. The rotor of an electric motor has rotational inertia $I_m = 2.0 \times 10^{-3}$ kg·m^2 about its central axis. The motor is used to change the orientation of the space probe in which it is mounted.

The motor axis is mounted parallel to the axis of the probe, which has rotational inertia $I_p = 12$ kg·m² about its axis. Calculate the number of revolutions of the rotor required to turn the probe through 30° about its axis.

41E. A wheel is rotating freely at angular speed 800 rev/min on a shaft whose rotational inertia is negligible. A second wheel, initially at rest and with twice the rotational inertia of the first, is suddenly coupled to the same shaft. (a) What is the angular speed of the resultant combination of the shaft and two wheels? (b) What fraction of the original rotational kinetic energy is lost? ssm ilw

42E. Two disks are mounted on low-friction bearings on the same axle and can be brought together so that they couple and rotate as one unit. (a) The first disk, with rotational inertia 3.3 kg·m² about its central axis, is set spinning at 450 rev/min. The second disk, with rotational inertia 6.6 kg·m² about its central axis, is set spinning at 900 rev/min in the same direction as the first. They then couple together. What is their angular speed after coupling? (b) If instead the second disk is set spinning at 900 rev/min in the direction opposite the first disk's rotation, what is their angular speed and direction of rotation after coupling?

43E. In a playground, there is a small merry-go-round of radius 1.20 m and mass 180 kg. Its radius of gyration (see Problem 43 of Chapter 11) is 91.0 cm. A child of mass 44.0 kg runs at a speed of 3.00 m/s along a path that is tangent to the rim of the initially stationary merry-go-round and then jumps on. Neglect friction between the bearings and the shaft of the merry-go-round. Calculate (a) the rotational inertia of the merry-go-round about its axis of rotation, (b) the magnitude of the angular momentum of the running child about the axis of rotation of the merry-go-round, and (c) the angular speed of the merry-go-round and child after the child has jumped on. ssm

44E. The rotational inertia of a collapsing spinning star changes to $\frac{1}{3}$ its initial value. What is the ratio of the new rotational kinetic energy to the initial rotational kinetic energy?

45P. A track is mounted on a large wheel that is free to turn with negligible friction about a vertical axis (Fig. 12-40). A toy train of mass m is placed on the track and, with the system initially at rest, the electrical power is turned on. The train reaches a steady speed v with respect to the track. What is the angular speed of the wheel if its mass is M and its radius is R? (Treat the wheel as a hoop, and neglect the mass of the spokes and hub.) ssm www

Fig. 12-40 Problem 45.

46P. In Fig. 12-41, two skaters, each of mass 50 kg, approach each other along parallel paths separated by 3.0 m. They have opposite velocities of 1.4 m/s each. One skater carries one end of a long pole with negligible mass, and the other skater grabs the other end of it as she passes. Assume frictionless ice. (a) Describe quantitatively the motion of the skaters after they have become connected by the pole. (b) What is the kinetic energy of the two-skater system?

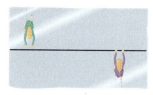

Fig. 12-41 Problem 46.

Next, the skaters each pull along the pole so as to reduce their separation to 1.0 m. What then are (c) their angular speed and (d) the kinetic energy of the system? (e) Explain the source of the increased kinetic energy.

47P. A cockroach of mass m runs counterclockwise around the rim of a lazy Susan (a circular dish mounted on a vertical axle) of radius R and rotational inertia I and having frictionless bearings. The cockroach's speed (relative to the ground) is v, whereas the lazy Susan turns clockwise with angular speed ω_0. The cockroach finds a bread crumb on the rim and, of course, stops. (a) What is the angular speed of the lazy Susan after the cockroach stops? (b) Is mechanical energy conserved as they stop?

48P. A girl of mass M stands on the rim of a frictionless merry-go-round of radius R and rotational inertia I that is not moving. She throws a rock of mass m horizontally in a direction that is tangent to the outer edge of the merry-go-round. The speed of the rock, relative to the ground, is v. Afterward, what are (a) the angular speed of the merry-go-round and (b) the linear speed of the girl?

49P. A horizontal vinyl record of mass 0.10 kg and radius 0.10 m rotates freely about a vertical axis through its center with an angular speed of 4.7 rad/s. The rotational inertia of the record about its axis of rotation is 5.0×10^{-4} kg·m². A wad of wet putty of mass 0.020 kg drops vertically onto the record from above and sticks to the edge of the record. What is the angular speed of the record immediately after the putty sticks to it?

50P. A uniform thin rod of length 0.50 m and mass 4.0 kg can rotate in a horizontal plane about a vertical axis through its center. The rod is at rest when a 3.0 g bullet traveling in the horizontal plane of the rod is fired into one end of the rod. As viewed from above, the direction of the bullet's velocity makes an angle of 60° with

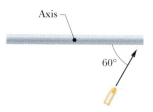

Fig. 12-42 Problem 50.

the rod (Fig. 12-42). If the bullet lodges in the rod and the angular velocity of the rod is 10 rad/s immediately after the collision, what is the bullet's speed just before impact?

51P*. Two 2.00 kg balls are attached to the ends of a thin rod of negligible mass, 50.0 cm long. The rod is free to rotate in a vertical plane without friction about a horizontal axis through its center. With the rod initially horizontal (Fig. 12-43), a 50.0 g wad of wet putty drops onto one of the balls, hitting it with a speed of 3.00 m/s and then sticking to it. (a) What is the angular speed of the system just after the putty wad hits? (b) What is the ratio of the kinetic

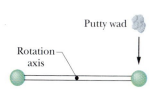

Fig. 12-43 Problem 51.

energy of the entire system after the collision to that of the putty wad just before? (c) Through what angle will the system rotate until it momentarily stops? ssm

52P. A cockroach of mass m lies on the rim of a uniform disk of mass $10.0m$ that can rotate freely about its center like a merry-go-round. Initially the cockroach and disk rotate together with an angular velocity of ω_0. Then the cockroach walks halfway to the center of the disk. (a) What is the change $\Delta\omega$ in the angular velocity of the cockroach–disk system? (b) What is the ratio K/K_0 of the new kinetic energy of the system to its initial kinetic energy? (c) What accounts for the change in the kinetic energy?

53P. If Earth's polar ice caps fully melted and the water returned to the oceans, the oceans would be deeper by about 30 m. What effect would this have on Earth's rotation? Make an estimate of the resulting change in the length of the day. (Concern has been expressed that warming of the atmosphere resulting from industrial pollution could cause the ice caps to melt.)

54P. A horizontal platform in the shape of a circular disk rotates on a frictionless bearing about a vertical axle through the center of the disk. The platform has a mass of 150 kg, a radius of 2.0 m, and a rotational inertia of $300 \text{ kg} \cdot \text{m}^2$ about the axis of rotation. A 60 kg student walks slowly from the rim of the platform toward the center. If the angular speed of the system is 1.5 rad/s when the student starts at the rim, what is the angular speed when she is 0.50 m from the center?

55P. A uniform disk of mass $10m$ and radius $3.0r$ can rotate freely about its fixed center like a merry-go-round. A smaller uniform disk of mass m and radius r lies on top of the larger disk, concentric with it. Initially the two disks rotate together with an angular velocity of 20 rad/s. Then a slight disturbance causes the smaller disk to slide outward across the larger disk, until the outer edge of the smaller disk catches on the outer edge of the larger disk. Afterward, the two disks again rotate together (without further sliding). (a) What then is their angular velocity about the center of the larger disk? (b) What is the ratio K/K_0 of the new kinetic energy of the two-disk system to the system's initial kinetic energy?

56P. A 30 kg child stands on the edge of a stationary merry-go-round of mass 100 kg and radius 2.0 m. The rotational inertia of the merry-go-round about its axis

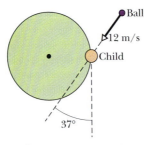

Fig. 12-44 Problem 56.

of rotation is $150 \text{ kg} \cdot \text{m}^2$. The child catches a ball of mass 1.0 kg thrown by a friend. Just before the ball is caught, it has a horizontal velocity of 12 m/s that makes an angle of 37° with a line tangent to the outer edge of the merry-go-round, as shown in the overhead view of Fig. 12-44. What is the angular speed of the merry-go-round just after the ball is caught?

57P. In Fig. 12-45, a 1.0 g bullet is fired into a 0.50 kg block that is mounted on the end of a 0.60 m nonuniform rod of mass 0.50 kg. The block–rod–bullet system then rotates about a fixed axis at point A. The rotational inertia of the rod alone about A is $0.060 \text{ kg} \cdot \text{m}^2$. Assume the block is small enough to treat as a particle on the end of the rod. (a) What is the rotational inertia of the block–rod–bullet system about point A? (b) If the angular speed of the system about A just after the bullet's impact is 4.5 rad/s, what is the speed of the bullet just before the impact?

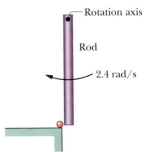

Fig. 12-45 Problem 57.

58P. In Fig. 12-46, a uniform rod (length $= 0.60$ m, mass $= 1.0$ kg) rotates about an axis through one end, with a rotational inertia of $0.12 \text{ kg} \cdot \text{m}^2$. As the rod swings through its lowest position, the end of the rod collides with a small 0.20 kg putty wad that sticks to the end of the rod. If the angular speed of the rod just before the collision is 2.4 rad/s, what is the angular speed of the rod–putty system immediately after the collision?

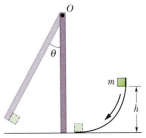

Fig. 12-46 Problem 58.

59P*. The particle of mass m in Fig. 12-47 slides down the frictionless surface through height h and collides with the uniform vertical rod (of mass M and length d), sticking to it. The rod pivots about point O through the angle θ before momentarily stopping. Find θ.

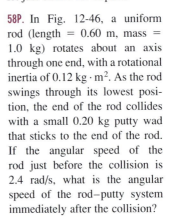

Fig. 12-44 Problem 56.

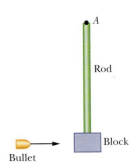

Fig. 12-47 Problem 59.

13 Equilibrium and Elasticity

Rock climbing may be the ultimate physics exam. Failure can mean death, and even "partial credit" can mean severe injury. For example, in a long chimney climb, in which your torso is pressed against one wall of a wide vertical fissure and your feet are pressed against the opposite wall, you need to rest occasionally or you will fall due to exhaustion. Here the exam consists of a single question: What can you do to relax your push on the walls in order to rest? If you relax without considering the physics, the walls will not hold you up.

What is the answer to this life-and-death, one-question exam?

The answer is in this chapter.

Fig. 13-1 A balancing rock near Petrified Forest National Park in Arizona. Although its perch seems precarious, the rock is in static equilibrium.

13-1 Equilibrium

Consider these objects: (1) a book resting on a table, (2) a hockey puck sliding across a frictionless surface with constant velocity, (3) the rotating blades of a ceiling fan, and (4) the wheel of a bicycle that is traveling along a straight path at constant speed. For each of these four objects:

1. The linear momentum $\vec{P}$ of its center of mass is constant.

2. Its angular momentum $\vec{L}$ about its center of mass, or about any other point, is also constant.

We say that such objects are in **equilibrium.** The two requirements for equilibrium are then

$$\vec{P} = \text{a constant} \quad \text{and} \quad \vec{L} = \text{a constant}. \tag{13-1}$$

Our concern in this chapter is with situations in which the constants in Eq. 13-1 are in fact zero; that is, we are concerned largely with objects that are not moving in any way—either in translation or in rotation—in the reference frame from which we observe them. Such objects are in **static equilibrium.** Of the four objects mentioned at the beginning of this section, only one—the book resting on the table—is in static equilibrium.

The balancing rock of Fig. 13-1 is another example of an object that, for the present at least, is in static equilibrium. It shares this property with countless other structures, such as cathedrals, houses, filing cabinets, and taco stands, that remain stationary over time.

As we discussed in Section 8-5, if a body returns to a state of static equilibrium after having been displaced from it by a force, the body is said to be in *stable* static equilibrium. A marble placed at the bottom of a hemispherical bowl is an example. However, if a small force can displace the body and end the equilibrium, the body is in *unstable* static equilibrium.

For example, suppose we balance a domino with the domino's center of mass vertically above the supporting edge, as in Fig. 13-2a. The torque about the supporting edge due to the gravitational force $\vec{F_g}$ on the domino is zero, because the line of action of $\vec{F_g}$ is through that edge. Thus, the domino is in equilibrium. Of course, even a slight force on it due to some chance disturbance ends the equilibrium. As the line of action of $\vec{F_g}$ moves to one side of the supporting edge (as in Fig. 13-2b), the torque due to $\vec{F_g}$ increases the rotation of the domino. Therefore, the domino in Fig. 13-2a is in unstable static equilibrium.

The domino in Fig. 13-2c is not quite as unstable. To topple this domino, a force would have to rotate it through and then beyond the balance position of Fig. 13-2a, in which the center of mass is above a supporting edge. A slight force will not topple this domino, but a vigorous flick of the finger against the domino certainly will. (If we arrange a chain of such upright dominos, a finger flick against the first can cause the whole chain to fall.)

Fig. 13-2 (a) A domino balanced on one edge, with its center of mass vertically above that edge. The gravitational force $\vec{F_g}$ on the domino is directed through the supporting edge. (b) If the domino is rotated even slightly from the balanced orientation, then $\vec{F_g}$ causes a torque that increases the rotation. (c) A domino upright on a narrow side is somewhat more stable than the domino in (a). (d) A square block is even more stable.

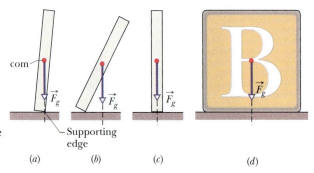

Fig. 13-3 A construction worker balanced above New York City is in static equilibrium but is more stable parallel to the beam than perpendicular to it.

The child's square block in Fig. 13-2d is even more stable because its center of mass would have to be moved even farther to get it to pass above a supporting edge. A flick of the finger may not topple the block. (This is why you never see a chain of toppling square blocks.) The worker in Fig. 13-3 is like both the domino and the square block: Parallel to the beam, his stance is wide and he is stable; perpendicular to the beam, his stance is narrow and he is unstable (and at the mercy of a chance gust of wind).

The analysis of static equilibrium is very important in engineering practice. The design engineer must isolate and identify all the external forces and torques that may act on a structure and, by good design and wise choice of materials, ensure that the structure will remain stable under these loads. Such analysis is necessary to ensure, for example, that bridges do not collapse under their traffic and wind loads, and that the landing gear of aircraft will survive the shock of rough landings.

13-2 The Requirements of Equilibrium

The translational motion of a body is governed by Newton's second law in its linear momentum form, given by Eq. 9-27 as

$$\vec{F}_{net} = \frac{d\vec{P}}{dt}.$$ (13-2)

If the body is in translational equilibrium—that is, if $\vec{P}$ is a constant—then $d\vec{P}/dt = 0$ and we must have

$$\vec{F}_{net} = 0 \qquad \text{(balance of forces)}.$$ (13-3)

The rotational motion of a body is governed by Newton's second law in its angular momentum form, given by Eq. 12-29 as

$$\vec{\tau}_{net} = \frac{d\vec{L}}{dt}.$$ (13-4)

If the body is in rotational equilibrium—that is, if $\vec{L}$ is a constant—then $d\vec{L}/dt = 0$ and we must have

$$\vec{\tau}_{net} = 0 \qquad \text{(balance of torques)}.$$ (13-5)

Thus, the two requirements for a body to be in equilibrium are as follows:

> **1.** The vector sum of all the external forces that act on the body must be zero.
>
> **2.** The vector sum of all the external torques that act on the body, measured about *any* possible point, must also be zero.

These requirements obviously hold for *static* equilibrium. They also hold for the more general equilibrium in which $\vec{P}$ and $\vec{L}$ are constant but not zero.

Equations 13-3 and 13-5, as vector equations, are each equivalent to three independent component equations, one for each direction of the coordinate axes:

Balance of forces	Balance of torques	
$F_{net,x} = 0$	$\tau_{net,x} = 0$	
$F_{net,y} = 0$	$\tau_{net,y} = 0$	(13-6)
$F_{net,z} = 0$	$\tau_{net,z} = 0$	

We shall simplify matters by considering only situations in which the forces that act on the body lie in the xy plane. This means that the only torques that can act on the body must tend to cause rotation around an axis parallel to the z axis. With this assumption, we eliminate one force equation and two torque equations from Eqs. 13-6, leaving

$$F_{net,x} = 0 \qquad \text{(balance of forces)}, \qquad (13\text{-}7)$$

$$F_{net,y} = 0 \qquad \text{(balance of forces)}, \qquad (13\text{-}8)$$

$$\tau_{net,z} = 0 \qquad \text{(balance of torques)}. \qquad (13\text{-}9)$$

Here, $\tau_{net,z}$ is the net torque that the external forces produce either about the z axis or about *any* axis parallel to it.

A hockey puck sliding at constant velocity over ice satisfies Eqs. 13-7, 13-8, and 13-9 and is thus in equilibrium *but not in static equilibrium*. For static equilibrium, the linear momentum $\vec{P}$ of the puck must be not only constant but also zero; the puck must be at rest on the ice. Thus, there is another requirement for static equilibrium:

> **3.** The linear momentum $\vec{P}$ of the body must be zero.

✔**CHECKPOINT 1:**   The figure gives six overhead views of a uniform rod on which two or more forces act perpendicularly to the rod. If the magnitudes of the forces are adjusted properly (but kept nonzero), in which situations can the rod be in static equilibrium?

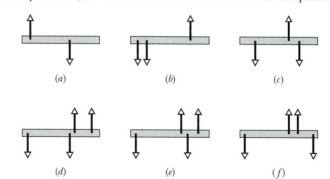

13-3   The Center of Gravity

The gravitational force on an extended body is the vector sum of the gravitational forces acting on the individual elements (the atoms) of the body. Instead of considering all those individual elements, we can say:

> The gravitational force $\vec{F}_g$ on a body effectively acts at a single point, called the **center of gravity** (cog) of the body.

Here the word "effectively" means that if the forces on the individual elements were somehow turned off and force $\vec{F}_g$ at the center of gravity were turned on, the net force and the net torque (about any point) acting on the body would not change.

Until now, we have assumed that the gravitational force $\vec{F}_g$ acts at the center of mass (com) of the body. This is equivalent to assuming that the center of gravity is

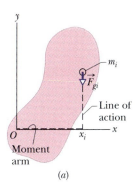

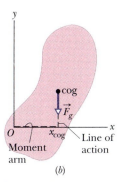

Fig. 13-4 (a) An element of mass m_i in an extended body. The gravitational force $\vec{F}_{gi}$ on it has moment arm x_i about the origin O of the coordinate system. (b) The gravitational force $\vec{F}_g$ on a body is said to act at the center of gravity (cog) of the body. Here it has moment arm x_{cog} about origin O.

at the center of mass. Recall that, for a body of mass M, the force $\vec{F}_g$ is equal to $M\vec{g}$, where $\vec{g}$ is the acceleration that the force would produce if the body were to fall freely. In the proof that follows, we show that

> If $\vec{g}$ is the same for all elements of a body, then the body's center of gravity (cog) is coincident with the body's center of mass (com).

This is approximately true for everyday objects because $\vec{g}$ varies only a little along Earth's surface and decreases in magnitude only slightly with altitude. Thus, for objects like a mouse or a moose, we have been justified in assuming that the gravitational force acts at the center of mass. After the following proof, we shall resume that assumption.

Proof

First, we consider the individual elements of the body. Figure 13-4a shows an extended body, of mass M, and one of its elements, of mass m_i. A gravitational force $\vec{F}_{gi}$ acts on each such element and is equal to $m_i\vec{g}_i$. The subscript on $\vec{g}_i$ means $\vec{g}_i$ is the gravitational acceleration *at the location of the element* (it can be different for other elements).

In Fig. 13-4a, each force $\vec{F}_{gi}$ produces a torque τ_i on the element about the origin O, with moment arm x_i. Using Eq. 11-33 ($\tau = r_\perp F$), we can write torque τ_i as

$$\tau_i = x_i F_{gi}. \tag{13-10}$$

The net torque on all the elements of the body is then

$$\tau_{net} = \sum \tau_i = \sum x_i F_{gi}. \tag{13-11}$$

Next, we consider the body as a whole. Figure 13-4b shows the gravitational force $\vec{F}_g$ acting at the body's center of gravity. This force produces a torque τ on the body about O, with moment arm x_{cog}. Again using Eq. 11-33, we can write this torque as

$$\tau = x_{cog} F_g. \tag{13-12}$$

The gravitational force $\vec{F}_g$ on the body is equal to the sum of the gravitational forces $\vec{F}_{gi}$ on all its elements, so we can substitute $\sum F_{gi}$ for F_g in Eq. 13-12 to write

$$\tau = x_{cog} \sum F_{gi}. \tag{13-13}$$

Now recall that the torque due to force $\vec{F}_g$ acting at the center of gravity is equal to the net torque due to all the forces $\vec{F}_{gi}$ acting on all the elements of the body. (That is how we defined the cog.) Therefore, τ in Eq. 13-13 is equal to τ_{net} in Eq. 13-11. Putting those two equations together, we can write

$$x_{cog} \sum F_{gi} = \sum x_i F_{gi}.$$

Substituting $m_i g_i$ for F_{gi} gives us

$$x_{cog} \sum m_i g_i = \sum x_i m_i g_i.$$

Now here is a key idea: If the accelerations g_i at all the locations of the elements are the same, we can cancel g_i from this equation to write

$$x_{cog} \sum m_i = \sum x_i m_i. \tag{13-14}$$

The sum $\sum m_i$ of the masses of all the elements is the mass M of the body. Therefore, we can rewrite Eq. 13-14 as

$$x_{cog} = \frac{1}{M} \sum x_i m_i. \tag{13-15}$$

The right side of this equation gives the coordinate x_{com} of the body's center of mass (Eq. 9-4). We now have what we sought to prove:

$$x_{cog} = x_{com}. \qquad (13\text{-}16)$$

✔ CHECKPOINT 2:   Suppose that you skewer an apple with a thin rod, missing the apple's center of gravity. When you hold the rod horizontally and allow the apple to rotate freely, where does the center of gravity end up and why?

13-4 Some Examples of Static Equilibrium

In this section we examine four sample problems involving static equilibrium. In each, we select a system of one or more objects to which we apply the equations of equilibrium (Eqs. 13-7, 13-8, and 13-9). The forces involved in the equilibrium are all in the xy plane, which means that the torques involved are parallel to the z axis. Thus, in applying Eq. 13-9, the balance of torques, we select an axis parallel to the z axis about which to calculate the torques. Although Eq. 13-9 is satisfied for *any* such choice of axis, you will see that certain choices simplify the application of Eq. 13-9 by eliminating one or more unknown force terms.

Sample Problem 13-1

In Fig. 13-5a, a uniform beam, of length L and mass $m = 1.8$ kg, is at rest with its ends on two scales. A uniform block, with mass $M = 2.7$ kg, is at rest on the beam, with its center a distance $L/4$ from the beam's left end. What do the scales read?

SOLUTION: The first steps in the solution of *any* problem about static equilibrium are these: Clearly define the system to be analyzed and then draw a free-body diagram of it, indicating all the forces on the system. Here, let us choose the system as the beam and block taken together. Then the forces on the system are shown in the free-body diagram of Fig. 13-5b. (Choosing the system takes experience and often there can be more than one good choice; see Problem Solving Tactic 1 below.)

The normal forces on the beam from the scales are $\vec{F}_l$ on the left and $\vec{F}_r$ on the right. The scale readings that we want are equal to the magnitudes of those forces. The gravitational force $\vec{F}_{g,beam}$ on the beam acts at the beam's center of mass and is equal to $m\vec{g}$. Similarly, the gravitational force $\vec{F}_{g,block}$ on the block acts at the block's center of mass and is equal to $M\vec{g}$. However, to simplify Fig. 13-5b, the block is represented by a dot within the boundary of the beam and the vector $\vec{F}_{g,block}$ is drawn with its tail on that dot. (This downward shift of vector $\vec{F}_{g,block}$ along its line of action does not alter the torque due to $\vec{F}_{g,block}$ about any axis perpendicular to the figure.)

The Key Idea here is that, because the system is in static equilibrium, we can apply the balance of forces equations (Eqs. 13-7 and 13-8) and the balance of torques equation (13-9) to it. The forces have no x components, so Eq. 13-7 ($F_{net,x} = 0$) provides no information. For the y components, Eq. 13-8 ($F_{net,y} = 0$) gives us

$$F_l + F_r - Mg - mg = 0. \qquad (13\text{-}17)$$

This equation contains two unknowns, the forces F_l and F_r, so we also need to use Eq. 13-9, the balance of torques equation.

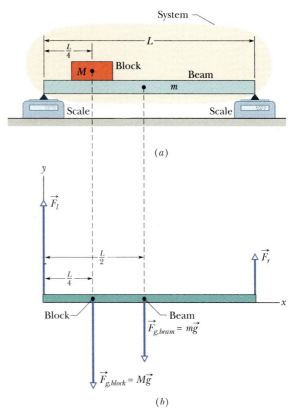

(a)

(b)

Fig. 13-5 Sample Problem 13-1. (a) A beam of mass m supports a block of mass M. (b) A free-body diagram, showing the forces that act on the system *beam + block*.

We can apply it to *any* rotation axis perpendicular to the plane of Fig. 13-5. Let us choose a rotation axis through the left end of the beam. We shall also use our general rule for assigning signs to torques: If a torque would cause an initially stationary body to rotate clockwise about the rotation axis, the torque is negative. If the rotation would be counterclockwise, the torque is positive. Finally, we shall write the torques in the form $r_\perp F$, where the moment arm $r_\perp$ is 0 for $\vec{F_l}$, $L/4$ for $M\vec{g}$, $L/2$ for $m\vec{g}$, and L for $\vec{F_r}$.

We now can write the balancing equation ($\tau_{\text{net},z} = 0$) as

$$(0)(F_l) - (L/4)(Mg) - (L/2)(mg) + (L)(F_r) = 0,$$

which gives us

$$F_r = \tfrac{1}{4}Mg + \tfrac{1}{2}mg$$
$$= \tfrac{1}{4}(2.7 \text{ kg})(9.8 \text{ m/s}^2) + \tfrac{1}{2}(1.8 \text{ kg})(9.8 \text{ m/s}^2)$$
$$= 15.44 \text{ N} \approx 15 \text{ N.} \qquad \text{(Answer)}$$

Now, solving Eq. 13-17 for F_l and substituting this result, we find

$$F_l = (M + m)g - F_r$$
$$= (2.7 \text{ kg} + 1.8 \text{ kg})(9.8 \text{ m/s}^2) - 15.44 \text{ N}$$
$$= 28.66 \text{ N} \approx 29 \text{ N.} \qquad \text{(Answer)}$$

Notice the strategy in the solution: When we wrote an equation for the balance of force components, we got stuck with two unknowns. If we had written an equation for the balance of torques around some *arbitrary* axis, we would have again gotten stuck with those two unknowns. However, because we chose the axis to pass through the point of application of one of the unknown forces, here $\vec{F_l}$, we did not get stuck. Our choice neatly eliminated that force from the torque equation, allowing us to solve for the other unknown force magnitude F_r. Then we returned to the equation for the balance of force components to find the remaining unknown force magnitude.

✔**CHECKPOINT 3:** The figure gives an overhead view of a uniform rod in static equilibrium. (a) Can you find the magnitudes of unknown forces $\vec{F_1}$ and $\vec{F_2}$ by balancing the forces? (b) If you wish to find the magnitude of force $\vec{F_2}$ by using a single equation, where should you place a rotational axis? (c) The magnitude of $\vec{F_2}$ turns out to be 65 N. What then is the magnitude of $\vec{F_1}$?

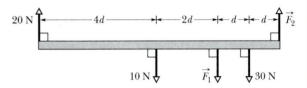

Sample Problem 13-2

In Fig. 13-6a, a ladder of length $L = 12$ m and mass $m = 45$ kg leans against a slick (frictionless) wall. Its upper end is at height $h = 9.3$ m above the pavement on which the lower end rests (the pavement is not frictionless). The ladder's center of mass is $L/3$ from the lower end. A firefighter of mass $M = 72$ kg climbs the ladder until her center of mass is $L/2$ from the lower end. What then are the magnitudes of the forces on the ladder from the wall and the pavement?

SOLUTION: First, we choose our system as being the firefighter and ladder, together, and then we draw the free-body diagram of Fig. 13-6b. The firefighter is represented with a dot within the boundary of the ladder. The gravitational force on her is represented with its equivalent $M\vec{g}$, and that vector has been shifted along its line of action, so that its tail is on the dot. (The shift does not alter a torque due to $M\vec{g}$ about any axis perpendicular to the figure.)

The only force on the ladder from the wall is the horizontal force $\vec{F_w}$ (there cannot be a frictional force along a frictionless wall). The force $\vec{F_p}$ on the ladder from the pavement has a horizontal component $\vec{F_{px}}$ that is a static frictional force and a vertical component $\vec{F_{py}}$ that is a normal force.

A **Key Idea** here is that the system is in static equilibrium, so

the balancing equations (Eqs. 13-7 through 13-9) apply to it. Let us start with Eq. 13-9 ($\tau_{\text{net},z} = 0$). To choose an axis about which to calculate the torques, note that we have unknown forces ($\vec{F_w}$ and $\vec{F_p}$) at the two ends of the ladder. To eliminate, say, $\vec{F_p}$ from the calculation, we place the axis at point O, perpendicular to the figure. We also place the origin of an xy coordinate system at O. We can find torques about O with any of Eqs. 11-31 through 11-33, but Eq. 11-33 ($\tau = r_\perp F$) is easiest to use here.

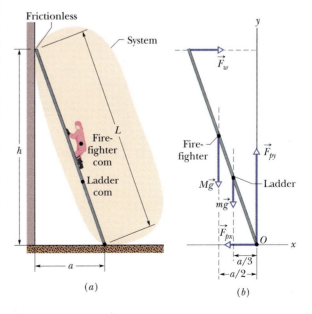

Fig. 13-6 Sample Problem 13-2 (a) A firefighter climbs halfway up a ladder that is leaning against a frictionless wall. The pavement beneath the ladder is not frictionless. (b) A free-body diagram, showing the forces that act on the firefighter–ladder system. The origin O of a coordinate system is placed at the point of application of the unknown force $\vec{F_p}$ (whose vector components $\vec{F_{px}}$ and $\vec{F_{py}}$ are shown).

To find the moment arm $r_\perp$ of $\vec{F}_w$, we draw a line of action through that vector (Fig. 13-6b). Then $r_\perp$ is the perpendicular distance between O and the line of action. In Fig. 13-6b, it extends along the y axis and is equal to the height h. We similarly draw lines of action for $M\vec{g}$ and $m\vec{g}$ and see that their moment arms extend along the x axis. For the distance a shown in Fig. 13-6a, the moment arms are $a/2$ (the firefighter is halfway up the ladder) and $a/3$ (the ladder's center of mass is one-third of the way up the ladder), respectively. The moment arms for $\vec{F}_{px}$ and $\vec{F}_{py}$ are zero.

Now, with torques written in the form $r_\perp F$, the balancing equation $\tau_{\mathrm{net},z} = 0$ becomes

$$-(h)(F_w) + (a/2)(Mg) + (a/3)(mg)$$
$$+ (0)(F_{px}) + (0)(F_{py}) = 0. \quad (13\text{-}18)$$

(Recall our rule: A positive torque corresponds to counterclockwise rotation and a negative torque corresponds to clockwise rotation.)

Using the Pythagorean theorem, we find that

$$a = \sqrt{L^2 - h^2} = 7.58 \text{ m.}$$

Then Eq. 13-18 gives us

$$F_w = \frac{ga(M/2 + m/3)}{h}$$
$$= \frac{(9.8 \text{ m/s}^2)(7.58 \text{ m})(72/2 \text{ kg} + 45/3 \text{ kg})}{9.3 \text{ m}}$$
$$= 407 \text{ N} \approx 410 \text{ N.} \qquad \text{(Answer)}$$

Now we need to use the force balancing equations. The equation $F_{\mathrm{net},x} = 0$ gives us

$$F_w - F_{px} = 0, \qquad (13\text{-}19)$$

so

$$F_{px} = F_w = 410 \text{ N.} \qquad \text{(Answer)}$$

The equation $F_{\mathrm{net},y} = 0$ gives us

$$F_{py} - Mg - mg = 0, \qquad (13\text{-}20)$$

so

$$F_{py} = (M + m)g = (72 \text{ kg} + 45 \text{ kg})(9.8 \text{ m/s}^2)$$
$$= 1146.6 \text{ N} \approx 1100 \text{ N.} \qquad \text{(Answer)}$$

Sample Problem 13-3

Figure 13-7a shows a safe, of mass $M = 430$ kg, hanging by a rope from a boom with dimensions $a = 1.9$ m and $b = 2.5$ m. The boom consists of a hinged beam and a horizontal cable that connects the beam to a wall. The uniform beam has a mass m of 85 kg; the mass of the cable and rope are negligible.

(a) What is the tension T_c in the cable? In other words, what is the magnitude of the force $\vec{T}_c$ on the beam from the cable?

SOLUTION: The system here is the beam alone, and the forces on it are shown in the free-body diagram of Fig. 13-7b. The force from the cable is $\vec{T}_c$. The gravitational force on the beam acts at the beam's center of mass (at the beam's center) and is represented by its equivalent $m\vec{g}$. The vertical component of the force on the beam from the hinge is $\vec{F}_v$, and the horizontal component of the force from the hinge is $\vec{F}_h$. The force from the rope supporting the safe

is $\vec{T}_r$. Because beam, rope, and safe are stationary, the magnitude of $\vec{T}_r$ is equal to the weight of the safe: $T_r = Mg$. We place the origin O of an xy coordinate system at the hinge.

One Key Idea here is that our system is in static equilibrium, so the balancing equations apply to it. Let us start with Eq. 13-9 ($\tau_{\mathrm{net},z} = 0$). Note that we are asked for the magnitude of force $\vec{T}_c$ and not of forces $\vec{F}_h$ and $\vec{F}_v$ acting at the hinge, at point O. Thus, a second Key Idea is that, to eliminate $\vec{F}_h$ and $\vec{F}_v$ from the torque calculation, we should calculate torques about an axis that is perpendicular to the figure at point O. Then $\vec{F}_h$ and $\vec{F}_v$ will have moment arms of zero. The lines of action for $\vec{T}_c$, $\vec{T}_r$, and $m\vec{g}$ are dashed in Fig. 13-7b. The corresponding moment arms are a, b, and $b/2$.

Writing torques in the form of $r_\perp F$ and using our rule about signs for torques, the balancing equation $\tau_{\mathrm{net},z} = 0$ becomes

$$(a)(T_c) - (b)(T_r) - (\tfrac{1}{2}b)(mg) = 0.$$

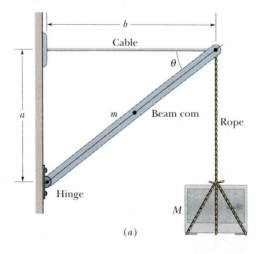

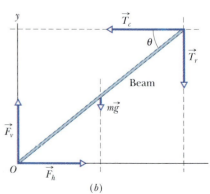

Fig. 13-7 Sample Problem 13-3. (a) A heavy safe is hung from a boom consisting of a horizontal steel cable and a uniform beam. (b) A free-body diagram for the beam.

(a)

(b)

Substituting Mg for T_r and solving for T_c, we find that

$$T_c = \frac{gb(M + \frac{1}{2}m)}{a}$$

$$= \frac{(9.8 \text{ m/s}^2)(2.5 \text{ m})(430 \text{ kg} + 85/2 \text{ kg})}{1.9 \text{ m}}$$

$$= 6093 \text{ N} \approx 6100 \text{ N}. \qquad \text{(Answer)}$$

(b) Find the magnitude F of the net force on the beam from the hinge.

SOLUTION: Now we want F_h and F_v so we can combine them to get F. Because we know T_c, our Key Idea here is to apply the force balancing equations to the beam. For the horizontal balance, we write $F_{\text{net},x} = 0$ as

$$F_h - T_c = 0,$$

and so

$$F_h = T_c = 6093 \text{ N}.$$

For the vertical balance, we write $F_{\text{net},y} = 0$ as

$$F_v - mg - T_r = 0.$$

Substituting Mg for T_r and solving for F_v, we find that

$$F_v = (m + M)g = (85 \text{ kg} + 430 \text{ kg})(9.8 \text{ m/s}^2)$$

$$= 5047 \text{ N}.$$

From the Pythagorean theorem, we now have

$$F = \sqrt{F_h^2 + F_v^2}$$

$$= \sqrt{(6093 \text{ N})^2 + (5047 \text{ N})^2} \approx 7900 \text{ N}. \qquad \text{(Answer)}$$

Note that F is substantially greater than either the combined weights of the safe and the beam, 5000 N, or the tension in the horizontal wire, 6100 N.

✔CHECKPOINT 4:   In the figure, a stationary 5 kg rod AC is held against a wall by a rope and friction between rod and wall. The uniform rod is 1 m long, and angle $\theta = 30°$. (a) If you are to find the magnitude of the force $\vec{T}$ on the rod from the rope with a single equation, at what labeled point should a rotational axis be placed? With that choice of axis and counterclockwise torques positive, what is the sign of (b) the torque τ_w due to the rod's weight and (c) the torque τ_r due to the pull on the rod by the rope? (d) Is the magnitude of τ_r greater than, less than, or equal to the magnitude of τ_w?

Sample Problem 13-4

In Fig. 13-8, a rock climber with mass $m = 55$ kg rests during a "chimney climb," pressing only with her shoulders and feet against the walls of a fissure of width $w = 1.0$ m. Her center of mass is a horizontal distance $d = 0.20$ m from the wall against which her shoulders are pressed. The coefficient of static friction between her shoes and the wall is $\mu_1 = 1.1$, and between her shoulders and the wall it is $\mu_2 = 0.70$. To rest, the climber wants to minimize her horizontal push on the walls. The minimum occurs when her feet and her shoulders are both on the verge of sliding.

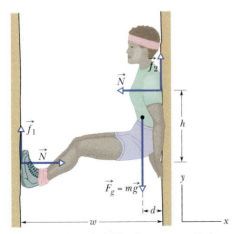

Fig. 13-8 Sample Problem 13-4. The forces on a climber resting in a rock chimney. The push of the climber on the chimney walls results in rise to the normal forces $\vec{N}$ and the static frictional forces $\vec{f}_1$ and $\vec{f}_2$.

(a) What is that minimum horizontal push on the walls?

SOLUTION: Our system is the climber, and Fig. 13-8 shows the forces that act on her. The only horizontal forces are the normal forces $\vec{N}$ on her from the walls, at her feet and shoulders. The static frictional forces on her are $\vec{f}_1$ and $\vec{f}_2$, directed upward. The gravitational force $\vec{F}_g = m\vec{g}$ acts at her center of mass.

A Key Idea is that, because the system is in static equilibrium, we can apply the force balancing equations (Eqs. 13-7 and 13-8) to it. The equation $F_{\text{net},x} = 0$ tells us that the two normal forces on her must be equal in magnitude and opposite in direction. We seek the magnitude N of these two forces, which is also the magnitude of her push against either wall.

The balancing equation $F_{\text{net},y} = 0$ gives us

$$f_1 + f_2 - mg = 0. \qquad (13\text{-}21)$$

We want the climber to be on the verge of sliding at both her feet and her shoulders. That means we want the static frictional forces there to be at their maximum values. Those maximum values are, from Eq. 6-1 ($f_{s,\text{max}} = \mu_s N$),

$$f_1 = \mu_1 N \quad \text{and} \quad f_2 = \mu_2 N. \qquad (13\text{-}22)$$

Substituting these expressions into Eq. 13-21 and solving for N give us

$$N = \frac{mg}{\mu_1 + \mu_2} = \frac{(55 \text{ kg})(9.8 \text{ m/s}^2)}{1.1 + 0.70} = 299 \text{ N} \approx 300 \text{ N}.$$

Thus, her minimum horizontal push must be about 300 N.

(b) For that push, what must be the vertical distance h between her feet and her shoulders if she is to be stable?

SOLUTION: A Key Idea here is that the climber will be stable if the torque balancing equation ($\tau_{net,z} = 0$) applies to her. This means that the forces on her must not produce a net torque about *any* rotation axis. Another Key Idea is that we are free to choose a rotation axis that helps simplify the calculation. We shall write the torques in the form $r_\perp F$, where $r_\perp$ is the moment arm of force F. In Fig. 13-8, we choose a rotation axis at her shoulders, perpendicular to the figure's plane. Then the moment arms of the forces acting there ($\vec{N}$ and $\vec{f}_2$) are zero. Frictional force $\vec{f}_1$, the normal force $\vec{N}$ at her feet, and the gravitational force $\vec{F}_g = m\vec{g}$ have the corresponding moment arms w, h, and d.

Recalling our rule about the signs of torques and the corresponding directions, we can now write $\tau_{net,z} = 0$ as

$$-(w)(f_1) + (h)(N) + (d)(mg) + (0)(f_2) + (0)(N) = 0. \quad (13\text{-}23)$$

(Note how the choice of rotation axis neatly eliminates f_2 from the calculation.) Next, solving Eq. 13-23 for h, setting $f_1 = \mu_1 N$, and substituting $N = 299$ N and other known values, we find that

$$h = \frac{f_1 w - mgd}{N} = \frac{\mu_1 N w - mgd}{N} = \mu_1 w - \frac{mgd}{N}$$

$$= (1.1)(1.0 \text{ m}) - \frac{(55 \text{ kg})(9.8 \text{ m/s}^2)(0.20 \text{ m})}{299 \text{ N}}$$

$$= 0.739 \text{ m} \approx 0.74 \text{ m}. \quad \text{(Answer)}$$

We would find the same required value of h if we wrote the torques about any other rotation axis perpendicular to the page, such as one at her feet.

If h is more than *or* less than 0.74 m, she must exert a force greater than 299 N on the walls to be stable. Here, then, is the advantage of knowing the physics before you climb a chimney. When you need to rest, you will avoid the (dire) error of novice climbers who place their feet too high or too low. Instead, you will know that there is a "best" distance between shoulders and feet, requiring the least push, and giving you a good chance to rest.

PROBLEM-SOLVING TACTICS

Tactic 1: *Static Equilibrium Problems*
Here is a list of steps for solving static equilibrium problems:

1. Draw a *sketch* of the problem.

2. Select the *system* to which you will apply the laws of equilibrium, drawing a closed curve around it on your sketch to fix it clearly in your mind. In some situations you can select a single object as the system; it is the object you wish to be in equilibrium (such as the rock climber in Sample Problem 13-4). In other situations, you might include additional objects in the system *if* their inclusion simplifies the calculations for equilibrium. For example, suppose in Sample Problem 13-2 you select only the ladder as the system. Then in Fig. 13-6*b* you will have to account for additional unknown forces exerted on the ladder by the hands and feet of the firefighter. These additional unknowns complicate the equilibrium calculations. The system of Fig. 13-6 was chosen to include the firefighter so that those unknown forces are *internal* to the system and thus need not be found in order to solve Sample Problem 13-2.

3. Draw a *free-body diagram* of the system. Show all the forces that act on the system, labeling them and making sure that their points of application and lines of action are correctly shown.

4. Draw in the *x and y axes* of a coordinate system. Choose them so that at least one axis is parallel to one or more unknown force. Resolve into components the forces that do not lie along one of the axes. In all our sample problems it made sense to choose the x axis horizontal and the y axis vertical.

5. Write the two *balance of forces equations,* using symbols throughout.

6. Choose one or more rotation axes perpendicular to the plane of the figure and write the *balance of torques equation* for each axis. If you choose an axis that passes through the line of action of an unknown force, the equation will be simplified because that force will not appear in it.

7. *Solve* your equations *algebraically* for the unknowns. Some students feel more confident in substituting numbers with units in the independent equations at this stage, especially if the algebra is particularly involved. However, experienced problem solvers prefer the algebraic approach, which reveals the dependence of solutions on the various variables.

8. Finally, *substitute numbers* with units in your algebraic solutions, obtaining numerical values for the unknowns.

9. Look at your answer—does it make sense? Is it obviously too large or too small? Is the sign correct? Are the units appropriate?

13-5 Indeterminate Structures

For the problems of this chapter, we have only three independent equations at our disposal, usually two balance of forces equations and one balance of torques equation about a given rotation axis. Thus, if a problem has more than three unknowns, we cannot solve it.

It is easy to find such problems. In Sample Problem 13-2, for example, we could have assumed that there is friction between the wall and the top of the ladder. Then

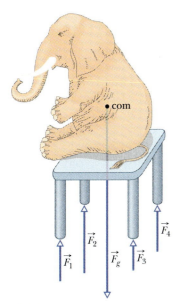

Fig. 13-9 The table is an indeterminate structure. The four forces on the table legs are different in magnitude and cannot be found from the laws of static equilibrium alone.

there would have been a vertical frictional force acting where the ladder touches the wall, making a total of four unknown forces. With only three equations, we could not have solved this problem.

Consider also an unsymmetrically loaded car. What are the forces—all different—on the four tires? Again, we cannot find them because we have only three independent equations with which to work. Similarly, we can solve an equilibrium problem for a table with three legs but not for one with four legs. Problems like these, in which there are more unknowns than equations, are called **indeterminate.**

Yet solutions to indeterminate problems exist in the real world. If you rest the tires of the car on four platform scales, each scale will register a definite reading, the sum of the readings being the weight of the car. What is eluding us in our efforts to find the individual forces by solving equations?

The problem is that we have assumed—without making a great point of it—that the bodies to which we apply the equations of static equilibrium are perfectly rigid. By this we mean that they do not deform when forces are applied to them. Strictly, there are no such bodies. The tires of the car, for example, deform easily under load until the car settles into a position of static equilibrium.

We have all had experience with a wobbly restaurant table, which we usually level by putting folded paper under one of the legs. If a big enough elephant sat on such a table, however, you may be sure that if the table did not collapse, it would deform just like the tires of a car. Its legs would all touch the floor, the forces acting upward on the table legs would all assume definite (and different) values as in Fig. 13-9, and the table would no longer wobble. How do we find the values of those forces acting on the legs?

To solve such indeterminate equilibrium problems, we must supplement equilibrium equations with some knowledge of *elasticity,* the branch of physics and engineering that describes how real bodies deform when forces are applied to them. The next section provides an introduction to this subject.

✔**CHECKPOINT 5:**   A horizontal uniform bar of weight 10 N is to hang from a ceiling by two wires that exert upward forces F_1 and F_2 on the bar. The figure shows four arrangements for the wires. Which arrangements, if any, are indeterminate (so that we cannot solve for numerical values of F_1 and F_2)?

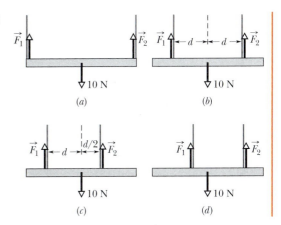

13-6 Elasticity

When a large number of atoms come together to form a metallic solid, such as an iron nail, they settle into equilibrium positions in a three-dimensional *lattice,* a repetitive arrangement in which each atom has a well-defined equilibrium distance from its nearest neighbors. The atoms are held together by interatomic forces that are modeled as tiny springs in Fig. 13-10. The lattice is remarkably rigid, which is another way of saying that the "interatomic springs" are extremely stiff. It is for this

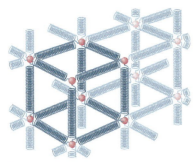

Fig. 13-10 The atoms of a metallic solid are distributed on a repetitive three-dimensional lattice. The springs represent interatomic forces.

reason that we perceive many ordinary objects such as metal ladders, tables, and spoons as perfectly rigid. Of course, some ordinary objects, such as garden hoses or rubber gloves, do not strike us as rigid at all. The atoms that make up these objects *do not* form a rigid lattice like that of Fig. 13-10 but are aligned in long, flexible molecular chains, each chain being only loosely bound to its neighbors.

All real "rigid" bodies are to some extent **elastic,** which means that we can change their dimensions slightly by pulling, pushing, twisting, or compressing them. To get a feeling for the orders of magnitude involved, consider a vertical steel rod 1 m long and 1 cm in diameter. If you hang a subcompact car from the end of such a rod, the rod will stretch, but only by about 0.5 mm, or 0.05%. Furthermore, the rod will return to its original length when the car is removed.

If you hang two cars from the rod, the rod will be permanently stretched and will not recover its original length when you remove the load. If you hang three cars from the rod, the rod will break. Just before rupture, the elongation of the rod will be less than 0.2%. Although deformations of this size seem small, they are important in engineering practice. (Whether a wing under load will stay on an airplane is obviously important.)

Figure 13-11 shows three ways in which a solid might change its dimensions when forces act on it. In Fig. 13-11a, a cylinder is stretched. In Fig. 13-11b, a cylinder is deformed by a force perpendicular to its axis, much as we might deform a pack of cards or a book. In Fig. 13-11c, a solid object, placed in a fluid under high pressure, is compressed uniformly on all sides. What the three deformation types have in common is that a **stress,** or deforming force per unit area, produces a **strain,** or unit deformation. In Fig. 13-11, *tensile stress* (associated with stretching) is illustrated in (a), *shearing stress* in (b), and *hydraulic stress* in (c).

The stresses and the strains take different forms in the three situations of Fig. 13-11, but—over the range of engineering usefulness—stress and strain are proportional to each other. The constant of proportionality is called a **modulus of elasticity,** so that

$$\text{stress} = \text{modulus} \times \text{strain}. \tag{13-24}$$

In a standard test of tensile properties, the tensile stress on a test cylinder (like that in Fig. 13-12) is slowly increased from zero to the point at which the cylinder fractures, and the strain is carefully measured and plotted. The result is a graph of stress versus strain like that in Fig. 13-13. For a substantial range of applied stresses, the stress–strain relation is linear, and the specimen recovers its original dimensions when the stress is removed; it is here that Eq. 13-24 applies. If the stress is increased beyond the **yield strength** S_y of the specimen, the specimen becomes permanently deformed. If the stress continues to increase, the specimen eventually ruptures, at a stress called the **ultimate strength** S_u.

Fig. 13-11 (a) A cylinder subject to *tensile stress* stretches by an amount ΔL. (b) A cylinder subject to *shearing stress* deforms by an amount Δx, somewhat like a pack of playing cards would. (c) A solid sphere subject to uniform *hydraulic stress* from a fluid shrinks in volume by an amount ΔV. All the deformations shown are greatly exaggerated.

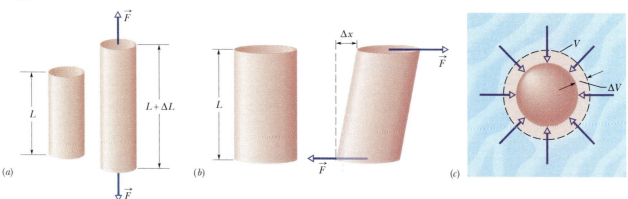

(a)   (b)   (c)

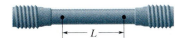

Fig. 13-12 A test specimen, used to determine a stress–strain curve such as that of Fig. 13-13. The change ΔL that occurs in a certain length L is measured in a tensile stress–strain test.

Tension and Compression

For simple tension or compression, the stress on an object is defined as F/A, where F is the magnitude of the force applied perpendicularly to the area A on the object. The strain, or unit deformation, is then the dimensionless quantity $\Delta L/L$, the fractional (or sometimes percentage) change in the length of the specimen. If the specimen is a long rod and the stress does not exceed the yield strength, then not only the entire rod but also every section of it experiences the same strain when a given stress is applied. Because the strain is dimensionless, the modulus in Eq. 13-24 has the same dimensions as the stress, namely, force per unit area.

The modulus for tensile and compressive stresses is called the **Young's modulus** and is represented in engineering practice by the symbol E. Equation 13-24 becomes

$$\frac{F}{A} = E \frac{\Delta L}{L} \qquad (13\text{-}25)$$

The strain $\Delta L/L$ in a specimen can often be measured conveniently with a *strain gauge* (Fig. 13-14). This simple and useful device, which can be attached directly to operating machinery with an adhesive, is based on the principle that its electrical properties are dependent on the strain it undergoes.

Although the Young's modulus for an object may be almost the same for tension and compression, the object's ultimate strength may well be different for the two types of stress. Concrete, for example, is very strong in compression but is so weak in tension that it is almost never used in that manner. Table 13-1 shows the Young's modulus and other elastic properties for some materials of engineering interest.

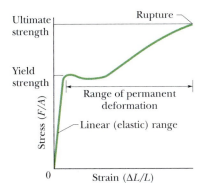

Fig. 13-13 A stress–strain curve for a steel test specimen such as that of Fig. 13-12. The specimen deforms permanently when the stress is equal to the *yield strength* of the material. It ruptures when the stress is equal to the *ultimate strength* of the material.

Shearing

In the case of shearing, the stress is also a force per unit area, but the force vector lies in the plane of the area rather than perpendicular to it. The strain is the dimensionless ratio $\Delta x/L$, with the quantities defined as shown in Fig. 13-11*b*. The corresponding modulus, which is given the symbol G in engineering practice, is called the **shear modulus.** For shearing, Eq. 13-24 is written as

$$\frac{F}{A} = G \frac{\Delta x}{L}. \qquad (13\text{-}26)$$

Shearing stresses play a critical role in the buckling of shafts that rotate under load and in bone fractures caused by bending.

Fig. 13-14 A strain gauge of overall dimensions 9.8 mm by 4.6 mm. The gauge is fastened with adhesive to the object whose strain is to be measured; it experiences the same strain as the object. The electrical resistance of the gauge varies with the strain, permitting strains up to 3% to be measured.

TABLE 13-1 Some Elastic Properties of Selected Materials of Engineering Interest

Material	Density ρ (kg/m^3)	Young's Modulus E (10^9 N/m^2)	Ultimate Strength S_u (10^6 N/m^2)	Yield Strength S_y (10^6 N/m^2)
Steel[a]	7860	200	400	250
Aluminum	2710	70	110	95
Glass	2190	65	50[b]	—
Concrete[c]	2320	30	40[b]	—
Wood[d]	525	13	50[b]	—
Bone	1900	9[b]	170[b]	—
Polystyrene	1050	3	48	—

[a]Structural steel (ASTM-A36).   [b]In compression.   [c]High strength.   [d]Douglas fir.

Hydraulic Stress

In Fig. 13-11c, the stress is the fluid pressure p on the object, which, as you will see in Chapter 15, is a force per unit area. The strain is $\Delta V/V$, where V is the original volume of the specimen and ΔV is the absolute value of the change in volume. The corresponding modulus, with symbol B, is called the **bulk modulus** of the material. The object is said to be under *hydraulic compression,* and the pressure can be called the *hydraulic stress.* For this situation, we write Eq. 13-24 as

$$p = B\,\frac{\Delta V}{V}. \tag{13-27}$$

The bulk modulus is 2.2×10^9 N/m^2 for water and 16×10^{10} N/m^2 for steel. The pressure at the bottom of the Pacific Ocean, at its average depth of about 4000 m, is 4.0×10^7 N/m^2. The fractional compression $\Delta V/V$ of a volume of water due to this pressure is 1.8%; that for a steel object is only about 0.025%. In general, solids—with their rigid atomic lattices—are less compressible than liquids, in which the atoms or molecules are less tightly coupled to their neighbors.

Sample Problem 13-5

A structural steel rod has a radius R of 9.5 mm and a length L of 81 cm. A 62 kN force $\vec{F}$ stretches it along its length. What are the stress on the rod and the elongation and strain of the rod?

SOLUTION: The first Key Idea here has to do with what is meant by the second sentence in the problem statement. We assume the rod is held stationary by, say, a clamp or vise at one end. Then force $\vec{F}$ is applied at the other end, parallel to the length of the rod and thus perpendicular to the end face there. Therefore, the situation is like that in Fig. 13-11a.

The next Key Idea is that we assume the force is applied uniformly across the end face and thus over an area $A = \pi R^2$. Then the stress on the rod is given by the left side of Eq. 13-25:

$$\text{stress} = \frac{F}{A} = \frac{F}{\pi R^2} = \frac{6.2 \times 10^4 \text{ N}}{(\pi)(9.5 \times 10^{-3}\text{ m})^2}$$
$$= 2.2 \times 10^8 \text{ N/m}^2. \qquad \text{(Answer)}$$

The yield strength for structural steel is 2.5×10^8 N/m^2, so this rod is dangerously close to its yield strength.

Another Key Idea is that the elongation of the rod depends on the stress, the original length L, and the type of material in the rod. The last determines which value we use for Young's modulus E (from Table 13-1). Using the value for steel, Eq. 13-25 gives us

$$\Delta L = \frac{(F/A)L}{E} = \frac{(2.2 \times 10^8 \text{ N/m}^2)(0.81 \text{ m})}{2.0 \times 10^{11} \text{ N/m}^2}$$
$$= 8.9 \times 10^{-4} \text{ m} = 0.89 \text{ mm}. \qquad \text{(Answer)}$$

The last Key Idea we need here is that strain is the ratio of the change in length to the original length, so we have

$$\frac{\Delta L}{L} = \frac{8.9 \times 10^{-4} \text{ m}}{0.81 \text{ m}}$$
$$= 1.1 \times 10^{-3} = 0.11\%. \qquad \text{(Answer)}$$

Sample Problem 13-6

A table has three legs that are 1.00 m in length and a fourth leg that is longer by $d = 0.50$ mm, so that the table wobbles slightly. A heavy steel cylinder with mass $M = 290$ kg is placed upright on the table (with a mass much less than M) so that all four legs are compressed and the table no longer wobbles. The legs are wooden cylinders with cross-sectional area $A = 1.0$ cm^2. The Young's modulus E for the wood is 1.3×10^{10} N/m^2. Assume that the tabletop remains level and that the legs do not buckle. What are the magnitudes of the forces on the legs from the floor?

SOLUTION: We take the table plus steel cylinder as our system. The situation is like that in Fig. 13-9, except now we have a steel cylinder on the table. One Key Idea is that if the tabletop remains level, the legs must be compressed in the following ways: Each of the short legs must be compressed by the same amount (call it ΔL_3) and thus by the same force of magnitude F_3. The single long leg

must be compressed by a larger amount ΔL_4 and thus by a force with a larger magnitude F_4. In other words, for a level tabletop, we must have

$$\Delta L_4 = \Delta L_3 + d. \tag{13-28}$$

A second Key Idea is that, from Eq. 13-25, we can relate a change in length to the force causing the change with $\Delta L = FL/AE$, where L is the original length of a leg. We can use this relation to replace ΔL_4 and ΔL_3 in Eq. 13-28. However, note that we can approximate the original length L as being the same for all four legs. The replacements then give us

$$\frac{F_4 L}{AE} = \frac{F_3 L}{AE} + d. \tag{13-29}$$

We cannot solve this equation because it has two unknowns, F_4 and F_3.

To get a second equation containing F_4 and F_3, we can use a vertical y axis and then write the balance of vertical forces ($F_{net,y} = 0$) as

$$3F_3 + F_4 - Mg = 0, \qquad (13\text{-}30)$$

where Mg is equal to the magnitude of the gravitational force on the system. (*Three* legs have force $\vec{F}_3$ on them.) To solve the simultaneous equations 13-29 and 13-30 for, say, F_3, we first use Eq. 13-30 to find that $F_4 = Mg - 3F_3$. Substituting that into Eq. 13-29 then yields, after some algebra,

$$F_3 = \frac{Mg}{4} - \frac{dAE}{4L}$$

$$= \frac{(290 \text{ kg})(9.8 \text{ m/s}^2)}{4}$$

$$- \frac{(5.0 \times 10^{-4} \text{ m})(10^{-4} \text{ m}^2)(1.3 \times 10^{10} \text{ N/m}^2)}{(4)(1.00 \text{ m})}$$

$$= 548 \text{ N} \approx 550 \text{ N}. \qquad \text{(Answer)}$$

From Eq. 13-30 we then find

$$F_4 = Mg - 3F_3 = (290 \text{ kg})(9.8 \text{ m/s}^2) - 3(548 \text{ N})$$

$$\approx 1200 \text{ N}. \qquad \text{(Answer)}$$

You can show that to reach their equilibrium configuration, the three short legs are each compressed by 0.42 mm and the single long leg by 0.92 mm.

✓**CHECKPOINT 6:** The figure shows a horizontal block that is suspended by two wires, A and B, which are identical except for their original lengths. The center of mass of the block is closer to wire B than to wire A. (a) Measuring torques about the block's center of mass, state whether the magnitude of the torque due to wire A is greater than, less than, or equal to the magnitude of the torque due to wire B. (b) Which wire exerts more force on the block? (c) If the wires are now equal in length, which one was originally shorter?

REVIEW & SUMMARY

Static Equilibrium A rigid body at rest is said to be in **static equilibrium.** For such a body, the vector sum of the external forces acting on it is zero:

$$\vec{F}_{net} = 0 \qquad \text{(balance of forces)}. \qquad (13\text{-}3)$$

If all the forces lie in the xy plane, this vector equation is equivalent to two component equations:

$$F_{net,x} = 0 \quad \text{and} \quad F_{net,y} = 0 \quad \text{(balance of forces)}. \qquad (13\text{-}7, 13\text{-}8)$$

Static equilibrium also implies that the vector sum of the external torques acting on the body about *any* point is zero, or

$$\vec{\tau}_{net} = 0 \qquad \text{(balance of torques)}. \qquad (13\text{-}5)$$

If the forces lie in the xy plane, all torque vectors are parallel to the z axis, and Eq. 13-5 is equivalent to the single component equation

$$\tau_{net,z} = 0 \qquad \text{(balance of torques)}. \qquad (13\text{-}9)$$

Center of Gravity The gravitational force acts individually on each element of a body. The net effect of all individual actions may be found by imagining an equivalent total gravitational force $\vec{F}_g$ acting at a specific point called the **center of gravity.** If the gravitational acceleration $\vec{g}$ is the same for all the elements of the body, the center of gravity is at the center of mass.

Elastic Moduli Three **elastic moduli** are used to describe the elastic behavior (deformations) of objects as they respond to forces that act on them. The **strain** (fractional change in length) is linearly related to the applied **stress** (force per unit area) by the proper modulus, according to the general relation

$$\text{stress} = \text{modulus} \times \text{strain}. \qquad (13\text{-}24)$$

Tension and Compression When an object is under tension or compression, Eq. 13-24 is written as

$$\frac{F}{A} = E \frac{\Delta L}{L}, \qquad (13\text{-}25)$$

where $\Delta L/L$ is the tensile or compressive strain of the object, F is the magnitude of the applied force $\vec{F}$ causing the strain, A is the cross-sectional area over which $\vec{F}$ is applied (perpendicular to A, as in Fig. 13-11a), and E is the **Young's modulus** for the object. The stress is F/A.

Shearing When an object is under a shearing stress, Eq. 13-24 is written as

$$\frac{F}{A} = G \frac{\Delta x}{L}, \qquad (13\text{-}26)$$

where $\Delta x/L$ is the shearing strain of the object, Δx is the displacement of one end of the object in the direction of the applied force $\vec{F}$ (as in Fig. 13-11b), and G is the **shear modulus** of the object. The stress is F/A.

Hydraulic Stress When an object undergoes *hydraulic compression* due to a stress exerted by a surrounding fluid, Eq. 13-24 is written as

$$p = B \frac{\Delta V}{V}, \qquad (13\text{-}27)$$

where p is the pressure (*hydraulic stress*) on the object due to the fluid, $\Delta V/V$ (the strain) is the absolute value of the fractional change in the object's volume due to that pressure, and B is the **bulk modulus** of the object.

QUESTIONS

1. Figure 13-15 shows an overhead view of a uniform stick on which four forces act. Suppose we choose a rotational axis through point O, calculate the torques about that axis due to the forces, and find that these torques balance. Will the torques balance if, instead, the rotational axis is chosen to be at (a) point A, (b) point B, or (c) point C? (d) Suppose, instead, that we find that the torques about point O do not balance. Is there another point about which the torques will balance?

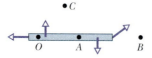

Fig. 13-15 Question 1.

2. In Fig. 13-16, a rigid beam is attached to two posts that are fastened to a floor. A small but heavy safe is placed at the six positions indicated, in turn. Assume that the mass of the beam is negligible compared to that of the safe. (a) Rank the positions according to the force on post A due to the safe, greatest compression first, greatest tension last, and indicate where, if anywhere, the force is zero. (b) Now rank them according to the force on post B.

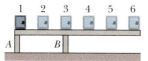

Fig. 13-16 Question 2.

3. Figure 13-17 shows four overhead views of rotating uniform disks that are sliding across a frictionless floor. Three forces, of magnitude F, $2F$, or $3F$, act on each disk, either at the rim, at the center, or halfway between rim and center. The force vectors rotate along with the disks, and, in the "snapshots" of Fig. 13-17, point left or right. Which disks are in equilibrium?

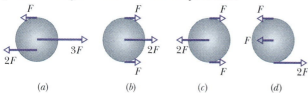

Fig. 13-17 Question 3.

4. Figure 13-18 shows overhead views of two structures on which three forces act. The directions of the forces are as indicated. If the magnitudes of the forces are adjusted properly (but kept nonzero), which structure can be in static equilibrium?

5. Figure 13-19 shows a mobile of toy penguins hanging from a

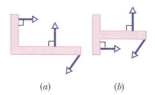

Fig. 13-18 Question 4.

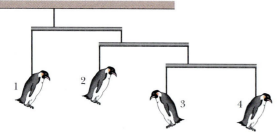

Fig. 13-19 Question 5.

ceiling. Each crossbar is horizontal, has negligible mass, and extends three times as far to the right of the wire supporting it as to the left. Penguin 1 has mass $m_1 = 48$ kg. What are the masses of the other penguins?

6. A ladder leans against a frictionless wall but is prevented from falling because of friction between it and the ground. Suppose you shift the base of the ladder toward the wall. Determine whether the following become larger, smaller, or stay the same (in magnitude): (a) the normal force on the ladder from the ground, (b) the force on the ladder from the wall, (c) the static frictional force on the ladder from the ground, and (d) the maximum value $f_{s,\text{max}}$ of the static frictional force.

7. Three piñatas hang from the (stationary) assembly of massless pulleys and cords seen in Fig. 13-20. One long cord runs from the ceiling at the right to the lower pulley at the left. Several shorter cords suspend pulleys from the ceiling or piñatas from the pulleys. The weights (in newtons) of two piñatas are given. (a) What is the weight of the third piñata? (*Hint:* When a cord loops half-way around a pulley, it pulls on the pulley with a net force that is twice the tension in the cord.) (b) What is the tension in the short cord labeled with T?

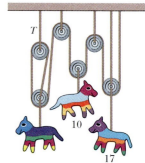

Fig. 13-20 Question 7.

8. (a) In Checkpoint 4, to express τ_r in terms of T, should you use $\sin \theta$ or $\cos \theta$? (b) If angle θ is decreased (by shortening the rope but still keeping the rod horizontal), does the torque τ_r required for equilibrium become larger, smaller, or stay the same? (c) Does the corresponding force magnitude T become larger, smaller, or stay the same?

9. The table gives the areas of three surfaces and the magnitude of a force that is applied perpendicular to the surface and uniformly across it. Rank the surfaces according to the stress on them, greatest first.

	Area	Force
Surface A	$0.5A_0$	$2F_0$
Surface B	$2A_0$	$4F_0$
Surface C	$3A_0$	$6F_0$

10. Four cylindrical rods are stretched as in Fig. 13-11a. The force magnitudes, the areas of the end faces, the changes in length, and the initial lengths are given here. Rank the rods according to their Young's moduli, greatest first.

Rod	Force	Area	Length Change	Initial Length
1	F	A	ΔL	L
2	$2F$	$2A$	$2\Delta L$	L
3	F	$2A$	$2\Delta L$	$2L$
4	$2F$	A	ΔL	$2L$

EXERCISES & PROBLEMS

ssm Solution is in the Student Solutions Manual.
www Solution is available on the World Wide Web at:
        http://www.wiley.com/college/hrw
ilw Solution is available on the Interactive LearningWare.

SEC. 13-4   Some Examples of Static Equilibrium

1E. A physics Brady Bunch, whose weights in newtons are indicated in Fig. 13-21, is balanced on a seesaw. What is the number of the person who causes the largest torque, about the rotation axis at *fulcrum f,* directed (a) out of the page and (b) into the page?

Fig. 13-21 Exercise 1.

2E. The leaning Tower of Pisa (Fig. 13-22) is 55 m high and 7.0 m in diameter. The top of the tower is displaced 4.5 m from the vertical. Treat the tower as a uniform, circular cylinder. (a) What additional displacement, measured at the top, would bring the tower to the verge of toppling? (b) What angle would the tower then make with the vertical?

Fig. 13-22 Exercise 2.

3E. A particle is acted on by forces given, in newtons, by $\vec{F}_1 = 10\hat{i} - 4\hat{j}$ and $\vec{F}_2 = 17\hat{i} + 2\hat{j}$. (a) What force $\vec{F}_3$ balances these forces? (b) What direction does $\vec{F}_3$ have relative to the *x* axis? ssm

4E. A bow is drawn at its midpoint until the tension in the string

is equal to the force exerted by the archer. What is the angle between the two halves of the string?

5E. A rope of negligible mass is stretched horizontally between two supports that are 3.44 m apart. When an object of weight 3160 N is hung at the center of the rope, the rope is observed to sag by 35.0 cm. What is the tension in the rope? ilw

6E. A scaffold of mass 60 kg and length 5.0 m is supported in a horizontal position by a vertical cable at each end. A window washer of mass 80 kg stands at a point 1.5 m from one end. What is the tension in (a) the nearer cable and (b) the farther cable?

7E. In Fig. 13-23, a uniform sphere of mass *m* and radius *r* is held in place by a massless rope attached to a frictionless wall a distance *L* above the center of the sphere. Find (a) the tension in the rope and (b) the force on the sphere from the wall. ssm

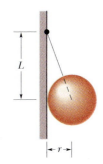

Fig. 13-23 Exercise 7.

8E. An automobile with a mass of 1360 kg has 3.05 m between the front and rear axles. Its center of gravity is located 1.78 m behind the front axle. With the automobile on level ground, determine the magnitude of the force from the ground on (a) each front wheel (assuming equal forces on the front wheels) and (b) each rear wheel (assuming equal forces on the rear wheels).

9E. A diver of weight 580 N stands at the end of a 4.5 m diving board of negligible mass (Fig. 13-24). The board is attached to two pedestals 1.5 m apart. What are the magnitude and direction of the force on the board from (a) the left pedestal and (b) the right pedestal? (c) Which pedestal is being stretched, and (d) which compressed? ssm

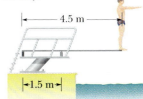

Fig. 13-24 Exercise 9.

10E. In Fig. 13-25, a man is trying to get his car out of mud on the shoulder of a road. He ties one end of a rope tightly around the front bumper and the other end tightly around a utility pole 18 m away. He then pushes sideways on the rope at its midpoint with a force of 550 N, displacing the center of the rope 0.30 m from its previous position, and the car barely moves. What is the magnitude of the force on the car from the rope? (The rope stretches somewhat.)

Fig. 13-25 Exercise 10.

11E. A meter stick balances horizontally on a knife-edge at the 50.0 cm mark. With two 5.0 g coins stacked over the 12.0 cm mark, the stick is found to balance at the 45.5 cm mark. What is the mass of the meter stick?   ssm

12E. A uniform cubical crate is 0.750 m on each side and weighs 500 N. It rests on a floor with one edge against a very small, fixed obstruction. At what least height above the floor must a horizontal force of magnitude 350 N be applied to the crate to tip it?

13E. A 75 kg window cleaner uses a 10 kg ladder that is 5.0 m long. He places one end on the ground 2.5 m from a wall, rests the upper end against a cracked window, and climbs the ladder. He is 3.0 m up along the ladder when the window breaks. Neglecting friction between the ladder and window and assuming that the base of the ladder does not slip, find (a) the magnitude of the force on the window from the ladder just before the window breaks and (b) the magnitude and direction of the force on the ladder from the ground just before the window breaks.   ssm

14E. Figure 13-26 shows the anatomical structures in the lower leg and foot that are involved in standing tiptoe with the heel raised off the floor so the foot effectively contacts the floor at only one point, shown as P in the figure. Calculate, in terms of a person's weight W, the forces on the foot from (a) the calf muscle (at A) and (b) the lower-leg bones (at B) when the person stands tiptoe on one foot. Assume that $a = 5.0$ cm and $b = 15$ cm.

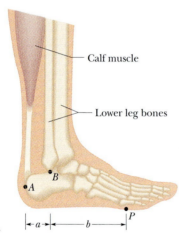

Calf muscle

Lower leg bones

B

A

P

$\leftarrow a \rightarrow \leftarrow b \rightarrow$

Fig. 13-26 Exercise 14.

15P. In Fig. 13-27, an 817 kg construction bucket is suspended by a cable A that is attached at O to two other cables B and C, making angles of 51.0° and 66.0° with the horizontal. Find the tensions in (a) cable A, (b) cable B, and (c) cable C. (*Hint:* To avoid solving two equations in two unknowns, position the axes as shown in the figure.)   ssm

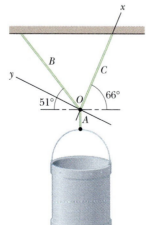

Fig. 13-27 Problem 15.

16P. The system in Fig. 13-28 is in equilibrium, with the string in the center exactly horizontal. Find (a) tension T_1, (b) tension T_2, (c) tension T_3, and (d) angle θ.

17P. The force $\vec{F}$ in Fig. 13-29 keeps the 6.40 kg block and the pulleys in equilibrium. The pulleys have negligible mass and friction. Calculate the tension T in the upper cable. (*Hint:* When a cable wraps halfway around a pulley as here, the magnitude of its net force on the pulley is twice the tension in the cable.)   ssm   ilw

18P. A 15 kg block is being lifted by the pulley system shown in Fig. 13-30. The upper arm is vertical, whereas the forearm makes an angle of 30° with the horizontal. What are the forces on the forearm from (a) the triceps muscle and (b) the upper-arm bone (the humerus)? The forearm and hand together have a mass of 2.0 kg with a center of mass 15 cm (measured along the arm) from the point where the forearm and upper-arm bones are in contact. The triceps muscle pulls vertically upward at a point 2.5 cm behind that contact point.

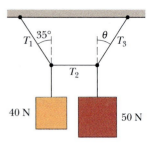

T_1 35° θ T_3

T_2

40 N   50 N

Fig. 13-28 Problem 16.

T

$\vec{F}$

Fig. 13-29 Problem 17.

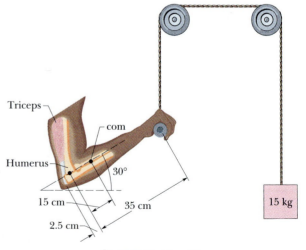

Triceps

com

Humerus

30°

15 cm   35 cm

2.5 cm

15 kg

Fig. 13-30 Problem 18.

19P. Forces $\vec{F}_1$, $\vec{F}_2$, and $\vec{F}_3$ act on the structure of Fig. 13-31, shown in an overhead view. We wish to put the structure in equilibrium by applying a fourth force, at a point such as P. The fourth force has vector components $\vec{F}_h$ and $\vec{F}_v$. We are given that $a = 2.0$ m,

$b = 3.0$ m, $c = 1.0$ m, $F_1 = 20$ N, $F_2 = 10$ N, and $F_3 = 5.0$ N. Find (a) F_h, (b) F_v, and (c) d. **ilw**

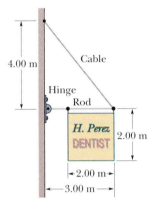

Fig. 13-31 Problem 19.

20P. In Fig. 13-32, a 50.0 kg uniform square sign, 2.00 m on a side, is hung from a 3.00 m horizontal rod of negligible mass. A cable is attached to the end of the rod and to a point on the wall 4.00 m above the point where the rod is hinged to the wall. (a) What is the tension in the cable? What are the magnitudes and directions of the (b) horizontal and (c) vertical components of the force on the rod from the wall?

21P. In Fig. 13-33, what magnitude of force $\vec{F}$ applied horizontally at the axle of the wheel is necessary to raise the wheel over an obstacle of height h? The wheel's radius is r and its mass is m. **ssm   www**

22P. In Fig. 13-34, a 55 kg rock climber is in a lie-back climb along a fissure, with hands pulling on one side of the fissure and feet pressed against the opposite side. The fissure has width $w = 0.20$ m, and the center of mass of the climber is a horizontal distance $d = 0.40$ m from the fissure. The coefficient of static friction between hands and rock is $\mu_1 = 0.40$, and between boots and rock it is $\mu_2 = 1.2$. (a) What is the least horizontal pull by the hands and push by the feet that will keep the climber stable? (b) For the horizontal pull of (a), what must be the vertical distance h between hands and feet? (c) If the climber encounters wet rock, so that μ_1 and μ_2 are reduced, what happens to the answers to (a) and (b), respectively?

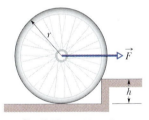

Fig. 13-32 Problem 20.

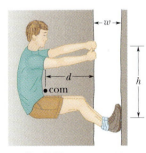

Fig. 13-33 Problem 21.

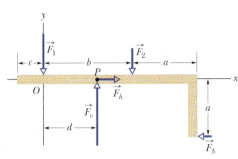

Fig. 13-34 Problem 22.

23P. In Fig. 13-35, one end of a uniform beam that weighs 222 N is attached to a wall with a hinge. The other end is supported by a wire. (a) Find the tension in the wire. What are the (b) horizontal and (c) vertical components of the force of the hinge on the beam? **ssm**

24P. Four bricks of length L, identical and uniform, are stacked on top of one another (Fig. 13-36) in such a way that part of each extends beyond the one beneath. Find, in terms of L, the maximum values of (a) a_1, (b) a_2, (c) a_3, (d) a_4, and (e) h, such that the stack is in equilibrium.

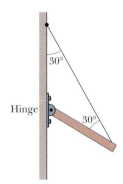

Fig. 13-35 Problem 23.

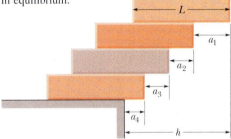

Fig. 13-36 Problem 24.

25P. The system in Fig. 13-37 is in equilibrium. A concrete block of mass 225 kg hangs from the end of the uniform strut whose mass is 45.0 kg. Find (a) the tension T in the cable and the (b) horizontal and (c) vertical force components on the strut from the hinge. **ilw**

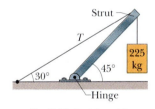

Fig. 13-37 Problem 25.

26P. A door 2.1 m high and 0.91 m wide has a mass of 27 kg. A hinge 0.30 m from the top and another 0.30 m from the bottom each support half the door's mass. Assume that the center of gravity is at the geometrical center of the door, and determine the (a) vertical and (b) horizontal components of the force from each hinge on the door.

27P. A nonuniform bar is suspended at rest in a horizontal position by two massless cords as shown in Fig. 13-38. One cord makes the angle $\theta = 36.9°$ with the vertical; the other makes the angle $\phi = 53.1°$ with the vertical. If the length L of the bar is 6.10 m, compute the distance x from the left-hand end of the bar to its center of mass. **ssm**

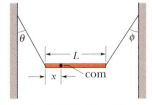

Fig. 13-38 Problem 27.

28P. In Fig. 13-39, a thin horizontal bar AB of negligible weight and length L is hinged to a vertical wall at A and supported at B

by a thin wire *BC* that makes an angle θ with the horizontal. A load of weight *W* can be moved anywhere along the bar; its position is defined by the distance *x* from the wall to its center of mass. As a function of *x*, find (a) the tension in the wire, and the (b) horizontal and (c) vertical components of the force on the bar from the hinge at *A*.

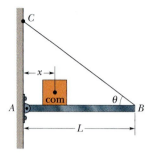

Fig. 13-39 Problems 28 and 30.

29P. In Fig. 13-40, a uniform plank, with a length *L* of 6.10 m and a weight of 445 N, rests on the ground and against a frictionless roller at the top of a wall of height *h* = 3.05 m. The plank remains in equilibrium for any value of $\theta \geq 70°$ but slips if $\theta <$ 70°. Find the coefficient of static friction between the plank and the ground. ssm

30P. In Fig. 13-39, suppose the length *L* of the uniform bar is 3.0 m and its weight is 200 N. Also, let the load's weight *W* = 300 N and the angle θ = 30°. The wire can withstand a maximum tension of 500 N. (a) What is the maximum possible distance *x* before the wire breaks? With the load placed at this maximum *x*, what are the (b) horizontal and (c) vertical components of the force on the bar from the hinge at *A*?

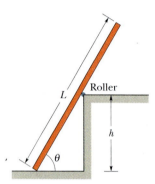

Fig. 13-40 Problem 29.

31P. For the stepladder shown in Fig. 13-41, sides *AC* and *CE* are each 2.44 m long and hinged at *C*. Bar *BD* is a tie-rod 0.762 m long, halfway up. A man weighing 854 N climbs 1.80 m along the ladder. Assuming that the floor is frictionless and neglecting the mass of the ladder, find (a) the tension in the tie-rod and the magnitudes of the forces on the ladder from the floor at (b) *A* and (c) *E*. (*Hint:* It will help to isolate parts of the ladder in applying the equilibrium conditions.) ssm

32P. Two uniform beams, *A* and *B*, are attached to a wall with hinges and then loosely bolted together as in Fig. 13-42. Find the

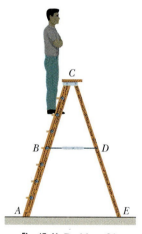

Fig. 13-41 Problem 31.

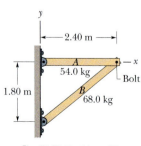

Fig. 13-42 Problem 32.

x and *y* components of the force on (a) beam *A* due to its hinge, (b) beam *A* due to the bolt, (c) beam *B* due to its hinge, and (d) beam *B* due to the bolt.

33P. A cubical box is filled with sand and weighs 890 N. We wish to "roll" the box by pushing horizontally on one of the upper edges. (a) What minimum force is required? (b) What minimum coefficient of static friction between box and floor is required? (c) Is there a more efficient way to roll the box? If so, find the smallest possible force that would have to be applied directly to the box to roll it. (*Hint:* At the onset of tipping, where is the normal force located?) ssm

34P. Four bricks of length *L*, identical and uniform, are stacked on a table in two ways, as shown in Fig. 13-43 (compare with Problem 24). We seek to maximize the overhang distance *h* in both arrangements. Find the optimum distances a_1, a_2, b_1, and b_2, and calculate *h* for the two arrangements. (See "The Amateur Scientist," *Scientific American,* June 1985, pp. 133–134, for a discussion and an even better version of arrangement (*b*).)

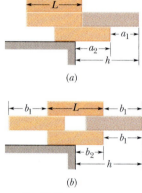

Fig. 13-43 Problem 34.

35P. A crate, in the form of a cube with edge lengths of 1.2 m, contains a piece of machinery; the center of mass of the crate and its contents is located 0.30 m above the crate's geometrical center. The crate rests on a ramp that makes an angle θ with the horizontal. As θ is increased from zero, an angle will be reached at which the crate will either start to slide down the ramp or tip over. Which event will occur (a) when the coefficient of static friction between ramp and crate is 0.60 and (b) when it is 0.70? In each case, give the angle at which the event occurs. (*Hint:* At the onset of tipping, where is the normal force located?) ssm www

SEC. 13-6 Elasticity

36E. Figure 13-44 shows the stress–strain curve for quartzite. What

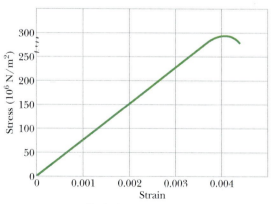

Fig. 13-44 Exercise 36.

are (a) the Young's modulus and (b) the approximate yield strength for this material?

37E. A horizontal aluminum rod 4.8 cm in diameter projects 5.3 cm from a wall. A 1200 kg object is suspended from the end of the rod. The shear modulus of aluminum is 3.0×10^{10} N/m². Neglecting the rod's mass, find (a) the shear stress on the rod and (b) the vertical deflection of the end of the rod. ssm ilw

38P. In Fig. 13-45, a lead brick rests horizontally on cylinders A and B. The areas of the top faces of the cylinders are related by $A_A = 2A_B$; the Young's moduli of the cylinders are related by $E_A = 2E_B$. The cylinders had identical lengths before the brick was placed on them. What fraction of the brick's mass is supported

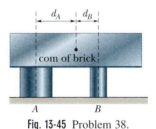

Fig. 13-45 Problem 38.

(a) by cylinder A and (b) by cylinder B? The horizontal distances between the center of mass of the brick and the centerlines of the cylinders are d_A for cylinder A and d_B for cylinder B. (c) What is the ratio d_A/d_B?

39P. In Fig. 13-46, a 103 kg uniform log hangs by two steel wires, A and B, both of radius 1.20 mm. Initially, wire A was 2.50 m long and 2.00 mm shorter than wire B. The log is now horizontal. What are the magnitudes of the forces on it from (a) wire A and (b) wire B? (c) What is the ratio d_A/d_B? ssm www

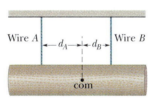

Fig. 13-46 Problem 39.

40P. A tunnel 150 m long, 7.2 m high, and 5.8 m wide (with a flat roof) is to be constructed 60 m beneath the ground. (See Fig. 13-47.) The tunnel roof is to be supported entirely by square steel columns, each with a cross-sectional area of 960 cm². The density of the ground material is 2.8 g/cm³. (a) What is the total mass of the material that the columns must support? (b) How many columns are needed to keep the compressive stress on each column at one-half its ultimate strength?

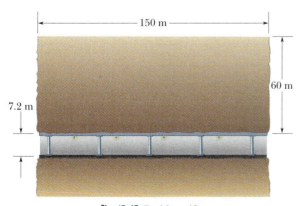

Fig. 13-47 Problem 40.

Additional Problems

41. Here is a way to move a heavy log through a tropical forest. Find a young tree in the general direction of travel; find a vine that hangs from the top of the tree down to ground level; pull the vine over to the log; wrap the vine around a limb on the log; pull hard enough on the vine to bend the tree over; and then tie off the vine on the limb. Repeat this procedure with several trees;

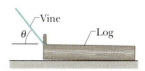

Fig. 13-48 Problem 41.

eventually the net force of the vines on the log moves the log forward. Although tedious, this technique allowed workers to move heavy logs long before modern machinery was available. Figure 13-48 shows the essentials of the technique. There, a single vine is shown attached to a branch at one end of a uniform log of mass M. The coefficient of static friction between the log and the ground is 0.80. If the log is on the verge of sliding, with the left end raised slightly by the vine, what are (a) the angle θ and (b) the magnitude T of the force on the log from the vine?

42. You have been hired to build a large sand mound in an indoor playground and must be careful about the stress that the sand will put on the floor. Consulting research literature, you are surprised to find that the greatest stress occurs, not directly beneath the apex (top) of the mound, but at points that are a distance r_m from that central point (Fig. 13-49a). This outward displacement of the maximum stress is presumably due to the sand grains form-

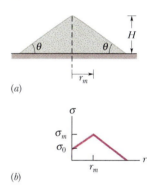

Fig. 13-49 Problem 42.

ing arches within the mound. For a mound of height $H = 3.00$ m and angle $\theta = 33°$, and with sand of density $\rho = 1800$ kg/m³, Fig. 13-49b gives the stress σ as a function of radius r from the central point of the mound's base. In that figure, $\sigma_0 = 40\,000$ N/m², $\sigma_m = 40\,024$ N/m², and $r_m = 1.82$ m.

(a) What is the volume of sand contained in the mound for $r \leq r_m/2$? (*Hint:* The volume is that of a vertical cylinder plus a cone on top of the cylinder. The volume of the cone is $\pi R^2 h/3$, where R is the cone's radius and h is the cone's height.) (b) What is the weight W of that volume of sand? (c) Use Fig. 13-49b to write an expression for the stress σ on the floor as a function of radius r, for $r \leq r_m$. (d) On the floor, what is the area dA of a thin ring of radius r centered on the mound's central axis and with radial width dr? (e) What then is the magnitude dF of the downward force on the ring due to the sand? (f) What is the magnitude F of the net downward force on the floor due to all the sand contained in the mound for $r \leq r_m/2$? (*Hint:* Integrate the expression of (e) from $r = 0$ to $r = r_m/2$.) Now note the surprise: This force magnitude F on the floor is less than the weight W of the sand above the floor, as found in (b). (g) By what fraction is F reduced from W; that is, what is $(F - W)/W$?

14 Gravitation

The Milky Way galaxy is a disk-shaped collection of dust, planets, and billions of stars, including our Sun and solar system. The force that binds it or any other galaxy together is the same force that holds the Moon in orbit and you on Earth — the gravitational force. That force is also responsible for one of nature's strangest objects, the black hole, a star that has completely collapsed onto itself. The gravitational force near a black hole is so strong that not even light can escape it.

If that is the case, how can a black hole be detected?

The answer is in this chapter.

Hydra

Centaurus

Sagittarius

Andromeda Galaxy

Large Magellanic C

14-1 The World and the Gravitational Force

The drawing that opens this chapter shows our view of the Milky Way galaxy. We are near the edge of the disk of the galaxy, about 26 000 light-years (2.5×10^{20} m) from its center, which in the drawing lies in the star collection known as Sagittarius. Our galaxy is a member of the Local Group of galaxies, which includes the Andromeda galaxy (Fig. 14-1) at a distance of 2.3×10^{6} light-years, and several closer dwarf galaxies, such as the Large Magellanic Cloud shown in the opening drawing.

The Local Group is part of the Local Supercluster of galaxies. Measurements taken during and since the 1980s suggest that the Local Supercluster and the supercluster consisting of the clusters Hydra and Centaurus are all moving toward an exceptionally massive region called the Great Attractor. This region appears to be about 300 million light-years away, on the opposite side of the Milky Way from us, past the clusters Hydra and Centaurus.

The force that binds together these progressively larger structures, from star to galaxy to supercluster, and may be drawing them all toward the Great Attractor, is the gravitational force. That force not only holds you on Earth but also reaches out across intergalactic space.

Fig. 14-1 The Andromeda galaxy. Located 2.3×10^{6} light-years from us, and faintly visible to the naked eye, it is very similar to our home galaxy, the Milky Way.

14-2 Newton's Law of Gravitation

Physicists like to study seemingly unrelated phenomena to show that a relationship can be found if they are examined closely enough. This search for unification has been going on for centuries. In 1665, the 23-year-old Isaac Newton made a basic contribution to physics when he showed that the force that holds the Moon in its orbit is the same force that makes an apple fall. We take this so much for granted now that it is not easy for us to comprehend the ancient belief that the motions of earthbound bodies and heavenly bodies were different in kind and were governed by different laws.

Newton concluded that not only does Earth attract an apple and the Moon but every body in the universe attracts every other body; this tendency of bodies to move toward each other is called **gravitation.** Newton's conclusion takes a little getting used to, because the familiar attraction of Earth for earthbound bodies is so great that it overwhelms the attraction that earthbound bodies have for each other. For example, Earth attracts an apple with a force magnitude of about 0.8 N. You also attract a nearby apple (and it attracts you), but the force of attraction has less magnitude than the weight of a speck of dust.

Quantitatively, Newton proposed a *force law* that we call **Newton's law of gravitation:** Every particle attracts any other particle with a **gravitational force** whose magnitude is given by

$$F = G \frac{m_1 m_2}{r^2} \qquad \text{(Newton's law of gravitation).} \qquad (14\text{-}1)$$

Here m_1 and m_2 are the masses of the particles, r is the distance between them, and G is the **gravitational constant,** with a value that is now known to be

$$G = 6.67 \times 10^{-11} \, \text{N} \cdot \text{m}^2/\text{kg}^2$$
$$= 6.67 \times 10^{-11} \, \text{m}^3/\text{kg} \cdot \text{s}^2. \qquad (14\text{-}2)$$

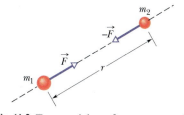

Fig. 14-2 Two particles, of masses m_1 and m_2 and with separation r, attract each other according to Newton's law of gravitation, Eq. 14-1. The forces of attraction, $\vec{F}$ and $-\vec{F}$, are equal in magnitude and in opposite directions.

As Fig. 14-2 shows, a particle m_2 attracts a particle m_1 with a gravitational force $\vec{F}$ that is directed toward particle m_2, and particle m_1 attracts particle m_2 with a gravitational force $-\vec{F}$ that is directed toward m_1. The forces $\vec{F}$ and $-\vec{F}$ form a third-law force pair; they are opposite in direction but equal in magnitude. They depend on the separation of the two particles, but not on their location: the particles could be in a deep cave or in deep space. Also, forces $\vec{F}$ and $-\vec{F}$ are not altered by the presence of other bodies, even if those bodies lie between the two particles we are considering.

The strength of the gravitational force—that is, how strongly two particles with given masses at a given separation attract each other—depends on the value of the gravitational constant G. If G—by some miracle—were suddenly multiplied by a factor of 10, you would be crushed to the floor by Earth's attraction. If G were divided by this factor, Earth's attraction would be weak enough that you could jump over a building.

Although Newton's law of gravitation applies strictly to particles, we can also apply it to real objects as long as the sizes of the objects are small compared to the distance between them. The Moon and Earth are far enough apart so that, to a good approximation, we can treat them both as particles—but what about an apple and Earth? From the point of view of the apple, the broad and level Earth, stretching out to the horizon beneath the apple, certainly does not look like a particle.

Newton solved the apple–Earth problem by proving an important theorem called the *shell theorem:*

> A uniform spherical shell of matter attracts a particle that is outside the shell as if all the shell's mass were concentrated at its center.

Earth can be thought of as a nest of such shells, one within another, and each attracting a particle outside Earth's surface as if the mass of that shell were at the center of the shell. Thus, from the apple's point of view, Earth *does* behave like a particle, one that is located at the center of Earth and has a mass equal to that of Earth.

Suppose, as in Fig. 14-3, that Earth pulls down on an apple with a force of magnitude 0.80 N. The apple must then pull up on Earth with a force of magnitude 0.80 N, which we take to act at the center of Earth. Although the forces are matched in magnitude, they produce different accelerations when the apple is released. For the apple, the acceleration is about 9.8 m/s^2, the familiar acceleration of a falling body near Earth's surface. For Earth, the acceleration measured in a reference frame attached to the center of mass of the apple–Earth system is only about 1×10^{-25} m/s^2.

Fig. 14-3 The apple pulls up on Earth just as much as Earth pulls down on the apple.

✓CHECKPOINT 1:   A particle is to be placed, in turn, outside four objects, each of mass m: (1) a large uniform solid sphere, (2) a large uniform spherical shell, (3) a small uniform solid sphere, and (4) a small uniform shell. In each situation, the distance between the particle and the center of the object is d. Rank the objects according to the magnitude of the gravitational force they exert on the particle, greatest first.

14-3   Gravitation and the Principle of Superposition

Given a group of particles, we find the net (or resultant) gravitational force on any one of them from the others by using the **principle of superposition.** This is a general principle that says a net effect is the sum of the individual effects. Here, the principle means that we first compute the gravitational force that acts on our selected particle due to each of the other particles, in turn. We then find the net force by adding these forces vectorially, as usual.

For n interacting particles, we can write the principle of superposition for gravitational forces as

$$\vec{F}_{1,\text{net}} = \vec{F}_{12} + \vec{F}_{13} + \vec{F}_{14} + \vec{F}_{15} + \cdots + \vec{F}_{1n}. \qquad (14\text{-}3)$$

Here $\vec{F}_{1,\text{net}}$ is the net force on particle 1 and, for example, $\vec{F}_{13}$ is the force on particle 1 from particle 3. We can express this equation more compactly as a vector sum:

$$\vec{F}_{1,\text{net}} = \sum_{i=2}^{n} \vec{F}_{1i}. \qquad (14\text{-}4)$$

What about the gravitational force on a particle from a real extended object? The force is found by dividing the object into parts small enough to treat as particles and then using Eq. 14-4 to find the vector sum of the forces on the particle from all the parts. In the limiting case, we can divide the extended object into differential parts of mass dm, each of which produces only a differential force $d\vec{F}$ on the particle. In this limit, the sum of Eq. 14-4 becomes an integral and we have

$$\vec{F}_1 = \int d\vec{F}, \qquad (14\text{-}5)$$

in which the integral is taken over the entire extended object and we drop the subscript "net." If the object is a uniform sphere or a spherical shell, we can avoid the integration of Eq. 14-5 by assuming that the object's mass is concentrated at the object's center and using Eq. 14-1.

Sample Problem 14-1

Figure 14-4a shows an arrangement of three particles, particle 1 having mass $m_1 = 6.0$ kg and particles 2 and 3 having mass $m_2 = m_3 = 4.0$ kg, and with distance $a = 2.0$ cm. What is the net gravitational force $\vec{F}_1$ that acts on particle 1 due to the other particles?

SOLUTION: One Key Idea here is that, because we have particles, the magnitude of the gravitational force on particle 1 due to either of the other particles is given by Eq. 14-1 ($F = Gm_1m_2/r^2$). Thus, the magnitude of the force $\vec{F}_{12}$ on particle 1 from particle 2 is

$$F_{12} = \frac{Gm_1m_2}{a^2}$$

$$= \frac{(6.67 \times 10^{-11}\ \text{m}^3/\text{kg} \cdot \text{s}^2)(6.0\ \text{kg})(4.0\ \text{kg})}{(0.020\ \text{m})^2}$$

$$= 4.00 \times 10^{-6}\ \text{N}.$$

Similarly, the magnitude of force $\vec{F}_{13}$ on particle 1 from particle 3 is

$$F_{13} = \frac{Gm_1m_3}{(2a)^2}$$

$$= \frac{(6.67 \times 10^{-11}\ \text{m}^3/\text{kg} \cdot \text{s}^2)(6.0\ \text{kg})(4.0\ \text{kg})}{(0.040\ \text{m})^2}$$

$$= 1.00 \times 10^{-6}\ \text{N}.$$

To determine the directions of $\vec{F}_{12}$ and $\vec{F}_{13}$, we use this Key Idea: Each force on particle 1 is directed toward the particle responsible for that force. Thus, $\vec{F}_{12}$ is directed in the positive direction of y (Fig. 14-4b) and has only the y component F_{12}. Similarly,

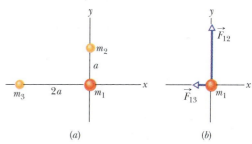

Fig. 14-4 Sample Problem 14-1. (a) An arrangement of three particles. (b) The forces acting on the particle of mass m_1 due to the other particles.

$\vec{F}_{13}$ is directed in the negative direction of x and has only the x component $-F_{13}$.

To find the net force $\vec{F}_{1,\text{net}}$ on particle 1, we first use this very important Key Idea: Because the forces are not directed along the same line, we *cannot* simply add or subtract their magnitudes or their components to get their net force. Instead, we must add them as vectors.

We can do so on a vector-capable calculator. However, here we note that $-F_{13}$ and F_{12} are actually the x and y components of $\vec{F}_{1,\text{net}}$. Therefore, we shall follow the guide of Eq. 3-6 to find first the magnitude and then the direction of $\vec{F}_{1,\text{net}}$. The magnitude is

$$F_{1,\text{net}} = \sqrt{(F_{12})^2 + (-F_{13})^2}$$

$$= \sqrt{(4.00 \times 10^{-6}\ \text{N})^2 + (-1.00 \times 10^{-6}\ \text{N})^2}$$

$$= 4.1 \times 10^{-6}\ \text{N}. \qquad \text{(Answer)}$$

Relative to the positive direction of the x axis, Eq. 3-6 gives the direction of $\vec{F}_{1,\text{net}}$ as

$$\theta = \tan^{-1}\frac{F_{12}}{-F_{13}} = \tan^{-1}\frac{4.00 \times 10^{-6}\ \text{N}}{-1.00 \times 10^{-6}\ \text{N}} = -76°.$$

Is this a reasonable direction? No, the direction of $\vec{F}_{1,\text{net}}$ must be between the directions of $\vec{F}_{12}$ and $\vec{F}_{13}$. Recall from Chapter 3 (Tactic 3) that a calculator displays only one of the two possible answers to a $\tan^{-1}$ function. We find the other answer by adding 180°. That gives us

$$-76° + 180° = 104°, \qquad \text{(Answer)}$$

which *is* a reasonable direction for $\vec{F}_{1,\text{net}}$.

✔**CHECKPOINT 2:**   The figure shows four arrangements of three particles of equal masses. (a) Rank the arrangements according to the magnitude of the net gravitational force on the particle labeled m, greatest first. (b) In arrangement 2, is the direction of the net force closer to the line of length d or to the line of length D?

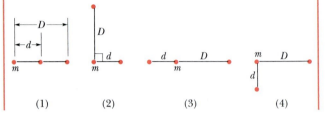

(1)   (2)   (3)   (4)

Sample Problem 14-2

Figure 14-5*a* shows an arrangement of five particles, with masses $m_1 = 8.0$ kg, $m_2 = m_3 = m_4 = m_5 = 2.0$ kg, and with $a = 2.0$ cm and $\theta = 30°$. What is the net gravitational force $\vec{F}_{1,\text{net}}$ on particle 1 due to the other particles?

SOLUTION: Our Key Ideas are the same as in Sample Problem 14-1. However, this problem has a lot of symmetry that can help simplify the solution.

For the magnitudes of the forces on particle 1, first note that particles 2 and 4 have equal masses and equal distances of $r = 2a$ from particle 1. Thus, from Eq. 14-1, we find

$$F_{12} = F_{14} = \frac{Gm_1m_2}{(2a)^2}. \qquad (14\text{-}6)$$

Similarly, since particles 3 and 5 have equal masses and are both a distance $r = a$ from particle 1, we find

$$F_{13} = F_{15} = \frac{Gm_1m_3}{a^2}. \qquad (14\text{-}7)$$

We can now substitute known data into these two equations to evaluate the magnitudes of the forces. Then we can indicate the directions of the forces on the free-body diagram of Fig. 14-5*b*, and find the net force in either of two basic ways: We could resolve the vectors into x and y components, find the net x component and the net y component, and then vectorially combine those net components. We could also add the vectors directly on a vector-capable calculator.

Instead, however, we shall make further use of the symmetry of the problem. First, we note that $\vec{F}_{12}$ and $\vec{F}_{14}$ are equal in magnitude but opposite in direction; thus, those forces *cancel*. Inspec-

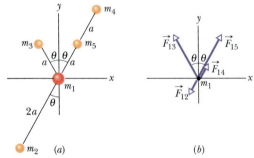

Fig. 14-5 Sample Problem 14-2. (*a*) An arrangement of five particles. (*b*) The forces acting on the particle of mass m_1 due to the other four particles.

tion of Fig. 14-5*b* and Eq. 14-7 reveals that the x components of $\vec{F}_{13}$ and $\vec{F}_{15}$ also *cancel*, and that their y components are identical in magnitude and both act in the positive direction of y. Thus, $\vec{F}_{1,\text{net}}$ acts in that same direction, and its magnitude is twice the y component of $\vec{F}_{13}$:

$$F_{1,\text{net}} = 2F_{13}\cos\theta = 2\frac{Gm_1m_3}{a^2}\cos\theta$$

$$= 2\frac{(6.67 \times 10^{-11}\ \text{m}^3/\text{kg}\cdot\text{s}^2)(8.0\ \text{kg})(2.0\ \text{kg})}{(0.020\ \text{m})^2}\cos 30°$$

$$= 4.6 \times 10^{-6}\ \text{N}. \qquad \text{(Answer)}$$

Note that the presence of particle 5 along the line between particles 1 and 4 does not alter the gravitational force on particle 1 from particle 4.

✔**CHECKPOINT 3:**   In the figure here, what is the direction of the net gravitational force on the particle of mass m_1 due to the other particles, each of mass m, that are arranged symmetrically relative to the y axis?

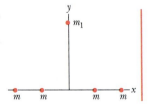

Tactic 1: *Drawing Gravitational Force Vectors*

When you are given a diagram of particles, such as Fig. 14-4a, and asked to find the net gravitational force on one of them, you should usually draw a free-body diagram showing only the particle of concern and the forces on *it alone*, as in Fig. 14-4b. If, instead, you choose to superimpose the force vectors on the given diagram, be sure to draw the vectors with either their tails (preferably) or their heads on the particle experiencing those forces. If you draw the vectors elsewhere, you invite confusion—and confusion is guar-

anteed if you draw the vectors on the particles *causing* the forces on the particle of concern.

Tactic 2: *Simplifying a Sum of Forces with Symmetry*

In Sample Problem 14-2 we used the symmetry of the situation: By realizing that particles 2 and 4 are positioned symmetrically about particle 1, and thus that $\vec{F}_{12}$ and $\vec{F}_{14}$ cancel, we avoided calculating either force. By realizing that the x components of $\vec{F}_{13}$ and $\vec{F}_{15}$ cancel and that their y components are identical and add, we saved even more effort.

14-4 Gravitation Near Earth's Surface

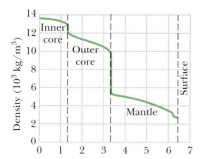

Fig. 14-6 The density of Earth as a function of distance from the center. The limits of the solid inner core, the largely liquid outer core, and the solid mantle are shown, but the crust of Earth is too thin to show clearly on this plot.

Let us assume that Earth is a uniform sphere of mass M. The magnitude of the gravitational force from Earth on a particle of mass m, located outside Earth a distance r from Earth's center, is then given by Eq. 14-1 as

$$F = G\frac{Mm}{r^2}. \tag{14-8}$$

If the particle is released, it will fall toward the center of Earth, as a result of the gravitational force $\vec{F}$, with an acceleration we shall call the **gravitational acceleration** $\vec{a}_g$. Newton's second law tells us that magnitudes F and a_g are related by

$$F = ma_g. \tag{14-9}$$

Now, substituting F from Eq. 14-8 into Eq. 14-9 and solving for a_g, we find

$$a_g = \frac{GM}{r^2}. \tag{14-10}$$

Table 14-1 shows values of a_g computed for various altitudes above Earth's surface.

Since Section 5-6, we have assumed that Earth is an inertial frame by neglecting its actual rotation. This simplification has allowed us to assume that the actual free-fall acceleration g of a particle is the same as the gravitational acceleration (which we now call a_g). Furthermore, we assumed that g has the constant value of 9.8 m/s² over Earth's surface. However, the g we would measure differs from the a_g we would calculate with Eq. 14-10 for three reasons: (1) Earth is not uniform, (2) it is not a perfect sphere, and (3) it rotates. Moreover, because g differs from a_g, the measured weight mg of the particle differs from the magnitude of the gravitational force on the particle as given by Eq. 14-8 for the same three reasons. Let us now examine those reasons.

1. **Earth is not uniform.** The density (mass per unit volume) of Earth varies radially as shown in Fig. 14-6, and the density of the crust (or outer section) of Earth varies from region to region over Earth's surface. Thus, g varies from region to region over the surface.

TABLE 14-1   Variation of a_g with Altitude

Altitude (km)	a_g (m/s²)	Altitude Example
0	9.83	Mean Earth surface
8.8	9.80	Mt. Everest
36.6	9.71	Highest manned balloon
400	8.70	Space shuttle orbit
35 700	0.225	Communications satellite

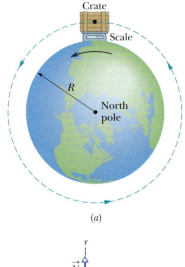

(a)

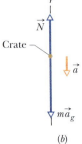

(b)

Fig. 14-7 (*a*) A crate lies on a scale at Earth's equator, as seen along Earth's rotation axis from above the north pole. (*b*) A free-body diagram for the crate, with a radially outward *r* axis. The gravitational force on the crate is represented with its equivalent $m\vec{a}_g$. The normal force on the crate from the scale is $\vec{N}$. Because of Earth's rotation, the crate has a centripetal acceleration $\vec{a}$ that is directed toward Earth's center.

2. Earth is not a sphere. Earth is approximately an ellipsoid, flattened at the poles and bulging at the equator. Its equatorial radius is greater than its polar radius by 21 km. Thus, a point at the poles is closer to the dense core of Earth than is a point on the equator. This is one reason the free-fall acceleration *g* increases as one proceeds, at sea level, from the equator toward either pole.

3. Earth is rotating. The rotation axis runs through the north and south poles of Earth. An object located on Earth's surface anywhere except at those poles must rotate in a circle about the rotation axis and thus must have a centripetal acceleration directed toward the center of the circle. This centripetal acceleration requires a centripetal net force that is also directed toward that center.

To see how Earth's rotation causes *g* to differ from a_g, let us analyze a simple situation in which a crate of mass *m* is on a scale at the equator. Figure 14-7*a* shows this situation as viewed from a point in space above the north pole.

Figure 14-7*b*, a free-body diagram for the crate, shows the two forces on the crate, both acting along a radial axis *r* that extends from Earth's center. The normal force $\vec{N}$ on the crate from the scale is directed outward, in the positive direction of axis *r*. The gravitational force, represented with its equivalent $m\vec{a}_g$, is directed inward. Because the crate travels in a circle about the center of Earth as Earth turns, it has a centripetal acceleration $\vec{a}$ directed inward. From Eq. 11-23, we know this acceleration is equal to $\omega^2 R$, where ω is Earth's angular speed and *R* is the circle's radius (approximately Earth's radius). Thus, we can write Newton's second law for the *r* axis ($F_{\text{net},r} = ma_r$) as

$$N - ma_g = m(-\omega^2 R). \qquad (14\text{-}11)$$

The magnitude *N* of the normal force is equal to the weight *mg* read on the scale. With *mg* substituted for *N*, Eq. 14-11 gives us

$$mg = ma_g - m(\omega^2 R), \qquad (14\text{-}12)$$

which says

$$\left(\begin{array}{c}\text{measured}\\\text{weight}\end{array}\right) = \left(\begin{array}{c}\text{magnitude of}\\\text{gravitational force}\end{array}\right) - \left(\begin{array}{c}\text{mass times}\\\text{centripetal acceleration}\end{array}\right).$$

Thus, the measured weight is actually less than the magnitude of the gravitational force on the crate, because of Earth's rotation.

To find a corresponding expression for *g* and a_g, we cancel *m* from Eq. 14-12 to write

$$g = a_g - \omega^2 R, \qquad (14\text{-}13)$$

which says

$$\left(\begin{array}{c}\text{free-fall}\\\text{acceleration}\end{array}\right) = \left(\begin{array}{c}\text{gravitational}\\\text{acceleration}\end{array}\right) - \left(\begin{array}{c}\text{centripetal}\\\text{acceleration}\end{array}\right).$$

Thus, the measured free-fall acceleration is actually less than the gravitational acceleration, because of Earth's rotation.

The difference between accelerations *g* and a_g is equal to $\omega^2 R$ and is greatest on the equator (for one reason, the radius of the circle traveled by the crate is greatest there). To find the difference, we can use Eq. 11-5 ($\omega = \Delta\theta/\Delta t$) and Earth's radius $R = 6.37 \times 10^6$ m. For one rotation of Earth, θ is 2π rad and the time period Δt is about 24 h. Using these values (and converting hours to seconds), we find that *g* is less than a_g by only about 0.034 m/s² (compared to 9.8 m/s²). Therefore, neglecting the difference in accelerations *g* and a_g is often justified. Similarly, neglecting the difference between weight and the magnitude of the gravitational acceleration is also often justified.

Sample Problem 14-3

(a) An astronaut whose height h is 1.70 m floats "feet down" in an orbiting space shuttle at a distance $r = 6.77 \times 10^6$ m from the center of Earth. What is the difference between the gravitational acceleration at her feet and that at her head?

SOLUTION: One Key Idea here is that we can approximate Earth as a uniform sphere of mass M_E. Then, from Eq. 14-10, the gravitational acceleration at any distance r from the center of Earth is

$$a_g = \frac{GM_E}{r^2}. \qquad (14\text{-}14)$$

We might simply apply Eq. 14-14 twice, first with, say, $r = 6.77 \times 10^6$ m for the feet and then with $r = 6.77 \times 10^6$ m + 1.70 m for the head. However, a calculator may give us the same value for a_g twice, and thus a difference of zero, because h is so small compared to r. A second Key Idea helps here: Because we have a differential change dr in r between the astronaut's feet and head, let us differentiate Eq. 14-14 with respect to r. That gives us

$$da_g = -2\frac{GM_E}{r^3} dr, \qquad (14\text{-}15)$$

where da_g is the differential change in the gravitational acceleration due to the differential change dr in r. For the astronaut, $dr = h$ and $r = 6.77 \times 10^6$ m. Substituting data into Eq. 14-15, we find

$$da_g = -2\frac{(6.67 \times 10^{-11}\ \text{m}^3/\text{kg} \cdot \text{s}^2)(5.98 \times 10^{24}\ \text{kg})}{(6.77 \times 10^6\ \text{m})^3}(1.70\ \text{m})$$

$$= -4.37 \times 10^{-6}\ \text{m/s}^2. \qquad \text{(Answer)}$$

This result means that the gravitational acceleration of the astronaut's feet toward Earth is slightly greater than the gravitational acceleration of her head toward Earth. This difference in acceleration tends to stretch her body, but the difference is so small that the stretching is unnoticeable.

(b) If the astronaut is now "feet down" at the same orbital radius r of 6.77×10^6 m about a black hole of mass $M_h = 1.99 \times 10^{31}$ kg (which is 10 times our Sun's mass), what is now the difference between the gravitational acceleration at her feet and that at her head? The black hole has a surface (called its *event horizon*) of radius $R_h = 2.95 \times 10^4$ m. Nothing, not even light, can escape from that surface or anywhere inside it. Note that the astronaut is (wisely) well outside the surface (at $r = 229R_h$).

SOLUTION: The Key Idea here is that we again have a differential change dr in r between the astronaut's feet and head, so we can again use Eq. 14-15. However, now we substitute $M_h = 1.99 \times 10^{31}$ kg for M_E. We find

$$da_g = -2\frac{(6.67 \times 10^{-11}\ \text{m}^3/\text{kg} \cdot \text{s}^2)(1.99 \times 10^{31}\ \text{kg})}{(6.77 \times 10^6\ \text{m})^3}(1.70\ \text{m})$$

$$= -14.5\ \text{m/s}^2. \qquad \text{(Answer)}$$

This means that the gravitational acceleration of the astronaut's feet toward the black hole is noticeably larger than that of her head. The resulting tendency to stretch her body would be bearable but quite painful. If she drifted closer to the black hole, the stretching tendency would increase drastically.

14-5 Gravitation Inside Earth

Newton's shell theorem can also be applied to a situation in which a particle is located *inside* a uniform shell, to show the following:

> A uniform shell of matter exerts no *net* gravitational force on a particle located inside it.

Caution: This statement does *not* mean that the gravitational forces on the particle from the various elements of the shell magically disappear. Rather, it means that the *sum* of the force vectors on the particle from all the elements is zero.

If the density of Earth were uniform, the gravitational force acting on a particle would be a maximum at Earth's surface and would decrease as the particle moved outward. If the particle were to move inward, perhaps down a deep mine shaft, the gravitational force would change for two reasons. (1) It would tend to increase because the particle would be moving closer to the center of Earth. (2) It would tend to decrease because the thickening shell of material lying outside the particle's radial position would not exert any net force on the particle.

For a uniform Earth, the second influence would prevail and the force on the particle would steadily decrease to zero as the particle approached the center of Earth. However, for the real (nonuniform) Earth, the force on the particle actually increases as the particle begins to descend. The force reaches a maximum at a certain depth; only then does it begin to decrease as the particle descends farther.

Sample Problem 14-4

In *Pole to Pole,* an early science fiction story by George Griffith, three explorers attempt to travel by capsule through a naturally formed (and, of course, fictional) tunnel directly from the south pole to the north pole (Fig. 14-8). According to the story, as the capsule approaches Earth's center, the gravitational force on the explorers becomes alarmingly large and then, exactly at the center, it suddenly but only momentarily disappears. Then the capsule travels through the second half of the tunnel, to the north pole.

Check Griffith's description by finding the gravitational force on the capsule of mass m when it reaches a distance r from Earth's center. Assume that Earth is a sphere of uniform density ρ (mass per unit volume).

SOLUTION: Newton's shell theorem gives us three Key Ideas here:

1. When the capsule is at a radius r from Earth's center, the portion of Earth that lies outside a sphere of radius r does *not* produce a net gravitational force on the capsule.

2. The portion that lies inside that sphere *does* produce a net gravitational force on the capsule.

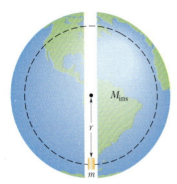

Fig. 14-8 Sample Problem 14-4. A capsule of mass m falls from rest through a tunnel that connects Earth's south and north poles. When the capsule is at distance r from Earth's center, the portion of Earth's mass that is contained in a sphere of that radius is M_{ins}.

3. We can treat the mass M_{ins} of that inside portion of Earth as being the mass of a particle located at Earth's center.

All three ideas tell us that we can write Eq. 14-1, for the magnitude of the gravitational force on the capsule, as

$$F = \frac{GmM_{\text{ins}}}{r^2}. \qquad (14\text{-}16)$$

To write the mass M_{ins} in terms of the radius r, we note that the volume V_{ins} containing this mass is $\frac{4}{3}\pi r^3$. Also, its density is Earth's density ρ. Thus, we have

$$M_{\text{ins}} = \rho V_{\text{ins}} = \rho \frac{4\pi r^3}{3}. \qquad (14\text{-}17)$$

Then, after substituting this expression into Eq. 14-16 and canceling, we have

$$F = \frac{4\pi Gm\rho}{3} r. \qquad \text{(Answer)} \quad (14\text{-}18)$$

This equation tells us that the force magnitude F depends linearly on the capsule's distance r from Earth's center. Thus, as r decreases, F also decreases (opposite of Griffith's description), until it is zero at Earth's center. At least Griffith got zero-at-the-center correct.

Equation 14-18 can also be written in terms of the force vector $\vec{F}$ and the capsule's position vector $\vec{r}$ along a radial axis extending from Earth's center. Let K represent the collection of constants $4\pi Gm\rho/3$. Then, Eq. 14-18 becomes

$$\vec{F} = -K\vec{r}, \qquad (14\text{-}19)$$

in which we have inserted a minus sign to indicate that $\vec{F}$ and $\vec{r}$ have opposite directions. Equation 14-19 has the form of Hooke's law (Eq. 7-20). Thus, under the idealized conditions of the story, the capsule would oscillate like a block on a spring, with the center of the oscillation at Earth's center. After the capsule had fallen from the south pole to Earth's center, it would travel from the center to the north pole (as Griffith said) and then back again.

14-6 Gravitational Potential Energy

In Section 8-3, we discussed the gravitational potential energy of a particle–Earth system. We were careful to keep the particle near Earth's surface, so that we could regard the gravitational force as constant. We then chose some reference configuration of the system as having a gravitational potential energy of zero. Often, in this configuration the particle was on Earth's surface. For particles not on Earth's surface, the gravitational potential energy decreased when the separation between the particle and Earth decreased.

Here, we broaden our view and consider the gravitational potential energy U of two particles, of masses m and M, separated by a distance r. We again choose a reference configuration with U equal to zero. However, to simplify the equations, the separation distance r in the reference configuration is now large enough to be approximated as *infinite*. As before, the gravitational potential energy decreases when

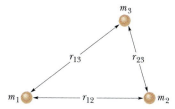

Fig. 14-9 Three particles form a system. (The separation for each pair of particles is labeled with a double subscript to indicate the particles.) The gravitational potential energy *of the system* is the sum of the gravitational potential energies of all three pairs of particles.

the separation decreases. Since $U = 0$ for $r = \infty$, the potential energy is negative for any finite separation and becomes progressively more negative as the particles move closer together.

With these facts in mind and as we shall justify next, we take the gravitational potential energy of the two-particle system to be

$$U = -\frac{GMm}{r} \qquad \text{(gravitational potential energy).} \qquad (14\text{-}20)$$

Note that $U(r)$ approaches zero as r approaches infinity and that for any finite value of r, the value of $U(r)$ is negative.

The potential energy given by Eq. 14-20 is a property of the system of two particles rather than of either particle alone. There is no way to divide this energy and say that so much belongs to one particle and so much to the other. However, if $M \gg m$, as is true for Earth (mass M) and a baseball (mass m), we often speak of "the potential energy of the baseball." We can get away with this because, when a baseball moves in the vicinity of Earth, changes in the potential energy of the baseball–Earth system appear almost entirely as changes in the kinetic energy of the baseball, since changes in the kinetic energy of Earth are too small to be measured. Similarly, in Section 14-8 we shall speak of "the potential energy of an artificial satellite" orbiting Earth, because the satellite's mass is so much smaller than Earth's mass. When we speak of the potential energy of bodies of comparable mass, however, we have to be careful to treat them as a system.

If our system contains more than two particles, we consider each pair of particles in turn, calculate the gravitational potential energy of that pair with Eq. 14-20 as if the other particles were not there, and then algebraically sum the results. Applying Eq. 14-20 to each of the three pairs of Fig. 14-9, for example, gives the potential energy of the system as

$$U = -\left(\frac{Gm_1m_2}{r_{12}} + \frac{Gm_1m_3}{r_{13}} + \frac{Gm_2m_3}{r_{23}}\right). \qquad (14\text{-}21)$$

Proof of Equation 14-20

Let us shoot a baseball directly away from Earth along the path in Fig. 14-10. We want to find an expression for the gravitational potential energy U of the ball at point P along its path, at radial distance R from Earth's center. To do so, we first find the work W done on the ball by the gravitational force as the ball travels from point P to a great (infinite) distance from Earth. Because the gravitational force $\vec{F}(r)$ is a variable force (its magnitude depends on r), we must use the techniques of Section 7-6 to find the work. In vector notation, we can write

$$W = \int_R^\infty \vec{F}(r) \cdot d\vec{r}. \qquad (14\text{-}22)$$

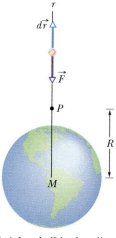

Fig. 14-10 A baseball is shot directly away from Earth, through point P at radial distance R from Earth's center. The gravitational force $\vec{F}$ on the ball and a differential displacement vector $d\vec{r}$ are shown, both directed along a radial r axis.

The integral contains the scalar (or dot) product of the force $\vec{F}(r)$ and the differential displacement vector $d\vec{r}$ along the ball's path. We can expand that product as

$$\vec{F}(r) \cdot d\vec{r} = F(r)\, dr \cos \phi, \qquad (14\text{-}23)$$

where ϕ is the angle between the directions of $\vec{F}(r)$ and $d\vec{r}$. When we substitute $180°$ for ϕ and Eq. 14-1 for $F(r)$, Eq. 14-23 becomes

$$\vec{F}(r) \cdot d\vec{r} = -\frac{GMm}{r^2}\, dr,$$

where M is Earth's mass and m is the mass of the ball.

Substituting this into Eq. 14-22 and integrating gives us

$$W = -GMm \int_R^\infty \frac{1}{r^2} \, dr = \left[\frac{GMm}{r} \right]_R^\infty$$

$$= 0 - \frac{GMm}{R} = -\frac{GMm}{R}. \quad (14\text{-}24)$$

W in Eq. 14-24 is the work required to move the ball from point P (at distance R) to infinity. Equation 8-1 ($\Delta U = -W$) tells us that we can also write that work in terms of potential energies as

$$U_\infty - U = -W.$$

The potential energy U_∞ at infinity is zero, and U is the potential energy at P. Thus, with Eq. 14-24 substituted for W, the previous equation becomes

$$U = W = -\frac{GMm}{R}.$$

Switching R to r gives us Eq. 14-20, which we set out to prove.

Path Independence

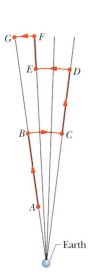

Fig. 14-11 Near Earth, a baseball is moved from point A to point G along a path consisting of radial lengths and circular arcs.

In Fig. 14-11, we move a baseball from point A to point G along a path consisting of three radial lengths and three circular arcs (centered on Earth). We are interested in the total work W done by Earth's gravitational force $\vec{F}$ on the ball as it moves from A to G. The work done along each circular arc is zero, because the direction of $\vec{F}$ is perpendicular to the arc at every point. Thus, the only works done by $\vec{F}$ are along the three radial lengths, and the total work W is the sum of those works.

Now, suppose we mentally shrink the arcs to zero. We would then be moving the ball directly from A to G along a single radial length. Does that change W? No. Because no work was done along the arcs, eliminating them does not change the work. The path taken from A to G now is clearly different, but the work done by $\vec{F}$ is the same.

We discussed such a result in a general way in Section 8-2. Here is the point: The gravitational force is a conservative force. Thus, the work done by the gravitational force on a particle moving from an initial point i to a final point f is independent of the actual path taken between the points. From Eq. 8-1, the change ΔU in the gravitational potential energy from point i to point f is given by

$$\Delta U = U_f - U_i = -W. \quad (14\text{-}25)$$

Since the work W done by a conservative force is independent of the actual path taken, the change ΔU in gravitational potential energy is *also independent* of the actual path taken.

Potential Energy and Force

In the proof of Eq. 14-20, we derived the potential energy function $U(r)$ from the force function $\vec{F}(r)$. We should be able to go the other way—that is, to start from the potential energy function and derive the force function. Guided by Eq. 8-20, we can write

$$F = -\frac{dU}{dr} = -\frac{d}{dr}\left(-\frac{GMm}{r} \right)$$

$$= -\frac{GMm}{r^2}. \quad (14\text{-}26)$$

This is Newton's law of gravitation (Eq. 14-1). The minus sign indicates that the force on mass m points radially inward, toward mass M.

Escape Speed

If you fire a projectile upward, usually it will slow, stop momentarily, and return to Earth. There is, however, a certain minimum initial speed that will cause it to move upward forever, theoretically coming to rest only at infinity. This initial speed is called the (Earth) **escape speed.**

Consider a projectile of mass m, leaving the surface of a planet (or some other astronomical body or system) with escape speed v. It has a kinetic energy K given by $\frac{1}{2}mv^2$ and a potential energy U given by Eq. 14-20:

$$U = -\frac{GMm}{R},$$

in which M is the mass of the planet, and R is its radius.

When the projectile reaches infinity, it stops and thus has no kinetic energy. It also has no potential energy because this is our zero-potential-energy configuration. Its total energy at infinity is therefore zero. From the principle of conservation of energy, its total energy at the planet's surface must also have been zero, so

$$K + U = \frac{1}{2}mv^2 + \left(-\frac{GMm}{R}\right) = 0.$$

This yields
$$v = \sqrt{\frac{2GM}{R}}. \qquad (14\text{-}27)$$

The escape speed v does not depend on the direction in which a projectile is fired from a planet. However, attaining that speed is easier if the projectile is fired in the direction the launch site is moving as the planet rotates about its axis. For example, rockets are launched eastward at Cape Canaveral to take advantage of the Cape's eastward speed of 1500 km/h due to Earth's rotation.

Equation 14-27 can be applied to find the escape speed of a projectile from any astronomical body, provided we substitute the mass of the body for M and the radius of the body for R. Table 14-2 shows escape speeds from some astronomical bodies.

TABLE 14-2   Some Escape Speeds

Body	Mass (kg)	Radius (m)	Escape Speed (km/s)
Ceres[a]	1.17×10^{21}	3.8×10^5	0.64
Earth's moon[a]	7.36×10^{22}	1.74×10^6	2.38
Earth	5.98×10^{24}	6.37×10^6	11.2
Jupiter	1.90×10^{27}	7.15×10^7	59.5
Sun	1.99×10^{30}	6.96×10^8	618
Sirius B[b]	2×10^{30}	1×10^7	5200
Neutron star[c]	2×10^{30}	1×10^4	2×10^5

[a] The most massive of the asteroids.

[b] A *white dwarf* (a star in a final stage of evolution) that is a companion of the bright star Sirius.

[c] The collapsed core of a star that remains after that star has exploded in a *supernova* event.

CHECKPOINT 4:   You move a ball of mass m away from a sphere of mass M. (a) Does the gravitational potential energy of the ball–sphere system increase or decrease? (b) Is positive or negative work done by the gravitational force between the ball and the sphere?

Sample Problem 14-5

An asteroid, headed directly toward Earth, has a speed of 12 km/s relative to the planet when it is at a distance of 10 Earth radii from Earth's center. Neglecting the effects of Earth's atmosphere on the asteroid, find the asteroid's speed v_f when it reaches Earth's surface.

SOLUTION: One Key Idea is that, because we are to neglect the effects of the atmosphere on the asteroid, the mechanical energy of the asteroid–Earth system is conserved during the fall. Thus, the final mechanical energy (when the asteroid reaches Earth's surface) is equal to the initial mechanical energy. We can write this as

$$K_f + U_f = K_i + U_i, \qquad (14\text{-}28)$$

where K is kinetic energy and U is gravitational potential energy.

A second Key Idea is that, if we assume the system is isolated, the system's linear momentum must be conserved during the fall. Therefore, the momentum change of the asteroid and that of Earth must be equal in magnitude and opposite in sign. However, because Earth's mass is so great relative to the asteroid's mass, the change in Earth's speed is negligible relative to the change in the asteroid's speed. So, the change in Earth's kinetic energy is also negligible. Thus, we can assume that the kinetic energies in Eq. 14-28 are those of the asteroid alone.

Let m represent the asteroid's mass and M represent Earth's mass (5.98×10^{24} kg). The asteroid is initially at the distance $10R_E$ and finally at the distance R_E, where R_E is Earth's radius (6.37×10^6 m). Substituting Eq. 14-20 for U and $\frac{1}{2}mv^2$ for K, we

rewrite Eq. 14-28 as

$$\tfrac{1}{2}mv_f^2 - \frac{GMm}{R_E} = \tfrac{1}{2}mv_i^2 - \frac{GMm}{10R_E}.$$

Rearranging and substituting known values, we find

$$
\begin{aligned}
v_f^2 &= v_i^2 + \frac{2GM}{R_E}\left(1 - \frac{1}{10}\right) \\
&= (12 \times 10^3 \text{ m/s})^2 \\
&\quad + \frac{2(6.67 \times 10^{-11} \text{ m}^3/\text{kg} \cdot \text{s}^2)(5.98 \times 10^{24} \text{ kg})}{6.37 \times 10^6 \text{ m}} 0.9 \\
&= 2.567 \times 10^8 \text{ m}^2/\text{s}^2,
\end{aligned}
$$

and thus

$$v_f = 1.60 \times 10^4 \text{ m/s} = 16 \text{ km/s}. \qquad \text{(Answer)}$$

At this speed, the asteroid would not have to be particularly large to do considerable damage at impact. As an example, if it were only 5 m across, the impact could release about as much energy as the nuclear explosion at Hiroshima. Alarmingly, about 500 million asteroids of this size are near Earth's orbit, and in 1994 one of them apparently penetrated Earth's atmosphere and exploded at an altitude of 20 km near a remote South Pacific island (setting off nuclear-explosion warnings on six military satellites). The impact of an asteroid 500 m across (there may be a million of them near Earth's orbit) could end modern civilization and almost eliminate humans worldwide.

Fig. 14-12 The path of the planet Mars as it moved against a background of the constellation Capricorn during 1971. Its position on four selected days is marked. Both Mars and Earth are moving in orbits around the Sun so that we see the position of Mars relative to us; this sometimes results in an apparent loop in the path of Mars.

14-7 Planets and Satellites: Kepler's Laws

The motions of the planets, as they seemingly wander against the background of the stars, have been a puzzle since the dawn of history. The "loop-the-loop" motion of Mars, shown in Fig. 14-12, was particularly baffling. Johannes Kepler (1571–1630), after a lifetime of study, worked out the empirical laws that govern these motions. Tycho Brahe (1546–1601), the last of the great astronomers to make observations without the help of a telescope, compiled the extensive data from which Kepler was able to derive the three laws of planetary motion that now bear his name. Later, Newton (1642–1727) showed that his law of gravitation leads to Kepler's laws.

In this section we discuss each of Kepler's laws in turn. Although here we apply the laws to planets orbiting the Sun, they hold equally well for satellites, either natural or artificial, orbiting Earth or any other massive central body.

▶ 1. THE LAW OF ORBITS: All planets move in elliptical orbits, with the Sun at one focus.

Figure 14-13 shows a planet of mass m moving in such an orbit around the Sun, whose mass is M. We assume that $M \gg m$, so that the center of mass of the planet–Sun system is approximately at the center of the Sun.

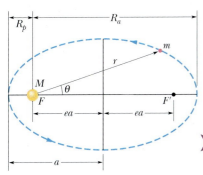

Fig. 14-13 A planet of mass m moving in an elliptical orbit around the Sun. The Sun, of mass M, is at one focus F of the ellipse. The other focus is F', which is located in empty space. Each focus is a distance ea from the ellipse's center, with e being the eccentricity of the ellipse. The semimajor axis a of the ellipse, the perihelion (nearest the Sun) distance R_p, and the aphelion (farthest from the Sun) distance R_a are also shown.

The orbit in Fig. 14-13 is described by giving its **semimajor axis** a and its **eccentricity** e, the latter defined so that ea is the distance from the center of the ellipse to either focus F or F'. *An eccentricity of zero corresponds to a circle,* in which the two foci merge to a single central point. The eccentricities of the planetary orbits are not large, so—sketched on paper—the orbits look circular. The eccentricity of the ellipse of Fig. 14-13, which has been exaggerated for clarity, is 0.74. The eccentricity of Earth's orbit is only 0.0167.

▶ **2. THE LAW OF AREAS:** A line that connects a planet to the Sun sweeps out equal areas in the plane of the planet's orbit in equal times; that is, the rate dA/dt at which it sweeps out area A is constant.

Qualitatively, this second law tells us that the planet will move most slowly when it is farthest from the Sun and most rapidly when it is nearest to the Sun. As it turns out, Kepler's second law is totally equivalent to the law of conservation of angular momentum. Let us prove it.

The area of the shaded wedge in Fig. 14-14a closely approximates the area swept out in time Δt by a line connecting the Sun and the planet, which are separated by a distance r. The area ΔA of the wedge is approximately the area of a triangle with base $r\,\Delta\theta$ and height r. Since the area of a triangle is one-half of the base times the height, $\Delta A \approx \frac{1}{2}r^2\,\Delta\theta$. This expression for ΔA becomes more exact as Δt (hence $\Delta\theta$) approaches zero. The instantaneous rate at which area is being swept out is then

$$\frac{dA}{dt} = \tfrac{1}{2}r^2\frac{d\theta}{dt} = \tfrac{1}{2}r^2\omega, \tag{14-29}$$

in which ω is the angular speed of the rotating line connecting Sun and planet.

Figure 14-14b shows the linear momentum $\vec{p}$ of the planet, along with its radial and perpendicular components. From Eq. 12-20 ($L = rp_\perp$), the magnitude of the angular momentum $\vec{L}$ of the planet about the Sun is given by the product of r and $p_\perp$, the component of $\vec{p}$ perpendicular to r. Here, for a planet of mass m,

$$L = rp_\perp = (r)(mv_\perp) = (r)(m\omega r)$$
$$= mr^2\omega, \tag{14-30}$$

where we have replaced $v_\perp$ with its equivalent ωr (Eq. 11-18). Eliminating $r^2\omega$ between Eqs. 14-29 and 14-30 leads to

$$\frac{dA}{dt} = \frac{L}{2m}. \tag{14-31}$$

If dA/dt is constant, as Kepler said it is, then Eq. 14-31 means that L must also be constant—angular momentum is conserved. Kepler's second law is indeed equivalent to the law of conservation of angular momentum.

Fig. 14-14 (a) In time Δt, the line r connecting the planet to the Sun (of mass M) sweeps through an angle $\Delta\theta$, sweeping out an area ΔA (shaded). (b) The linear momentum $\vec{p}$ of the planet and its components.

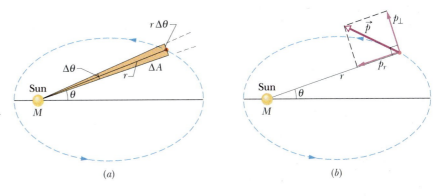

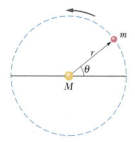

Fig. 14-15 A planet of mass m moving around the Sun in a circular orbit of radius r.

3. THE LAW OF PERIODS: The square of the period of any planet is proportional to the cube of the semimajor axis of its orbit.

To see this, consider the circular orbit of Fig. 14-15, with radius r (the radius of a circle is equivalent to the semimajor axis of an ellipse). Applying Newton's second law ($F = ma$) to the orbiting planet in Fig. 14-15 yields

$$\frac{GMm}{r^2} = (m)(\omega^2 r). \qquad (14\text{-}32)$$

Here we have substituted from Eq. 14-1 for the force magnitude F and used Eq. 11-23 to substitute $\omega^2 r$ for the centripetal acceleration. If we use Eq. 11-20 to replace ω with $2\pi/T$, where T is the period of the motion, we obtain Kepler's third law:

$$T^2 = \left(\frac{4\pi^2}{GM}\right) r^3 \qquad \text{(law of periods).} \qquad (14\text{-}33)$$

The quantity in parentheses is a constant that depends only on the mass M of the central body about which the planet orbits.

Equation 14-33 holds also for elliptical orbits, provided we replace r with a, the semimajor axis of the ellipse. This law predicts that the ratio T^2/a^3 has essentially the same value for every planetary orbit around a given massive body. Table 14-3 shows how well it holds for the orbits of the planets of the solar system.

TABLE 14-3   Kepler's Law of Periods for the Solar System

Planet	Semimajor Axis a (10^{10} m)	Period T (y)	T^2/a^3 (10^{-34} y²/m³)
Mercury	5.79	0.241	2.99
Venus	10.8	0.615	3.00
Earth	15.0	1.00	2.96
Mars	22.8	1.88	2.98
Jupiter	77.8	11.9	3.01
Saturn	143	29.5	2.98
Uranus	287	84.0	2.98
Neptune	450	165	2.99
Pluto	590	248	2.99

CHECKPOINT 5:   Satellite 1 is in a certain circular orbit about a planet, while satellite 2 is in a larger circular orbit. Which satellite has (a) the longer period and (b) the greater speed?

Sample Problem 14-6

Comet Halley orbits about the Sun with a period of 76 years and, in 1986, had a distance of closest approach to the Sun, its *perihelion distance* R_p, of 8.9×10^{10} m. Table 14-3 shows that this is between the orbits of Mercury and Venus.

(a) What is the comet's farthest distance from the Sun, its *aphelion distance* R_a?

SOLUTION: One Key Idea comes from Fig. 14-13, in which we see that $R_a + R_p = 2a$, where a is the semimajor axis of the orbit of comet

Halley. Thus, we can find R_a if we first find a. A second Key Idea is that we can relate a to the given period via the law of periods (Eq. 14-33) if we simply substitute the semimajor axis a for r. Doing so and then solving for a, we have

$$a = \left(\frac{GMT^2}{4\pi^2}\right)^{1/3}. \qquad (14\text{-}34)$$

If we substitute the mass M of the Sun, 1.99×10^{30} kg, and the period T of the comet, 76 years or 2.4×10^9 s, into Eq. 14-34, we find that $a = 2.7 \times 10^{12}$ m. Now we have

$$R_a = 2a - R_p$$
$$= (2)(2.7 \times 10^{12} \text{ m}) - 8.9 \times 10^{10} \text{ m}$$
$$= 5.3 \times 10^{12} \text{ m}. \qquad \text{(Answer)}$$

Table 14-3 shows that this is a little less than the semimajor axis of the orbit of Pluto. Thus, the comet does not get farther from the Sun than Pluto.

(b) What is the eccentricity e of the orbit of comet Halley?

SOLUTION: The Key Idea here is that we can relate e, a, and R_p via Fig. 14-13. We see there that $ea = a - R_p$, or

$$e = \frac{a - R_p}{a} = 1 - \frac{R_p}{a}$$

$$= 1 - \frac{8.9 \times 10^{10} \text{ m}}{2.7 \times 10^{12} \text{ m}} = 0.97. \qquad \text{(Answer)}$$

This cometary orbit, with an eccentricity approaching unity, is a long thin ellipse.

Sample Problem 14-7

Hunting a black hole. Observations of the light from a certain star indicate that it is part of a binary (two-star) system. This visible star has orbital speed $v = 270$ km/s, orbital period $T = 1.70$ days, and approximate mass $m_1 = 6M_s$, where M_s is the Sun's mass, 1.99×10^{30} kg. Assuming that the visible star and its companion star, which is dark and unseen, are both in circular orbits (see Fig. 14-16), determine the approximate mass m_2 of the dark star.

SOLUTION: Some of the Key Ideas in this challenging problem are as follows:

1. The two stars are in circular orbits, not about each other, but about the center of mass of this two-star system.

2. As with the two-particle systems of Section 9-2, the center of mass of the two-star system must lie along a line connecting the centers of the stars—that is, at point O in Fig. 14-16. The visible star orbits at radius r_1, the dark star at radius r_2.

3. The center of mass of the system is not even approximately at the center of a central, massive object (like the Sun). Therefore, Kepler's law of periods, Eq. 14-33, does *not* apply here and we cannot easily find mass m_2 with it.

4. The centripetal force causing each star to move in a circle is the gravitational force due to the other star. The magnitude of the force is Gm_1m_2/r^2, where r is the distance between the centers of the stars.

5. From Eq. 4-32, the centripetal acceleration a of the visible star is v^2/r_1.

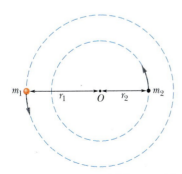

Fig. 14-16 Sample Problem 14-7. A visible star with mass m_1 and a dark, unseen star with mass m_2 orbit around the center of mass of the two-star system at O.

These ideas lead us to write Newton's second law ($F = ma$) for the visible star as

$$\frac{Gm_1m_2}{r^2} = m_1 \frac{v^2}{r_1}. \qquad (14\text{-}35)$$

This equation contains the required mass m_2, but to find it we need expressions for r and r_1. (Note that m_1 cancels out.)

We start by locating the center of mass relative to the visible star, using Eq. 9-1. That star is at distance zero relative to itself, the center of mass is at distance r_1, and the dark star is at distance r. Equation 9-2 then becomes

$$r_1 = \frac{m_1(0) + m_2 r}{m_1 + m_2}, \qquad (14\text{-}36)$$

which yields

$$r = r_1 \frac{m_1 + m_2}{m_2}. \qquad (14\text{-}37)$$

To find an expression for r_1, we note that the visible star is moving in a circle of radius r_1, at speed v, and with period T. Thus, from Eq. 4-33, we have $v = 2\pi r_1/T$ or

$$r_1 = \frac{vT}{2\pi}. \qquad (14\text{-}38)$$

Substituting this for r_1 in Eq. 14-37 results in

$$r = \frac{vT}{2\pi} \frac{m_1 + m_2}{m_2}. \qquad (14\text{-}39)$$

Now we return to Eq. 14-35 and substitute for r with Eq. 14-39, for r_1 with Eq. 14-38, and for m_1 with the given $6M_s$. Rearranging the result and substituting known data then give us

$$\frac{m_2^3}{(6M_s + m_2)^2} = \frac{v^3 T}{2\pi G}$$

$$= \frac{(2.7 \times 10^5 \text{ m/s})^3 (1.70 \text{ days})(86\,400 \text{ s/day})}{(2\pi)(6.67 \times 10^{-11} \text{ N} \cdot \text{m}^2/\text{kg}^2)}$$

$$= 6.90 \times 10^{30} \text{ kg},$$

or

$$\frac{m_2^3}{(6M_s + m_2)^2} = 3.47 M_s. \qquad (14\text{-}40)$$

We can solve this cubic equation for m_2 with a polynomial solver on a calculator. Instead, since we are working with approximate masses anyway, we can substitute integer multiples of M_s for m_2 until we find one that makes Eq. 14-40 nearly true. This occurs for

$$m_2 \approx 9M_s. \qquad \text{(Answer)}$$

The data here approximate those for the binary system LMC X-3 in the Large Magellanic Cloud (shown in the figure that begins

this chapter). From other data, the dark object is known to be especially compact: It may be a star that collapsed under its own gravitational pull to become a neutron star or a black hole. Since a neutron star cannot have a mass larger than about $2M_s$, the result $m_2 \approx 9M_s$ strongly suggests that the dark object is a black hole.

Thus, we can detect the presence of a black hole provided it is part of a binary system with a visible star whose mass, orbital speed, and orbital period can be measured.

14-8 Satellites: Orbits and Energy

As a satellite orbits Earth on its elliptical path, both its speed, which fixes its kinetic energy K, and its distance from the center of Earth, which fixes its gravitational potential energy U, fluctuate with fixed periods. However, the mechanical energy E of the satellite remains constant. (Since the satellite's mass is so much smaller than Earth's mass, we assign U and E for the Earth–satellite system to the satellite alone.)

The potential energy of the system is given by Eq. 14-20 and is

$$U = -\frac{GMm}{r}$$

(with $U = 0$ for infinite separation). Here r is the radius of the orbit, assumed for the time being to be circular, and M and m are the masses of Earth and the satellite, respectively.

To find the kinetic energy of a satellite in a circular orbit, we write Newton's second law ($F = ma$) as

$$\frac{GMm}{r^2} = m\frac{v^2}{r}, \qquad (14\text{-}41)$$

where v^2/r is the centripetal acceleration of the satellite. Then, from Eq. 14-41, the kinetic energy is

$$K = \tfrac{1}{2}mv^2 = \frac{GMm}{2r}, \qquad (14\text{-}42)$$

which shows us that for a satellite in a circular orbit,

$$K = -\frac{U}{2} \qquad \text{(circular orbit).} \qquad (14\text{-}43)$$

The total mechanical energy of the orbiting satellite is

$$E = K + U = \frac{GMm}{2r} - \frac{GMm}{r}$$

or

$$E = -\frac{GMm}{2r} \qquad \text{(circular orbit).} \qquad (14\text{-}44)$$

This tells us that for a satellite in a circular orbit, the total energy E is the negative of the kinetic energy K:

$$E = -K \qquad \text{(circular orbit).} \qquad (14\text{-}45)$$

For a satellite in an elliptical orbit of semimajor axis a, we can substitute a for r in Eq. 14-44 to find the mechanical energy as

$$E = -\frac{GMm}{2a} \qquad \text{(elliptical orbit).} \qquad (14\text{-}46)$$

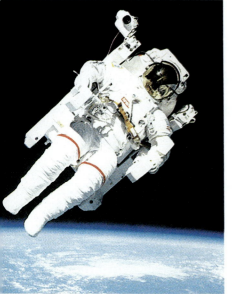

On February 7, 1984, at a height of 102 km above Hawaii and with a speed of about 29 000 km/h, Bruce McCandless stepped (untethered) into space from a space shuttle and became the first human satellite.

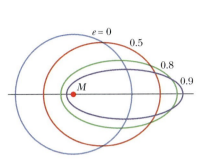

Fig. 14-17 Four orbits about an object of mass M. All four orbits have the same semimajor axis a and thus correspond to the same total mechanical energy E. Their eccentricities e are marked.

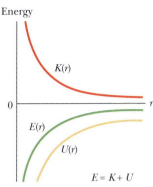

Fig. 14-18 The variation of kinetic energy K, potential energy U, and total energy E with radius r for a satellite in a circular orbit. For any value of r, the values of U and E are negative, the value of K is positive, and $E = -K$. As $r \rightarrow \infty$, all three energy curves approach a value of zero.

Equation 14-46 tells us that the total energy of an orbiting satellite depends only on the semimajor axis of its orbit and not on its eccentricity e. For example, four orbits with the same semimajor axis are shown in Fig. 14-17; the same satellite would have the same total mechanical energy E in all four orbits. Figure 14-18 shows the variation of K, U, and E with r for a satellite moving in a circular orbit about a massive central body.

✔**CHECKPOINT 6:**   In the figure here, a space shuttle is initially in a circular orbit of radius r about Earth. At point P, the pilot briefly fires a forward-pointing thruster to decrease the shuttle's kinetic energy K and mechanical energy E. (a) Which of the dashed elliptical orbits shown in the figure will the shuttle then take? (b) Is the orbital period T of the shuttle (the time to return to P) then greater than, less than, or the same as in the circular orbit?

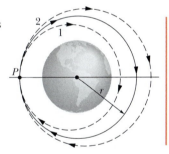

Sample Problem 14-8

A playful astronaut releases a bowling ball, of mass $m = 7.20$ kg, into circular orbit about Earth at an altitude h of 350 km.

(a) What is the mechanical energy E of the ball in its orbit?

SOLUTION: The Key Idea here is that we can get E from the orbital energy, given by Eq. 14-44 ($E = -GMm/2r$), if we first find the orbital radius r. That radius must be

$$r = R + h = 6370 \text{ km} + 350 \text{ km} = 6.72 \times 10^6 \text{ m},$$

in which R is the radius of Earth. Then, from Eq. 14-44, the mechanical energy is

$$E = -\frac{GMm}{2r}$$

$$= -\frac{(6.67 \times 10^{-11} \text{ N} \cdot \text{m}^2/\text{kg}^2)(5.98 \times 10^{24} \text{ kg})(7.20 \text{ kg})}{(2)(6.72 \times 10^6 \text{ m})}$$

$$= -2.14 \times 10^8 \text{ J} = -214 \text{ MJ}. \quad \text{(Answer)}$$

(b) What is the mechanical energy E_0 of the ball on the launchpad at Cape Canaveral? From there to the orbit, what is the change ΔE in the ball's mechanical energy?

SOLUTION: The Key Idea here is that, on the launchpad, the ball is *not* in orbit and thus Eq. 14-44 does *not* apply. Instead, we must find $E_0 = K_0 + U_0$, where K_0 is the ball's kinetic energy and U_0 is the gravitational potential energy of the ball–Earth system. To find U_0, we use Eq. 14-20 to write

$$U_0 = -\frac{GMm}{R}$$

$$= -\frac{(6.67 \times 10^{-11} \text{ N} \cdot \text{m}^2/\text{kg}^2)(5.98 \times 10^{24} \text{ kg})(7.20 \text{ kg})}{6.37 \times 10^6 \text{ m}}$$

$$= -4.51 \times 10^8 \text{ J} = -451 \text{ MJ}.$$

The kinetic energy K_0 of the ball is due to the ball's motion with

Earth's rotation. You can show that K_0 is less than 1 MJ, which is negligible relative to U_0. Thus, the mechanical energy of the ball on the launchpad is

$$E_0 = K_0 + U_0 \approx 0 - 451 \text{ MJ} = -451 \text{ MJ}. \quad \text{(Answer)}$$

The *increase* in the mechanical energy of the ball from launch-pad to orbit is

$$\Delta E = E - E_0 = (-214 \text{ MJ}) - (-451 \text{ MJ})$$
$$= 237 \text{ MJ}. \quad \text{(Answer)}$$

You can buy this amount of energy from your utility company for a few dollars. Obviously the high cost of placing objects into orbit is not due to the mechanical energy those objects require.

14-9   Einstein and Gravitation

Principle of Equivalence

Albert Einstein once said: "I was . . . in the patent office at Bern when all of a sudden a thought occurred to me: 'If a person falls freely, he will not feel his own weight.' I was startled. This simple thought made a deep impression on me. It impelled me toward a theory of gravitation."

Thus Einstein tells us how he began to form his **general theory of relativity.** The fundamental postulate of this theory about gravitation (the gravitating of objects toward each other) is called the **principle of equivalence,** which says that gravitation and acceleration are equivalent. If a physicist were locked up in a small box as in Fig. 14-19, he would not be able to tell whether the box was at rest on Earth (and subject only to Earth's gravitational force), as in Fig. 14-19a, or accelerating through interstellar space at 9.8 m/s² (and subject only to the force producing that acceleration), as in Fig. 14-19b. In both situations he would feel the same and would read the same value for his weight on a scale. Moreover, if he watched an object fall past him, the object would have the same acceleration relative to him in both situations.

Curvature of Space

We have thus far explained gravitation as due to a force between masses. Einstein showed that, instead, gravitation is due to a curvature (or shape) of space that is caused by the masses. (As is discussed later in this book, space and time are entangled, so the curvature of which Einstein spoke is really a curvature of *spacetime*, the combined four dimensions of our universe.)

Picturing how space (such as vacuum) can have curvature is difficult. An analogy might help: Suppose that from orbit we watch a race in which two boats begin on the equator with a separation of 20 km and head due south (Fig. 14-20a). To the sailors, the boats travel along flat, parallel paths. However, with time the boats draw together until, nearer the south pole, they touch. The sailors in the boats can interpret this drawing together in terms of a force acting on the boats. However, we can see that the boats draw together simply because of the curvature of Earth's surface. We can see this because we are viewing the race from "outside" that surface.

Figure 14-20b shows a similar race: Two horizontally separated apples are dropped from the same height above Earth. Although the apples may appear to travel along parallel paths, they actually move toward each other because they both fall toward Earth's center. We can interpret the motion of the apples in terms of the gravitational force on the apples from Earth. We can also interpret the motion in terms of a curvature of the space near Earth, due to the presence of Earth's mass. This time we cannot see the curvature because we cannot get "outside" the curved space, as we got "outside" the curved Earth in the boat example. However, we can depict the curvature with a drawing like Fig. 14-20c; there the apples would move along a surface that curves toward Earth because of Earth's mass.

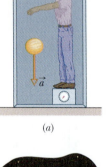

(a)

(b)

Fig. 14-19 (a) A physicist in a box resting on Earth sees a cantaloupe falling with acceleration $a = 9.8$ m/s². (b) If he and the box accelerate in deep space at 9.8 m/s², the cantaloupe has the same acceleration relative to him. It is not possible, by doing experiments within the box, for the physicist to tell which situation he is in. For example, the platform scale on which he stands reads the same weight in both situations.

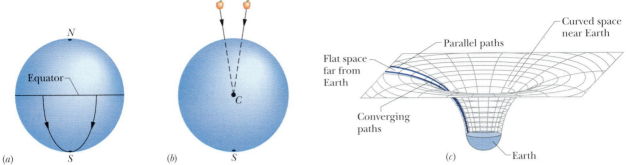

Fig. 14-20 (*a*) Two objects moving along lines of longitude toward the south pole converge because of the curvature of Earth's surface. (*b*) Two objects falling freely near Earth move along lines that converge toward the center of Earth because of the curvature of space near Earth. (*c*) Far from Earth (and other masses), space is flat and parallel paths remain parallel. Close to Earth, the parallel paths begin to converge because space is curved by Earth's mass.

When light passes near Earth, its path bends slightly because of the curvature of space there, an effect called *gravitational lensing.* When it passes a more massive structure, like a galaxy or a black hole having large mass, its path can be bent more. If such a massive structure is between us and a quasar (an extremely bright, extremely distant source of light), the light from the quasar can bend around the massive structure and toward us (Fig. 14-21*a*). Then, because the light seems to be coming to us from a number of slightly different directions in the sky, we see the same quasar in all those different directions. In some situations, the quasars we see blend together to form a giant luminous arc, which is called an *Einstein ring* (Fig. 14-21*b*).

Should we attribute gravitation to the curvature of spacetime due to the presence of masses or to a force between masses? Or should we attribute it to the actions of a type of fundamental particle called a *graviton,* as conjectured in some modern physics theories? We do not know.

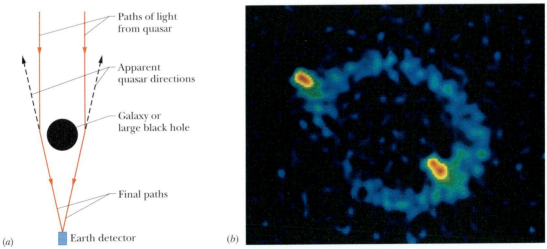

Fig. 14-21 (*a*) Light from a distant quasar follows curved paths around a galaxy or a large black hole because the mass of the galaxy or black hole has curved the adjacent space. If the light is detected, it appears to have originated along the backward extensions of the final paths (dashed lines). (*b*) The Einstein ring known as MG1131+0456 on the computer screen of a telescope. The source of the light (actually, radio waves, which are a form of invisible light) is far behind the large, unseen galaxy that produces the ring; a portion of the source appears as the two bright spots seen along the ring.

REVIEW & SUMMARY

The Law of Gravitation Any particle in the universe attracts any other particle with a **gravitational force** whose magnitude is

$$F = G \frac{m_1 m_2}{r^2} \qquad \text{(Newton's law of gravitation)}, \qquad (14\text{-}1)$$

where m_1 and m_2 are the masses of the particles, r is their separation, and G ($= 6.67 \times 10^{-11}$ N·m²/kg²) is the *gravitational constant.*

Gravitational Behavior of Uniform Spherical Shells Equation 14-1 holds only for particles. The gravitational force between extended bodies must generally be found by adding (integrating) the individual forces on individual particles within the bodies. However, if either of the bodies is a uniform spherical shell or a spherically symmetric solid, the net gravitational force it exerts on an *external* object may be computed as if all the mass of the shell or body were located at its center.

Superposition Gravitational forces obey the **principle of superposition;** that is, if n particles interact, the net force $\vec{F}_{1,\text{net}}$ on a particle labeled as particle 1 is the sum of the forces on it from all the other particles taken one at a time:

$$\vec{F}_{1,\text{net}} = \sum_{i=2}^{n} \vec{F}_{1i}, \qquad (14\text{-}4)$$

in which the sum is a vector sum of the forces $\vec{F}_{1i}$ on particle 1 from particles 2, 3, $\cdots$, n. The gravitational force $\vec{F}_1$ on a particle from an extended body is found by dividing the body into units of differential mass dm, each of which produces a differential force $d\vec{F}$ on the particle, and then integrating to find the sum of those forces:

$$\vec{F}_1 = \int d\vec{F}. \qquad (14\text{-}5)$$

Gravitational Acceleration The *gravitational acceleration* a_g of a particle (of mass m) is due solely to the gravitational force acting on it. When the particle is at distance r from the center of a uniform, spherical body of mass M, the magnitude F of the gravitational force on the particle is given by Eq. 14-1. Thus, by Newton's second law,

$$F = ma_g, \qquad (14\text{-}9)$$

which gives

$$a_g = \frac{GM}{r^2}. \qquad (14\text{-}10)$$

Free-Fall Acceleration and Weight The actual free-fall acceleration $\vec{g}$ of a particle near Earth differs slightly from the gravitational acceleration $\vec{a}_g$, and the particle's weight (equal to mg) differs from the magnitude of the gravitational force acting on the particle as computed with Eq. 14-1, because Earth is not uniform or spherical and because Earth rotates.

Gravitation Within a Spherical Shell A uniform shell of matter exerts no net gravitational force on a particle located inside it. This means that if a particle is located inside a uniform solid sphere at distance r from its center, the gravitational force exerted on the particle is due only to the mass M_{ins} that lies inside a sphere of radius r. This mass is given by

$$M_{\text{ins}} = \rho \frac{4\pi r^3}{3}, \qquad (14\text{-}17)$$

where ρ is the density of the sphere.

Gravitational Potential Energy The gravitational potential energy $U(r)$ of a system of two particles, with masses M and m and separated by a distance r, is the negative of the work that would be done by the gravitational force of either particle acting on the other if the separation between the particles were changed from infinite (very large) to r. This energy is

$$U = -\frac{GMm}{r} \qquad \text{(gravitational potential energy)}. \qquad (14\text{-}20)$$

Potential Energy of a System If a system contains more than two particles, its total gravitational potential energy U is the sum of terms representing the potential energies of all the pairs. As an example, for three particles, of masses m_1, m_2, and m_3,

$$U = -\left(\frac{Gm_1 m_2}{r_{12}} + \frac{Gm_1 m_3}{r_{13}} + \frac{Gm_2 m_3}{r_{23}} \right). \qquad (14\text{-}21)$$

Escape Speed An object will escape the gravitational pull of an astronomical body of mass M and radius R (to reach an infinite distance) if the object's speed near the body's surface is at least equal to the **escape speed,** given by

$$v = \sqrt{\frac{2GM}{R}}. \qquad (14\text{-}27)$$

Kepler's Laws Gravitational attraction holds the solar system together and makes possible orbiting Earth satellites, both natural and artificial. Such motions are governed by Kepler's three laws of planetary motion, all of which are direct consequences of Newton's laws of motion and gravitation:

1. ***The law of orbits.*** All planets move in elliptical orbits with the Sun at one focus.

2. ***The law of areas.*** A line joining any planet to the Sun sweeps out equal areas in equal times as the planet orbits the Sun. (This statement is equivalent to conservation of angular momentum.)

3. ***The law of periods.*** The square of the period T of any planet about the Sun is proportional to the cube of the semimajor axis

a of the orbit. For circular orbits with radius *r*, the semimajor axis *a* is replaced by *r* and the law is written as

$$T^2 = \left(\frac{4\pi^2}{GM}\right)r^3 \qquad \text{(law of periods)}, \qquad (14\text{-}33)$$

where *M* is the mass of the attracting body—the Sun in the case of the solar system. This equation is generally valid for elliptical planetary orbits, when the semimajor axis *a* is inserted in place of the circular radius *r*.

Energy in Planetary Motion When a planet or satellite with mass *m* moves in a circular orbit with radius *r*, its potential energy *U* and kinetic energy *K* are given by

$$U = -\frac{GMm}{r} \quad \text{and} \quad K = \frac{GMm}{2r}. \qquad (14\text{-}20, 14\text{-}42)$$

The mechanical energy $E = K + U$ is then

$$E = -\frac{GMm}{2r}. \qquad (14\text{-}44)$$

For an elliptical orbit of semimajor axis *a*,

$$E = -\frac{GMm}{2a}. \qquad (14\text{-}46)$$

Einstein's View of Gravitation Einstein pointed out that gravitation and acceleration are equivalent. This **principle of equivalence** led him to a theory of gravitation (the **general theory of relativity**) that explains gravitational effects in terms of a curvature of space.

QUESTIONS

1. In Fig. 14-22, two particles, of masses *m* and 2*m*, are fixed in place on an axis. (a) Where on the axis can a third particle of mass 3*m* be placed (other than at infinity) so that the net gravitational force on it from the first two particles is zero: to the left of the first two particles, to their right, between them but closer to the more massive particle, or between them but closer to the less massive particle? (b) Does the answer change if the third particle has, instead, a mass of 16*m*? (c) Is there a point off the axis at which the net force on the third particle would be zero?

Fig. 14-22 Question 1.

2. In Fig. 14-23, a central particle is surrounded by two circular rings of particles, at radii *r* and *R*, with $R > r$. All the particles have mass *m*. What are the magnitude and direction of the net gravitational force on the central particle due to the particles in the rings?

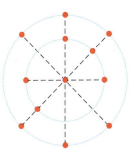

Fig. 14-23 Question 2.

3. In Fig. 14-24, a central particle of mass *M* is surrounded by a square array of other particles, separated by either distance *d* or distance *d*/2 along the perimeter of the square. What are the magnitude and direction of the net gravitational force on the central particle due to the other particles?

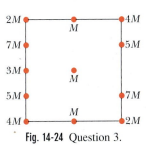

Fig. 14-24 Question 3.

4. Figure 14-25 shows four arrangements of a particle of mass *m* and one or more uniform rods of mass *M* and length *L*, each a distance *d* from the particle. Rank the arrangements according to

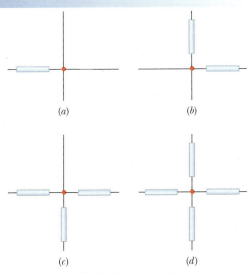

Fig. 14-25 Question 4.

the magnitude of the net gravitational force on the particle from the rods, greatest first.

5. Figure 14-26 gives the gravitational acceleration a_g for four planets as a function of the radial distance *r* from the center of the

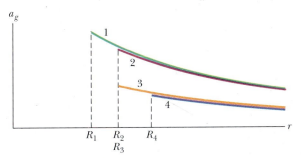

Fig. 14-26 Question 5.

planet, starting at the surface of the planet (at radius $R_1, R_2, R_3,$ or R_4). Plots 1 and 2 coincide for $r \geq R_2$; plots 3 and 4 coincide for $r \geq R_4$. Rank the four planets according to (a) their mass and (b) their density, greatest first.

6. Figure 14-27 shows three uniform spherical planets that are identical in size and mass. The periods of rotation T for the planets are given, and six lettered points are indicated—three points are on the equators of the planets and three points are on the north poles. Rank the points according to the value of the free-fall acceleration g at them, greatest first.

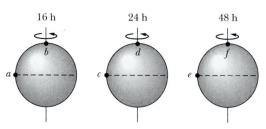

Fig. 14-27 Question 6.

7. From an inertial frame in space, we watch two identical uniform spheres fall toward one another owing to their mutual gravitational attraction. Approximate their initial speed as zero and take the initial gravitational potential energy of the two-sphere system as U_i. When the separation between the two spheres is half the initial separation, what is the kinetic energy of each sphere?

8. Rank the four systems of equal-mass particles in Checkpoint 2 according to the absolute value of the gravitational potential energy of the system, greatest first.

9. Figure 14-28 shows six paths by which a rocket orbiting a moon might move from point a to point b. Rank the paths according to (a) the corresponding change in the gravitational potential energy of the rocket–moon system and (b) the net work done on the rocket by the gravitational force from the moon, greatest first.

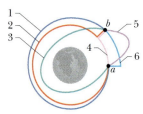

Fig. 14-28 Question 9.

10. In Fig. 14-29, a particle of mass m (not shown) is to be moved from an infinite distance to one of the three possible locations a, b, and c. Two other particles, of masses m and $2m$, are fixed in place. Rank the three possible locations according to the work done by the net gravitational force on the moving particle due to the fixed particles, greatest first.

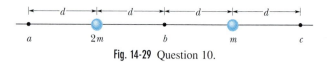

Fig. 14-29 Question 10.

11. In Fig. 14-30, a particle of mass m is initially at point A, at distance d from the center of one uniform sphere and distance $4d$ from the center of another uniform sphere, both of mass $M \gg m$. State whether, if you moved the particle to point D, the following would be positive, negative, or zero: (a) the change in the gravitational potential energy of the particle, (b) the work done by the net gravitational force on the particle, (c) the work done by your force. (d) What are the answers if, instead, the move were from point B to point C?

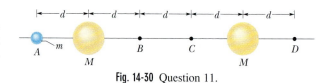

Fig. 14-30 Question 11.

12. Figure 14-31 gives the masses and separations for three pairs of stars that each form a binary star system. (a) Locate the point about which the stars of each pair orbit. (b) Rank the pairs according to the magnitude of the centripetal acceleration of the stars, greatest first.

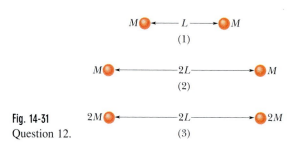

Fig. 14-31
Question 12.

EXERCISES & PROBLEMS

ssm   Solution is in the Student Solutions Manual.
www   Solution is available on the World Wide Web at:
          http://www.wiley.com/college/hrw
ilw   Solution is available on the Interactive LearningWare.

SEC. 14-2   Newton's Law of Gravitation

1E. What must the separation be between a 5.2 kg particle and a 2.4 kg particle for their gravitational attraction to have a magnitude of 2.3×10^{-12} N? ssm

2E. Some believe that the positions of the planets at the time of birth influence the newborn. Others deride this belief and claim that the gravitational force exerted on a baby by the obstetrician is greater than that exerted by the planets. To check this claim, calculate and compare the magnitude of the gravitational force exerted on a 3 kg baby (a) by a 70 kg obstetrician who is 1 m away and roughly approximated as a point mass, (b) by the massive planet Jupiter ($m = 2 \times 10^{27}$ kg) at its closest approach to Earth ($= 6 \times 10^{11}$ m), and (c) by Jupiter at its greatest distance from Earth ($= 9 \times 10^{11}$ m). (d) Is the claim correct?

3E. One of the *Echo* satellites consisted of an inflated spherical aluminum balloon 30 m in diameter and of mass 20 kg. Suppose a meteor having a mass of 7.0 kg passes within 3.0 m of the surface of the satellite. What is the magnitude of the gravitational force on the meteor from the satellite at the closest approach? **ssm**

4E. The Sun and Earth each exert a gravitational force on the Moon. What is the ratio F_{Sun}/F_{Earth} of these two forces? (The average Sun–Moon distance is equal to the Sun–Earth distance.)

5E. A mass M is split into two parts, m and $M - m$, which are then separated by a certain distance. What ratio m/M maximizes the magnitude of the gravitational force between the parts? **ilw**

SEC. 14-3 Gravitation and the Principle of Superposition

6E. A spaceship is on a straight-line path between Earth and its moon. At what distance from Earth is the net gravitational force on the spaceship zero?

7E. How far from Earth must a space probe be along a line toward the Sun so that the Sun's gravitational pull on the probe balances Earth's pull? **ssm**

8P. Three 5.0 kg spheres are located in the *xy* plane as shown in Fig. 14-32. What is the magnitude of the net gravitational force on the sphere at the origin due to the other two spheres?

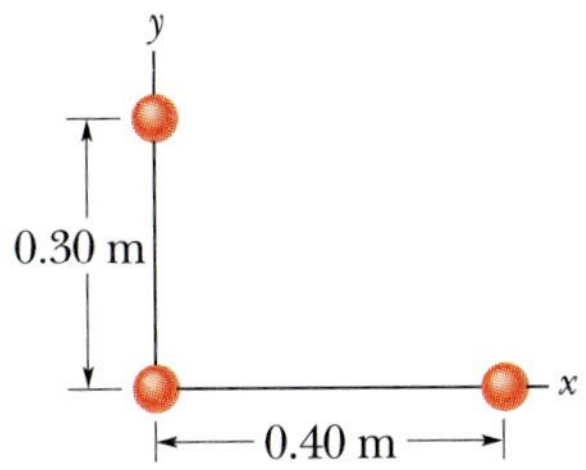

9P. In Fig. 14-33*a*, four spheres form the corners of a square whose side is 2.0 cm long. What are the magnitude and direction of the net gravitational force from them on a central sphere with mass $m_5 = 250$ kg?

10P. In Fig. 14-33*b*, two spheres of mass m and a third sphere of mass M form an equilateral triangle, and a fourth sphere of mass m_4 is at the center of the triangle. The net gravitational force on that central sphere from the three other spheres is zero. (a) What is M in terms of m? (b) If we double the value of m_4, what then is the magnitude of the net gravitational force on the central sphere?

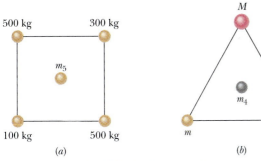

Fig. 14-33 Problems 9 and 10.

11P. The masses and coordinates of three spheres are as follows: 20 kg, $x = 0.50$ m, $y = 1.0$ m; 40 kg, $x = -1.0$ m, $y = -1.0$ m; 60 kg, $x = 0$ m, $y = -0.50$ m. What is the magnitude of the gravitational force on a 20 kg sphere located at the origin due to the other spheres? **ilw**

12P. Four uniform spheres, with masses $m_A = 400$ kg, $m_B = 350$ kg, $m_C = 2000$ kg, and $m_D = 500$ kg, have (x, y) coordinates of (0, 50 cm), (0, 0), (−80 cm, 0), and (40 cm, 0), respectively. What is the net gravitational force on sphere B due to the other spheres?

13P. Figure 14-34 shows a spherical hollow inside a lead sphere of radius R; the surface of the hollow passes through the center of the sphere and "touches" the right side of the sphere. The mass of the sphere before hollowing was M. With what gravitational force does the hollowed-out lead sphere attract a small sphere of mass m that lies at a distance d from the center of the lead sphere, on the straight line connecting the centers of the spheres and of the hollow? **ssm**

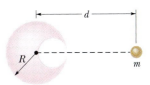

Fig. 14-34 Problem 13.

SEC. 14-4 Gravitation Near Earth's Surface

14E. You weigh 530 N at sidewalk level outside the World Trade Center in New York City. Suppose that you ride from this level to the top of one of its 410 m towers. Ignoring Earth's rotation, how much less would you weigh there (because you are slightly farther from the center of Earth)?

15E. At what altitude above Earth's surface would the gravitational acceleration be 4.9 m/s²? **ssm**

16E. (a) What will an object weigh on the Moon's surface if it weighs 100 N on Earth's surface? (b) How many Earth radii must this same object be from the center of Earth if it is to weigh the same as it does on the Moon?

17P. The fastest possible rate of rotation of a planet is that for which the gravitational force on material at the equator just barely provides the centripetal force needed for the rotation. (Why?) (a) Show that the corresponding shortest period of rotation is

$$T = \sqrt{\frac{3\pi}{G\rho}},$$

where ρ is the uniform density of the spherical planet. (b) Calculate the rotation period assuming a density of 3.0 g/cm³, typical of many planets, satellites, and asteroids. No astronomical object has ever been found to be spinning with a period shorter than that determined by this analysis. **ssm**

18P. One model for a certain planet has a core of radius R and mass M surrounded by an outer shell of inner radius R, outer radius $2R$, and mass $4M$. If $M = 4.1 \times 10^{24}$ kg and $R = 6.0 \times 10^6$ m, what is the gravitational acceleration of a particle at points (a) R and (b) $3R$ from the center of the planet?

19P. A body is suspended from a spring scale in a ship sailing along the equator with speed v. (a) Show that the scale reading will be very close to $W_0(1 \pm 2\omega v/g)$, where ω is the angular speed of Earth and W_0 is the scale reading when the ship is at rest. (b) Explain the $\pm$ sign. **ssm** **www**

20P. The radius R_h and mass M_h of a black hole are related by $R_h = 2GM_h/c^2$, where c is the speed of light. Assume that the gravitational acceleration a_g of an object at a distance $r_o =$

Fig. 14-32 Problem 8.

$1.001R_h$ from the center of a black hole is given by Eq. 14-10 (it is, for large black holes). (a) Find an expression for a_g at r_o in terms of M_h. (b) Does a_g at r_o increase or decrease with an increase of M_h? (c) What is a_g at r_o for a very large black hole whose mass is 1.55×10^{12} times the solar mass of 1.99×10^{30} kg? (d) If the astronaut of Sample Problem 14-3 is at r_o with her feet toward this black hole, what is the difference in gravitational acceleration between her head and her feet? (e) Is the tendency to stretch the astronaut severe?

21P. Certain neutron stars (extremely dense stars) are believed to be rotating at about 1 rev/s. If such a star has a radius of 20 km, what must be its minimum mass so that material on its surface remains in place during the rapid rotation? **ilw**

SEC. 14-5 Gravitation Inside Earth

22E. Two concentric shells of uniform density having masses M_1 and M_2 are situated as shown in Fig. 14-35. Find the magnitude of the net gravitational force on a particle of mass m, due to the shells, when the particle is located at (a) point A, at distance $r = a$ from the center, (b) point B at $r = b$, and (c) point C at $r = c$. The distance r is measured from the center of the shells.

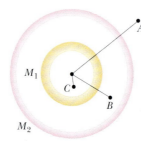

Fig. 14-35 Exercise 22.

23P. A solid sphere of uniform density has a mass of 1.0×10^4 kg and a radius of 1.0 m. What is the magnitude of the gravitational force due to the sphere on a particle of mass m located at a distance of (a) 1.5 m and (b) 0.50 m from the center of the sphere? (c) Write a general expression for the magnitude of the gravitational force on the particle at a distance $r \leq 1.0$ m from the center of the sphere.

24P. A uniform solid sphere of radius R produces a gravitational acceleration of a_g on its surface. At what two distances from the center of the sphere is the gravitational acceleration $a_g/3$? (*Hint:* Consider distances both inside and outside the sphere.)

25P. Figure 14-36 shows, not to scale, a cross section through the interior of Earth. Rather than being uniform throughout, Earth is divided into three zones: an outer *crust,* a *mantle,* and an

inner *core*. The dimensions of these zones and the masses contained within them are shown on the figure. Earth has a total mass of 5.98×10^{24} kg and a radius of 6370 km. Ignore rotation and assume that Earth is spherical. (a) Calculate a_g at the surface. (b) Suppose that a bore hole (the *Mohole*) is driven to the crust–mantle interface at a depth of 25 km; what would be the value of a_g at the bottom of the hole? (c) Suppose that Earth were a uniform sphere with the same total mass and size. What would be the value of a_g at a depth of 25 km? (Precise measurements of a_g are sensitive probes of the interior structure of Earth, although results can be clouded by local density variations.) **ssm**

SEC. 14-6 Gravitational Potential Energy

26E. (a) What is the gravitational potential energy of the two-particle system in Exercise 1? If you triple the separation between the particles, how much work is done (b) by the gravitational force between the particles and (c) by you?

27E. (a) In Problem 12, remove sphere A and calculate the gravitational potential energy of the remaining three-particle system. (b) If A is then put back in place, is the potential energy of the four-particle system more or less than that of the system in (a)? (c) In (a), is the work done by you to remove A positive or negative? (d) In (b), is the work done by you to replace A positive or negative?

28E. In Problem 5, what ratio m/M gives the least gravitational potential energy for the system?

29E. The mean diameters of Mars and Earth are 6.9×10^3 km and 1.3×10^4 km, respectively. The mass of Mars is 0.11 times Earth's mass. (a) What is the ratio of the mean density of Mars to that of Earth? (b) What is the value of the gravitational acceleration on Mars? (c) What is the escape speed on Mars? **ssm**

30E. Calculate the amount of energy required to escape from (a) Earth's moon and (b) Jupiter relative to that required to escape from Earth.

31P. The three spheres in Fig. 14-37, with masses $m_A = 800$ g, $m_B = 100$ g, and $m_C = 200$ g, have their centers on a common line, with $L = 12$ cm and $d = 4.0$ cm. You move sphere B along the line until its center-to-center separation from C is $d = 4.0$ cm. How much work is done on sphere B (a) by you and (b) by the net gravitational force on B due to spheres A and C? **ssm**

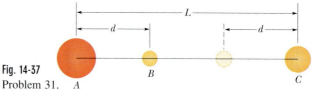

Fig. 14-37
Problem 31. A B C

32P. Zero, a hypothetical planet, has a mass of 5.0×10^{23} kg, a radius of 3.0×10^6 m, and no atmosphere. A 10 kg space probe is to be launched vertically from its surface. (a) If the probe is launched with an initial energy of 5.0×10^7 J, what will be its kinetic energy when it is 4.0×10^6 m from the center of Zero? (b) If the probe is to achieve a maximum distance of 8.0×10^6 m from the center of Zero, with what initial kinetic energy must it be launched from the surface of Zero?

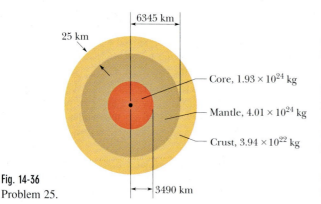

Fig. 14-36
Problem 25.

33P. A rocket is accelerated to speed $v = 2\sqrt{gR_e}$ near Earth's surface (where Earth's radius is R_e), and it then coasts upward. (a) Show that it will escape from Earth. (b) Show that very far from Earth its speed will be $v = \sqrt{2gR_e}$. ssm

34P. Planet Roton, with a mass of 7.0×10^{24} kg and a radius of 1600 km, gravitationally attracts a meteorite that is initially at rest relative to the planet, at a great enough distance to take as infinite. The meteorite falls toward the planet. Assuming the planet is airless, find the speed of the meteorite when it reaches the planet's surface.

35P. (a) What is the escape speed on a spherical asteroid whose radius is 500 km and whose gravitational acceleration at the surface is 3.0 m/s²? (b) How far from the surface will a particle go if it leaves the asteroid's surface with a radial speed of 1000 m/s? (c) With what speed will an object hit the asteroid if it is dropped from 1000 km above the surface? ssm www

36P. A 150.0 kg rocket moving radially outward from Earth has a speed of 3.70 km/s when its engine shuts off 200 km above Earth's surface. (a) Assuming negligible air drag, find the rocket's kinetic energy when the rocket is 1000 km above Earth's surface. (b) What maximum height above the surface is reached by the rocket?

37P. Two neutron stars are separated by a distance of 10^{10} m. They each have a mass of 10^{30} kg and a radius of 10^5 m. They are initially at rest with respect to each other. As measured from that rest frame, how fast are they moving when (a) their separation has decreased to one-half its initial value and (b) they are about to collide? ssm www

38P. In deep space, sphere A of mass 20 kg is located at the origin of an x axis and sphere B of mass 10 kg is located on the axis at $x = 0.80$ m. Sphere B is released from rest while sphere A is held at the origin. (a) What is the gravitational potential energy of the two-sphere system as B is released? (b) What is the kinetic energy of B when it has moved 0.20 m toward A?

39P. A projectile is fired vertically from Earth's surface with an initial speed of 10 km/s. Neglecting air drag, how far above the surface of Earth will it go? ilw

SEC. 14-7 Planets and Satellites: Kepler's Laws

40E. The mean distance of Mars from the Sun is 1.52 times that of Earth from the Sun. From Kepler's law of periods, calculate the number of years required for Mars to make one revolution about the Sun; compare your answer with the value given in Appendix C.

41E. The Martian satellite Phobos travels in an approximately circular orbit of radius 9.4×10^6 m with a period of 7 h 39 min. Calculate the mass of Mars from this information. ssm

42E. Determine the mass of Earth from the period T (27.3 days) and the radius r (3.82×10^5 km) of the Moon's orbit about Earth. Assume the Moon orbits the center of Earth rather than the center of mass of the Earth–Moon system.

43E. Our Sun, with mass 2.0×10^{30} kg, revolves about the center of the Milky Way galaxy, which is 2.2×10^{20} m away, once every 2.5×10^8 years. Assuming that each of the stars in the galaxy has a mass equal to that of our Sun, that the stars are distributed uniformly in a sphere about the galactic center, and that our Sun is essentially at the edge of that sphere, estimate roughly the number of stars in the galaxy. ssm

44E. A satellite is placed in a circular orbit about Earth with a radius equal to one-half the radius of the Moon's orbit. What is its period of revolution in lunar months? (A lunar month is the period of revolution of the Moon.)

45E. (a) What linear speed must an Earth satellite have to be in a circular orbit at an altitude of 160 km? (b) What is the period of revolution? ssm

46E. The Sun's center is at one focus of Earth's orbit. How far from this focus is the other focus, (a) in meters and (b) in terms of the solar radius, 6.96×10^8 m? The eccentricity of Earth's orbit is 0.0167, and the semimajor axis is 1.50×10^{11} m.

47E. A satellite, moving in an elliptical orbit, is 360 km above Earth's surface at its farthest point and 180 km above at its closest point. Calculate (a) the semimajor axis and (b) the eccentricity of the orbit. (*Hint:* See Sample Problem 14-6.) ssm

48E. A satellite hovers over a certain spot on the equator of (rotating) Earth. What is the altitude of its orbit (called a *geosynchronous orbit*)?

49E. A comet that was seen in April 574 by Chinese astronomers on a day known by them as the Woo Woo day was spotted again in May 1994. Assume the time between observations is the period of the Woo Woo day comet and take its eccentricity as 0.11. What are (a) the semimajor axis of the comet's orbit and (b) its greatest distance from the Sun in terms of the mean orbital radius R_P of Pluto?

50E. In 1993 the spacecraft *Galileo* sent home an image (Fig. 14-38) of asteroid 243 Ida and a tiny orbiting moon (now known as Dactyl), the first confirmed example of an asteroid–moon system. In the image, the moon, which is 1.5 km wide, is 100 km from the center of the asteroid, which is 55 km long. The shape of the moon's orbit is not well known; assume it is circular with a period of 27 h. (a) What is the mass of the asteroid? (b) The volume of the asteroid, measured from the *Galileo* images, is 14 100 km³. What is the density of the asteroid?

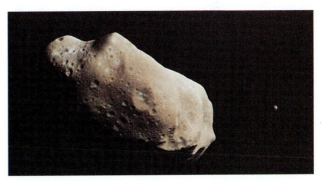

Fig. 14-38 Exercise 50. An image from the spacecraft *Galileo* shows a tiny moon orbiting asteroid 243 Ida.

51P. In 1610, Galileo used his telescope to discover four prominent moons around Jupiter. Their mean orbital radii a and periods T are as follows:

Name	a (10^8 m)	T (days)
Io	4.22	1.77
Europa	6.71	3.55
Ganymede	10.7	7.16
Callisto	18.8	16.7

(a) Plot log a (y axis) against log T (x axis) and show that you get a straight line. (b) Measure the slope of the line and compare it with the value that you expect from Kepler's third law. (c) Find the mass of Jupiter from the intercept of this line with the y axis.

52P. A 20 kg satellite has a circular orbit with a period of 2.4 h and a radius of 8.0×10^6 m around a planet of unknown mass. If the magnitude of the gravitational acceleration on the surface of the planet is 8.0 m/s², what is the radius of the planet?

53P. In a certain binary-star system, each star has the same mass as our Sun, and they revolve about their center of mass. The distance between them is the same as the distance between Earth and the Sun. What is their period of revolution in years? **ilw**

54P. A certain triple-star system consists of two stars, each of mass m, revolving about a central star of mass M in the same circular orbit of radius r (Fig. 14-39). The two stars are always at opposite ends of a diameter of the circular orbit. Derive an expression for the period of revolution of the stars.

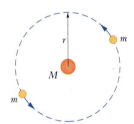

Fig. 14-39 Problem 54.

55P*. Three identical stars of mass M are located at the vertices of an equilateral triangle with side L. At what speed must they move if they all revolve under the influence of one another's gravitational force in a circular orbit circumscribing the triangle while still preserving the equilateral triangle? **ssm www**

SEC. 14-8 Satellites: Orbits and Energy

56E. Consider two satellites, A and B, both of mass m, moving in the same circular orbit of radius r around Earth, of mass M_E, but in opposite senses of rotation and therefore on a collision course (see Fig. 14-40). (a) In terms of G, M_E, m, and r, find the total mechanical energy $E_A + E_B$ of the two-satellite-plus-Earth system before collision. (b) If the collision is completely inelastic so that the wreckage remains as one piece of tangled material (mass $= 2m$), find the total mechanical energy immediately after collision. (c) Describe the subsequent motion of the wreckage.

Fig. 14-40 Exercise 56.

57E. An asteroid, whose mass is 2.0×10^{-4} times the mass of Earth, revolves in a circular orbit around the Sun at a distance that is twice Earth's distance from the Sun. (a) Calculate the period of revolution of the asteroid in years. (b) What is the ratio of the kinetic energy of the asteroid to that of Earth? **ssm**

58P. Two Earth satellites, A and B, each of mass m, are to be launched into circular orbits about Earth's center. Satellite A is to orbit at an altitude of 6370 km. Satellite B is to orbit at an altitude of 19 110 km. The radius of Earth R_E is 6370 km. (a) What is the ratio of the potential energy of satellite B to that of satellite A, in orbit? (b) What is the ratio of the kinetic energy of satellite B to that of satellite A, in orbit? (c) Which satellite has the greater total energy if each has a mass of 14.6 kg? By how much?

59P. Show that if an object is in an elliptical orbit with semimajor axis a about a planet of mass M, then its distance r from the planet and speed v are related by

$$v^2 = GM \left(\frac{2}{r} - \frac{1}{a} \right).$$

(*Hint:* Use the law of conservation of mechanical energy and Eq. 14-46.) **ssm**

60P. Use the result of Problem 59 and data contained in Sample Problem 14-6 to calculate (a) the speed v_p of comet Halley at perihelion and (b) its speed v_a at aphelion. (c) Using the law of conservation of angular momentum relative to the Sun, find the ratio of the comet's perihelion distance R_p to its aphelion distance R_a in terms of v_p and v_a.

61P. (a) Does it take more energy to get a satellite up to 1500 km above Earth than to put it in circular orbit once it is there? (Take Earth's radius to be 6370 km.) (b) What about 3185 km? (c) What about 4500 km?

62P. One way to attack a satellite in Earth orbit is to launch a swarm of pellets in the same orbit as the satellite but in the opposite direction. Suppose a satellite in a circular orbit 500 km above Earth's surface collides with a pellet having mass 4.0 g. (a) What is the kinetic energy of the pellet in the reference frame of the satellite just before the collision? (b) What is the ratio of this kinetic energy to the kinetic energy of a 4.0 g bullet from a modern army rifle with a muzzle speed of 950 m/s?

63P. What are (a) the speed and (b) the period of a 220 kg satellite in an approximately circular orbit 640 km above the surface of Earth? Suppose the satellite loses mechanical energy at the average rate of 1.4×10^5 J per orbital revolution. Adopting the reasonable approximation that the satellite's orbit becomes a "circle of slowly diminishing radius," determine the satellite's (c) altitude, (d) speed, and (e) period at the end of its 1500th revolution. (f) What is the magnitude of the average retarding force on the satellite? Is angular momentum around Earth's center conserved for (g) the satellite and (h) the satellite–Earth system? **ssm**

SEC. 14-9 Einstein and Gravitation

64E. In Fig. 14-19b, the scale on which the 60 kg physicist stands reads 220 N. How long will the cantaloupe take to reach the floor if the physicist drops it from rest (relative to himself), 2.1 m from the floor?

15 Fluids

The force exerted by water on the body of a descending diver increases noticeably, even for a relatively shallow descent to the bottom of a swimming pool. However, in 1975, using scuba gear with a special gas mixture for breathing, William Rhodes emerged from a chamber that had been lowered 300 m into the Gulf of Mexico, and he then swam to a record depth of 350 m. Strangely, a novice scuba diver practicing in a swimming pool might be in more danger from the force exerted by the water than was Rhodes. Occasionally, novice scuba divers die because they have neglected that danger.

What is this potentially lethal risk?

The answer is in this chapter.

15-1   Fluids and the World Around Us

Fluids—which include both liquids and gases—play a central role in our daily lives. We breathe and drink them, and a rather vital fluid circulates in the human cardiovascular system. There are the fluid ocean and the fluid atmosphere.

In a car, there are fluids in the tires, the gas tank, the radiator, the combustion chambers of the engine, the exhaust manifold, the battery, the air-conditioning system, the windshield wiper reservoir, the lubrication system, and the hydraulic system. (*Hydraulic* means operated via a liquid.) The next time you see a large piece of earthmoving machinery, count the hydraulic cylinders that permit the machine to do its work. Large jet planes have scores of them.

We use the kinetic energy of a moving fluid in windmills, and the potential energy of another fluid in hydroelectric power plants. Given time, fluids carve the landscape. We often travel great distances just to watch fluids move. Perhaps it is time to see what physics can tell us about fluids.

15-2   What Is a Fluid?

A **fluid,** in contrast to a solid, is a substance that can flow. Fluids conform to the boundaries of any container in which we put them. They do so because a fluid cannot sustain a force that is tangential to its surface. (In the more formal language of Section 13-6, a fluid is a substance that flows because it cannot withstand a shearing stress. It can, however, exert a force in the direction perpendicular to its surface.) Some materials, such as pitch, take a long time to conform to the boundaries of a container, but they do so eventually; thus, we classify them as fluids.

You may wonder why we lump liquids and gases together and call them fluids. After all (you may say), liquid water is as different from steam as it is from ice. Actually, it is not. Ice, like other crystalline solids, has its constituent atoms organized in a fairly rigid three-dimensional array called a crystalline lattice. In neither steam nor liquid water, however, is there any such orderly long-range arrangement.

15-3   Density and Pressure

When we discuss rigid bodies, we are concerned with particular lumps of matter, such as wooden blocks, baseballs, or metal rods. Physical quantities that we find useful, and in whose terms we express Newton's laws, are *mass* and *force*. We might speak, for example, of a 3.6 kg block acted on by a 25 N force.

With fluids, we are more interested in the extended substance, and in properties that can vary from point to point in that substance. It is more useful to speak of **density** and **pressure** than of mass and force.

Density

To find the density ρ of a fluid at any point, we isolate a small volume element ΔV around that point and measure the mass Δm of the fluid contained within that element. The **density** is then

$$\rho = \frac{\Delta m}{\Delta V}. \tag{15-1}$$

In theory, the density at any point in a fluid is the limit of this ratio as the volume element ΔV at that point is made smaller and smaller. In practice, we assume that a fluid sample is large compared to atomic dimensions and thus is "smooth" (with uniform density), rather than "lumpy" with atoms. This assumption allows us to

TABLE 15-1 Some Densities

Material or Object	Density (kg/m^3)
Interstellar space	10^{-20}
Best laboratory vacuum	10^{-17}
Air: 20°C and 1 atm pressure	1.21
20°C and 50 atm	60.5
Styrofoam	1×10^2
Ice	0.917×10^3
Water: 20°C and 1 atm	0.998×10^3
20°C and 50 atm	1.000×10^3
Seawater: 20°C and 1 atm	1.024×10^3
Whole blood	1.060×10^3
Iron	7.9×10^3
Mercury (the metal)	13.6×10^3
Earth: average	5.5×10^3
core	9.5×10^3
crust	2.8×10^3
Sun: average	1.4×10^3
core	1.6×10^5
White dwarf star (core)	10^{10}
Uranium nucleus	3×10^{17}
Neutron star (core)	10^{18}
Black hole (1 solar mass)	10^{19}

write Eq. 15-1 as

$$\rho = \frac{m}{V} \qquad \text{(uniform density)}, \qquad (15\text{-}2)$$

where m and V are the mass and volume of the sample.

Density is a scalar property; its SI unit is the kilogram per cubic meter. Table 15-1 shows the densities of some substances and the average densities of some objects. Note that the density of a gas (see Air in the table) varies considerably with pressure, but the density of a liquid (see Water) does not; that is, gases are readily *compressible* but liquids are not.

Pressure

Let a small pressure-sensing device be suspended inside a fluid-filled vessel, as in Fig. 15-1a. The sensor (Fig. 15-1b) consists of a piston of area ΔA riding in a close-fitting cylinder and resting against a spring. A readout arrangement allows us to record the amount by which the (calibrated) spring is compressed by the surrounding fluid, thus indicating the magnitude ΔF of the force that acts normal to the piston. We define the **pressure** on the piston from the fluid as

$$p = \frac{\Delta F}{\Delta A}. \qquad (15\text{-}3)$$

In theory, the pressure at any point in the fluid is the limit of this ratio as the area ΔA of the piston, centered on that point, is made smaller and smaller. However, if the force is uniform over a flat area A, we can write Eq. 15-3 as

$$p = \frac{F}{A} \qquad \text{(pressure of uniform force on flat area)}, \qquad (15\text{-}4)$$

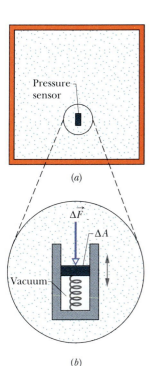

(a)

(b)

Fig. 15-1 (a) A fluid-filled vessel containing a small pressure sensor, shown in (b). The pressure is measured by the relative position of the movable piston in the sensor.

TABLE 15-2   Some Pressures

	Pressure (Pa)
Center of the Sun	2×10^{16}
Center of Earth	4×10^{11}
Highest sustained laboratory pressure	1.5×10^{10}
Deepest ocean trench (bottom)	1.1×10^8
Spike heels on a dance floor	1×10^6
Automobile tire[a]	2×10^5
Atmosphere at sea level	1.0×10^5
Normal blood pressure[a,b]	1.6×10^4
Best laboratory vacuum	10^{-12}

[a]Pressure in excess of atmospheric pressure.
[b]The systolic pressure, corresponding to 120 torr on the physician's pressure gauge.

where F is the magnitude of the normal force on area A. (When we say a force is uniform over an area, we mean that it is evenly distributed over every point of the area.)

We find by experiment that at a given point in a fluid at rest, the pressure p defined by Eq. 15-3 has the same value no matter how the pressure sensor is oriented. Pressure is a scalar, having no directional properties. It is true that the force acting on the piston of our pressure sensor is a vector, but Eq. 15-3 involves only the *magnitude* of that force, a scalar quantity.

The SI unit of pressure is the newton per square meter, which is given a special name, the **pascal** (Pa). In metric countries, tire pressure gauges are calibrated in kilopascals. The pascal is related to some other common (non-SI) pressure units as follows:

$$1 \text{ atm} = 1.01 \times 10^5 \text{ Pa} = 760 \text{ torr} = 14.7 \text{ lb/in.}^2.$$

The *atmosphere* (atm) is, as the name suggests, the approximate average pressure of the atmosphere at sea level. The *torr* (named for Evangelista Torricelli, who invented the mercury barometer in 1674) was formerly called the *millimeter of mercury* (mm Hg). The pound per square inch is often abbreviated psi. Table 15-2 shows some pressures.

Sample Problem 15-1

A living room has floor dimensions of 3.5 m and 4.2 m and a height of 2.4 m.

(a) What does the air in the room weigh when the air pressure is 1.0 atm?

SOLUTION: The Key Ideas here are these: (1) The air's weight is equal to mg, where m is its mass. (2) Mass m is related to the air density ρ and the air's volume V by Eq. 15-2 ($\rho = m/V$). Putting these two ideas together and taking the density of air at 1.0 atm from Table 15-1, we find

$$mg = (\rho V)g$$
$$= (1.21 \text{ kg/m}^3)(3.5 \text{ m} \times 4.2 \text{ m} \times 2.4 \text{ m})(9.8 \text{ m/s}^2)$$
$$= 418 \text{ N} \approx 420 \text{ N}. \qquad \text{(Answer)}$$

This is the weight of about 110 cans of Pepsi.

(b) What is the magnitude of the atmosphere's force on the floor of the room?

SOLUTION: The Key Idea here is that the atmosphere pushes down on the floor with a force of magnitude F that is uniform over the floor. Thus, it produces a pressure that is related to F and the flat area A of the floor by Eq. 15-4 ($p = F/A$), which gives us

$$F = pA = (1.0 \text{ atm})\left(\frac{1.01 \times 10^5 \text{ N/m}^2}{1.0 \text{ atm}}\right)(3.5 \text{ m})(4.2 \text{ m})$$
$$= 1.5 \times 10^6 \text{ N}. \qquad \text{(Answer)}$$

This enormous force is equal to the weight of the column of air that covers the floor and extends all the way to the top of the atmosphere.

15-4   Fluids at Rest

Figure 15-2a shows a tank of water—or other liquid—open to the atmosphere. As every diver knows, the pressure *increases* with depth below the air–water interface. The diver's depth gauge, in fact, is a pressure sensor much like that of Fig. 15-1b. As every mountaineer knows, the pressure *decreases* with altitude as one ascends into the atmosphere. The pressures encountered by the diver and the mountaineer are usually called *hydrostatic pressures,* because they are due to fluids that are static (at rest). Here we want to find an expression for hydrostatic pressure as a function of depth or altitude.

Let us look first at the increase in pressure with depth below the water's surface. We set up a vertical y axis in the tank, with its origin at the air–water interface and the positive direction upward. We next consider a water sample contained in an imaginary right circular cylinder of horizontal base (or face) area A, such that y_1 and

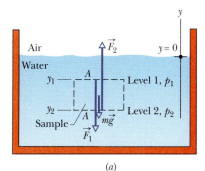

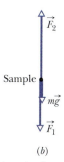

Fig. 15-2 (a) A tank of water in which a sample of water is contained in an imaginary cylinder of horizontal base area A. Force $\vec{F_1}$ acts at the top surface of the cylinder; force $\vec{F_2}$ acts at the bottom surface of the cylinder; the gravitational force on the water in the cylinder is represented by $m\vec{g}$. (b) A free-body diagram of the water sample.

y_2 (both of which are *negative* numbers) are the depths below the surface of the upper and lower cylinder faces, respectively.

Figure 15-2b shows a free-body diagram for the water in the cylinder. The water is in *static equilibrium;* that is, it is stationary and the forces on it balance. Three forces act on it vertically: Force $\vec{F_1}$ acts at the top surface of the cylinder and is due to the water above the cylinder. Similarly, force $\vec{F_2}$ acts at the bottom surface of the cylinder and is due to the water below the cylinder. The gravitational force on the water in the cylinder is represented by $m\vec{g}$, where m is the mass of the water in the cylinder. The balance of these forces is written as

$$F_2 = F_1 + mg. \tag{15-5}$$

We want to transform Eq. 15-5 into an equation involving pressures. From Eq. 15-4, we know that

$$F_1 = p_1 A \quad \text{and} \quad F_2 = p_2 A. \tag{15-6}$$

The mass m of the water in the cylinder is, from Eq. 15-2, $m = \rho V$, where the cylinder's volume V is the product of its face area A and its height $y_1 - y_2$. Thus, m is equal to $\rho A(y_1 - y_2)$. Substituting this and Eq. 15-6 into Eq. 15-5, we find

$$p_2 A = p_1 A + \rho A g(y_1 - y_2)$$

or

$$p_2 = p_1 + \rho g(y_1 - y_2). \tag{15-7}$$

This equation can be used to find pressure both in a liquid (as a function of depth) and in the atmosphere (as a function of altitude or height). For the former, suppose we seek the pressure p at a depth h below the liquid surface. Then we choose level 1 to be the surface, level 2 to be a distance h below it (as in Fig. 15-3), and p_0 to represent the atmospheric pressure on the surface. We then substitute

$$y_1 = 0, \quad p_1 = p_0 \quad \text{and} \quad y_2 = -h, \quad p_2 = p$$

into Eq. 15-7, which becomes

$$p = p_0 + \rho g h \qquad \text{(pressure at depth } h\text{).} \tag{15-8}$$

Note that the pressure at a given depth in the liquid depends on that depth but not on any horizontal dimension.

▶ The pressure at a point in a fluid in static equilibrium depends on the depth of that point but not on any horizontal dimension of the fluid or its container.

Thus, Eq. 15-8 holds no matter what the shape of the container. If the bottom surface of the container is at depth h, then Eq. 15-8 gives the pressure p there.

In Eq. 15-8, p is said to be the total pressure or **absolute pressure** at level 2. To see why, note in Fig. 15-3 that the pressure p at level 2 consists of two contributions: (1) p_0, the pressure due to the atmosphere, which bears down on the liquid, and (2) $\rho g h$, the pressure due to the liquid above level 2, which bears down on level 2. In general, the difference between an absolute pressure and an atmospheric pressure is called the **gauge pressure.** (The name comes from the use of a gauge to measure this difference in pressures.) For the situation of Fig. 15-3, the gauge pressure is $\rho g h$.

Equation 15-7 also holds above the liquid surface: It gives the atmospheric pressure at a given distance above level 1 in terms of the atmospheric pressure p_1 at level 1 (*assuming* that the atmospheric density is uniform over that distance). For example, to find the atmospheric pressure at a distance d above level 1 in

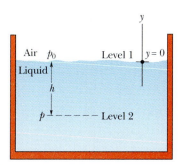

Fig. 15-3 The pressure p increases with depth h below the liquid surface according to Eq. 15-8.

Fig. 15-3, we substitute

$$y_1 = 0, \quad p_1 = p_0 \quad \text{and} \quad y_2 = d, \quad p_2 = p.$$

Then with $\rho = \rho_{\text{air}}$, we obtain

$$p = p_0 - \rho_{\text{air}}gd.$$

✔**CHECKPOINT 1:**
The figure shows four containers of olive oil. Rank them according to the pressure at depth h, greatest first.

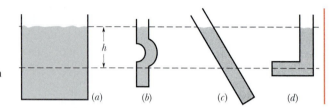

(a)   (b)   (c)   (d)

Sample Problem 15-2

A novice scuba diver practicing in a swimming pool takes enough air from his tank to fully expand his lungs before abandoning the tank at depth L and swimming to the surface. He ignores instructions and fails to exhale during his ascent. When he reaches the surface, the difference between the external pressure on him and the air pressure in his lungs is 9.3 kPa. From what depth does he start? What potentially lethal danger does he face?

SOLUTION: The **Key Idea** here is that when the diver fills his lungs at depth L, the external pressure on him (and thus the air pressure within his lungs) is greater than normal and given by Eq. 15-8 as

$$p = p_0 + \rho g L,$$

where p_0 is atmospheric pressure and ρ is the water's density (998 kg/m³, from Table 15-1). As he ascends, the external pressure on him decreases, until it is atmospheric pressure p_0 at the surface. His blood pressure also decreases, until it is normal. However, because he does not exhale, the air pressure in his lungs remains

at the value it had at depth L. At the surface, the pressure difference between the higher pressure in his lungs and the lower pressure on his chest is

$$\Delta p = p - p_0 = \rho g L,$$

from which we find

$$L = \frac{\Delta p}{\rho g} = \frac{9300 \text{ Pa}}{(998 \text{ kg/m}^3)(9.8 \text{ m/s}^2)}$$
$$= 0.95 \text{ m.} \qquad \text{(Answer)}$$

This is not deep! Yet, the pressure difference of 9.3 kPa (about 9% of atmospheric pressure) is sufficient to rupture the diver's lungs and force air from them into the depressurized blood, which then carries the air to the heart, killing the diver. If the diver follows instructions and gradually exhales as he ascends, he allows the pressure in his lungs to equalize with the external pressure, and then there is no danger.

Sample Problem 15-3

The U-tube in Fig. 15-4 contains two liquids in static equilibrium: Water of density ρ_w (= 998 kg/m³) is in the right arm, and oil of unknown density ρ_x is in the left. Measurement gives $l = 135$ mm and $d = 12.3$ mm. What is the density of the oil?

SOLUTION: One **Key Idea** here is that the pressure p_{int} at the oil–water interface in the left arm depends on the density ρ_x and height of the oil above the interface. A second **Key Idea** is that the water in the right arm *at the same level* must be at the same pressure p_{int}. The reason is that, because the water is in static equilibrium, pressures at points in the water at the same level must be the same even if the points are separated horizontally.

In the right arm, the interface is a distance l below the free surface of the *water* and we have, from Eq. 15-8,

$$p_{\text{int}} = p_0 + \rho_w g l \qquad \text{(right arm).}$$

In the left arm, the interface is a distance $l + d$ below the free surface of the *oil* and we have, again from Eq. 15-8,

$$p_{\text{int}} = p_0 + \rho_x g (l + d) \qquad \text{(left arm).}$$

Equating these two expressions and solving for the unknown density yield

$$\rho_x = \rho_w \frac{l}{l + d} = (998 \text{ kg/m}^3) \frac{135 \text{ mm}}{135 \text{ mm} + 12.3 \text{ mm}}$$
$$= 915 \text{ kg/m}^3. \qquad \text{(Answer)}$$

Note that the answer does not depend on the atmospheric pressure p_0 or the free-fall acceleration g.

Fig. 15-4 Sample Problem 15-3. The oil in the left arm stands higher than the water in the right arm because the oil is less dense than the water. Both fluid columns produce the same pressure p_{int} at the level of the interface.

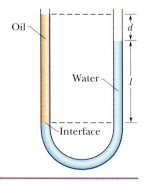

Oil

Water

Interface

15-5 Measuring Pressure

The Mercury Barometer

Figure 15-5a shows a very basic *mercury barometer,* a device used to measure the pressure of the atmosphere. The long glass tube is filled with mercury and inverted with its open end in a dish of mercury, as the figure shows. The space above the mercury column contains only mercury vapor, whose pressure is so small at ordinary temperatures that it can be neglected.

We can use Eq. 15-7 to find the atmospheric pressure p_0 in terms of the height h of the mercury column. We choose level 1 of Fig. 15-2 to be that of the air–mercury interface and level 2 to be that of the top of the mercury column, as labeled in Fig. 15-5a. We then substitute

$$y_1 = 0, \quad p_1 = p_0 \quad \text{and} \quad y_2 = h, \quad p_2 = 0$$

into Eq. 15-7, finding that

$$p_0 = \rho g h, \tag{15-9}$$

where ρ is the density of the mercury.

For a given pressure, the height h of the mercury column does not depend on the cross-sectional area of the vertical tube. The fanciful mercury barometer of Fig. 15-5b gives the same reading as that of Fig. 15-5a; all that counts is the vertical distance h between the mercury levels.

Equation 15-9 shows that, for a given pressure, the height of the column of mercury depends on the value of g at the location of the barometer and on the density of mercury, which varies with temperature. The column height (in millimeters) is numerically equal to the pressure (in torr) *only* if the barometer is at a place where g has its accepted standard value of 9.80665 m/s^2 *and* the temperature of the mercury is 0°C. If these conditions do not prevail (and they rarely do), small corrections must be made before the height of the mercury column can be transformed into a pressure.

The Open-Tube Manometer

An *open-tube manometer* (Fig. 15-6) measures the gauge pressure p_g of a gas. It consists of a U-tube containing a liquid, with one end of the tube connected to the

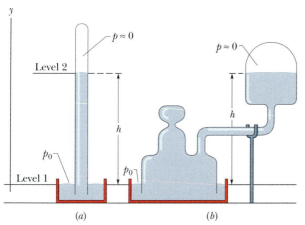

(a)   (b)

Fig. 15-5 (a) A mercury barometer. (b) Another mercury barometer. The distance h is the same in both cases.

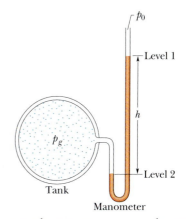

Fig. 15-6 An open-tube manometer, connected to measure the gauge pressure of the gas in the tank on the left. The right arm of the U-tube is open to the atmosphere.

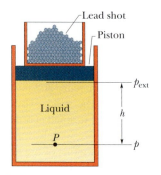

Fig. 15-7 Lead shot (small balls of lead) loaded onto the piston create a pressure p_{ext} at the top of the enclosed (incompressible) liquid. If p_{ext} is increased, by adding more lead shot, the pressure increases by the same amount at all points within the liquid.

vessel whose gauge pressure we wish to measure and the other end open to the atmosphere. We can use Eq. 15-7 to find the gauge pressure in terms of the height h shown in Fig. 15-6. Let us choose levels 1 and 2 as shown in Fig. 15-6. We then substitute

$$y_1 = 0, \quad p_1 = p_0 \quad \text{and} \quad y_2 = -h, \quad p_2 = p$$

into Eq. 15-7, finding that

$$p_g = p - p_0 = \rho g h, \tag{15-10}$$

where ρ is the density of the liquid in the tube. The gauge pressure p_g is directly proportional to h.

The gauge pressure can be positive or negative, depending on whether $p > p_0$ or $p < p_0$. In inflated tires or the human circulatory system, the (absolute) pressure is greater than atmospheric pressure, so the gauge pressure is a positive quantity, sometimes called the *overpressure*. If you suck on a straw to pull fluid up the straw, the (absolute) pressure in your lungs is actually less than atmospheric pressure. The gauge pressure in your lungs is then a negative quantity.

15-6   Pascal's Principle

When you squeeze one end of a tube to get toothpaste out the other end, you are watching **Pascal's principle** in action. This principle is also the basis for the Heimlich maneuver, in which a sharp pressure increase properly applied to the abdomen is transmitted to the throat, forcefully ejecting food lodged there. The principle was first stated clearly in 1652 by Blaise Pascal (for whom the unit of pressure is named):

> A change in the pressure applied to an enclosed incompressible fluid is transmitted undiminished to every portion of the fluid and to the walls of its container.

Demonstrating Pascal's Principle

Consider the case in which the incompressible fluid is a liquid contained in a tall cylinder, as in Fig. 15-7. The cylinder is fitted with a piston on which a container of lead shot rests. The atmosphere, container, and shot put pressure p_{ext} on the piston and thus on the liquid. The pressure p at any point P in the liquid is then

$$p = p_{ext} + \rho g h. \tag{15-11}$$

Let us add a little more lead shot to the container to increase p_{ext} by an amount Δp_{ext}. The quantities ρ, g, and h in Eq. 15-11 are unchanged, so the pressure change at P is

$$\Delta p = \Delta p_{ext}. \tag{15-12}$$

This pressure change is independent of h, so it must hold for all points within the liquid, as Pascal's principle states.

Pascal's Principle and the Hydraulic Lever

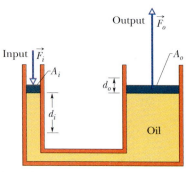

Fig. 15-8 A hydraulic arrangement that can be used to magnify a force $\vec{F}_i$. The work done is, however, not magnified and is the same for both the input and output forces.

Figure 15-8 shows how Pascal's principle can be made the basis of a hydraulic lever. In operation, let an external force of magnitude F_i be directed downward on the left-hand (or input) piston, whose area is A_i. An incompressible liquid in the device then produces an upward force of magnitude F_o on the righthand (or output) piston, whose area is A_o. To keep the system in equilibrium, there must be a downward force of magnitude F_o on the output piston from an external load (not shown). The force $\vec{F}_i$ applied on the left and the downward force $\vec{F}_o$ from the load on the right produce a change Δp in the pressure of the liquid that is given by

$$\Delta p = \frac{F_i}{A_i} = \frac{F_o}{A_o},$$

so

$$F_o = F_i \frac{A_o}{A_i}. \qquad (15\text{-}13)$$

Equation 15-13 shows that the output force F_o on the load must be greater than the input force F_i if $A_o > A_i$, as is the case in Fig. 15-8.

If we move the input piston downward a distance d_i, the output piston moves upward a distance d_o, such that the same volume V of the incompressible liquid is displaced at both pistons. Then

$$V = A_i d_i = A_o d_o,$$

which we can write as

$$d_o = d_i \frac{A_i}{A_o}. \qquad (15\text{-}14)$$

This shows that, if $A_o > A_i$ (as in Fig. 15-8), the output piston moves a smaller distance than the input piston moves.

From Eqs. 15-13 and 15-14 we can write the output work as

$$W = F_o d_o = \left(F_i \frac{A_o}{A_i} \right) \left(d_i \frac{A_i}{A_o} \right) = F_i d_i, \qquad (15\text{-}15)$$

which shows that the work W done *on* the input piston by the applied force is equal to the work W done *by* the output piston in lifting the load placed on it.

The advantage of a hydraulic lever is this:

> ➤ With a hydraulic lever, a given force applied over a given distance can be transformed to a greater force applied over a smaller distance.

The product of force and distance remains unchanged so that the same work is done. However, there is often tremendous advantage in being able to exert the larger force. Most of us, for example, cannot lift an automobile directly but can with a hydraulic jack, even though we have to pump the handle farther than the automobile rises. In this device, the displacement d_i is accomplished not in a single stroke but over a series of small strokes.

15-7 Archimedes' Principle

Figure 15-9 shows a student in a swimming pool, manipulating a very thin plastic sack (of negligible mass) that is filled with water. She finds that the sack and its contained water are in static equilibrium, tending neither to rise nor to sink. The downward gravitational force $\vec{F}_g$ on the contained water must be balanced by a net upward force from the water surrounding the sack.

This net upward force is a **buoyant force** $\vec{F}_b$. It exists because the pressure in the surrounding water increases with depth below the surface. Thus, the pressure near the bottom of the sack is greater than the pressure near the top. Then the forces on the sack due to this pressure are greater in magnitude near the bottom of the sack than near the top. Some of the forces are represented in Fig. 15-10a, where the space occupied by the sack has been left empty. Note that the force vectors drawn near the bottom of that space (with upward components) have longer lengths than those drawn near the top of the sack (with downward components). If we vectorially add all the forces on the sack from the water, the horizontal components cancel and the

Fig. 15-9 A thin-walled plastic sack of water is in static equilibrium in the pool. The gravitational force on it must be balanced by a net upward force on the sack from the surrounding water.

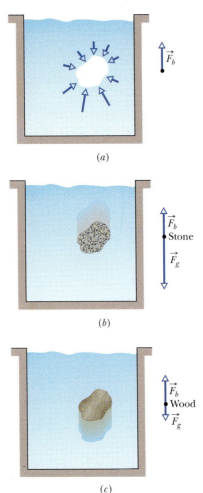

(a)

(b)

(c)

Fig. 15-10 (a) The water surrounding the hole in the water produces a net upward buoyant force on whatever fills the hole. (b) For a stone of the same volume as the hole, the gravitational force exceeds the buoyant force in magnitude. (c) For a lump of wood of the same volume, the gravitational force is less than the buoyant force in magnitude.

vertical components add to yield the upward buoyant force $\vec{F}_b$ on the sack. (Force $\vec{F}_b$ is shown to the right of the pool in Fig. 15-10a.)

Because the sack of water is in static equilibrium, the magnitude of $\vec{F}_b$ is equal to the magnitude $m_f g$ of the gravitational force $\vec{F}_g$ on the sack of water: $F_b = m_f g$. (Subscript f refers to *fluid*, here the water.) In words, the magnitude of the buoyant force is equal to the weight of the water in the sack.

In Fig. 15-10b, we have replaced the sack of water with a stone that exactly fills the hole in Fig. 15-10a. The stone is said to *displace* the water, meaning that it occupies space that would otherwise be occupied by water. We have changed nothing about the shape of the hole, so the forces at the hole's surface must be the same as when the water-filled sack was in place. Thus, the same upward buoyant force that acted on the water-filled sack now acts on the stone; that is, the magnitude F_b of the buoyant force is equal to $m_f g$, the weight of the water displaced by the stone.

Unlike the water-filled sack, the stone is not in static equilibrium. The downward gravitational force $\vec{F}_g$ on the stone is greater in magnitude than the upward buoyant force, as is shown in the free-body diagram to the right of the pool in Fig. 15-10b. The stone thus accelerates downward, sinking to the bottom of the pool.

Let us next exactly fill the hole in Fig. 15-10a with a block of light-weight wood, as in Fig. 15-10c. Again, nothing has changed about the forces at the hole's surface, so the magnitude F_b of the buoyant force is still equal to $m_f g$, the weight of the displaced water. Like the stone, the block is not in static equilibrium. However, this time the gravitational force $\vec{F}_g$ is lesser in magnitude than the buoyant force (as shown to the right of the pool), and so the block accelerates upward, rising to the top surface of the water.

Our results with the sack, stone, and block apply to all fluids and are summarized in **Archimedes' principle:**

> When a body is fully or partially submerged in a fluid, a buoyant force $\vec{F}_b$ from the surrounding fluid acts on the body. The force is directed upward and has a magnitude equal to the weight $m_f g$ of the fluid that has been displaced by the body.

The buoyant force on a body in a fluid has the magnitude

$$F_b = m_f g \qquad \text{(buoyant force)}, \qquad (15\text{-}16)$$

where m_f is the mass of the fluid that is displaced by the body.

In the late evening of August 21, 1986, something (possibly a volcanic tremor) disturbed Cameroon's Lake Nyos, which has a high concentration of dissolved carbon dioxide. The disturbance caused that gas to form bubbles. Being lighter than the surrounding fluid (the water), those bubbles were buoyed to the surface, where they released the carbon dioxide. The gas, being heavier than the surrounding fluid (now the air), rushed down the mountainside like a river, asphyxiating 1700 persons and the scores of animals seen here.

Floating

When we release a block of light-weight wood just above the water in a pool, it moves into the water because the gravitational force on it pulls it downward. As the block displaces more and more water, the magnitude F_b of the upward buoyant force acting on it increases. Eventually, F_b is large enough to equal the magnitude F_g of the downward gravitational force on the block, and the block comes to rest. The block is then in static equilibrium and is said to be *floating* in the water. In general,

> ▶ When a body floats in a fluid, the magnitude F_b of the buoyant force on the body is equal to the magnitude F_g of the gravitational force on the body.

We can write this statement as

$$F_b = F_g \qquad \text{(floating).} \tag{15-17}$$

From Eq. 15-16, we know that $F_b = m_f g$. Thus,

> ▶ When a body floats in a fluid, the magnitude F_g of the gravitational force on the body is equal to the weight $m_f g$ of the fluid that has been displaced by the body.

We can write this statement as

$$F_g = m_f g \qquad \text{(floating).} \tag{15-18}$$

In other words, a floating body displaces its own weight of fluid.

Apparent Weight in a Fluid

If we place a stone on a scale that is calibrated to measure weight, then the reading on the scale is the stone's weight. However, if we do this underwater, the upward buoyant force on the stone from the water decreases the reading. That reading is then an apparent weight. In general, an apparent weight is related to the actual weight of a body and the buoyant force on the body by

$$\begin{pmatrix} \text{apparent} \\ \text{weight} \end{pmatrix} = \begin{pmatrix} \text{actual} \\ \text{weight} \end{pmatrix} - \begin{pmatrix} \text{magnitude of} \\ \text{buoyant force} \end{pmatrix},$$

which we can write as

$$\text{weight}_{\text{app}} = \text{weight} - F_b \qquad \text{(apparent weight).} \tag{15-19}$$

If, in some strange test of strength, you had to lift a heavy stone, you could do it more easily with the stone underwater. Then your applied force would need to exceed only the stone's apparent weight, not its larger actual weight, because the upward buoyant force would help you lift the stone.

The magnitude of the buoyant force on a floating body is equal to the body's weight. Equation 15-19 thus tells us that a floating body has an apparent weight of zero—the body would produce a reading of zero on a scale. (When astronauts prepare to perform a complex task in space, they practice the task floating underwater, where their apparent weight is zero as it is in space.)

✔**CHECKPOINT 2:**   A penguin floats first in a fluid of density ρ_0, then in a fluid of density $0.95\rho_0$, and then in a fluid of density $1.1\rho_0$. (a) Rank the densities according to the magnitude of the buoyant force on the penguin, greatest first. (b) Rank the densities according to the amount of fluid displaced by the penguin, greatest first.

Sample Problem 15-4

What fraction of the volume of an iceberg floating in seawater is visible?

SOLUTION: Let V_i be the total volume of the iceberg. The nonvisible portion is below water and thus is equal to the volume V_f of the fluid (the seawater) displaced by the iceberg. We seek the fraction (call it frac)

$$\text{frac} = \frac{V_i - V_f}{V_i} = 1 - \frac{V_f}{V_i}, \qquad (15\text{-}20)$$

but we know neither volume. A **Key Idea** here is that, because the iceberg is floating, Eq. 15-18 ($F_g = m_f g$) applies. We can write that equation as

$$m_i g = m_f g,$$

from which we see that $m_i = m_f$. Thus, the mass of the iceberg is equal to the mass of the displaced fluid (seawater). Although we know neither mass, we can relate them to the densities of ice and seawater given in Table 15-1 by using Eq. 15-2 ($\rho = m/V$). Because $m_i = m_f$, we can write

$$\rho_i V_i = \rho_f V_f$$

or

$$\frac{V_f}{V_i} = \frac{\rho_i}{\rho_f}.$$

Substituting this into Eq. 15-20 and then using the known densities, we find

$$\text{frac} = 1 - \frac{\rho_i}{\rho_f} = 1 - \frac{917 \text{ kg/m}^3}{1024 \text{ kg/m}^3}$$

$$= 0.10 \text{ or } 10\%. \qquad \text{(Answer)}$$

Sample Problem 15-5

A spherical, helium-filled balloon has a radius R of 12.0 m. The balloon, support cables, and basket have a mass m of 196 kg. What maximum load M can the balloon support while it floats at an altitude at which the helium density ρ_{He} is 0.160 kg/m^3 and the air density ρ_{air} is 1.25 kg/m^3? Assume that the volume of air displaced by the load, support cables, and basket is negligible.

SOLUTION: The **Key Idea** here is that the balloon, cables, basket, load, *and* the contained helium form a floating body, with a total mass of $m + M + m_{He}$, where m_{He} is the mass of the contained helium. Then the magnitude of the total gravitational force on this body must be equal to the weight of the air displaced by the body (the air is the fluid in which this body floats). Let m_{air} be the mass of the air displaced by the body. From Eq. 15-18 ($F_g = m_f g$), we have

$$(m + M + m_{He})g = m_{air}g$$

or

$$M = m_{air} - m_{He} - m. \qquad (15\text{-}21)$$

We do not know m_{He} and m_{air} but we do know the corresponding densities, so we can use Eq. 15-2 ($\rho = m/V$) to rewrite Eq. 15-21 in terms of those densities. First we note that, because the load, support cables, and basket displace a negligible amount of air, the volume of the displaced air is equal to the volume V ($= \frac{4}{3}\pi R^3$) of the spherical balloon. Then Eq. 15-21 becomes

$$M = \rho_{air}V - \rho_{He}V - m$$

$$= (\tfrac{4}{3}\pi R^3)(\rho_{air} - \rho_{He}) - m$$

$$= (\tfrac{4}{3}\pi)(12.0 \text{ m})^3(1.25 \text{ kg/m}^3 - 0.160 \text{ kg/m}^3) - 196 \text{ kg}$$

$$= 7694 \text{ kg} \approx 7690 \text{ kg}. \qquad \text{(Answer)}$$

15-8   Ideal Fluids in Motion

The motion of *real fluids* is very complicated and not yet fully understood. Instead, we shall discuss the motion of an **ideal fluid,** which is simpler to handle mathematically and yet provides useful results. Here are four assumptions that we make about our ideal fluid; they all are concerned with *flow:*

1. *Steady flow* In *steady* (or *laminar*) flow, the velocity of the moving fluid at any fixed point does not change with time, either in magnitude or in direction. The gentle flow of water near the center of a quiet stream is steady; that in a chain of rapids is not. Figure 15-11 shows a transition from steady flow to *nonsteady* (or *turbulent*) flow for a rising stream of smoke. The speed of the smoke particles increases as they rise and, at a certain critical speed, the flow changes from steady to nonsteady (that is, from laminar to *nonlaminar* flow).

2. *Incompressible flow* We assume, as we have already done for fluids at rest, that our ideal fluid is incompressible; that is, its density has a constant, uniform value.

3. *Nonviscous flow* Roughly speaking, the viscosity of a fluid is a measure of how resistive the fluid is to flow. For example, thick honey is more resistive to flow than water, and so honey is said to be more viscous than water. Viscosity is the

Fig. 15-11 At a certain point, the rising flow of smoke and heated gas changes from steady to turbulent.

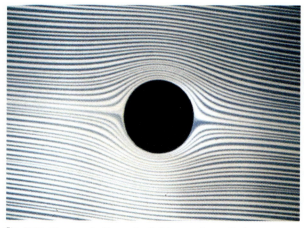

Fig. 15-12 The steady flow of a fluid around a cylinder, as revealed by a dye tracer that was injected into the fluid upstream of the cylinder.

Fig. 15-13 Smoke reveals streamlines in airflow past a car in a wind-tunnel test.

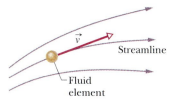

Fig. 15-14 A fluid element *P* traces out a streamline as it moves. The velocity vector of the element is tangent to the streamline at every point.

fluid analog of friction between solids; both are mechanisms by which the kinetic energy of moving objects can be transferred to thermal energy. In the absence of friction, a block could glide at constant speed along a horizontal surface. In the same way, an object moving through a nonviscous fluid would experience no *viscous drag force*—that is, no resistive force due to viscosity; it could move at constant speed through the fluid. The British scientist Lord Rayleigh noted that in an ideal fluid a ship's propeller would not work but, on the other hand, a ship (once set into motion) would not need a propeller!

4. Irrotational flow Although it need not concern us further, we also assume that the flow is *irrotational*. To test for this property, let a tiny grain of dust move with the fluid. Although this test body may (or may not) move in a circular path, in irrotational flow the test body will not rotate about an axis through its own center of mass. For a loose analogy, the motion of a Ferris wheel is rotational; that of its passengers is irrotational.

We can make the flow of a fluid visible by adding a *tracer*. This might be a dye injected into many points across a liquid stream (Fig. 15-12) or smoke particles added to a gas flow (Figs. 15-11 and 15-13). Each bit of a tracer follows a *streamline*, which is the path that a tiny element of the fluid would take as the fluid flows. Recall from Chapter 2 that the velocity of a particle is always tangent to the path taken by the particle. Here the particle is the fluid element, and its velocity $\vec{v}$ is always tangent to a streamline (Fig. 15-14). For this reason, two streamlines can never intersect; if they did, then an element arriving at their intersection would have two different velocities simultaneously, an impossibility.

15-9 The Equation of Continuity

You may have noticed that you can increase the speed of the water emerging from a garden hose by partially closing the hose opening with your thumb. Apparently the speed v of the water depends on the cross-sectional area A through which the water flows.

Here we wish to derive an expression that relates v and A for the steady flow of an ideal fluid through a tube with varying cross section, like that in Fig. 15-15. The flow there is toward the right, and the tube segment shown (part of a longer

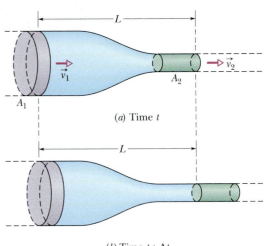

Fig. 15-15 Fluid flows from left to right at a steady rate through a tube segment of length L. The fluid's speed is v_1 at the left side and v_2 at the right side. The tube's cross-sectional area is A_1 at the left side and A_2 at the right side. From time t in (*a*) to time $t + \Delta t$ in (*b*), the amount of fluid shown in purple enters at the left side and the equal amount of fluid shown in green emerges at the right side.

(*a*) Time t

(*b*) Time $t + \Delta t$

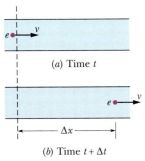

(*a*) Time t

(*b*) Time $t + \Delta t$

Fig. 15-16 Fluid flows at a constant speed v through a tube. (*a*) At time t, fluid element e is about to pass the dashed line. (*b*) At time $t + \Delta t$, element e is a distance $\Delta x = v \Delta t$ from the dashed line.

tube) has length L. The fluid has speeds v_1 at the left end of the segment and v_2 at the right end. The tube has cross-sectional areas A_1 at the left end and A_2 at the right end. Suppose that in a time interval Δt a volume ΔV of fluid enters the tube segment at its left end (that volume is colored purple in Fig. 15-15*a*). Then, because the fluid is incompressible, an identical volume ΔV must emerge from the right end of the segment (it is colored green in Fig. 15-15*b*).

We can use this common volume ΔV to relate the speeds and areas. To do so, we first consider Fig. 15-16, which shows a side view of a tube of *uniform* cross-sectional area A. In Fig. 15-16*a*, a fluid element e is about to pass through the dashed line drawn across the tube width. The element's speed is v, so during a time interval Δt, the element moves along the tube a distance $\Delta x = v \Delta t$. The volume ΔV of fluid that has passed through the dashed line in that time interval Δt is

$$\Delta V = A \, \Delta x = Av \, \Delta t. \tag{15-22}$$

Applying Eq. 15-22 to both the left and right ends of the tube segment in Fig. 15-15, we have

$$\Delta V = A_1 v_1 \, \Delta t = A_2 v_2 \, \Delta t$$

or $\qquad\qquad A_1 v_1 = A_2 v_2 \qquad$ (equation of continuity). $\qquad$ (15-23)

This relation between speed and cross-sectional area is called the **equation of continuity** for the flow of an ideal fluid. It tells us that the flow speed increases when we decrease the cross-sectional area through which the fluid flows (as when we partially close off a garden hose with a thumb).

Equation 15-23 applies not only to an actual tube but also to any so-called *tube of flow,* or imaginary tube whose boundary consists of streamlines. Such a tube acts like a real tube because no fluid element can cross a streamline; thus, all the fluid within a tube of flow must remain within its boundary. Figure 15-17 shows a tube of flow in which the cross-sectional area increases from area A_1 to area A_2 along the flow direction. From Eq. 15-23 we know that, with the increase in area, the speed must decrease, as is indicated by the greater spacing between streamlines at the right in Fig. 15-17. Similarly, you can see that in Fig. 15-12 the speed of the flow is greatest just above and just below the cylinder.

We can rewrite Eq. 15-23 as

$$R_V = Av = \text{a constant} \qquad \text{(volume flow rate, equation of continuity),} \qquad (15\text{-}24)$$

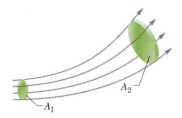

Fig. 15-17 A tube of flow is defined by the streamlines that form the boundary. The volume flow rate must be the same for all cross sections of the tube of flow.

in which R_V is the **volume flow rate** of the fluid (volume per unit time). Its SI unit is the cubic meter per second (m³/s). If the density ρ of the fluid is uniform, we can multiply Eq. 15-24 by that density to get the **mass flow rate** R_m (mass per unit time):

$$R_m = \rho R_V = \rho A v = \text{a constant} \qquad \text{(mass flow rate)}. \qquad (15\text{-}25)$$

The SI unit of mass flow rate is the kilogram per second (kg/s). Equation 15-25 says that the mass that flows into the tube segment of Fig. 15-15 each second must be equal to the mass that flows out of that segment each second.

✔**CHECKPOINT 3:** The figure shows a pipe and gives the volume flow rate (in cm³/s) and the direction of flow for all but one section. What are the volume flow rate and the direction of flow for that section?

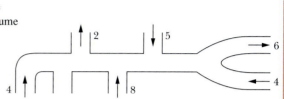

Sample Problem 15-6

The cross-sectional area A_0 of the aorta (the major blood vessel emerging from the heart) of a normal resting person is 3 cm², and the speed v_0 of the blood through it is 30 cm/s. A typical capillary (diameter $\approx$ 6 μm) has a cross-sectional area A of 3×10^{-7} cm² and a flow speed v of 0.05 cm/s. How many capillaries does such a person have?

SOLUTION: The **Key Idea** here is that all the blood that passes through the capillaries must have passed through the aorta. Therefore, the volume flow rate through the aorta must equal the total volume flow rate through the capillaries. Let us assume that the capillaries

are identical, with the given cross-sectional area A and flow speed v. Then, from Eq. 15-24 we have

$$A_0 v_0 = nAv,$$

where n is the number of capillaries. Solving for n yields

$$n = \frac{A_0 v_0}{Av} = \frac{(3 \text{ cm}^2)(30 \text{ cm/s})}{(3 \times 10^{-7} \text{ cm}^2)(0.05 \text{ cm/s})}$$

$$= 6 \times 10^9 \text{ or 6 billion}. \qquad \text{(Answer)}$$

You can easily show that the combined cross-sectional area of the capillaries is about 600 times the cross-sectional area of the aorta.

Sample Problem 15-7

Figure 15-18 shows how the stream of water emerging from a faucet "necks down" as it falls. The indicated cross-sectional areas are $A_0 = 1.2$ cm² and $A = 0.35$ cm². The two levels are separated by a vertical distance $h = 45$ mm. What is the volume flow rate from the tap?

SOLUTION: The **Key Idea** here is simply that the volume flow rate through the higher cross section must be the same as that through

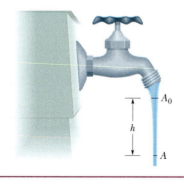

Fig. 15-18 Sample Problem 15-7. As water falls from a tap, its speed increases. Because the flow rate must be the same at all cross sections, the stream must "neck down."

the lower cross section. Thus, from Eq. 15-24, we have

$$A_0 v_0 = Av, \qquad (15\text{-}26)$$

where v_0 and v are the water speeds at the levels corresponding to A_0 and A. From Eq. 2-16 we can also write, because the water is falling freely with acceleration g,

$$v^2 = v_0^2 + 2gh. \qquad (15\text{-}27)$$

Eliminating v between Eqs. 15-26 and 15-27 and solving for v_0, we obtain

$$v_0 = \sqrt{\frac{2ghA^2}{A_0^2 - A^2}}$$

$$= \sqrt{\frac{(2)(9.8 \text{ m/s}^2)(0.045 \text{ m})(0.35 \text{ cm}^2)^2}{(1.2 \text{ cm}^2)^2 - (0.35 \text{ cm}^2)^2}}$$

$$= 0.286 \text{ m/s} = 28.6 \text{ cm/s}.$$

From Eq. 15-24, the volume flow rate R_V is then

$$R_V = A_0 v_0 = (1.2 \text{ cm}^2)(28.6 \text{ cm/s})$$

$$= 34 \text{ cm}^3/\text{s}. \qquad \text{(Answer)}$$

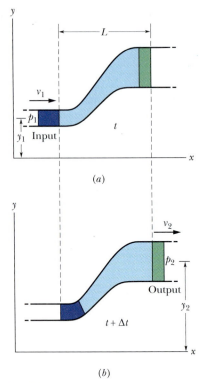

Fig. 15-19 Fluid flows at a steady rate through a length L of a tube, from the input end at the left to the output end at the right. From time t in (a) to time $t + \Delta t$ in (b), the amount of fluid shown in purple enters the input end and the equal amount shown in green emerges from the output end.

15-10 Bernoulli's Equation

Figure 15-19 represents a tube through which an ideal fluid is flowing at a steady rate. In a time interval Δt, suppose that a volume of fluid ΔV, colored purple in Fig. 15-19a, enters the tube at the left (or input) end and an identical volume, colored green in Fig. 15-19b, emerges at the right (or output) end. The emerging volume must be the same as the entering volume because the fluid is incompressible, with an assumed constant density ρ.

Let y_1, v_1, and p_1 be the elevation, speed, and pressure of the fluid entering at the left, and y_2, v_2, and p_2 be the corresponding quantities for the fluid emerging at the right. By applying the principle of conservation of energy to the fluid, we shall show that these quantities are related by

$$p_1 + \tfrac{1}{2}\rho v_1^2 + \rho g y_1 = p_2 + \tfrac{1}{2}\rho v_2^2 + \rho g y_2. \qquad (15\text{-}28)$$

We can also write this equation as

$$p + \tfrac{1}{2}\rho v^2 + \rho g y = \text{a constant} \qquad \text{(Bernoulli's equation).} \qquad (15\text{-}29)$$

Equations 15-28 and 15-29 are equivalent forms of **Bernoulli's equation,** after Daniel Bernoulli, who studied fluid flow in the 1700s.* Like the equation of continuity (Eq. 15-24), Bernoulli's equation is not a new principle but simply the reformulation of a familiar principle in a form more suitable to fluid mechanics. As a check, let us apply Bernoulli's equation to fluids at rest, by putting $v_1 = v_2 = 0$ in Eq. 15-28. The result is

$$p_2 = p_1 + \rho g(y_1 - y_2),$$

which is Eq. 15-7 with a slight change in notation.

A major prediction of Bernoulli's equation emerges if we take y to be a constant ($y = 0$, say) so that the fluid does not change elevation as it flows. Equation 15-28 then becomes

$$p_1 + \tfrac{1}{2}\rho v_1^2 = p_2 + \tfrac{1}{2}\rho v_2^2, \qquad (15\text{-}30)$$

which tells us that:

▶ If the speed of a fluid element increases as it travels along a horizontal streamline, the pressure of the fluid must decrease, and conversely.

Put another way, where the streamlines are relatively close together (that is, where the velocity is relatively great), the pressure is relatively low, and conversely.

The link between a change in speed and a change in pressure makes sense if you consider a fluid element. When the element nears a narrow region, the higher pressure behind it accelerates it so that it then has a greater speed in the narrow region. When it nears a wide region, the higher pressure ahead of it decelerates it so that it then has a lesser speed in the wide region.

Bernoulli's equation is strictly valid only to the extent that the fluid is ideal. If viscous forces are present, thermal energy will be involved. We take no account of this in the derivation that follows.

Proof of Bernoulli's Equation

Let us take as our system the entire volume of the (ideal) fluid shown in Fig. 15-19. We shall apply the principle of conservation of energy to this system as it moves

*For irrotational flow (which we assume), the constant in Eq. 15-29 has the same value for all points within the tube of flow; the points do not have to lie along the same streamline. Similarly, the points 1 and 2 in Eq. 15-28 can lie anywhere within the tube of flow.

from its initial state (Fig. 15-19a) to its final state (Fig. 15-19b). The fluid lying between the two vertical planes separated by a distance L in Fig. 15-19 does not change its properties during this process; we need be concerned only with changes that take place at the input and output ends.

We apply energy conservation in the form of the work–kinetic energy theorem,

$$W = \Delta K, \tag{15-31}$$

which tells us that the change in the kinetic energy of our system must equal the net work done on the system. The change in kinetic energy results from the change in speed between the ends of the tube and is

$$\begin{aligned} \Delta K &= \tfrac{1}{2}\Delta m\, v_2^2 - \tfrac{1}{2}\Delta m\, v_1^2 \\ &= \tfrac{1}{2}\rho\, \Delta V\,(v_2^2 - v_1^2), \end{aligned} \tag{15-32}$$

in which $\Delta m\,(=\rho\,\Delta V)$ is the mass of the fluid that enters at the input end and leaves at the output end during a small time interval Δt.

The work done on the system arises from two sources. The work W_g done by the gravitational force $(\Delta m\,\vec{g})$ on the fluid of mass Δm during the vertical lift of the mass from the input level to the output level is

$$\begin{aligned} W_g &= -\Delta m\, g(y_2 - y_1) \\ &= -\rho g\, \Delta V\,(y_2 - y_1). \end{aligned} \tag{15-33}$$

This work is negative because the upward displacement and the downward gravitational force have opposite directions.

Work must also be done on the system (at the input end) to push the entering fluid into the tube and by the system (at the output end) to push forward the fluid that is located ahead of the emerging fluid. In general, the work done by a force of magnitude F, acting on a fluid sample contained in a tube of area A to move the fluid through a distance Δx, is

$$F\,\Delta x = (pA)(\Delta x) = p(A\,\Delta x) = p\,\Delta V.$$

The work done on the system is then $p_1\,\Delta V$, and the work done by the system is $-p_2\,\Delta V$. Their sum W_p is

$$\begin{aligned} W_p &= -p_2\,\Delta V + p_1\,\Delta V \\ &= -(p_2 - p_1)\,\Delta V. \end{aligned} \tag{15-34}$$

The work–kinetic energy theorem of Eq. 15-31 now becomes

$$W = W_g + W_p = \Delta K.$$

Substituting from Eqs. 15-32, 15-33, and 15-34 yields

$$-\rho g\, \Delta V\,(y_2 - y_1) - \Delta V\,(p_2 - p_1) = \tfrac{1}{2}\rho\, \Delta V\,(v_2^2 - v_1^2).$$

This, after a slight rearrangement, matches Eq. 15-28, which we set out to prove.

✔**CHECKPOINT 4:** Water flows smoothly through the pipe shown in the figure, descending in the process. Rank the four numbered sections of pipe according to (a) the volume flow rate R_V through them, (b) the flow speed v through them, and (c) the water pressure p within them, greatest first.

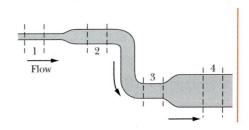

Sample Problem 15-8

Ethanol of density $\rho = 791$ kg/m^3 flows smoothly through a horizontal pipe that tapers (as in Fig. 15-15) in cross-sectional area from $A_1 = 1.20 \times 10^{-3}$ m^2 to $A_2 = A_1/2$. The pressure difference between the wide and narrow sections of pipe is 4120 Pa. What is the volume flow rate R_V of the ethanol?

SOLUTION: One Key Idea here is that, because the fluid flowing through the wide section of pipe must entirely pass through the narrow section, the volume flow rate R_V must be the same in the two sections. Thus, from Eq. 15-24,

$$R_V = v_1A_1 = v_2A_2. \tag{15-35}$$

However, with two unknown speeds, we cannot evaluate this equation for R_V.

A second Key Idea is that, because the flow is smooth, we can apply Bernoulli's equation. From Eq. 15-28, we can write

$$p_1 + \tfrac{1}{2}\rho v_1^2 + \rho gy = p_2 + \tfrac{1}{2}\rho v_2^2 + \rho gy, \tag{15-36}$$

where subscripts 1 and 2 refer to the wide and narrow sections of pipe, respectively, and y is their common elevation. This equation hardly seems to help because it does not contain the desired volume flow R_V and it contains the unknown speeds v_1 and v_2.

However, there is a neat way to make it work for us: First, we can use Eq. 15-35 and the fact that $A_2 = A_1/2$ to write

$$v_1 = \frac{R_V}{A_1} \quad \text{and} \quad v_2 = \frac{R_V}{A_2} = \frac{2R_V}{A_1}. \tag{15-37}$$

Then we can substitute these expressions into Eq. 15-36 to eliminate the unknown speeds and introduce the desired volume flow rate. Doing this and solving for R_V yield

$$R_V = A_1\sqrt{\frac{2(p_1 - p_2)}{3\rho}}. \tag{15-38}$$

We still have a decision to make: We know that the pressure difference between the two sections is 4120 Pa, but does that mean that $p_1 - p_2$ is 4120 Pa or -4120 Pa? We could guess the former is true, or otherwise the square root in Eq. 15-38 would give us an imaginary number. Instead of guessing, however, let's try some reasoning. From Eq. 15-35 we see that speed v_2 in the narrow section (small A_2) must be greater than speed v_1 in the wider section (larger A_1). Recall that if the speed of a fluid increases as it travels along a horizontal path (as here), the pressure of the fluid must decrease. Thus, p_1 is greater than p_2, and $p_1 - p_2 = 4120$ Pa. Inserting this and known data into Eq. 15-38 gives

$$R_V = 1.20 \times 10^{-3} \text{ m}^2 \sqrt{\frac{(2)(4120 \text{ Pa})}{(3)(791 \text{ kg/m}^3)}}$$
$$= 2.24 \times 10^{-3} \text{ m}^3/\text{s}. \tag{Answer}$$

Sample Problem 15-9

In the old West, a desperado fires a bullet into an open water tank (Fig. 15-20), creating a hole a distance h below the water surface. What is the speed v of the water emerging from the hole?

SOLUTION: A Key Idea is that this situation is essentially that of water moving (downward) with speed v_0 through a wide pipe (the tank) of cross-sectional area A and then moving (horizontally) with speed v through a narrow pipe (the hole) of cross-sectional area a. Another Key Idea is that, because the water flowing through the wide pipe must entirely pass through the narrow pipe, the volume flow rate R_V must be the same in the two "pipes." Then from Eq. 15-24,

$$R_V = av = Av_0$$

and thus

$$v_0 = \frac{a}{A}v.$$

Because $a \ll A$, we see that $v_0 \ll v$.

A third Key Idea is that we can also relate v to v_0 (and to h) through Bernoulli's equation (Eq. 15-28). We take the level of the hole as our reference level for measuring elevations (and thus gravitational potential energy). Noting that the pressure at the top of the tank and at the bullet hole is the atmospheric pressure p_0 (because both places are exposed to the atmosphere), we write Eq. 15-28 as

$$p_0 + \tfrac{1}{2}\rho v_0^2 + \rho gh = p_0 + \tfrac{1}{2}\rho v^2 + \rho g(0). \tag{15-39}$$

(Here the top of the tank is represented by the left side of the equation, and the hole by the right side. The zero on the right indicates that the hole is at our reference level.) Before we solve Eq. 15-39 for v, we can use our result that $v_0 \ll v$ to simplify it: We assume that v_0^2, and thus the term $\tfrac{1}{2}\rho v_0^2$ in Eq. 15-39, is negligible compared to the other terms, and we drop it. Solving the remaining equation for v then yields

$$v = \sqrt{2gh}. \tag{Answer}$$

This is the same speed that an object would have when falling a height h from rest.

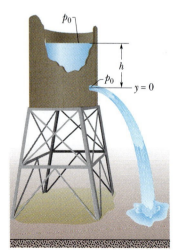

Fig. 15-20 Sample Problem 15-9. Water pours through a hole in a water tank, at a distance h below the water surface. The pressure at the water surface and at the hole is atmospheric pressure p_0.

REVIEW & SUMMARY

Density   The **density** ρ of any material is defined as its mass per unit volume:

$$\rho = \frac{\Delta m}{\Delta V}. \qquad (15\text{-}1)$$

Usually, where a material sample is large compared with atomic dimensions, we can write Eq. 15-1 as

$$\rho = \frac{m}{V}. \qquad (15\text{-}2)$$

Fluid Pressure   A **fluid** is a substance that can flow; it conforms to the boundaries of its container because it cannot withstand shearing stress. It can, however, exert a force perpendicular to its surface. That force is described in terms of **pressure** p:

$$p = \frac{\Delta F}{\Delta A}, \qquad (15\text{-}3)$$

in which ΔF is the force acting on a surface element of area ΔA. If the force is uniform over a flat area, Eq. 15-3 can be written as

$$p = \frac{F}{A}. \qquad (15\text{-}4)$$

The force resulting from fluid pressure at a particular point in a fluid has the same magnitude in all directions. *Gauge pressure* is the difference between the actual pressure (or *absolute pressure*) at a point and the atmospheric pressure.

Pressure Variation with Height and Depth   Pressure in a fluid at rest varies with vertical position y. For y measured positive upward,

$$p_2 = p_1 + \rho g(y_1 - y_2). \qquad (15\text{-}7)$$

The pressure in a fluid is the same for all points at the same level. If h is the *depth* of a fluid sample below some reference level at which the pressure is p_0, Eq. 15-7 becomes

$$p = p_0 + \rho g h, \qquad (15\text{-}8)$$

where p is the pressure in the sample.

Pascal's Principle   *Pascal's principle,* which can be derived from Eq. 15-7, states that a change in the pressure applied to an enclosed fluid is transmitted undiminished to every portion of the fluid and to the walls of the containing vessel.

Archimedes' Principle   When a body is fully or partially submerged in a fluid, a buoyant force $\vec{F}_b$ from the surrounding fluid acts on the body. The force is directed upward and has a magnitude given by

$$F_b = m_f g, \qquad (15\text{-}16)$$

where m_f is the mass of the fluid that has been displaced by the body.

When a body floats in a fluid, the magnitude F_b of the (upward) buoyant force on the body is equal to the magnitude F_g of the (downward) gravitational force on the body. The apparent weight of a body on which a buoyant force acts is related to its actual weight by

$$\text{weight}_{app} = \text{weight} - F_b. \qquad (15\text{-}19)$$

Flow of Ideal Fluids   An *ideal fluid* is incompressible and lacks viscosity, and its flow is steady and irrotational. A **streamline** is the path followed by an individual fluid particle. A *tube of flow* is a bundle of streamlines. The flow within any tube of flow obeys the **equation of continuity:**

$$R_V = Av = \text{a constant}, \qquad (15\text{-}24)$$

in which R_V is the **volume flow rate**, A is the cross-sectional area of the tube of flow at any point, and v is the speed of the fluid at that point, assumed to be constant across A. The **mass flow rate** R_m is

$$R_m = \rho R_V = \rho Av = \text{a constant}. \qquad (15\text{-}25)$$

Bernoulli's Equation   Applying the principle of conservation of mechanical energy to the flow of an ideal fluid leads to **Bernoulli's equation:**

$$p + \tfrac{1}{2}\rho v^2 + \rho g y = \text{a constant} \qquad (15\text{-}29)$$

along any tube of flow.

QUESTIONS

1. Figure 15-21 shows a tank filled with water. Five horizontal floors and ceilings are indicated; all have the same area and are located at distances L, $2L$, or $3L$ below the top of the tank. Rank the floors and ceilings according to the force on them due to the water, greatest first.

2. *The Teapot Effect:* When water is poured slowly from a teapot

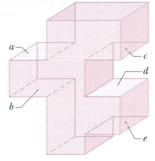

Fig. 15-21 Question 1.

spout, it can double back under the spout for a considerable distance before detaching and falling. (The water layer is held against the underside of the spout by atmospheric pressure.) In Fig. 15-22, within the water layer in the spout, point a is at the top of the layer

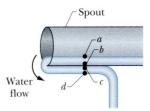

Fig. 15-22 Question 2.

and point b is at the bottom of the layer; within the water layer below the spout, point c is at the top of the layer and point d is at the bottom of the layer. Rank those four points according to the gauge pressure in the water there, most positive first, most negative last.

3. Figure 15-23 shows four situations in which a red liquid and a gray liquid are in a **U**-tube. In one situation the liquids cannot be in static equilibrium. (a) Which situation is that? (b) For the other three situations, assume static equilibrium. For each, is the density of the red liquid greater than, less than, or equal to the density of the gray liquid?

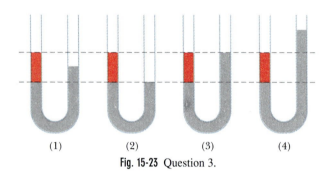

(1)   (2)   (3)   (4)

Fig. 15-23 Question 3.

4. Three hydraulic levers such as that in Fig. 15-8 are used to lift identical loads (at the output side) by identical distances. The levels are identical on the input side but differ in the area of the output piston: Lever 1 has a piston of area A, lever 2 has a piston of area $2A$, and lever 3 has a piston of area $3A$. For the lift, rank the levers according to (a) the required work at the input side, (b) the required magnitude of the force (assumed to be constant) at the input side, and (c) the displacement of the piston at the input side, greatest first.

5. We fully submerge an irregular 3 kg lump of material in a certain fluid. The fluid that would have been in the space now occupied by the lump has a mass of 2 kg. (a) When we release the lump, does it move upward, move downward, or remain in place? (b) If we next fully submerge the lump in a less dense fluid and again release it, what does it do?

6. Figure 15-24 shows four solid objects floating in corn syrup. Rank the objects according to their density, greatest first.

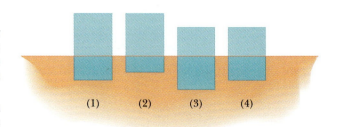

(1)   (2)   (3)   (4)

Fig. 15-24 Question 6.

7. Figure 15-25 shows three identical open-top containers filled to the brim with water; toy ducks float in two of them. Rank the containers and contents according to their weight, greatest first.

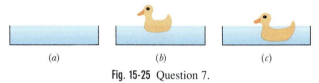

(a)   (b)   (c)

Fig. 15-25 Question 7.

8. A block floats in a pail of water in a stationary elevator. Does the block float higher, lower, or at the same level when the elevator cab (a) moves upward at constant speed, (b) moves downward at constant speed, (c) accelerates upward, and (d) accelerates downward with an acceleration magnitude less than g?

9. A boat with an anchor on board floats in a swimming pool that is somewhat wider than the boat. Does the water level in the pool move upward, move downward, or remain the same if the anchor is (a) dropped into the water or (b) thrown onto the surrounding ground? (c) Does the water level in the pool move upward, move downward, or remain the same if, instead, a cork is dropped from the boat into the water, where it floats?

10. Figure 15-26 shows three straight pipes through which water flows. The figure gives the speed of the water in each pipe and the cross-sectional area of each pipe. Rank the pipes according to the volume of water that passes through the cross-sectional area per minute, greatest first.

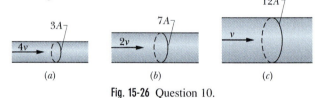

(a)   (b)   (c)

Fig. 15-26 Question 10.

EXERCISES & PROBLEMS

SEC. 15-3   Density and Pressure

1E. Find the pressure increase in the fluid in a syringe when a nurse applies a force of 42 N to the syringe's circular piston, which has a radius of 1.1 cm. **ssm**

2E. Three liquids that will not mix are poured into a cylindrical container. The volumes and densities of the liquids are 0.50 L, 2.6 g/cm³; 0.25 L, 1.0 g/cm³; and 0.40 L, 0.80 g/cm³. What is the force on the bottom of the container due to these liquids? One liter = 1 L = 1000 cm³. (Ignore the contribution due to the atmosphere.)

3E. An office window has dimensions 3.4 m by 2.1 m. As a result of the passage of a storm, the outside air pressure drops to 0.96 atm, but inside the pressure is held at 1.0 atm. What net force pushes out on the window? **ssm**

4E. You inflate the front tires on your car to 28 psi. Later, you measure your blood pressure, obtaining a reading of 120/80, the readings being in mm Hg. In metric countries (which is to say, most of the world), these pressures are customarily reported in kilopascals (kPa). In kilopascals, what are (a) your tire pressure and (b) your blood pressure?

5E. A fish maintains its depth in fresh water by adjusting the air content of porous bone or air sacs to make its average density the same as that of the water. Suppose that with its air sacs collapsed, a fish has a density of 1.08 g/cm³. To what fraction of its expanded body volume must the fish inflate the air sacs to reduce its density to that of water? ilw

6P. An airtight container having a lid with negligible mass and an area of 77 cm² is partially evacuated. If a 480 N force is required to pull the lid off the container and the atmospheric pressure is 1.0×10^5 Pa, what is the air pressure in the container before it is opened? ssm

7P. In 1654 Otto von Guericke, inventor of the air pump, gave a demonstration before the noblemen of the Holy Roman Empire in which two teams of eight horses could not pull apart two evacuated brass hemispheres. (a) Assuming that the hemispheres have thin walls, so that R in Fig. 15-27 may be considered both the inside and outside radius, show that the force $\vec{F}$ required to pull apart the hemispheres has magnitude $F = \pi R^2 \, \Delta p$, where Δp is the difference between the pressures outside and inside the sphere. (b) Taking R as 30 cm, the inside pressure as 0.10 atm, and the outside pressure as 1.00 atm, find the force magnitude the teams of horses would have had to exert to pull apart the hemispheres. (c) Explain why one team of horses could have proved the point just as well if the hemispheres were attached to a sturdy wall. ssm
www

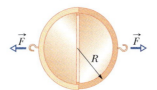

Fig. 15-27 Problem 7.

SEC. 15-4   Fluids at Rest

8E. Calculate the hydrostatic difference in blood pressure between the brain and the foot in a person of height 1.83 m. The density of blood is 1.06×10^3 kg/m³.

9E. The sewage outlet of a house constructed on a slope is 8.2 m below street level. If the sewer is 2.1 m below street level, find the minimum pressure difference that must be created by the sewage pump to transfer waste of average density 900 kg/m³ from outlet to sewer. ssm

10E. Figure 15-28 displays the *phase diagram* of carbon, showing the ranges of temperature and pressure in which carbon will crystallize either as diamond or graphite. What is the minimum depth at which diamonds can form if the temperature at that

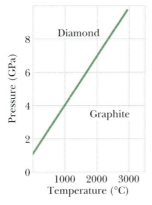

Fig. 15-28 Exercise 10.

depth is 1000°C and the rocks there have density 3.1 g/cm³? Assume that, as in a fluid, the pressure at any level is due to the gravitational force on the material lying above that level.

11E. A swimming pool has the dimensions 24 m × 9.0 m × 2.5 m. When it is filled with water, what is the force (resulting from the water alone) on (a) the bottom, (b) each short side, and (c) each long side? (d) If you are concerned with the possibility that the concrete walls and floor will collapse, is it appropriate to take the atmospheric pressure into account? Why? ilw

12E. (a) Assuming the density of seawater is 1.03 g/cm³, find the total weight of water on top of a nuclear submarine at a depth of 200 m if its (horizontal cross-sectional) hull area is 3000 m². (b) In atmospheres, what water pressure would a diver experience at this depth? Do you think that occupants of a damaged submarine at this depth could escape without special equipment?

13E. Crew members attempt to escape from a damaged submarine 100 m below the surface. What force must be applied to a pop-out hatch, which is 1.2 m by 0.60 m, to push it out at that depth? Assume that the density of the ocean water is 1025 kg/m³. ssm

14E. A cylindrical barrel has a narrow tube fixed to the top, as shown (with dimensions) in Fig. 15-29. The vessel is filled with water to the top of the tube. Calculate the ratio of the hydrostatic force on the bottom of the barrel to the gravitational force on the water contained inside the barrel. Why is that ratio not equal to one? (You need not consider the atmospheric pressure.)

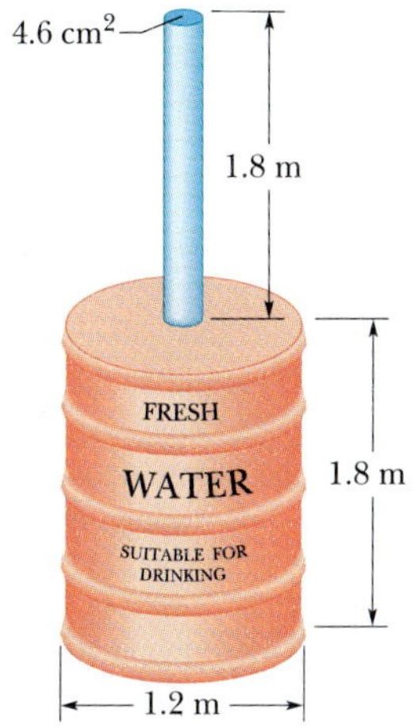

15P. Two identical cylindrical vessels with their bases at the same level each contain a liquid of density ρ. The area of each base is A, but in one vessel the liquid height is h_1, and in the other it is h_2. Find the work done by the gravitational force in equalizing the levels when the two vessels are connected. ssm

16P. In analyzing certain geological features, it is often appropriate to assume that the pressure at some horizontal *level of compensation*, deep inside Earth, is the same over a large region and is equal to the pressure due to the gravitational force on the overlying material. Thus, the pressure on the level of compensation is given by the fluid pressure formula. This model requires, for one thing, that mountains have *roots* of continental rock extending into the denser mantle (Fig. 15-30). Consider a mountain

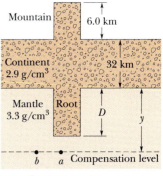

Fig. 15-30 Problem 16.

Fig. 15-29 Exercise 14.

6.0 km high. The continental rocks have a density of 2.9 g/cm³, and beneath the continent the mantle has a density of 3.3 g/cm³. Calculate the depth D of the root. (*Hint:* Set the pressure at points a and b equal; the depth y of the level of compensation will cancel out.)

17P. Figure 15-31 shows the juncture of ocean and continent. Find the depth h of the ocean using the level-of-compensation technique presented in Problem 16. ssm www

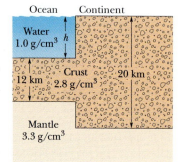

Fig. 15-31 Problem 17.

18P. The **L**-shaped tank shown in Fig. 15-32 is filled with water and is open at the top. If $d = 5.0$ m, what are (a) the force on face A and (b) the force on face B due to the water?

19P. Water stands at a depth D behind the vertical upstream face of a dam, as shown in Fig. 15-33. Let W be the width of the dam. Find (a) the net horizontal force on the dam from the gauge pressure of the water and (b) the net torque due to that force (and thus gauge pressure) about a line through O parallel to the width of the dam. (c) Find the moment arm of the net horizontal force about the line through O. ssm

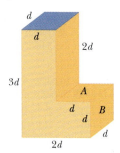

Fig. 15-32 Problem 18.

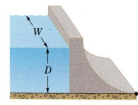

Fig. 15-33 Problem 19.

SEC. 15-5   Measuring Pressure

20E. To suck lemonade of density 1000 kg/m³ up a straw to a maximum height of 4.0 cm, what minimum gauge pressure (in atmospheres) must you produce in your lungs?

21E. What would be the height of the atmosphere if the air density (a) were uniform and (b) decreased linearly to zero with height? Assume that at sea level the air pressure is 1.0 atm and the air density is 1.3 kg/m³. ssm

SEC. 15-6   Pascal's Principle

22E. A piston of small cross-sectional area a is used in a hydraulic press to exert a small force $\vec{f}$ on the enclosed liquid. A connecting pipe leads to a larger piston of cross-sectional area A (Fig. 15-34). (a) What force magnitude F will the larger piston sustain without moving? (b) If the small piston has a diameter of 3.80 cm and the large piston one of 53.0 cm, what force magnitude on the small piston will balance a 20.0 kN force on the large piston? ssm

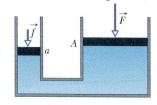

Fig. 15-34 Exercises 22 and 23.

23E. In the hydraulic press of Exercise 22, through what distance must the large piston be moved to raise the small piston a distance of 0.85 m? ssm

SEC. 15-7   Archimedes' Principle

24E. A boat floating in fresh water displaces water weighing 35.6 kN. (a) What is the weight of the water that this boat would displace if it were floating in salt water with a density of 1.10×10^3 kg/m³? (b) Would the volume of the displaced water change? If so, by how much?

25E. An iron anchor of density 7870 kg/m³ appears 200 N lighter in water than in air. (a) What is the volume of the anchor? (b) How much does it weigh in air? ssm

26E. In Fig. 15-35, a cubical object of dimensions $L = 0.600$ m on a side and with a mass of 450 kg is suspended by a rope in an open tank of liquid of density 1030 kg/m³. (a) Find the magnitude of the total downward force on the top of the object from the liquid and the atmosphere, assuming that atmospheric pressure is 1.00 atm. (b) Find the magnitude of the total upward force on the bottom of the object. (c) Find the tension in the rope. (d) Calculate the magnitude of the buoyant force on the object using Archimedes' principle. What relation exists among all these quantities?

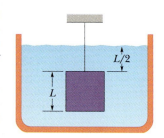

Fig. 15-35 Exercise 26.

27E. A block of wood floats in fresh water with two-thirds of its volume submerged. In oil the block floats with 0.90 of its volume submerged. Find the density of (a) the wood and (b) the oil. ssm

28E. A blimp is cruising slowly at low altitude, filled as usual with helium gas. Its maximum useful payload, including crew and cargo, is 1280 kg. The volume of the helium-filled interior space is 5000 m³. The density of helium gas is 0.16 kg/m³, and the density of hydrogen is 0.081 kg/m³. How much more payload could the blimp carry if you replaced the helium with hydrogen? (Why not do it?)

29P. A hollow sphere of inner radius 8.0 cm and outer radius 9.0 cm floats half-submerged in a liquid of density 800 kg/m³. (a) What is the mass of the sphere? (b) Calculate the density of the material of which the sphere is made. ssm www

30P. About one-third of the body of a person floating in the Dead Sea will be above the water line. Assuming that the human body density is 0.98 g/cm³, find the density of the water in the Dead Sea. (Why is it so much greater than 1.0 g/cm³?)

31P. A hollow spherical iron shell floats almost completely submerged in water. The outer diameter is 60.0 cm, and the density of iron is 7.87 g/cm³. Find the inner diameter. ilw

32P. A block of wood has a mass of 3.67 kg and a density of 600 kg/m³. It is to be loaded with lead so that it will float in water with 0.90 of its volume submerged. What mass of lead is needed (a) if the lead is attached to the top of the wood and (b) if the lead is attached to the bottom of the wood? The density of lead is 1.13×10^4 kg/m³.

33P. An iron casting containing a number of cavities weighs 6000 N in air and 4000 N in water. What is the total volume of all the cavities in the casting? The density of iron (that is, a sample with no cavities) is 7.87 g/cm^3. ssm

34P. Assume the density of brass weights to be 8.0 g/cm^3 and that of air to be 0.0012 g/cm^3. What percent error arises from neglecting the buoyancy of air in weighing an object of mass m and density ρ on a beam balance, as in Fig. 5-6?

35P. (a) What is the minimum area of the top surface of a slab of ice 0.30 m thick floating on fresh water that will hold up an automobile of mass 1100 kg? (b) Does it matter where the car is placed on the block of ice? ssm

36P. Three children, each of weight 356 N, make a log raft by lashing together logs of diameter 0.30 m and length 1.80 m. How many logs will be needed to keep them afloat in fresh water? Take the density of the logs to be 800 kg/m^3.

37P. A metal rod of length 80 cm and mass 1.6 kg has a uniform cross-sectional area of 6.0 cm^2. Due to a nonuniform density, the center of mass of the rod is 20 cm from one end of the rod. The rod is suspended in a horizontal position in water by ropes attached to both ends (Fig. 15-36). (a) What is the tension in the rope closer to the center of mass? (b) What is the tension in the rope farther from center of mass? (*Hint:* The buoyancy force on the rod effectively acts at the rod's center.) ssm

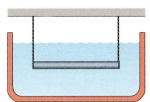

Fig. 15-36 Problem 37.

38P. A car has a total mass of 1800 kg. The volume of air space in the passenger compartment is 5.00 m^3. The volume of the motor and front wheels is 0.750 m^3, and the volume of the rear wheels, gas tank, and trunk is 0.800 m^3; water cannot enter these areas. The car is parked on a hill; the handbrake cable snaps and the car rolls down the hill into a lake (Fig. 15-37). (a) At first, no water enters the passenger compartment. How much of the car, in cubic meters, is below the water surface with the car floating as shown? (b) As water slowly enters, the car sinks. How many cubic meters of water are in the car as it disappears below the water surface? (The car, with a heavy load in the trunk, remains horizontal.)

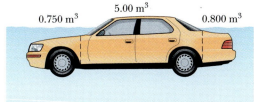

Fig. 15-37
Problem 38.

SEC. 15-9 The Equation of Continuity

39E. A garden hose with an internal diameter of 1.9 cm is connected to a (stationary) lawn sprinkler that consists merely of an enclosure with 24 holes, each 0.13 cm in diameter. If the water in the hose has a speed of 0.91 m/s, at what speed does it leave the sprinkler holes? ssm

40E. Figure 15-38 shows the merging of two streams to form a river. One stream has a width of 8.2 m, depth of 3.4 m, and current speed of 2.3 m/s. The other stream is 6.8 m wide and 3.2 m deep, and flows at 2.6 m/s. The width of the river is 10.5 m, and the current speed is 2.9 m/s. What is its depth?

Fig. 15-38 Exercise 40.

41P. Water is pumped steadily out of a flooded basement at a speed of 5.0 m/s through a uniform hose of radius 1.0 cm. The hose passes out through a window 3.0 m above the waterline. What is the power of the pump? ssm

42E. The water flowing through a 1.9 cm (inside diameter) pipe flows out through three 1.3 cm pipes. (a) If the flow rates in the three smaller pipes are 26, 19, and 11 L/min, what is the flow rate in the 1.9 cm pipe? (b) What is the ratio of the speed of water in the 1.9 cm pipe to that in the pipe carrying 26 L/min?

SEC. 15-10 Bernoulli's Equation

43E. Water is moving with a speed of 5.0 m/s through a pipe with a cross-sectional area of 4.0 cm^2. The water gradually descends 10 m as the pipe increases in area to 8.0 cm^2. (a) What is the speed at the lower level? (b) If the pressure at the upper level is 1.5×10^5 Pa, what is the pressure at the lower level? ssm

44E. Models of torpedoes are sometimes tested in a horizontal pipe of flowing water, much as a wind tunnel is used to test model airplanes. Consider a circular pipe of internal diameter 25.0 cm and a torpedo model, aligned along the axis of the pipe, with a diameter of 5.00 cm. The model is to be tested with water flowing past it at 2.50 m/s. (a) With what speed must the water flow in the part of the pipe that is unconstricted by the model? (b) What will the pressure difference be between the constricted and unconstricted parts of the pipe?

45E. A water pipe having a 2.5 cm inside diameter carries water into the basement of a house at a speed of 0.90 m/s and a pressure of 170 kPa. If the pipe tapers to 1.2 cm and rises to the second floor 7.6 m above the input point, what are (a) the speed and (b) the water pressure at the second floor? ilw

46E. A water intake at a pump storage reservoir (Fig. 15-39) has a cross-sectional area of 0.74 m². The water flows in at a speed of 0.40 m/s. At the generator building 180 m below the intake point, the cross-sectional area is smaller than at the intake and the water flows out at 9.5 m/s. What is the difference in pressure, in megapascals, between inlet and outlet?

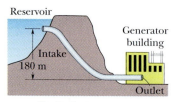

Fig. 15-39 Exercise 46.

47E. A tank of large area is filled with water to a depth $D = 0.30$ m. A hole of cross-sectional area $A = 6.5$ cm² in the bottom of the tank allows water to drain out. (a) What is the rate at which water flows out, in cubic meters per second? (b) At what distance below the bottom of the tank is the cross-sectional area of the stream equal to one-half the area of the hole? **ssm**

48E. Air flows over the top of an airplane wing of area A with speed v_t and past the underside of the wing (also of area A) with speed v_u. Show that in this simplified situation Bernoulli's equation predicts that the magnitude L of the upward lift force on the wing will be

$$L = \tfrac{1}{2}\rho A(v_t^2 - v_u^2),$$

where ρ is the density of the air.

49E. If the speed of flow past the lower surface of an airplane wing is 110 m/s, what speed of flow over the upper surface will give a pressure difference of 900 Pa between upper and lower surfaces? Take the density of air to be 1.30×10^{-3} g/cm³, and see Exercise 48. **ssm**

50E. Suppose that two tanks, 1 and 2, each with a large opening at the top, contain different liquids. A small hole is made in the side of each tank at the same depth h below the liquid surface, but the hole in tank 1 has half the cross-sectional area of the hole in tank 2. (a) What is the ratio ρ_1/ρ_2 of the densities of the liquids if the mass flow rate is the same for the two holes? (b) What is the ratio of the volume flow rates from the two tanks? (c) To what height above the hole in the second tank should liquid be added or drained to equalize the volume flow rates?

51P. In Fig. 15-40, water flows through a horizontal pipe, and then out into the atmosphere at a speed of 15 m/s. The diameters of the left and right sections of the pipe are 5.0 cm and 3.0 cm, respectively. (a) What volume of water flows into the atmosphere during a 10 min period? In the left section of the pipe, what are (b) the speed v_2, and (c) the gauge pressure?

Fig. 15-40 Problem 51.

52P. An opening of area 0.25 cm² in an otherwise closed beverage keg is 50 cm below the level of the liquid (of density 1.0 g/cm³) in the keg. What is the speed of the liquid flowing through the opening if the gauge pressure in the air space above the liquid is (a) zero and (b) 0.40 atm?

53P. The fresh water behind a reservoir dam is 15 m deep. A horizontal pipe 4.0 cm in diameter passes through the dam 6.0 m below the water surface, as shown in Fig. 15-41. A plug secures the pipe opening. (a) Find the magnitude of the frictional force between plug and pipe wall. (b) The plug is removed. What volume of water flows out of the pipe in 3.0 h? **ilw**

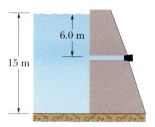

Fig. 15-41 Problem 53.

54P. A tank is filled with water to a height H. A hole is punched in one of the walls at a depth h below the water surface (Fig. 15-42). (a) Show that the distance x from the base of the tank to the point at which the resulting stream strikes the floor is given by $x = 2\sqrt{h(H - h)}$. (b) Could a

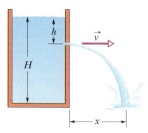

Fig. 15-42 Problem 54.

hole be punched at another depth to produce a second stream that would have the same range? If so, at what depth? (c) At what depth should the hole be placed to make the emerging stream strike the ground at the maximum distance from the base of the tank?

55P. A *venturi meter* is used to measure the flow speed of a fluid in a pipe. The meter is connected between two sections of the pipe (Fig. 15-43); the cross-sectional area A of the entrance and exit of the meter matches the pipe's cross-sectional area. Between the entrance and exit, the fluid flows from the pipe with speed V and then through a narrow "throat" of cross-sectional area a with speed v. A manometer connects the wider portion of the meter to the narrower portion. The change in the fluid's speed is accompanied by a change Δp in the fluid's pressure, which causes a height difference h of the liquid in the two arms of the manometer. (Here Δp means pressure in the throat minus pressure in the pipe.) (a) By applying Bernoulli's equation and the equation of continuity to points 1 and 2 in Fig. 15-43, show that

$$V = \sqrt{\frac{2a^2\,\Delta p}{\rho(a^2 - A^2)}},$$

where ρ is the density of the fluid. (b) Suppose that the fluid is fresh water, that the cross-sectional areas are 64 cm² in the pipe

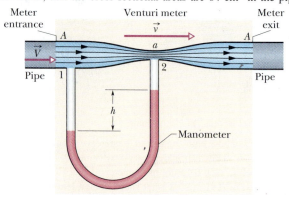

Fig. 15-43 Problems 55 and 56.

and 32 cm^2 in the throat, and that the pressure is 55 kPa in the pipe and 41 kPa in the throat. What is the rate of water flow in cubic meters per second? ssm www

56P. Consider the venturi tube of Problem 55 and Fig. 15-43 without the manometer. Let A equal $5a$. Suppose that the pressure p_1 at A is 2.0 atm. Compute the values of (a) V at A and (b) v at a that would make the pressure p_2 at a equal to zero. (c) Compute the corresponding volume flow rate if the diameter at A is 5.0 cm. The phenomenon that occurs at a when p_2 falls to nearly zero is known as cavitation. The water vaporizes into small bubbles.

57P. A pitot tube (Fig. 15-44) is used to determine the airspeed of an airplane. It consists of an outer tube with a number of small holes B (four are shown) that allow air into the tube; that tube is connected to one arm of a U-tube. The other arm of the U-tube is connected to hole A at the front end of the device, which points in the direction the plane is headed. At A the air becomes stagnant so that $v_A = 0$. At B, however, the speed of the air presumably equals the airspeed v of the aircraft. (a) Use Bernoulli's equation to show that

$$v = \sqrt{\frac{2\rho g h}{\rho_{air}}},$$

where ρ is the density of the liquid in the U-tube and h is the difference in the fluid levels in that tube. (b) Suppose that the tube contains alcohol and indicates a level difference h of 26.0 cm. What is the plane's speed relative to the air? The density of the air is 1.03 kg/m^3 and that of alcohol is 810 kg/m^3.

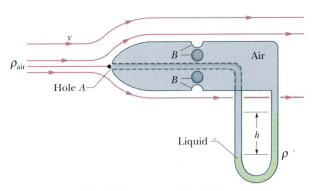

Fig. 15-44 Problems 57 and 58.

58P. A pitot tube (see Problem 57) on a high-altitude aircraft measures a differential pressure of 180 Pa. What is the airspeed if the density of the air is 0.031 kg/m^3?

Additional Problems

59. The dinosaur *Diplodocus* was enormous, with long neck and tail and a mass that was great enough to test its leg strength. According to conjecture, *Diplodocus* waded in water, perhaps up to its head, so that buoyancy could offset its weight and lighten the load on its legs. To check the conjecture, take the density of *Diplodocus* to be 0.90 that of water, and assume that its mass was the published estimate of 1.85×10^4 kg. (a) What then would be its actual weight? Find its apparent weight when it had the follow-

ing fractions of its volume submerged: (b) 0.50, (c) 0.80, and (d) 0.90. When almost fully submerged, with only its head above water, its lungs would have been about 8.0 m below the water surface. (e) At that depth, what would be the difference between the (external) water pressure and the pressure of the air in its lungs? For the dinosaur to breathe in, its lung muscles would have had to expand its lungs against this pressure difference. It probably could not do so against a pressure difference of more than 8 kPa. (f) Did *Diplodocus* wade as conjectured?

60. When you cough, you expel air at high speed through the trachea and upper bronchi so that the air will remove excess mucus lining the pathway. You produce the high speed by this procedure: You breathe in a large amount of air, trap it by closing the glottis (the narrow opening in the larynx), increase the air pressure by contracting the lungs, partially collapse the trachea and upper bronchi to narrow the pathway, and then expel the air through the pathway by suddenly reopening the glottis. Assume that during the expulsion the volume flow rate is 7.0×10^{-3} m^3/s. In terms of the speed of sound $v_s = 343$ m/s, what is the air speed through the trachea if the trachea diameter (a) remains its normal diameter of 14 mm and (b) contracts to a diameter of 5.2 mm?

61. Suppose that your body has a uniform density of 0.95 times that of water. (a) If you float in a swimming pool, what fraction of your body's volume is above the water surface?

Quicksand is a fluid that is produced when water is forced up into sand, moving the sand grains away from one another so they are no longer locked together by friction. Pools of quicksand can occur when water drains underground from hills into valleys with sand pockets. (b) If you step into a deep pool of quicksand with a density 1.6 times that of water, what fraction of your body's volume is then above the quicksand surface? (c) In particular, are you submerged enough to be unable to breathe? The viscosity of quicksand dramatically increases during any quick movements (the fluid is said to be *thixotropic*). Thus, if you struggle to escape from quicksand, the quicksand holds you tighter. (d) How might you escape without someone helping?

62. In a sink with a flat bottom, turn on a sink faucet so that a smoothly flowing (laminar) stream strikes the bottom. The water spreads from the impact point in a shallow layer but then, at a certain radius r_J from the impact point, it suddenly increases in depth. This depth change, called a *hydraulic jump*, forms a prominent circle around the impact point. Inside the circle, the speed v_1 of the spreading water is constant and is equal to its speed in the falling stream just before impact.

In a certain experiment, the radius of the falling stream is 1.3 mm just before impact, the volume flow rate R_V is 7.9 cm^3/s, the jump radius r_J is 2.0 cm, and the depth just after the jump is 2.0 mm. (a) What is speed v_1? (b) For $r < r_J$, express the water depth d as a function of the radial distance r from the impact point. (c) Does the depth of the water increase or decrease with r? (d) What is the depth just before the water reaches the hydraulic jump? (e) What is the speed v_2 of the water just after the jump? What are the kinetic energy densities (f) just before and (g) just after the jump? (h) What is the change in pressure on the sink due to the jump? (i) Does Bernoulli's equation apply to a streamline through the jump?

16 Oscillations

On September 19, 1985, seismic waves from an earthquake that originated along the west coast of Mexico caused terrible and widespread damage in Mexico City, about 400 km from the origin.

Why did the seismic waves cause such extensive damage in Mexico City but relatively little on the way there?

The answer is in this chapter.

16-1 Oscillations

We are surrounded by oscillations—motions that repeat themselves. There are swinging chandeliers, boats bobbing at anchor, and the surging pistons in the engines of cars. There are oscillating guitar strings, drums, bells, diaphragms in telephones and speaker systems, and quartz crystals in wristwatches. Less evident are the oscillations of the air molecules that transmit the sensation of sound, the oscillations of the atoms in a solid that convey the sensation of temperature, and the oscillations of the electrons in the antennas of radio and TV transmitters that convey information.

Oscillations in the real world are usually *damped;* that is, the motion dies out gradually, transferring mechanical energy to thermal energy by the action of frictional forces. Although we cannot totally eliminate such loss of mechanical energy, we can replenish the energy from some source. As an example, you know that by swinging your legs or torso you can "pump" a swing to maintain or increase the oscillations. In doing this, you transfer biochemical energy to mechanical energy of the oscillating system.

16-2 Simple Harmonic Motion

Figure 16-1a shows a sequence of "snapshots" of a simple oscillating system, a particle moving repeatedly back and forth about the origin of an x axis. In this section we simply describe the motion. Later, we shall discuss how to attain such motion.

One important property of oscillatory motion is its **frequency,** or number of oscillations that are completed each second. The symbol for frequency is f, and its SI unit is the **hertz** (abbreviated Hz), where

$$1 \text{ hertz} = 1 \text{ Hz} = 1 \text{ oscillation per second} = 1 \text{ s}^{-1}. \qquad (16\text{-}1)$$

Related to the frequency is the **period** T of the motion, which is the time for one complete oscillation (or **cycle**); that is,

$$T = \frac{1}{f}. \qquad (16\text{-}2)$$

Any motion that repeats itself at regular intervals is called **periodic motion** or **harmonic motion.** We are interested here in motion that repeats itself in a particular way—namely, like that in Fig. 16-1a. For such motion the displacement x of the

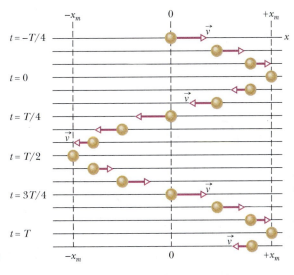

Fig. 16-1 (a) A sequence of "snapshots" (taken at equal time intervals) showing the position of a particle as it oscillates back and forth about the origin along an x axis, between the limits $+x_m$ and $-x_m$. The vector arrows are scaled to indicate the speed of the particle. The speed is maximum when the particle is at the origin and zero when it is at $\pm x_m$. If the time t is chosen to be zero when the particle is at $+x_m$, then the particle returns to $+x_m$ at $t = T$, where T is the period of the motion. The motion is then repeated. (b) A graph of x as a function of time for the motion of (a).

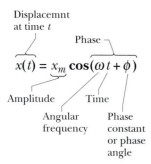

Fig. 16-2 A handy reference to the quantities in Eq. 16-3 for simple harmonic motion.

particle from the origin is given as a function of time by

$$x(t) = x_m \cos(\omega t + \phi) \qquad \text{(displacement)}, \qquad (16\text{-}3)$$

in which x_m, ω, and ϕ are constants. This motion is called **simple harmonic motion** (SHM), a term that means the periodic motion is a sinusoidal function of time. Equation 16-3, in which the sinusoidal function is a cosine function, is graphed in Fig. 16-1b. (You can get that graph by rotating Fig. 16-1a counterclockwise by 90° and then connecting the successive locations of the particle with a curve.) The quantities that determine the shape of the graph are displayed in Fig. 16-2 with their names. We now shall define those quantities.

The quantity x_m, called the **amplitude** of the motion, is a positive constant whose value depends on how the motion was started. The subscript m stands for *maximum* because the amplitude is the magnitude of the maximum displacement of the particle in either direction. The cosine function in Eq. 16-3 varies between the limits ± 1, so the displacement $x(t)$ varies between the limits $\pm x_m$.

The time-varying quantity $(\omega t + \phi)$ in Eq. 16-3 is called the **phase** of the motion, and the constant ϕ is called the **phase constant** (or **phase angle**). The value of ϕ depends on the displacement and velocity of the particle at time $t = 0$. For the $x(t)$ plots of Fig. 16-3a, the phase constant ϕ is zero.

To interpret the constant ω, called the **angular frequency** of the motion, we first note that the displacement $x(t)$ must return to its initial value after one period T of the motion; that is, $x(t)$ must equal $x(t + T)$ for all t. To simplify this analysis, let us put $\phi = 0$ in Eq. 16-3. From that equation we then can write

$$x_m \cos \omega t = x_m \cos \omega(t + T). \qquad (16\text{-}4)$$

The cosine function first repeats itself when its argument (the phase) has increased by 2π rad, so Eq. 16-4 gives us

$$\omega(t + T) = \omega t + 2\pi$$

or

$$\omega T = 2\pi.$$

Thus, from Eq. 16-2 the angular frequency is

$$\omega = \frac{2\pi}{T} = 2\pi f. \qquad (16\text{-}5)$$

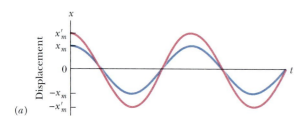

(a)

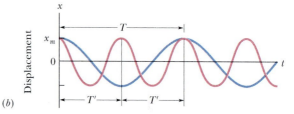

(b)

Fig. 16-3 In each case, the blue curve is obtained from Eq. 16-3 with $\phi = 0$. (a) The red curve differs from the blue curve *only* in that its amplitude x'_m is greater (the red-curve extremes of displacement are higher and lower). (b) The red curve differs from the blue curve *only* in that its period is $T' = T/2$ (the red curve is compressed horizontally). (c) The red curve differs from the blue curve *only* in that $\phi = -\pi/4$ rad rather than zero (the negative value of ϕ shifts the red curve to the right).

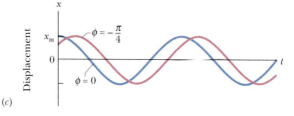

(c)

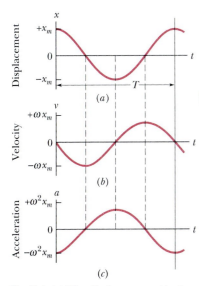

Fig. 16-4 (*a*) The displacement *x*(*t*) of a particle oscillating in SHM with phase angle ϕ equal to zero. The period *T* marks one complete oscillation. (*b*) The velocity *v*(*t*) of the particle. (*c*) The acceleration *a*(*t*) of the particle.

The SI unit of angular frequency is the radian per second. (To be consistent, then, ϕ must be in radians.) Figure 16-3 compares *x*(*t*) for two simple harmonic motions that differ either in amplitude, in period (and thus in frequency and angular frequency), or in phase constant.

✔**CHECKPOINT 1:** A particle undergoing simple harmonic oscillation of period *T* (like that in Fig. 16-1) is at $-x_m$ at time *t* = 0. Is it at $-x_m$, at $+x_m$, at 0, between $-x_m$ and 0, or between 0 and $+x_m$ when (a) *t* = 2.00*T*, (b) *t* = 3.50*T*, and (c) *t* = 5.25*T*?

The Velocity of SHM

By differentiating Eq. 16-3, we can find an expression for the velocity of a particle moving with simple harmonic motion; that is,

$$v(t) = \frac{dx(t)}{dt} = \frac{d}{dt}[x_m \cos(\omega t + \phi)]$$

or $$v(t) = -\omega x_m \sin(\omega t + \phi) \quad \text{(velocity)}. \quad (16\text{-}6)$$

Figure 16-4*a* is a plot of Eq. 16-3 with $\phi = 0$. Figure 16-4*b* shows Eq. 16-6, also with $\phi = 0$. Analogous to the amplitude x_m in Eq. 16-3, the positive quantity ωx_m in Eq. 16-6 is called the **velocity amplitude** v_m. As you can see in Fig. 16-4*b*, the velocity of the oscillating particle varies between the limits $\pm v_m = \pm \omega x_m$. Note also in that figure that the curve of *v*(*t*) is *shifted* (to the left) from the curve of *x*(*t*) by one-quarter period; when the magnitude of the displacement is greatest (that is, $x(t) = x_m$), the magnitude of the velocity is least (that is, *v*(*t*) = 0). When the magnitude of the displacement is least (that is, zero), the magnitude of the velocity is greatest (that is, $v_m = \omega x_m$).

The Acceleration of SHM

Knowing the velocity *v*(*t*) for simple harmonic motion, we can find an expression for the acceleration of the oscillating particle by differentiating once more. Thus, we have, from Eq. 16-6,

$$a(t) = \frac{dv(t)}{dt} = \frac{d}{dt}[-\omega x_m \sin(\omega t + \phi)]$$

or $$a(t) = -\omega^2 x_m \cos(\omega t + \phi) \quad \text{(acceleration)}. \quad (16\text{-}7)$$

Figure 16-4*c* is a plot of Eq. 16-7 for the case $\phi = 0$. The positive quantity $\omega^2 x_m$ in Eq. 16-7 is called the **acceleration amplitude** a_m; that is, the acceleration of the particle varies between the limits $\pm a_m = \pm \omega^2 x_m$, as Fig. 16-4*c* shows. Note also that the curve of *a*(*t*) is shifted (to the left) by $\frac{1}{4}T$ relative to the curve of *v*(*t*).

We can combine Eqs. 16-3 and 16-7 to yield

$$a(t) = -\omega^2 x(t), \quad (16\text{-}8)$$

which is the hallmark of simple harmonic motion:

➤ In SHM, the acceleration is proportional to the displacement but opposite in sign, and the two quantities are related by the square of the angular frequency.

Thus, as Fig. 16-4 shows, when the displacement has its greatest positive value, the acceleration has its greatest negative value, and conversely. When the displacement is zero, the acceleration is also zero.

Tactic 1: *Phase Angles*

Note the effect of the phase angle ϕ on a plot of $x(t)$. When $\phi = 0$, $x(t)$ has a graph like that in Fig. 16-4a, a typical cosine curve. Increasing ϕ shifts the curve leftward along the t axis. (You might remember this with the symbol ⤹ϕ, where the up arrow indicates an increase in ϕ and the left arrow indicates the resulting shift in the curve.) Decreasing ϕ shifts the curve rightward, as in Fig. 16-3c for $\phi = -\pi/4$.

Two plots of SHM with different phase angles are said to have a *phase difference*; each is said to be *phase-shifted* from the other,

or *out of phase* with the other. The curves in Fig. 16-3c, for example, have a phase difference of $\pi/4$ rad.

Because SHM repeats after each period T and the cosine function repeats after each 2π rad, one period T represents a phase difference of 2π rad. In Fig. 16-4, $x(t)$ is phase-shifted to the right from $v(t)$ by one-quarter period, or $-\pi/2$ rad; it is shifted to the right from $a(t)$ by one-half period, or $-\pi$ rad. A phase shift of 2π rad causes a curve of SHM to coincide with itself; that is, it looks unchanged.

16-3   The Force Law for Simple Harmonic Motion

Once we know how the acceleration of a particle varies with time, we can use Newton's second law to learn what force must act on the particle to give it that acceleration. If we combine Newton's second law and Eq. 16-8, we find, for simple harmonic motion,

$$F = ma = -(m\omega^2)x. \qquad (16\text{-}9)$$

This result—a restoring force that is proportional to the displacement but opposite in sign—is familiar. It is Hooke's law,

$$F = -kx, \qquad (16\text{-}10)$$

for a spring, the spring constant here being

$$k = m\omega^2. \qquad (16\text{-}11)$$

We can in fact take Eq. 16-10 as an alternative definition of simple harmonic motion. It says:

> Simple harmonic motion is the motion executed by a particle of mass m subject to a force that is proportional to the displacement of the particle but opposite in sign.

The block–spring system of Fig. 16-5 forms a **linear simple harmonic oscillator** (linear oscillator, for short), where "linear" indicates that F is proportional to x rather than to some other power of x. The angular frequency ω of the simple harmonic motion of the block is related to the spring constant k and the mass m of the block by Eq. 16-11, which yields

$$\omega = \sqrt{\frac{k}{m}} \qquad \text{(angular frequency).} \qquad (16\text{-}12)$$

By combining Eqs. 16-5 and 16-12, we can write, for the **period** of the linear oscillator of Fig. 16-5,

$$T = 2\pi\sqrt{\frac{m}{k}} \qquad \text{(period).} \qquad (16\text{-}13)$$

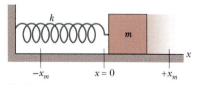

Fig. 16-5 A linear simple harmonic oscillator. The surface is frictionless. Like the particle of Fig. 16-2, the block moves in simple harmonic motion once it has been pulled to the side and released. Its displacement is then given by Eq. 16-3.

Equations 16-12 and 16-13 tell us that a large angular frequency (and thus a small period) goes with a stiff spring (large k) and a light block (small m).

Every oscillating system, be it the linear oscillator of Fig. 16-5, a diving board, or a violin string, has some element of "springiness" and some element of "inertia" or mass, and thus resembles a linear oscillator. In the linear oscillator of Fig. 16-5, these elements are located in separate parts of the system: The springiness is entirely in the spring, which we assume to be massless, and the inertia is entirely in the block, which we assume to be rigid. In a violin string, however, the two elements are both within the string itself, as you will see in Chapter 17.

✔CHECKPOINT 2: Which of the following relationships between the force F on a particle and the particle's position x implies simple harmonic oscillation: (a) $F = -5x$, (b) $F = -400x^2$, (c) $F = 10x$, (d) $F = 3x^2$?

Sample Problem 16-1

A block whose mass m is 680 g is fastened to a spring whose spring constant k is 65 N/m. The block is pulled a distance $x = 11$ cm from its equilibrium position at $x = 0$ on a frictionless surface and released from rest at $t = 0$.

(a) What are the angular frequency, the frequency, and the period of the resulting motion?

SOLUTION: The Key Idea here is that the block–spring system forms a linear simple harmonic oscillator, with the block undergoing SHM. Then the angular frequency is given by Eq. 16-12:

$$\omega = \sqrt{\frac{k}{m}} = \sqrt{\frac{65 \text{ N/m}}{0.68 \text{ kg}}} = 9.78 \text{ rad/s}$$

$$\approx 9.8 \text{ rad/s}. \qquad \text{(Answer)}$$

The frequency follows from Eq. 16-5, which yields

$$f = \frac{\omega}{2\pi} = \frac{9.78 \text{ rad/s}}{2\pi \text{ rad}} = 1.56 \text{ Hz} \approx 1.6 \text{ Hz}. \quad \text{(Answer)}$$

The period follows from Eq. 16-2, which yields

$$T = \frac{1}{f} = \frac{1}{1.56 \text{ Hz}} = 0.64 \text{ s} = 640 \text{ ms}. \qquad \text{(Answer)}$$

(b) What is the amplitude of the oscillation?

SOLUTION: The Key Idea here is that, with no friction involved, the mechanical energy of the spring–block system is conserved. The block is released from rest 11 cm from its equilibrium position, with zero kinetic energy and the elastic potential energy of the system at a maximum. Thus, the block will have zero kinetic energy whenever it is again 11 cm from its equilibrium position, which means it will never be farther than 11 cm from that position. Its maximum displacement is 11 cm:

$$x_m = 11 \text{ cm}. \qquad \text{(Answer)}$$

(c) What is the maximum speed v_m of the oscillating block, and where is the block when it occurs?

SOLUTION: The Key Idea here is that the maximum speed v_m is the velocity amplitude ωx_m in Eq. 16-6; that is,

$$v_m = \omega x_m = (9.78 \text{ rad/s})(0.11 \text{ m})$$

$$= 1.1 \text{ m/s}. \qquad \text{(Answer)}$$

This maximum speed occurs when the oscillating block is rushing through the origin; compare Figs. 16-4a and 16-4b, where you can see that the speed is a maximum whenever $x = 0$.

(d) What is the magnitude a_m of the maximum acceleration of the block?

SOLUTION: The Key Idea this time is that the magnitude a_m of the maximum acceleration is the acceleration amplitude $\omega^2 x_m$ in Eq. 16-7; that is,

$$a_m = \omega^2 x_m = (9.78 \text{ rad/s})^2 (0.11 \text{ m})$$

$$= 11 \text{ m/s}^2. \qquad \text{(Answer)}$$

This maximum acceleration occurs when the block is at the ends of its path. At those points, the force acting on the block has its maximum magnitude; compare Figs. 16-4a and 16-4c, where you can see that the magnitudes of the displacement and acceleration are maximum at the same times.

(e) What is the phase constant ϕ for the motion?

SOLUTION: Here the Key Idea is that Eq. 16-3 gives the displacement of the block as a function of time. We know that at time $t = 0$, the block is located at $x = x_m$. Substituting these *initial conditions,* as they are called, into Eq. 16-3 and canceling x_m give us

$$1 = \cos \phi. \qquad (16\text{-}14)$$

Taking the inverse cosine then yields

$$\phi = 0 \text{ rad}. \qquad \text{(Answer)}$$

(Any angle that is an integer multiple of 2π rad also satisfies Eq. 16-14; we chose the smallest angle.)

(f) What is the displacement function $x(t)$ for the spring–block system?

SOLUTION: The Key Idea here is that $x(t)$ is given in general form by Eq. 16-3. Substituting known quantities into that equation gives us

$$x(t) = x_m \cos(\omega t + \phi)$$

$$= (0.11 \text{ m}) \cos[(9.8 \text{ rad/s})t + 0]$$

$$= 0.11 \cos(9.8t), \qquad \text{(Answer)}$$

where x is in meters and t is in seconds.

Sample Problem 16-2

At $t = 0$, the displacement $x(0)$ of the block in a linear oscillator like that of Fig. 16-5 is -8.50 cm. (Read $x(0)$ as "x at time zero.") The block's velocity $v(0)$ then is -0.920 m/s, and its acceleration $a(0)$ is $+47.0$ m/s^2.

(a) What is the angular frequency ω of this system?

SOLUTION: A Key Idea here is that, with the block in SHM, Eqs. 16-3, 16-6, and 16-7 give its displacement, velocity, and acceleration, respectively, and each contains ω. Let's substitute $t = 0$ into each to see whether we can solve any one of them for ω. We find

$$x(0) = x_m \cos \phi, \tag{16-15}$$

$$v(0) = -\omega x_m \sin \phi, \tag{16-16}$$

and
$$a(0) = -\omega^2 x_m \cos \phi. \tag{16-17}$$

In Eq. 16-15, ω has disappeared. In Eqs. 16-16 and 16-17, we know values for the left sides, but we do not know x_m and ϕ. However, if we divide Eq. 16-17 by Eq. 16-15, we neatly eliminate both x_m and ϕ and can then solve for ω as

$$\omega = \sqrt{-\frac{a(0)}{x(0)}} = \sqrt{-\frac{47.0 \text{ m/s}^2}{-0.0850 \text{ m}}}$$

$$= 23.5 \text{ rad/s.} \qquad \text{(Answer)}$$

(b) What are the phase constant ϕ and amplitude x_m?

SOLUTION: The same Key Idea as in part (a) also applies here, as do Eqs. 16-15 through 16-17. Now, however, we know ω and want ϕ and x_m. If we divide Eq. 16-16 by Eq. 16-15, we find

$$\frac{v(0)}{x(0)} = \frac{-\omega x_m \sin \phi}{x_m \cos \phi} = -\omega \tan \phi.$$

Solving for $\tan \phi$, we find

$$\tan \phi = -\frac{v(0)}{\omega x(0)} = -\frac{-0.920 \text{ m/s}}{(23.5 \text{ rad/s})(-0.0850 \text{ m})}$$

$$= -0.461.$$

This equation has two solutions:

$$\phi = -25° \quad \text{and} \quad \phi = 180° + (-25°) = 155°.$$

(Normally only the first solution here is displayed by a calculator.) A Key Idea in choosing the proper solution is to test them both by using them to compute values for the amplitude x_m. From Eq. 16-15, we find that if $\phi = -25°$, then

$$x_m = \frac{x(0)}{\cos \phi} = \frac{-0.0850 \text{ m}}{\cos(-25°)} = -0.094 \text{ m.}$$

We find similarly that if $\phi = 155°$, then $x_m = 0.094$ m. Because the amplitude of SHM must be a positive constant, the correct phase constant and amplitude here are

$$\phi = 155° \quad \text{and} \quad x_m = 0.094 \text{ m} = 9.4 \text{ cm.} \quad \text{(Answer)}$$

PROBLEM-SOLVING TACTICS

Tactic 2: *Identifying SHM*
In linear SHM the acceleration a and displacement x of the system are related by an equation of the form

$$a = -(\text{a positive constant})x,$$

which says that the acceleration is proportional to the displacement from the equilibrium position but is in the opposite direction. Once you find such an expression for an oscillating system, you can immediately compare it to Eq. 16-8, identify the positive constant as being equal to ω^2, and so quickly get an expression for the angular frequency of the motion. With Eq. 16-5 you then can find the period T and the frequency f.

In some problems you might derive an expression for the force F as a function of displacement x. If the motion is linear SHM, the force and displacement are related by

$$F = -(\text{a positive constant})x,$$

which says that the force is proportional to the displacement but is in the opposite direction. Once you have found such an expression, you can immediately compare it to Eq. 16-10 and identify the positive constant as being k. If you know the mass that is involved, you can then use Eqs. 16-12, 16-13, and 16-5 to find the angular frequency ω, period T, and frequency f.

16-4   Energy in Simple Harmonic Motion

In Chapter 8 we saw that the energy of a linear oscillator transfers back and forth between kinetic energy and potential energy, while the sum of the two—the mechanical energy E of the oscillator—remains constant. We now consider this situation quantitatively.

The potential energy of a linear oscillator like that of Fig. 16-5 is associated entirely with the spring. Its value depends on how much the spring is stretched or compressed—that is, on $x(t)$. We can use Eqs. 8-11 and 16-3 to find

$$U(t) = \tfrac{1}{2}kx^2 = \tfrac{1}{2}kx_m^2 \cos^2(\omega t + \phi). \tag{16-18}$$

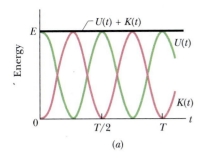

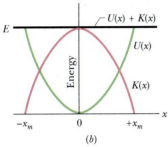

Fig. 16-6 (*a*) Potential energy $U(t)$, kinetic energy $K(t)$, and mechanical energy E as functions of time t for a linear harmonic oscillator. Note that all energies are positive and that the potential energy and the kinetic energy peak twice during every period. (*b*) Potential energy $U(x)$, kinetic energy $K(x)$, and mechanical energy E as functions of position x for a linear harmonic oscillator with amplitude x. For $x = 0$ the energy is all kinetic, and for $x = \pm x_m$ it is all potential.

Note carefully that a function written in the form $\cos^2 A$ (as here) means $(\cos A)^2$ and is *not* the same as one written $\cos A^2$, which means $\cos (A^2)$.

The kinetic energy of the system of Fig. 16-5 is associated entirely with the block. Its value depends on how fast the block is moving—that is, on $v(t)$. We can use Eq. 16-6 to find

$$K(t) = \tfrac{1}{2}mv^2 = \tfrac{1}{2}m\omega^2 x_m^2 \sin^2(\omega t + \phi). \qquad (16\text{-}19)$$

If we use Eq. 16-12 to substitute k/m for ω^2, we can write Eq. 16-19 as

$$K(t) = \tfrac{1}{2}mv^2 = \tfrac{1}{2}kx_m^2 \sin^2(\omega t + \phi). \qquad (16\text{-}20)$$

The mechanical energy follows from Eqs. 16-18 and 16-20 and is

$$\begin{aligned} E &= U + K \\ &= \tfrac{1}{2}kx_m^2 \cos^2(\omega t + \phi) + \tfrac{1}{2}kx_m^2 \sin^2(\omega t + \phi) \\ &= \tfrac{1}{2}kx_m^2[\cos^2(\omega t + \phi) + \sin^2(\omega t + \phi)]. \end{aligned}$$

For any angle α,

$$\cos^2 \alpha + \sin^2 \alpha = 1.$$

Thus, the quantity in the square brackets above is unity and we have

$$E = U + K = \tfrac{1}{2}kx_m^2. \qquad (16\text{-}21)$$

The mechanical energy of a linear oscillator is indeed constant and independent of time. The potential energy and kinetic energy of a linear oscillator are shown as functions of time t in Fig. 16-6*a*, and as functions of displacement x in Fig. 16-6*b*.

You might now understand why an oscillating system normally contains an element of springiness and an element of inertia: The former stores its potential energy and the latter stores its kinetic energy.

✔**CHECKPOINT 3:** In Fig. 16-5, the block has a kinetic energy of 3 J and the spring has an elastic potential energy of 2 J when the block is at $x = +2.0$ cm. (a) What is the kinetic energy when the block is at $x = 0$? What are the elastic potential energies when the block is at (b) $x = -2.0$ cm and (c) $x = -x_m$?

Sample Problem 16-3

(a) What is the mechanical energy E of the linear oscillator of Sample Problem 16-1? (Initially, the block's position is $x = 11$ cm and its speed is $v = 0$. Spring constant k is 65 N/m.)

SOLUTION: The **Key Idea** here is that the mechanical energy E (the sum of the kinetic energy $K = \tfrac{1}{2}mv^2$ of the block and the potential energy $U = \tfrac{1}{2}kx^2$ of the spring) is constant throughout the motion of the oscillator. Thus, we can evaluate E at any point during the motion. Because we are given the initial conditions of the oscillator as $x = 11$ cm and $v = 0$, let us evaluate E for those conditions. We find

$$E = K + U = \tfrac{1}{2}mv^2 + \tfrac{1}{2}kx^2 = 0 + \tfrac{1}{2}(65 \text{ N/m})(0.11 \text{ m})^2$$
$$= 0.393 \text{ J} \approx 0.39 \text{ J}. \qquad \text{(Answer)}$$

(b) What are the potential energy U and kinetic energy K of the oscillator when the block is at $x = \tfrac{1}{2}x_m$? What are they when the block is at $x = -\tfrac{1}{2}x_m$?

SOLUTION: The **Key Idea** here is that, because we are given the location of the block, we can easily find the spring's potential energy with $U = \tfrac{1}{2}kx^2$. For $x = \tfrac{1}{2}x_m$, we have

$$U = \tfrac{1}{2}kx^2 = \tfrac{1}{2}k(\tfrac{1}{2}x_m)^2 = (\tfrac{1}{2})(\tfrac{1}{4})kx_m^2.$$

We can substitute for k and x_m, or we can use the **Key Idea** that the total mechanical energy, which we know from part (a), is $\tfrac{1}{2}kx_m^2$. That idea allows us to write, from the above equation,

$$U = \tfrac{1}{4}(\tfrac{1}{2}kx_m^2) = \tfrac{1}{4}E = \tfrac{1}{4}(0.393 \text{ J}) = 0.098 \text{ J}. \quad \text{(Answer)}$$

Now, using the **Key Idea** of (a) (namely, $E = K + U$), we can write

$$K = E - U = 0.393 \text{ J} - 0.098 \text{ J} \approx 0.30 \text{ J}. \quad \text{(Answer)}$$

By repeating these calculations for $x = -\tfrac{1}{2}x_m$, we would find the same answers for that displacement, consistent with the left–right symmetry of Fig. 16-6*b*.

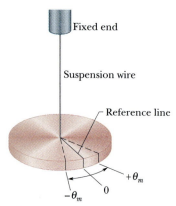

Fig. 16-7 An angular simple harmonic oscillator, or torsion pendulum, is an angular version of the linear simple harmonic oscillator of Fig. 16-5. The disk oscillates in a horizontal plane; the reference line oscillates with angular amplitude θ_m. The twist in the suspension wire stores potential energy as a spring does and provides the restoring torque.

16-5 An Angular Simple Harmonic Oscillator

Figure 16-7 shows an angular version of a simple harmonic oscillator; the element of springiness or elasticity is associated with the twisting of a suspension wire rather than the extension and compression of a spring as we previously had. The device is called a **torsion pendulum,** with *torsion* referring to the twisting.

If we rotate the disk in Fig. 16-7 by some angular displacement θ from its rest position (where the reference line is at $\theta = 0$) and release it, it will oscillate about that position in **angular simple harmonic motion.** Rotating the disk through an angle θ in either direction introduces a restoring torque given by

$$\tau = -\kappa\theta. \tag{16-22}$$

Here κ (Greek *kappa*) is a constant, called the **torsion constant,** that depends on the length, diameter, and material of the suspension wire.

Comparison of Eq. 16-22 with Eq. 16-10 leads us to suspect that Eq. 16-22 is the angular form of Hooke's law, and that we can transform Eq. 16-13, which gives the period of linear SHM, into an equation for the period of angular SHM: We replace the spring constant k in Eq. 16-13 with its equivalent, the constant κ of Eq. 16-22, and we replace the mass m in Eq. 16-13 with *its* equivalent, the rotational inertia I of the oscillating disk. These replacements lead to

$$T = 2\pi\sqrt{\frac{I}{\kappa}} \qquad \text{(torsion pendulum)}, \tag{16-23}$$

which is the correct equation for the period of an angular simple harmonic oscillator, or torsion pendulum.

Tactic 3: *Identifying Angular SHM*
When a system undergoes angular simple harmonic motion, its angular acceleration α and angular displacement θ are related by an equation of the form

$$\alpha = -\text{(a positive constant)}\theta.$$

This equation is the angular equivalent of Eq. 16-8 ($a = -\omega^2 x$). It says that the angular acceleration α is proportional to the angular displacement θ from the equilibrium position but tends to rotate the system in the direction opposite the displacement. If you have an expression of this form, you can identify the positive constant as being ω^2, and then you can determine ω, f, and T.

You can also identify angular SHM if you have an expression for the torque τ in terms of the angular displacement, because that expression must be in the form of Eq. 16-22 ($\tau = -\kappa\theta$) or

$$\tau = -\text{(a positive constant)}\theta.$$

This equation is the angular equivalent of Eq. 16-10 ($F = -kx$). It says that the torque τ is proportional to the angular displacement θ from the equilibrium position but tends to rotate the system in the opposite direction. If you have an expression of this form, then you can identify the positive constant as being the system's torsion constant κ. If you know the rotational inertia I of the system, you can then determine T with Eq. 16-23.

Sample Problem 16-4

Figure 16-8a shows a thin rod whose length L is 12.4 cm and whose mass m is 135 g, suspended at its midpoint from a long wire. Its period T_a of angular SHM is measured to be 2.53 s. An irregularly shaped object, which we call object X, is then hung from the same wire, as in Fig. 16-8b, and its period T_b is found to be 4.76 s. What is the rotational inertia of object X about its suspension axis?

SOLUTION: The Key Idea here is that the rotational inertia of either the rod or object X is related to the measured period by Eq. 16-23. In Table 11-2e, the rotational inertia of a thin rod about a perpendic-

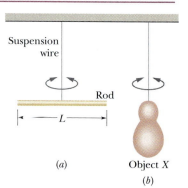

Fig. 16-8 Sample Problem 16-4. Two torsion pendulums, consisting of (a) a wire and a rod and (b) the same wire and an irregularly shaped object.

ular axis through its midpoint is given as $\frac{1}{12}mL^2$. Thus, we have, for the rod in Fig. 16-8a,

$$I_a = \tfrac{1}{12}mL^2 = (\tfrac{1}{12})(0.135 \text{ kg})(0.124 \text{ m})^2$$
$$= 1.73 \times 10^{-4} \text{ kg} \cdot \text{m}^2.$$

Now let us write Eq. 16-23 twice, once for the rod and once for object X:

$$T_a = 2\pi\sqrt{\frac{I_a}{\kappa}} \quad \text{and} \quad T_b = 2\pi\sqrt{\frac{I_b}{\kappa}}.$$

The constant κ, which is a property of the wire, is the same for both figures; only the periods and the rotational inertias differ.

Let us square each of these equations, divide the second by the first, and solve the resulting equation for I_b. The result is

$$I_b = I_a\frac{T_b^2}{T_a^2} = (1.73 \times 10^{-4} \text{ kg} \cdot \text{m}^2)\frac{(4.76 \text{ s})^2}{(2.53 \text{ s})^2}$$
$$= 6.12 \times 10^{-4} \text{ kg} \cdot \text{m}^2. \qquad \text{(Answer)}$$

16-6 Pendulums

We turn now to a class of simple harmonic oscillators in which the springiness is associated with the gravitational force rather than with the elastic properties of a twisted wire or a compressed or stretched spring.

The Simple Pendulum

If you hang an apple at the end of a long thread fixed at its upper end, and then set the apple swinging back and forth a small distance, you easily see that the apple's motion is periodic. Is it, in fact, simple harmonic motion? 'f so, what is the period T? To answer, we consider a **simple pendulum,** which consists of a particle of mass m (called the *bob* of the pendulum) suspended from one end of an unstretchable, massless string of length L that is fixed at the other end, as in Fig. 16-9a. The bob is free to swing back and forth in the plane of the page, to the left and right of a vertical line through the pendulum's pivot point.

The forces acting on the bob are the force $\vec{T}$ from the string and the gravitational force $\vec{F}_g$, as shown in Fig. 16-9b where the string makes an angle θ with the vertical. We resolve $\vec{F}_g$ into a radial component $F_g \cos\theta$ and a component $F_g \sin\theta$ that is tangent to the path taken by the bob. This tangential component produces a restoring torque about the pendulum's pivot point, because it always acts opposite the displacement of the bob so as to bring the bob back toward its central location. That location is called the *equilibrium position* ($\theta = 0$), because the pendulum would be at rest there were it not swinging.

From Eq. 11-33 ($\tau = r_\perp F$), we can write this restoring torque as

$$\tau = -L(F_g \sin\theta), \qquad (16\text{-}24)$$

where the minus sign indicates that the torque acts to reduce θ, and L is the moment arm of the force component $F_g \sin\theta$ about the pivot point. Substituting Eq. 16-24 into Eq. 11-36 ($\tau = I\alpha$) and then substituting mg as the magnitude of F_g, we obtain

$$-L(mg \sin\theta) = I\alpha, \qquad (16\text{-}25)$$

where I is the pendulum's rotational inertia about the pivot point and α is its angular acceleration about that point.

We can simplify Eq. 16-25 if we assume the angle θ is small, for then we can approximate $\sin\theta$ with θ (expressed in radian measure). (As an example, if $\theta = 5.00° = 0.0873$ rad, then $\sin\theta = 0.0872$, a difference of only about 0.1%.) With that approximation and some rearranging, we then have

$$\alpha = -\frac{mgL}{I}\theta. \qquad (16\text{-}26)$$

This equation is the angular equivalent of Eq. 16-8, the hallmark of SHM. It tells

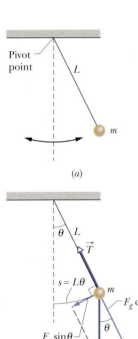

Fig. 16-9 (a) A simple pendulum. (b) The forces acting on the bob are the gravitational force $\vec{F}_g$ and the force $\vec{T}$ from the string. The tangential component $F_g \sin\theta$ of the gravitational force is a restoring force that tends to bring the pendulum back to its central position.

us that the angular acceleration α of the pendulum is proportional to the angular displacement θ but opposite in sign. Thus, as the pendulum bob moves to, say, the right as in Fig. 16-9a, its acceleration *to the left* increases until it stops and begins moving to the left. Then, when it is on the left, its acceleration to the right tends to return it to the right, and so on, as it swings back and forth in SHM. More precisely, the motion of a *simple pendulum swinging through only small angles* is approximately SHM. We can state this restriction to small angles another way: The **angular amplitude** θ_m of the motion (the maximum angle of swing) must be small.

Comparing Eqs. 16-26 and Eq. 16-8, we see that the angular frequency of the pendulum is $\omega = \sqrt{mgL/I}$. Next, if we substitute this expression for ω into Eq. 16-5 ($\omega = 2\pi/T$), we see that the period of the pendulum may be written as

$$T = 2\pi\sqrt{\frac{I}{mgL}}. \tag{16-27}$$

All the mass of a simple pendulum is concentrated in the mass m of the particle-like bob, which is at radius L from the pivot point. Thus, we can use Eq. 11-26 ($I = mr^2$) to write $I = mL^2$ for the rotational inertia of the pendulum. Substituting this into Eq. 16-27 and simplifying then yield

$$T = 2\pi\sqrt{\frac{L}{g}} \qquad \text{(simple pendulum, small amplitude)} \tag{16-28}$$

as a simpler expression for the period of a simple pendulum swinging through only small angles. (We assume small-angle swinging in the problems of this chapter.)

The Physical Pendulum

A real pendulum, usually called a **physical pendulum,** can have a complicated distribution of mass, much different from that of a simple pendulum. Does a physical pendulum also undergo SHM? If so, what is its period?

Figure 16-10 shows an arbitrary physical pendulum displaced to one side by angle θ. The gravitational force $\vec{F_g}$ acts at its center of mass C, at a distance h from the pivot point O. In spite of their shapes, comparison of Figs. 16-10 and 16-9b reveals only one important difference between an arbitrary physical pendulum and a simple pendulum. For a physical pendulum the restoring component $F_g \sin \theta$ of the gravitational force has a moment arm of distance h about the pivot point, rather than of string length L. In all other respects, an analysis of the physical pendulum would duplicate our analysis of the simple pendulum up through Eq. 16-27. Again, (for small θ_m) we would find that the motion is approximately SHM.

If we replace L with h in Eq. 16-27, we can write the period of a physical pendulum as

$$T = 2\pi\sqrt{\frac{I}{mgh}} \qquad \text{(physical pendulum, small amplitude).} \tag{16-29}$$

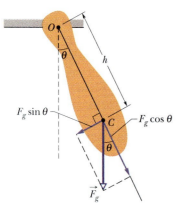

Fig. 16-10 A physical pendulum. The restoring torque is $hF_g \sin \theta$. When $\theta = 0$, center of mass C hangs directly below pivot point O.

As with the simple pendulum, I is the rotational inertia of the pendulum about O. However, now I is not simply mL^2 (it depends on the shape of the physical pendulum), but it is still proportional to m.

A physical pendulum will not swing if it pivots at its center of mass. Formally, this corresponds to putting $h = 0$ in Eq. 16-29. That equation then predicts $T \rightarrow \infty$, which implies that such a pendulum will never complete one swing.

Corresponding to any physical pendulum that oscillates about a given pivot point O with period T is a simple pendulum of length L_0 with the same period T. We can find L_0 with Eq. 16-28. The point along the physical pendulum at distance L_0 from point O is called the *center of oscillation* of the physical pendulum for the given suspension point.

Measuring g

We can use a physical pendulum to measure the free-fall acceleration g at a particular location on Earth's surface. (Countless thousands of such measurements have been made during geophysical prospecting.)

To analyze a simple case, take the pendulum to be a uniform rod of length L, suspended from one end. For such a pendulum, h in Eq. 16-29, the distance between the pivot point and the center of mass, is $\frac{1}{2}L$. Table 11-2e tells us that the rotational inertia of this pendulum about a perpendicular axis through its center of mass is $\frac{1}{12}mL^2$. From the parallel-axis theorem of Eq. 11-29 ($I = I_{com} + Mh^2$), we then find that the rotational inertia about a perpendicular axis through one end of the rod is

$$I = I_{com} + mh^2 = \tfrac{1}{12}mL^2 + m(\tfrac{1}{2}L)^2 = \tfrac{1}{3}mL^2. \tag{16-30}$$

If we put $h = \frac{1}{2}L$ and $I = \frac{1}{3}mL^2$ in Eq. 16-29 and solve for g, we find

$$g = \frac{8\pi^2 L}{3T^2}. \tag{16-31}$$

Thus, by measuring L and the period T, we can find the value of g at the pendulum's location. (If precise measurements are to be made, a number of refinements are needed, such as swinging the pendulum in an evacuated chamber.)

Sample Problem 16-5

In Fig. 16-11a, a meter stick swings about a pivot point at one end, at distance h from its center of mass.

(a) What is its period of oscillation T?

SOLUTION: One **Key Idea** here is that the stick is not a simple pendulum because its mass is not concentrated in a bob at the end opposite the pivot point—so the stick is a physical pendulum. Then its period is given by Eq. 16-29, for which we need the rotational inertia I of the stick about the pivot point. We can treat the stick as a uniform rod of length L and mass m. Then Eq. 16-30 tells us that $I = \frac{1}{3}mL^2$, and the distance h in Eq. 16-29 is $\frac{1}{2}L$. Substituting these quantities into Eq. 16-29, we find

$$T = 2\pi\sqrt{\frac{I}{mgh}} = 2\pi\sqrt{\frac{\frac{1}{3}mL^2}{mg(\frac{1}{2}L)}} = 2\pi\sqrt{\frac{2L}{3g}} \tag{16-32}$$

$$= 2\pi\sqrt{\frac{(2)(1.00\text{ m})}{(3)(9.8\text{ m/s}^2)}} = 1.64\text{ s}. \quad\text{(Answer)}$$

Note that the result is independent of the pendulum's mass m.

(b) What is the distance L_0 between the pivot point O of the stick and the center of oscillation of the stick?

SOLUTION: The **Key Idea** here is that we want the length L_0 of the simple pendulum (drawn in Fig. 16-11b) that has the same period as the physical pendulum (the stick) of Fig. 16-11a. Setting

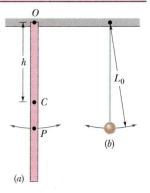

Fig. 16-11 Sample Problem 16-5. (a) A meter stick suspended from one end as a physical pendulum. (b) A simple pendulum whose length L_0 is chosen so that the periods of the two pendulums are equal. Point P on the pendulum of (a) marks the center of oscillation.

Eqs. 16-28 and 16-32 equal yields

$$T = 2\pi\sqrt{\frac{L_0}{g}} = 2\pi\sqrt{\frac{2L}{3g}}.$$

You can see by inspection that

$$L_0 = \tfrac{2}{3}L = (\tfrac{2}{3})(100\text{ cm}) = 66.7\text{ cm}. \quad\text{(Answer)}$$

In Fig. 16-11a, point P marks this distance from suspension point O. Thus, point P is the stick's center of oscillation for the given suspension point.

✓CHECKPOINT 4: Three physical pendulums, of masses m_0, $2m_0$, and $3m_0$, have the same shape and size and are suspended at the same point. Rank the masses according to the periods of the pendulums, greatest period first.

Sample Problem 16-6

In Fig. 16-12, a penguin (obviously skilled in aquatic sports) dives from a uniform board that is hinged at the left and attached to a spring at the right. The board has length $L = 2.0$ m and mass $m = 12$ kg; the spring constant k is 1300 N/m. When the penguin dives, it leaves the board and spring oscillating with a small amplitude. Assume that the board is stiff enough not to bend, and find the period T of the oscillations.

SOLUTION: Since a spring is involved, we might guess that the oscillations are SHM, but we shall not assume that. Instead, we use the following **Key Idea**: If the board is in SHM, then the acceleration and displacement of the oscillating end of the board must be related by an expression in the form of Eq. 16-8 ($a = -\omega^2 x$). If so, we shall be able to find ω and then the desired T from the expression. Let us check by finding the relation between the acceleration and displacement of the board's right end.

Because the board rotates about the hinge as one end oscillates, we are concerned with a torque $\vec{\tau}$ on the board about the hinge. That torque is due to the force $\vec{F}$ on the board from the spring. Because $\vec{F}$ varies with time, $\vec{\tau}$ must also. However, at any given instant we can relate the magnitudes of $\vec{\tau}$ and $\vec{F}$ with Eq. 11-31 ($\tau = rF \sin \phi$). Here we have

$$\tau = LF \sin 90°, \qquad (16\text{-}33)$$

where L is the moment arm of force $\vec{F}$ and 90° is the angle between the moment arm and the force's line of action. Combining Eq. 16-33 with Eq. 11-36 ($\tau = I\alpha$) gives us

$$LF = I\alpha, \qquad (16\text{-}34)$$

where I is the board's rotational inertia about the hinge, and α is its angular acceleration about that point. We may treat the board as a thin rod pivoted about one end. Then, from Eq. 16-30, the board's rotational inertia I is $\frac{1}{3}mL^2$.

Now let us mentally erect a vertical x through the oscillating right end of the board, with the positive direction upward. Then the force on the right end of the board from the spring is $F = -kx$, where x is the vertical displacement of the right end.

Substituting these expressions for I and F into Eq. 16-34 gives us

$$-Lkx = \frac{mL^2\alpha}{3}. \qquad (16\text{-}35)$$

We now have a mixture of linear displacement x (vertically) and rotational acceleration α (about the hinge). We can replace α in Eq. 16-35 with the (linear) acceleration a along the x axis by substituting according to Eq. 11-22 ($a_t = \alpha r$) for tangential acceleration. Here the tangential acceleration is a and the radius of rotation r is L, so $\alpha = a/L$. With that substitution, Eq. 16-35 becomes

$$-Lkx = \frac{mL^2 a}{3L},$$

which yields

$$a = -\frac{3k}{m}x. \qquad (16\text{-}36)$$

Equation 16-36 is, in fact, of the same form as Eq. 16-8 ($a = -\omega^2 x$). Therefore, the board does indeed undergo SHM, and comparison of Eqs. 16-36 and 16-8 shows that

$$\omega^2 = \frac{3k}{m},$$

which gives $\omega = \sqrt{3k/m}$. Using Eq. 16-5 ($\omega = 2\pi/T$) to find T then gives us

$$T = 2\pi\sqrt{\frac{m}{3k}} = 2\pi\sqrt{\frac{12 \text{ kg}}{3(1300 \text{ N/m})}}$$

$$= 0.35 \text{ s}. \qquad \text{(Answer)}$$

Perhaps surprisingly, the period is independent of the board's length L.

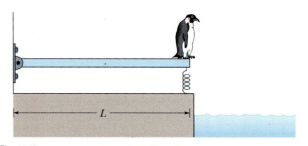

Fig. 16-12 Sample Problem 16-6. The dive by the penguin causes the board and spring to oscillate; the board pivots about the hinge at the left.

16-7 Simple Harmonic Motion and Uniform Circular Motion

In 1610, Galileo, using his newly constructed telescope, discovered the four principal moons of Jupiter. Over weeks of observation, each moon seemed to him to be moving back and forth relative to the planet in what today we would call simple harmonic motion; the disk of the planet was the midpoint of the motion. The record of Galileo's observations, written in his own hand, is still available. A. P. French of MIT used Galileo's data to work out the position of the moon Callisto relative to

Fig. 16-13 The angle between Jupiter and its moon Callisto as seen from Earth. The circles are based on Galileo's 1610 measurements. The curve is a best fit, strongly suggesting simple harmonic motion. At Jupiter's mean distance, 10 minutes of arc corresponds to about 2×10^6 km. (Adapted from A. P. French, *Newtonian Mechanics*, W. W. Norton & Company, New York, 1971, p. 288.)

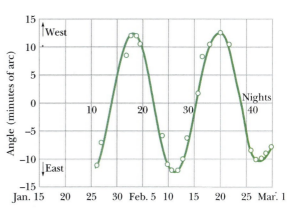

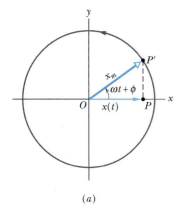

(a)

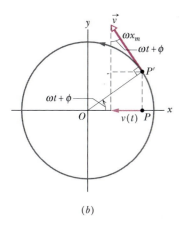

(b)

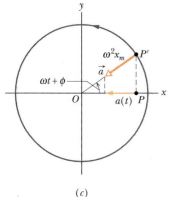

(c)

Jupiter. In the results shown in Fig. 16-13, the circles are based on Galileo's observations and the curve is a best fit to the data. The curve strongly suggests Eq. 16-3, the displacement function for SHM. A period of about 16.8 days can be measured from the plot.

Actually, Callisto moves with essentially constant speed in an essentially circular orbit around Jupiter. Its true motion—far from being simple harmonic—is uniform circular motion. What Galileo saw—and what you can see with a good pair of binoculars and a little patience—is the projection of this uniform circular motion on a line in the plane of the motion. We are led by Galileo's remarkable observations to the conclusion that simple harmonic motion is uniform circular motion viewed edge-on. In more formal language:

> Simple harmonic motion is the projection of uniform circular motion on a diameter of the circle in which the latter motion occurs.

Figure 16-14a gives an example. It shows a *reference particle P'* moving in uniform circular motion with (constant) angular speed ω in a *reference circle.* The radius x_m of the circle is the magnitude of the particle's position vector. At any time t, the angular position of the particle is $\omega t + \phi$, where ϕ is its angular position at $t = 0$.

The projection of particle P' onto the x axis is a point P, which we take to be a second particle. The projection of the position vector of particle P' onto the x axis gives the location $x(t)$ of P. Thus, we find

$$x(t) = x_m \cos(\omega t + \phi),$$

which is precisely Eq. 16-3. Our conclusion is correct. If reference particle P' moves in uniform circular motion, its projection particle P moves in simple harmonic motion along a diameter of the circle.

Figure 16-14b shows the velocity $\vec{v}$ of the reference particle. From Eq. 11-18 ($v = \omega r$), the magnitude of the velocity vector is ωx_m; its projection on the x axis is

$$v(t) = -\omega x_m \sin(\omega t + \phi),$$

Fig. 16-14 (a) A reference particle P' moving with uniform circular motion in a reference circle of radius x_m. Its projection P on the x axis executes simple harmonic motion. (b) The projection of the velocity $\vec{v}$ of the reference particle is the velocity of SHM. (c) The projection of the radial acceleration $\vec{a}$ of the reference particle is the acceleration of SHM.

which is exactly Eq. 16-6. The minus sign appears because the velocity component of P in Fig. 16-14b is directed to the left, in the negative direction of x.

Figure 16-14c shows the radial acceleration $\vec{a}$ of the reference particle. From Eq. 11-23 ($a_r = \omega^2 r$), the magnitude of the radial acceleration vector is $\omega^2 x_m$; its projection on the x axis is

$$a(t) = -\omega^2 x_m \cos(\omega t + \phi),$$

which is exactly Eq. 16-7. Thus, whether we look at the displacement, the velocity, or the acceleration, the projection of uniform circular motion is indeed simple harmonic motion.

16-8 Damped Simple Harmonic Motion

A pendulum will swing only briefly under water, because the water exerts a drag force on the pendulum that quickly eliminates the motion. A pendulum swinging in air does better, but still the motion dies out eventually, because the air exerts a drag force on the pendulum (and friction acts at its support), transferring energy from the pendulum's motion.

When the motion of an oscillator is reduced by an external force, the oscillator and its motion are said to be **damped**. An idealized example of a damped oscillator is shown in Fig. 16-15, where a block with mass m oscillates vertically on a spring with spring constant k. From the block, a rod extends to a vane (both assumed massless) that is submerged in a liquid. As the vane moves up and down, the liquid exerts an inhibiting drag force on it and thus on the entire oscillating system. With time, the mechanical energy of the block–spring system decreases, as energy is transferred to thermal energy of the liquid and vane.

Let us assume the liquid exerts a **damping force** $\vec{F}_d$ that is proportional in magnitude to the velocity $\vec{v}$ of the vane and block (an assumption that is accurate if the vane moves slowly). Then, for components along the x axis in Fig. 16-15, we have

$$F_d = -bv, \tag{16-37}$$

where b is a **damping constant** that depends on the characteristics of both the vane and the liquid and has the SI unit of kilogram per second. The minus sign indicates that $\vec{F}_d$ opposes the motion.

The force on the block from the spring is $F_s = -kx$. Let us assume that the gravitational force on the block is negligible compared to F_d and F_s. Then we can write Newton's second law for components along the x axis ($F_{net,x} = ma_x$) as

$$-bv - kx = ma. \tag{16-38}$$

Substituting dx/dt for v and d^2x/dt^2 for a and rearranging give us the differential equation

$$m\frac{d^2x}{dt^2} + b\frac{dx}{dt} + kx = 0. \tag{16-39}$$

The solution of this equation is

$$x(t) = x_m e^{-bt/2m} \cos(\omega' t + \phi), \tag{16-40}$$

where x_m is the amplitude and ω' is the angular frequency of the damped oscillator. This angular frequency is given by

$$\omega' = \sqrt{\frac{k}{m} - \frac{b^2}{4m^2}}. \tag{16-41}$$

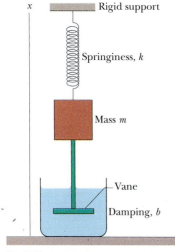

Fig. 16-15 An idealized damped simple harmonic oscillator. A vane immersed in a liquid exerts a damping force on the block as the block oscillates parallel to the x axis.

If $b = 0$ (there is no damping), then Eq. 16-41 reduces to Eq. 16-12 ($\omega = \sqrt{k/m}$) for the angular frequency of an undamped oscillator, and Eq. 16-40 reduces to Eq. 16-3 for the displacement of an undamped oscillator. If the damping constant is small but not zero (so that $b \ll \sqrt{km}$), then $\omega' \approx \omega$.

We can regard Eq. 16-40 as a cosine function whose amplitude, which is $x_m e^{-bt/2m}$, gradually decreases with time, as Fig. 16-16 suggests. For an undamped oscillator, the mechanical energy is constant and is given by Eq. 16-21 ($E = \frac{1}{2}kx_m^2$). If the oscillator is damped, the mechanical energy is not constant but decreases with time. If the damping is small, we can find $E(t)$ by replacing x_m in Eq. 16-21 with $x_m e^{-bt/2m}$, the amplitude of the damped oscillations. By doing so, we find that

$$E(t) \approx \tfrac{1}{2}kx_m^2 e^{-bt/m}, \tag{16-42}$$

which tells us that, like the amplitude, the mechanical energy decreases exponentially with time.

✔**CHECKPOINT 5:**   Here are three sets of values for the spring constant, damping constant, and mass for the damped oscillator of Fig. 16-15. Rank the sets according to the time required for the mechanical energy to decrease to one-fourth of its initial value, greatest first.

Set 1	$2k_0$	b_0	m_0
Set 2	k_0	$6b_0$	$4m_0$
Set 3	$3k_0$	$3b_0$	m_0

Sample Problem 16-7

For the damped oscillator of Fig. 16-15, $m = 250$ g, $k = 85$ N/m, and $b = 70$ g/s.

(a) What is the period of the motion?

SOLUTION: The **Key Idea** here is that because $b \ll \sqrt{km} = 4.6$ kg/s, the period is approximately that of the undamped oscillator. From Eq. 16-13, we then have

$$T = 2\pi\sqrt{\frac{m}{k}} = 2\pi\sqrt{\frac{0.25 \text{ kg}}{85 \text{ N/m}}} = 0.34 \text{ s}. \quad \text{(Answer)}$$

(b) How long does it take for the amplitude of the damped oscillations to drop to half its initial value?

SOLUTION: Now the **Key Idea** is that the amplitude at time t is displayed in Eq. 16-40 as $x_m e^{-bt/2m}$. It has the value x_m at $t = 0$. Thus, we must find the value of t for which

$$x_m e^{-bt/2m} = \tfrac{1}{2}x_m.$$

Canceling x_m and taking the natural logarithm of the equation that remains, we have $\ln \frac{1}{2}$ on the right side and

$$\ln(e^{-bt/2m}) = -bt/2m$$

on the left side. Thus,

$$t = \frac{-2m \ln \frac{1}{2}}{b} = \frac{-(2)(0.25 \text{ kg})(\ln \frac{1}{2})}{0.070 \text{ kg/s}}$$

$$= 5.0 \text{ s}. \quad \text{(Answer)}$$

Because $T = 0.34$ s, this is about 15 periods of oscillation.

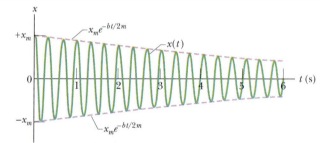

Fig. 16-16 The displacement function $x(t)$ for the damped oscillator of Fig. 16-15, with values given in Sample Problem 16-7. The amplitude, which is $x_m e^{-bt/2m}$, decreases exponentially with time.

(c) How long does it take for the mechanical energy to drop to one-half its initial value?

SOLUTION: Here the **Key Idea** is that, from Eq. 16-42, the mechanical energy at time t is $\frac{1}{2}kx_m^2 e^{-bt/m}$. It has the value $\frac{1}{2}kx_m^2$ at $t = 0$. Thus, we must find the value of t for which

$$\tfrac{1}{2}kx_m^2 e^{-bt/m} = \tfrac{1}{2}(\tfrac{1}{2}kx_m^2).$$

If we divide both sides of this equation by $\frac{1}{2}kx_m^2$ and solve for t as we did above, we find

$$t = \frac{-m \ln \frac{1}{2}}{b} = \frac{-(0.25 \text{ kg})(\ln \frac{1}{2})}{0.070 \text{ kg/s}} = 2.5 \text{ s}. \quad \text{(Answer)}$$

This is exactly half the time we calculated in (b), or about 7.5 periods of oscillation. Figure 16-16 was drawn to illustrate this sample problem.

16-9 Forced Oscillations and Resonance

A person swinging in a swing without anyone pushing it is an example of *free oscillation.* However, if someone pushes the swing periodically, the swing has *forced,* or *driven, oscillations. Two* angular frequencies are associated with a system undergoing driven oscillations: (1) the *natural* angular frequency ω of the system, which is the angular frequency at which it would oscillate if it were suddenly disturbed and then left to oscillate freely, and (2) the angular frequency ω_d of the external driving force causing the driven oscillations.

We can use Fig. 16-15 to represent an idealized forced simple harmonic oscillator if we allow the structure marked "rigid support" to move up and down at a variable angular frequency ω_d. Such a forced oscillator oscillates at the angular frequency ω_d of the driving force, and its displacement $x(t)$ is given by

$$x(t) = x_m \cos(\omega_d t + \phi), \tag{16-43}$$

where x_m is the amplitude of the oscillations.

How large the displacement amplitude x_m is depends on a complicated function of ω_d and ω. The velocity amplitude v_m of the oscillations is easier to describe: it is greatest when

$$\omega_d = \omega \qquad \text{(resonance)}, \tag{16-44}$$

a condition called **resonance.** Equation 16-44 is also *approximately* the condition at which the displacement amplitude x_m of the oscillations is greatest. Thus, if you push a swing at its natural angular frequency, the displacement and velocity amplitudes will increase to large values, a fact that children learn quickly by trial and error. If you push at other angular frequencies, either higher or lower, the displacement and velocity amplitudes will be smaller.

Figure 16-17 shows how the displacement amplitude of an oscillator depends on the angular frequency ω_d of the driving force, for three values of the damping coefficient b. Note that for all three the amplitude is approximately greatest when $\omega_d/\omega = 1$—that is, when the resonance condition of Eq. 16-44 is satisfied. The curves of Fig. 16-17 show that less damping gives a taller and narrower *resonance peak.*

All mechanical structures have one or more natural angular frequencies, and if a structure is subjected to a strong external driving force that matches one of these

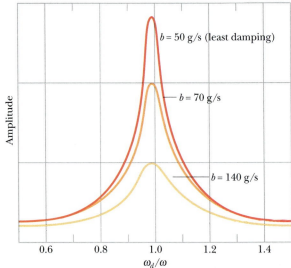

Fig. **16-17** The displacement amplitude x_m of a forced oscillator varies as the angular frequency ω_d of the driving force is varied. The amplitude is greatest approximately at $\omega_d/\omega = 1$, the resonance condition. The curves here correspond to three values of the damping constant b.

angular frequencies, the resulting oscillations of the structure may rupture it. Thus, for example, aircraft designers must make sure that none of the natural angular frequencies at which a wing can oscillate matches the angular frequency of the engines in flight. A wing that flaps violently at certain engine speeds would obviously be dangerous.

Mexico's earthquake in September 1985 was a major earthquake (8.1 on the Richter scale), but the seismic waves from it should have been too weak to cause extensive damage when they reached Mexico City about 400 km away. However, Mexico City is largely built on an ancient lake bed, where the soil is still soft with water. Although the amplitude of the seismic waves was weak in the firmer ground en route to Mexico City, their amplitude substantially increased in the loose soil of the city. Acceleration amplitudes of the waves were as much as $0.20g$, and the angular frequency was (surprisingly) concentrated around 3 rad/s. Not only was the ground severely oscillated, but many of the buildings with intermediate height had resonant angular frequencies of about 3 rad/s. Most of those intermediate-height buildings collapsed during the shaking, while shorter buildings (with higher resonant angular frequencies) and taller buildings (with lower resonant angular frequencies) remained standing.

REVIEW & SUMMARY

Frequency   The *frequency f* of periodic or oscillatory motion is the number of oscillations per second. In the SI system, it is measured in hertz:

$$1 \text{ hertz} = 1 \text{ Hz} = 1 \text{ oscillation per second} = 1 \text{ s}^{-1}. \quad (16\text{-}1)$$

Period   The *period T* is the time required for one complete oscillation, or **cycle.** It is related to the frequency by

$$T = \frac{1}{f}. \quad (16\text{-}2)$$

Simple Harmonic Motion   In *simple harmonic motion* (SHM), the displacement $x(t)$ of a particle from its equilibrium position is described by the equation

$$x = x_m \cos(\omega t + \phi) \quad \text{(displacement)}, \quad (16\text{-}3)$$

in which x_m is the **amplitude** of the displacement, the quantity $(\omega t + \phi)$ is the **phase** of the motion, and ϕ is the **phase constant.** The **angular frequency** ω is related to the period and frequency of the motion by

$$\omega = \frac{2\pi}{T} = 2\pi f \quad \text{(angular frequency)}. \quad (16\text{-}5)$$

Differentiating Eq. 16-3 leads to equations for the particle's velocity and acceleration during SHM as functions of time:

$$v = -\omega x_m \sin(\omega t + \phi) \quad \text{(velocity)} \quad (16\text{-}6)$$

and $\quad a = -\omega^2 x_m \cos(\omega t + \phi) \quad \text{(acceleration)}. \quad (16\text{-}7)$

In Eq. 16-6, the positive quantity ωx_m is the **velocity amplitude** v_m of the motion. In Eq. 16-7, the positive quantity $\omega^2 x_m$ is the **acceleration amplitude** a_m of the motion.

The Linear Oscillator   A particle with mass m that moves under the influence of a Hooke's law restoring force given by $F = -kx$ exhibits simple harmonic motion with

$$\omega = \sqrt{\frac{k}{m}} \quad \text{(angular frequency)} \quad (16\text{-}12)$$

and $\quad\quad T = 2\pi\sqrt{\frac{m}{k}} \quad \text{(period)}. \quad (16\text{-}13)$

Such a system is called a **linear simple harmonic oscillator.**

Energy   A particle in simple harmonic motion has, at any time, kinetic energy $K = \frac{1}{2}mv^2$ and potential energy $U = \frac{1}{2}kx^2$. If no friction is present, the mechanical energy $E = K + U$ remains constant even though K and U change.

Pendulums   Examples of devices that undergo simple harmonic motion are the **torsion pendulum** of Fig. 16-7, the **simple pendulum** of Fig. 16-9, and the **physical pendulum** of Fig. 16-10. Their periods of oscillation for small oscillations are, respectively,

$$T = 2\pi\sqrt{I/\kappa}, \quad (16\text{-}23)$$

$$T = 2\pi\sqrt{L/g}, \quad (16\text{-}28)$$

and $\quad\quad T = 2\pi\sqrt{I/mgh}. \quad (16\text{-}29)$

Simple Harmonic Motion and Uniform Circular Motion
Simple harmonic motion is the projection of uniform circular motion onto the diameter of the circle in which the latter motion occurs. Figure 16-14 shows that all parameters of circular motion (position, velocity, and acceleration) project to the corresponding values for simple harmonic motion.

Damped Harmonic Motion The mechanical energy E in a real oscillating system decreases during the oscillations because external forces, such as a drag force, inhibit the oscillations and transfer mechanical energy to thermal energy. The real oscillator and its motion are then said to be **damped.** If the **damping force** is given by $\vec{F}_d = -b\vec{v}$, where v is the velocity of the oscillator and b is a **damping constant,** then the displacement of the oscillator is given by

$$x(t) = x_m e^{-bt/2m} \cos(\omega' t + \phi), \qquad (16\text{-}40)$$

where ω', the angular frequency of the damped oscillator, is given by

$$\omega' = \sqrt{\frac{k}{m} - \frac{b^2}{4m^2}}. \qquad (16\text{-}41)$$

If the damping constant is small ($b \ll \sqrt{km}$), then $\omega' \approx \omega$, where ω is the angular frequency of the undamped oscillator. For small b, the mechanical energy E of the oscillator is given by

$$E(t) \approx \tfrac{1}{2}kx_m^2\, e^{-bt/m}. \qquad (16\text{-}42)$$

Forced Oscillations and Resonance If an external driving force with angular frequency ω_d acts on an oscillating system with *natural* angular frequency ω, the system oscillates with angular frequency ω_d. The velocity amplitude v_m of the system is greatest when

$$\omega_d = \omega, \qquad (16\text{-}44)$$

a condition called **resonance.** The amplitude x_m of the system is (approximately) greatest under the same condition.

QUESTIONS

1. Which of the following relationships between the acceleration a and the displacement x of a particle involve SHM: (a) $a = 0.5x$, (b) $a = 400x^2$, (c) $a = -20x$, (d) $a = -3x^2$?

2. Given $x = (2.0\text{ m})\cos(5t)$ for SHM and needing to find the velocity at $t = 2$ s, should you substitute for t and then differentiate with respect to t or vice versa?

3. The acceleration $a(t)$ of a particle undergoing SHM is graphed in Fig. 16-18. (a) Which of the labeled points corresponds to the particle at $-x_m$? (b) At point 4, is the velocity of the particle positive, negative, or zero? (c) At point 5, is the particle at $-x_m$, at $+x_m$, at 0, between $-x_m$ and 0, or between 0 and $+x_m$?

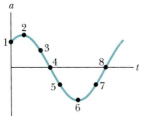

Fig. 16-18 Question 3.

4. Which of the following describe ϕ for the SHM of Fig. 16-19a:

(a) $-\pi < \phi < -\pi/2,$ (b) $\pi < \phi < 3\pi/2,$
(c) $-3\pi/2 < \phi < -\pi?$

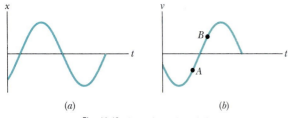

Fig. 16-19 Questions 4 and 5.

5. The velocity $v(t)$ of a particle undergoing SHM is graphed in Fig. 16-19b. Is the particle momentarily stationary, headed toward $-x_m$, or headed toward $+x_m$ at (a) point A on the graph and (b) point B? Is the particle at $-x_m$, at $+x_m$, at 0, between $-x_m$ and 0, or between 0 and $+x_m$ when its velocity is represented by (c) point A and (d) point B? Is the speed of the particle increasing or decreasing at (e) point A and (f) point B?

6. Figure 16-20 gives, for three situations, the displacements $x(t)$ of a pair of simple harmonic oscillators (A and B) that are identical except for phase. For each pair, what phase shift (in radians and in degrees) is needed to shift the curve for A to coincide with the curve for B? Of the many possible answers, choose the shift with the smallest absolute magnitude.

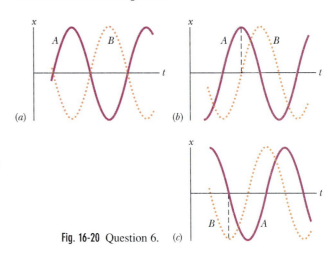

Fig. 16-20 Question 6.

7. Figures 16-21a and b show the positions of four linear oscillators with identical masses and spring constants, in snapshots at the same instant. What is the phase difference of the two linear oscil-

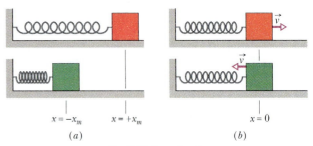

Fig. 16-21 Question 7.

lators in (a) Fig. 16-21a and (b) Fig. 16-21b? (c) What is the phase difference between the red oscillator in Fig. 16-21a and the green oscillator in Fig. 16-21b?

8. (a) Which curve in Fig. 16-22a gives the acceleration $a(t)$ versus displacement $x(t)$ of a simple harmonic oscillator? (b) Which curve in Fig. 16-22b gives the velocity $v(t)$ versus $x(t)$?

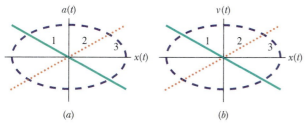

Fig. 16-22 Question 8.

9. In Fig. 16-23, a small block A sits on a large block B with a certain nonzero coefficient of static friction between the two blocks. Block B, which lies on a frictionless surface, is initially at $x = 0$, with the spring at its relaxed length; then we pull the block a distance d to the right and release it. As the spring–blocks system undergoes SHM, with amplitude x_m, block A is on the verge of slipping over B.

(a) Is the acceleration of block A constant or does it vary? (b) Is the magnitude of the frictional force accelerating A constant or does it vary? (c) Is A more likely to slip at $x = 0$ or at $x = \pm x_m$? (d) If the SHM began with an initial displacement that was greater than d, would slippage then be more likely or less likely? (Warm-up for Problem 16)

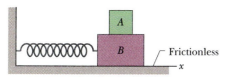

Fig. 16-23 Question 9.

10. In Fig. 16-24, a spring–block system is put into SHM in two experiments. In the first, the block is pulled from the equilibrium position through a displacement d_1 and then released. In the second,

it is pulled from the equilibrium position through a greater displacement d_2 and then released. Are the (a) amplitude, (b) period, (c) frequency, (d) maximum kinetic energy, and (e) maximum potential energy in the second experiment greater than, less than, or the same as those in the first experiment?

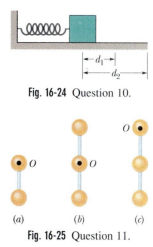

Fig. 16-24 Question 10.

11. Figure 16-25 shows three physical pendulums consisting of identical uniform spheres of the same mass that are rigidly connected by identical rods of negligible mass. Each pendulum is vertical and can pivot about suspension point O. Rank the pendulums according to period of oscillation, greatest first.

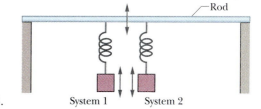

Fig. 16-25 Question 11.

12. Follow-up to Exercise 36: If the speed of the bullet were greater, would the following quantities of the resulting SHM then be greater, less, or the same: (a) amplitude, (b) period, (c) maximum potential energy?

13. You are to build the oscillation transfer device shown in Fig. 16-26. It consists of two spring–block systems hanging from a flexible rod. When the spring of system 1 is stretched and then released, the resulting SHM of system 1 at frequency f_1 oscillates the rod. The rod then provides a driving force on system 2, at the same frequency f_1. You can choose from four springs with spring constants k of 1600, 1500, 1400, and 1200 N/m, and four blocks with masses m of 800, 500, 400, and 200 kg. Mentally determine which spring should go with which block in each of the two systems to maximize the amplitude of oscillations in system 2.

Fig. 16-26
Question 13.

System 1   System 2

EXERCISES & PROBLEMS

SEC. 16-3   The Force Law for Simple Harmonic Motion

1E. An object undergoing simple harmonic motion takes 0.25 s to travel from one point of zero velocity to the next such point. The distance between those points is 36 cm. Calculate (a) the period, (b) the frequency, and (c) the amplitude of the motion.

2E. An oscillating block–spring system takes 0.75 s to begin repeating its motion. Find (a) the period, (b) the frequency in hertz, and (c) the angular frequency in radians per second.

3E. An oscillator consists of a block of mass 0.500 kg connected to a spring. When set into oscillation with amplitude 35.0 cm, the oscillator repeats its motion every 0.500 s. Find (a) the period, (b) the frequency, (c) the angular frequency, (d) the spring constant, (e) the maximum speed, and (f) the magnitude of the maximum force on the block from the spring. ssm

4E. What is the maximum acceleration of a platform that oscillates with an amplitude of 2.20 cm at a frequency of 6.60 Hz?

5E. A loudspeaker produces a musical sound by means of the oscillation of a diaphragm. If the amplitude of oscillation is limited to 1.0×10^{-3} mm, what frequencies will result in the magnitude of the diaphragm's acceleration exceeding g? ssm

6E. The scale of a spring balance that reads from 0 to 15.0 kg is 12.0 cm long. A package suspended from the balance is found to oscillate vertically with a frequency of 2.00 Hz. (a) What is the spring constant? (b) How much does the package weigh?

7E. A particle with a mass of 1.00×10^{-20} kg is oscillating with simple harmonic motion with a period of 1.00×10^{-5} s and a maximum speed of 1.00×10^3 m/s. Calculate (a) the angular frequency and (b) the maximum displacement of the particle. ssm

8E. A small body of mass 0.12 kg is undergoing simple harmonic motion of amplitude 8.5 cm and period 0.20 s. (a) What is the magnitude of the maximum force acting on it? (b) If the oscillations are produced by a spring, what is the spring constant?

9E. In an electric shaver, the blade moves back and forth over a distance of 2.0 mm in simple harmonic motion, with frequency 120 Hz. Find (a) the amplitude, (b) the maximum blade speed, and (c) the magnitude of the maximum blade acceleration. ssm

10E. A loudspeaker diaphragm is oscillating in simple harmonic motion with a frequency of 440 Hz and a maximum displacement of 0.75 mm. What are (a) the angular frequency, (b) the maximum speed, and (c) the magnitude of the maximum acceleration?

11E. An automobile can be considered to be mounted on four identical springs as far as vertical oscillations are concerned. The springs of a certain car are adjusted so that the oscillations have a frequency of 3.00 Hz. (a) What is the spring constant of each spring if the mass of the car is 1450 kg and the mass is evenly distributed over the springs? (b) What will be the oscillation frequency if five passengers, averaging 73.0 kg each, ride in the car? (Again, consider an even distribution of mass.)

12E. A body oscillates with simple harmonic motion according to the equation

$$x = (6.0 \text{ m}) \cos[(3\pi \text{ rad/s})t + \pi/3 \text{ rad}].$$

At $t = 2.0$ s, what are (a) the displacement, (b) the velocity, (c) the acceleration, and (d) the phase of the motion? Also, what are (e) the frequency and (f) the period of the motion?

13E. The piston in the cylinder head of a locomotive has a stroke (twice the amplitude) of 0.76 m. If the piston moves with simple harmonic motion with an angular frequency of 180 rev/min, what is its maximum speed? ssm

14P. Figure 16-27 shows an astronaut on a body-mass measuring device (BMMD). Designed for use on orbiting space vehicles, its purpose is to allow astronauts to measure their mass in the "weightless" conditions in Earth orbit. The BMMD is a spring-mounted chair; an astronaut measures his or her period of oscillation in the chair; the mass follows from the formula for the period of an oscillating block–spring system. (a) If M is the mass of the astronaut and m the effective mass of that part of the BMMD that also oscillates, show that

$$M = (k/4\pi^2)T^2 - m,$$

where T is the period of oscillation and k is the spring constant. (b) The spring constant was $k = 605.6$ N/m for the BMMD on

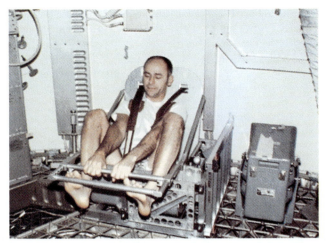

Fig. 16-27 Problem 14.

Skylab Mission Two; the period of oscillation of the empty chair was 0.90149 s. Calculate the effective mass of the chair. (c) With an astronaut in the chair, the period of oscillation became 2.08832 s. Calculate the mass of the astronaut.

15P. At a certain harbor, the tides cause the ocean surface to rise and fall a distance d (from highest level to lowest level) in simple harmonic motion, with a period of 12.5 h. How long does it take for the water to fall a distance $d/4$ from its highest level?

16P. In Fig. 16-28, two blocks ($m = 1.0$ kg and $M = 10$ kg) and a spring ($k = 200$ N/m) are arranged on a horizontal, frictionless surface. The coefficient of static friction between the two blocks is 0.40. What amplitude of simple harmonic motion of the spring–blocks system puts the smaller block on the verge of slipping over the larger block?

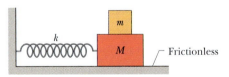

Fig. 16-28 Problem 16.

17P. A block is on a horizontal surface (a shake table) that is moving back and forth horizontally with simple harmonic motion of frequency 2.0 Hz. The coefficient of static friction between block and surface is 0.50. How great can the amplitude of the SHM be if the block is not to slip along the surface? ssm www

18P. A block rides on a piston that is moving vertically with simple harmonic motion. (a) If the SHM has period 1.0 s, at what amplitude of motion will the block and piston separate? (b) If the piston has an amplitude of 5.0 cm, what is the maximum frequency for which the block and piston will be in contact continuously?

19P. An oscillator consists of a block attached to a spring ($k = 400$ N/m). At some time t, the position (measured from the system's equilibrium location), velocity, and acceleration of the block are $x = 0.100$ m, $v = -13.6$ m/s, and $a = -123$ m/s². Calculate (a) the frequency of oscillation, (b) the mass of the block, and (c) the amplitude of the motion. ilw

20P. A simple harmonic oscillator consists of a block of mass 2.00 kg attached to a spring of spring constant 100 N/m. When $t = 1.00$ s, the position and velocity of the block are $x = 0.129$ m and $v = 3.415$ m/s. (a) What is the amplitude of the oscillations? What were the (b) position and (c) velocity of the block at $t = 0$ s?

21P. A massless spring hangs from the ceiling with a small object attached to its lower end. The object is initially held at rest in a position y_i such that the spring is at its rest length. The object is then released from y_i and oscillates up and down, with its lowest position being 10 cm below y_i. (a) What is the frequency of the oscillation? (b) What is the speed of the object when it is 8.0 cm below the initial position? (c) An object of mass 300 g is attached to the first object, after which the system oscillates with half the original frequency. What is the mass of the first object? (d) Relative to y_i, where is the new equilibrium (rest) position with both objects attached to the spring? ssm

22P. Two particles execute simple harmonic motion of the same amplitude and frequency along close parallel lines. They pass each other moving in opposite directions each time their displacement is half their amplitude. What is their phase difference?

23P. Two particles oscillate in simple harmonic motion along a common straight-line segment of length A. Each particle has a period of 1.5 s, but they differ in phase by $\pi/6$ rad. (a) How far apart are they (in terms of A) 0.50 s after the lagging particle leaves one end of the path? (b) Are they then moving in the same direction, toward each other, or away from each other? ssm

24P. In Fig. 16-29, two identical springs of spring constant k are attached to a block of mass m and to fixed supports. Show that the block's frequency of oscillation on the frictionless surface is

$$f = \frac{1}{2\pi}\sqrt{\frac{2k}{m}}.$$

25P. Suppose that the two springs in Fig. 16-29 have different spring constants k_1 and k_2. Show that the frequency f of oscillation of the block is then given by

$$f = \sqrt{f_1^2 + f_2^2},$$

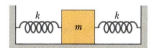

Fig. 16-29 Problems 24 and 25.

where f_1 and f_2 are the frequencies at which the block would oscillate if connected only to spring 1 or only to spring 2. ilw

26P. The end of one of the prongs of a tuning fork that executes simple harmonic motion of frequency 1000 Hz has an amplitude of 0.40 mm. Find (a) the magnitude of the maximum acceleration and (b) the maximum speed of the end of the prong. Find (c) the magnitude of the acceleration and (d) the speed of the end of the prong when the end has a displacement of 0.20 mm.

27P. In Fig. 16-30, two springs are joined and connected to a block of mass m. The surface is frictionless. If the springs both have spring constant k, show that

$$f = \frac{1}{2\pi}\sqrt{\frac{k}{2m}}$$

gives the block's frequency of oscillation. ssm

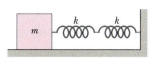

Fig. 16-30 Problem 27.

28P. In Fig. 16-31, a block weighing 14.0 N, which slides without friction on a 40.0° incline, is connected to the top of the incline by a massless spring of unstretched length 0.450 m and spring constant 120 N/m. (a) How far from the top of the incline does the block stop? (b) If the block is pulled slightly down the incline and released, what is the period of the resulting oscillations?

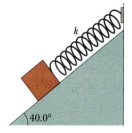

Fig. 16-31 Problem 28.

29P. A uniform spring with unstretched length L and spring constant k is cut into two pieces of unstretched lengths L_1 and L_2, with $L_1 = nL_2$. What are the corresponding spring constants (a) k_1 and (b) k_2 in terms of n and k? If a block is attached to the original spring, as in Fig. 16-5, it oscillates with frequency f. If the spring is replaced with the piece L_1 or L_2, the corresponding frequency is f_1 or f_2. Find (c) f_1 and (d) f_2 in terms of f. ssm www

30P. In Fig. 16-32, three 10000 kg ore cars are held at rest on a 30° incline on a mine railway using a cable that is parallel to the incline. The cable stretches 15 cm just before the coupling between the two lower cars breaks, detaching the lowest car. Assuming that the cable obeys Hooke's law, find (a) the frequency and (b) the amplitude of the resulting oscillations of the remaining two cars.

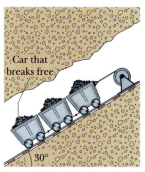

Fig. 16-32 Problem 30.

SEC. 16-4 Energy in Simple Harmonic Motion

31E. Find the mechanical energy of a block–spring system having a spring constant of 1.3 N/cm and an oscillation amplitude of 2.4 cm. ssm

32E. An oscillating block–spring system has a mechanical energy of 1.00 J, an amplitude of 10.0 cm, and a maximum speed of 1.20 m/s. Find (a) the spring constant, (b) the mass of the block and (c) the frequency of oscillation.

33E. A 5.00 kg object on a horizontal frictionless surface is attached to a spring with spring constant 1000 N/m. The object is displaced from equilibrium 50.0 cm horizontally and given an initial velocity of 10.0 m/s back toward the equilibrium position. (a) What is the frequency of the motion? What are (b) the initial potential energy of the block–spring system, (c) the initial kinetic energy, and (d) the amplitude of the oscillation? ilw

34E. A (hypothetical) large slingshot is stretched 1.50 m to launch a 130 g projectile with speed sufficient to escape from Earth (11.2 km/s). Assume the elastic bands of the slingshot obey Hooke's law. (a) What is the spring constant of the device, if all the elastic potential energy is converted to kinetic energy? (b) Assume that an average person can exert a force of 220 N. How many people are required to stretch the elastic bands?

35E. A vertical spring stretches 9.6 cm when a 1.3 kg block is hung from its end. (a) Calculate the spring constant. This block is then displaced an additional 5.0 cm downward and released from rest. Find (b) the period, (c) the frequency, (d) the amplitude, and (e) the maximum speed of the resulting SHM. ssm

36E. A block of mass M, at rest on a horizontal frictionless table, is attached to a rigid support by a spring of constant k. A bullet of mass m and velocity $\vec{v}$ strikes the block as shown in Fig. 16-33. The bullet is embedded in the

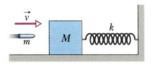

Fig. 16-33 Exercise 36.

block. Determine (a) the speed of the block immediately after the collision and (b) the amplitude of the resulting simple harmonic motion.

37E. When the displacement in SHM is one-half the amplitude x_m, what fraction of the total energy is (a) kinetic energy and (b) potential energy? (c) At what displacement, in terms of the amplitude, is the energy of the system half kinetic energy and half potential energy? ssm

38P. A 10 g particle is undergoing simple harmonic motion with an amplitude of 2.0×10^{-3} m and a maximum acceleration of magnitude 8.0×10^3 m/s². The phase constant is $-\pi/3$ rad. (a) Write an equation for the force on the particle as a function of time. (b) What is the period of the motion? (c) What is the maximum speed of the particle? (d) What is the total mechanical energy of this simple harmonic oscillator?

39P*. A 4.0 kg block is suspended from a spring with a spring constant of 500 N/m. A 50 g bullet is fired into the block from directly below with a speed of 150 m/s and becomes embedded in the block. (a) Find the amplitude of the resulting simple harmonic motion. (b) What fraction of the original kinetic energy of the bullet is transferred to mechanical energy of the harmonic oscillator? ssm www

SEC. 16-5 An Angular Simple Harmonic Oscillator

40E. A flat uniform circular disk has a mass of 3.00 kg and a radius of 70.0 cm. It is suspended in a horizontal plane by a vertical wire attached to its center. If the disk is rotated 2.50 rad about the wire, a torque of 0.0600 N·m is required to maintain that orientation. Calculate (a) the rotational inertia of the disk about the wire, (b) the torsion constant, and (c) the angular frequency of this torsion pendulum when it is set oscillating.

41P. The balance wheel of a watch oscillates with an angular amplitude of π rad and a period of 0.500 s. Find (a) the maximum angular speed of the wheel, (b) the angular speed of the wheel when its displacement is $\pi/2$ rad, and (c) the magnitude of the angular acceleration of the wheel when its displacement is $\pi/4$ rad. ssm

SEC. 16-6 Pendulums

42E. In Fig. 16-34, a 2500 kg demolition ball swings from the end of a crane. The length of the swinging segment of cable is 17 m. (a) Find the period of the swinging, assuming that the system can be treated as a simple pendulum. (b) Does the period depend on the ball's mass?

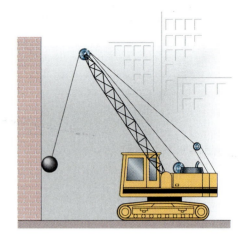

Fig. 16-34
Exercise 42.

43E. What is the length of a simple pendulum that marks seconds by completing a full swing from left to right and then back again every 2.0 s? ssm

44E. A performer seated on a trapeze is swinging back and forth with a period of 8.85 s. If she stands up, thus raising the center of mass of the *trapeze + performer* system by 35.0 cm, what will be the new period of the system? Treat *trapeze + performer* as a simple pendulum.

45E. A physical pendulum consists of a meter stick that is pivoted at a small hole drilled through the stick a distance d from the 50 cm mark. The period of oscillation is 2.5 s. Find d. ilw

46E. In Fig. 16-35, a physical pendulum consists of a uniform solid disk (of mass M and radius R) supported in a vertical plane by a pivot located a distance d from the center of the disk. The disk is displaced by a small angle and released. Find an expression for the period of the resulting simple harmonic motion.

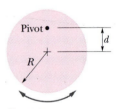

Fig. 16-35 Exercise 46.

47E. A pendulum is formed by pivoting a long thin rod of length L and mass m about a point on the rod that is a distance d above the center of the rod. (a) Find the period of this pendulum in terms of d, L, m, and g, assuming small-amplitude swinging. What happens to the period if (b) d is decreased, (c) L is increased, or (d) m is increased? ssm

48E. A uniform circular disk whose radius R is 12.5 cm is suspended as a physical pendulum from a point on its rim. (a) What is its period? (b) At what radial distance $r < R$ is there a pivot point that gives the same period?

49E. The pendulum in Fig. 16-36 consists of a uniform disk with

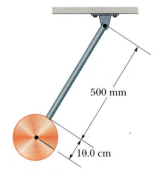

Fig. 16-36 Exercise 49.

radius 10.0 cm and mass 500 g attached to a uniform rod with length 500 mm and mass 270 g. (a) Calculate the rotational inertia of the pendulum about the pivot point. (b) What is the distance between the pivot point and the center of mass of the pendulum? (c) Calculate the period of oscillation. ssm

50E. (a) If the physical pendulum of Sample Problem 16-5 is inverted and suspended at point P, what is its period of oscillation? (b) Is the period now greater than, less than, or equal to its previous value?

51E. In Sample Problem 16-5, we saw that a physical pendulum has a center of oscillation at distance $2L/3$ from its point of suspension. Show that the distance between the point of suspension and the center of oscillation for a physical pendulum of any form is I/mh, where I and h have the meanings assigned to them in Eq. 16-29, and m is the mass of the pendulum.

52P. A stick with length L oscillates as a physical pendulum, pivoted about point O in Fig. 16-37. (a) Derive an expression for the period of the pendulum in terms of L and x, the distance from the pivot point to the center of mass of the pendulum. (b) For what value of x/L is the period a minimum? (c) Show that if $L = 1.00$ m and $g = 9.80$ m/s², this minimum is 1.53 s.

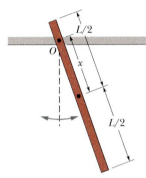

Fig. 16-37 Problem 52.

53P. In the overhead view of Fig. 16-38, a long uniform rod of length L and mass m is free to rotate in a horizontal plane about a vertical axis through its center. A spring with force constant k is connected horizontally between one end of the rod and a fixed wall. When the rod is in equilibrium, it is parallel to the wall. What is the period of the small oscillations that result when the rod is rotated slightly and released? ssm ilw www

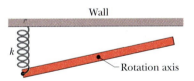

Fig. 16-38 Problem 53.

54P. A simple pendulum of length L and mass m is suspended in a car that is traveling with constant speed v around a circle of radius R. If the pendulum undergoes small oscillations in a radial direction about its equilibrium position, what will be its frequency of oscillation?

55P. What is the frequency of a simple pendulum 2.0 m long (a) in a room, (b) in an elevator accelerating upward at a rate of 2.0 m/s², and (c) in free fall? ssm

56P. For a simple pendulum, find the angular amplitude θ_m at which the restoring torque required for simple harmonic motion deviates from the actual restoring torque by 1.0%. (See "Trigonometric Expansions" in Appendix E.)

57P. The bob on a simple pendulum of length R moves in an arc of a circle. (a) By considering that the radial acceleration of the

bob as it moves through its equilibrium position is that for uniform circular motion (v^2/R), show that the tension in the string at that position is $mg(1 + \theta_m^2)$ if the angular amplitude θ_m is small. (See "Trigonometric Expansions" in Appendix E.) (b) Is the tension at other positions of the bob greater, smaller, or the same?

58P. A wheel is free to rotate about its fixed axle. A spring is attached to one of its spokes a distance r from the axle, as shown in Fig. 16-39. (a) Assuming that the wheel is a hoop of mass m and radius R, obtain the angular frequency of small oscillations of this system in terms of $m, R, r,$ and the spring constant k. How does the result change if (b) $r = R$ and (c) $r = 0$?

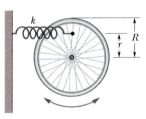

Fig. 16-39 Problem 58.

SEC. 16-8 Damped Simple Harmonic Motion

59E. In Sample Problem 16-7, what is the ratio of the amplitude of the damped oscillations to the initial amplitude when 20 full oscillations have elapsed? ssm

60E. The amplitude of a lightly damped oscillator decreases by 3.0% during each cycle. What fraction of the mechanical energy of the oscillator is lost in each full oscillation?

61E. For the system shown in Fig. 16-15, the block has a mass of 1.50 kg and the spring constant is 8.00 N/m. The damping force is given by $-b(dx/dt)$, where $b = 230$ g/s. Suppose that the block is initially pulled down a distance 12.0 cm and released. (a) Calculate the time required for the amplitude of the resulting oscillations to fall to one-third of its initial value. (b) How many oscillations are made by the block in this time? ssm

62P. Assume that you are examining the oscillation characteristics of the suspension system of a 2000 kg automobile. The suspension "sags" 10 cm when the entire automobile is placed on it. Also, the amplitude of oscillation decreases by 50% during one complete oscillation. Estimate the values of (a) the spring constant k and (b) the damping constant b for the spring and shock absorber system of one wheel, assuming each wheel supports 500 kg.

SEC. 16-9 Forced Oscillations and Resonance

63E. For Eq. 16-43, suppose the amplitude x_m is given by

$$x_m = \frac{F_m}{[m^2(\omega_d^2 - \omega^2)^2 + b^2\omega_d^2]^{1/2}},$$

where F_m is the (constant) amplitude of the external oscillating force exerted on the spring by the rigid support in Fig. 16-15. At resonance, what are (a) the amplitude and (b) the velocity amplitude of the oscillating object?

64P. A 1000 kg car carrying four 82 kg people travels over a rough "washboard" dirt road with corrugations 4.0 m apart, which cause the car to bounce on its spring suspension. The car bounces with maximum amplitude when its speed is 16 km/h. The car now stops, and the four people get out. By how much does the car body rise on its suspension owing to this decrease in mass?

17 Waves—I

When a beetle moves along the sand within a few tens of centimeters of this sand scorpion, the scorpion immediately turns toward the beetle and dashes to it (for lunch). The scorpion can do this without seeing (it is nocturnal) or hearing the beetle.

How can the scorpion so precisely locate its prey?

The answer is in this chapter.

17-1 Waves and Particles

Two ways to get in touch with a friend in a distant city are to write a letter and to use the telephone.

The first choice (the letter) involves the concept of "particle": A material object moves from one point to another, carrying with it information and energy. Most of the preceding chapters deal with particles or with systems of particles.

The second choice (the telephone) involves the concept of "wave," the subject of this chapter and the next. In a wave, information and energy move from one point to another but no material object makes that journey. In your telephone call, a sound wave carries your message from your vocal cords to the telephone. There, an electromagnetic wave takes over, passing along a copper wire or an optical fiber or through the atmosphere, possibly by way of a communications satellite. At the receiving end there is another sound wave, from a telephone to your friend's ear. Although the message is passed, nothing that you have touched reaches your friend. Leonardo da Vinci understood about waves when he wrote of water waves: "It often happens that the wave flees the place of its creation, while the water does not; like the waves made in a field of grain by the wind, where we see the waves running across the field while the grain remains in place."

Particle and *wave* are the two great concepts in classical physics, in the sense that we seem able to associate almost every branch of the subject with one or the other. The two concepts are quite different. The word *particle* suggests a tiny concentration of matter capable of transmitting energy. The word *wave* suggests just the opposite—namely, a broad distribution of energy, filling the space through which it passes. The job at hand is to put aside particles for a while and to learn something about waves.

17-2 Types of Waves

Waves are of three main types:

1. *Mechanical waves.* These waves are most familiar because we encounter them almost constantly; common examples include water waves, sound waves, and seismic waves. All these waves have certain central features: They are governed by Newton's laws, and they can exist only within a material medium, such as water, air, and rock.

2. *Electromagnetic waves.* These waves are less familiar, but you use them constantly; common examples include visible and ultraviolet light, radio and television waves, microwaves, x rays, and radar waves. These waves require no material medium to exist. Light waves from stars, for example, travel through the vacuum of space to reach us. All electromagnetic waves travel through a vacuum at the same speed c, given by

$$c = 299\ 792\ 458 \text{ m/s} \qquad \text{(speed of light).} \qquad (17\text{-}1)$$

3. *Matter waves.* Although these waves are commonly used in modern technology, their type is probably very unfamiliar to you. These waves are associated with electrons, protons, and other fundamental particles, and even atoms and molecules. Because we commonly think of these things as constituting matter, such waves are called matter waves.

Much of what we discuss in this chapter applies to waves of all kinds. However, for specific examples we shall refer to mechanical waves.

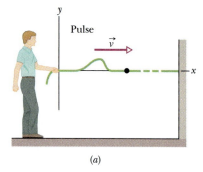

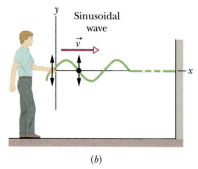

Fig. 17-1 (a) A single pulse is sent along a stretched string. A typical string element (marked with a dot) moves up once and then down as the pulse passes. The element's motion is perpendicular to the wave's direction of travel, so the pulse is a *transverse wave*. (b) A sinusoidal wave is sent along the string. A typical string element moves up and down continuously as the wave passes. This too is a transverse wave.

17-3 Transverse and Longitudinal Waves

A wave sent along a stretched, taut string is the simplest mechanical wave. If you give one end of a stretched string a single up-and-down jerk, a wave in the form of a single *pulse* travels along the string, as in Fig. 17-1a. This pulse and its motion can occur because the string is under tension. When you pull your end of the string upward, it begins to pull upward on the adjacent section of the string via tension between the two sections. As the adjacent section moves upward, it begins to pull the next section upward, and so on. Meanwhile, you have pulled down on your end of the string. As each section moves upward in turn, it begins to be pulled back downward by neighboring sections that are already on the way down. The net result is that a distortion in the string's shape (the pulse) moves along the string at some velocity $\vec{v}$.

If you move your hand up and down in continuous simple harmonic motion, a continuous wave travels along the string at velocity $\vec{v}$. Because the motion of your hand is a sinusoidal function of time, the wave has a sinusoidal shape at any given instant, as in Fig. 17-1b; that is, the wave has the shape of a sine curve or a cosine curve.

We consider here only an "ideal" string, in which no frictionlike forces within the string cause the wave to die out as it travels along the string. In addition, we assume that the string is so long that we need not consider a wave rebounding from the far end.

One way to study the waves of Fig. 17-1 is to monitor the **wave forms** (shapes of the waves) as they move to the right. Alternatively, we could monitor the motion of an element of the string as the element oscillates up and down while a wave passes through it. We would find that the displacement of every such oscillating string element is *perpendicular* to the direction of travel of the wave, as indicated in Fig. 17-1. This motion is said to be **transverse**, and the wave is said to be a **transverse wave**.

Figure 17-2 shows how a sound wave can be produced by a piston in a long, air-filled pipe. If you suddenly move the piston rightward and then leftward, you can send a pulse of sound along the pipe. The rightward motion of the piston moves the elements of air next to it rightward, changing the air pressure there. The increased air pressure then pushes rightward on the elements of air somewhat farther along the pipe. Moving the piston leftward reduces the air pressure next to it. Once they have moved rightward, the nearest elements, and then farther elements, move back leftward. Thus, the motion of the air and the change in air pressure travel rightward along the pipe as a pulse.

If you push and pull on the piston in simple harmonic motion, as is being done in Fig. 17-2, a sinusoidal wave travels along the pipe. Because the motion of the elements of air is parallel to the direction of the wave's travel, the motion is said to be **longitudinal**, and the wave is said to be a **longitudinal wave**. In this chapter we concentrate on transverse waves, and string waves in particular; in Chapter 18 we shall concentrate on longitudinal waves, and sound waves in particular.

Both a transverse wave and a longitudinal wave are said to be **traveling waves** because they both travel from one point to another, as from one end of the string to the other end in Fig. 17-1 or from one end of the pipe to the other end in Fig. 17-2. Note that it is the wave that moves from end to end, not the material (string or air) through which the wave moves.

The sand scorpion shown in the photograph opening this chapter uses waves of both transverse and longitudinal motion to locate its prey. When a beetle even slightly disturbs the sand, it sends pulses along the sand's surface (Fig. 17-3). One

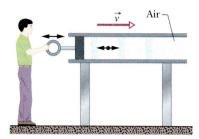

Fig. 17-2 A sound wave is set up in an air-filled pipe by moving a piston back and forth. Because the oscillations of an element of the air (represented by the black dot) are parallel to the direction in which the wave travels, the wave is a *longitudinal wave*.

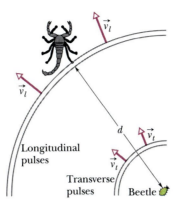

Fig. 17-3 A beetle's motion sends fast longitudinal pulses and slower transverse pulses along the sand's surface. The sand scorpion first intercepts the longitudinal pulses; here, it is the rearmost right leg that senses the pulses earliest.

set of pulses is longitudinal, traveling with speed $v_l = 150$ m/s. A second set is transverse, traveling with speed $v_t = 50$ m/s.

The scorpion, with its eight legs spread roughly in a circle about 5 cm in diameter, intercepts the faster longitudinal pulses first and learns the direction of the beetle; it is in the direction of whichever leg is disturbed earliest by the pulses. The scorpion then senses the time interval Δt between that first interception and the interception of the slower transverse waves and uses it to determine the distance d to the beetle. This distance is given by

$$\Delta t = \frac{d}{v_t} - \frac{d}{v_l},$$

and it turns out to be

$$d = (75 \text{ m/s}) \, \Delta t.$$

For example, if $\Delta t = 4.0$ ms, then $d = 30$ cm, which gives the scorpion a perfect fix on the beetle.

17-4 Wavelength and Frequency

To completely describe a wave on a string (and the motion of any element along its length), we need a function that gives the shape of the wave. This means that we need a relation in the form $y = h(x, t)$, in which y is the transverse displacement of any string element as a function h of the time t and the position x of the element along the string. In general, a sinusoidal shape like the wave in Fig. 17-1b can be described with h being either a sine function or a cosine function; both give the same general shape for the wave. In this chapter we use the sine function.

Imagine a sinusoidal wave like that of Fig. 17-1b traveling in the positive direction of an x axis. As the wave sweeps through succeeding elements (that is, very short sections) of the string, the elements oscillate parallel to the y axis. At time t, the displacement y of the element located at position x is given by

$$y(x, t) = y_m \sin(kx - \omega t). \tag{17-2}$$

Because this equation is written in terms of position x, it can be used to find the displacements of all the elements of the string as a function of time. Thus, it can tell us the shape of the wave at any given time and how that shape changes as the wave moves along the string. The names of the quantities in Eq. 17-2 are displayed in Fig. 17-4 and defined next.

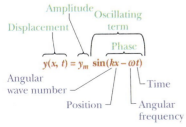

Fig. 17-4 The names of the quantities in Eq. 17-2, for a transverse sinusoidal wave.

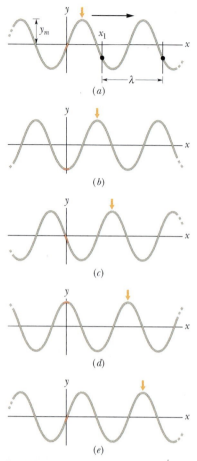

Fig. 17-5 Five "snapshots" of a string wave traveling in the positive direction of an x axis. The amplitude y_m is indicated. A typical wavelength λ, measured from an arbitrary position x_1, is also indicated.

Before we discuss them, however, let us examine Fig. 17-5, which shows five "snapshots" of a sinusoidal wave traveling in the positive direction of an x axis. The movement of the wave is indicated by the rightward progress of the short arrow pointing to a high point of the wave. From snapshot to snapshot, the short arrow moves to the right with the wave shape, but the string moves *only* parallel to the y axis. To see that, let us follow the motion of the red-dyed string element at $x = 0$. In the first snapshot (Fig. 17-5a), it is at displacement $y = 0$. In the next snapshot, it is at its extreme downward displacement because a *valley* (or extreme low point) of the wave is passing through it. It then moves back up through $y = 0$. In the fourth snapshot, it is at its extreme upward displacement because a *peak* (or extreme high point) of the wave is passing through it. In the fifth snapshot, it is again at $y = 0$, having completed one full oscillation.

Amplitude and Phase

The **amplitude** y_m of a wave, such as that in Fig. 17-5, is the magnitude of the maximum displacement of the elements from their equilibrium positions as the wave passes through them. (The subscript m stands for maximum.) Because y_m is a magnitude, it is always a positive quantity, even if it is measured downward in Fig. 17-5a, instead of upward as drawn.

The **phase** of the wave is the *argument* $kx - \omega t$ of the sine in Eq. 17-2. As the wave sweeps through a string element at a particular position x, the phase changes linearly with time t. This means that the sine also changes, oscillating between $+1$ and -1. Its extreme positive value ($+1$) corresponds to a peak of the wave moving through the element; then, the value of y at position x is y_m. Its extreme negative value (-1) corresponds to a valley of the wave moving through the element; then, the value of y at position x is $-y_m$. Thus, the sine function and the time-dependent phase of a wave correspond to the oscillation of a string element, and the amplitude of the wave determines the extremes of the element's displacement.

Wavelength and Angular Wave Number

The **wavelength** λ of a wave is the distance (parallel to the direction of the wave's travel) between repetitions of the shape of the wave (or *wave shape*). A typical wavelength is marked in Fig. 17-5a, which is a snapshot of the wave at time $t = 0$. At that time, Eq. 17-2 gives, for the description of the wave shape,

$$y(x, 0) = y_m \sin kx. \tag{17-3}$$

By definition, the displacement y is the same at both ends of this wavelength—that is, at $x = x_1$ and $x = x_1 + \lambda$. Thus, by Eq. 17-3,

$$y_m \sin kx_1 = y_m \sin k(x_1 + \lambda)$$
$$= y_m \sin(kx_1 + k\lambda). \tag{17-4}$$

A sine function begins to repeat itself when its angle (or argument) is increased by 2π rad, so in Eq. 17-4 we must have $k\lambda = 2\pi$, or

$$k = \frac{2\pi}{\lambda} \quad \text{(angular wave number).} \tag{17-5}$$

We call k the **angular wave number** of the wave; its SI unit is the radian per meter, or the inverse meter. (Note that the symbol k here does *not* represent a spring constant as previously.)

Notice that the wave in Fig. 17-5 moves to the right by $\frac{1}{4}\lambda$ from one snapshot to the next. Thus, by the fifth snapshot, it has moved to the right by 1λ.

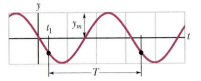

Fig. 17-6 A graph of the displacement of the string element at $x = 0$ as a function of time, as the sinusoidal wave of Fig. 17-5 passes through it. The amplitude y_m is indicated. A typical period T, measured from an arbitrary time t_1, is also indicated.

Period, Angular Frequency, and Frequency

Figure 17-6 shows a graph of the displacement y of Eq. 17-2 versus time t at a certain position along the string, taken to be $x = 0$. If you were to monitor the string, you would see that the single element of the string at that position moves up and down in simple harmonic motion given by Eq. 17-2 with $x = 0$:

$$y(0, t) = y_m \sin(-\omega t)$$
$$= -y_m \sin \omega t \qquad (x = 0). \qquad (17\text{-}6)$$

Here we have made use of the fact that $\sin(-\alpha) = -\sin \alpha$, where α is any angle. Figure 17-6 is a graph of this equation; it *does not* show the shape of the wave.

We define the **period** of oscillation T of a wave to be the time any string element takes to move through one full oscillation. A typical period is marked on the graph of Fig. 17-6. Applying Eq. 17-6 to both ends of this time interval and equating the results yield

$$-y_m \sin \omega t_1 = -y_m \sin \omega(t_1 + T)$$
$$= -y_m \sin(\omega t_1 + \omega T). \qquad (17\text{-}7)$$

This can be true only if $\omega T = 2\pi$, or if

$$\omega = \frac{2\pi}{T} \qquad \text{(angular frequency)}. \qquad (17\text{-}8)$$

We call ω the **angular frequency** of the wave; its SI unit is the radian per second.

Look back at the five snapshots of a traveling wave in Fig. 17-5. The time between snapshots is $\frac{1}{4}T$. Thus, by the fifth snapshot, every string element has made one full oscillation.

The **frequency** f of a wave is defined as $1/T$ and is related to the angular frequency ω by

$$f = \frac{1}{T} = \frac{\omega}{2\pi} \qquad \text{(frequency)}. \qquad (17\text{-}9)$$

Like the frequency of simple harmonic motion in Chapter 16, this frequency f is a number of oscillations per unit time—here, the number made by a string element as the wave moves through it. As in Chapter 16, f is usually measured in hertz or its multiples, such as kilohertz.

> ✔**CHECKPOINT 1:** The figure is a composite of three snapshots, each of a wave traveling along a particular string. The phases for the waves are given by (a) $2x - 4t$, (b) $4x - 8t$, and (c) $8x - 16t$. Which phase corresponds to which wave in the figure?

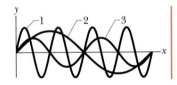

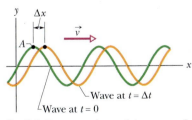

Fig. 17-7 Two snapshots of the wave of Fig. 17-5, at time $t = 0$ and then at time $t = \Delta t$. As the wave moves to the right at velocity $\vec{v}$, the entire curve shifts a distance Δx during Δt. Point A "rides" with the wave form but the string elements move only up and down.

17-5 The Speed of a Traveling Wave

Figure 17-7 shows two snapshots of the wave of Eq. 17-2, taken a small time interval Δt apart. The wave is traveling in the positive direction of x (to the right in Fig. 17-7), the entire wave pattern moving a distance Δx in that direction during the interval Δt. The ratio $\Delta x/\Delta t$ (or, in the differential limit, dx/dt) is the **wave speed** v. How can we find its value?

As the wave in Fig. 17-7 moves, each point of the moving wave form, such as point A marked on a peak, retains its displacement y. (Points on the string do not retain their displacement, but points on the wave *form* do.) If point A retains its displacement as it moves, the phase in Eq. 17-2 giving it that displacement must remain a constant:

$$kx - \omega t = \text{a constant.} \tag{17-10}$$

Note that although this argument is constant, both x and t are changing. In fact, as t increases, x must also, to keep the argument constant. This confirms that the wave pattern is moving in the positive direction of x.

To find the wave speed v, we take the derivative of Eq. 17-10, getting

$$k\frac{dx}{dt} - \omega = 0$$

or

$$\frac{dx}{dt} = v = \frac{\omega}{k}. \tag{17-11}$$

Using Eq. 17-5 ($k = 2\pi/\lambda$) and Eq. 17-8 ($\omega = 2\pi/T$), we can rewrite the wave speed as

$$v = \frac{\omega}{k} = \frac{\lambda}{T} = \lambda f \qquad \text{(wave speed).} \tag{17-12}$$

The equation $v = \lambda/T$ tells us that the wave speed is one wavelength per period; the wave moves a distance of one wavelength in one period of oscillation.

Equation 17-2 describes a wave moving in the positive direction of x. We can find the equation of a wave traveling in the opposite direction by replacing t in Eq. 17-2 with $-t$. This corresponds to the condition

$$kx + \omega t = \text{a constant,} \tag{17-13}$$

which (compare Eq. 17-10) requires that x *decrease* with time. Thus, a wave traveling in the negative direction of x is described by the equation

$$y(x, t) = y_m \sin(kx + \omega t). \tag{17-14}$$

If you analyze the wave of Eq. 17-14 as we have just done for the wave of Eq. 17-2, you will find for its velocity

$$\frac{dx}{dt} = -\frac{\omega}{k}. \tag{17-15}$$

The minus sign (compare Eq. 17-11) verifies that the wave is indeed moving in the negative direction of x and justifies our switching the sign of the time variable.

Consider now a wave of arbitrary shape, given by

$$y(x, t) = h(kx \pm \omega t), \tag{17-16}$$

where h represents *any* function, the sine function being one possibility. Our previous analysis shows that all waves in which the variables x and t enter in the combination $kx \pm \omega t$ are traveling waves. Furthermore, all traveling waves *must* be of the form of Eq. 17-16. Thus, $y(x, t) = \sqrt{ax + bt}$ represents a possible (though perhaps physically a little bizarre) traveling wave. The function $y(x, t) = \sin(ax^2 - bt)$, on the other hand, does *not* represent a traveling wave.

Sample Problem 17-1

A wave traveling along a string is described by

$$y(x, t) = 0.00327 \sin(72.1x - 2.72t), \tag{17-17}$$

in which the numerical constants are in SI units (0.00327 m, 72.1 rad/m, and 2.72 rad/s).

(a) What is the amplitude of this wave?

SOLUTION: The Key Idea is that Eq. 17-17 is of the same form as Eq. 17-2,

$$y = y_m \sin(kx - \omega t), \tag{17-18}$$

so we have a sinusoidal wave. By comparing the two equations, we see that the amplitude is

$$y_m = 0.00327 \text{ m} = 3.27 \text{ mm}. \qquad \text{(Answer)}$$

(b) What are the wavelength, period, and frequency of this wave?

SOLUTION: By comparing Eqs. 17-17 and 17-18, we see that the angular wave number and angular frequency are

$$k = 72.1 \text{ rad/m} \quad \text{and} \quad \omega = 2.72 \text{ rad/s}.$$

We then relate wavelength λ to k via Eq. 17-5:

$$\lambda = \frac{2\pi}{k} = \frac{2\pi \text{ rad}}{72.1 \text{ rad/m}}$$
$$= 0.0871 \text{ m} = 8.71 \text{ cm}. \qquad \text{(Answer)}$$

Next, we relate T to ω with Eq. 17-8:

$$T = \frac{2\pi}{\omega} = \frac{2\pi \text{ rad}}{2.72 \text{ rad/s}} = 2.31 \text{ s}, \qquad \text{(Answer)}$$

and from Eq. 17-9 we have

$$f = \frac{1}{T} = \frac{1}{2.31 \text{ s}} = 0.433 \text{ Hz}. \qquad \text{(Answer)}$$

(c) What is the velocity of this wave?

SOLUTION: The speed of the wave is given by Eq. 17-12:

$$v = \frac{\omega}{k} = \frac{2.72 \text{ rad/s}}{72.1 \text{ rad/m}} = 0.0377 \text{ m/s}$$
$$= 3.77 \text{ cm/s}. \qquad \text{(Answer)}$$

Because the phase in Eq. 17-17 contains the position variable x, the wave is moving along the x axis. Also, because the wave equation is written in the form of Eq. 17-2, the *minus* sign in front of the ωt term indicates that the wave is moving in the *positive* direction of the x axis. (Note that the quantities calculated in (b) and (c) are independent of the amplitude of the wave.)

(d) What is the displacement y at $x = 22.5$ cm and $t = 18.9$ s?

SOLUTION: The Key Idea here is that Eq. 17-17 gives the displacement as a function of position x and time t. Substituting the given values into the equation yields

$$y = 0.00327 \sin(72.1 \times 0.225 - 2.72 \times 18.9)$$
$$= (0.00327 \text{ m}) \sin(-35.1855 \text{ rad})$$
$$= (0.00327 \text{ m})(0.588)$$
$$= 0.00192 \text{ m} = 1.92 \text{ mm}. \qquad \text{(Answer)}$$

Thus, the displacement is positive. (Be sure to change your calculator mode to radians before evaluating the sine.)

Sample Problem 17-2

In Sample Problem 17-1d, we showed that at $t = 18.9$ s the transverse displacement y of the element of the string at $x = 0.255$ m due to the wave of Eq. 17-17 is 1.92 mm.

(a) What is u, the transverse velocity of the same element of the string, at that time? (This speed, which is associated with the transverse oscillation of an element of the string, is in the y direction. Do not confuse it with v, the constant velocity at which the *wave form* travels along the x axis.)

SOLUTION: The Key Idea here is that the transverse velocity u is the rate at which the displacement y of the element is changing. In general, that displacement is given by

$$y(x, t) = y_m \sin(kx - \omega t). \tag{17-19}$$

For an element at a certain location x, we find the rate of change of y by taking the derivative of Eq. 17-19 with respect to t while treating x as a constant. A derivative taken while one (or more) of the variables is treated as a constant is called a *partial derivative* and is represented by the symbol $\partial/\partial x$ rather than d/dx. Here we have

$$u = \frac{\partial y}{\partial t} = -\omega y_m \cos(kx - \omega t). \tag{17-20}$$

Next, substituting numerical values from Sample Problem 17-1, we obtain

$$u = (-2.72 \text{ rad/s})(3.27 \text{ mm}) \cos(-35.1855 \text{ rad})$$
$$= 7.20 \text{ mm/s}. \qquad \text{(Answer)}$$

Thus, at $t = 18.9$ s, the element of string at $x = 22.5$ cm is moving in the positive direction of y, with a velocity of 7.20 mm/s.

(b) What is the transverse acceleration a_y of the same element at that time?

SOLUTION: The Key Idea here is that the transverse acceleration a_y is the rate at which the transverse velocity of the element is changing. From Eq. 17-20, again treating x as a constant but allowing t to vary, we find

$$a_y = \frac{\partial u}{\partial t} = -\omega^2 y_m \sin(kx - \omega t).$$

Comparison with Eq. 17-19 shows that we can write this as

$$a_y = -\omega^2 y.$$

We see that the transverse acceleration of an oscillating string element is proportional to its transverse displacement but opposite in sign. This is completely consistent with the action of the element itself—namely, that it is moving transversely in simple harmonic motion. Substituting numerical values yields

$$a_y = -(2.72 \text{ rad/s})^2(1.92 \text{ mm})$$
$$= -14.2 \text{ mm/s}^2. \qquad \text{(Answer)}$$

Thus, at $t = 18.9$ s, the element of string at $x = 22.5$ cm is displaced from its equilibrium position by 1.92 mm in the positive y direction and has an acceleration of magnitude 14.2 mm/s^2 in the negative y direction.

✔**CHECKPOINT 2:**   Here are the equations of three waves:
(1) $y(x, t) = 2 \sin(4x - 2t)$,   (2) $y(x, t) = \sin(3x - 4t)$,   (3) $y(x, t) = 2 \sin(3x - 3t)$.
Rank the waves according to their (a) wave speed and (b) maximum transverse speed, greatest first.

PROBLEM-SOLVING TACTICS

Tactic 1: *Evaluating Large Phases*

Sometimes, as in Sample Problems 17-1d and 17-2, an angle much greater than 2π rad (or 360°) crops up and you are asked to find its sine or cosine. Adding or subtracting an integral multiple of 2π rad to such an angle does not change the value of any of its trigonometric functions. In Sample Problem 17-1d, for example, the angle is -35.1855 rad. Adding $(6)(2\pi \text{ rad})$ to this angle yields

$$-35.1855 \text{ rad} + (6)(2\pi \text{ rad}) = 2.51361 \text{ rad},$$

an angle of less than 2π rad that has the same trigonometric functions as -35.1855 rad (Fig. 17-8). As an example, the sine of both 2.51361 rad and -35.1855 rad is 0.588.

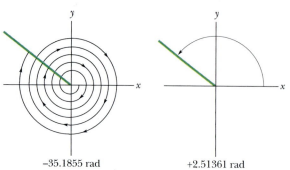

−35.1855 rad   +2.51361 rad

Fig. 17-8 These two angles are different but all their trigonometric functions are identical.

Your calculator will reduce such large angles for you automatically. *Caution:* Do not round off large angles if you intend to take their sines or cosines. In taking the sine of a very large angle, you are throwing away most of the angle and taking the sine of what is left over. If, for example, you were to round -35.1855 rad to -35 rad (a change of 0.5% and normally a reasonable step), you would be changing the sine of the angle by 27%. Also, if you change a large angle from degrees to radians, be sure to use an exact conversion factor (such as 180° $= \pi$ rad) rather than an approximate one (such as 57.3° ≈ 1 rad).

17-6 Wave Speed on a Stretched String

The speed of a wave is related to the wave's wavelength and frequency by Eq. 17-12, but *it is set by the properties of the medium*. If a wave is to travel through a medium such as water, air, steel, or a stretched string, it must cause the particles of that medium to oscillate as it passes. For that to happen, the medium must possess both mass (so that there can be kinetic energy) and elasticity (so that there can be potential energy). Thus, the medium's mass and elasticity properties determine how fast the wave can travel in the medium. Conversely, it should be possible to calculate the speed of the wave through the medium in terms of these properties. We do so now for a stretched string, in two ways.

Dimensional Analysis

In dimensional analysis we carefully examine the dimensions of all the physical quantities that enter into a given situation to determine the quantities they produce. In this case, we examine mass and elasticity to find a speed v, which has the dimension of length divided by time, or LT^{-1}.

For the mass, we use the mass of a string element, which is represented by the mass m of the string divided by the length l of the string. We call this ratio the *linear density* μ of the string. Thus, $\mu = m/l$, its dimension being mass divided by length, ML^{-1}.

You cannot send a wave along a string unless the string is under tension, which means that it has been stretched and pulled taut by forces at its two ends. The tension τ in the string is equal to the common magnitude of those two forces. As a wave travels along the string, it displaces elements of the string by causing additional stretching, with adjacent sections of string pulling on each other because of the tension. Thus, we can associate the tension in the string with the stretching (elasticity) of the string. The tension and the stretching forces it produces have the dimension of a force—namely, MLT^{-2} (from $F = ma$).

The goal here is to combine μ (dimension ML^{-1}) and τ (dimension MLT^{-2}) in such a way as to generate v (dimension LT^{-1}). A little juggling of various combinations suggests

$$v = C\sqrt{\frac{\tau}{\mu}}, \qquad (17\text{-}21)$$

in which C is a dimensionless constant that cannot be determined with dimensional analysis. In our second approach to determining wave speed, you will see that Eq. 17-21 is indeed correct and that $C = 1$.

Derivation from Newton's Second Law

Instead of the sinusoidal wave of Fig. 17-1b, let us consider a single symmetrical pulse such as that of Fig. 17-9, moving from left to right along a string with speed v. For convenience, we choose a reference frame in which the pulse remains stationary; that is, we run along with the pulse, keeping it constantly in view. In this frame, the string appears to move past us, from right to left in Fig. 17-9, with speed v.

Consider a small string element within the pulse, of length Δl, forming an arc of a circle of radius R and subtending an angle 2θ at the center of that circle. A force $\vec{\tau}$ with a magnitude equal to the tension in the string pulls tangentially on this element at each end. The horizontal components of these forces cancel, but the vertical components add to form a radial restoring force $\vec{F}$. In magnitude,

$$F = 2(\tau \sin \theta) \approx \tau (2\theta) = \tau \frac{\Delta l}{R} \qquad \text{(force)}, \qquad (17\text{-}22)$$

where we have approximated $\sin \theta$ as θ for the small angles θ in Fig. 17-9. From that figure, we have also used $2\theta = \Delta l/R$.

The mass of the element is given by

$$\Delta m = \mu \, \Delta l \qquad \text{(mass)}, \qquad (17\text{-}23)$$

where μ is the string's linear density.

At the moment shown in Fig. 17-9, the string element Δl is moving in an arc of a circle. Thus, it has a centripetal acceleration toward the center of that circle, given by

$$a = \frac{v^2}{R} \qquad \text{(acceleration)}. \qquad (17\text{-}24)$$

Equations 17-22, 17-23, and 17-24 contain the elements of Newton's second law. Combining them in the form

$$\text{force} = \text{mass} \times \text{acceleration}$$

gives

$$\frac{\tau \, \Delta l}{R} = (\mu \, \Delta l) \frac{v^2}{R}.$$

Solving this equation for the speed v yields

$$v = \sqrt{\frac{\tau}{\mu}} \qquad \text{(speed)}. \qquad (17\text{-}25)$$

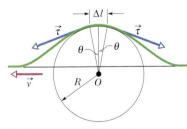

Fig. 17-9 A symmetrical pulse, viewed from a reference frame in which the pulse is stationary and the string appears to move right to left with speed v. We find speed v by applying Newton's second law to a string element of length Δl, located at the top of the pulse.

in exact agreement with Eq. 17-21 if the constant C in that equation is given the value unity. Equation 17-25 gives the speed of the pulse in Fig. 17-9 and the speed of *any* other wave on the same string under the same tension.

Equation 17-25 tells us:

> The speed of a wave along a stretched ideal string depends only on the tension and linear density of the string and not on the frequency of the wave.

The *frequency* of the wave is fixed entirely by whatever generates the wave (for example, the person in Fig. 17-1*b*). The *wavelength* of the wave is then fixed by Eq. 17-12 in the form $\lambda = v/f$.

✓**CHECKPOINT 3:**   You send a traveling wave along a particular string by oscillating one end. If you increase the frequency of the oscillations, do (a) the speed of the wave and (b) the wavelength of the wave increase, decrease, or remain the same? If, instead, you increase the tension in the string, do (c) the speed of the wave and (d) the wavelength of the wave increase, decrease, or remain the same?

Sample Problem 17-3

In Fig. 17-10, two strings have been tied together with a knot and then stretched between two rigid supports. The strings have linear densities $\mu_1 = 1.4 \times 10^{-4}$ kg/m and $\mu_2 = 2.8 \times 10^{-4}$ kg/m. Their lengths are $L_1 = 3.0$ m and $L_2 = 2.0$ m, and string 1 is under a tension of 400 N. Simultaneously, on each string a pulse is sent from the rigid support end, toward the knot. Which pulse reaches the knot first?

SOLUTION: We need several Key Ideas here:

1. The time t taken by a pulse to travel a length L is $t = L/v$, where v is the constant speed of the pulse.

2. The speed of a pulse on a stretched string depends on the string's

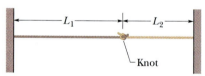

Fig. 17-10 Sample Problem 17-3. Two strings, of lengths L_1 and L_2, tied together with a knot and stretched between two rigid supports.

tension τ and linear density μ, and is given by Eq. 17-25 $(v = \sqrt{\tau/\mu})$.

3. Because the two strings are stretched together, they must both be under the same tension τ (= 400 N).

Putting these three ideas together gives us, as the time for the pulse on string 1 to reach the knot,

$$t_1 = \frac{L_1}{v_1} = L_1\sqrt{\frac{\mu_1}{\tau}} = (3.0 \text{ m})\sqrt{\frac{1.4 \times 10^{-4} \text{ kg/m}}{400 \text{ N}}}$$
$$= 1.77 \times 10^{-3} \text{ s}.$$

Similarly, the data for the pulse on string 2 give us

$$t_2 = L_2\sqrt{\frac{\mu_2}{\tau}} = 1.67 \times 10^{-3} \text{ s}.$$

Thus, the pulse on string 2 reaches the knot first.

Now look back at Key Idea 2. The linear density of string 2 is greater than that of string 1, so the pulse on string 2 must be slower than that on string 1. Could we have guessed the answer from that fact alone? No, because from the first Key Idea we see that the distance traveled by a pulse also matters.

17-7 Energy and Power of a Traveling String Wave

When we set up a wave on a stretched string, we provide energy for the motion of the string. As the wave moves away from us, it transports that energy as both kinetic energy and elastic potential energy. Let us consider each form in turn.

Kinetic Energy

An element of the string of mass dm, oscillating transversely in simple harmonic motion as the wave passes through it, has kinetic energy associated with its transverse

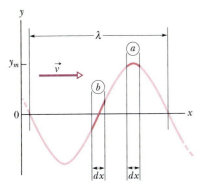

Fig. 17-11 A snapshot of a traveling wave on a string at time $t = 0$. String element a is at displacement $y = y_m$, and string element b is at displacement $y = 0$. The kinetic energy of the string element at each position depends on the transverse velocity of the element. The potential energy depends on the amount by which the string element is stretched as the wave passes through it.

velocity $\vec{u}$. When the element is rushing through its $y = 0$ position (element b in Fig. 17-11), its transverse velocity—and thus its kinetic energy—is a maximum. When the element is at its extreme position $y = y_m$ (as is element a), its transverse velocity—and thus its kinetic energy—is zero.

Elastic Potential Energy

To send a sinusoidal wave along a previously straight string, the wave must necessarily stretch the string. As a string element of length dx oscillates transversely, its length must increase and decrease in a periodic way if the string element is to fit the sinusoidal wave form. Elastic potential energy is associated with these length changes, just as for a spring.

When the string element is at its $y = y_m$ position (element a in Fig. 17-11), its length has its normal undisturbed value dx, so its elastic potential energy is zero. However, when the element is rushing through its $y = 0$ position, it is stretched to its maximum extent, and its elastic potential energy then is a maximum.

Energy Transport

The oscillating string element thus has both its maximum kinetic energy and its maximum elastic potential energy at $y = 0$. In the snapshot of Fig. 17-11, the regions of the string at maximum displacement have no energy, and the regions at zero displacement have maximum energy. As the wave travels along the string, forces due to the tension in the string continuously do work to transfer energy from regions with energy to regions with no energy.

Suppose we set up a wave on a string stretched along a horizontal x axis so that Eq. 17-2 describes the string's displacement. We might send a wave along the string by continuously oscillating one end of the string, as in Fig. 17-1b. In doing so, we continuously provide energy for the motion and stretching of the string—as the string sections oscillate perpendicularly to the x axis, they have kinetic energy and elastic potential energy. As the wave moves into sections that were previously at rest, energy is transferred into those new sections. Thus, we say that the wave *transports* the energy along the string.

The Rate of Energy Transmission

The kinetic energy dK associated with a string element of mass dm is given by

$$dK = \tfrac{1}{2}\, dm\, u^2,\tag{17-26}$$

where u is the transverse speed of the oscillating string element. To find u, we differentiate Eq. 17-2 with respect to time while holding x constant:

$$u = \frac{\partial y}{\partial t} = -\omega y_m \cos(kx - \omega t).\tag{17-27}$$

Using this relation and putting $dm = \mu\, dx$, we rewrite Eq. 17-26 as

$$dK = \tfrac{1}{2}(\mu\, dx)(-\omega y_m)^2 \cos^2(kx - \omega t).\tag{17-28}$$

Dividing Eq. 17-28 by dt gives the rate at which the kinetic energy of a string element changes, and thus the rate at which kinetic energy is carried along by the wave. The ratio dx/dt that then appears on the right of Eq. 17-28 is the wave speed v, so we obtain

$$\frac{dK}{dt} = \tfrac{1}{2}\mu v\omega^2 y_m^2 \cos^2(kx - \omega t).\tag{17-29}$$

The *average* rate at which kinetic energy is transported is

$$\left(\frac{dK}{dt}\right)_{avg} = \tfrac{1}{2}\mu v\omega^2 y_m^2 [\cos^2(kx - \omega t)]_{avg}$$

$$= \tfrac{1}{4}\mu v\omega^2 y_m^2. \tag{17-30}$$

Here we have taken the average over an integer number of wavelengths and have used the fact that the average value of the square of a cosine function over an integer number of periods is $\tfrac{1}{2}$.

Elastic potential energy is also carried along with the wave, and at the same average rate given by Eq. 17-30. Although we shall not examine the proof, you should recall that, in an oscillating system such as a pendulum or a spring–block system, the average kinetic energy and the average potential energy are indeed equal.

The **average power,** which is the average rate at which energy of both kinds is transmitted by the wave, is then

$$P_{avg} = 2\left(\frac{dK}{dt}\right)_{avg} \tag{17-31}$$

or, from Eq. 17-30,

$$P_{avg} = \tfrac{1}{2}\mu v\omega^2 y_m^2 \qquad \text{(average power).} \tag{17-32}$$

The factors μ and v in this equation depend on the material and tension of the string. The factors ω and y_m depend on the process that generates the wave. The dependence of the average power of a wave on the square of its amplitude and also on the square of its angular frequency is a general result, true for waves of all types.

Sample Problem 17-4

A stretched string has linear density $\mu = 525$ g/m and is under tension $\tau = 45$ N. We send a sinusoidal wave with frequency $f = 120$ Hz and amplitude $y_m = 8.5$ mm along the string from one end. At what average rate does the wave transport energy?

SOLUTION: The **Key Idea** here is that the average rate of energy transport is the average power P_{avg} as given by Eq. 17-32. To use that equation, however, we first must calculate the angular frequency ω and the wave speed v. From Eq. 17-9,

$$\omega = 2\pi f = (2\pi)(120 \text{ Hz}) = 754 \text{ rad/s.}$$

From Eq. 17-25 we have

$$v = \sqrt{\frac{\tau}{\mu}} = \sqrt{\frac{45 \text{ N}}{0.525 \text{ kg/m}}} = 9.26 \text{ m/s.}$$

Equation 17-32 then yields

$$P_{avg} = \tfrac{1}{2}\mu v\omega^2 y_m^2$$

$$= (\tfrac{1}{2})(0.525 \text{ kg/m})(9.26 \text{ m/s})(754 \text{ rad/s})^2(0.0085 \text{ m})^2$$

$$\approx 100 \text{ W.} \qquad \text{(Answer)}$$

✓**CHECKPOINT 4:** For the string and wave of this sample problem, we can adjust three parameters: the tension in the string, the frequency of the wave, and the amplitude of the wave. Does the average rate at which the wave transports energy along the string increase, decrease, or remain the same if we increase (a) the tension, (b) the frequency, and (c) the amplitude?

17-8 The Principle of Superposition for Waves

It often happens that two or more waves pass simultaneously through the same region. When we listen to a concert, for example, sound waves from many instruments fall simultaneously on our eardrums. The electrons in the antennas of our radio and television receivers are set in motion by the net effect of many electromagnetic waves from many different broadcasting centers. The water of a lake or harbor may be churned up by waves in the wakes of many boats.

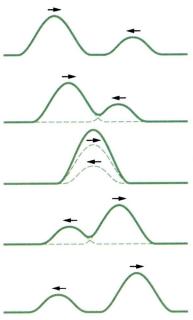

Fig. 17-12 A series of snapshots that show two pulses traveling in opposite directions along a stretched string. The superposition principle applies as the pulses move through each other.

Suppose that two waves travel simultaneously along the same stretched string. Let $y_1(x, t)$ and $y_2(x, t)$ be the displacements that the string would experience if each wave traveled alone. The displacement of the string when the waves overlap is then the algebraic sum

$$y'(x, t) = y_1(x, t) + y_2(x, t). \qquad (17\text{-}33)$$

This summation of displacements along the string means that

► Overlapping waves algebraically add to produce a **resultant wave** (or **net wave**).

This is another example of the **principle of superposition,** which says that when several effects occur simultaneously, their net effect is the sum of the individual effects.

Figure 17-12 shows a sequence of snapshots of two pulses traveling in opposite directions on the same stretched string. When the pulses overlap, the resultant pulse is their sum. Moreover, each pulse moves through the other, as if the other were not present:

► Overlapping waves do not in any way alter the travel of each other.

17-9 Interference of Waves

Suppose we send two sinusoidal waves of the same wavelength and amplitude in the same direction along a stretched string. The superposition principle applies. What resultant wave does it predict for the string?

The resultant wave depends on the extent to which the waves are *in phase* (in step) with respect to each other—that is, how much one wave form is shifted from the other wave form. If the waves are exactly in phase (so that the peaks and valleys of one are exactly aligned with those of the other), they combine to double the displacement of either wave acting alone. If they are exactly out of phase (the peaks of one are exactly aligned with the valleys of the other), they combine to cancel everywhere, and the string remains straight. We call this phenomenon of combining waves **interference,** and the waves are said to **interfere.** (These terms refer only to the displacements of the waves; the travel of the waves is unaffected.)

Let one wave traveling along a stretched string be given by

$$y_1(x, t) = y_m \sin(kx - \omega t) \qquad (17\text{-}34)$$

and another, shifted from the first, by

$$y_2(x, t) = y_m \sin(kx - \omega t + \phi). \qquad (17\text{-}35)$$

These waves have the same angular frequency ω (and thus the same frequency f), the same angular wave number k (and thus the same wavelength λ), and the same amplitude y_m. They both travel in the positive direction of the x axis, with the same speed, given by Eq. 17-25. They differ only by a constant angle ϕ, which we call the **phase constant.** These waves are said to be *out of phase* by ϕ or to have a *phase difference* of ϕ, or one wave is said to be *phase-shifted* from the other by ϕ.

From the principle of superposition (Eq. 17-33), the resultant wave is the algebraic sum of the two interfering waves and has displacement

$$y'(x, t) = y_1(x, t) + y_2(x, t)$$
$$= y_m \sin(kx - \omega t) + y_m \sin(kx - \omega t + \phi). \qquad (17\text{-}36)$$

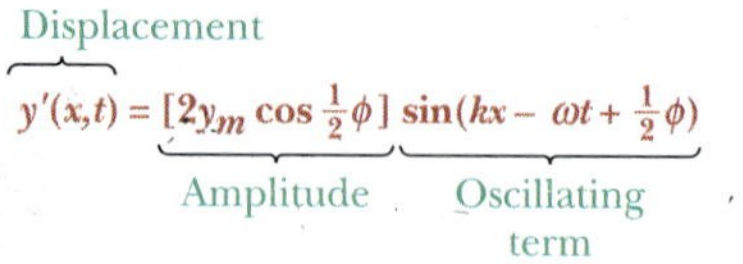

Displacement

$$y'(x,t) = [\underbrace{2y_m \cos \tfrac{1}{2}\phi}_{\text{Amplitude}}]\, \underbrace{\sin(kx - \omega t + \tfrac{1}{2}\phi)}_{\substack{\text{Oscillating}\\\text{term}}}$$

Fig. 17-13 The resultant wave of Eq. 17-38, due to the interference of two sinusoidal transverse waves, is also a sinusoidal transverse wave, with an amplitude and an oscillating term.

In Appendix E we see that we can write the sum of the sines of two angles α and β as

$$\sin \alpha + \sin \beta = 2 \sin \tfrac{1}{2}(\alpha + \beta) \cos \tfrac{1}{2}(\alpha - \beta). \qquad (17\text{-}37)$$

Applying this relation to Eq. 17-36 leads to

$$y'(x, t) = [2y_m \cos \tfrac{1}{2}\phi] \sin(kx - \omega t + \tfrac{1}{2}\phi). \qquad (17\text{-}38)$$

As Fig. 17-13 shows, the resultant wave is also a sinusoidal wave traveling in the direction of increasing x. It is the only wave you would actually see on the string (you would *not* see the two interfering waves of Eqs. 17-34 and 17-35).

> If two sinusoidal waves of the same amplitude and wavelength travel in the *same* direction along a stretched string, they interfere to produce a resultant sinusoidal wave traveling in that direction.

The resultant wave differs from the interfering waves in two respects: (1) its phase constant is $\tfrac{1}{2}\phi$, and (2) its amplitude y'_m is the quantity in the brackets in Eq. 17-38:

$$y'_m = 2y_m \cos \tfrac{1}{2}\phi \qquad \text{(amplitude)}. \qquad (17\text{-}39)$$

If $\phi = 0$ rad (or $0°$), the two interfering waves are exactly in phase, as in Fig. 17-14a. Then Eq. 17-38 reduces to

$$y'(x, t) = 2y_m \sin(kx - \omega t) \qquad (\phi = 0). \qquad (17\text{-}40)$$

This resultant wave is plotted in Fig. 17-14d. Note from both that figure and Eq. 17-40 that the amplitude of the resultant wave is twice the amplitude of either interfering wave. That is the greatest amplitude the resultant wave can have, because the cosine term in Eqs. 17-38 and 17-39 has its greatest value (unity) when $\phi = 0$. Interference that produces the greatest possible amplitude is called *fully constructive interference*.

If $\phi = \pi$ rad (or $180°$), the interfering waves are exactly out of phase as in Fig. 17-14b. Then $\cos \tfrac{1}{2}\phi$ becomes $\cos \pi/2 = 0$, and the amplitude of the resultant

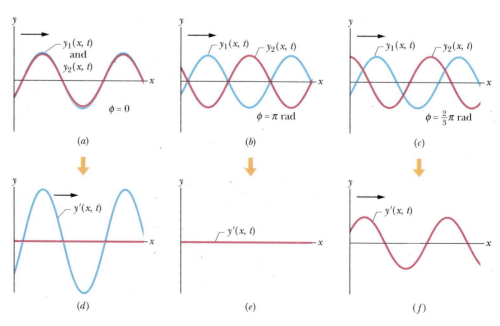

Fig. 17-14 Two identical sinusoidal waves, $y_1(x, t)$ and $y_2(x, t)$, travel along a string in the positive direction of an x axis. They interfere to give a resultant wave $y'(x, t)$. The resultant wave is what is actually seen on the string. The phase difference ϕ between the two interfering waves is (a) 0 rad or $0°$, (b) π rad or $180°$, and (c) $\tfrac{2}{3}\pi$ rad or $120°$. The corresponding resultant waves are shown in (d), (e), and (f).

TABLE 17-1   Phase Differences and Resulting Interference Types[a]

Phase Difference, in			Amplitude of Resultant Wave	Type of Interference
Degrees	Radians	Wavelengths		
0	0	0	$2y_m$	Fully constructive
120	$\frac{2}{3}\pi$	0.33	y_m	Intermediate
180	π	0.50	0	Fully destructive
240	$\frac{4}{3}\pi$	0.67	y_m	Intermediate
360	2π	1.00	$2y_m$	Fully constructive
865	15.1	2.40	$0.60y_m$	Intermediate

[a]The phase difference is between two otherwise identical waves, with amplitude y_m, moving in the same direction.

wave as given by Eq. 17-39 is zero. We then have, for all values of x and t,

$$y'(x, t) = 0 \qquad (\phi = \pi \text{ rad}). \qquad (17\text{-}41)$$

The resultant wave is plotted in Fig. 17-14e. Although we sent two waves along the string, we see no motion of the string. This type of interference is called *fully destructive interference*.

Because a sinusoidal wave repeats its shape every 2π rad, a phase difference $\phi = 2\pi$ rad (or 360°) corresponds to a shift of one wave relative to the other wave by a distance equivalent to one wavelength. Thus, phase differences can be described in terms of wavelengths as well as angles. For example, in Fig. 17-14b the waves may be said to be 0.50 wavelength out of phase. Table 17-1 shows some other examples of phase differences and the interference they produce. Note that when interference is neither fully constructive nor fully destructive, it is called *intermediate interference*. The amplitude of the resultant wave is then intermediate between 0 and $2y_m$. For example, from Table 17-1, if the interfering waves have a phase difference of 120° ($\phi = \frac{2}{3}\pi$ rad = 0.33 wavelength), then the resultant wave has an amplitude of y_m, the same as the interfering waves (see Figs. 17-14c and f).

Two waves with the same wavelength are in phase if their phase difference is zero or any integer number of wavelengths. Thus, the integer part of any phase difference *expressed in wavelengths* may be discarded. For example, a phase difference of 0.40 wavelength is equivalent in every way to one of 2.40 wavelengths, and so the simpler of the two numbers can be used in computations.

Sample Problem 17-5

Two identical sinusoidal waves, moving in the same direction along a stretched string, interfere with each other. The amplitude y_m of each wave is 9.8 mm, and the phase difference ϕ between them is 100°.

(a) What is the amplitude y'_m of the resultant wave due to the interference of these two waves, and what type of interference occurs?

SOLUTION: The Key Idea here is that these are identical sinusoidal waves traveling in the *same direction* along a string, so they interfere to produce a sinusoidal traveling wave. Because they are identical, they have the *same amplitude*. Thus, the amplitude y'_m of the

resultant wave is given by Eq. 17-39:

$$y'_m = 2y_m \cos \tfrac{1}{2}\phi = (2)(9.8 \text{ mm}) \cos(100°/2)$$
$$= 13 \text{ mm}. \qquad \text{(Answer)}$$

We can tell that the interference is *intermediate* in two ways. The phase difference is between 0 and 180° and, correspondingly, amplitude y'_m is between 0 and $2y_m$ (= 19.6 mm).

(b) What phase difference, in radians and wavelengths, will give the resultant wave an amplitude of 4.9 mm?

SOLUTION: The same Key Idea applies here as in part (a), but now we are given y'_m and seek ϕ. From Eq. 17-39,

$$y'_m = 2y_m \cos \tfrac{1}{2}\phi.$$

We now have

$$4.9 \text{ mm} = (2)(9.8 \text{ mm}) \cos \tfrac{1}{2}\phi,$$

which gives us (with a calculator in the radian mode)

$$\phi = 2 \cos^{-1} \frac{4.9 \text{ mm}}{(2)(9.8 \text{ mm})}$$

$$= \pm 2.636 \text{ rad} \approx \pm 2.6 \text{ rad}. \qquad \text{(Answer)}$$

There are two solutions because we can obtain the same resultant wave by letting the first wave *lead* (travel ahead of) or *lag* (travel behind) the second wave by 2.6 rad. In wavelengths, the phase difference is

$$\frac{\phi}{2\pi \text{ rad/wavelength}} = \frac{\pm 2.636 \text{ rad}}{2\pi \text{ rad/wavelength}}$$

$$= \pm 0.42 \text{ wavelength.} \qquad \text{(Answer)}$$

✔ **CHECKPOINT 5:** Here are four other possible phase differences between the two waves of this sample problem, expressed in wavelengths: 0.20, 0.45, 0.60, and 0.80. Rank them according to the amplitude of the resultant wave, greatest first.

17-10 Phasors

We can represent a string wave (or any other type of wave) vectorially with a **phasor.** In essence, a phasor is a vector that has a magnitude equal to the amplitude of the wave and that rotates around an origin; the angular speed of the phasor is equal to the angular frequency ω of the wave. For example, the wave

$$y_1(x, t) = y_{m1} \sin(kx - \omega t) \qquad (17\text{-}42)$$

is represented by the phasor shown in Fig. 17-15a. The magnitude of the phasor is the amplitude y_{m1} of the wave. As the phasor rotates around the origin at angular speed ω, its projection y_1 on the vertical axis varies sinusoidally, from a maximum of y_{m1} through zero to a minimum of $-y_{m1}$ and then back to y_{m1}. This variation corresponds to the sinusoidal variation in the displacement y_1 of any point along the string as the wave passes through it.

When two waves travel along the same string in the same direction, we can represent them and their resultant wave in a *phasor diagram.* The phasors in Fig. 17-15b represent the wave of Eq. 17-42 and a second wave given by

$$y_2(x, t) = y_{m2} \sin(kx - \omega t + \phi). \qquad (17\text{-}43)$$

This second wave is phase shifted from the first wave by phase constant ϕ. Because the phasors rotate at the same angular speed ω, the angle between the two phasors is always ϕ. If ϕ is a *positive* quantity, then the phasor for wave 2 *lags* the phasor for wave 1 as they rotate, as drawn in Fig. 17-15b. If ϕ is a negative quantity, then the phasor for wave 2 *leads* the phasor for wave 1.

Because waves y_1 and y_2 have the same angular wave number k and angular frequency ω, we know from Eq. 17-38 that their resultant is of the form

$$y'(x, t) = y'_m \sin(kx - \omega t + \beta), \qquad (17\text{-}44)$$

where y'_m is the amplitude of the resultant wave and β is its phase constant. To find the values of y'_m and β, we would have to sum the two combining waves, as we did to obtain Eq. 17-38.

To do this on a phasor diagram, we vectorially add the two phasors at any instant during their rotation, as in Fig. 17-15c where phasor y_{m2} has been shifted to the

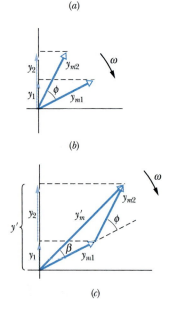

Fig. 17-15 (a) A phasor of magnitude y_{m1} rotating about an origin at angular speed ω represents a sinusoidal wave. The phasor's projection y_1 on the vertical axis represents the displacement of a point through which the wave passes. (b) A second phasor, also of angular speed ω but of magnitude y_{m2} and rotating at a constant angle ϕ from the first phasor, represents a second wave, with a phase constant ϕ. (c) The resultant wave of the two waves is represented by the vector sum y'_m of the two phasors. The projection y' on the vertical axis represents the displacement of a point as that resultant wave passes through it.

head of phasor y_{m1}. The magnitude of the vector sum equals the amplitude y'_m in Eq. 17-44. The angle between the vector sum and the phasor for y_1 equals the phase constant β in Eq. 17-44.

Note that, in contrast to the method of Section 17-9:

> We can use phasors to combine waves *even if their amplitudes are different.*

Sample Problem 17-6

Two sinusoidal waves $y_1(x, t)$ and $y_2(x, t)$ have the same wavelength and travel together in the same direction along a string. Their amplitudes are $y_{m1} = 4.0$ mm and $y_{m2} = 3.0$ mm, and their phase constants are 0 and $\pi/3$ rad, respectively. What are the amplitude y'_m and phase constant β of the resultant wave? Write the resultant wave in the form of Eq. 17-44.

SOLUTION: One **Key Idea** here is that the two waves have a number of properties in common: Because they travel along the same string, they must have the same speed v, as set by the tension and linear density of the string according to Eq. 17-25. With the same wavelength λ, they must have the same angular wave number $k\ (= 2\pi/\lambda)$. Also, with the same wave number k and speed v, they must have the same angular frequency $\omega\ (= kv)$.

A second **Key Idea** is that the waves (call them waves 1 and 2) can be represented by phasors rotating at the same angular speed ω about an origin. Because the phase constant for wave 2 is *greater* than that for wave 1 by $\pi/3$, phasor 2 must *lag* phasor 1 by $\pi/3$ rad in their clockwise rotation, as shown in Fig. 17-16a. The resultant wave due to the interference of waves 1 and 2 can then be represented by a phasor that is the vector sum of phasors 1 and 2.

To simplify the vector summation, we drew phasors 1 and 2 in Fig. 17-16a at the instant when phasor 1 lies along the horizontal axis. We then drew lagging phasor 2 at positive angle $\pi/3$ rad. In Fig. 17-16b we shifted phasor 2 so its tail is at the head of phasor 1. Then we can draw the phasor y'_m of the resultant wave from the tail of phasor 1 to the head of phasor 2. The phase constant β is the angle it makes with phasor 1.

To find values for y'_m and β, we can sum phasors 1 and 2 directly on a vector-capable calculator, by adding a vector of magnitude 4.0 and angle 0 rad to a vector of magnitude 3.0 and angle $\pi/3$ rad, or we can add the vectors by components. For the hori-

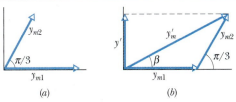

Fig. 17-16 Sample Problem 17-6. (a) Two phasors of magnitudes y_{m1} and y_{m2} and with phase difference $\pi/3$. (b) Vector addition of these phasors at any instant during their rotation gives the magnitude y'_m of the phasor for the resultant wave.

zontal components we have

$$y'_{mh} = y_{m1} \cos 0 + y_{m2} \cos \pi/3$$
$$= 4.0 \text{ mm} + (3.0 \text{ mm}) \cos \pi/3 = 5.50 \text{ mm}.$$

For the vertical components we have

$$y'_{mv} = y_{m1} \sin 0 + y_{m2} \sin \pi/3$$
$$= 0 + (3.0 \text{ mm}) \sin \pi/3 = 2.60 \text{ mm}.$$

Thus, the resultant wave has an amplitude of

$$y'_m = \sqrt{(5.50 \text{ mm})^2 + (2.60 \text{ mm})^2}$$
$$= 6.1 \text{ mm} \qquad \text{(Answer)}$$

and a phase constant of

$$\beta = \tan^{-1} \frac{2.60 \text{ mm}}{5.50 \text{ mm}} = 0.44 \text{ rad.} \qquad \text{(Answer)}$$

From Fig. 17-16b, phase constant β is a *positive* angle relative to phasor 1. Thus, the resultant wave *lags* wave 1 in their travel by phase constant $\beta = +0.44$ rad. From Eq. 17-44, we can write the resultant wave as

$$y'(x, t) = (6.1 \text{ mm}) \sin(kx - \omega t + 0.44 \text{ rad}). \quad \text{(Answer)}$$

17-11 Standing Waves

In the preceding two sections, we discussed two sinusoidal waves of the same wavelength and amplitude traveling *in the same direction* along a stretched string. What if they travel in opposite directions? We can again find the resultant wave by applying the superposition principle.

Figure 17-17 suggests the situation graphically. It shows the two combining waves, one traveling to the left in Fig. 17-17a, the other to the right in Fig. 17-17b. Figure 17-17c shows their sum, obtained by applying the superposition principle graphically. The outstanding feature of the resultant wave is that there are places

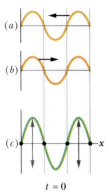

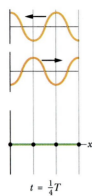

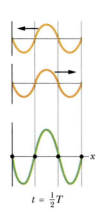

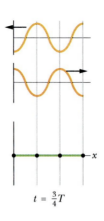

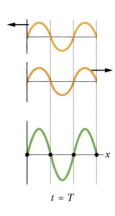

(a)

(b)

(c)

$t = 0$ $t = \frac{1}{4}T$ $t = \frac{1}{2}T$ $t = \frac{3}{4}T$ $t = T$

Fig. 17-17 (a) Five snapshots of a wave traveling to the left, at the times t indicated below part (c) (T is the period of oscillation). (b) Five snapshots of a wave identical to that in (a) but traveling to the right, at the same times t. (c) Corresponding snapshots for the superposition of the two waves on the same string. At $t = 0$, $\frac{1}{2}T$, and T, fully constructive interference occurs because of the alignment of peaks with peaks and valleys with valleys. At $t = \frac{1}{4}T$ and $\frac{3}{4}T$, fully destructive interference occurs because of the alignment of peaks with valleys. Some points (the nodes, marked with dots) never oscillate; some points (the antinodes) oscillate the most.

along the string, called **nodes**, where the string never moves. Four such nodes are marked by dots in Fig. 17-17c. Halfway between adjacent nodes are **antinodes,** where the amplitude of the resultant wave is a maximum. Wave patterns such as that of Fig. 17-17c are called **standing waves** because the wave patterns do not move left or right; the locations of the maxima and minima do not change.

> ▶ If two sinusoidal waves of the same amplitude and wavelength travel in *opposite* directions along a stretched string, their interference with each other produces a standing wave.

To analyze a standing wave, we represent the two combining waves with the equations

$$y_1(x, t) = y_m \sin(kx - \omega t) \tag{17-45}$$

and

$$y_2(x, t) = y_m \sin(kx + \omega t). \tag{17-46}$$

The principle of superposition gives, for the combined wave,

$$y'(x, t) = y_1(x, t) + y_2(x, t) = y_m \sin(kx - \omega t) + y_m \sin(kx + \omega t).$$

Applying the trigonometric relation of Eq. 17-37 leads to

$$y'(x, t) = [2y_m \sin kx] \cos \omega t, \tag{17-47}$$

which is displayed in Fig. 17-18. This equation does not describe a traveling wave because it is not of the form of Eq. 17-16. Instead, it describes a standing wave.

The quantity $2y_m \sin kx$ in the brackets of Eq. 17-47 can be viewed as the amplitude of oscillation of the string element that is located at position x. However, since an amplitude is always positive and $\sin kx$ can be negative, we take the absolute value of the quantity $2y_m \sin kx$ to be the amplitude at x.

In a traveling sinusoidal wave, the amplitude of the wave is the same for all string elements. That is not true for a standing wave, in which the amplitude *varies with position*. In the standing wave of Eq. 17-47, for example, the amplitude is zero for values of kx that give $\sin kx = 0$. Those values are

$$kx = n\pi, \quad \text{for } n = 0, 1, 2, \dots . \tag{17-48}$$

Substituting $k = 2\pi/\lambda$ in this equation and rearranging, we get

$$x = n\frac{\lambda}{2}, \quad \text{for } n = 0, 1, 2, \dots \quad \text{(nodes)}, \tag{17-49}$$

Displacement

$$y'(x,t) = [2y_m \sin kx]\cos \omega t$$

Amplitude   Oscillating
at position x term

Fig. 17-18 The resultant wave of Eq. 17-47 is a standing wave and is due to the interference of two sinusoidal waves of the same amplitude and wavelength that travel in opposite directions.

as the positions of zero amplitude—the nodes—for the standing wave of Eq. 17-47. Note that adjacent nodes are separated by $\lambda/2$, half a wavelength.

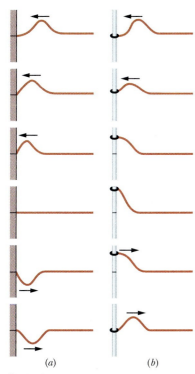

Fig. 17-19 (*a*) A pulse incident from the right is reflected at the left end of the string, which is tied to a wall. Note that the reflected pulse is inverted from the incident pulse. (*b*) Here the left end of the string is tied to a ring that can slide without friction up and down the rod. Now the pulse is not inverted by the reflection.

The amplitude of the standing wave of Eq. 17-47 has a maximum value of $2y$, which occurs for values of kx that give $|\sin kx| = 1$. Those values are

$$kx = \tfrac{1}{2}\pi, \tfrac{3}{2}\pi, \tfrac{5}{2}\pi, \ldots$$
$$= (n + \tfrac{1}{2})\pi, \qquad \text{for } n = 0, 1, 2, \ldots. \tag{17-50}$$

Substituting $k = 2\pi/\lambda$ in Eq. 17-50 and rearranging, we get

$$x = \left(n + \frac{1}{2}\right)\frac{\lambda}{2}, \qquad \text{for } n = 0, 1, 2, \ldots \quad \text{(antinodes)}, \tag{17-51}$$

as the positions of maximum amplitude—the antinodes—of the standing wave of Eq. 17-47. The antinodes are separated by $\lambda/2$ and are located halfway between pairs of nodes.

Reflections at a Boundary

We can set up a standing wave in a stretched string by allowing a traveling wave to be reflected from the far end of the string so that it travels back through itself. The incident (original) wave and the reflected wave can then be described by Eqs. 17-45 and 17-46, respectively, and they can combine to form a pattern of standing waves.

In Fig. 17-19, we use a single pulse to show how such reflections take place. In Fig. 17-19*a*, the string is fixed at its left end. When the pulse arrives at that end, it exerts an upward force on the support (the wall). By Newton's third law, the support exerts an opposite force of equal magnitude on the string. This second force generates a pulse at the support, which travels back along the string in the direction opposite that of the incident pulse. In a "hard" reflection of this kind, there must be a node at the support because the string is fixed there. The reflected and incident pulses must have opposite signs, so as to cancel each other at that point.

In Fig. 17-19*b*, the left end of the string is fastened to a light ring that is free to slide without friction along a rod. When the incident pulse arrives, the ring moves up the rod. As the ring moves, it pulls on the string, stretching the string and producing a reflected pulse with the same sign and amplitude as the incident pulse. Thus, in such a "soft" reflection, the incident and reflected pulses reinforce each other, creating an antinode at the end of the string; the maximum displacement of the ring is twice the amplitude of either of these pulses.

✓**CHECKPOINT 6:** Two waves with the same amplitude and wavelength interfere in three different situations to produce resultant waves with the following equations:
(1) $y'(x, t) = 4 \sin(5x - 4t)$
(2) $y'(x, t) = 4 \sin(5x) \cos(4t)$
(3) $y'(x, t) = 4 \sin(5x + 4t)$
In which situation are the two combining waves traveling (a) toward positive x, (b) toward negative x, and (c) in opposite directions?

17-12 Standing Waves and Resonance

Consider a string, such as a guitar string, that is stretched between two clamps. Suppose we send a continuous sinusoidal wave of a certain frequency along the string, say, toward the right. When the wave reaches the right end, it reflects and begins to travel back to the left. That left-going wave then overlaps the wave that is still traveling to the right. When the left-going wave reaches the left end, it reflects again and the newly reflected wave begins to travel to the right, overlapping the left-

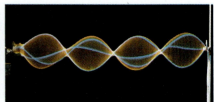

Fig. 17-20 Stroboscopic photographs reveal (imperfect) standing wave patterns on a string being made to oscillate by a vibrator at the left end. The patterns occur at certain frequencies of oscillation.

going and right-going waves. In short, we very soon have many overlapping traveling waves, which interfere with one another.

For certain frequencies, the interference produces a standing wave pattern (or **oscillation mode**) with nodes and large antinodes like those in Fig. 17-20. Such a standing wave is said to be produced at **resonance,** and the string is said to *resonate* at these certain frequencies, called **resonant frequencies.** If the string is oscillated at some frequency other than a resonant frequency, a standing wave is not set up. Then the interference of the right-going and left-going traveling waves results in only small (perhaps imperceptible) oscillations of the string.

Let a string be stretched between two clamps separated by a fixed distance L. To find expressions for the resonant frequencies of the string, we note that a node must exist at each of its ends, because each end is fixed and cannot oscillate. The simplest pattern that meets this key requirement is that in Fig. 17-21a, which shows the string at both its extreme displacements (one solid and one dashed, together forming a single "loop"). There is only one antinode, which is at the center of the string. Note that half a wavelength spans the length L, which we take to be the string's length. Thus, for this pattern, $\lambda/2 = L$. This condition tells us that if the left-going and right-going traveling waves are to set up this pattern by their interference, they must have the wavelength $\lambda = 2L$.

A second simple pattern meeting the requirement of nodes at the fixed ends is shown in Fig. 17-21b. This pattern has three nodes and two antinodes and is said to be a two-loop pattern. For the left-going and right-going waves to set it up, they must have a wavelength $\lambda = L$. A third pattern is shown in Fig. 17-21c. It has four nodes, three antinodes, and three loops, and the wavelength is $\lambda = \frac{2}{3}L$. We could continue this progression by drawing increasingly more complicated patterns. In each step of the progression, the pattern would have one more node and one more antinode than the preceding step, and an additional $\lambda/2$ would be fitted into the distance L.

Thus, a standing wave can be set up on a string of length L by a wave with a wavelength equal to one of the values

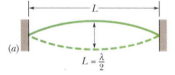

(a)

$$L = \frac{\lambda}{2}$$

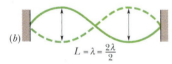

(b)

$$L = \lambda = \frac{2\lambda}{2}$$

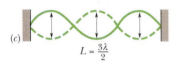

(c)

$$L = \frac{3\lambda}{2}$$

Fig. 17-21 A string, stretched between two clamps, is made to oscillate in standing wave patterns. (a) The simplest possible pattern consists of one *loop,* which refers to the composite shape formed by the string in its extreme displacements (the solid and dashed lines). (b) The next simplest pattern has two loops. (c) The next has three loops.

$$\lambda = \frac{2L}{n}, \qquad \text{for } n = 1, 2, 3, \ldots. \qquad (17\text{-}52)$$

The resonant frequencies that correspond to these wavelengths follow from Eq. 17-12:

$$f = \frac{v}{\lambda} = n\frac{v}{2L}, \qquad \text{for } n = 1, 2, 3, \ldots. \qquad (17\text{-}53)$$

Here v is the speed of traveling waves on the string.

Equation 17-53 tells us that the resonant frequencies are integer multiples of the lowest resonant frequency, $f = v/2L$, which corresponds to $n = 1$. The oscillation mode with that lowest frequency is called the *fundamental mode* or the *first harmonic.* The *second harmonic* is the oscillation mode with $n = 2$, the *third harmonic* is that with $n = 3$, and so on. The frequencies associated with these modes are often

Fig. 17-22 One of many possible standing wave patterns for a kettledrum head, made visible by dark powder sprinkled on the drumhead. As the head is set into oscillation at a single frequency by a mechanical vibrator at the upper left of the photograph, the powder collects at the nodes, which are circles and straight lines in this two-dimensional example.

labeled f_1, f_2, f_3, and so on. The collection of all possible oscillation modes is called the **harmonic series**, and n is called the **harmonic number** of the nth harmonic.

The phenomenon of resonance is common to all oscillating systems and can occur in two and three dimensions. For example, Fig. 17-22 shows a two-dimensional standing wave pattern on the oscillating head of a kettledrum.

✔**CHECKPOINT 7:**   In the following series of resonant frequencies, one frequency (lower than 400 Hz) is missing: 150, 225, 300, 375 Hz. (a) What is the missing frequency? (b) What is the frequency of the seventh harmonic?

Sample Problem 17-7

In Fig. 17-23, a string, tied to a sinusoidal vibrator at P and running over a support at Q, is stretched by a block of mass m. The separation L between P and Q is 1.2 m, the linear density of the string is 1.6 g/m, and the frequency f of the vibrator is fixed at 120 Hz. The amplitude of the motion at P is small enough for that point to be considered a node. A node also exists at Q.

(a) What mass m allows the vibrator to set up the fourth harmonic on the string?

SOLUTION: One **Key Idea** here is that the string will resonate at only certain frequencies, determined by the wave speed v on the string and the length L of the string. From Eq. 17-53, these resonant frequencies are

$$f = n\frac{v}{2L}, \qquad \text{for } n = 1, 2, 3, \ldots . \qquad (17\text{-}54)$$

To set up the fourth harmonic (for which $n = 4$), we need to adjust the right side of this equation, with $n = 4$, so that the left side equals the frequency of the vibrator (120 Hz).

We cannot adjust L in Eq. 17-54; it is set. However, a second **Key Idea** is that we *can* adjust v, because it depends on how much mass m we hang on the string. According to Eq. 17-25, wave speed $v = \sqrt{\tau/\mu}$. Here the tension τ in the string is equal to the weight mg of the block. Thus,

$$v = \sqrt{\frac{\tau}{\mu}} = \sqrt{\frac{mg}{\mu}}. \qquad (17\text{-}55)$$

Fig. 17-23 Sample Problem 17-7. A string under tension connected to a vibrator. For a fixed vibrator frequency, standing wave patterns will occur for certain values of the string tension.

Substituting v from Eq. 17-55 into Eq. 17-54, setting $n = 4$ for the fourth harmonic, and solving for m give us

$$m = \frac{4L^2 f^2 \mu}{n^2 g} \qquad (17\text{-}56)$$

$$= \frac{(4)(1.2 \text{ m})^2(120 \text{ Hz})^2(0.0016 \text{ kg/m})}{(4)^2(9.8 \text{ m/s}^2)}$$

$$= 0.846 \text{ kg} \approx 0.85 \text{ kg}. \qquad \text{(Answer)}$$

(b) What standing wave mode is set up if $m = 1.00$ kg?

SOLUTION: If we insert this value of m into Eq. 17-56 and solve for n, we find that $n = 3.7$. A **Key Idea** here is that n must be an integer, so $n = 3.7$ is impossible. Thus, with $m = 1.00$ kg, the vibrator cannot set up a standing wave on the string, and any oscillation of the string will be small, perhaps even imperceptible.

Tactic 2: *Harmonics on a String*

When you need to obtain information about a certain harmonic on a stretched string of given length L, first draw that harmonic (as in Fig. 17-21). If you are asked about, say, the fifth harmonic, you need to draw five loops between the fixed support points. That would mean that five loops, each of length $\lambda/2$, occupy the length

L of the string. Thus, $5(\lambda/2) = L$, and $\lambda = 2L/5$. You can then use Eq. 17-12 ($f = v/\lambda$) to find the frequency of the harmonic.

Keep in mind that the wavelength of a harmonic is set only by the length L of the string, but the frequency depends also on the wave speed v, which is set by the tension and the linear density of the string via Eq. 17-25.

REVIEW & SUMMARY

Transverse and Longitudinal Waves   Mechanical waves can exist only in material media and are governed by Newton's laws. **Transverse** mechanical waves, like those on a stretched string, are waves in which the particles of the medium oscillate perpendicular to the wave's direction of travel. Waves in which the particles of the medium oscillate parallel to the wave's direction of travel are **longitudinal** waves.

Sinusoidal Waves   A sinusoidal wave moving in the positive x direction has the mathematical form

$$y(x, t) = y_m \sin(kx - \omega t), \qquad (17\text{-}2)$$

where y_m is the **amplitude** of the wave, k is the **angular wave number,** ω is the **angular frequency,** and $kx - \omega t$ is the **phase.** The **wavelength** λ is related to k by

$$k = \frac{2\pi}{\lambda}. \qquad (17\text{-}5)$$

The **period** T and **frequency** f of the wave are related to ω by

$$\frac{\omega}{2\pi} = f = \frac{1}{T}. \qquad (17\text{-}9)$$

Finally, the **wave speed** v is related to these other parameters by

$$v = \frac{\omega}{k} = \frac{\lambda}{T} = \lambda f. \qquad (17\text{-}12)$$

Equation of a Traveling Wave   Any function of the form

$$y(x, t) = h(kx \pm \omega t) \qquad (17\text{-}16)$$

can represent a **traveling wave** with a wave speed given by Eq. 17-12 and a wave shape given by the mathematical form of h. The plus sign denotes a wave traveling in the negative x direction, and the minus sign a wave traveling in the positive x direction.

Wave Speed on Stretched String   The speed of a wave on a stretched string is set by properties of the string. The speed on a string with tension τ and linear density μ is

$$v = \sqrt{\frac{\tau}{\mu}}. \qquad (17\text{-}25)$$

Power   The **average power,** or average rate at which energy is transmitted by a sinusoidal wave on a stretched string, is given by

$$P_{\text{avg}} = \tfrac{1}{2}\mu v \omega^2 y_m^2. \qquad (17\text{-}32)$$

Superposition of Waves   When two or more waves traverse the same medium, the displacement of any particle of the medium is the sum of the displacements that the individual waves would give it.

Interference of Waves   Two sinusoidal waves on the same string exhibit **interference,** adding or canceling according to the principle of superposition. If the two are traveling in the same direction and have the same amplitude y_m and frequency (hence the same wavelength) but differ in phase by a **phase constant** ϕ, the result is a single wave with this same frequency:

$$y'(x, t) = [2y_m \cos \tfrac{1}{2}\phi] \sin(kx - \omega t + \tfrac{1}{2}\phi). \qquad (17\text{-}38)$$

If $\phi = 0$, the waves are exactly in phase and their interference is fully constructive; if $\phi = \pi$ rad, they are exactly out of phase and their interference is fully destructive.

Phasors   A wave $y(x, t)$ can be represented with a *phasor*. This is a vector that has a magnitude equal to the amplitude y_m of the wave and that rotates about an origin with an angular speed equal to the angular frequency ω of the wave. The projection of the rotating phasor on a vertical axis gives the displacement y of a point along the wave's travel.

Standing Waves   The interference of two identical sinusoidal waves moving in opposite directions produces **standing waves.** For a string with fixed ends, the standing wave is given by

$$y'(x, t) = [2y_m \sin kx] \cos \omega t. \qquad (17\text{-}47)$$

Standing waves are characterized by fixed locations of zero displacement called **nodes** and fixed locations of maximum displacement called **antinodes.**

Resonance   Standing waves on a string can be set up by reflection of traveling waves from the ends of the string. If an end is fixed, it must be the position of a node. This limits the frequencies at which standing waves will occur on a given string. Each possible frequency is a **resonant frequency,** and the corresponding standing wave pattern is an **oscillation mode.** For a stretched string of length L with fixed ends, the resonant frequencies are

$$f = \frac{v}{\lambda} = n\frac{v}{2L}, \qquad \text{for } n = 1, 2, 3, \ldots. \qquad (17\text{-}53)$$

The oscillation mode corresponding to $n = 1$ is called the *fundamental mode* or the *first harmonic*; the mode corresponding to $n = 2$ is the *second harmonic*; and so on.

QUESTIONS

1. What is the wavelength of the (strange) wave in Fig. 17-24, where each segment of the wave has length d?

Fig. 17-24 Question 1.

2. Figure 17-25a gives a snapshot of a wave traveling in the direction of positive x along a string under tension. Four string elements are indicated by the lettered points. For each of those elements, determine whether, at the instant of the snapshot, the element is moving upward or downward or is momentarily at rest. (*Hint:* Imagine the wave as it moves through the four string elements.)

.Figure 17-25b gives the displacement of a string element

located at, say, $x = 0$ as a function of time. At the lettered times, is the element moving upward or downward or is it momentarily at rest?

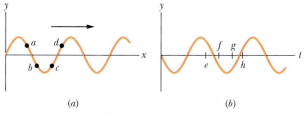

(a)   (b)

Fig. 17-25 Question 2.

3. In Fig. 17-26, five points are indicated on a snapshot of a sinusoidal wave. What is the phase difference between point 1 and (a) point 2, (b) point 3, (c) point 4, and (d) point 5? Answer in radians and in terms of the wavelength of the wave. The snapshot shows a point of zero displacement at $x = 0$. In terms of the period T of the wave, when will (e) a peak and (f) the next point of zero displacement reach $x = 0$?

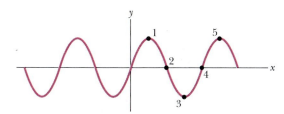

Fig. 17-26 Question 3.

4. The following four waves are sent along strings with the same linear densities (x is in meters and t is in seconds). Rank the waves according to (a) their wave speed and (b) the tension in the strings along which they travel, greatest first:

(1) $y_1 = (3 \text{ mm}) \sin(x - 3t)$,   (3) $y_3 = (1 \text{ mm}) \sin(4x - t)$,

(2) $y_2 = (6 \text{ mm}) \sin(2x - t)$,   (4) $y_4 = (2 \text{ mm}) \sin(x - 2t)$.

5. In Fig. 17-27, wave 1 consists of a rectangular peak of height 4 units and width d, and a rectangular valley of depth 2 units and width d. The wave travels rightward along an x axis. Choices 2, 3, and 4 are similar waves, with the same heights, depths, and widths, that will travel leftward along that axis and through wave 1. With

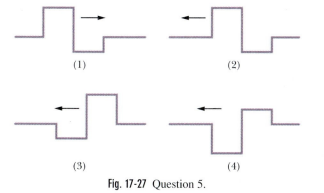

Fig. 17-27 Question 5.

which choice will the interference give, for an instant, (a) the deepest valley, (b) a flat line, and (c) a level peak $2d$ wide?

6. If you start with two sinusoidal waves of the same amplitude traveling in phase on a string and then somehow phase-shift one of them by 5.4 wavelengths, what type of interference will occur on the string?

7. The amplitudes and phase differences for four pairs of waves of equal wavelengths are (a) 2 mm, 6 mm, and π rad; (b) 3 mm, 5 mm, and π rad; (c) 7 mm, 9 mm, and π rad; (d) 2 mm, 2 mm, and 0 rad. Each pair travels in the same direction along the same string. Without written calculation, rank the four pairs according to the amplitude of their resultant wave, greatest first. (*Hint:* Construct phasor diagrams.)

8. If you set up the seventh harmonic on a string, (a) how many nodes are present, and (b) is there a node, antinode, or some intermediate state at the midpoint? If you next set up the sixth harmonic, (c) is its resonant wavelength longer or shorter than that for the seventh harmonic, and (d) is the resonant frequency higher or lower?

9. Strings A and B have identical lengths and linear densities, but string B is under greater tension than string A. Figure 17-28 shows four situations, (a) through (d), in which standing wave patterns exist on the two strings. In which situations is there the possibility that strings A and B are oscillating at the same resonant frequency?

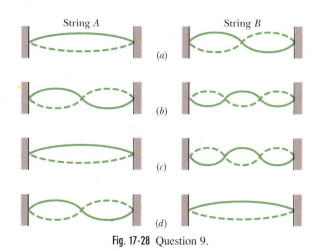

Fig. 17-28 Question 9.

10. (a) If a standing wave on a string is given by

$$y'(t) = (3 \text{ mm}) \sin(5x) \cos(4t),$$

is there a node or an antinode of the oscillations of the string at $x = 0$? (b) If the standing wave is given by

$$y'(t) = (3 \text{ mm}) \sin(5x + \pi/2) \cos(4t),$$

is there a node or an antinode at $x = 0$?

11. (a) In Sample Problem 17-7 and Fig. 17-23, if we gradually increase the mass of the block (the frequency remains fixed), new resonant modes appear. Do the harmonic numbers of the new resonant modes increase or decrease from one to the next? (b) Is the shift from one resonant mode to the next gradual, or does each resonant mode disappear well before the next one appears?

EXERCISES & PROBLEMS

ssm   Solution is in the Student Solutions Manual.
www   Solution is available on the World Wide Web at:
          http://www.wiley.com/college/hrw
ilw   Solution is available on the Interactive LearningWare.

SEC. 17-5   The Speed of a Traveling Wave

1E. A wave has an angular frequency of 110 rad/s and a wavelength of 1.80 m. Calculate (a) the angular wave number and (b) the speed of the wave.

2E. The speed of electromagnetic waves (which include visible light, radio, and x rays) in vacuum is 3.0×10^8 m/s. (a) Wavelengths of visible light waves range from about 400 nm in the violet to about 700 nm in the red. What is the range of frequencies of these waves? (b) The range of frequencies for shortwave radio (for example, FM radio and VHF television) is 1.5 to 300 MHz. What is the corresponding wavelength range? (c) X ray wavelengths range from about 5.0 nm to about 1.0×10^{-2} nm. What is the frequency range for x rays?

3E. A sinusoidal wave travels along a string. The time for a particular point to move from maximum displacement to zero is 0.170 s. What are the (a) period and (b) frequency? (c) The wavelength is 1.40 m; what is the wave speed?   ssm

4E. Write the equation for a sinusoidal wave traveling in the negative direction along an x axis and having an amplitude of 0.010 m, a frequency of 550 Hz, and a speed of 330 m/s.

5E. Show that

$$y = y_m \sin k(x - vt), \qquad y = y_m \sin 2\pi\left(\frac{x}{\lambda} - ft\right),$$

$$y = y_m \sin \omega\left(\frac{x}{v} - t\right), \qquad y = y_m \sin 2\pi\left(\frac{x}{\lambda} - \frac{t}{T}\right)$$

are all equivalent to $y = y_m \sin(kx - \omega t)$.   ssm

6P. The equation of a transverse wave traveling along a very long string is $y = 6.0 \sin(0.020\pi x + 4.0\pi t)$, where x and y are expressed in centimeters and t is in seconds. Determine (a) the amplitude, (b) the wavelength, (c) the frequency, (d) the speed, (e) the direction of propagation of the wave, and (f) the maximum transverse speed of a particle in the string. (g) What is the transverse displacement at $x = 3.5$ cm when $t = 0.26$ s?

7P. (a) Write an equation describing a sinusoidal transverse wave traveling on a cord in the $+x$ direction with a wavelength of 10 cm, a frequency of 400 Hz, and an amplitude of 2.0 cm. (b) What is the maximum speed of a point on the cord? (c) What is the speed of the wave?   ssm

8P. A transverse sinusoidal wave of wavelength 20 cm is moving along a string in the positive x direction. The transverse displacement of the string particle at $x = 0$ as a function of time is shown in Fig. 17-29. (a) Make a rough sketch of one wavelength

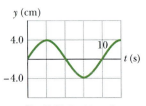

Fig. 17-29 Problem 8.

of the wave (the portion between $x = 0$ and $x = 20$ cm) at time $t = 0$. (b) What is the speed of the wave? (c) Write the equation for the wave with all the constants evaluated. (d) What is the transverse velocity of the particle at $x = 0$ at $t = 5.0$ s?

9P. A sinusoidal wave of frequency 500 Hz has a speed of 350 m/s. (a) How far apart are two points that differ in phase by $\pi/3$ rad? (b) What is the phase difference between two displacements at a certain point at times 1.00 ms apart?   ilw

SEC. 17-6   Wave Speed on a Stretched String

10E. The heaviest and lightest strings on a certain violin have linear densities of 3.0 and 0.29 g/m. What is the ratio of the diameter of the heaviest string to that of the lightest string, assuming that the strings are of the same material?

11E. What is the speed of a transverse wave in a rope of length 2.00 m and mass 60.0 g under a tension of 500 N?   ssm

12E. The tension in a wire clamped at both ends is doubled without appreciably changing the wire's length between the clamps. What is the ratio of the new to the old wave speed for transverse waves traveling along this wire?

13E. The linear density of a string is 1.6×10^{-4} kg/m. A transverse wave on the string is described by the equation

$$y = (0.021 \text{ m}) \sin[(2.0 \text{ m}^{-1})x + (30 \text{ s}^{-1})t].$$

What is (a) the wave speed and (b) the tension in the string?   ssm

14E. The equation of a transverse wave on a string is

$$y = (2.0 \text{ mm}) \sin[(20 \text{ m}^{-1})x - (600 \text{ s}^{-1})t].$$

The tension in the string is 15 N. (a) What is the wave speed? (b) Find the linear density of this string in grams per meter.

15P. A stretched string has a mass per unit length of 5.0 g/cm and a tension of 10 N. A sinusoidal wave on this string has an amplitude of 0.12 mm and a frequency of 100 Hz and is traveling in the negative direction of x. Write an equation for this wave.   ssm

16P. What is the fastest transverse wave that can be sent along a steel wire? For safety reasons, the maximum tensile stress to which steel wires should be subjected is 7.0×10^8 N/m². The density of steel is 7800 kg/m³. Show that your answer does not depend on the diameter of the wire.

17P. A sinusoidal transverse wave of amplitude y_m and wavelength λ travels on a stretched cord. (a) Find the ratio of the maximum particle speed (the speed with which a single particle in the cord moves transverse to the wave) to the wave speed. (b) If a wave having a certain wavelength and amplitude is sent along a cord, would this speed ratio depend on the material of which the cord is made, such as wire or nylon?   ssm

18P. A sinusoidal wave is traveling on a string with speed 40 cm/s. The displacement of the particles of the string at $x = 10$ cm is found to vary with time according to the equation $y = (5.0 \text{ cm}) \sin[1.0 - (4.0 \text{ s}^{-1})t]$. The linear density of the string is 4.0 g/cm. What are (a) the frequency and (b) the wavelength of the wave? (c) Write the general equation giving the transverse displacement of the particles of the string as a function of position and time. (d) Calculate the tension in the string.

19P. A sinusoidal transverse wave is traveling along a string in the negative direction of an x axis. Figure 17-30 shows a plot of the displacement as a function of position at time $t = 0$; the y intercept is 4.0 cm. The string tension is 3.6 N, and its linear density is 25 g/m. Find (a) the amplitude, (b) the wavelength, (c) the wave speed, and (d) the period of the wave. (e) Find the maximum transverse speed of a particle in the string. (f) Write an equation describing the traveling wave. ssm ilw

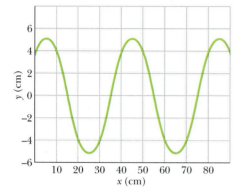

Fig. 17-30
Problem 19.

20P. In Fig. 17-31a, string 1 has a linear density of 3.00 g/m, and string 2 has a linear density of 5.00 g/m. They are under tension owing to the hanging block of mass $M = 500$ g. Calculate the wave speed on (a) string 1 and (b) string 2. (*Hint:* When a string loops halfway around a pulley, it pulls on the pulley with a net force that is twice the tension in the string.) Next the block is divided into two blocks (with $M_1 + M_2 = M$) and the apparatus is rearranged as shown in Fig. 17-31b. Find (c) M_1 and (d) M_2 such that the wave speeds in the two strings are equal.

21P. A wire 10.0 m long and having a mass of 100 g is stretched under a tension of 250 N. If two pulses, separated in time by 30.0 ms, are generated, one at each end of the wire, where will the pulses first meet? ssm ilw

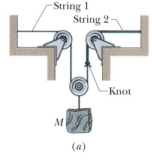
(a)

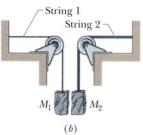
(b)

Fig. 17-31 Problem 20.

22P. The type of rubber band used inside some baseballs and golf balls obeys Hooke's law over a wide range of elongation of the band. A segment of this material has an unstretched length ℓ and a mass m. When a force F is applied, the band stretches an additional length $\Delta\ell$. (a) What is the speed (in terms of m, $\Delta\ell$, and the spring constant k) of transverse waves on this stretched rubber band? (b) Using your answer to (a), show that the time required for a transverse pulse to travel the length of the rubber band is proportional to $1/\sqrt{\Delta\ell}$ if $\Delta\ell \ll \ell$ and is constant if $\Delta\ell \gg \ell$.

23P*. A uniform rope of mass m and length L hangs from a ceiling. (a) Show that the speed of a transverse wave on the rope is a function of y, the distance from the lower end, and is given by $v = \sqrt{gy}$. (b) Show that the time a transverse wave takes to travel the length of the rope is given by $t = 2\sqrt{L/g}$. ssm

SEC. 17-7   Energy and Power of a Traveling String Wave

24E. A string along which waves can travel is 2.70 m long and has a mass of 260 g. The tension in the string is 36.0 N. What must be the frequency of traveling waves of amplitude 7.70 mm for the average power to be 85.0 W?

25P. A transverse sinusoidal wave is generated at one end of a long, horizontal string by a bar that moves up and down through a distance of 1.00 cm. The motion is continuous and is repeated regularly 120 times per second. The string has linear density 120 g/m and is kept under a tension of 90.0 N. Find the maximum value of (a) the transverse speed u and (b) the transverse component of the tension τ. (*Hint:* That component is $\tau \sin\theta$, where θ is the angle the string makes with the horizontal. You will need to relate angle θ to dy/dx.)

(c) Show that the two maximum values calculated above occur at the same phase values for the wave. What is the transverse displacement y of the string at these phases? (d) What is the maximum rate of energy transfer along the string? (e) What is the transverse displacement y when this maximum transfer occurs? (f) What is the minimum rate of energy transfer along the string? (g) What is the transverse displacement y when this minimum transfer occurs? ssm www

SEC. 17-9   Interference of Waves

26E. What phase difference between two otherwise identical traveling waves, moving in the same direction along a stretched string, will result in the combined wave having an amplitude 1.50 times that of the common amplitude of the two combining waves? Express your answer in (a) degrees, (b) radians, and (c) wavelengths.

27E. Two identical traveling waves, moving in the same direction, are out of phase by $\pi/2$ rad. What is the amplitude of the resultant wave in terms of the common amplitude y_m of the two combining waves? ssm

28P. Two sinusoidal waves, identical except for phase, travel in the same direction along a string and interfere to produce a resultant wave given by $y'(x, t) = (3.0 \text{ mm}) \sin(20x - 4.0t + 0.820 \text{ rad})$, with x in meters and t in seconds. What are (a) the wavelength λ of the two waves, (b) the phase difference between them, and (c) their amplitude y_m?

SEC. 17-10   Phasors

29E. Determine the amplitude of the resultant wave when two sinusoidal string waves having the same frequency and traveling in the same direction on the same string are combined, if their amplitudes are 3.0 cm and 4.0 cm and they have phase constants of 0 and $\pi/2$ rad, respectively. ssm

30P. Two sinusoidal waves of the same period, with amplitudes of 5.0 and 7.0 mm, travel in the same direction along a stretched string; they produce a resultant wave with an amplitude of 9.0 mm. The phase constant of the 5.0 mm wave is 0. What is the phase constant of the 7.0 mm wave?

31P. Three sinusoidal waves of the same frequency travel along a string in the positive direction of an x axis. Their amplitudes are y_1, $y_1/2$, and $y_1/3$, and their phase constants are 0, $\pi/2$, and π, respectively. What are (a) the amplitude and (b) the phase constant of the resultant wave? (c) Plot the wave form of the resultant wave at $t = 0$, and discuss its behavior as t increases. ssm www

SEC. 17-12   Standing Waves and Resonance

32E. A string under tension τ_i oscillates in the third harmonic at frequency f_3, and the waves on the string have wavelength λ_3. If the tension is increased to $\tau_f = 4\tau_i$ and the string is again made to oscillate in the third harmonic, what then are (a) the frequency of oscillation in terms of f_3 and (b) the wavelength of the waves in terms of λ_3?

33E. A nylon guitar string has a linear density of 7.2 g/m and is under a tension of 150 N. The fixed supports are 90 cm apart.

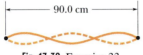
Fig. 17-32 Exercise 33.

The string is oscillating in the standing wave pattern shown in Fig. 17-32. Calculate the (a) speed, (b) wavelength, and (c) frequency of the traveling waves whose superposition gives this standing wave. ilw

34E. Two sinusoidal waves with identical wavelengths and amplitudes travel in opposite directions along a string with a speed of 10 cm/s. If the time interval between instants when the string is flat is 0.50 s, what is the wavelength of the waves?

35E. A string fixed at both ends is 8.40 m long and has a mass of 0.120 kg. It is subjected to a tension of 96.0 N and set oscillating. (a) What is the speed of the waves on the string? (b) What is the longest possible wavelength for a standing wave? (c) Give the frequency of that wave. ssm

36E. A 125 cm length of string has a mass of 2.00 g. It is stretched with a tension of 7.00 N between fixed supports. (a) What is the wave speed for this string? (b) What is the lowest resonant frequency of this string?

37E. What are the three lowest frequencies for standing waves on a wire 10.0 m long having a mass of 100 g, which is stretched under a tension of 250 N? ssm

38E. String A is stretched between two clamps separated by distance L. String B, with the same linear density and under the same tension as string A, is stretched between two clamps separated by distance $4L$. Consider the first eight harmonics of string B. Which, if any, has a resonant frequency that matches a resonant frequency of string A?

39P. A string that is stretched between fixed supports separated by 75.0 cm has resonant frequencies of 420 and 315 Hz, with no intermediate resonant frequencies. What are (a) the lowest resonant frequency and (b) the wave speed? ssm ilw www

40P. In Fig. 17-33, two pulses travel along a string in opposite

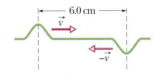
Fig. 17-33 Problem 40.

directions. The wave speed v is 2.0 m/s and the pulses are 6.0 cm apart at $t = 0$. (a) Sketch the wave patterns when t is equal to 5.0, 10, 15, 20, and 25 ms. (b) In what form (or type) is the energy of the pulses at $t = 15$ ms?

41P. A string oscillates according to the equation

$$y' = (0.50 \text{ cm}) \sin\left[\left(\frac{\pi}{3} \text{ cm}^{-1}\right)x\right] \cos[(40\pi \text{ s}^{-1})t].$$

What are (a) the amplitude and (b) the speed of the two waves (identical except for direction of travel) whose superposition gives this oscillation? (c) What is the distance between nodes? (d) What is the speed of a particle of the string at the position $x = 1.5$ cm when $t = \frac{9}{8}$ s? ssm

42P. A standing wave results from the sum of two transverse traveling waves given by

$$y_1 = 0.050 \cos(\pi x - 4\pi t)$$

and
$$y_2 = 0.050 \cos(\pi x + 4\pi t),$$

where x, y_1, and y_2 are in meters and t is in seconds. (a) What is the smallest positive value of x that corresponds to a node? (b) At what times during the interval $0 \le t \le 0.50$ s will the particle at $x = 0$ have zero velocity?

43P. A string 3.0 m long is oscillating as a three-loop standing wave with an amplitude of 1.0 cm. The wave speed is 100 m/s. (a) What is the frequency? (b) Write equations for two waves that, when combined, will result in this standing wave. ssm

44P. In an experiment on standing waves, a string 90 cm long is attached to the prong of an electrically driven tuning fork that oscillates perpendicular to the length of the string at a frequency of 60 Hz. The mass of the string is 0.044 kg. What tension must the string be under (weights are attached to the other end) if it is to oscillate in four loops?

45P. Oscillation of a 600 Hz tuning fork sets up standing waves in a string clamped at both ends. The wave speed for the string is 400 m/s. The standing wave has four loops and an amplitude of 2.0 mm. (a) What is the length of the string? (b) Write an equation for the displacement of the string as a function of position and time. ssm

46P. A rope, under a tension of 200 N and fixed at both ends, oscillates in a second-harmonic standing wave pattern. The displacement of the rope is given by

$$y = (0.10 \text{ m})(\sin \pi x/2) \sin 12\pi t,$$

where $x = 0$ at one end of the rope, x is in meters, and t is in seconds. What are (a) the length of the rope, (b) the speed of the waves on the rope, and (c) the mass of the rope? (d) If the rope oscillates in a third-harmonic standing wave pattern, what will be the period of oscillation?

47P. A generator at one end of a very long string creates a wave given by

$$y = (6.0 \text{ cm}) \cos \frac{\pi}{2} [(2.0 \text{ m}^{-1})x + (8.0 \text{ s}^{-1})t],$$

and one at the other end creates the wave

$$y = (6.0 \text{ cm}) \cos \frac{\pi}{2} [(2.0 \text{ m}^{-1})x - (8.0 \text{ s}^{-1})t].$$

Calculate (a) the frequency, (b) the wavelength, and (c) the speed of each wave. At what *x* values are (d) the nodes and (e) the antinodes? ssm

48P. A standing wave pattern on a string is described by

$$y(x, t) = 0.040 \sin 5\pi x \cos 40\pi t,$$

where *x* and *y* are in meters and *t* is in seconds. (a) Determine the location of all nodes for $0 \le x \le 0.40$ m. (b) What is the period of the oscillatory motion of any (nonnode) point on the string? What are (c) the speed and (d) the amplitude of the two traveling waves that interfere to produce this wave? (e) At what times for $0 \le t \le 0.050$ s will all the points on the string have zero transverse velocity?

49P. Show that the maximum kinetic energy in each loop of a standing wave produced by two traveling waves of identical amplitudes is $2\pi^2\mu y_m^2 fv$. ssm www

50P. For a certain transverse standing wave on a long string, an antinode is at $x = 0$ and a node is at $x = 0.10$ m. The displacement $y(t)$ of the string particle at $x = 0$ is shown in Fig. 17-34. When $t = 0.50$ s, what are the displacements of the string particles at (a) $x = 0.20$ m and (b) $x = 0.30$ m? At $x = 0.20$ m, what are the transverse velocities of the string particles at (c) $t = 0.50$ s and (d) $t = 1.0$ s? (e) Sketch the standing wave at $t = 0.50$ s for the range $x = 0$ to $x = 0.40$ m.

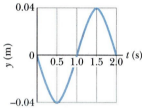

Fig. 17-34 Problem 50.

51P. In Fig. 17-35, an aluminum wire, of length $L_1 = 60.0$ cm, cross-sectional area 1.00×10^{-2} cm², and density 2.60 g/cm³, is joined to a steel wire, of density 7.80 g/cm³ and the same cross-sectional area. The compound wire, loaded with a block of mass $m = 10.0$ kg, is arranged so that the distance L_2 from the joint to the supporting pulley is 86.6 cm. Transverse waves are set up in the wire by using an external source of variable frequency; a node is located at the pulley. (a) Find the lowest frequency of excitation for which standing waves are observed such that the joint in the wire is one of the nodes. (b) How many nodes are observed at this frequency? ssm

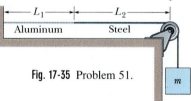

Fig. 17-35 Problem 51.

Additional Problem

52. *Body armor.* When a high-speed projectile such as a bullet or bomb fragment strikes modern body armor, the fabric of the armor stops the projectile and prevents penetration by quickly spreading the projectile's energy over a large area. This spreading is done by longitudinal and transverse pulses that move *radially* from the impact point, where the projectile pushes a cone-shaped dent into the fabric. The longitudinal pulse, racing along the fibers of the fabric at speed v_l ahead of the denting, causes the fibers to thin and stretch, with material flowing radially inward into the dent. One such radial fiber is shown in Fig. 17-36a. Part of the projectile's energy goes into this motion and stretching. The transverse pulse, moving at a slower speed v_t, is due to the denting. As the projectile increases the dent's depth, the dent increases in radius, causing the material in the fibers to move in the same direction as the projectile (perpendicular to the transverse pulse's direction of travel). The rest of the projectile's energy goes into this motion. All the energy that does not eventually go into permanently deforming the fibers ends up as thermal energy.

Figure 17-36b is a graph of speed v versus time t for a bullet of mass 10.2 g fired from a .38 Special revolver directly into body armor. Take $v_l = 2000$ m/s, and assume that the half-angle θ of the conical dent is 60°. At the end of the collision, what are the radii of (a) the thinned region and (b) the dent (assuming that the person wearing the armor remains stationary)?

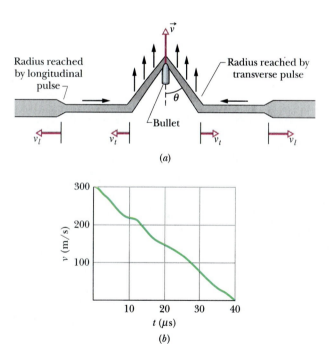

Fig. 17-36 Problem 52.

18 Waves—II

This horseshoe bat not only can locate a moth flying in total darkness but can also determine the moth's relative speed, to home in on the insect.

How does the bat's detection system work, and how can a moth "jam" the system or otherwise reduce its effectiveness?

The answer is in this chapter.

Fig. 18-1 An image of a fetus searching for a thumb to suck; the image is made with ultrasound (which has a frequency above your hearing range).

18-1 Sound Waves

As we saw in Chapter 17, mechanical waves are waves that require a material medium to exist. There are two types of mechanical waves: *Transverse waves* involve oscillations perpendicular to the direction in which the wave travels; *longitudinal waves* involve oscillations parallel to the direction of wave travel.

In this book, a **sound wave** is defined roughly as any longitudinal wave. Seismic prospecting teams use such waves to probe Earth's crust for oil. Ships carry sound-ranging gear (sonar) to detect underwater obstacles. Submarines use sound waves to stalk other submarines, largely by listening for the characteristic noises produced by the propulsion system. Figure 18-1, a computer-processed image of a fetal head and arm, shows how sound waves can be used to explore the soft tissues of the human body. In this chapter we shall focus on sound waves that travel through the air and that are audible to people.

Figure 18-2 illustrates several ideas that we shall use in our discussions. Point S represents a tiny sound source, called a *point source*, that emits sound waves in all directions. The *wavefronts* and *rays* indicate the direction of travel and the spread of the sound waves. **Wavefronts** are surfaces over which the oscillations of the air due to the sound wave have the same value; such surfaces are represented by whole or partial circles in a two-dimensional drawing for a point source. **Rays** are directed lines perpendicular to the wavefronts that indicate the direction of travel of the wavefronts. The short double arrows superimposed on the rays of Fig. 18-2 indicate that the longitudinal oscillations of the air are parallel to the rays.

Near a point source like that of Fig. 18-2, the wavefronts are spherical and spread out in three dimensions, and there the waves are said to be *spherical*. As the wavefronts move outward and their radii become larger, their curvature decreases. Far from the source, we approximate the wavefronts as planes (or lines on two-dimensional drawings), and the waves are said to be *planar*.

18-2 The Speed of Sound

The speed of any mechanical wave, transverse or longitudinal, depends on both an inertial property of the medium (to store kinetic energy) and an elastic property of the medium (to store potential energy). Thus, we can generalize Eq. 17-25, which gives the speed of a transverse wave along a stretched string, by writing

$$v = \sqrt{\frac{\tau}{\mu}} = \sqrt{\frac{\text{elastic property}}{\text{inertial property}}}, \qquad (18\text{-}1)$$

where (for transverse waves) τ is the tension in the string and μ is the string's linear density. If the medium is air and the wave is longitudinal, we can guess that the inertial property, corresponding to μ, is the volume density ρ of air. What shall we put for the elastic property?

In a stretched string, potential energy is associated with the periodic stretching of the string elements as the wave passes through them. As a sound wave passes through air, potential energy is associated with periodic compressions and expansions of small volume elements of the air. The property that determines the extent to which an element of a medium changes in volume when the pressure (force per unit area) on it changes is the **bulk modulus** B, defined (from Eq. 13-27) as

$$B = -\frac{\Delta p}{\Delta V/V} \qquad \text{(definition of bulk modulus).} \qquad (18\text{-}2)$$

Here $\Delta V/V$ is the fractional change in volume produced by a change in pressure Δp. As explained in Section 15-3, the SI unit for pressure is the newton per square meter,

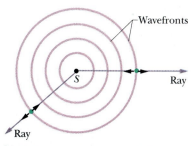

Fig. 18-2 A sound wave travels from a point source S through a three-dimensional medium. The wavefronts form spheres centered on S; the rays are radial to S. The short, double-headed arrows indicate that elements of the medium oscillate parallel to the rays.

TABLE 18-1   The Speed of Sound[a]

Medium	Speed (m/s)
Gases	
Air (0°C)	331
Air (20°C)	343
Helium	965
Hydrogen	1284
Liquids	
Water (0°C)	1402
Water (20°C)	1482
Seawater[b]	1522
Solids	
Aluminum	6420
Steel	5941
Granite	6000

[a]At 0°C and 1 atm pressure, except where noted.

[b]At 20°C and 3.5% salinity.

which is given a special name, the *pascal* (Pa). From Eq. 18-2 we see that the unit for B is also the pascal. The signs of Δp and ΔV are always opposite: When we increase the pressure on an element (Δp is positive), its volume decreases (ΔV is negative). We include a minus sign in Eq. 18-2 so that B is always a positive quantity. Now substituting B for τ and ρ for μ in Eq. 18-1 yields

$$v = \sqrt{\frac{B}{\rho}} \qquad \text{(speed of sound)} \qquad (18\text{-}3)$$

as the speed of sound in a medium with bulk modulus B and density ρ. This is actually the correct equation, as we shall derive shortly. Table 18-1 lists the speed of sound in various media.

The density of water is almost 1000 times greater than the density of air. If this were the only relevant factor, we would expect from Eq. 18-3 that the speed of sound in water would be considerably less than the speed of sound in air. However, Table 18-1 shows us that the reverse is true. We conclude (again from Eq. 18-3) that the bulk modulus of water must be more than 1000 times greater than that of air. This is indeed the case. Water is much more incompressible than air, which (see Eq. 18-2) is another way of saying that its bulk modulus is much greater.

Formal Derivation of Eq. 18-3

We now derive Eq. 18-3 by direct application of Newton's laws. Let a single pulse in which air is compressed travel (from right to left) with speed v through the air in a long tube, like that in Fig. 17-2. Let us run along with the pulse at that speed, so that the pulse appears to stand still in our reference frame. Figure 18-3a shows the situation as it is viewed from that frame. The pulse is standing still, and air is moving at speed v through it from left to right.

Let the pressure of the undisturbed air be p and the pressure inside the pulse be $p + \Delta p$, where Δp is positive owing to the compression. Consider a slice of air of thickness Δx and face area A, moving toward the pulse at speed v. As this element of air enters the pulse, its leading face encounters a region of higher pressure, which slows it to speed $v + \Delta v$, in which Δv is negative. This slowing is complete when the rear face reaches the pulse, which requires time interval

$$\Delta t = \frac{\Delta x}{v}. \qquad (18\text{-}4)$$

Fig. 18-3 A compression pulse is sent down a long air-filled tube. The reference frame of the figure is chosen so that the pulse is at rest and the air moves from left to right. (*a*) A slice of air of width Δx moves toward the pulse with speed v. (*b*) The leading face of the slice enters the pulse. The forces acting on the leading and trailing faces (due to air pressure) are shown.

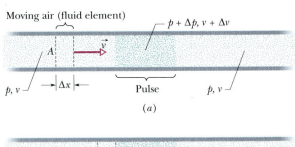

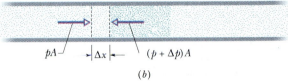

Let us apply Newton's second law to the element. During Δt, the average force on the element's trailing face is pA toward the right, and the average force on the leading face is $(p + \Delta p)A$ toward the left (Fig. 18-3b). Therefore, the average net force on the element during Δt is

$$F = pA - (p + \Delta p)A$$
$$= -\Delta p\, A \quad \text{(net force)}. \quad (18\text{-}5)$$

The minus sign indicates that the net force on the fluid element is directed to the left in Fig. 18-3b. The volume of the element is $A\, \Delta x$, so with the aid of Eq. 18-4, we can write its mass as

$$\Delta m = \rho A\, \Delta x = \rho A v\, \Delta t \quad \text{(mass)}. \quad (18\text{-}6)$$

Then the average acceleration of the element during Δt is

$$a = \frac{\Delta v}{\Delta t} \quad \text{(acceleration)}. \quad (18\text{-}7)$$

From Newton's second law ($F = ma$), we have, from Eqs. 18-5, 18-6, and 18-7,

$$-\Delta p\, A = (\rho A v\, \Delta t)\frac{\Delta v}{\Delta t},$$

which we can write as

$$\rho v^2 = -\frac{\Delta p}{\Delta v/v}. \quad (18\text{-}8)$$

The air that occupies a volume $V\ (= Av\, \Delta t)$ outside the pulse is compressed by an amount $\Delta V\ (= A\, \Delta v\, \Delta t)$ as it enters the pulse. Thus,

$$\frac{\Delta V}{V} = \frac{A\, \Delta v\, \Delta t}{Av\, \Delta t} = \frac{\Delta v}{v}. \quad (18\text{-}9)$$

Substituting Eq. 18-9 and then Eq. 18-2 into Eq. 18-8 leads to

$$\rho v^2 = -\frac{\Delta p}{\Delta v/v} = -\frac{\Delta p}{\Delta V/V} = B.$$

Solving for v yields Eq. 18-3 for the speed of the air toward the right in Fig. 18-3, and thus for the actual speed of the pulse toward the left.

Sample Problem 18-1

One clue used by your brain to determine the direction of a source of sound is the time delay Δt between the arrival of the sound at the ear closer to the source and the arrival at the farther ear. Assume that the source is distant so that a wavefront from it is approximately planar when it reaches you, and let D represent the separation between your ears.

(a) Find an expression that gives Δt in terms of D and the angle θ between the direction of the source and the forward direction.

SOLUTION: The situation is shown (from an overhead view) in Fig. 18-4, where wavefronts approach you from a source that is located in front of you and to your right. The **Key Idea** here is that

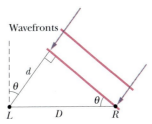

Fig. 18-4 Sample Problem 18-1. A wavefront travels a distance d ($= D \sin \theta$) farther to reach the left ear (L) than to reach the right ear (R).

the time delay Δt is due to the distance d that each wavefront must travel to reach your left ear (L) after it reaches your right ear (R). From Fig. 18-4, we find

$$\Delta t = \frac{d}{v} = \frac{D \sin \theta}{v}, \qquad \text{(Answer)} \quad (18\text{-}10)$$

where v is the speed of sound in air. Based on a lifetime of experience, your brain correlates each detected value of Δt (from zero to the maximum value) with a value of θ (from zero to 90°) for the direction of the sound source.

(b) Suppose that you are submerged in water at 20°C when a wavefront arrives from directly to your right. Based on the time-delay clue, at what angle θ from the forward direction does the source seem to be?

SOLUTION: The Key Idea here is that the speed of the sound is now the speed v_w in water, so in Eq. 18-10 we substitute v_w for v and

90° for θ, finding that

$$\Delta t_w = \frac{D \sin 90°}{v_w} = \frac{D}{v_w}. \qquad (18\text{-}11)$$

Since v_w is about four times v, delay Δt_w is about one-fourth the maximum time delay in air. Based on experience, your brain will process the water time delay as if it occurred in air. Thus, the sound source appears to be at an angle θ smaller than 90°. To find that apparent angle, we substitute the time delay D/v_w from Eq. 18-11 for Δt in Eq. 18-10, obtaining

$$\frac{D}{v_w} = \frac{D \sin \theta}{v}. \qquad (18\text{-}12)$$

Then, to solve for θ we substitute $v = 343$ m/s and $v_w = 1482$ m/s (from Table 18-1) into Eq. 18-12, finding

$$\sin \theta = \frac{v}{v_w} = \frac{343 \text{ m/s}}{1482 \text{ m/s}} = 0.231$$

and thus

$$\theta = 13°. \qquad \text{(Answer)}$$

18-3 Traveling Sound Waves

Here we examine the displacements and pressure variations associated with a sinusoidal sound wave traveling through air. Figure 18-5a displays such a wave traveling rightward through a long air-filled tube. Recall from Chapter 17 that we can produce such a wave by sinusoidally moving a piston at the left end of the tube (as in Fig. 17-2). The piston's rightward motion moves the element of air next to it and compresses that air; the piston's leftward motion allows the element of air to move back to the left and the pressure to decrease. As each element of air pushes on the next element in turn, the right–left motion of the air and the change in its pressure travel along the tube as a sound wave.

Consider a thin element of air of thickness Δx, located at a position x along the tube. As the wave travels through x, the element of air oscillates left and right in simple harmonic motion about its equilibrium position (Fig. 18-5b). Thus, the oscillations of each air element due to the traveling sound wave are like those of a string element due to a transverse wave, except that the air element oscillates *longitudinally* rather than *transversely*. Because string elements oscillate parallel to the y axis, we write their displacements in the form $y(x, t)$. Similarly, because air ele-

Fig. 18-5 (a) A sound wave, traveling through a long air-filled tube with speed v, consists of a moving, periodic pattern of expansions and compressions of the air. The wave is shown at an arbitrary instant. (b) A horizontally expanded view of a short piece of the tube. As the wave passes, a fluid element of thickness Δx oscillates left and right in simple harmonic motion about its equilibrium position. At the instant shown in (b), the element happens to be displaced a distance s to the right of its equilibrium position. Its maximum displacement, either right or left, is s_m.

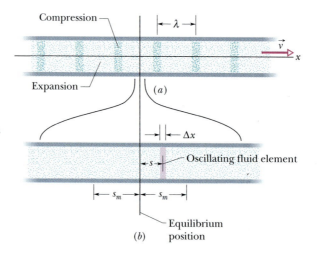

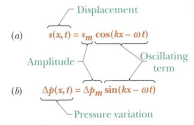

Fig. 18-6 (*a*) The displacement function and (*b*) the pressure-variation function of a traveling sound wave consist of an amplitude and an oscillating term.

ments oscillate parallel to the x axis, we could write their displacements in the form $x(x, t)$. However, we shall avoid that confusing notation and use $s(x, t)$ instead.

To show that the displacements $s(x, t)$ are sinusoidal functions of x and t, we can use either a sine function or a cosine function. In this chapter we use a cosine function, writing

$$s(x, t) = s_m \cos(kx - \omega t). \qquad (18\text{-}13)$$

Figure 18-6*a* shows the major parts of this equation. In it, s_m is the **displacement amplitude**—that is, the maximum displacement of the air element to either side of its equilibrium position (see Fig. 18-5*b*). The angular wave number k, angular frequency ω, frequency f, wavelength λ, speed v, and period T for a sound (longitudinal) wave are defined and interrelated exactly as for a transverse wave, except that λ is now the distance (again along the direction of travel) in which the pattern of compression and expansion due to the wave begins to repeat itself (see Fig. 18-5*a*). (We assume s_m is much less than λ.)

As the wave moves, the air pressure at any position x in Fig. 18-5*a* varies sinusoidally, as we prove next. To describe this variation we write

$$\Delta p(x, t) = \Delta p_m \sin(kx - \omega t). \qquad (18\text{-}14)$$

Figure 18-6*b* shows the major parts of this equation. A negative value of Δp in Eq. 18-14 corresponds to an expansion of the air, and a positive value to a compression. Here Δp_m is the **pressure amplitude,** which is the maximum increase or decrease in pressure due to the wave; Δp_m is normally very much less than the pressure p present when there is no wave. As we shall prove, the pressure amplitude Δp_m is related to the displacement amplitude s_m in Eq. 18-13 by

$$\Delta p_m = (v\rho\omega)s_m. \qquad (18\text{-}15)$$

Figure 18-7 shows plots of Eqs. 18-13 and 18-14 at $t = 0$; with time, the two curves would move rightward along the horizontal axes. Note that the displacement and pressure variation are $\pi/2$ rad (or 90°) out of phase. Thus, for example, the pressure variation Δp at any point along the wave is zero when the displacement there is a maximum.

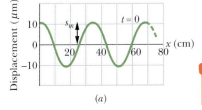

(*a*)

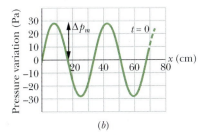

(*b*)

Fig. 18-7 (*a*) A plot of the displacement function (Eq. 18-13) for $t = 0$. (*b*) A similar plot of the pressure-variation function (Eq. 18-14). Both plots are for a 1000 Hz sound wave whose pressure amplitude is at the threshold of pain; see Sample Problem 18-2.

✓**CHECKPOINT 1:**   When the oscillating fluid element in Fig. 18-5*b* is moving rightward through the point of zero displacement, is the pressure in the element at its equilibrium value, just beginning to increase, or just beginning to decrease?

Derivation of Eqs. 18-14 and 18-15

Figure 18-5*b* shows an oscillating element of air of cross-sectional area A and thickness Δx, with its center displaced from its equilibrium position by distance s.

From Eq. 18-2 we can write, for the pressure variation in the displaced element,

$$\Delta p = -B \frac{\Delta V}{V}. \qquad (18\text{-}16)$$

The quantity V in Eq. 18-16 is the volume of the element, given by

$$V = A \, \Delta x. \qquad (18\text{-}17)$$

The quantity ΔV in Eq. 18-16 is the change in volume that occurs when the element is displaced. This volume change comes about because the displacements of the two faces of the element are not quite the same, differing by some amount Δs. Thus, we can write the change in volume as

$$\Delta V = A \, \Delta s. \qquad (18\text{-}18)$$

Substituting Eqs. 18-17 and 18-18 into Eq. 18-16 and passing to the differential limit yield

$$\Delta p = -B \frac{\Delta s}{\Delta x} = -B \frac{\partial s}{\partial x}. \qquad (18\text{-}19)$$

The symbols ∂ indicate that the derivative in Eq. 18-19 is a *partial derivative*, which tells us how s changes with x when the time t is fixed. From Eq. 18-13 we then have, treating t as a constant,

$$\frac{\partial s}{\partial x} = \frac{\partial}{\partial x} [s_m \cos(kx - \omega t)] = -ks_m \sin(kx - \omega t).$$

Substituting this quantity for the partial derivative in Eq. 18-19 yields

$$\Delta p = Bks_m \sin(kx - \omega t).$$

Setting $\Delta p_m = Bks_m$, this yields Eq. 18-14, which we set out to prove.

Using Eq. 18-3, we can now write

$$\Delta p_m = (Bk)s_m = (v^2 \rho k)s_m.$$

Equation 18-15, which we also promised to prove, follows at once if we substitute ω/v for k from Eq. 17-12.

Sample Problem 18-2

The maximum pressure amplitude Δp_m that the human ear can tolerate in loud sounds is about 28 Pa (which is very much less than the normal air pressure of about 10^5 Pa). What is the displacement amplitude s_m for such a sound in air of density $\rho = 1.21$ kg/m^3, at a frequency of 1000 Hz and a speed of 343 m/s?

SOLUTION: The **Key Idea** is that the displacement amplitude s_m of a sound wave is related to the pressure amplitude Δp_m of the wave according to Eq. 18-15. Solving that equation for s_m yields

$$s_m = \frac{\Delta p_m}{v\rho\omega} = \frac{\Delta p_m}{v\rho(2\pi f)}.$$

Substituting known data then gives us

$$s_m = \frac{28 \text{ Pa}}{(343 \text{ m/s})(1.21 \text{ kg/m}^3)(2\pi)(1000 \text{ Hz})}$$
$$= 1.1 \times 10^{-5} \text{ m} = 11 \ \mu\text{m}. \qquad \text{(Answer)}$$

That is only about one-seventh the thickness of this page. Obviously, the displacement amplitude of even the loudest sound that the ear can tolerate is very small.

The pressure amplitude Δp_m for the *faintest* detectable sound at 1000 Hz is 2.8×10^{-5} Pa. Proceeding as above leads to $s_m = 1.1 \times 10^{-11}$ m or 11 pm, which is about one-tenth the radius of a typical atom. The ear is indeed a sensitive detector of sound waves.

18-4   Interference

Like transverse waves, sound waves can undergo interference. Let us consider, in particular, the interference between two identical sound waves traveling in the same direction. Figure 18-8 shows how we can set up such a situation: Two point sources S_1 and S_2 emit sound waves that are in phase and of identical wavelength λ. Thus, the sources themselves are said to be in phase; that is, as the waves emerge from the sources, their displacements are always identical. We are interested in the waves that then travel through point P in Fig. 18-8. We assume that the distance to P is much greater than the distance between the sources so that we can approximate the waves as traveling in the same direction at P.

If the waves traveled along paths with identical lengths to reach point P, they would be in phase there. As with transverse waves, this means that they would undergo fully constructive interference there. However, in Fig. 18-8, path L_2 traveled

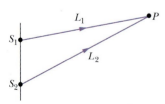

Fig. 18-8 Two point sources S_1 and S_2 emit spherical sound waves in phase. The rays indicate that the waves pass through a common point P.

by the wave from S_2 is longer than path L_1 traveled by the wave from S_1. The difference in path lengths means that the waves may not be in phase at point P. In other words, their phase difference ϕ at P depends on their **path length difference** $\Delta L = |L_2 - L_1|$.

To relate phase difference ϕ to path length difference ΔL, we recall (from Section 17-4) that a phase difference of 2π rad corresponds to one wavelength. Thus, we can write the proportion

$$\frac{\phi}{2\pi} = \frac{\Delta L}{\lambda}, \tag{18-20}$$

from which

$$\phi = \frac{\Delta L}{\lambda} 2\pi. \tag{18-21}$$

Fully constructive interference occurs when ϕ is zero, 2π, or any integer multiple of 2π. We can write this condition as

$$\phi = m(2\pi), \qquad \text{for } m = 0, 1, 2, \ldots \qquad \text{(fully constructive interference).} \tag{18-22}$$

From Eq. 18-21, this occurs when the ratio $\Delta L/\lambda$ is

$$\frac{\Delta L}{\lambda} = 0, 1, 2, \ldots \qquad \text{(fully constructive interference).} \tag{18-23}$$

For example, if the path length difference $\Delta L = |L_2 - L_1|$ in Fig. 18-8 is equal to 2λ, then $\Delta L/\lambda = 2$ and the waves undergo fully constructive interference at point P. The interference is fully constructive because the wave from S_2 is phase-shifted relative to the wave from S_1 by 2λ, putting the two waves *exactly in phase* at P.

Fully destructive interference occurs when ϕ is an odd multiple of π, a condition we can write as

$$\phi = (2m + 1)\pi, \qquad \text{for } m = 0, 1, 2, \ldots \qquad \text{(fully destructive interference).} \tag{18-24}$$

From Eq. 18-21, this occurs when the ratio $\Delta L/\lambda$ is

$$\frac{\Delta L}{\lambda} = 0.5, 1.5, 2.5, \ldots \qquad \text{(fully destructive interference).} \tag{18-25}$$

For example, if the path length difference $\Delta L = |L_2 - L_1|$ in Fig. 18-8 is equal to 2.5λ, then $\Delta L/\lambda = 2.5$ and the waves undergo fully destructive interference at point P. The interference is fully destructive because the wave from S_2 is phase-shifted relative to the wave from S_1 by 2.5 wavelengths, which puts the two waves *exactly out of phase* at P.

Of course, two waves could produce intermediate interference as, say, when $\Delta L/\lambda = 1.2$. This would be closer to fully constructive interference ($\Delta L/\lambda = 1.0$) than to fully destructive interference ($\Delta L/\lambda = 1.5$).

Sample Problem 18-3

In Fig. 18-9a, two point sources S_1 and S_2, which are in phase and separated by distance $D = 1.5\lambda$, emit identical sound waves of wavelength λ.

(a) What is the path length difference of the waves from S_1 and S_2 at point P_1, which lies on the perpendicular bisector of distance

D, at a distance greater than D from the sources? What type of interference occurs at P_1?

SOLUTION: The Key Idea here is that, because the waves travel identical distances to reach P_1, their path length difference is

$$\Delta L = 0. \qquad \text{(Answer)}$$

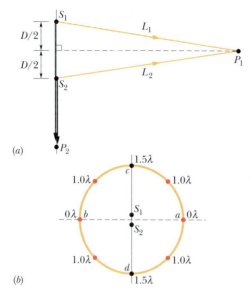

(a)

(b)

Fig. 18-9 Sample Problem 18-3. (a) Two point sources S_1 and S_2, separated by distance D, emit spherical sound waves in phase. The waves travel equal distances to reach point P_1. Point P_2 is on the line extending through S_1 and S_2. (b) The path length difference (in terms of wavelength) between the waves from S_1 and S_2, at eight points on a large circle around the sources.

From Eq. 18-23, this means that the waves undergo fully constructive interference at P_1.

(b) What are the path length difference and type of interference at point P_2 in Fig. 18-9a?

SOLUTION: Now the **Key Idea** is that the wave from S_1 travels the extra distance D (= 1.5λ) to reach P_2. Thus, the path length difference is

$$\Delta L = 1.5\lambda. \qquad \text{(Answer)}$$

From Eq. 18-25, this means that the waves are exactly out of phase at P_2 and undergo fully destructive interference there.

(c) Figure 18-9b shows a circle with a radius much greater than D, centered on the midpoint between sources S_1 and S_2. What is the number of points N around this circle at which the interference is fully constructive?

SOLUTION: Imagine that, starting at point a, we move clockwise along the circle to point d. One **Key Idea** here is that as we move to point d, the path length difference ΔL increases and so the type of interference changes. From (a), we know that the path length difference is $\Delta L = 0\lambda$ at point a. From (b), we know that $\Delta L = 1.5\lambda$ at point d. Thus, there must be one point along the circle between a and d at which $\Delta L = \lambda$, as indicated in Fig. 18-9b. From Eq. 18-23, fully constructive interference occurs at that point. Also, there can be no other point along the way from point a to point d at which fully constructive interference occurs, because there is no other integer than 1 between 0 and 1.5.

Another **Key Idea** here is to use symmetry to locate the other points of fully constructive interference along the rest of the circle. Symmetry about line cd gives us point b, at which $\Delta L = 0\lambda$. Also, there are three more points at which $\Delta L = \lambda$. In all we have

$$N = 6. \qquad \text{(Answer)}$$

✔**CHECKPOINT 2:** In this sample problem, if the distance D between sources S_1 and S_2 were, instead, equal to 4λ, what would be the path length difference and what type of interference would occur at (a) point P_1 and (b) point P_2?

18-5 Intensity and Sound Level

If you have ever tried to sleep while someone played loud music nearby, you are well aware that there is more to sound than frequency, wavelength, and speed. There is also intensity. The **intensity** I of a sound wave at a surface is the average rate per unit area at which energy is transferred by the wave through or onto the surface. We can write this as

$$I = \frac{P}{A}, \qquad (18\text{-}26)$$

where P is the time rate of energy transfer (the power) of the sound wave, and A is the area of the surface intercepting the sound. As we shall derive shortly, the intensity I is related to the displacement amplitude s_m of the sound wave by

$$I = \tfrac{1}{2}\rho v \omega^2 s_m^2. \qquad (18\text{-}27)$$

Variation of Intensity with Distance

How intensity varies with distance from a real sound source is often complex. Some real sources (like loudspeakers) may transmit sound only in particular directions,

Sound can cause the wall of a drinking glass to oscillate. If the sound produces a standing wave of oscillations and if the intensity of the sound is large enough, the glass will shatter.

and the environment usually produces echoes (reflected sound waves) that overlap the direct sound waves. In some situations, however, we can ignore echoes and assume that the sound source is a point source that emits the sound *isotropically*—that is, with equal intensity in all directions. The wavefronts spreading from such an isotropic point source S at a particular instant are shown in Fig. 18-10.

Let us assume that the mechanical energy of the sound waves is conserved as they spread from this source. Let us also center an imaginary sphere of radius r on the source, as shown in Fig. 18-10. All the energy emitted by the source must pass through the surface of the sphere. Thus, the time rate at which energy is transferred through the surface by the sound waves must equal the time rate at which energy is emitted by the source (that is, the power P_s of the source). From Eq. 18-26, the intensity I at the sphere must then be

$$I = \frac{P_s}{4\pi r^2},\tag{18-28}$$

where $4\pi r^2$ is the area of the sphere. Equation 18-28 tells us that the intensity of sound from an isotropic point source decreases with the square of the distance r from the source.

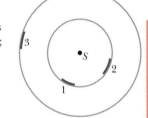

CHECKPOINT 3: The figure indicates three small patches 1, 2, and 3 that lie on the surfaces of two imaginary spheres; the spheres are centered on an isotropic point source S of sound. The rates at which energy is transmitted through the three patches by the sound waves are equal. Rank the patches according to (a) the intensity of the sound on them and (b) their area, greatest first.

The Decibel Scale

You saw in Sample Problem 18-2 that the displacement amplitude at the human ear ranges from about 10^{-5} m for the loudest tolerable sound to about 10^{-11} m for the faintest detectable sound, a ratio of 10^6. From Eq. 18-27 we see that the intensity of a sound varies as the *square* of its amplitude, so the ratio of intensities at these two limits of the human auditory system is 10^{12}. Humans can hear over an enormous range of intensities.

We deal with such an enormous range of values by using logarithms. Consider the relation

$$y = \log x,$$

in which x and y are variables. It is a property of this equation that if we *multiply x* by 10, then y increases by 1. To see this, we write

$$y' = \log(10x) = \log 10 + \log x = 1 + y.$$

Similarly, if we multiply x by 10^{12}, y increases by only 12.

Thus, instead of speaking of the intensity I of a sound wave, it is much more convenient to speak of its **sound level** β, defined as

$$\beta = (10 \text{ dB}) \log \frac{I}{I_0}.\tag{18-29}$$

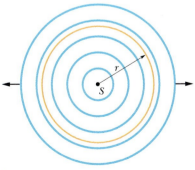

Fig. 18-10 A point source S emits sound waves uniformly in all directions. The waves pass through an imaginary sphere of radius r that is centered on S.

Here dB is the abbreviation for **decibel**, the unit of sound level, a name that was chosen to recognize the work of Alexander Graham Bell. I_0 in Eq. 18-29 is a standard reference intensity ($= 10^{-12}$ W/m^2), chosen because it is near the lower limit of the human range of hearing. For $I = I_0$, Eq. 18-29 gives $\beta = 10 \log 1 = 0$, so our standard reference level corresponds to zero decibels. Then β increases by 10 dB every time the sound intensity increases by an order of magnitude (a factor of 10). Thus, $\beta = 40$ corresponds to an intensity that is 10^4 times the standard reference level. Table 18-2 lists the sound levels for a variety of environments.

Derivation of Eq. 18-27

Consider, in Fig. 18-5a, a thin slice of air of thickness dx, area A, and mass dm, oscillating back and forth as the sound wave of Eq. 18-13 passes through it. The kinetic energy dK of the slice of air is

$$dK = \tfrac{1}{2}dm \, v_s^2. \tag{18-30}$$

Here v_s is not the speed of the wave but the speed of the oscillating element of air, obtained from Eq. 18-13 as

$$v_s = \frac{\partial s}{\partial t} = -\omega s_m \sin(kx - \omega t).$$

Using this relation and putting $dm = \rho A \, dx$ allow us to rewrite Eq. 18-30 as

$$dK = \tfrac{1}{2}(\rho A \, dx)(-\omega s_m)^2 \sin^2(kx - \omega t). \tag{18-31}$$

Dividing Eq. 18-31 by dt gives the rate at which kinetic energy moves along with the wave. As we saw in Chapter 17 for transverse waves, dx/dt is the wave speed v, so we have

$$\frac{dK}{dt} = \tfrac{1}{2}\rho A v \omega^2 s_m^2 \sin^2(kx - \omega t). \tag{18-32}$$

The *average* rate at which kinetic energy is transported is

$$\left(\frac{dK}{dt}\right)_{\text{avg}} = \tfrac{1}{2}\rho A v \omega^2 s_m^2 [\sin^2(kx - \omega t)]_{\text{avg}}$$

$$= \tfrac{1}{4}\rho A v \omega^2 s_m^2. \tag{18-33}$$

To obtain this equation, we have used the fact that the average value of the square of a sine (or a cosine) function over one full oscillation is $\tfrac{1}{2}$.

We assume that *potential* energy is carried along with the wave at this same average rate. The wave intensity I, which is the average rate per unit area at which energy of both kinds is transmitted by the wave, is then, from Eq. 18-33,

$$I = \frac{2(dK/dt)_{\text{avg}}}{A} = \tfrac{1}{2}\rho v \omega^2 s_m^2,$$

which is Eq. 18-27, the equation we set out to derive.

Sample Problem 18-4

An electric spark jumps along a straight line of length $L = 10$ m, emitting a pulse of sound that travels radially outward from the spark. (The spark is said to be a *line source* of sound.) The power of the emission is $P_s = 1.6 \times 10^4$ W.

(a) What is the intensity I of the sound when it reaches a distance $r = 12$ m from the spark?

SOLUTION: Let us center an imaginary cylinder of radius $r = 12$ m and length $L = 10$ m (open at both ends) on the spark, as shown in Fig. 18-11. One **Key Idea** here is that the intensity I at the cylin-

drical surface is the ratio P/A of the time rate P at which sound energy passes through the surface to the surface area A. Another **Key Idea** is to assume that the principle of conservation of energy applies to the sound energy. This means that the rate P at which energy is transferred through the cylinder must equal the rate P_s at which energy is emitted by the source. Putting these ideas together and noting that the area of the cylindrical surface is $A = 2\pi r L$, we have

$$I = \frac{P}{A} = \frac{P_s}{2\pi r L}. \tag{18-34}$$

This tells us that the intensity of the sound from a line source

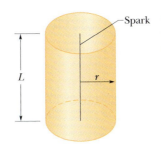

Fig. 18-11 Sample Problem 18-4. A spark along a straight line of length L emits sound waves radially outward. The waves pass through an imaginary cylinder of radius r and length L that is centered on the spark.

(b) At what time rate P_d is sound energy intercepted by an acoustic detector of area $A_d = 2.0 \text{ cm}^2$, aimed at the spark and located a distance $r = 12$ m from the spark?

SOLUTION: Applying the first Key Idea of part (a), we know that the intensity of sound at the detector is the ratio of the energy transfer rate P_d there to the detector's area A_d:

$$I = \frac{P_d}{A_d}. \quad (18\text{-}35)$$

We can imagine that the detector lies on the cylindrical surface of (a). Then the sound intensity at the detector is the intensity $I \; (= 21.2 \text{ W/m}^2)$ at the cylindrical surface. Solving Eq. 18-35 for P_d gives us

$$P_d = (21.2 \text{ W/m}^2)(2.0 \times 10^{-4} \text{ m}^2) = 4.2 \text{ mW}. \quad \text{(Answer)}$$

decreases with distance r (and not with the square of distance r as for a point source). Substituting the given data, we find

$$I = \frac{1.6 \times 10^4 \text{ W}}{2\pi(12 \text{ m})(10 \text{ m})} = 21.2 \text{ W/m}^2 \approx 21 \text{ W/m}^2. \quad \text{(Answer)}$$

Sample Problem 18-5

In 1976, the Who set a record for the loudest concert—the sound level 46 m in front of the speaker systems was $\beta_2 = 120$ dB. What is the ratio of the intensity I_2 of the band at that spot to the intensity I_1 of a jackhammer operating at sound level $\beta_1 = 92$ dB?

SOLUTION: The Key Idea here is that for both the Who and the jackhammer, the sound level β is related to the intensity by the definition of sound level in Eq. 18-29. For the Who, we have

$$\beta_2 = (10 \text{ dB}) \log \frac{I_2}{I_0},$$

and for the jackhammer, we have

$$\beta_1 = (10 \text{ dB}) \log \frac{I_1}{I_0}.$$

The difference in the sound levels is

$$\beta_2 - \beta_1 = (10 \text{ dB})\left(\log \frac{I_2}{I_0} - \log \frac{I_1}{I_0}\right). \quad (18\text{-}36)$$

Using the identity

$$\log \frac{a}{b} - \log \frac{c}{d} = \log \frac{ad}{bc},$$

we can rewrite Eq. 18-36 as

$$\beta_2 - \beta_1 = (10 \text{ dB}) \log \frac{I_2}{I_1}. \quad (18\text{-}37)$$

Rearranging and substituting the known sound levels now yield

$$\log \frac{I_2}{I_1} = \frac{\beta_2 - \beta_1}{10 \text{ dB}} = \frac{120 \text{ dB} - 92 \text{ dB}}{10 \text{ dB}} = 2.8.$$

Taking the antilog of the far left and far right sides of this equation (the antilog key on your calculator is probably marked as 10^x), we find

$$\frac{I_2}{I_1} = \log^{-1} 2.8 = 630. \quad \text{(Answer)}$$

Thus, the Who was *very* loud.

Temporary exposure to sound intensities as great as those of a jackhammer and the 1976 Who concert results in a temporary

Fig. 18-12 Sample Problem 18-5. Peter Townshend of the Who, playing in front of a speaker system. He suffered a permanent reduction in his hearing ability due to his exposure to high-intensity sound, not so much during on-stage performances as from wearing headphones in recording studios and at home.

reduction of hearing. Repeated or prolonged exposure can result in permanent reduction of hearing (Fig. 18-12). Loss of hearing is a clear risk for anyone continually listening to, say, heavy metal at high volume, especially on headphones.

18-6 Sources of Musical Sound

Musical sounds can be set up by oscillating strings (guitar, piano, violin), membranes (kettledrum, snare drum), air columns (flute, oboe, pipe organ, and the fujara of Fig. 18-13), wooden blocks or steel bars (marimba, xylophone), and many other oscillating bodies. Most instruments involve more than a single oscillating part. In the violin, for example, both the strings and the body of the instrument participate in producing the music.

Recall from Chapter 17 that standing waves can be set up on a stretched string that is fixed at both ends. They arise because waves traveling along the string are reflected back onto the string at each end. If the wavelength of the waves is suitably matched to the length of the string, the superposition of waves traveling in opposite directions produces a standing wave pattern (or oscillation mode). The wavelength required of the waves for such a match is one that corresponds to a *resonant frequency* of the string. The advantage of setting up standing waves is that the string then oscillates with a large, sustained amplitude, pushing back and forth against the surrounding air and thus generating a noticeable sound wave with the same frequency as the oscillations of the string. This production of sound is of obvious importance to, say, a guitarist.

We can set up standing waves of sound in an air-filled pipe in a similar way. As sound waves travel through the air in the pipe, they are reflected at each end and travel back through the pipe. (The reflection occurs even if an end is open, but the reflection is not as complete as when the end is closed.) If the wavelength of the sound waves is suitably matched to the length of the pipe, the superposition of waves traveling in opposite directions through the pipe sets up a standing wave pattern. The wavelength required of the sound waves for such a match is one that corresponds to a resonant frequency of the pipe. The advantage of such a standing wave is that the air in the pipe oscillates with a large, sustained amplitude, emitting at any open end a sound wave that has the same frequency as the oscillations in the pipe. This emission of sound is of obvious importance to, say, an organist.

Many other aspects of standing sound wave patterns are similar to those of string waves: The closed end of a pipe is like the fixed end of a string in that there must be a node (zero displacement) there, and the open end of a pipe is like the end of a string attached to a freely moving ring, as in Fig. 17-19*b*, in that there must be an antinode there. (Actually, the antinode for the open end of a pipe is located slightly beyond the end, but we shall not dwell on that detail here.)

The simplest standing wave pattern that can be set up in a pipe with two open ends is shown in Fig. 18-14*a*. There is an antinode across each open end, as required. There is also a node across the middle of the pipe. An easier way of representing this standing longitudinal sound wave is shown in Fig. 18-14*b*—by drawing it as a standing transverse string wave.

The standing wave pattern of Fig. 18-14*a* is called the *fundamental mode* or *first harmonic*. For it to be set up, the sound waves in a pipe of length L must have a wavelength given by $L = \lambda/2$, so that $\lambda = 2L$. Several more standing sound wave patterns for a pipe with two open ends are shown in Fig. 18-15*a* using string wave representations. The *second harmonic* requires sound waves of wavelength $\lambda = L$, the *third harmonic* requires wavelength $\lambda = 2L/3$, and so on.

More generally, the resonant frequencies for a pipe of length L with two open ends correspond to the wavelengths

$$\lambda = \frac{2L}{n}, \qquad \text{for } n = 1, 2, 3, \ldots, \tag{18-38}$$

where n is called the *harmonic number*. The resonant frequencies for a pipe with

Fig. 18-13 The air column within a fujara oscillates when that traditional Slovakian instrument is played.

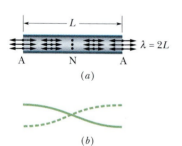

Fig. 18-14 (*a*) The simplest standing wave pattern of displacement for (longitudinal) sound waves in a pipe with both ends open has an antinode (A) across each end and a node (N) across the middle. (The longitudinal displacements represented by the double arrows are greatly exaggerated.) (*b*) The corresponding standing wave pattern for (transverse) string waves.

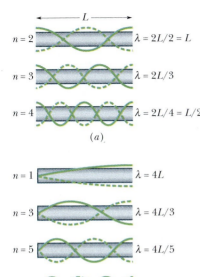

Fig. 18-15 Standing wave patterns for string waves superimposed on pipes to represent standing sound wave patterns in the pipes. (a) With *both* ends of the pipe open, any harmonic can be set up in the pipe. (b) With only *one* end open, only odd harmonics can be set up.

two open ends are then given by

$$f = \frac{v}{\lambda} = \frac{nv}{2L}, \qquad \text{for } n = 1, 2, 3, \ldots \qquad \text{(pipe, two open ends),} \quad (18\text{-}39)$$

where v is the speed of sound.

Figure 18-15*b* shows (using string wave representations) some of the standing sound wave patterns that can be set up in a pipe with only one open end. As required, across the open end there is an antinode and across the closed end there is a node. The simplest pattern requires sound waves having a wavelength given by $L = \lambda/4$, so that $\lambda = 4L$. The next simplest pattern requires a wavelength given by $L = 3\lambda/4$, so that $\lambda = 4L/3$, and so on.

More generally, the resonant frequencies for a pipe of length L with only one open end correspond to the wavelengths

$$\lambda = \frac{4L}{n}, \qquad \text{for } n = 1, 3, 5, \ldots, \qquad (18\text{-}40)$$

in which the harmonic number n *must be an odd number*. The resonant frequencies are then given by

$$f = \frac{v}{\lambda} = \frac{nv}{4L}, \qquad \text{for } n = 1, 3, 5, \ldots \qquad \text{(pipe, one open end).} \quad (18\text{-}41)$$

Note again that only odd harmonics can exist in a pipe with one open end. For example, the second harmonic, with $n = 2$, cannot be set up in such a pipe. Note also that for such a pipe the adjective in a phrase such as "the third harmonic" still refers to the harmonic number n (and not to, say, the third possible harmonic).

The length of a musical instrument reflects the range of frequencies over which the instrument is designed to function, and smaller length implies higher frequencies. Figure 18-16, for example, shows the saxophone and violin families, with their frequency ranges suggested by the piano keyboard. Note that, for every instrument, there is overlap with its higher- and lower-frequency neighbors.

In any oscillating system that gives rise to a musical sound, whether it is a violin string or the air in an organ pipe, the fundamental and one or more of the higher

Fig. 18-16 The saxophone and violin families, showing the relations between instrument length and frequency range. The frequency range of each instrument is indicated by a horizontal bar along a frequency scale suggested by the keyboard at the bottom; the frequency increases toward the right.

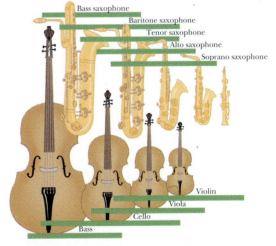

(a)

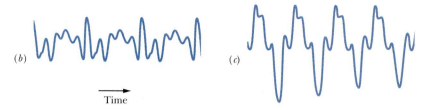

(b)

Time

(c)

Fig. 18-17 The wave forms produced by (a) a flute, (b) an oboe, and (c) a saxophone when they all play the same note, with the same first harmonic frequency.

harmonics are usually generated simultaneously. Thus, you hear them together—that is, superimposed as a net wave. When different instruments are played at the same note, they produce the same fundamental frequency but different intensities for the higher harmonics. For example, the fourth harmonic of middle C might be relatively loud on one instrument and relatively quiet or even missing on another. Thus, because different instruments produce different net waves, they sound different to you even when they are played at the same note. That would be the case for the three net waves shown in Fig. 18-17, which were produced at the same note by different instruments.

CHECKPOINT 4: Pipe A, with length L, and pipe B, with length 2L, both have two open ends. Which harmonic of pipe B has the same frequency as the fundamental of pipe A?

Sample Problem 18-6

Weak background noises from a room set up the fundamental standing wave in a cardboard tube of length $L = 67.0$ cm with two open ends. Assume that the speed of sound in the air within the tube is 343 m/s.

(a) What frequency do you hear from the tube?

SOLUTION: The Key Idea here is that, with both pipe ends open, we have a symmetric situation in which the standing wave has an antinode at each end of the tube. The standing wave pattern (in string wave style) is that of Fig. 18-14b. The frequency is given by Eq. 18-39 with $n = 1$ for the fundamental mode:

$$f = \frac{nv}{2L} = \frac{(1)(343 \text{ m/s})}{2(0.670 \text{ m})} = 256 \text{ Hz}. \qquad \text{(Answer)}$$

If the background noises set up any higher harmonics, such as the

second harmonic, you must also hear frequencies that are *integer* multiples of 256 Hz.

(b) If you jam your ear against one end of the tube, what fundamental frequency do you hear from the tube?

SOLUTION: The Key Idea now is that, with your ear effectively closing one end of the tube, we have an asymmetric situation—an antinode still exists at the open end but a node is now at the other (closed) end. The standing wave pattern is the top one in Fig. 18-15b. The frequency is given by Eq. 18-41 with $n = 1$ for the fundamental mode:

$$f = \frac{nv}{4L} = \frac{(1)(343 \text{ m/s})}{4(0.670 \text{ m})} = 128 \text{ Hz}. \qquad \text{(Answer)}$$

If the background noises set up any higher harmonics, they will be *odd* multiples of 128 Hz. That means that the frequency of 256 Hz (which is an even multiple) cannot now occur.

18-7 Beats

If we listen, a few minutes apart, to two sounds whose frequencies are, say, 552 and 564 Hz, most of us cannot tell one from the other. However, if the sounds reach our ears simultaneously, what we hear is a sound whose frequency is 558 Hz, the *average* of the two combining frequencies. We also hear a striking variation in the intensity of this sound—it increases and decreases in slow, wavering **beats** that repeat at a frequency of 12 Hz, the *difference* between the two combining frequencies. Figure 18-18 shows this beat phenomenon.

Let the time-dependent variations of the displacements due to two sound waves at a particular location be

$$s_1 = s_m \cos \omega_1 t \quad \text{and} \quad s_2 = s_m \cos \omega_2 t, \qquad (18\text{-}42)$$

where $\omega_1 > \omega_2$. We have assumed, for simplicity, that the waves have the same

(a) (b) (c)

Fig. 18-18 (a, b) The pressure variations Δp of two sound waves as they would be detected separately. The frequencies of the waves are nearly equal. (c) The resultant pressure variation if the two waves are detected simultaneously.

Time

amplitude. According to the superposition principle, the resultant displacement is

$$s = s_1 + s_2 = s_m(\cos \omega_1 t + \cos \omega_2 t).$$

Using the trigonometric identity (see Appendix E)

$$\cos \alpha + \cos \beta = 2 \cos \tfrac{1}{2}(\alpha - \beta) \cos \tfrac{1}{2}(\alpha + \beta)$$

allows us to write the resultant displacement as

$$s = 2s_m \cos \tfrac{1}{2}(\omega_1 - \omega_2)t \cos \tfrac{1}{2}(\omega_1 + \omega_2)t. \tag{18-43}$$

If we write

$$\omega' = \tfrac{1}{2}(\omega_1 - \omega_2) \quad \text{and} \quad \omega = \tfrac{1}{2}(\omega_1 + \omega_2), \tag{18-44}$$

we can then write Eq. 18-43 as

$$s(t) = [2s_m \cos \omega' t] \cos \omega t. \tag{18-45}$$

We now assume that the angular frequencies ω_1 and ω_2 of the combining waves are almost equal, which means that $\omega \gg \omega'$ in Eq. 18-44. We can then regard Eq. 18-45 as a cosine function whose angular frequency is ω and whose amplitude (which is not constant but varies with angular frequency ω') is the quantity in the brackets.

A maximum amplitude will occur whenever $\cos \omega' t$ in Eq. 18-45 has the value $+1$ or -1, which happens twice in each repetition of the cosine function. Because $\cos \omega' t$ has angular frequency ω', the angular frequency ω_{beat} at which beats occur is $\omega_{beat} = 2\omega'$. Then, with the aid of Eq. 18-44, we can write

$$\omega_{beat} = 2\omega' = (2)(\tfrac{1}{2})(\omega_1 - \omega_2) = \omega_1 - \omega_2.$$

Because $\omega = 2\pi f$, we can recast this as

$$f_{beat} = f_1 - f_2 \quad \text{(beat frequency).} \tag{18-46}$$

Musicians use the beat phenomenon in tuning their instruments. If an instrument is sounded against a standard frequency (for example, the lead oboe's reference A) and tuned until the beat disappears, then the instrument is in tune with that standard. In musical Vienna, concert A (440 Hz) is available as a telephone service for the benefit of the city's many professional and amateur musicians.

Sample Problem 18-7

You wish to tune the note A_3 on a piano to its proper frequency of 220 Hz. You have available a tuning fork whose frequency is 440 Hz. How should you proceed?

SOLUTION: We need two **Key Ideas** here: (1) The two frequencies are too far apart to produce beats. (2) However, the piano string will oscillate not only in its fundamental mode (at 220 Hz when tuned) but also in its second harmonic mode (at 440 Hz when in tune). Thus, with the string somewhat out of tune, the frequency of its

second harmonic will beat against the 440 Hz of the tuning fork. To tune the string, you can listen for those beats and then either tighten or loosen the string to decrease the beat frequency until the beating disappears.

✔CHECKPOINT 5: In this sample problem, you tighten the string and the beat frequency increases from 6 Hz. Should you continue to tighten the string or should you loosen the string to put the string in tune?

18-8 The Doppler Effect

A police car is parked by the side of the highway, sounding its 1000 Hz siren. If you are also parked by the highway, you will hear that same frequency. However, if there is relative motion between you and the police car, either toward or away from each other, you will hear a different frequency. For example, if you are driving *toward* the police car at 120 km/h (about 75 mi/h), you will hear a *higher* frequency (1096 Hz, an *increase* of 96 Hz). If you are driving *away from* the police car at that same speed, you will hear a *lower* frequency (904 Hz, a *decrease* of 96 Hz).

These motion-related frequency changes are examples of the **Doppler effect.** The effect was proposed (although not fully worked out) in 1842 by Austrian physicist Johann Christian Doppler. It was tested experimentally in 1845 by Buys Ballot in Holland, "using a locomotive drawing an open car with several trumpeters."

The Doppler effect holds not only for sound waves but also for electromagnetic waves, including microwaves, radio waves, and visible light. Here, however, we shall consider only sound waves, and we shall take as a reference frame the body of air through which these waves travel. This means that we shall measure the speeds of a source S of sound waves and a detector D of those waves *relative to that body of air*. (Unless otherwise stated, the body of air is stationary relative to the ground, so the speeds can also be measured relative to the ground.) We shall assume that S and D move either directly toward or directly away from each other, at speeds less than the speed of sound.

If either the detector or the source is moving, or both are moving, the emitted frequency f and the detected frequency f' are related by

$$f' = f\frac{v \pm v_D}{v \pm v_S} \qquad \text{(general Doppler effect)}, \qquad (18\text{-}47)$$

where v is the speed of sound through the air, v_D is the detector's speed relative to the air, and v_S is the source's speed relative to the air. The choice of plus or minus signs is set by this rule:

▶ When the motion of detector or source is toward the other, the sign on its speed must give an upward shift in frequency. When the motion of detector or source is away from the other, the sign on its speed must give a downward shift in frequency.

In short, *toward* means *shift up,* and *away* means *shift down.*

Here are some examples of the rule. If the detector moves toward the source, use the plus sign in the numerator of Eq. 18-47 to get a shift up in the frequency. If it moves away, use the minus sign in the numerator to get a shift down. If it is stationary, substitute 0 for v_D. If the source moves toward the detector, use the minus sign in the denominator of Eq. 18-47 to get a shift up in the frequency. If it moves away, use the plus sign in the denominator to get a shift down. If the source is stationary, substitute 0 for v_S.

Next, we derive equations for the Doppler effect for the following two specific situations and then derive Eq. 18-47 for the general situation.

1. When the detector moves relative to the air and the source is stationary relative to the air, the motion changes the frequency at which the detector intercepts wavefronts and thus the detected frequency of the sound wave.

2. When the source moves relative to the air and the detector is stationary relative to the air, the motion changes the wavelength of the sound wave and thus the detected frequency (recall that frequency is related to wavelength).

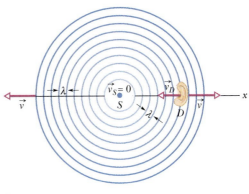

Fig. 18-19 A stationary source of sound *S* emits spherical wavefronts, shown one wavelength apart, that expand outward at speed *v*. A sound detector *D*, represented by an ear, moves with velocity $\vec{v}_D$ toward the source. The detector senses a higher frequency because of its motion.

Detector Moving; Source Stationary

In Fig. 18-19, a detector *D* (represented by an ear) is moving at speed v_D toward a stationary source *S* that emits spherical wavefronts, of wavelength λ and frequency *f*, moving at the speed *v* of sound in air. The wavefronts are drawn one wavelength apart. The frequency detected by detector *D* is the rate at which *D* intercepts wavefronts (or individual wavelengths). If *D* were stationary, that rate would be *f*, but since *D* is moving into the wavefronts, the rate of interception is greater, and thus the detected frequency f' is greater than *f*.

Let us for the moment consider the situation in which *D* is stationary (Fig. 18-20). In time *t*, the wavefronts move to the right a distance *vt*. The number of wavelengths in that distance *vt* is the number of wavelengths intercepted by *D* in time *t*, and that number is vt/λ. The rate at which *D* intercepts wavelengths, which is the frequency *f* detected by *D*, is

$$f = \frac{vt/\lambda}{t} = \frac{v}{\lambda}. \tag{18-48}$$

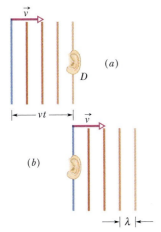

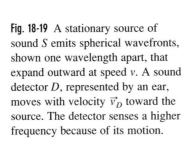

Fig. 18-20 Wavefronts of Fig. 18-19, assumed planar, (*a*) reach and (*b*) pass a stationary detector *D*; they move a distance *vt* to the right in time *t*.

In this situation, with *D* stationary, there is no Doppler effect—the frequency detected by *D* is the frequency emitted by *S*.

Now let us again consider the situation in which *D* moves opposite the wavefronts (Fig. 18-21). In time *t*, the wavefronts move to the right a distance *vt* as previously, but now *D* moves to the left a distance $v_D t$. Thus, in this time *t*, the distance moved by the wavefronts relative to *D* is $vt + v_D t$. The number of wavelengths in this relative distance $vt + v_D t$ is the number of wavelengths intercepted by *D* in time *t*, and is $(vt + v_D t)/\lambda$. The *rate* at which *D* intercepts wavelengths in this situation is the frequency f', given by

$$f' = \frac{(vt + v_D t)/\lambda}{t} = \frac{v + v_D}{\lambda}. \tag{18-49}$$

From Eq. 18-48, we have $\lambda = v/f$. Then Eq. 18-49 becomes

$$f' = \frac{v + v_D}{v/f} = f\frac{v + v_D}{v}. \tag{18-50}$$

Note that in Eq. 18-50, f' must be greater than *f* unless $v_D = 0$ (the detector is stationary).

Similarly, we can find the frequency detected by *D* if *D* moves away from the source. In this situation, the wavefronts move a distance $vt - v_D t$ relative to *D* in time *t*, and f' is given by

$$f' = f\frac{v - v_D}{v}. \tag{18-51}$$

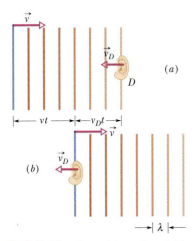

Fig. 18-21 Wavefronts (*a*) reach and (*b*) pass detector *D*, which moves opposite the wavefronts. In time *t*, the wavefronts move a distance *vt* to the right and *D* moves a distance $v_D t$ to the left.

In Eq. 18-51, f' must be less than f unless $v_D = 0$.

We can summarize Eqs. 18-50 and 18-51 with

$$f' = f\frac{v \pm v_D}{v} \qquad \text{(detector moving; source stationary).} \qquad (18\text{-}52)$$

Source Moving; Detector Stationary

Let detector D be stationary with respect to the body of air, and let source S move toward D at speed v_S (Fig. 18-22). The motion of S changes the wavelength of the sound waves it emits, and thus the frequency detected by D.

To see this change, let T ($= 1/f$) be the time between the emission of any pair of successive wavefronts W_1 and W_2. During T, wavefront W_1 moves a distance vT and the source moves a distance $v_S T$. At the end of T, wavefront W_2 is emitted. In the direction in which S moves, the distance between W_1 and W_2, which is the wavelength λ' of the waves moving in that direction, is $vT - v_S T$. If D detects those waves, it detects frequency f' given by

$$f' = \frac{v}{\lambda'} = \frac{v}{vT - v_S T} = \frac{v}{v/f - v_S/f}$$

$$= f\frac{v}{v - v_S}. \qquad (18\text{-}53)$$

Note that f' must be greater than f unless $v_S = 0$.

In the direction opposite that taken by S, the wavelength λ' of the waves is $vT + v_S T$. If D detects those waves, it detects frequency f' given by

$$f' = f\frac{v}{v + v_S}. \qquad (18\text{-}54)$$

Now f' must be less than f unless $v_S = 0$.

We can summarize Eqs. 18-53 and 18-54 with

$$f' = f\frac{v}{v \pm v_S} \qquad \text{(source moving; detector stationary).} \qquad (18\text{-}55)$$

General Doppler Effect Equation

We can now derive the general Doppler effect equation by replacing f in Eq. 18-55 (the frequency of the source) with f' of Eq. 18-52 (the frequency associated with motion of the detector). The result is Eq. 18-47 for the general Doppler effect.

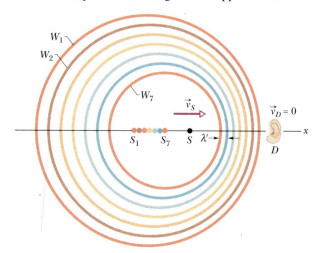

Fig. 18-22 A detector D is stationary, and a source S is moving toward it at speed v_S. Wavefront W_1 was emitted when the source was at S_1, wavefront W_7 when it was at S_7. At the moment depicted, the source is at S. The detector senses a higher frequency because the moving source, chasing its own wavefronts, emits a reduced wavelength λ' in the direction of its motion.

That general equation holds not only when both detector and source are moving but also in the two specific situations we just discussed. For the situation in which the detector is moving and the source is stationary, substitution of $v_S = 0$ into Eq. 18-47 gives us Eq. 18-52, which we previously found. For the situation in which the source is moving and the detector is stationary, substitution of $v_D = 0$ into Eq. 18-47 gives us Eq. 18-55, which we previously found. Thus, Eq. 18-47 is the equation to remember.

Bat Navigation

Bats navigate and search out prey by emitting, and then detecting reflections of, ultrasonic waves. These are sound waves with frequencies greater than can be heard by a human. For example, a horseshoe bat emits ultrasonic waves at 83 kHz, well above the 20 kHz limit of human hearing.

After the sound is emitted through the bat's nostrils, it might reflect (echo) from a moth, and then return to the bat's ears. The motions of the bat and the moth relative to the air cause the frequency heard by the bat to differ by a few kilohertz from the frequency it emitted. The bat automatically translates this difference into a relative speed between itself and the moth, so it can zero in on the moth.

Some moths evade capture by flying away from the direction in which they hear ultrasonic waves. That choice of flight path reduces the frequency difference between what the bat emits and what it hears, and then the bat may not notice the echo. Some moths avoid capture by clicking to produce their own ultrasonic waves, thus "jamming" the detection system and confusing the bat. (Surprisingly, moths and bats do all this without first studying physics.)

✔CHECKPOINT 6:   The figure indicates the directions of motion of a sound source and a detector for six situations in stationary air. For each situation, is the detected frequency greater than or less than the emitted frequency, or can't we tell without more information about the actual speeds?

Sample Problem 18-8

A rocket moves at a speed of 242 m/s directly toward a stationary pole (through stationary air) while emitting sound waves at frequency $f = 1250$ Hz.

(a) What frequency f' is measured by a detector that is attached to the pole?

SOLUTION: We can find f' with Eq. 18-47 for the general Doppler effect. The Key Idea here is that, because the sound source (the rocket) moves through the air *toward* the stationary detector on the pole, we need to choose the sign on v_S that gives a *shift up* in the frequency of the sound. Thus, in Eq. 18-47 we use the minus sign in the denominator. We then substitute 0 for the detector speed v_D, 242 m/s for the source speed v_S, 343 m/s for the speed of sound v (from Table 18-1), and 1250 Hz for the emitted frequency f.

We find

$$f' = f\frac{v \pm v_D}{v \pm v_S} = (1250 \text{ Hz})\frac{343 \text{ m/s} \pm 0}{343 \text{ m/s} - 242 \text{ m/s}}$$

$$= 4245 \text{ Hz} \approx 4250 \text{ Hz}, \qquad \text{(Answer)}$$

which, indeed, is a greater frequency than the emitted frequency.

(b) Some of the sound reaching the pole reflects back to the rocket as an echo. What frequency f'' does a detector on the rocket detect for the echo?

SOLUTION: Two Key Ideas here are the following:

1. The pole is now the source of sound (because it is the source of the echo), and the rocket's detector is now the detector (because it detects the echo).

2. The frequency of the sound emitted by the source (the pole) is equal to f', the frequency of the sound the pole intercepts and reflects.

We can rewrite Eq. 18-47 in terms of the source frequency f' and the detected frequency f'' as

$$f'' = f' \frac{v \pm v_D}{v \pm v_S}. \tag{18-56}$$

A third **Key Idea** here is that, because the detector (on the rocket) moves through the air *toward* the stationary source, we need to use the sign on v_D that gives a *shift up* in the frequency of the sound. Thus, we use the plus sign in the numerator of Eq. 18-56. Also, we substitute $v_D = 242$ m/s, $v_S = 0$, $v = 343$ m/s, and

$f' = 4245$ Hz. We find

$$f'' = (4245 \text{ Hz}) \frac{343 \text{ m/s} + 242 \text{ m/s}}{343 \text{ m/s} \pm 0}$$

$$= 7240 \text{ Hz}, \tag{Answer}$$

which, indeed, is greater than the frequency of the sound reflected by the pole.

✔CHECKPOINT 7:   If the air in this sample problem is moving toward the pole at speed 20 m/s, (a) what value for the source speed v_S should be used in the solution of part (a), and (b) what value for the detector speed v_D should be used in the solution of part (b)?

18-9 Supersonic Speeds; Shock Waves

If a source is moving toward a stationary detector at a speed equal to the speed of sound—that is, if $v_S = v$—Eqs. 18-47 and 18-55 predict that the detected frequency f' will be infinitely great. This means that the source is moving so fast that it keeps pace with its own spherical wavefronts, as Fig. 18-23a suggests. What happens when the speed of the source *exceeds* the speed of sound?

For such *supersonic* speeds, Eqs. 18-47 and 18-55 no longer apply. Figure 18-23b depicts the spherical wavefronts that originated at various positions of the source. The radius of any wavefront in this figure is vt, where v is the speed of sound and t is the time that has elapsed since the source emitted that wavefront. Note that all the wavefronts bunch along a V-shaped envelope in the two-dimensional drawing of Fig. 18-23b. The wavefronts actually extend in three dimensions, and the bunching actually forms a cone called the *Mach cone*. A *shock wave* is said to exist along the surface of this cone, because the bunching of wavefronts causes an abrupt rise and fall of air pressure as the surface passes through any point. From Fig. 18-23b, we see that the half-angle θ of the cone, called the *Mach cone angle,* is given by

$$\sin \theta = \frac{vt}{v_S t} = \frac{v}{v_S} \qquad \text{(Mach cone angle).} \tag{18-57}$$

The ratio v_S/v is called the *Mach number*. When you hear that a particular plane has flown at Mach 2.3, it means that its speed was 2.3 times the speed of sound in the air through which the plane was flying. The shock wave generated by a super-

Fig. 18-23 (*a*) A source of sound S moves at speed v_S equal to the speed of sound and thus as fast as the wavefronts it generates. (*b*) A source S moves at speed v_S faster than the speed of sound and thus faster than the wavefronts. When the source was at position S_1 it generated wavefront W_1, and at position S_6 it generated W_6. All the spherical wavefronts expand at the speed of sound v and bunch along the surface of a cone called the Mach cone, forming a shock wave. The surface of the cone has half-angle θ and is tangent to all the wavefronts.

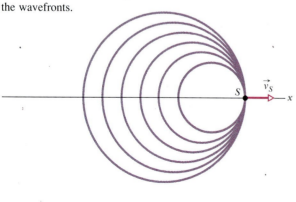

(a)

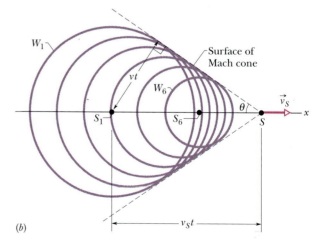

(b)

sonic aircraft (Fig. 18-24) or projectile produces a burst of sound, called a *sonic boom,* in which the air pressure first suddenly increases and then suddenly decreases below normal before returning to normal. Part of the sound that is heard when a rifle is fired is the sonic boom produced by the bullet. A sonic boom can also be heard from a long bullwhip when it is snapped quickly: Near the end of the whip's motion, its tip is moving faster than sound and produces a small sonic boom— the *crack* of the whip.

Fig. 18-24 Shock waves produced by the wings of a Navy FA 18 jet. They are visible because the sudden decrease in air pressure in the shock waves caused water molecules in the air to condense, forming a fog.

REVIEW & SUMMARY

Sound Waves   Sound waves are longitudinal mechanical waves that can travel through solids, liquids, or gases. The speed v of a sound wave in a medium having **bulk modulus** B and density ρ is

$$v = \sqrt{\frac{B}{\rho}} \qquad \text{(speed of sound).} \qquad (18\text{-}3)$$

In air at 20°C, the speed of sound is 343 m/s.

A sound wave causes a longitudinal displacement s of a mass element in a medium as given by

$$s = s_m \cos(kx - \omega t), \qquad (18\text{-}13)$$

where s_m is the **displacement amplitude** (maximum displacement) from equilibrium, $k = 2\pi/\lambda$, and $\omega = 2\pi f$, λ and f being the wavelength and frequency, respectively, of the sound wave. The sound wave also causes a pressure change Δp of the medium from the equilibrium pressure:

$$\Delta p = \Delta p_m \sin(kx - \omega t), \qquad (18\text{-}14)$$

where the **pressure amplitude** is

$$\Delta p_m = (v\rho\omega)s_m. \qquad (18\text{-}15)$$

Interference   The interference of two sound waves with identical wavelengths passing through a common point depends on their phase difference ϕ there. If the sound waves were emitted in phase and are traveling in approximately the same direction, ϕ is given by

$$\phi = \frac{\Delta L}{\lambda} 2\pi, \qquad (18\text{-}21)$$

where ΔL is their **path length difference** (the difference in the distances traveled by the waves to reach the common point). Fully constructive interference occurs when ϕ is an integer multiple of 2π,

$$\phi = m(2\pi), \qquad \text{for } m = 0, 1, 2, \ldots, \qquad (18\text{-}22)$$

and, equivalently, when ΔL is related to wavelength λ by

$$\frac{\Delta L}{\lambda} = 0, 1, 2, \ldots. \qquad (18\text{-}23)$$

Fully destructive interference occurs when ϕ is an odd multiple of π,

$$\phi = (2m + 1)\pi, \qquad \text{for } m = 0, 1, 2, \ldots, \qquad (18\text{-}24)$$

and, equivalently, when ΔL is related to λ by

$$\frac{\Delta L}{\lambda} = 0.5, 1.5, 2.5, \ldots. \qquad (18\text{-}25)$$

Sound Intensity   The **intensity** I of a sound wave at a surface is the average rate per unit area at which energy is transferred by the wave through or onto the surface:

$$I = \frac{P}{A}, \qquad (18\text{-}26)$$

where P is the time rate of energy transfer (power) of the sound wave and A is the area of the surface intercepting the sound. The intensity I is related to the displacement amplitude s_m of the sound wave by

$$I = \tfrac{1}{2}\rho v\omega^2 s_m^2. \qquad (18\text{-}27)$$

The intensity at a distance r from a point source that emits sound waves of power P_s is

$$I = \frac{P_s}{4\pi r^2}. \qquad (18\text{-}28)$$

Sound Level in Decibels   The *sound level* β in *decibels* (dB) is defined as

$$\beta = (10 \text{ dB}) \log \frac{I}{I_0}, \qquad (18\text{-}29)$$

where I_0 (= 10^{-12} W/m²) is a reference intensity level to which all intensities are compared. For every factor-of-10 increase in intensity, 10 dB is added to the sound level.

Standing Wave Patterns in Pipes   Standing sound wave patterns can be set up in pipes. A pipe open at both ends will resonate at frequencies

$$f = \frac{v}{\lambda} = \frac{nv}{2L}, \qquad n = 1, 2, 3, \ldots, \qquad (18\text{-}39)$$

where v is the speed of sound in the air in the pipe. For a pipe closed at one end and open at the other, the resonant frequencies are

$$f = \frac{v}{\lambda} = \frac{nv}{4L}, \qquad n = 1, 3, 5, \ldots . \qquad (18\text{-}41)$$

Beats   *Beats* arise when two waves having slightly different frequencies, f_1 and f_2, are detected together. The beat frequency is

$$f_{\text{beat}} = f_1 - f_2. \qquad (18\text{-}46)$$

The Doppler Effect   The *Doppler effect* is a change in the observed frequency of a wave when the source or the detector moves relative to the transmitting medium (such as air). For sound the observed frequency f' is given in terms of the source frequency f by

$$f' = f \frac{v \pm v_D}{v \pm v_S} \qquad \text{(general Doppler effect)}, \qquad (18\text{-}47)$$

where v_D is the speed of the detector relative to the medium, v_S is that of the source, and v is the speed of sound in the medium. The signs are chosen such that f' tends to be *greater* for motion (of detector or source) "toward" and *less* for motion "away."

Shock Wave   If the speed of a source relative to the medium exceeds the speed of sound in the medium, the Doppler equation no longer applies. In such a case, shock waves result. The half angle θ of the Mach cone is given by

$$\sin \theta = \frac{v}{v_S} \qquad \text{(Mach cone angle)}. \qquad (18\text{-}57)$$

QUESTIONS

1. Figure 18-25 shows the paths taken by two pulses of sound that begin simultaneously and then race each other through equal distances in air. The only difference between the paths is that a region of hot (low density) air lies along path 2. Which pulse wins the race?

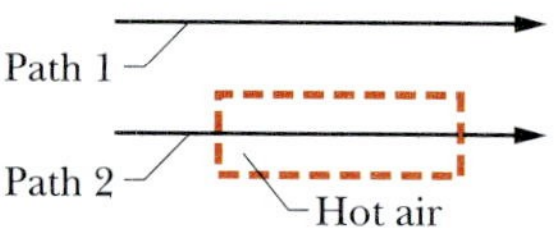

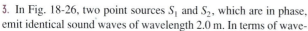
Path 1
Path 2
Hot air

Fig. 18-25 Question 1.

2. A sound wave of wavelength λ and displacement amplitude s_m begins to travel down a passageway (a tube, the opening of the ear, etc.). When a small device in the passageway detects this wave, it issues a second sound wave (said to be *antisound*) that is able to cancel the first wave, so that nothing is heard at the far end of the passageway. For such cancellation, what must be (a) the direction of travel, (b) the wavelength, and (c) the displacement amplitude of the second wave? (d) What must be the phase difference between the two waves? (Such antisound devices are used to eliminate unwanted sound in a noisy environment.)

3. In Fig. 18-26, two point sources S_1 and S_2, which are in phase, emit identical sound waves of wavelength 2.0 m. In terms of wavelengths, what is the phase difference between the waves arriving at point P if (a) L_1 = 38 m and L_2 = 34 m, and (b) L_1 = 39 m and L_2 = 36 m? (c) Assuming that the source separation is much smaller than L_1 and L_2, what type of interference occurs at P in situations (a) and (b), respectively?

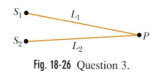
S_1 L_1
S_2 L_2 P

Fig. 18-26 Question 3.

4. In Fig. 18-27, sound waves of wavelength λ are emitted by a point source S and travel to a detector D directly along path 1 and via reflection from a panel along path 2. Initially, the panel is al-

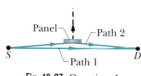
Panel   Path 2
S Path 1 D

Fig. 18-27 Question 4.

most along path 1 and the waves arriving at D along the two paths are almost exactly in phase. Then the panel is moved away from path 1 as shown until the waves arriving at D are exactly out of phase. What then is the path length difference ΔL of the waves along the two paths?

5. In Fig. 18-28, two point sources S_1 and S_2, which are in phase, emit identical sound waves of wavelength λ, and point P is at equal distances from them. Then S_2 is moved directly away from P by a distance equal to $\lambda/4$. Are the waves at P then exactly in phase, exactly out of phase, or do they have some intermediate phase relation if (a) S_1 is moved directly toward P by a distance equal to $\lambda/4$ and (b) S_1 is moved directly away from P by a distance equal to $3\lambda/4$?

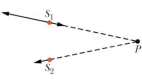
S_1
P
S_2

Fig. 18-28 Question 5.

6. In Sample Problem 18-3 and Fig. 18-9a, the waves arriving at point P_1 on the perpendicular bisector are exactly in phase; that is, the waves from S_1 and S_2 always tend to move an element of air at P_1 in the same direction. Let the intersection of the perpendicular bisector and the line through S_1 and S_2 be point P_3. (a) Are the waves arriving at P_3 exactly in phase, exactly out of phase, or do they have some intermediate relation? (b) What is the answer if we increase the separation between the sources to 1.7λ?

7. A standing sound wave in a pipe has five nodes and five antinodes. (a) How many open ends does the pipe have? (b) What is the harmonic number n for this standing wave?

8. The sixth harmonic is set up in a pipe. (a) How many open ends does the pipe have (it has at least one)? (b) Is there a node, antinode, or some intermediate state at the midpoint?

9. (a) When an orchestra warms up, the players' warm breath increases the temperature of the air within the wind instruments (and thus decreases the density of that air). Do the resonant frequencies of those instruments increase or decrease? (b) When the slide of a slide trombone is pushed outward, do the resonant frequencies of the instrument increase or decrease?

10. For a particular tube, here are four of the six harmonic frequencies below 1000 Hz: 300, 600, 750, and 900 Hz. What two frequencies are missing from the list?

11. Pipe A has length L and one open end. Pipe B has length $2L$ and two open ends. Which harmonics of pipe B have a frequency that matches a resonant frequency of pipe A?

12. Figure 18-29 shows a stretched string of length L and pipes a, b, c, and d of lengths L, $2L$, $L/2$, and $L/2$, respectively. The string's tension is adjusted until the speed of waves on the string equals the speed of sound waves in the air. The fundamental mode of oscillation is then set up on the string. In which pipe will the sound produced by the string cause resonance, and what oscillation mode will that sound set up?

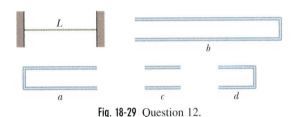

Fig. 18-29 Question 12.

13. Sound waves of frequency f are reflected by a fluid moving through a narrow tube along an x axis (Fig. 18-30a). The tube's inside diameter varies with x. The change in frequency Δf of the sound, due to the Doppler effect, also varies with x, as shown in Fig. 18-30b. Rank the five indicated regions in terms of the tube's inside diameter, greatest first. (*Hint:* See Section 15-10.)

14. A friend rides, in turn, the rims of three fast merry-go-rounds while holding a sound source that emits isotropically at a certain frequency. You stand far from each merry-go-round. The frequency you hear for each of your friend's three rides varies as the merry-go-round rotates. The variations in frequency for the three rides are given by the three curves in Fig. 18-31. Rank the curves according to (a) the linear speed v of the sound source, (b) the angular speeds ω of the merry-go-rounds, and (c) the radii r of the merry-go-rounds, greatest first.

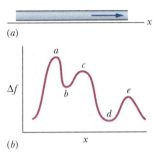

Fig. 18-30 Question 13.

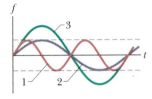

Fig. 18-31 Question 14.

EXERCISES & PROBLEMS

ssm   Solution is in the Student Solutions Manual.
www   Solution is available on the World Wide Web at:
        http://www.wiley.com/college/hrw
ilw   Solution is available on the Interactive LearningWare.

Where needed in the problems, use

$$\text{speed of sound in air} = 343 \text{ m/s}$$

and $$\text{density of air} = 1.21 \text{ kg/m}^3$$

unless otherwise specified.

SEC. 18-2   The Speed of Sound

1E. Devise a rule for finding your distance in kilometers from a lightning flash by counting the seconds from the time you see the flash until you hear the thunder. Assume that the sound travels to you along a straight line.

2E. You are at a large outdoor concert, seated 300 m from the speaker system. The concert is also being broadcast live via satellite (at the speed of light, 3.0×10^8 m/s). Consider a listener 5000 km away who receives the broadcast. Who hears the music first, you or the listener, and by what time difference?

3E. Two spectators at a soccer game in Montjuic Stadium see, and a moment later hear, the ball being kicked on the playing field. The time delay for one spectator is 0.23 s and for the other 0.12 s. Sight lines from the two spectators to the player kicking the ball meet at an angle of 90°. (a) How far is each spectator from the player? (b) How far are the spectators from each other? ssm

4E. A column of soldiers, marching at 120 paces per minute, keep in step with the beat of a drummer at the head of the column. It is observed that the soldiers in the rear end of the column are striding forward with the left foot when the drummer is advancing with the right. What is the approximate length of the column?

5P. Earthquakes generate sound waves inside Earth. Unlike a gas, Earth can experience both transverse (S) and longitudinal (P) sound waves. Typically, the speed of S waves is about 4.5 km/s, and that of P waves 8.0 km/s. A seismograph records P and S waves from an earthquake. The first P waves arrive 3.0 min before the first S waves (Fig. 18-32). Assuming the waves travel in a straight line, how far away does the earthquake occur? ssm ilw

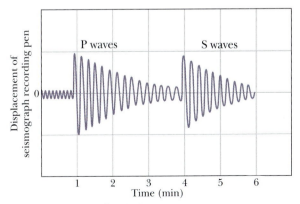

Fig. 18-32 Problem 5.

6P. The speed of sound in a certain metal is V. One end of a long pipe of that metal of length L is struck a hard blow. A listener at the other end hears two sounds, one from the wave that travels along the pipe and the other from the wave that travels through the air. (a) If v is the speed of sound in air, what time interval t elapses between the arrivals of the two sounds? (b) Suppose that $t = 1.00$ s and the metal is steel. Find the length L.

7P. A stone is dropped into a well. The sound of the splash is heard 3.00 s later. What is the depth of the well? ssm

SEC. 18-3   Traveling Sound Waves

8E. The audible frequency range for normal hearing is from about 20 Hz to 20 kHz. What are the wavelengths of sound waves at these frequencies?

9E. Diagnostic ultrasound of frequency 4.50 MHz is used to examine tumors in soft tissue. (a) What is the wavelength in air of such a sound wave? (b) If the speed of sound in tissue is 1500 m/s, what is the wavelength of this wave in tissue? ssm

10P. (a) A continuous sinusoidal longitudinal wave is sent along a very long coiled spring from an oscillating source attached to it. The frequency of the source is 25 Hz, and at any time the distance between successive points of maximum expansion in the spring is 24 cm. Find the wave speed. (b) If the maximum longitudinal displacement of a particle in the spring is 0.30 cm and the wave moves in the negative direction of an x axis, write the equation for the wave. Place $x = 0$ at the source and take the displacement there to be zero when $t = 0$.

11P. The pressure in a traveling sound wave is given by the equation

$$\Delta p = (1.50 \text{ Pa}) \sin \pi[(0.900 \text{ m}^{-1})x - (315 \text{ s}^{-1})t].$$

Find (a) the pressure amplitude, (b) the frequency, (c) the wavelength, and (d) the speed of the wave.

SEC. 18-4   Interference

12P. Two point sources of sound waves of identical wavelength λ and amplitude are separated by distance $D = 2.0\lambda$. The sources are in phase. (a) How many points of maximum signal (that is, maximum constructive interference) lie along a large circle around the sources? (b) How many points of minimum signal (destructive interference) lie around the circle?

13P. In Fig. 18-33, two loudspeakers, separated by a distance of 2.00 m, are in phase. Assume the amplitudes of the sound from the speakers are approximately the same at the position of a listener, who is 3.75 m directly in front of one of the speakers. (a) For what frequencies in the audible range (20 Hz to 20 kHz) does the listener hear a minimum signal? (b) For what frequencies is the signal a maximum? ssm www

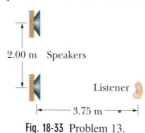

Fig. 18-33 Problem 13.

14P. Two sound waves, from two different sources with the same frequency, 540 Hz, travel in the same direction at 330 m/s. The sources are in phase. What is the phase difference of the waves at a point that is 4.40 m from one source and 4.00 m from the other?

15P. Two loudspeakers are located 3.35 m apart on an outdoor stage. A listener is 18.3 m from one and 19.5 m from the other. During the sound check, a signal generator drives the two speakers in phase with the same amplitude and frequency. The transmitted frequency is swept through the audible range (20 Hz to 20 kHz). (a) What are the three lowest frequencies at which the listener will hear a minimum signal because of destructive interference? (b) What are the three lowest frequencies at which the listener will hear a maximum signal? itw

16P. In Fig. 18-34, sound with a 40.0 cm wavelength travels rightward from a source and through a tube that consists of a straight portion and a half-circle. Part of the sound wave travels through the half-circle and then rejoins the rest of the wave, which goes directly through the straight portion. This rejoining results in interference. What is the smallest radius r that results in an intensity minimum at the detector?

Fig. 18-34 Problem 16.

SEC. 18-5   Intensity and Sound Level

17E. A source emits sound waves isotropically. The intensity of the waves 2.50 m from the source is 1.91×10^{-4} W/m². Assuming that the energy of the waves is conserved, find the power of the source. ssm

18E. A 1.0 W point source emits sound waves isotropically. Assuming that the energy of the waves is conserved, find the intensity (a) 1.0 m from the source and (b) 2.5 m from the source.

19E. A sound wave of frequency 300 Hz has an intensity of 1.00 μW/m². What is the amplitude of the air oscillations caused by this wave? ssm

20E. Two sounds differ in sound level by 1.00 dB. What is the ratio of the greater intensity to the smaller intensity?

21E. A certain sound source is increased in sound level by 30 dB. By what multiple is (a) its intensity increased and (b) its pressure amplitude increased? ssm

22E. The source of a sound wave has a power of 1.00 μW. If it is a point source, (a) what is the intensity 3.00 m away and (b) what is the sound level in decibels at that distance?

23E. (a) If two sound waves, one in air and one in (fresh) water, are equal in intensity, what is the ratio of the pressure amplitude of the wave in water to that of the wave in air? Assume the water and the air are at 20°C. (See Table 15-1.) (b) If the pressure amplitudes are equal instead, what is the ratio of the intensities of the waves? ssm

24P. Assume that a noisy freight train on a straight track emits a cylindrical, expanding sound wave, and that the air absorbs no energy. How does the amplitude s_m of the wave depend on the perpendicular distance r from the source?

25P. (a) Show that the intensity I of a wave is the product of the wave's energy per unit volume u and its speed v. (b) Radio waves travel at a speed of 3.00×10^8 m/s. Find u for a radio wave 480 km from a 50 000 W source, assuming the wavefronts are spherical. ssm

26P. Find the ratios (greater to smaller) of (a) the intensities,

(b) the pressure amplitudes, and (c) the particle displacement amplitudes for two sounds whose sound levels differ by 37 dB.

27P. A sound wave travels out uniformly in all directions from a point source. (a) Justify the following expression for the displacement s of the transmitting medium at any distance r from the source:

$$s = \frac{b}{r} \sin k(r - vt),$$

where b is a constant. Consider the speed, direction of propagation, periodicity, and intensity of the wave. (b) What is the dimension of the constant b? ssm www

28P. A point source emits 30.0 W of sound isotropically. A small microphone intercepts the sound in an area of 0.750 cm², 200 m from the source. Calculate (a) the sound intensity there and (b) the power intercepted by the microphone.

29P*. Figure 18-35 shows an air-filled, acoustic interferometer, used to demonstrate the interference of sound waves. Sound source S is an oscillating diaphragm; D is a sound detector, such as the ear or a microphone. Path SBD can be varied in length, but path SAD is fixed. At D, the sound wave coming along path SBD interferes with that coming along path SAD. In one demonstration, the sound intensity at D has a minimum value of 100 units at one position of the movable arm and continuously climbs to a maximum value of 900 units when that arm is shifted by 1.65 cm. Find (a) the frequency of the sound emitted by the source and (b) the ratio of the amplitude at D of the SAD wave to that of the SBD wave. (c) How can it happen that these waves have different amplitudes, considering that they originate at the same source? ssm

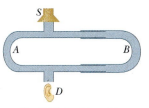

Fig. 18-35 Problem 29.

SEC. 18-6 Sources of Musical Sound

30E. A violin string 15.0 cm long and fixed at both ends oscillates in its $n = 1$ mode. The speed of waves on the string is 250 m/s, and the speed of sound in air is 348 m/s. What are (a) the frequency and (b) the wavelength of the emitted sound wave?

31E. Organ pipe A, with both ends open, has a fundamental frequency of 300 Hz. The third harmonic of organ pipe B, with one end open, has the same frequency as the second harmonic of pipe A. How long are (a) pipe A and (b) pipe B?

32E. The water level in a vertical glass tube 1.00 m long can be adjusted to any position in the tube. A tuning fork vibrating at 686 Hz is held just over the open top end of the tube, to set up a standing wave of sound in the air-filled top portion of the tube. (That air-filled top portion acts as a tube with one end closed and the other end open.) At what positions of the water level is there resonance?

33E. (a) Find the speed of waves on a violin string of mass 800 mg and length 22.0 cm if the fundamental frequency is 920 Hz. (b) What is the tension in the string? For the fundamental, what is the wavelength of (c) the waves on the string and (d) the sound waves emitted by the string? ssm ilw

34P. A certain violin string is 30 cm long between its fixed ends and has a mass of 2.0 g. The "open" string (no applied finger) sounds an A note (440 Hz). (a) To play a C note (523 Hz), how far down the string must one place a finger? (b) What is the ratio of the wavelength of the string waves required for an A note to that required for a C note? (c) What is the ratio of the wavelength of the sound wave for an A note to that for a C note?

35P. In Fig. 18-36, S is a small loudspeaker driven by an audio oscillator and amplifier, adjustable in frequency from 1000 to 2000 Hz only. Tube D is a piece of cylindrical sheet-metal pipe 45.7 cm long and open at both ends. (a) If the speed of sound in air is 344 m/s at the existing temperature, at what frequencies will resonance occur in the pipe when the frequency emitted by the speaker is varied from 1000 Hz to 2000 Hz? (b) Sketch the standing wave (using the style of Fig. 18-14b) for each resonant frequency. ssm www

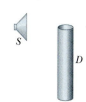

Fig. 18-36 Problem 35.

36P. A string on a cello has length L, for which the fundamental frequency is f. (a) By what length l must the string be shortened by fingering to change the fundamental frequency to rf? (b) What is l if $L = 0.80$ m and $r = 1.2$? (c) For $r = 1.2$, what is the ratio of the wavelength of the new sound wave emitted by the string to that of the wave emitted before fingering?

37P. A well with vertical sides and water at the bottom resonates at 7.00 Hz and at no lower frequency. (The air-filled portion of the well acts as a tube with one closed end and one open end.) The air in the well has a density of 1.10 kg/m³ and a bulk modulus of 1.33×10^5 Pa. How far down in the well is the water surface? ssm

38P. A tube 1.20 m long is closed at one end. A stretched wire is placed near the open end. The wire is 0.330 m long and has a mass of 9.60 g. It is fixed at both ends and oscillates in its fundamental mode. By resonance, it sets the air column in the tube into oscillation at that column's fundamental frequency. Find (a) that frequency and (b) the tension in the wire.

39P. The period of a pulsating variable star may be estimated by considering the star to be executing *radial* longitudinal pulsations in the fundamental standing wave mode; that is, the star's radius varies periodically with time, with a displacement antinode at the star's surface. (a) Would you expect the center of the star to be a displacement node or antinode? (b) By analogy with a pipe with one open end, show that the period of pulsation T is given by

$$T = \frac{4R}{v},$$

where R is the equilibrium radius of the star and v is the average sound speed in the material of the star. (c) Typical white dwarf stars are composed of material with a bulk modulus of 1.33×10^{22} Pa and a density of 10^{10} kg/m³. They have radii equal to 9.0×10^{-3} solar radius. What is the approximate pulsation period of a white dwarf? ssm

40P. Pipe A, which is 1.2 m long and open at both ends, oscillates at its third lowest harmonic frequency. It is filled with air for which the speed of sound is 343 m/s. Pipe B, which is closed at one end, oscillates at its second lowest harmonic frequency. The frequencies of pipes A and B happen to match. (a) If an x axis extends along the interior of pipe A, with $x = 0$ at one end, where along the axis are the displacement nodes? (b) How long is pipe B? (c) What is the lowest harmonic frequency of pipe A?

41P. A violin string 30.0 cm long with linear density 0.650 g/m is placed near a loudspeaker that is fed by an audio oscillator of variable frequency. It is found that the string is set into oscillation only at the frequencies 880 and 1320 Hz as the frequency of the oscillator is varied over the range 500–1500 Hz. What is the tension in the string? ssm

SEC. 18-7   Beats

42E. The A string of a violin is a little too tightly stretched. Four beats per second are heard when the string is sounded together with a tuning fork that is oscillating accurately at concert A (440 Hz). What is the period of the violin string oscillation?

43E. A tuning fork of unknown frequency makes three beats per second with a standard fork of frequency 384 Hz. The beat frequency decreases when a small piece of wax is put on a prong of the first fork. What is the frequency of this fork? ssm

44P. You have five tuning forks that oscillate at close but different frequencies. What are the (a) maximum and (b) minimum number of different beat frequencies you can produce by sounding the forks two at a time, depending on how the frequencies differ?

45P. Two identical piano wires have a fundamental frequency of 600 Hz when kept under the same tension. What fractional increase in the tension of one wire will lead to the occurrence of 6 beats/s when both wires oscillate simultaneously? ssm

SEC. 18-8   The Doppler Effect

46E. Trooper B is chasing speeder A along a straight stretch of road. Both are moving at a speed of 160 km/h. Trooper B, failing to catch up, sounds his siren again. Take the speed of sound in air to be 343 m/s and the frequency of the source to be 500 Hz. What is the Doppler shift in the frequency heard by speeder A?

47E. The 16 000 Hz whine of the turbines in the jet engines of an aircraft moving with speed 200 m/s is heard at what frequency by the pilot of a second craft trying to overtake the first at a speed of 250 m/s? ssm

48E. An ambulance with a siren emitting a whine at 1600 Hz overtakes and passes a cyclist pedaling a bike at 2.44 m/s. After being passed, the cyclist hears a frequency of 1590 Hz. How fast is the ambulance moving?

49P. A whistle of frequency 540 Hz moves in a circle of radius 60.0 cm at an angular speed of 15.0 rad/s. What are (a) the lowest and (b) the highest frequencies heard by a listener a long distance away, at rest with respect to the center of the circle? ilw

50P. A stationary motion detector sends sound waves of frequency 0.150 MHz toward a truck approaching at a speed of 45.0 m/s. What is the frequency of the waves reflected back to the detector?

51P. A French submarine and a U.S. submarine move toward each other during maneuvers in motionless water in the North Atlantic (Fig. 18-37). The French sub moves at 50.0 km/h, and the U.S. sub at 70.0 km/h. The French sub sends out a sonar signal (sound wave in water) at 1000 Hz. Sonar waves travel at 5470 km/h. (a) What is the signal's frequency as detected by the U.S. sub? (b) What

frequency is detected by the French sub in the signal reflected back to it by the U.S. sub?

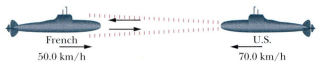

Fig. 18-37 Problem 51.

52P. A sound source A and a reflecting surface B move directly toward each other. Relative to the air, the speed of source A is 29.9 m/s, the speed of surface B is 65.8 m/s, and the speed of sound is 329 m/s. The source emits waves at frequency 1200 Hz as measured in the source frame. In the reflector frame, what are (a) the frequency and (b) the wavelength of the arriving sound waves? In the source frame, what are (c) the frequency and (d) the wavelength of the sound waves reflected back to the source?

53P. An acoustic burglar alarm consists of a source emitting waves of frequency 28.0 kHz. What is the beat frequency between the source waves and the waves reflected from an intruder walking at an average speed of 0.950 m/s directly away from the alarm? ilw

54P. A bat is flitting about in a cave, navigating via ultrasonic bleeps. Assume that the sound emission frequency of the bat is 39 000 Hz. During one fast swoop directly toward a flat wall surface, the bat is moving at 0.025 times the speed of sound in air. What frequency does the bat hear reflected off the wall?

55P. A girl is sitting near the open window of a train that is moving at a velocity of 10.00 m/s to the east. The girl's uncle stands near the tracks and watches the train move away. The locomotive whistle emits sound at frequency 500.0 Hz. The air is still. (a) What frequency does the uncle hear? (b) What frequency does the girl hear? A wind begins to blow from the east at 10.00 m/s. (c) What frequency does the uncle now hear? (d) What frequency does the girl now hear? ssm www

56P. A 2000 Hz siren and a civil defense official are both at rest with respect to the ground. What frequency does the official hear if the wind is blowing at 12 m/s (a) from source to official and (b) from official to source?

57P. Two trains are traveling toward each other at 30.5 m/s relative to the ground. One train is blowing a whistle at 500 Hz. (a) What frequency is heard on the other train in still air? (b) What frequency is heard on the other train if the wind is blowing at 30.5 m/s toward the whistle and away from the listener? (c) What frequency is heard if the wind direction is reversed?

SEC. 18-9   Supersonic Speeds; Shock Waves

58E. A bullet is fired with a speed of 685 m/s. Find the angle made by the shock cone with the line of motion of the bullet.

59P. A jet plane passes over you at a height of 5000 m and a speed of Mach 1.5. (a) Find the Mach cone angle. (b) How long after the jet passes directly overhead does the shock wave reach you? Use 331 m/s for the speed of sound. ssm

60P. A plane flies at 1.25 times the speed of sound. Its sonic boom reaches a man on the ground 1.00 min after the plane passes directly overhead. What is the altitude of the plane? Assume the speed of sound to be 330 m/s.

19 Temperature, Heat, and the First Law of Thermodynamics

The giant hornet *Vespa mandarinia japonica* preys on Japanese bees. However, if one of the hornets attempts to invade a bee hive, several hundred of the bees quickly form a compact ball around the hornet to stop it. After about 20 minutes the hornet is dead, although the bees do not sting, bite, crush, or suffocate it.

Why, then, does the hornet die?

The answer is in this chapter.

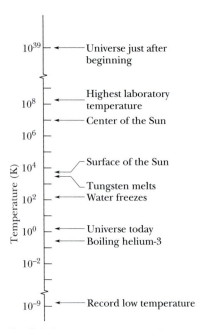

Fig. 19-1 Some temperatures on the Kelvin scale. Temperature $T = 0$ corresponds to $10^{-\infty}$ and cannot be plotted on this logarithmic scale.

19-1   Thermodynamics

In this and the next two chapters we focus on a new subject—**thermodynamics,** the study of the *thermal energy* (often called the *internal energy*) of systems. The central concept of thermodynamics is temperature. This word is so familiar that most of us—because of our built-in sense of hot and cold—tend to be overconfident in our understanding of it. Our "temperature sense" is in fact not always reliable. On a cold winter day, for example, an iron railing seems much colder to the touch than a wooden fence post, yet both are at the same temperature. This error in our perception comes about because iron removes energy from our fingers more quickly than wood does. Here, we shall develop the concept of temperature from its foundations, without relying in any way on our temperature sense.

Temperature is one of the seven SI base quantities. Physicists measure temperature on the **Kelvin scale,** which is marked in units called *kelvins.* Although the temperature of a body apparently has no upper limit, it does have a lower limit; this limiting low temperature is taken as the zero of the Kelvin temperature scale. Room temperature is about 290 kelvins, or 290 K as we write it, above this *absolute zero.* Figure 19-1 shows a wide range of temperatures, either measured or conjectured.

When the universe began, some 10 to 20 billion years ago, its temperature was about 10^{39} K. As the universe expanded it cooled, and it has now reached an average temperature of about 3 K. We on Earth are a little warmer than that because we happen to live near a star. Without our Sun, we too would be at 3 K (or, rather, we could not exist).

19-2   The Zeroth Law of Thermodynamics

The properties of many bodies change as we alter their temperature, perhaps by moving them from a refrigerator to a warm oven. To give a few examples: As their temperatures increase, the volume of a liquid increases, a metal rod grows a little longer, and the electrical resistance of a wire increases, as does the pressure exerted by a confined gas. We can use any one of these properties as the basis of an instrument that will help us to pin down the concept of temperature.

Figure 19-2 shows such an instrument. Any resourceful engineer could design and construct it, using any one of the properties listed above. The instrument is fitted with a digital readout display and has the following properties: If you heat it (say, with a Bunsen burner), the displayed number starts to increase; if you then put it into a refrigerator, the displayed number starts to decrease. The instrument is not calibrated in any way, and the numbers have (as yet) no physical meaning. The device is a *thermoscope* but not (as yet) a *thermometer.*

Suppose that, as in Fig. 19-3a, we put the thermoscope (which we shall call body T) into intimate contact with another body (body A). The entire system is confined within a thick-walled insulating box. The numbers displayed by the thermoscope roll by until, eventually, they come to rest (let us say the reading is "137.04") and no further change takes place. In fact, we suppose that every measurable property of body T and of body A has assumed a stable, unchanging value. Then we say that the two bodies are in *thermal equilibrium* with each other. Even though the displayed readings for body T have not been calibrated, we conclude that bodies T and A must be at the same (unknown) temperature.

Suppose that we next put body T in intimate contact with body B (Fig. 19-3b) and find that the two bodies come to thermal equilibrium *at the same reading of the thermoscope.* Then bodies T and B must be at the same (still unknown) temperature. If we now put bodies A and B into intimate contact (Fig. 19-3c), are they immediately in thermal equilibrium with each other? Experimentally, we find that they are.

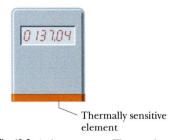

Fig. 19-2 A thermoscope. The numbers increase when the device is heated and decrease when it is cooled. The thermally sensitive element could be— among many possibilities—a coil of wire whose electrical resistance is measured and displayed.

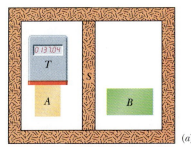

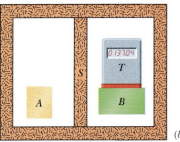

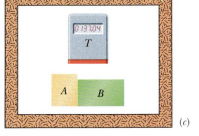

(a)

(b)

(c)

Fig. 19-3 (a) Body T (a thermoscope) and body A are in thermal equilibrium. (Body S is a thermally insulating screen.) (b) Body T and body B are also in thermal equilibrium, at the same reading of the thermoscope. (c) If (a) and (b) are true, the zeroth law of thermodynamics states that body A and body B are also in thermal equilibrium.

The experimental fact shown in Fig. 19-3 is summed up in the **zeroth law of thermodynamics:**

> ▶ If bodies A and B are each in thermal equilibrium with a third body T, then they are in thermal equilibrium with each other.

In less formal language, the message of the zeroth law is: "Every body has a property called **temperature.** When two bodies are in thermal equilibrium, their temperatures are equal. And vice versa." We can now make our thermoscope (the third body T) into a thermometer, confident that its readings will have physical meaning. All we have to do is calibrate it.

We use the zeroth law constantly in the laboratory. If we want to know whether the liquids in two beakers are at the same temperature, we measure the temperature of each with a thermometer. We do not need to bring the two liquids into intimate contact and observe whether they are or are not in thermal equilibrium.

The zeroth law, which has been called a logical afterthought, came to light only in the 1930s, long after the first and second laws of thermodynamics had been discovered and numbered. Because the concept of temperature is fundamental to those two laws, the law that establishes temperature as a valid concept should have the lowest number — hence the zero.

19-3 Measuring Temperature

Here we first define and measure temperatures on the Kelvin scale. Then we calibrate a thermoscope so as to make it a thermometer.

The Triple Point of Water

To set up a temperature scale, we pick some reproducible thermal phenomenon and, quite arbitrarily, assign a certain Kelvin temperature to its environment; that is, we select a *standard fixed point* and give it a standard fixed-point *temperature*. We could, for example, select the freezing point or the boiling point of water but, for various technical reasons, we select instead the **triple point of water.**

Liquid water, solid ice, and water vapor (gaseous water) can coexist, in thermal equilibrium, at only one set of values of pressure and temperature. Figure 19-4 shows a triple-point cell, in which this so-called triple point of water can be achieved in the laboratory. By international agreement, the triple point of water has been assigned a value of 273.16 K as the standard fixed-point temperature for the calibration of thermometers; that is,

$$T_3 = 273.16 \text{ K} \qquad \text{(triple-point temperature)}, \qquad (19\text{-}1)$$

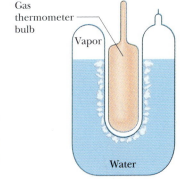

Fig. 19-4 A triple-point cell, in which solid ice, liquid water, and water vapor coexist in thermal equilibrium. By international agreement, the temperature of this mixture has been defined to be 273.16 K. The bulb of a constant-volume gas thermometer is shown inserted into the well of the cell.

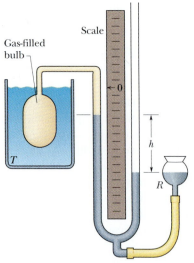

Fig. 19-5 A constant-volume gas thermometer, its bulb immersed in a liquid whose temperature T is to be measured.

in which the subscript 3 means "triple point." This agreement also sets the size of the kelvin as 1/273.16 of the difference between absolute zero and the triple-point temperature of water.

Note that we do not use a degree mark in reporting Kelvin temperatures. It is 300 K (not 300°K), and it is read "300 kelvins" (not "300 degrees Kelvin"). The usual SI prefixes apply. Thus, 0.0035 K is 3.5 mK. No distinction in nomenclature is made between Kelvin temperatures and temperature differences, so we can write, "the boiling point of sulfur is 717.8 K" and "the temperature of this water bath was raised by 8.5 K."

The Constant-Volume Gas Thermometer

The standard thermometer, against which all other thermometers are calibrated, is based on the pressure of a gas in a fixed volume. Figure 19-5 shows such a **constant-volume gas thermometer**; it consists of a gas-filled bulb connected by a tube to a mercury manometer. By raising and lowering reservoir R, the mercury level on the left can always be brought to the zero of the scale to keep the gas volume constant (variations in the gas volume can affect temperature measurements).

The temperature of any body in thermal contact with the bulb (like the liquid in Fig. 19-5) is then defined to be

$$T = Cp, \tag{19-2}$$

in which p is the pressure within the gas and C is a constant. From Eq. 15-10, the pressure p is

$$p = p_0 - \rho g h, \tag{19-3}$$

in which p_0 is the atmospheric pressure, ρ is the density of the mercury in the manometer, and h is the measured difference between the mercury levels in the two arms of the tube.*

If we next put the bulb in a triple-point cell (Fig. 19-4), the temperature now being measured is

$$T_3 = Cp_3, \tag{19-4}$$

in which p_3 is the gas pressure now. Eliminating C between Eqs. 19-2 and 19-4 gives us the temperature as

$$T = T_3\left(\frac{p}{p_3}\right) = (273.16 \text{ K})\left(\frac{p}{p_3}\right) \qquad \text{(provisional)}. \tag{19-5}$$

We still have a problem with this thermometer. If we use it to measure, say, the boiling point of water, we find that different gases in the bulb give slightly different results. However, as we use smaller and smaller amounts of gas to fill the bulb, the readings converge nicely to a single temperature, no matter what gas we use. Figure 19-6 shows this convergence for three gases.

Thus the recipe for measuring a temperature with a gas thermometer is

$$T = (273.16 \text{ K})\left(\lim_{\text{gas}\to 0} \frac{p}{p_3}\right). \tag{19-6}$$

The recipe instructs us to measure an unknown temperature T as follows: Fill the thermometer bulb with an arbitrary amount of *any* gas (for example, nitrogen) and measure p_3 (using a triple-point cell) and p, the gas pressure at the temperature being

---

*For pressure units, we shall use units introduced in Section 15-3. The SI unit for pressure is the newton per square meter, which is called the pascal (Pa). The pascal is related to other common pressure units by

$$1 \text{ atm} = 1.01 \times 10^5 \text{ Pa} = 760 \text{ torr} = 14.7 \text{ lb/in.}^2.$$

Fig. 19-6 Temperatures measured by a constant-volume gas thermometer, with its bulb immersed in boiling water. For temperature calculations using Eq. 19-5, pressure p_3 was measured at the triple point of water. Three different gases in the thermometer bulbs gave generally different results at different gas pressures, but as the amount of gas was decreased (decreasing p_3), all three curves converged to 373.125 K.

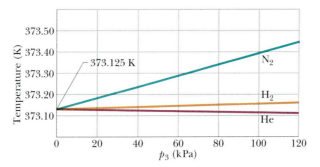

measured. (Keep the gas volume the same.) Calculate the ratio p/p_3. Then repeat both measurements with a smaller amount of gas in the bulb, and again calculate this ratio. Continue this way, using smaller and smaller amounts of gas, until you can extrapolate to the ratio p/p_3 that you would find if there were approximately no gas in the bulb. Calculate the temperature T by substituting that extrapolated ratio into Eq. 19-6. (The temperature is called the *ideal gas temperature*.)

19-4   The Celsius and Fahrenheit Scales

So far, we have discussed only the Kelvin scale, used in basic scientific work. In nearly all countries of the world, the Celsius scale (formerly called the centigrade scale) is the scale of choice for popular and commercial use and much scientific use. Celsius temperatures are measured in degrees, and the Celsius degree has the same size as the kelvin. However, the zero of the Celsius scale is shifted to a more convenient value than absolute zero. If T_C represents a Celsius temperature and T a Kelvin temperature, then

$$T_C = T - 273.15°. \qquad (19-7)$$

In expressing temperatures on the Celsius scale, the degree symbol is commonly used. Thus, we write 20.00°C for a Celsius reading but 293.15 K for a Kelvin reading.

The Fahrenheit scale, used in the United States, employs a smaller degree than the Celsius scale and a different zero of temperature. You can easily verify both these differences by examining an ordinary room thermometer on which both scales are marked. The relation between the Celsius and Fahrenheit scales is

$$T_F = \tfrac{9}{5}T_C + 32°, \qquad (19-8)$$

where T_F is Fahrenheit temperature. Transferring between these two scales can be done easily by remembering a few corresponding points, such as the freezing and boiling points of water (see Table 19-1). Figure 19-7 compares the Kelvin, Celsius, and Fahrenheit scales.

TABLE 19-1 Some Corresponding Temperatures

Temperature	°C	°F
Boiling point of water[a]	100	212
Normal body temperature	37.0	98.6
Accepted comfort level	20	68
Freezing point of water[a]	0	32
Zero of Fahrenheit scale	≈ -18	0
Scales coincide	-40	-40

[a]Strictly, the boiling point of water on the Celsius scale is 99.975°C, and the freezing point is 0.00°C. Thus, there is slightly less than 100 C° between those two points.

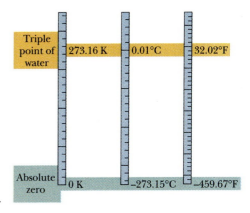

Fig. 19-7 The Kelvin, Celsius, and Fahrenheit temperature scales compared.

We use the letters C and F to distinguish measurements and degrees on the two scales. Thus,

$$0°C = 32°F$$

means that 0° on the Celsius scale measures the same temperature as 32° on the Fahrenheit scale, whereas

$$5 \; C° = 9 \; F°$$

means that a temperature difference of 5 Celsius degrees (note the degree symbol appears *after* C) is equivalent to a temperature difference of 9 Fahrenheit degrees.

Sample Problem 19-1

Suppose you come across old scientific notes that describe a temperature scale called Z on which the boiling point of water is 65.0°Z and the freezing point is −14.0°Z. To what temperature on the Fahrenheit scale would a temperature of $T = -98.0°Z$ correspond? Assume that the Z scale is linear; that is, the size of a Z degree is the same everywhere on the Z scale.

SOLUTION: One **Key Idea** here is to relate the given temperature T to *either* of the two known temperatures on the Z scale. Since $T = -98.0°Z$ is closer to the freezing point of −14.0°Z, we use that point for simplicity. Then we note that T is *below the freezing point* by the difference −14.0°Z − (−98.0°Z) = 84.0 Z° (Fig. 19-8). (Read this difference as "84.0 Z degrees.")

Another **Key Idea** is to set up a conversion factor between the Z and Fahrenheit scales to convert this difference. To do so, we use *both* known temperatures on the Z scale and the corresponding

temperatures on the Fahrenheit scale. On the Z scale, the difference between the boiling and freezing points is 65.0°Z − (−14.0°Z) = 79.0 Z°. On the Fahrenheit scale, it is 212°F − 32.0°F = 180 F°. Thus, a temperature difference of 79.0 Z° is equivalent to a temperature difference of 180 F° (Fig. 19-8), and we can use the ratio (180 F°)/(79.0 Z°) as our conversion factor.

Now, since T is below the freezing point by 84.0 Z°, it must also be below the freezing point by

$$(84.0 \; Z°) \frac{180 \; F°}{79.0 \; Z°} = 191 \; F°.$$

Because the freezing point is at 32.0°F, this means that

$$T = 32.0°F - 191 \; F° = -159°F. \qquad \text{(Answer)}$$

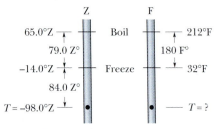

Z and F scale diagram: 65.0°Z / Boil / 212°F; 79.0 Z°; −14.0°Z / Freeze / 32°F; 180 F°; 84.0 Z°; $T = -98.0°Z$ / $T = ?$

Fig. 19-8 Sample Problem 19-1. An unknown temperature scale compared to the Fahrenheit temperature scale.

✔ **CHECKPOINT 1:** The figure here shows three temperature scales with the freezing and boiling points of water indicated. (a) Rank the degrees on these scales by size, greatest first. (b) Rank the following temperatures, highest first: 50°X, 50°W, and 50°Y.

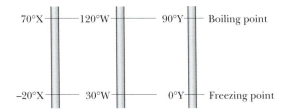

70°X — 120°W — 90°Y — Boiling point

−20°X — 30°W — 0°Y — Freezing point

PROBLEM-SOLVING TACTICS

Tactic 1: *Temperature Changes*

Between the boiling and freezing points of water, there are (approximately) 100 kelvins and 100 Celsius degrees. Thus, a kelvin is the same size as a Celsius degree. From this or from Eq. 19-7, we then know that any temperature change is the same number whether expressed in kelvins or Celsius degrees. For example, a temperature change of 10 K is exactly equivalent to a temperature change of 10 C°.

Between the boiling and freezing points of water, there are 180 Fahrenheit degrees. Thus, 180 F° = 100 K, and a Fahrenheit degree must be 100/180, or 5/9, the size of a kelvin or Celsius degree. From this or from Eq. 19-8, we then know that any temperature change expressed in Fahrenheit degrees must be $\frac{9}{5}$ times

that same temperature change expressed in either kelvins or Celsius degrees. For example, in Fahrenheit degrees, a temperature change of 10 K is (9/5)(10 K), or 18 F°.

You should take care not to confuse a *temperature* with a temperature *change* or *difference*. A temperature of 10 K is certainly not the same as one of 10°C or 18°F but, as above, a temperature *change* of 10 K is the same as one of 10 C° or 18 F°. This distinction is very important in an equation containing a temperature T instead of a temperature change or difference such as $T_2 - T_1$: A temperature T by itself should generally be in kelvins and not degrees Celsius or Fahrenheit. In short, beware the "bare T."

Fig. 19-9 Railroad tracks in Asbury Park, New Jersey, distorted because of thermal expansion on a very hot July day.

19-5 Thermal Expansion

You can often loosen a tight metal jar lid by holding it under a stream of hot water. Both the metal of the lid and the glass of the jar expand as the hot water adds energy to their atoms. (With the added energy, the atoms can move a bit farther from each other than usual, against the spring-like interatomic forces that hold every solid together.) However, because the atoms in the metal move farther apart than those in the glass, the lid expands more than the jar and thus is loosened.

Such **thermal expansion** is not always desirable, as Fig. 19-9 suggests. To preclude buckling, therefore, expansion slots are placed in bridges to accommodate roadway expansion on hot days. Dental materials used for fillings must be matched in their thermal expansion properties to those of tooth enamel (otherwise consuming hot coffee or cold ice cream would be quite painful). In aircraft manufacture, however, rivets and other fasteners are often cooled in dry ice before insertion and then allowed to expand to a tight fit.

Thermometers and thermostats may be based on the differences in expansion between the components of a *bimetal strip* (Fig. 19-10). Also, the familiar liquid-in-glass thermometers are based on the fact that liquids such as mercury and alcohol expand to a different (greater) extent than their glass containers.

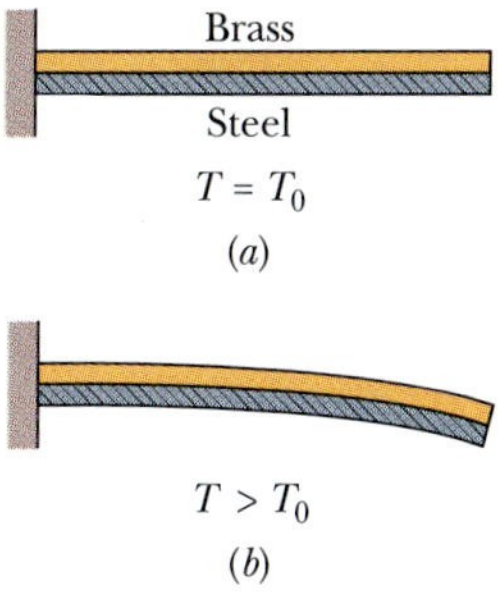

Linear Expansion

If the temperature of a metal rod of length L is raised by an amount ΔT, its length is found to increase by an amount

$$\Delta L = L\alpha\,\Delta T, \tag{19-9}$$

in which α is a constant called the **coefficient of linear expansion.** The coefficient α has the unit "per degree" or "per kelvin" and depends on the material. Although

Brass

Steel

$T = T_0$

(a)

$T > T_0$

(b)

Fig. 19-10 (a) A bimetal strip, consisting of a strip of brass and a strip of steel welded together, at temperature T_0. (b) The strip bends as shown at temperatures above this reference temperature. Below the reference temperature the strip bends the other way. Many thermostats operate on this principle, making and breaking an electrical contact as the temperature rises and falls.

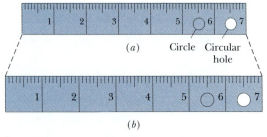

Fig. 19-11 The same steel ruler at two different temperatures. When it expands, the scale, the numbers, the thickness, and the diameters of the circle and circular hole are all increased by the same factor. (The expansion has been exaggerated for clarity.)

TABLE 19-2   Some Coefficients of Linear Expansion[a]

Substance	$\alpha\ (10^{-6}/\text{C}°)$	Substance	$\alpha\ (10^{-6}/\text{C}°)$
Ice (at 0°C)	51	Steel	11
Lead	29	Glass (ordinary)	9
Aluminum	23	Glass (Pyrex)	3.2
Brass	19	Diamond	1.2
Copper	17	Invar[b]	0.7
Concrete	12	Fused quartz	0.5

[a]Room temperature values except for the listing for ice.
[b]This alloy was designed to have a low coefficient of expansion. The word is a shortened form of "invariable."

α varies somewhat with temperature, for most practical purposes it can be taken as constant for a particular material. Table 19-2 shows some coefficients of linear expansion. Note that the unit C° there could be replaced with the unit K.

The thermal expansion of a solid is like (three-dimensional) photographic enlargement. Figure 19-11b shows the (exaggerated) expansion of a steel ruler after its temperature is increased from that of Fig. 19-11a. Equation 19-9 applies to every linear dimension of the ruler, including its edge, thickness, diagonals, and the diameters of the circle etched on it and the circular hole cut in it. If the disk cut from that hole originally fits snugly in the hole, it will continue to fit snugly if it undergoes the same temperature increase as the ruler.

Volume Expansion

If all dimensions of a solid expand with temperature, the volume of that solid must also expand. For liquids, volume expansion is the only meaningful expansion parameter. If the temperature of a solid or liquid whose volume is V is increased by an amount ΔT, the increase in volume is found to be

$$\Delta V = V\beta\ \Delta T, \tag{19-10}$$

where β is the **coefficient of volume expansion** of the solid or liquid. The coefficients of volume expansion and linear expansion for a solid are related by

$$\beta = 3\alpha. \tag{19-11}$$

The most common liquid, water, does not behave like other liquids. Above about 4°C, water expands as the temperature rises, as we would expect. Between 0 and about 4°C, however, water *contracts* with increasing temperature. Thus, at about 4°C, the density of water passes through a maximum. At all other temperatures, the density of water is less than this maximum value.

This behavior of water is the reason why lakes freeze from the top down rather than from the bottom up. As water on the surface is cooled from, say, 10°C toward the freezing point, it becomes denser ("heavier") than lower water and sinks to the bottom. Below 4°C, however, further cooling makes the water then on the surface *less* dense ("lighter") than the lower water, so it stays on the surface until it freezes. Thus the surface freezes while the lower water is still liquid. If lakes froze from the bottom up, the ice so formed would tend not to melt completely during the summer, because it would be insulated by the water above. After a few years, many bodies of open water in the temperate zones of Earth would be frozen solid all year round—and aquatic life as we know it could not exist.

✔**CHECKPOINT 2:** The figure here shows four rectangular metal plates, with sides of L, $2L$, or $3L$. They are all made of the same material, and their temperature is to be increased by the same amount. Rank the plates according to the expected increase in (a) their vertical heights and (b) their areas, greatest first.

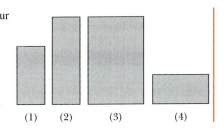

(1)   (2)   (3)   (4)

Sample Problem 19-2

On a hot day in Las Vegas, an oil trucker loaded 37,000 L of diesel fuel. He encountered cold weather on the way to Payson, Utah, where the temperature was 23.0 K lower than in Las Vegas, and where he delivered his entire load. How many liters did he deliver? The coefficient of volume expansion for diesel fuel is $9.50 \times 10^{-4}/C°$, and the coefficient of linear expansion for his steel truck tank is $11 \times 10^{-6}/C°$.

SOLUTION: The Key Idea here is that the volume of the diesel fuel depends directly on the temperature. Thus, because the temperature decreased, the volume of the fuel did also. From Eq. 19-10, the volume change is

$$\Delta V = V \beta \, \Delta T$$
$$= (37{,}000 \text{ L})(9.50 \times 10^{-4}/C°)(-23.0 \text{ K}) = -808 \text{ L}.$$

Thus, the amount delivered was

$$V_{\text{del}} = V + \Delta V = 37{,}000 \text{ L} - 808 \text{ L}$$
$$= 36{,}190 \text{ L}. \qquad \text{(Answer)}$$

Note that the thermal expansion of the steel tank has nothing to do with the problem. Question: Who paid for the "missing" diesel fuel?

19-6 Temperature and Heat

If you take a can of cola from the refrigerator and leave it on the kitchen table, its temperature will rise—rapidly at first but then more slowly—until the temperature of the cola equals that of the room (the two are then in thermal equilibrium). In the same way, the temperature of a cup of hot coffee, left sitting on the table, will fall until it also reaches room temperature.

In generalizing this situation, we describe the cola or the coffee as a *system* (with temperature T_S) and the relevant parts of the kitchen as the *environment* (with temperature T_E) of that system. Our observation is that if T_S is not equal to T_E, then T_S will change (T_E may also change some) until the two temperatures are equal and thus thermal equilibrium is reached.

Such a change in temperature is due to the transfer of energy between the thermal energy of the system and the system's environment. (*Thermal energy* is an internal energy that consists of the kinetic and potential energies associated with the random motions of the atoms, molecules, and other microscopic bodies within an object.) The transferred energy is called **heat** and is symbolized Q. Heat is *positive* when energy is transferred to a system's thermal energy from its environment (we say that heat is absorbed). Heat is *negative* when energy is transferred from a system's thermal energy to its environment (we say that heat is released or lost).

This transfer of energy is shown in Fig. 19-12. In the situation of Fig. 19-12a, in which $T_S > T_E$, energy is transferred from the system to the environment, so Q is negative. In Fig. 19-12b, in which $T_S = T_E$, there is no such transfer, Q is zero, and heat is neither released nor absorbed. In Fig. 19-12c, in which $T_S < T_E$, the transfer is to the system from the environment, so Q is positive.

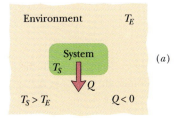

(a)

Environment $\quad T_E$

System T_S

Q

$T_S > T_E$ $\qquad Q < 0$

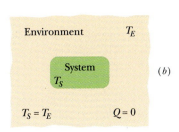

(b)

Environment $\quad T_E$

System T_S

$T_S = T_E$ $\qquad Q = 0$

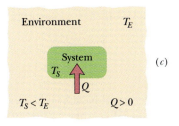

(c)

Environment $\quad T_E$

System T_S

Q

$T_S < T_E$ $\qquad Q > 0$

Fig. 19-12 If the temperature of a system exceeds that of its environment as in (a), heat Q is lost by the system to the environment until thermal equilibrium (b) is established. (c) If the temperature of the system is below that of the environment, heat is absorbed by the system until thermal equilibrium is established.

We are led then to this definition of heat:

> ► Heat is the energy that is transferred between a system and its environment because of a temperature difference that exists between them.

Recall that energy can also be transferred between a system and its environment as *work W* via a force acting on a system. Heat and work, unlike temperature, pressure, and volume, are not intrinsic properties of a system. They have meaning only as they describe the transfer of energy into or out of a system. Thus, it is proper to say: "During the last 3 min, 15 J of heat was transferred to the system from its environment" or "During the last minute, 12 J of work was done on the system by its environment." It is meaningless to say: "This system contains 450 J of heat" or "This system contains 385 J of work."

Before scientists realized that heat is transferred energy, heat was measured in terms of its ability to raise the temperature of water. Thus, the **calorie** (cal) was defined as the amount of heat that would raise the temperature of 1 g of water from 14.5°C to 15.5°C. In the British system, the corresponding unit of heat was the **British thermal unit** (Btu), defined as the amount of heat that would raise the temperature of 1 lb of water from 63°F to 64°F.

In 1948, the scientific community decided that since heat (like work) is transferred energy, the SI unit for heat should be the one we use for energy, namely, the **joule.** The calorie is now defined to be 4.1860 J (exactly), with no reference to the heating of water. (The "calorie" used in nutrition, sometimes called the Calorie (Cal), is really a kilocalorie.) The relations among the various heat units are

$$1 \text{ cal} = 3.969 \times 10^{-3} \text{ Btu} = 4.1860 \text{ J.} \tag{19-12}$$

19-7   The Absorption of Heat by Solids and Liquids

Heat Capacity

The **heat capacity** C of an object is the proportionality constant between the heat Q that the object absorbs or loses and the resulting temperature change ΔT of the object; that is,

$$Q = C \, \Delta T = C(T_f - T_i), \tag{19-13}$$

in which T_i and T_f are the initial and final temperatures of the object. Heat capacity C has the unit of energy per degree or energy per kelvin. The heat capacity C of, say, a marble slab used in a bun warmer might be 179 cal/C°, which we can also write as 179 cal/K or as 749 J/K.

The word "capacity" in this context is really misleading in that it suggests analogy with the capacity of a bucket to hold water. *That analogy is false,* and you should not think of the object as "containing" heat or being limited in its ability to absorb heat. Heat transfer can proceed without limit as long as the necessary temperature difference is maintained. The object may, of course, melt or vaporize during the process.

Specific Heat

Two objects made of the same material—say, marble—will have heat capacities proportional to their masses. It is therefore convenient to define a "heat capacity per

TABLE 19-3 Specific Heats of Some Substances at Room Temperature

Substance	Specific Heat		Molar Specific Heat
	cal / g·K	J / kg·K	J / mol·K
Elemental Solids			
Lead	0.0305	128	26.5
Tungsten	0.0321	134	24.8
Silver	0.0564	236	25.5
Copper	0.0923	386	24.5
Aluminum	0.215	900	24.4
Other Solids			
Brass	0.092	380	
Granite	0.19	790	
Glass	0.20	840	
Ice (−10°C)	0.530	2220	
Liquids			
Mercury	0.033	140	
Ethyl alcohol	0.58	2430	
Seawater	0.93	3900	
Water	1.00	4190	

unit mass" or **specific heat** c that refers not to an object but to a unit mass of the material of which the object is made. Equation 19-13 then becomes

$$Q = cm \, \Delta T = cm(T_f - T_i). \qquad (19\text{-}14)$$

Through experiment we would find that although the heat capacity of a particular marble slab might be 179 cal/C° (or 749 J/K), the specific heat of marble itself (in that slab or in any other marble object) is 0.21 cal/g · C° (or 880 J/kg · K).

From the way the calorie and the British thermal unit were initially defined, the specific heat of water is

$$c = 1 \text{ cal/g} \cdot \text{C}° = 1 \text{ Btu/lb} \cdot \text{F}° = 4190 \text{ J/kg} \cdot \text{K}. \qquad (19\text{-}15)$$

Table 19-3 shows the specific heats of some substances at room temperature. Note that the value for water is relatively high. The specific heat of any substance actually depends somewhat on temperature, but the values in Table 19-3 apply reasonably well in a range of temperatures near room temperature.

✔**CHECKPOINT 3:** A certain amount of heat Q will warm 1 g of material A by 3 C° and 1 g of material B by 4 C°. Which material has the greater specific heat?

Molar Specific Heat

In many instances the most convenient unit for specifying the amount of a substance is the mole (mol), where

$$1 \text{ mol} = 6.02 \times 10^{23} \text{ elementary units}$$

of *any* substance. Thus 1 mol of aluminum means 6.02×10^{23} atoms (the atom being the elementary unit), and 1 mol of aluminum oxide means 6.02×10^{23} molecules of the oxide (because the molecule is the elementary unit of a compound).

When quantities are expressed in moles, specific heats must also involve moles (rather than a mass unit); they are then called **molar specific heats.** Table 19-3 shows the values for some elemental solids (each consisting of a single element) at room temperature.

An Important Point

In determining and then using the specific heat of any substance, we need to know the conditions under which energy is transferred as heat. For solids and liquids, we usually assume that the sample is under constant pressure (usually atmospheric) during the transfer. It is also conceivable that the sample is held at constant volume while the heat is absorbed. This means that thermal expansion of the sample is prevented by applying external pressure. For solids and liquids, this is very hard to arrange experimentally but the effect can be calculated, and it turns out that the specific heats under constant pressure and constant volume for any solid or liquid differ usually by no more than a few percent. Gases, as you will see, have quite different values for their specific heats under constant-pressure conditions and under constant-volume conditions.

Heats of Transformation

When energy is absorbed as heat by a solid or liquid, the temperature of the sample does not necessarily rise. Instead, the sample may change from one *phase,* or *state,* to another. Matter can exist in three common states: In the *solid state,* the molecules of a sample are locked into a fairly rigid structure by their mutual attraction. In the

TABLE 19-4   Some Heats of Transformation

| Substance | Melting | | Boiling | |
	Melting Point (K)	Heat of Fusion L_F (kJ/kg)	Boiling Point (K)	Heat of Vaporization L_V (kJ/kg)
Hydrogen	14.0	58.0	20.3	455
Oxygen	54.8	13.9	90.2	213
Mercury	234	11.4	630	296
Water	273	333	373	2256
Lead	601	23.2	2017	858
Silver	1235	105	2323	2336
Copper	1356	207	2868	4730

liquid state, the molecules have more energy and move about more. They may form brief clusters, but the sample does not have a rigid structure and can flow or settle into a container. In the *gas* or *vapor state,* the molecules have even more energy, are free of one another, and can fill up the full volume of a container.

To *melt* a solid means to change it from the solid state to the liquid state. The process requires energy because the molecules of the solid must be freed from their rigid structure. Melting an ice cube to form liquid water is a common example. To *freeze* a liquid to form a solid is the reverse of melting and requires that energy be removed from the liquid, so that the molecules can settle into a rigid structure.

To *vaporize* a liquid means to change it from the liquid state to the vapor or gas state. This process, like melting, requires energy because the molecules must be freed from their clusters. Boiling liquid water to transfer it to water vapor (or steam — a gas of individual water molecules) is a common example. *Condensing* a gas to form a liquid is the reverse of vaporizing; it requires that energy be removed from the gas, so that the molecules can cluster instead of flying away from one another.

The amount of energy per unit mass that must be transferred as heat when a sample completely undergoes a phase change is called the **heat of transformation** *L*. Thus, when a sample of mass *m* completely undergoes a phase change, the total energy transferred is

$$Q = Lm. \qquad (19\text{-}16)$$

When the phase change is from liquid to gas (then the sample must absorb heat) or from gas to liquid (then the sample must release heat), the heat of transformation is called the **heat of vaporization** L_V. For water at its normal boiling or condensation temperature,

$$L_V = 539 \text{ cal/g} = 40.7 \text{ kJ/mol} = 2256 \text{ kJ/kg.} \qquad (19\text{-}17)$$

When the phase change is from solid to liquid (then the sample must absorb heat) or from liquid to solid (then the sample must release heat), the heat of transformation is called the **heat of fusion** L_F. For water at its normal freezing or melting temperature,

$$L_F = 79.5 \text{ cal/g} = 6.01 \text{ kJ/mol} = 333 \text{ kJ/kg.} \qquad (19\text{-}18)$$

Table 19-4 shows the heats of transformation for some substances.

Sample Problem 19-3

(a) How much heat must be absorbed by ice of mass *m* = 720 g at −10°C to take it to liquid state at 15°C?

SOLUTION: The first Key Idea is that the heating process is accomplished in three steps.

Step 1. The Key Idea here is that the ice cannot melt at a temperature below the freezing point—so initially, any energy transferred to the ice as heat can only increase the temperature of the ice. The heat Q_1 needed to increase that temperature from the initial value $T_i = -10°C$ to a final value $T_f = 0°C$ (so that the ice can then melt) is given by Eq. 19-14 ($Q = cm\,\Delta T$). Using the specific heat of ice c_{ice} in Table 19-3 gives us

$$Q_1 = c_{ice}m(T_f - T_i)$$
$$= (2220 \text{ J/kg}\cdot\text{K})(0.720 \text{ kg})[0°C - (-10°C)]$$
$$= 15{,}984 \text{ J} \approx 15.98 \text{ kJ}.$$

Step 2. The next Key Idea is that the temperature cannot increase from 0°C until all the ice melts—so any energy transferred to the ice as heat now can only change ice to liquid water. The heat Q_2 needed to melt all the ice is given by Eq. 19-16 ($Q = Lm$). Here L is the heat of fusion L_F, with the value given in Eq. 19-18 and Table 19-4. We find

$$Q_2 = L_F m = (333 \text{ kJ/kg})(0.720 \text{ kg}) \approx 239.8 \text{ kJ}.$$

Step 3. Now we have liquid water at 0°C. The next Key Idea is that the energy transferred to the liquid water as heat now can only increase the temperature of the liquid water. The heat Q_3 needed to increase the temperature of the water from the initial value $T_i = 0°C$ to the final value $T_f = 15°C$ is given by Eq. 19-14 (with the specific heat of liquid water c_{liq}):

$$Q_3 = c_{liq}m(T_f - T_i)$$
$$= (4190 \text{ J/kg}\cdot\text{K})(0.720 \text{ kg})(15°C - 0°C)$$
$$= 45{,}252 \text{ J} \approx 45.25 \text{ kJ}.$$

The total required heat Q_{tot} is the sum of the amounts required in the three steps:

$$Q_{tot} = Q_1 + Q_2 + Q_3$$
$$= 15.98 \text{ kJ} + 239.8 \text{ kJ} + 45.25 \text{ kJ}$$
$$\approx 300 \text{ kJ}. \qquad\text{(Answer)}$$

Note that the heat required to melt the ice is much greater than the heat required to raise the temperature of either the ice or the liquid water.

(b) If we supply the ice with a total energy of only 210 kJ (as heat), what then are the final state and temperature of the water?

SOLUTION: From step 1, we know that 15.98 kJ is needed to raise the temperature of the ice to the melting point. The remaining heat Q_{rem} is then 210 kJ − 15.98 kJ, or about 194 kJ. From step 2, we can see that this amount of heat is insufficient to melt all the ice. Then this Key Idea becomes important: Because the melting of the ice is incomplete, we must end up with a mixture of ice and liquid; the temperature of the mixture must be the freezing point, 0°C.

We can find the mass m of ice that is melted by the available energy Q_{rem} by using Eq. 19-16 with L_F:

$$m = \frac{Q_{rem}}{L_F} = \frac{194 \text{ kJ}}{333 \text{ kJ/kg}} = 0.583 \text{ kg} \approx 580 \text{ g}.$$

Thus, the mass of the ice that remains is 720 g − 580 g, or 140 g, and we have

$$580 \text{ g water} \quad\text{and}\quad 140 \text{ g ice}, \quad\text{at } 0°C. \qquad\text{(Answer)}$$

Sample Problem 19-4

A copper slug whose mass m_c is 75 g is heated in a laboratory oven to a temperature T of 312°C. The slug is then dropped into a glass beaker containing a mass $m_w = 220$ g of water. The heat capacity C_b of the beaker is 45 cal/K. The initial temperature T_i of the water and the beaker is 12°C. Assuming that the slug, beaker, and water are an isolated system and the water does not vaporize, find the final temperature T_f of the system at thermal equilibrium.

SOLUTION: One Key Idea here is that, with the system isolated, only internal transfers of energy can occur. There are three such transfers, all as heat. The slug loses energy, the water gains energy, and the beaker gains energy. Another Key Idea is that, because these transfers do not involve a phase change, the energy transfers can only change the temperatures. To relate the transfers to the temperature changes, we can use Eqs. 19-13 and 19-14 to write

$$\text{for the water: } \quad Q_w = c_w m_w(T_f - T_i); \qquad (19\text{-}19)$$
$$\text{for the beaker: } \quad Q_b = C_b(T_f - T_i); \qquad (19\text{-}20)$$
$$\text{for the copper: } \quad Q_c = c_c m_c(T_f - T). \qquad (19\text{-}21)$$

A third Key Idea is that, with the system isolated, the total energy of the system cannot change. This means that the sum of these three energy transfers is zero:

$$Q_w + Q_b + Q_c = 0. \qquad (19\text{-}22)$$

Substituting Eqs. 19-19 through 19-21 into Eq. 19-22 yields

$$c_w m_w(T_f - T_i) + C_b(T_f - T_i) + c_c m_c(T_f - T) = 0. \qquad (19\text{-}23)$$

Temperatures are contained in Eq. 19-23 only as differences. Thus, because the differences on the Celsius and Kelvin scales are identical, we can use either of those scales in this equation. Solving it for T_f, we obtain

$$T_f = \frac{c_c m_c T + C_b T_i + c_w m_w T_i}{c_w m_w + C_b + c_c m_c}.$$

Using Celsius temperatures and taking values for c_c and c_w from Table 19-3, we find the numerator to be

$$(0.0923 \text{ cal/g}\cdot\text{K})(75 \text{ g})(312°C) + (45 \text{ cal/K})(12°C)$$
$$+ (1.00 \text{ cal/g}\cdot\text{K})(220 \text{ g})(12°C) = 5339.8 \text{ cal},$$

and the denominator to be

$$(1.00 \text{ cal/g}\cdot\text{K})(220 \text{ g}) + 45 \text{ cal/K}$$
$$+ (0.0923 \text{ cal/g}\cdot\text{K})(75 \text{ g}) = 271.9 \text{ cal/C°}.$$

We then have

$$T_f = \frac{5339.8 \text{ cal}}{271.9 \text{ cal/C°}} = 19.6°C \approx 20°C. \qquad\text{(Answer)}$$

From the given data you can show that

$$Q_w \approx 1670 \text{ cal}, \qquad Q_b \approx 342 \text{ cal}, \qquad Q_c \approx -2020 \text{ cal}.$$

Apart from rounding errors, the algebraic sum of these three heat transfers is indeed zero, as Eq. 19-22 requires.

19-8 A Closer Look at Heat and Work

Here we look in some detail at how energy can be transferred as heat and work between a system and its environment. Let us take as our system a gas confined to a cylinder with a movable piston, as in Fig. 19-13. The upward force on the piston due to the pressure of the confined gas is equal to the weight of lead shot loaded onto the top of the piston. The walls of the cylinder are made of insulating material that does not allow any transfer of energy as heat. The bottom of the cylinder rests on a reservoir for thermal energy, a *thermal reservoir* (perhaps a hot plate) whose temperature T you can control by turning a knob.

The system (the gas) starts from an *initial state i*, described by a pressure p_i, a volume V_i, and a temperature T_i. You want to change the system to a *final state f*, described by a pressure p_f, a volume V_f, and a temperature T_f. The procedure by which you change the system from its initial state to its final state is called a *thermodynamic process*. During such a process, energy may be transferred into the system from the thermal reservoir (positive heat) or vice versa (negative heat). Also, work can be done by the system to raise the loaded piston (positive work) or lower it (negative work). We assume that all such changes occur slowly, with the result that the system is always in (approximate) thermal equilibrium (that is, every part of the system is always in thermal equilibrium with every other part).

Suppose that you remove a few lead shot from the piston of Fig. 19-13, allowing the gas to push the piston and remaining shot upward through a differential displacement $d\vec{s}$ with an upward force $\vec{F}$. Since the displacement is tiny, we can assume that $\vec{F}$ is constant during the displacement. Then $\vec{F}$ has a magnitude that is equal to pA, where p is the pressure of the gas and A is the face area of the piston. The differential work dW done by the gas during the displacement is

$$dW = \vec{F} \cdot d\vec{s} = (pA)(ds) = p(A\, ds)$$
$$= p\, dV, \tag{19-24}$$

in which dV is the differential change in the volume of the gas owing to the movement of the piston. When you have removed enough shot to allow the gas to change its volume from V_i to V_f, the total work done by the gas is

$$W = \int dW = \int_{V_i}^{V_f} p\, dV. \tag{19-25}$$

During the change in volume, the pressure and temperature of the gas may also change. To evaluate the integral in Eq. 19-25 directly, we would need to know how pressure varies with volume for the actual process by which the system changes from state i to state f.

There are actually many ways to take the gas from state i to state f. One way is shown in Fig. 19-14a, which is a plot of the pressure of the gas versus its volume and which is called a *p-V diagram*. In Fig. 19-14a, the curve indicates that the pressure decreases as the volume increases. The integral Eq. 19-25 (and thus the work W done by the gas) is represented by the shaded area under the curve between points i and f. Regardless of what exactly we do to take the gas along the curve, that work is positive, owing to the fact that the gas increases its volume by forcing the piston upward.

Another way to get from state i to state f is shown in Fig. 19-14b. There the change takes place in two steps—the first from state i to state a, and the second from state a to state f.

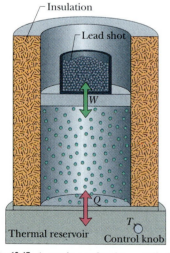

Fig. 19-13 A gas is confined to a cylinder with a movable piston. Heat Q can be added to, or withdrawn from, the gas by regulating the temperature T of the adjustable thermal reservoir. Work W can be done by the gas by raising or lowering the piston.

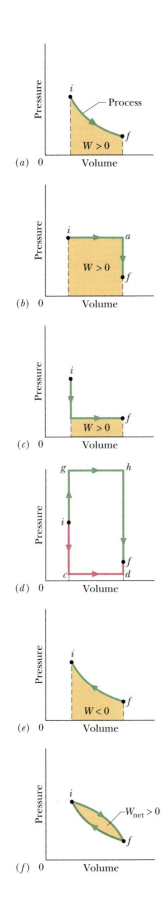

(a)

(b)

(c)

(d)

(e)

(f)

Fig. 19-14 (*a*) The shaded area represents the work *W* done by a system as it goes from an initial state *i* to a final state *f*. Work *W* is positive because the system's volume increases. (*b*) *W* is still positive, but now greater. (*c*) *W* is still positive, but now smaller. (*d*) *W* can be even smaller (path *icdf*) or larger (path *ighf*). (*e*) Here the system goes from state *f* to state *i*, as the gas is compressed to less volume by an external force. The work *W* done *by* the system is now negative. (*f*) The net work W_{net} done by the system during a complete cycle is represented by the shaded area.

Step *ia* of this process is carried out at constant pressure, which means that you leave undisturbed the lead shot that ride on top of the piston in Fig. 19-13. You cause the volume to increase (from V_i to V_f) by slowly turning up the temperature control knob, raising the temperature of the gas to some higher value T_a. (Increasing the temperature increases the force from the gas on the piston, moving it upward.) During this step, positive work is done by the expanding gas (to lift the loaded piston) and heat is absorbed by the system from the thermal reservoir (in response to the arbitrarily small temperature differences that you create as you turn up the temperature). This heat is positive because it is added to the system.

Step *af* of the process of Fig. 19-14*b* is carried out at constant volume, so you must wedge the piston, preventing it from moving. Then as you use the control knob to decrease the temperature, you find that the pressure drops from p_a to its final value p_f. During this step, heat is lost by the system to the thermal reservoir.

For the overall process *iaf*, the work *W*, which is positive and is carried out only during step *ia*, is represented by the shaded area under the curve. Energy is transferred as heat during both steps *ia* and *af*, with a net energy transfer *Q*.

Figure 19-14*c* shows a process in which the previous two steps are carried out in reverse order. The work *W* in this case is smaller than for Fig. 19-14*b*, as is the net heat absorbed. Figure 19-14*d* suggests that you can make the work done by the gas as small as you want (by following a path like *icdf*) or as large as you want (by following a path like *ighf*).

To sum up: A system can be taken from a given initial state to a given final state by an infinite number of processes. Heat may or may not be involved, and in general, the work *W* and the heat *Q* will have different values for different processes. We say that heat and work are *path-dependent* quantities.

Figure 19-14*e* shows an example in which negative work is done by a system as some external force compresses the system, reducing its volume. The absolute value of the work done is still equal to the area beneath the curve, but because the gas is *compressed,* the work done by the gas is negative.

Figure 19-14*f* shows a *thermodynamic cycle* in which the system is taken from some initial state *i* to some other state *f* and then back to *i*. The net work done by the system during the cycle is the sum of the *positive* work done during the expansion and the *negative* work done during the compression. In Fig. 19-14*f*, the net work is positive because the area under the expansion curve (*i* to *f*) is greater than the area under the compression curve (*f* to *i*).

✔**CHECKPOINT 4:** The *p-V* diagram here shows six curved paths (connected by vertical paths) that can be followed by a gas. Which two of them should be part of a closed cycle if the net work done by the gas is to be at its maximum positive value?

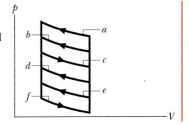

19-9   The First Law of Thermodynamics

You have just seen that when a system changes from a given initial state to a given final state, both the work W and the heat Q depend on the nature of the process. Experimentally, however, we find a surprising thing. *The quantity $Q - W$ is the same for all processes.* It depends only on the initial and final states and does not depend at all on how the system gets from one to the other. All other combinations of Q and W, including Q alone, W alone, $Q + W$, and $Q - 2W$, are *path dependent;* only the quantity $Q - W$ is not.

The quantity $Q - W$ must represent a change in some intrinsic property of the system. We call this property the *internal energy E_{int}* and we write

$$\Delta E_{int} = E_{int,f} - E_{int,i} = Q - W \qquad \text{(first law).} \qquad (19\text{-}26)$$

Equation 19-26 is the **first law of thermodynamics.** If the thermodynamic system undergoes only a differential change, we can write the first law as*

$$dE_{int} = dQ - dW \qquad \text{(first law).} \qquad (19\text{-}27)$$

> The internal energy E_{int} of a system tends to increase if energy is added as heat Q and tends to decrease if energy is lost as work W done by the system.

In Chapter 8, we discussed the principle of energy conservation as it applies to isolated systems—that is, to systems in which no energy enters or leaves the system. The first law of thermodynamics is an extension of that principle to systems that are *not* isolated. In such cases, energy may be transferred into or out of the system as either work W or heat Q. In our statement of the first law of thermodynamics above, we assume that there are no changes in the kinetic energy or the potential energy of the system as a whole; that is, $\Delta K = \Delta U = 0$.

Before this chapter, the term *work* and the symbol W always meant the work done *on* a system. However, starting with Eq. 19-24 and continuing through the next two chapters about thermodynamics, we focus on the work done *by* a system, such as the gas in Fig. 19-13.

The work done *on* a system is always the negative of the work done *by* the system, so if we rewrite Eq. 19-26 in terms of the work W_{on} done *on* the system, we have $\Delta E_{int} = Q + W_{on}$. This tells us the following: The internal energy of a system tends to increase if heat is absorbed by the system or if positive work is done *on* the system. Conversely, the internal energy tends to decrease if heat is lost by the system or if negative work is done *on* the system.

✔**CHECKPOINT 5:**   The figure here shows four paths on a p-V diagram along which a gas can be taken from state i to state f. Rank the paths according to (a) the change ΔE_{int}, (b) the work W done by the gas, and (c) the magnitude of the energy transferred as heat Q, greatest first.

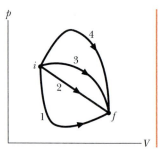

*Here dQ and dW, unlike dE_{int}, are not true differentials; that is, there are no such functions as $Q(p, V)$ and $W(p, V)$ that depend only on the state of the system. The quantities dQ and dW are called *inexact differentials* and are usually represented by the symbols $đQ$ and $đW$. For our purposes, we can treat them simply as infinitesimally small energy transfers.

19-10 Some Special Cases of the First Law of Thermodynamics

Here we look at four different thermodynamic processes, in each of which a certain restriction is imposed on the system. We then see what consequences follow when we apply the first law of thermodynamics to the process. The results are summarized in Table 19-5.

1. **Adiabatic processes.** An adiabatic process is one that occurs so rapidly or occurs in a system that is so well insulated that *no transfer of energy as heat* occurs between the system and its environment. Putting $Q = 0$ in the first law (Eq. 19-26) yields

$$\Delta E_{int} = -W \qquad \text{(adiabatic process).} \qquad (19\text{-}28)$$

This tells us that if work is done *by* the system (that is, if W is positive), the internal energy of the system decreases by the amount of work. Conversely, if work is done *on* the system (that is, if W is negative), the internal energy of the system increases by that amount.

Figure 19-15 shows an idealized adiabatic process. Heat cannot enter or leave the system because of the insulation. Thus, the only way energy can be transferred between the system and its environment is by work. If we remove shot from the piston and allow the gas to expand, the work done by the system (the gas) is positive and the internal energy of the gas decreases. If, instead, we add shot and compress the gas, the work done by the system is negative and the internal energy of the gas increases.

2. **Constant-volume processes.** If the volume of a system (such as a gas) is held constant, that system can do no work. Putting $W = 0$ in the first law (Eq. 19-26) yields

$$\Delta E_{int} = Q \qquad \text{(constant-volume process).} \qquad (19\text{-}29)$$

Thus, if heat is absorbed by a system (that is, if Q is positive), the internal energy of the system increases. Conversely, if heat is lost during the process (that is, if Q is negative), the internal energy of the system must decrease.

3. **Cyclical processes.** There are processes in which, after certain interchanges of heat and work, the system is restored to its initial state. In that case, no intrinsic property of the system—including its internal energy—can possibly change. Putting $\Delta E_{int} = 0$ in the first law (Eq. 19-26) yields

$$Q = W \qquad \text{(cyclical process).} \qquad (19\text{-}30)$$

Thus, the net work done during the process must exactly equal the net amount of energy transferred as heat; the store of internal energy of the system remains unchanged. Cyclical processes form a closed loop on a p-V plot, as shown in Fig. 19-14f. We shall discuss such processes in some detail in Chapter 21.

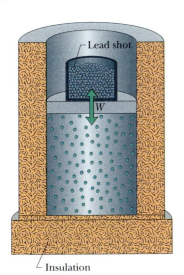

Fig. 19-15 An adiabatic expansion can be carried out by slowly removing lead shot from the top of the piston. Adding lead shot reverses the process at any stage.

TABLE 19-5 The First Law of Thermodynamics: Four Special Cases

The Law: $\Delta E_{int} = Q - W$ (Eq. 19-26)		
Process	Restriction	Consequence
Adiabatic	$Q = 0$	$\Delta E_{int} = -W$
Constant volume	$W = 0$	$\Delta E_{int} = Q$
Closed cycle	$\Delta E_{int} = 0$	$Q = W$
Free expansion	$Q = W = 0$	$\Delta E_{int} = 0$

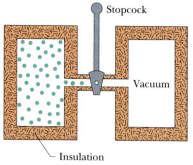

Fig. 19-16 The initial stage of a free-expansion process. After the stopcock is opened, the gas fills both chambers and eventually reaches an equilibrium state.

4. *Free expansions.* These are adiabatic processes in which no transfer of heat occurs between the system and its environment and no work is done on or by the system. Thus, $Q = W = 0$ and the first law requires that

$$\Delta E_{int} = 0 \qquad \text{(free expansion).} \qquad (19\text{-}31)$$

Figure 19-16 shows how such an expansion can be carried out. A gas, which is in thermal equilibrium within itself, is initially confined by a closed stopcock to one half of an insulated double chamber; the other half is evacuated. The stopcock is opened, and the gas expands freely to fill both halves of the chamber. No heat is transferred to or from the gas because of the insulation. No work is done by the gas because it rushes into a vacuum and thus does not meet any pressure.

A free expansion differs from all other processes we have considered because it cannot be done slowly and in a controlled way. As a result, at any given instant during the sudden expansion, the gas is not in thermal equilibrium and its pressure is not the same everywhere. Therefore, although we can plot the initial and final states on a p-V diagram, we cannot plot the expansion itself.

✔**CHECKPOINT 6:**   For one complete cycle as shown in the p-V diagram here, are (a) ΔE_{int} for the gas and (b) the net energy transferred as heat Q positive, negative, or zero?

Sample Problem 19-5

Let 1.00 kg of liquid water at 100°C be converted to steam at 100°C by boiling at standard atmospheric pressure (which is 1.00 atm or 1.01×10^5 Pa) in the arrangement of Fig. 19-17. The volume of that water changes from an initial value of 1.00×10^{-3} m³ as a liquid to 1.671 m³ as steam.

(a) How much work is done by the system during this process?

SOLUTION: The **Key Idea** here is that the system must do positive work because the volume increases. In the general case we would calculate the work W done by integrating the pressure with respect to the volume (Eq. 19-25). However, here the pressure is constant at 1.01×10^5 Pa, so we can take p outside the integral. We then have

$$W = \int_{V_i}^{V_f} p \, dV = p \int_{V_i}^{V_f} dV = p(V_f - V_i)$$
$$= (1.01 \times 10^5 \text{ Pa})(1.671 \text{ m}^3 - 1.00 \times 10^{-3} \text{ m}^3)$$
$$= 1.69 \times 10^5 \text{ J} = 169 \text{ kJ.} \qquad \text{(Answer)}$$

(b) How much energy is transferred as heat during the process?

SOLUTION: The **Key Idea** here is that the heat causes only a phase change and not a change in temperature, so it is given fully by Eq. 19-16 ($Q = Lm$). Because the change is from liquid to gaseous phase, L is the heat of vaporization L_V, with the value given in Eq. 19-17 and Table 19-4. We find

$$Q = L_V m = (2256 \text{ kJ/kg})(1.00 \text{ kg})$$
$$= 2256 \text{ kJ} \approx 2260 \text{ kJ.} \qquad \text{(Answer)}$$

(c) What is the change in the system's internal energy during the process?

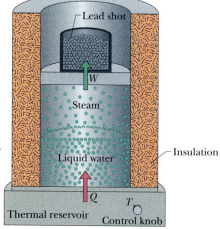

Fig. 19-17 Sample Problem 19-5. Water boiling at constant pressure. Energy is transferred from the thermal reservoir as heat until the liquid water has changed completely into steam. Work is done by the expanding gas as it lifts the loaded piston.

SOLUTION: The **Key Idea** here is that the change in the system's internal energy is related to the heat (here, this is energy transferred into the system) and the work (here, this is energy transferred out of the system) by the first law of thermodynamics (Eq. 19-26). Thus, we can write

$$\Delta E_{int} = Q - W = 2256 \text{ kJ} - 169 \text{ kJ}$$
$$\approx 2090 \text{ kJ} = 2.09 \text{ MJ.} \qquad \text{(Answer)}$$

This quantity is positive, indicating that the internal energy of the system has increased during the boiling process. This energy goes into separating the H_2O molecules, which strongly attract each other in the liquid state. We see that, when water is boiled, about 7.5% (= 169 kJ/2260 kJ) of the heat goes into the work of pushing back the atmosphere. The rest of the heat goes into the system's internal energy.

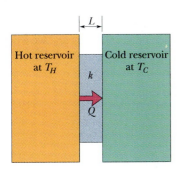

Fig. 19-18 Thermal conduction. Energy is transferred as heat from a reservoir at temperature T_H to a cooler reservoir at temperature T_C through a conducting slab of thickness L and thermal conductivity k.

19-11 Heat Transfer Mechanisms

We have discussed the transfer of energy as heat between a system and its environment, but we have not yet described how that transfer takes place. There are three transfer mechanisms: conduction, convection, and radiation.

Conduction

If you leave the end of a metal poker in a fire for enough time, its handle will get hot. Energy is transferred from the fire to the handle by (thermal) **conduction** along the length of the poker. The vibration amplitudes of the atoms and electrons of the metal at the fire end of the poker become relatively large because of the high temperature of their environment. These increased vibrational amplitudes, and thus the associated energy, are passed along the poker, from atom to atom, during collisions between adjacent atoms. In this way, a region of rising temperature extends itself along the poker to the handle.

Consider a slab of face area A and thickness L, whose faces are maintained at temperatures T_H and T_C by a hot reservoir and a cold reservoir, as in Fig. 19-18. Let Q be the energy that is transferred as heat through the slab, from its hot face to its cold face, in time t. Experiment shows that the *conduction rate* P_{cond} (the amount of energy transferred per unit time) is

$$P_{\text{cond}} = \frac{Q}{t} = kA\frac{T_H - T_C}{L}, \qquad (19\text{-}32)$$

in which k, called the *thermal conductivity*, is a constant that depends on the material of which the slab is made. A material that readily transfers energy by conduction is a *good thermal conductor* and has a high value of k. Table 19-6 gives the thermal conductivities of some common metals, gases, and building materials.

Thermal Resistance to Conduction (*R*-Value)

If you are interested in insulating your house or in keeping cola cans cold on a picnic, you are more concerned with poor heat conductors than with good ones. For this reason, the concept of *thermal resistance R* has been introduced into engineering practice. The *R*-value of a slab of thickness L is defined as

$$R = \frac{L}{k}. \qquad (19\text{-}33)$$

The lower the thermal conductivity of the material of which a slab is made, the higher the *R*-value of the slab, so something that has a high *R*-value is a *poor thermal conductor* and thus a *good thermal insulator*.

Note that R is a property attributed to a slab of a specified thickness, not to a material. The commonly used unit for R (which, in the United States at least, is almost never stated) is the square foot–Fahrenheit degree–hour per British thermal unit ($\text{ft}^2 \cdot \text{F}° \cdot \text{h/Btu}$). (Now you know why the unit is rarely stated.)

Conduction Through a Composite Slab

Figure 19-19 shows a composite slab, consisting of two materials having different thicknesses L_1 and L_2 and different thermal conductivities k_1 and k_2. The temperatures of the outer surfaces of the slab are T_H and T_C. Each face of the slab has area A. Let us derive an expression for the conduction rate through the slab under the assumption that the transfer is a *steady-state* process; that is, the temperatures everywhere in the slab and the rate of energy transfer do not change with time.

TABLE 19-6 Some Thermal Conductivities[a]

Substance	k (W/m · K)
Metals	
Stainless steel	14
Lead	35
Aluminum	235
Copper	401
Silver	428
Gases	
Air (dry)	0.026
Helium	0.15
Hydrogen	0.18
Building Materials	
Polyurethane form	0.024
Rock wool	0.043
Fiberglass	0.048
White pine	0.11
Window glass	1.0

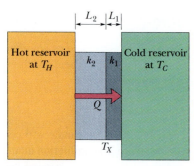

Fig. 19-19 Heat is transferred at a steady rate through a composite slab made up of two different materials with different thicknesses and different thermal conductivities. The steady-state temperature at the interface of the two materials is T_X.

In the steady state, the conduction rates through the two materials must be equal. This is the same as saying that the energy transferred through one material in a certain time must be equal to that transferred through the other material in the same time. If this were not true, temperatures in the slab would be changing and we would not have a steady-state situation. Letting T_X be the temperature of the interface between the two materials, we can now use Eq. 19-32 to write

$$P_{cond} = \frac{k_2 A(T_H - T_X)}{L_2} = \frac{k_1 A(T_X - T_C)}{L_1}. \tag{19-34}$$

Solving Eq. 19-34 for T_X yields, after a little algebra,

$$T_X = \frac{k_1 L_2 T_C + k_2 L_1 T_H}{k_1 L_2 + k_2 L_1}. \tag{19-35}$$

Substituting this expression for T_X into either equality of Eq. 19-34 yields

$$P_{cond} = \frac{A(T_H - T_C)}{L_1/k_1 + L_2/k_2}. \tag{19-36}$$

We can extend Eq. 19-36 to apply to any number n of materials making up a slab:

$$P_{cond} = \frac{A(T_H - T_C)}{\Sigma \,(L/k)}. \tag{19-37}$$

The summation sign in the denominator tells us to add the values of L/k for all the materials.

✔**CHECKPOINT 7:**   The figure shows the face and interface temperatures of a composite slab consisting of four materials, of identical thicknesses, through which the heat transfer is steady.

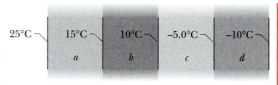

Rank the materials according to their thermal conductivities, greatest first.

Convection

When you look at the flame of a candle or a match, you are watching thermal energy being transported upward by **convection.** Such energy transfer occurs when a fluid, such as air or water, comes in contact with an object whose temperature is higher than that of the fluid. The temperature of the part of the fluid that is in contact with the hot object increases, and (in most cases) that fluid expands and thus becomes less dense. Because this expanded fluid is now lighter than the surrounding cooler fluid, buoyant forces cause it to rise. Some of the surrounding cooler fluid then flows so as to take the place of the rising warmer fluid, and the process can then continue.

Convection is part of many natural processes. Atmospheric convection plays a fundamental role in determining global climate patterns and daily weather variations. Glider pilots and birds alike seek rising thermals (convection currents of warm air) that keep them aloft. Huge energy transfers take place within the oceans by the same process. Finally, energy is transported to the surface of the Sun from the nuclear furnace at its core by enormous cells of convection, in which hot gas rises to the surface along the cell core and cooler gas around the core descends below the surface.

Radiation

The third method by which an object and its environment can exchange energy as heat is via electromagnetic waves (visible light is one kind of electromagnetic wave).

Fig. 19-20 A false-color thermogram reveals the rate at which energy is radiated by houses along a street. The rates, from largest to smallest, are color coded as white, red, pink, blue, and black. You can tell where there is insulation in the walls, a heavy curtain over a window, and a higher air temperature at the ceiling on the second floor.

Energy transferred in this way is often called **thermal radiation** to distinguish it from electromagnetic *signals* (as in, say, television broadcasts) and from nuclear radiation (energy and particles emitted by nuclei). (To "radiate" generally means to emit.) When you stand in front of a big fire, you are warmed by absorbing thermal radiation from the fire; that is, your thermal energy increases as the fire's thermal energy decreases. No medium is required for heat transfer via radiation—the radiation can travel through vacuum from, say, the Sun to you.

The rate P_{rad} at which an object emits energy via electromagnetic radiation depends on the object's surface area A and the temperature T of that area in kelvins and is given by

$$P_{rad} = \sigma \varepsilon A T^4. \qquad (19\text{-}38)$$

Here $\sigma = 5.6703 \times 10^{-8}$ W/m$^2 \cdot$ K^4 is called the *Stefan–Boltzmann constant* after Josef Stefan (who discovered Eq. 19-38 experimentally in 1879) and Ludwig Boltzmann (who derived it theoretically soon after). The symbol ε represents the *emissivity* of the object's surface, which has a value between 0 and 1, depending on the composition of the surface. A surface with the maximum emissivity of 1.0 is said to be a *blackbody radiator,* but such a surface is an ideal limit and does not occur in nature. Note again that the temperature in Eq. 19-38 must be in kelvins so that a temperature of absolute zero corresponds to no radiation. Note also that every object whose temperature is above 0 K—including you—emits thermal radiation. (See Fig. 19-20.)

The rate P_{abs} at which an object absorbs energy via thermal radiation from its environment, which we take to be at uniform temperature T_{env} (in kelvins), is

$$P_{abs} = \sigma \varepsilon A T_{env}^4. \qquad (19\text{-}39)$$

The emissivity ε in Eq. 19-39 is the same as that in Eq. 19-38. An idealized blackbody radiator, with $\varepsilon = 1$, will absorb all the radiated energy it intercepts (rather than sending a portion back away from itself through reflection or scattering).

Because an object will radiate energy to the environment while it absorbs energy from the environment, the object's net rate P_{net} of energy exchange due to thermal radiation is

$$P_{net} = P_{abs} - P_{rad} = \sigma \varepsilon A (T_{env}^4 - T^4). \qquad (19\text{-}40)$$

P_{net} is positive if net energy is being absorbed via radiation, and negative if it is being lost via radiation.

Sample Problem 19-6

Figure 19-21 shows the cross section of a wall made of white pine of thickness L_a and brick of thickness L_d $(= 2.0L_a)$, sandwiching two layers of unknown material with identical thicknesses and thermal conductivities. The thermal conductivity of the pine is k_a and that of the brick is k_d $(= 5.0k_a)$. The face area A of the wall is unknown. Thermal conduction through the wall has reached the steady state; the only known interface temperatures are $T_1 = 25°C$, $T_2 = 20°C$, and $T_5 = -10°C$. What is interface temperature T_4?

SOLUTION: One Key Idea here is that temperature T_4 helps determine the rate P_d at which energy is conducted through the brick, as given by Eq. 19-32. However, we lack enough data to solve Eq. 19-32 for T_4. A second Key Idea is that because the conduction is steady, the conduction rate P_d through the brick must equal the conduction rate P_a through the pine. From Eq. 19-32 and Fig. 19-21, we can write

$$P_a = k_a A \frac{T_1 - T_2}{L_a} \quad \text{and} \quad P_d = k_d A \frac{T_4 - T_5}{L_d}.$$

Setting $P_a = P_d$ and solving for T_4 yield

$$T_4 = \frac{k_a L_d}{k_d L_a}(T_1 - T_2) + T_5.$$

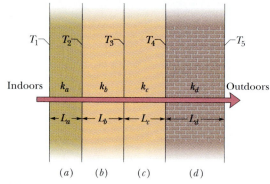

Fig. 19-21 Sample Problem 19-6. A wall of four layers through which there is steady-state heat transfer.

Letting $L_d = 2.0L_a$ and $k_d = 5.0k_a$, and inserting the known temperatures, we find

$$T_4 = \frac{k_a(2.0L_a)}{(5.0k_a)L_a}(25°C - 20°C) + (-10°C)$$
$$= -8.0°C. \qquad \text{(Answer)}$$

Sample Problem 19-7

When hundreds of Japanese bees form a compact ball around a giant hornet that attempts to invade their hive, they can quickly raise their body temperature from the normal 35°C to 47°C or 48°C. That higher temperature is lethal to the hornet but not to the bees (Fig. 19-22). Assume the following: 500 bees form a ball of radius $R = 2.0$ cm for a time $t = 20$ min, the primary loss of energy by the ball is by thermal radiation, the ball's surface has emissivity $\varepsilon = 0.80$, and the ball has a uniform temperature. On average, how much additional energy must each bee produce during the 20 min to maintain 47°C?

SOLUTION: The Key Idea here is that, because the surface temperature of the bee ball increases after the ball forms, the rate at which energy is radiated by the ball also increases. Thus, the bees lose an additional amount of energy to thermal radiation. We can relate the surface temperature to the rate of radiation (energy per unit time)

Fig. 19-22 Sample Problem 19-7. The bees were unharmed by their increased body temperature, which the hornet could not withstand.

with Eq. 19-38 ($P_{rad} = \sigma \varepsilon A T^4$), in which A is the ball's surface area and T is the ball's surface temperature in kelvins. This rate is an energy per unit time; that is,

$$P_{rad} = \frac{E}{t}.$$

Thus, the amount of energy E radiated in time t is $E = P_{rad}t$.

At the normal temperature $T_1 = 35°C$, the radiation rate would be P_{r1} and the amount of energy radiated in time t would be $E_1 = P_{r1}t$. At the increased temperature $T_2 = 47°C$, the (greater) radiation rate is P_{r2} and the (greater) amount of energy radiated in time t is $E_2 = P_{r2}t$. Thus, in maintaining the ball at T_2 for time t, the bees must (together) provide an additional energy of $\Delta E = E_2 - E_1$.

We can now write

$$\Delta E = E_2 - E_1 = P_{r2}t - P_{r1}t$$
$$= (\sigma \varepsilon A T_2^4)t - (\sigma \varepsilon A T_1^4)t = \sigma \varepsilon A t(T_2^4 - T_1^4). \quad (19\text{-}41)$$

The temperatures here *must* be in kelvins; thus, we write them as

$$T_2 = 47°C + 273 C° = 320 \text{ K}$$

and

$$T_1 = 35°C + 273 C° = 308 \text{ K}.$$

The surface area A of the ball is

$$A = 4\pi R^2 = (4\pi)(0.020 \text{ m})^2 = 5.027 \times 10^{-3} \text{ m}^2,$$

and the time t is 20 min = 1200 s. Substituting these and other known values into Eq. 19-41, we find

$$\Delta E = (5.6703 \times 10^{-8} \text{ W/m}^2 \cdot \text{K}^4)(0.80)(5.027 \times 10^{-3} \text{ m}^2)$$
$$\times (1200 \text{ s})[(320 \text{ K})^4 - (308 \text{ K})^4] = 406.8 \text{ J}.$$

Thus, with 500 bees in the ball, each bee must produce an additional energy of

$$\frac{\Delta E}{500} = \frac{406.8 \text{ J}}{500} = 0.81 \text{ J}. \qquad \text{(Answer)}$$

REVIEW & SUMMARY

Temperature; Thermometers   Temperature is an SI base quantity related to our sense of hot and cold. It is measured with a thermometer, which contains a working substance with a measurable property, such as length or pressure, that changes in a regular way as the substance becomes hotter or colder.

Zeroth Law of Thermodynamics   When a thermometer and some other object are placed in contact with each other, they eventually reach thermal equilibrium. The reading of the thermometer is then taken to be the temperature of the other object. The process provides consistent and useful temperature measurements because of the **zeroth law of thermodynamics:** If bodies *A* and *B* are each in thermal equilibrium with a third body *C* (the thermometer), then *A* and *B* are in thermal equilibrium with each other.

The Kelvin Temperature Scale   In the SI system, temperature is measured on the **Kelvin scale,** which is based on the *triple point* of water (273.16 K). Other temperatures are then defined by use of a *constant-volume gas thermometer,* in which a sample of gas is maintained at constant volume so its pressure is proportional to its temperature. We define the *temperature T* as measured with a gas thermometer to be

$$T = (273.16 \text{ K}) \left(\lim_{\text{gas} \to 0} \frac{p}{p_3} \right). \tag{19-6}$$

Here *T* is measured in kelvins, and p_3 and *p* are the pressures of the gas at 273.16 K and the measured temperature, respectively.

Celsius and Fahrenheit Scales   The Celsius temperature scale is defined by

$$T_\text{C} = T - 273.15°, \tag{19-7}$$

with *T* in kelvins. The Fahrenheit temperature scale is defined by

$$T_\text{F} = \tfrac{9}{5} T_\text{C} + 32°. \tag{19-8}$$

Thermal Expansion   All objects change size with changes in temperature. For a temperature change ΔT, a change ΔL in any linear dimension *L* is given by

$$\Delta L = L \alpha \, \Delta T, \tag{19-9}$$

in which α is the **coefficient of linear expansion.** The change ΔV in the volume *V* of a solid or liquid is

$$\Delta V = V \beta \, \Delta T. \tag{19-10}$$

Here $\beta = 3\alpha$ is the material's **coefficient of volume expansion.**

Heat   Heat *Q* is energy that is transferred between a system and its environment because of a temperature difference between them. It can be measured in **joules** (J), **calories** (cal), **kilocalories** (Cal or kcal), or **British thermal units** (Btu), with

$$1 \text{ cal} = 3.969 \times 10^{-3} \text{ Btu} = 4.1860 \text{ J}. \tag{19-12}$$

Heat Capacity and Specific Heat   If heat *Q* is absorbed by an object, the object's temperature change $T_f - T_i$ is related to *Q* by

$$Q = C(T_f - T_i), \tag{19-13}$$

in which *C* is the **heat capacity** of the object. If the object has mass *m*, then

$$Q = cm(T_f - T_i), \tag{19-14}$$

where *c* is the **specific heat** of the material making up the object. The **molar specific heat** of a material is the heat capacity per mole, or per 6.02×10^{23} elementary units of the material.

Heat of Transformation   Heat absorbed by a material may change the material's physical state or phase—for example, from solid to liquid or from liquid to gas. The amount of energy required per unit mass to change the phase (but not the temperature) of a particular material is its **heat of transformation** *L*. Thus,

$$Q = Lm. \tag{19-16}$$

The **heat of vaporization** L_V is the amount of energy per unit mass that must be added to vaporize a liquid or that must be removed to condense a gas. The **heat of fusion** L_F is the amount of energy per unit mass that must be added to melt a solid or that must be removed to freeze a liquid.

Work Associated with Volume Change   A gas may exchange energy with its surroundings through work. The amount of work *W* done *by* a gas as it expands or contracts from an initial volume V_i to a final volume V_f is given by

$$W = \int dW = \int_{V_i}^{V_f} p \, dV. \tag{19-25}$$

The integration is necessary because the pressure *p* may vary during the volume change.

First Law of Thermodynamics   The principle of conservation of energy for a thermodynamic process is expressed in the **first law of thermodynamics,** which may assume either of the forms

$$\Delta E_\text{int} = E_{\text{int},f} - E_{\text{int},i} = Q - W \quad \text{(first law)} \tag{19-26}$$

or

$$dE_\text{int} = dQ - dW \quad \text{(first law)}. \tag{19-27}$$

E_int represents the internal energy of the material, which depends only on its state (temperature, pressure, and volume). *Q* represents the energy exchanged as heat between the system and its surroundings; *Q* is positive if the system absorbs heat, and negative if the system loses heat. *W* is the work done *by* the system; *W* is positive if the system expands against some external force exerted by the surroundings, and negative if the system contracts because of some external force. *Both Q and W are path dependent;* ΔE_int *is path independent.*

Applications of the First Law   The first law of thermodynamics finds application in several special cases:

$$
\begin{aligned}
\text{adiabatic processes:} \quad & Q = 0, \quad \Delta E_\text{int} = -W \\
\text{constant-volume processes:} \quad & W = 0, \quad \Delta E_\text{int} = Q \\
\text{cyclical processes:} \quad & \Delta E_\text{int} = 0, \quad Q = W \\
\text{free expansions:} \quad & Q = W = \Delta E_\text{int} = 0
\end{aligned}
$$

Conduction, Convection, and Radiation   The rate P_{cond} at which energy is *conducted* through a slab whose faces are maintained at temperatures T_H and T_C is

$$P_{cond} = \frac{Q}{t} = kA\frac{T_H - T_C}{L}, \qquad (19\text{-}32)$$

in which A and L are the face area and length of the slab, and k is the thermal conductivity of the material.

Convection occurs when temperature differences cause an energy transfer by motion within a fluid. *Radiation* is an energy transfer via the emission of electromagnetic energy. The rate P_{rad} at which an object emits energy via thermal radiation is

$$P_{rad} = \sigma\varepsilon AT^4, \qquad (19\text{-}38)$$

where σ ($= 5.6703 \times 10^{-8}$ W/m$^2 \cdot$ K^4) is the Stefan–Boltzmann constant, ε is the emissivity of the object's surface, A is its surface area, and T is its surface temperature (in kelvins). The rate P_{abs} at which an object absorbs energy via thermal radiation from its environment, which is at the uniform temperature T_{env} (in kelvins), is

$$P_{abs} = \sigma\varepsilon AT_{env}^4. \qquad (19\text{-}39)$$

QUESTIONS

1. Figure 19-23 shows three linear temperature scales with the freezing and boiling points of water indicated. Rank changes of 25 R°, 25 S°, and 25 U° according to the corresponding changes in temperature, greatest first.

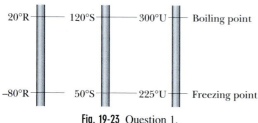

Fig. 19-23 Question 1.

2. The table gives the initial length L, change in temperature ΔT, and change in length ΔL of four rods. Rank the rods according to their coefficients of thermal expansion, greatest first.

Rod	L (m)	ΔT (C°)	ΔL (m)
a	2	10	4×10^{-4}
b	1	20	4×10^{-4}
c	2	10	8×10^{-4}
d	4	5	4×10^{-4}

3. In a thermally isolated container, material A of mass m is placed against material B, also of mass m but at a higher temperature. When thermal equilibrium is reached, the temperature changes ΔT_A and ΔT_B of A and B are recorded. Then the experiment is repeated, using A with other materials, all of the same mass m. The results are given in the table. Rank the four materials according to their specific heats, greatest first.

Experiment	Temperature Changes	
1	$\Delta T_A = +50$ C°	$\Delta T_B = -50$ C°
2	$\Delta T_A = +10$ C°	$\Delta T_C = -20$ C°
3	$\Delta T_A = +2$ C°	$\Delta T_D = -40$ C°

4. Materials A, B, and C are solids that are at their melting temperatures. Material A requires 200 J to melt 4 kg, material B requires 300 J to melt 5 kg, and material C requires 300 J to melt 6 kg. Rank the materials according to their heats of fusion, greatest first.

5. Figure 19-24 shows two closed cycles on p-V diagrams for a gas. The three parts of cycle 1 are of the same length and shape as those of cycle 2. For each cycle, should the cycle be traversed clockwise or counterclockwise if (a) the net work W done by the gas is to be positive and (b) the net energy transferred by the gas as heat Q is to be positive?

6. For which cycle in Fig. 19-24, traversed clockwise, is (a) W greater and (b) Q greater?

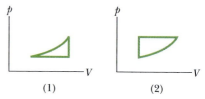

Fig. 19-24 Questions 5 and 6.

7. Figure 19-25 shows a composite slab of three different materials, a, b, and c, with identical thicknesses and with thermal conductivities $k_b > k_a > k_c$. The transfer of energy through them as heat is nonzero and steady. Rank the materials according to the temperature difference ΔT across them, greatest first.

Fig. 19-25 Question 7.

8. During an icicle's growth, its outer surface is covered with a thin sheath of liquid water that slowly seeps downward to form drops, one at a time, that hang at the icicle's tip (Fig. 19-26). Each drop straddles a thin tube of liquid water that extends up into the icicle, toward (but not all the way to) its *root* (its top). As the water at the top of the tube gradually freezes, energy is released. Is that energy conducted radially outward through the ice, downward through the water to the hanging

Fig. 19-26 Question 8.

drop, or upward toward the root? (Assume that the air temperature is below 0°C.)

9. The following solid objects, made of the same material, are maintained at a temperature of 300 K in an environment whose temperature is 350 K: a cube of edge length r, a sphere of radius r, and a hemisphere of radius r. Rank the objects according to the net rate at which thermal radiation is exchanged with the environment, greatest first.

10. Three different materials of identical masses are placed, in turn, in a special freezer that can extract energy from a material at a certain constant rate. During the cooling process, each material begins in the liquid state and ends in the solid state; Fig. 19-27 shows graphs of the temperature T versus time t for the three materials. (a) For material 1, is the specific heat for the liquid state greater than or less than that for the solid state? Rank the materials according to (b) their freezing-point temperatures, (c) their specific heats in the liquid state, (d) their specific heats in the solid state, and (e) their heats of fusion, all greatest first.

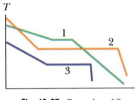

Fig. 19-27 Question 10.

11. A sample A of liquid water and a sample B of ice, of identical masses, are placed in a thermally isolated (insulated) container and allowed to come to thermal equilibrium. Figure 19-28a is a sketch of the temperature T of the samples versus time t. (a) Is the equilibrium temperature above, below, or at the freezing point of water? (b) In reaching equilibrium, does the liquid partly freeze or fully

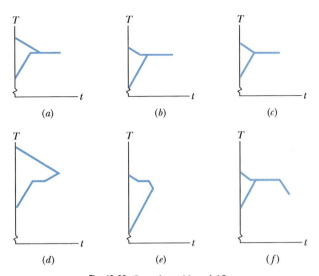

Fig. 19-28 Questions 11 and 12.

freeze, or does it undergo no freezing? (c) Does the ice partly melt or fully melt, or does it undergo no melting?

12. Question 11 continued: Figure 19-28 gives additional sketches of T versus t, of which one or more are impossible to produce. (a) Which sketch is impossible and why? (b) In the possible ones, is the equilibrium temperature above, below, or at the freezing point of water? (c) As the possible situations reach equilibrium, does the liquid partly freeze or fully freeze, or does it undergo no freezing? Does the ice partly melt or fully melt, or undergo no melting?

EXERCISES & PROBLEMS

ssm Solution is in the Student Solutions Manual.
www Solution is available on the World Wide Web at:
        http://www.wiley.com/college/hrw
ilw Solution is available on the Interactive LearningWare.

SEC. 19-3 Measuring Temperature

1E. Two constant-volume gas thermometers are assembled, one with nitrogen and the other hydrogen. Both contain enough gas so that $p_3 = 80$ kPa. What is the difference between the pressures in the two thermometers if both bulbs are inserted into boiling water? (*Hint:* See Fig. 19-6.) Which gas is at higher pressure? **ssm**

2E. Suppose the temperature of a gas at the boiling point of water is 373.15 K. What then is the limiting value of the ratio of the pressure of the gas at that boiling point to its pressure at the triple point of water? (Assume the volume of the gas is the same at both temperatures.)

3E. A particular gas thermometer is constructed of two gas-containing bulbs, each of which is put into a water bath, as shown in Fig. 19-29. The pressure difference between the two bulbs is

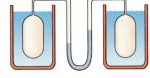

Fig. 19-29 Exercise 3.

measured by a mercury manometer as shown. Appropriate reservoirs, not shown in the diagram, maintain constant gas volume in the two bulbs. There is no difference in pressure when both baths are at the triple point of water. The pressure difference is 120 torr when one bath is at the triple point and the other is at the boiling point of water. It is 90.0 torr when one bath is at the triple point and the other is at an unknown temperature to be measured. What is the unknown temperature? **ssm**

SEC. 19-4 The Celsius and Fahrenheit Scales

4E. At what temperature is the Fahrenheit scale reading equal to (a) twice that of the Celsius and (b) half that of the Celsius?

5E. At what temperature do the following pairs of scales read the same, if ever: (a) Fahrenheit and Celsius (verify the listing in Table 19-1), (b) Fahrenheit and Kelvin, and (c) Celsius and Kelvin? **ssm**

6E. (a) In 1964, the temperature in the Siberian village of Oymyakon reached −71°C. What temperature is this on the Fahrenheit scale? (b) The highest officially recorded temperature in the continental United States was 134°F in Death Valley, California. What is this temperature on the Celsius scale?

7P. It is an everyday observation that hot and cold objects cool down or warm up to the temperature of their surroundings. If the temperature difference ΔT between an object and its surroundings

$(\Delta T = T_{\text{obj}} - T_{\text{sur}})$ is not too great, the rate of cooling or warming of the object is proportional, approximately, to this temperature difference; that is,

$$\frac{d\,\Delta T}{dt} = -A(\Delta T),$$

where A is a constant. (The minus sign appears because ΔT decreases with time if ΔT is positive and increases if ΔT is negative.) This is known as *Newton's law of cooling*. (a) On what factors does A depend? What are its dimensions? (b) If at some instant $t = 0$ the temperature difference is ΔT_0, show that it is

$$\Delta T = \Delta T_0 e^{-At}$$

at a later time t. ssm

8P. The heater of a house breaks down one day when the outside temperature is 7.0°C. As a result, the inside temperature drops from 22°C to 18°C in 1.0 h. The owner fixes the heater and adds insulation to the house. Now she finds that, on a similar day, the house takes twice as long to drop from 22°C to 18°C when the heater is not operating. What is the ratio of the new value of constant A in Newton's law of cooling (see Problem 7) to the previous value?

9P. Suppose that on a linear temperature scale X, water boils at $-53.5°X$ and freezes at $-170°X$. What is a temperature of 340 K on the X scale? ilw

SEC. 19-5 Thermal Expansion

10E. An aluminum flagpole is 33 m high. By how much does its length increase as the temperature increases by 15 C°?

11E. The Pyrex glass mirror in the telescope at the Mt. Palomar Observatory has a diameter of 200 in. The temperature ranges from $-10°C$ to 50°C on Mt. Palomar. In micrometers, what is the maximum change in the diameter of the mirror, assuming that the glass can freely expand and contract? ssm

12E. An aluminum-alloy rod has a length of 10.000 cm at 20.000°C and a length of 10.015 cm at the boiling point of water. (a) What is the length of the rod at the freezing point of water? (b) What is the temperature if the length of the rod is 10.009 cm?

13E. A circular hole in an aluminum plate is 2.725 cm in diameter at 0.000°C. What is its diameter when the temperature of the plate is raised to 100.0°C? ilw

14E. What is the volume of a lead ball at 30°C if the ball's volume at 60°C is 50 cm³?

15E. Find the change in volume of an aluminum sphere with an initial radius of 10 cm when the sphere is heated from 0.0°C to 100°C. ssm

16E. The area A of a rectangular plate is ab. Its coefficient of linear expansion is α. After a temperature rise ΔT, side a is longer by Δa and side b is longer by Δb (Fig. 19-30). Show that if the small quantity $(\Delta a\,\Delta b)/ab$ is neglected, then $\Delta A = 2\alpha A\,\Delta T$.

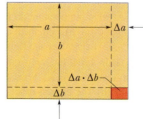

Fig. 19-30 Exercise 16.

17E. An aluminum cup of 100 cm³ capacity is completely filled with glycerin at 22°C. How much glycerin, if any, will spill out of the cup if the temperature of both

the cup and glycerin is increased to 28°C? (The coefficient of volume expansion of glycerin is $5.1 \times 10^{-4}/C°$.) ssm

18P. At 20°C, a rod is exactly 20.05 cm long on a steel ruler. Both the rod and the ruler are placed in an oven at 270°C, where the rod now measures 20.11 cm on the same ruler. What is the coefficient of thermal expansion for the material of which the rod is made?

19P. A steel rod is 3.000 cm in diameter at 25°C. A brass ring has an interior diameter of 2.992 cm at 25°C. At what common temperature will the ring just slide onto the rod? ssm ilw www

20P. When the temperature of a metal cylinder is raised from 0.0°C to 100°C, its length increases by 0.23%. (a) Find the percent change in density. (b) What is the metal? Use Table 19-2.

21P. Show that when the temperature of a liquid in a barometer changes by ΔT and the pressure is constant, the liquid's height h changes by $\Delta h = \beta h\,\Delta T$, where β is the coefficient of volume expansion. Neglect the expansion of the glass tube. ssm

22P. When the temperature of a copper coin is raised by 100 C°, its diameter increases by 0.18%. To two significant figures, give the percent increase in (a) the area of a face, (b) the thickness, (c) the volume, and (d) the mass of the coin. (e) Calculate the coefficient of linear expansion of the coin.

23P. A pendulum clock with a pendulum made of brass is designed to keep accurate time at 20°C. If the clock operates at 0.0°C, what is the magnitude of its error, in seconds per hour, and does the clock run fast or slow? ilw

24P. In a certain experiment, a small radioactive source must move at selected, extremely slow speeds. This motion is accomplished by fastening the source to one end of an aluminum rod and heating the central section of the rod in a controlled way. If the effective heated section of the rod in Fig. 19-31 is 2.00 cm, at what constant rate must the temperature of the rod be changed if the source is to move at a constant speed of 100 nm/s?

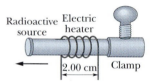

Fig. 19-31 Problem 24.

25P. As a result of a temperature rise of 32 C°, a bar with a crack at its center buckles upward (Fig. 19-32). If the fixed distance L_0 is 3.77 m and the coefficient of linear expansion of the bar is $25 \times 10^{-6}/C°$, find the rise x of the center. ssm ilw

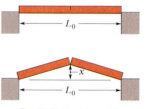

Fig. 19-32 Problem 25.

SEC. 19-7 The Absorption of Heat by Solids and Liquids

26E. A certain diet doctor encourages people to diet by drinking ice water. His theory is that the body must burn off enough fat to raise the temperature of the water from 0.00°C to the body temperature of 37.0°C. How many liters of ice water would have to be consumed to burn off 454 g (about 1 lb) of fat, assuming that this much fat burning requires 3500 Cal be transferred to the ice water? Why is it not advisable to follow this diet? (One liter $= 10^3$ cm³. The density of water is 1.00 g/cm³.)

27E. A certain substance has a mass per mole of 50 g/mol. When 314 J is added as heat to a 30.0 g sample, the sample's temperature rises from 25.0°C to 45.0°C. What are (a) the specific heat and (b) the molar specific heat of this substance? (c) How many moles are present? ssm

28E. How much water remains unfrozen after 50.2 kJ is transferred as heat from 260 g of liquid water initially at its freezing point?

29E. Calculate the minimum amount of energy, in joules, required to completely melt 130 g of silver initially at 15.0°C. ssm

30E. A room is lighted by four 100 W incandescent lightbulbs. (The power of 100 W is the rate at which a bulb converts electrical energy to heat and the energy of visible light.) Assuming that 90% of the energy is converted to heat, how much heat does the room receive in 1.00 h?

31E. An energetic athlete can use up all the energy from a diet of 4000 Cal/day. If he were to use up this energy at a steady rate, how would his rate of energy use compare with the power of a 100 W bulb? (The power of 100 W is the rate at which the bulb converts electrical energy to heat and the energy of visible light.) ilw

32E. How many grams of butter, which has a usable energy content of 6.0 Cal/g (= 6000 cal/g), would be equivalent to the change in gravitational potential energy of a 73.0 kg man who ascends from sea level to the top of Mt. Everest, at elevation 8.84 km? Assume that the average value of g is 9.80 m/s².

33E. A power of 0.400 hp is required for 2.00 min to drill a hole in a 1.60-lb copper block. (a) If the full power is the rate at which thermal energy is generated, how much is generated in Btu? (b) What is the rise in temperature of the copper if the copper absorbs 75.0% of this energy? (Use the energy conversion 1 ft·lb = 1.285 × 10⁻³ Btu.) ssm

34E. One way to keep the contents of a garage from becoming too cold on a night when a severe subfreezing temperature is forecast is to put a tub of water in the garage. If the mass of the water is 125 kg and its initial temperature is 20°C, (a) how much energy must the water transfer to its surroundings in order to freeze completely and (b) what is the lowest possible temperature of the water and its surroundings until that happens?

35E. A small electric immersion heater is used to heat 100 g of water for a cup of instant coffee. The heater is labeled "200 watts," which means that it converts electrical energy to thermal energy at this rate. Calculate the time required to bring all this water from 23°C to 100°C, ignoring any heat losses. ssm

36P. A 150 g copper bowl contains 220 g of water, both at 20.0°C. A very hot 300 g copper cylinder is dropped into the water, causing the water to boil, with 5.00 g being converted to steam. The final temperature of the system is 100°C. Neglect energy transfers with the environment. (a) How much energy (in calories) is transferred to the water as heat? (b) How much to the bowl? (c) What is the original temperature of the cylinder?

37P. A chef, on finding his stove out of order, decides to boil the water for his wife's coffee by shaking it in a thermos flask. Suppose that he uses tap water at 15°C and that the water falls 30 cm each shake, the chef making 30 shakes each minute. Neglecting any loss of thermal energy by the flask, how long must he shake the flask until the water reaches 100°C? ssm www

38P. *Nonmetric version:* How long does a 2.0 × 10⁵ Btu/h water heater take to raise the temperature of 40 gal of water from 70°F

to 100°F? *Metric version:* How long does a 59 kW water heater take to raise the temperature of 150 L of water from 21°C to 38°C?

39P. Ethyl alcohol has a boiling point of 78°C, a freezing point of −114°C, a heat of vaporization of 879 kJ/kg, a heat of fusion of 109 kJ/kg, and a specific heat of 2.43 kJ/kg · K. How much energy must be removed from 0.510 kg of ethyl alcohol that is initially a gas at 78°C so that it becomes a solid at −114°C?

40P. A 1500 kg Buick moving at 90 km/h brakes to a stop, at uniform deceleration and without skidding, over a distance of 80 m. At what average rate is mechanical energy transferred to thermal energy in the brake system?

41P. The specific heat of a substance varies with temperature according to $c = 0.20 + 0.14T + 0.023T^2$, with T in °C and c in cal/g · K. Find the energy required to raise the temperature of 2.0 g of this substance from 5.0°C to 15°C.

42P. In a solar water heater, energy from the Sun is gathered by water that circulates through tubes in a rooftop collector. The solar radiation enters the collector through a transparent cover and warms the water in the tubes; this water is pumped into a holding tank. Assume that the efficiency of the overall system is 20% (that is, 80% of the incident solar energy is lost from the system). What collector area is necessary to raise the temperature of 200 L of water in the tank from 20°C to 40°C in 1.0 h when the intensity of incident sunlight is 700 W/m²?

43P. What mass of steam at 100°C must be mixed with 150 g of ice at its melting point, in a thermally insulated container, to produce liquid water at 50°C? ilw

44P. A person makes a quantity of iced tea by mixing 500 g of hot tea (essentially water) with an equal mass of ice at its melting point. If the initial hot tea is at a temperature of (a) 90°C and (b) 70°C, what are the temperature and mass of the remaining ice when the tea and ice reach a common temperature? Neglect energy transfers with the environment.

45P. (a) Two 50 g ice cubes are dropped into 200 g of water in a thermally insulated container. If the water is initially at 25°C, and the ice comes directly from a freezer at −15°C, what is the final temperature of the drink when the drink reaches thermal equilibrium? (b) What is the final temperature if only one ice cube is used? ssm

46P. An insulated Thermos contains 130 cm³ of hot coffee, at a temperature of 80.0°C. You put in a 12.0 g ice cube at its melting point to cool the coffee. By how many degrees has your coffee cooled once the ice has melted? Treat the coffee as though it were pure water and neglect energy transfers with the environment.

47P. A 20.0 g copper ring has a diameter of 2.54000 cm at its temperature of 0.000°C. An aluminum sphere has a diameter of 2.54508 cm at its temperature of 100.0°C. The sphere is placed on top of the ring (Fig. 19-33), and the two are allowed to come to thermal equilibrium, with no heat lost to the surroundings. The sphere just passes through the ring at the equilibrium temperature. What is the mass of the sphere? ssm

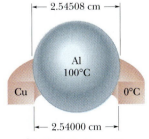

Fig. 19-33 Problem 47.

SEC. 19-10   Some Special Cases of the First Law of Thermodynamics

48E. Consider that 200 J of work is done on a system and 70.0 cal is extracted from the system as heat. In the sense of the first law of thermodynamics, what are the values (including algebraic signs) of (a) W, (b) Q, and (c) ΔE_{int}?

49E. A sample of gas expands from 1.0 m³ to 4.0 m³ while its pressure decreases from 40 Pa to 10 Pa. How much work is done by the gas if its pressure changes with volume via each of the three paths shown in the p-V diagram in Fig. 19-34? **ssm www**

50E. A thermodynamic system is taken from an initial state A to another state B and back again to A, via state C, as shown by path $ABCA$ in the p-V diagram of Fig. 19-35a. (a) Complete the table in Fig. 19-35b by filling in either + or − for the sign of each thermodynamic quantity associated with each step of the cycle. (b) Calculate the numerical value of the work done by the system for the complete cycle $ABCA$.

51E. Gas within a closed chamber undergoes the cycle shown in the p-V diagram of Fig. 19-36. Calculate the net energy added to the system as heat during one complete cycle. **ssm ilw**

52E. Gas within a chamber passes through the cycle shown in Fig. 19-37. Determine the energy transferred by the system as heat during process CA if the energy added as heat Q_{AB} during process AB is 20.0 J, no energy is transferred as heat during process BC, and the net work done during the cycle is 15.0 J.

53P. When a system is taken from state i to state f along path iaf in Fig. 19-38, $Q = 50$ cal and $W = 20$ cal. Along path ibf, $Q = 36$ cal. (a) What is W along path ibf? (b) If $W = -13$ cal for the return path fi, what is Q for this path? (c) Take $E_{int,i} = 10$ cal. What is $E_{int,f}$? (d) If $E_{int,b} = 22$ cal, what

are the values of Q for path ib and path bf? **ssm**

SEC. 19-11   Heat Transfer Mechanisms

54E. The average rate at which energy is conducted outward through the ground surface in North America is 54.0 mW/m², and the average thermal conductivity of the near-surface rocks is 2.50 W/m · K. Assuming a surface temperature of 10.0°C, find the temperature at a depth of 35.0 km (near the base of the crust). Ignore the heat generated by the presence of radioactive elements.

55E. The ceiling of a single-family dwelling in a cold climate should have an R-value of 30. To give such insulation, how thick would a layer of (a) polyurethane foam and (b) silver have to be?

56E. (a) Calculate the rate at which body heat is conducted through the clothing of a skier in a steady-state process, given the following data: the body surface area is 1.8 m² and the clothing is 1.0 cm thick; the skin surface temperature is 33°C and the outer surface of the clothing is at 1.0°C; the thermal conductivity of the clothing is 0.040 W/m · K. (b) How would the answer to (a) change if, after a fall, the skier's clothes became soaked with water of thermal conductivity 0.60 W/m · K?

57E. Consider the slab shown in Fig. 19-18. Suppose that $L = 25.0$ cm, $A = 90.0$ cm², and the material is copper. If $T_H = 125°C$, $T_C = 10.0°C$, and a steady state is reached, find the conduction rate through the slab. **ssm**

58E. If you were to walk briefly in space without a spacesuit while far from the Sun (as an astronaut does in the movie *2001*), you would feel the cold of space—while you radiated energy, you would absorb almost none from your environment. (a) At what rate would you lose energy? (b) How much energy would you lose in 30 s? Assume that your emissivity is 0.90, and estimate other data needed in the calculations.

59E. A cylindrical copper rod of length 1.2 m and cross-sectional area 4.8 cm² is insulated to prevent heat loss through its surface. The ends are maintained at a temperature difference of 100 C° by having one end in a water–ice mixture and the other in boiling water and steam. (a) Find the rate at which energy is conducted along the rod. (b) Find the rate at which ice melts at the cold end. **ilw**

60E. Four square pieces of insulation of two different materials, all with the same thickness and area A, are available to cover an opening of area $2A$. This can be done in either of the two ways shown in Fig. 19-39. Which arrangement, (a) or (b), gives the lower energy flow if $k_2 \neq k_1$?

61P. Two identical rectangular rods of metal are welded end to

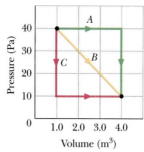

Fig. 19-34   Exercise 49.

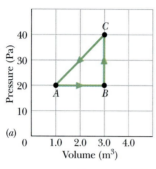

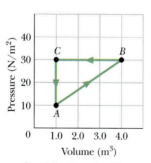

	Q	W	ΔE_{int}
$A \longrightarrow B$			+
$B \longrightarrow C$	+		
$C \longrightarrow A$			

Fig. 19-35   Exercise 50.

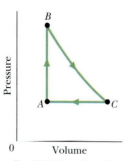

Fig. 19-36   Exercise 51.

Fig. 19-37   Exercise 52.

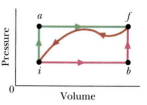

Fig. 19-38   Problem 53.

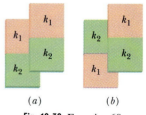

Fig. 19-39   Exercise 60.

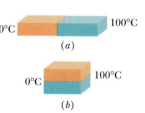

Fig. 19-40   Problem 61.

end as shown in Fig. 19-40*a*, and 10 J is conducted (in a steady-state process) through the rods as heat in 2.0 min. How long would it take for 10 J to be conducted through the rods if they were welded together as shown in Fig. 19-40*b*?

62P. A sphere of radius 0.500 m, temperature 27.0°C, and emissivity 0.850 is located in an environment of temperature 77.0°C. At what rate does the sphere (a) emit and (b) absorb thermal radiation? (c) What is the sphere's net rate of energy exchange?

63P. (a) What is the rate of energy loss in watts per square meter through a glass window 3.0 mm thick if the outside temperature is −20°F and the inside temperature is +72°F? (b) A storm window having the same thickness of glass is installed parallel to the first window, with an air gap of 7.5 cm between the two windows. What now is the rate of energy loss if conduction is the only important energy-loss mechanism? ilw

64P. Figure 19-41 shows (in cross section) a wall that consists of four layers. The thermal conductivities are $k_1 = 0.060$ W/m · K, $k_3 = 0.040$ W/m · K, and $k_4 = 0.12$ W/m · K (k_2 is not known). The layer thicknesses are $L_1 = 1.5$ cm, $L_3 = 2.8$ cm, and $L_4 = 3.5$ cm (L_2 is not known). Energy transfer through the wall is steady. What is the temperature of the interface indicated?

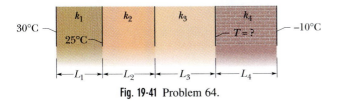

Fig. 19-41 Problem 64.

65P. A tank of water has been outdoors in cold weather, and a slab of ice 5.0 cm thick has formed on its surface (Fig. 19-42). The air above the ice is at −10°C. Calculate the rate of formation of ice (in centimeters per hour) on the ice slab. Take the thermal conductivity and density of ice to be 0.0040 cal/s · cm · C° and 0.92 g/cm³. Assume that energy is not transferred through the walls or bottom of the tank. ssm www

Fig. 19-42 Problem 65.

66P. Ice has formed on a shallow pond and a steady state has been reached, with the air above the ice at −5.0°C and the bottom of the pond at 4.0°C. If the total depth of *ice + water* is 1.4 m, how thick is the ice? (Assume that the thermal conductivities of ice and water are 0.40 and 0.12 cal/m · C° · s, respectively.)

Additional Problems

67. You can join the semi-secret "300 F" club at the Amundsen–Scott South Pole Station only when the outside temperature is below −70°C. On such a day, you first bask in a hot sauna and then run outside wearing only your shoes. (This is, of course, extremely dangerous, but the rite is effectively a protest against the constant danger of the winter cold at the south pole.)

Assume that when you step out of the sauna, your skin temperature is 102°F and the walls, ceiling, and floor of the sauna room have a temperature of 30°C. Estimate your surface area, and take your skin emissivity to be 0.80. (a) What is the approximate net rate P_{net} at which you lose energy via thermal radiation exchanges with the room? Next, assume that when you are outside half your surface area exchanges thermal radiation with the sky at a temperature of −25°C and the other half exchanges thermal radiation with the snow and ground at a temperature of −80°C. What is the approximate net rate at which you lose energy via thermal radiation exchanges with (b) the sky and (c) the snow and ground?

68. Emperor penguins (Fig. 19-43), those large penguins that resemble stuffy English butlers, breed and hatch their young even during severe Antarctic winters. Once an egg is laid, the father balances the egg on his feet to prevent the egg from freezing. He must do this for the full incubation period of 105 to 115 days, during which he cannot eat because his food is in the water. He can survive this long without food only if he can reduce his consumption of internal energy significantly. If he is alone, he consumes that energy too quickly to stay warm, and eventually abandons the egg in order to eat. To protect themselves and each other from the cold so as to reduce the consumption of internal energy, penguin fathers huddle closely together, in groups of perhaps several thousand. In addition to providing other benefits, the huddling reduces the rate at which the penguins thermally radiate energy to their surroundings.

Assume that a penguin father is a circular cylinder with top surface area *a*, height *h*, surface temperature *T*, and emissivity *ε*. (a) Find an expression for the rate P_i at which an individual father would radiate energy to the environment from his top surface and his side surface were he alone with his egg.

If *N* identical fathers were well apart from one another, the total rate of energy loss via radiation would be NP_i. Suppose, instead, that they huddle closely to form a *huddled cylinder* with top surface area *Na* and height *h*. (b) Find an expression for the rate P_h at which energy is radiated by the top surface and the side surface of the huddled cylinder.

(c) Assuming $a = 0.34$ m² and $h = 1.1$ m and using the expressions you obtained for P_i and P_h, graph the ratio P_h/NP_i versus *N*. Of course, the penguins know nothing about algebra or graphing, but their instinctive huddling reduces this ratio so that more of their eggs survive to the hatching stage. From the graphs (as you will see, you probably need more than one version), approximate how many penguins must huddle so that P_h/NP_i is reduced to (d) 0.5, (e) 0.4, (f) 0.3, (g) 0.2, and (h) 0.15. (i) For the assumed data, what is the lower limiting value for P_h/NP_i?

Fig. 19-43 Problem 68.

20 The Kinetic Theory of Gases

When a container of cold champagne, soda pop, or any other carbonated drink is opened, a slight fog forms around the opening and some of the liquid sprays outward. (In the photograph, the fog is the white cloud that surrounds the stopper, and the spray has formed streaks within the cloud.)

What causes the fog?

The answer is in this chapter.

20-1 A New Way to Look at Gases

Classical thermodynamics—the subject of the last chapter—has nothing to say about atoms. Its laws are concerned only with such macroscopic variables as pressure, volume, and temperature. However, we know that a gas is made up of moving atoms or molecules (groups of atoms bound together). The pressure exerted by a gas must surely be related to the collisions of its molecules with the walls of its container. The ability of a gas to fill the volume of its container must surely be due to the freedom of motion of its molecules, and the temperature and internal energy of a gas must surely be related to the kinetic energy of these molecules. Thus, we can learn something about gases by approaching the subject from this direction. We call this molecular approach the **kinetic theory of gases.** It is the subject of this chapter.

20-2 Avogadro's Number

When our thinking is slanted toward molecules, it makes sense to measure the sizes of our samples in moles. If we do so, we can be certain that we are comparing samples that contain the same number of atoms or molecules. The *mole* is one of the seven SI base units and is defined as follows:

> One mole is the number of atoms in a 12 g sample of carbon-12.

The obvious question now is: "How many atoms or molecules are there in a mole?" The answer is determined experimentally and, as you saw in Chapter 19, is

$$N_A = 6.02 \times 10^{23} \text{ mol}^{-1} \qquad \text{(Avogadro's number)}, \qquad (20\text{-}1)$$

where mol^{-1} represents the inverse mole or "per mole," and mol is the abbreviation for mole. The number N_A is called **Avogadro's number** after Italian scientist Amadeo Avogadro (1776–1856), who suggested that all gases contain the same number of atoms or molecules when they occupy the same volume under the same conditions of temperature and pressure.

The number of moles n contained in a sample of any substance is equal to the ratio of the number of molecules N in the sample to the number of molecules N_A in 1 mole:

$$n = \frac{N}{N_A}. \qquad (20\text{-}2)$$

(*Caution:* The three symbols in this equation can easily be confused with one another, so you should sort them with their meanings now, before you end in "N-confusion.") We can find the number of moles in a sample from the mass M_{sam} of the sample and either the *molar mass M* (the mass of 1 mol) or the molecular mass m (the mass of one molecule):

$$n = \frac{M_{sam}}{M} = \frac{M_{sam}}{mN_A}. \qquad (20\text{-}3)$$

In Eq. 20-3, we used the fact that the mass M of 1 mol is the product of the mass m of one molecule and the number of molecules N_A in 1 mol:

$$M = mN_A. \qquad (20\text{-}4)$$

Tactic 1: *Avogadro's Number of What?*

In Eq. 20-1, Avogadro's number is expressed in terms of mol^{-1}, which is the inverse mole, or 1/mol. We could instead explicitly state the elementary unit involved in a given situation. For example, we might write $N_A = 6.02 \times 10^{23}$ atoms/mole if the elementary unit is an atom. If, instead, the elementary unit is a molecule, then we might write $N_A = 6.02 \times 10^{23}$ molecules/mole.

20-3 Ideal Gases

Our goal in this chapter is to explain the macroscopic properties of a gas—such as its pressure and its temperature—in terms of the behavior of the molecules that make it up. However, there is an immediate problem: which gas? Should it be hydrogen or oxygen, or methane, or perhaps uranium hexafluoride? They are all different. Experimenters have found, though, that if we confine 1 mole samples of various gases in boxes of identical volume and hold the gases at the same temperature, then their measured pressures are nearly—though not exactly—the same. If we repeat the measurements at lower gas densities, then these small differences in the measured pressures tend to disappear. Further experiments show that, at low enough densities, all real gases tend to obey the relation

$$pV = nRT \qquad \text{(ideal gas law)}, \qquad (20\text{-}5)$$

in which p is the absolute (not gauge) pressure, n is the number of moles of gas present, and T is the temperature in kelvins. The symbol R is a constant called the **gas constant** that has the same value for all gases—namely,

$$R = 8.31 \text{ J/mol} \cdot \text{K}. \qquad (20\text{-}6)$$

Equation 20-5 is called the **ideal gas law.** Provided the gas density is low, this law holds for any single gas or for any mixture of different gases. (For a mixture, n is the total number of moles in the mixture.)

We can rewrite Eq. 20-5 in an alternative form, in terms of a constant called the **Boltzmann constant** k, which is defined as

$$k = \frac{R}{N_A} = \frac{8.31 \text{ J/mol} \cdot \text{K}}{6.02 \times 10^{23} \text{ mol}^{-1}} = 1.38 \times 10^{-23} \text{ J/K}. \qquad (20\text{-}7)$$

This allows us to write $R = kN_A$. Then, with Eq. 20-2 ($n = N/N_A$), we see that

$$nR = Nk. \qquad (20\text{-}8)$$

Substituting this into Eq. 20-5 gives a second expression for the ideal gas law:

$$pV = NkT \qquad \text{(ideal gas law)}. \qquad (20\text{-}9)$$

(*Caution:* Note the difference between the two expressions for the ideal gas law—Eq. 20-5 involves the number of moles n and Eq. 20-9 involves the number of molecules N.)

You may well ask, "What is an *ideal gas* and what is so 'ideal' about one?" The answer lies in the simplicity of the law (Eqs. 20-5 and 20-9) that governs its macroscopic properties. Using this law—as you will see—we can deduce many properties of the ideal gas in a simple way. Although there is no such thing in nature as a truly ideal gas, *all real* gases approach the ideal state at low enough densities—that is, under conditions in which their molecules are far enough apart that they do not interact with one another. Thus, the ideal gas concept allows us to gain useful insights into the limiting behavior of real gases.

Work Done by an Ideal Gas at Constant Temperature

Suppose we put an ideal gas in a piston–cylinder arrangement like those in Chapter 19. Suppose also that we allow the gas to expand from an initial volume V_i to a final volume V_f while we keep the temperature T of the gas constant. Such a process, at *constant temperature*, is called an **isothermal expansion** (and the reverse is called an **isothermal compression**).

On a p-V diagram, an *isotherm* is a curve that connects points that have the same temperature. Thus, it is a graph of pressure versus volume for a gas whose temperature T is held constant. For n moles of an ideal gas, it is a graph of the equation

$$p = nRT\,\frac{1}{V} = (\text{a constant})\,\frac{1}{V}. \tag{20-10}$$

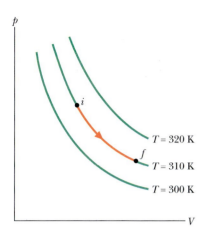

Fig. 20-1 Three isotherms on a p-V diagram. The path shown along the middle isotherm represents an isothermal expansion of a gas from an initial state i to a final state f. The path from f to i along the isotherm would represent the reverse process, that is, an isothermal compression.

Figure 20-1 shows three isotherms, each corresponding to a different (constant) value of T. (Note that the values of T for the isotherms increase upward to the right.) Superimposed on the middle isotherm is the path followed by a gas during an isothermal expansion from state i to state f at a constant temperature of 310 K.

To find the work done by an ideal gas during an isothermal expansion, we start with Eq. 19-25,

$$W = \int_{V_i}^{V_f} p\,dV. \tag{20-11}$$

This is a general expression for the work done during any change in volume of any gas. For an ideal gas, we can use Eq. 20-5 to substitute for p, obtaining

$$W = \int_{V_i}^{V_f} \frac{nRT}{V}\,dV. \tag{20-12}$$

Because we are considering an isothermal expansion, T is constant and we can move it in front of the integral sign to write

$$W = nRT \int_{V_i}^{V_f} \frac{dV}{V} = nRT\left[\ln V\right]_{V_i}^{V_f}. \tag{20-13}$$

By evaluating the expression in brackets at the limits and then using the relationship $\ln a - \ln b = \ln(a/b)$, we find that

$$W = nRT \ln \frac{V_f}{V_i} \qquad \text{(ideal gas, isothermal process).} \tag{20-14}$$

Recall that the symbol ln specifies a *natural* logarithm, which has base e.

For an expansion, V_f is greater than V_i, so the ratio V_f/V_i in Eq. 20-14 is greater than unity. The natural logarithm of a quantity greater than unity is positive, and so the work W done by an ideal gas during an isothermal expansion is positive, as we expect. For a compression, V_f is less than V_i, so the ratio of volumes in Eq. 20-14 is less than unity. The natural logarithm in that equation—hence the work W—is negative, again as we expect.

Work Done at Constant Volume and at Constant Pressure

Equation 20-14 does not give the work W done by an ideal gas during *every* thermodynamic process. Instead, it gives the work only for a process in which the temperature is held constant. If the temperature varies, then the symbol T in Eq. 20-12 cannot be moved in front of the integral symbol as in Eq. 20-13, and thus we do not end up with Eq. 20-14.

However, we can go back to Eq. 20-11 to find the work W done by an ideal gas (or any other gas) during two more processes—a constant-volume process and a constant-pressure process. If the volume of the gas is constant, then Eq. 20-11 yields

$$W = 0 \qquad \text{(constant-volume process)}. \qquad (20\text{-}15)$$

If, instead, the volume changes while the pressure p of the gas is held constant, then Eq. 20-11 becomes

$$W = p(V_f - V_i) = p\,\Delta V \qquad \text{(constant-pressure process)}. \qquad (20\text{-}16)$$

✔**CHECKPOINT 1:** An ideal gas has an initial pressure of 3 pressure units and an initial volume of 4 volume units. The table gives the final pressure and volume of the gas (in those same units) in five processes. Which processes start and end on the same isotherm?

	a	b	c	d	e
p	12	6	5	4	1
V	1	2	7	3	12

Sample Problem 20-1

A cylinder contains 12 L of oxygen at 20°C and 15 atm. The temperature is raised to 35°C, and the volume is reduced to 8.5 L. What is the final pressure of the gas in atmospheres? Assume that the gas is ideal.

SOLUTION: The Key Idea here is that, because the gas is ideal, its pressure, volume, temperature, and number of moles are related by the ideal gas law, both in the initial state i and in the final state f (after the changes). Thus, from Eq. 20-5 we can write

$$p_i V_i = nRT_i \quad \text{and} \quad p_f V_f = nRT_f.$$

Dividing the second equation by the first equation and solving for p_f yields

$$p_f = \frac{p_i T_f V_i}{T_i V_f}. \qquad (20\text{-}17)$$

Note here that if we converted the given initial and final volumes from liters to the proper units of cubic meters, the multiplying conversion factors would cancel out of Eq. 20-17. The same would be true for conversion factors that convert the pressures from atmospheres to the proper pascals. However, to convert the given temperatures to kelvins requires the addition of an amount that would not cancel and thus must be included. Hence, we must write

$$T_i = (273 + 20)\ \text{K} = 293\ \text{K}$$

and

$$T_f = (273 + 35)\ \text{K} = 308\ \text{K}.$$

Inserting the given data into Eq. 20-17 then yields

$$p_f = \frac{(15\ \text{atm})(308\ \text{K})(12\ \text{L})}{(293\ \text{K})(8.5\ \text{L})} = 22\ \text{atm}. \qquad \text{(Answer)}$$

Sample Problem 20-2

One mole of oxygen (assume it to be an ideal gas) expands at a constant temperature T of 310 K from an initial volume V_i of 12 L to a final volume V_f of 19 L. How much work is done by the gas during the expansion?

SOLUTION: The Key Idea is this: Generally we find the work by integrating the gas pressure with respect to the gas volume, using Eq. 20-11. However, because the gas here is ideal and the expansion is isothermal, that integration leads to Eq. 20-14. Therefore, we can write

$$W = nRT \ln \frac{V_f}{V_i}$$

$$= (1\ \text{mol})(8.31\ \text{J/mol} \cdot \text{K})(310\ \text{K}) \ln \frac{19\ \text{L}}{12\ \text{L}}$$

$$= 1180\ \text{J}. \qquad \text{(Answer)}$$

The expansion is graphed in the p-V diagram of Fig. 20-2. The work done by the gas during the expansion is represented by the area beneath the curve if.

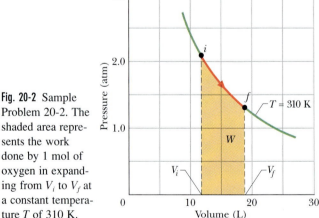

Fig. 20-2 Sample Problem 20-2. The shaded area represents the work done by 1 mol of oxygen in expanding from V_i to V_f at a constant temperature T of 310 K.

You can show that if the expansion is now reversed, with the gas undergoing an isothermal compression from 19 L to 12 L, the work done by the gas will be -1180 J. Thus, an external force would have to do 1180 J of work on the gas to compress it.

20-4 Pressure, Temperature, and RMS Speed

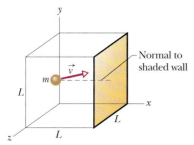

Fig. 20-3 A cubical box of edge L, containing n moles of an ideal gas. A molecule of mass m and velocity $\vec{v}$ is about to collide with the shaded wall of area L^2. A normal to that wall is shown.

Here is our first kinetic theory problem. Let n moles of an ideal gas be confined in a cubical box of volume V, as in Fig. 20-3. The walls of the box are held at temperature T. What is the connection between the pressure p exerted by the gas on the walls and the speeds of the molecules?

The molecules of gas in the box are moving in all directions and with various speeds, bumping into each other and bouncing from the walls of the box like balls in a racquetball court. We ignore (for the time being) collisions of the molecules with one another and consider only elastic collisions with the walls.

Figure 20-3 shows a typical gas molecule, of mass m and velocity $\vec{v}$, that is about to collide with the shaded wall. Because we assume that any collision of a molecule with a wall is elastic, when this molecule collides with the shaded wall, the only component of its velocity that is changed is the x component, and that component is reversed. This means that the only change in the particle's momentum is along the x axis, and that change is

$$\Delta p_x = (-mv_x) - (mv_x) = -2mv_x.$$

Hence, the momentum Δp_x delivered to the wall by the molecule during the collision is $+2mv_x$. (Because in this book the symbol p represents both momentum and pressure, we must be careful to note that here p represents momentum and is a vector quantity.)

The molecule of Fig. 20-3 will hit the shaded wall repeatedly. The time Δt between collisions is the time the molecule takes to travel to the opposite wall and back again (a distance of $2L$) at speed v_x. Thus, Δt is equal to $2L/v_x$. (Note that this result holds even if the molecule bounces off any of the other walls along the way, because those walls are parallel to x and so cannot change v_x.) Therefore, the average rate at which momentum is delivered to the shaded wall by this single molecule is

$$\frac{\Delta p_x}{\Delta t} = \frac{2mv_x}{2L/v_x} = \frac{mv_x^2}{L}.$$

From Newton's second law ($\vec{F} = d\vec{p}/dt$), the rate at which momentum is delivered to the wall is the force acting on that wall. To find the total force, we must add up the contributions of all the molecules that strike the wall, allowing for the possibility that they all have different speeds. Dividing the magnitude of the total force F_x by the area of the wall ($= L^2$) then gives the pressure p on that wall, where now and in the rest of this discussion, p represents pressure. Thus, using the expression for $\Delta p_x/\Delta t$, we can write this pressure as

$$p = \frac{F_x}{L^2} = \frac{mv_{x1}^2/L + mv_{x2}^2/L + \cdots + mv_{xN}^2/L}{L^2}$$

$$= \left(\frac{m}{L^3}\right)(v_{x1}^2 + v_{x2}^2 + \cdots + v_{xN}^2), \tag{20-18}$$

where N is the number of molecules in the box.

Since $N = nN_A$, there are nN_A terms in the second parentheses of Eq. 20-18. We can replace that quantity by $nN_A(v_x^2)_{avg}$, where $(v_x^2)_{avg}$ is the average value of the square of the x components of all the molecular speeds. Equation 20-18 then becomes

$$p = \frac{nmN_A}{L^3}(v_x^2)_{avg}.$$

However, mN_A is the molar mass M of the gas (that is, the mass of 1 mol of the gas). Also, L^3 is the volume of the box, so

$$p = \frac{nM(v_x^2)_{avg}}{V}. \qquad (20\text{-}19)$$

For any molecule, $v^2 = v_x^2 + v_y^2 + v_z^2$. Because there are many molecules and because they are all moving in random directions, the average values of the squares of their velocity components are equal, so that $v_x^2 = \frac{1}{3}v^2$. Thus, Eq. 20-19 becomes

$$p = \frac{nM(v^2)_{avg}}{3V}. \qquad (20\text{-}20)$$

The square root of $(v^2)_{avg}$ is a kind of average speed, called the **root-mean-square speed** of the molecules and symbolized by v_{rms}. Its name describes it rather well: You *square* each speed, you find the *mean* (that is, the average) of all these squared speeds, and then you take the square *root* of that mean. With $\sqrt{(v^2)_{avg}} = v_{rms}$, we can then write Eq. 20-20 as

$$p = \frac{nMv_{rms}^2}{3V}. \qquad (20\text{-}21)$$

Equation 20-21 is very much in the spirit of kinetic theory. It tells us how the pressure of the gas (a purely macroscopic quantity) depends on the speed of the molecules (a purely microscopic quantity).

We can turn Eq. 20-21 around and use it to calculate v_{rms}. Combining Eq. 20-21 with the ideal gas law ($pV = nRT$) leads to

$$v_{rms} = \sqrt{\frac{3RT}{M}}. \qquad (20\text{-}22)$$

Table 20-1 shows some rms speeds calculated from Eq. 20-22. The speeds are surprisingly high. For hydrogen molecules at room temperature (300 K), the rms speed is 1920 m/s or 4300 mi/h—faster than a speeding bullet! On the surface of the Sun, where the temperature is 2×10^6 K, the rms speed of hydrogen molecules would be 82 times greater than at room temperature were it not for the fact that at such high speeds, the molecules cannot survive collisions among themselves. Remember too that the rms speed is only a kind of average speed; many molecules move much faster than this, and some much slower.

TABLE 20-1   Some Molecular Speeds at Room Temperature ($T = 300$ K)[a]

Gas	Molar Mass (10^{-3} kg/mol)	v_{rms} (m/s)
Hydrogen (H_2)	2.02	1920
Helium (He)	4.0	1370
Water vapor (H_2O)	18.0	645
Nitrogen (N_2)	28.0	517
Oxygen (O_2)	32.0	483
Carbon dioxide (CO_2)	44.0	412
Sulfur dioxide (SO_2)	64.1	342

[a]For convenience, we often set room temperature — 300 K even though (at 27°C or 81°F) that represents a fairly warm room.

The speed of sound in a gas is closely related to the rms speed of the molecules of that gas. In a sound wave, the disturbance is passed on from molecule to molecule by means of collisions. The wave cannot move any faster than the "average" speed of the molecules. In fact, the speed of sound must be somewhat less than this "average" molecular speed because not all molecules are moving in exactly the same direction as the wave. As examples, at room temperature, the rms speeds of hydrogen and nitrogen molecules are 1920 m/s and 517 m/s, respectively. The speeds of sound in these two gases at this temperature are 1350 m/s and 350 m/s, respectively.

A question often arises: If molecules move so fast, why does it take as long as a minute or so before you can smell perfume when someone opens a bottle across a room? The answer is that, as we shall discuss in Section 20-6, each perfume molecule moves away from the bottle only very slowly because its repeated collisions with other molecules prevent it from moving directly across the room to you.

Sample Problem 20-3

Here are five numbers: 5, 11, 32, 67, and 89.

(a) What is the average value n_{avg} of these numbers?

SOLUTION: We find this from

$$n_{avg} = \frac{5 + 11 + 32 + 67 + 89}{5} = 40.8. \quad \text{(Answer)}$$

(b) What is the rms value n_{rms} of these numbers?

SOLUTION: We find this from

$$n_{rms} = \sqrt{\frac{5^2 + 11^2 + 32^2 + 67^2 + 89^2}{5}} = 52.1. \quad \text{(Answer)}$$

The rms value is greater than the average value because the larger numbers—being squared—are relatively more important in forming the rms value. To test this, let us replace 89 in our set of five numbers by 300. The average value of the new set of five numbers (as you should show) is 2.0 times the previous average value. The rms value, however, is 2.7 times the previous rms value.

20-5 Translational Kinetic Energy

We again consider a single molecule of an ideal gas as it moves around in the box of Fig. 20-3, but we now assume that its speed changes when it collides with other molecules. Its translational kinetic energy at any instant is $\frac{1}{2}mv^2$. Its *average* translational kinetic energy over the time that we watch it is

$$K_{avg} = (\tfrac{1}{2}mv^2)_{avg} = \tfrac{1}{2}m(v^2)_{avg} = \tfrac{1}{2}mv_{rms}^2, \quad (20\text{-}23)$$

in which we make the assumption that the average speed of the molecule during our observation is the same as the average speed of all the molecules at any given time. (Provided the total energy of the gas is not changing and we observe our molecule for long enough, this assumption is appropriate.) Substituting for v_{rms} from Eq. 20-22 leads to

$$K_{avg} = (\tfrac{1}{2}m)\frac{3RT}{M}.$$

However, M/m, the molar mass divided by the mass of a molecule, is simply Avogadro's number. Thus,

$$K_{avg} = \frac{3RT}{2N_A}.$$

Using Eq. 20-7 ($k = R/N_A$), we can then write

$$K_{avg} = \tfrac{3}{2}kT. \quad (20\text{-}24)$$

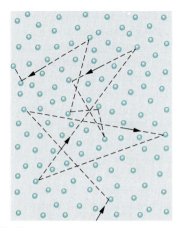

Fig. 20-4 A molecule traveling through a gas, colliding with other gas molecules in its path. Although the other molecules are shown as stationary, they are also moving in a similar fashion.

This equation tells us something unexpected:

> At a given temperature T, all ideal gas molecules—no matter what their mass—have the same average translational kinetic energy, namely, $\frac{3}{2}kT$. When we measure the temperature of a gas, we are also measuring the average translational kinetic energy of its molecules.

✓**CHECKPOINT 2:**   A gas mixture consists of molecules of types 1, 2, and 3, with molecular masses $m_1 > m_2 > m_3$. Rank the three types according to (a) average kinetic energy and (b) rms speed, greatest first.

20-6 Mean Free Path

We continue to examine the motion of molecules in an ideal gas. Figure 20-4 shows the path of a typical molecule as it moves through the gas, changing both speed and direction abruptly as it collides elastically with other molecules. Between collisions, our typical molecule moves in a straight line at constant speed. Although the figure shows all the other molecules as stationary, they are moving similarly.

One useful parameter to describe this random motion is the **mean free path** λ of the molecules. As its name implies, λ is the average distance traversed by a molecule between collisions. We expect λ to vary inversely with N/V, the number of molecules per unit volume (or density of molecules). The larger N/V is, the more collisions there should be and the smaller the mean free path. We also expect λ to vary inversely with the size of the molecules, say, with their diameter d. (If the molecules were points, as we have assumed them to be, they would never collide and the mean free path would be infinite.) Thus, the larger the molecules are, the smaller the mean free path. We can even predict that λ should vary (inversely) as the *square* of the molecular diameter because the cross section of a molecule—not its diameter—determines its effective target area.

The expression for the mean free path does, in fact, turn out to be

$$\lambda = \frac{1}{\sqrt{2}\,\pi d^2\,N/V} \qquad \text{(mean free path).} \qquad (20\text{-}25)$$

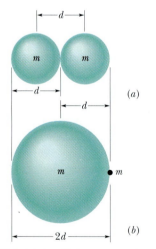

Fig. 20-5 (a) A collision occurs when the centers of two molecules come within a distance d of each other, d being the molecular diameter. (b) An equivalent but more convenient representation is to think of the moving molecule as having a *radius* d and all other molecules as being points. The condition for a collision is unchanged.

To justify Eq. 20-25, we focus attention on a single molecule and assume—as Fig. 20-4 suggests—that our molecule is traveling with a constant speed v and that all the other molecules are at rest. Later, we shall relax this assumption.

We assume further that the molecules are spheres of diameter d. A collision will then take place if the centers of the molecules come within a distance d of each other, as in Fig. 20-5a. Another, more helpful way to look at the situation is to consider our single molecule to have a *radius* of d and all the other molecules to be *points,* as in Fig. 20-5b. This does not change our criterion for a collision.

As our single molecule zigzags through the gas, it sweeps out a short cylinder of cross-sectional area πd^2 between successive collisions. If we watch this molecule for a time interval Δt, it moves a distance $v\,\Delta t$, where v is its assumed speed. Thus, if we align all the short cylinders swept out in interval Δt, we form a composite cylinder (Fig. 20-6) of length $v\,\Delta t$ and volume $(\pi d^2)(v\,\Delta t)$. The number of collisions that occur in time Δt is then equal to the number of (point) molecules that lie within this cylinder.

Since N/V is the number of molecules per unit volume, the number of molecules in the cylinder is N/V times the volume of the cylinder, or $(N/V)(\pi d^2 v\,\Delta t)$. This is also the number of collisions in time Δt. The mean free path is the length of the

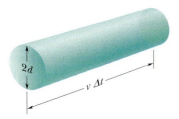

Fig. 20-6 In time Δt the moving molecule effectively sweeps out a cylinder of length $v\,\Delta t$ and radius d.

path (and of the cylinder) divided by this number:

$$\lambda = \frac{\text{length of path during } \Delta t}{\text{number of collisions in } \Delta t} \approx \frac{v\,\Delta t}{\pi d^2 v\,\Delta t\, N/V}$$

$$= \frac{1}{\pi d^2\, N/V}. \tag{20-26}$$

This equation is only approximate because it is based on the assumption that all the molecules except one are at rest. In fact, *all* the molecules are moving; when this is taken properly into account, Eq. 20-25 results. Note that it differs from the (approximate) Eq. 20-26 only by a factor of $1/\sqrt{2}$.

We can even get a glimpse of what is "approximate" about Eq. 20-26. The v in the numerator and that in the denominator are—strictly—not the same. The v in the numerator is v_{avg}, the mean speed of the molecule *relative to the container*. The v in the denominator is v_{rel}, the mean speed of our single molecule *relative to the other molecules*, which are moving. It is this latter average speed that determines the number of collisions. A detailed calculation, taking into account the actual speed distribution of the molecules, gives $v_{\text{rel}} = \sqrt{2}\, v_{\text{avg}}$ and thus the factor $\sqrt{2}$.

The mean free path of air molecules at sea level is about 0.1 μm. At an altitude of 100 km, the density of air has dropped to such an extent that the mean free path rises to about 16 cm. At 300 km, the mean free path is about 20 km. A problem faced by those who would study the physics and chemistry of the upper atmosphere in the laboratory is the unavailability of containers large enough to hold gas samples that simulate upper atmospheric conditions. Yet studies of the concentrations of Freon, carbon dioxide, and ozone in the upper atmosphere are of vital public concern.

Sample Problem 20-4

(a) What is the mean free path λ for oxygen molecules at temperature $T = 300$ K and pressure $p = 1.0$ atm? Assume that the molecular diameter is $d = 290$ pm and the gas is ideal.

SOLUTION: The **Key Idea** here is that each oxygen molecule moves among other *moving* oxygen molecules in a zigzag path due to the resulting collisions. Thus, we use Eq. 20-25 for the mean free path, for which we need the number of molecules per unit volume, N/V. Because we assume the gas is ideal, we can use the ideal gas law of Eq. 20-9 ($pV = NkT$) to write $N/V = p/kT$. Substituting this into Eq. 20-25, we find

$$\lambda = \frac{1}{\sqrt{2}\,\pi d^2\, N/V} = \frac{kT}{\sqrt{2}\,\pi d^2 p}$$

$$= \frac{(1.38 \times 10^{-23}\ \text{J/K})(300\ \text{K})}{\sqrt{2}\,\pi (2.9 \times 10^{-10}\ \text{m})^2 (1.01 \times 10^5\ \text{Pa})}$$

$$= 1.1 \times 10^{-7}\ \text{m}. \qquad \text{(Answer)}$$

This is about 380 molecular diameters.

(b) Assume the average speed of the oxygen molecules is $v = 450$ m/s. What is the average time t between successive collisions for any given molecule? At what rate does the molecule collide; that is, what is the frequency f of its collisions?

SOLUTION: To find the time t between collisions, we use this **Key Idea**: Between collisions, the molecule travels, on average, the mean free path λ at speed v. Thus, the average time between collisions is

$$t = \frac{\text{distance}}{\text{speed}} = \frac{\lambda}{v} = \frac{1.1 \times 10^{-7}\ \text{m}}{450\ \text{m/s}}$$

$$= 2.44 \times 10^{-10}\ \text{s} \approx 0.24\ \text{ns.} \qquad \text{(Answer)}$$

This tells us that, on average, any given oxygen molecule has less than a nanosecond between collisions.

To find the frequency f of the collisions, we use this **Key Idea**: The average rate or frequency at which the collisions occur is the inverse of the time t between collisions. Thus,

$$f = \frac{1}{t} = \frac{1}{2.44 \times 10^{-10}\ \text{s}} = 4.1 \times 10^9\ \text{s}^{-1}. \qquad \text{(Answer)}$$

This tells us that, on average, any given oxygen molecule makes about 4 billion collisions per second.

✓CHECKPOINT 3: One mole of gas A, with molecular diameter $2d_0$ and average molecular speed v_0, is placed inside a certain container. One mole of gas B, with molecular diameter d_0 and average molecular speed $2v_0$ (the molecules of B are smaller but faster), is placed in an identical container. Which gas has the greater average collision rate within its container?

20-7 | The Distribution of Molecular Speeds

The root-mean-square speed v_{rms} gives us a general idea of molecular speeds in a gas at a given temperature. We often want to know more. For example, what fraction of the molecules have speeds greater than the rms value? Greater than twice the rms value? To answer such questions, we need to know how the possible values of speed are distributed among the molecules. Figure 20-7a shows this distribution for oxygen molecules at room temperature ($T = 300$ K); Fig. 20-7b compares it with the distribution at $T = 80$ K.

In 1852, Scottish physicist James Clerk Maxwell first solved the problem of finding the speed distribution of gas molecules. His result, known as **Maxwell's speed distribution law,** is

$$P(v) = 4\pi\left(\frac{M}{2\pi RT}\right)^{3/2} v^2 e^{-Mv^2/2RT}. \qquad (20\text{-}27)$$

Here v is the molecular speed, T is the gas temperature, M is the molar mass of the gas, and R is the gas constant. It is this equation that is plotted in Fig. 20-7a,b. The quantity $P(v)$ in Eq. 20-27 and Fig. 20-7 is a *probability distribution function:* For any speed v, the product $P(v)\,dv$ (a dimensionless quantity) is the fraction of molecules whose speeds lie in the interval of width dv centered on speed v.

As Fig. 20-7a shows, this fraction is equal to the area of a strip with height $P(v)$ and width dv. The total area under the distribution curve corresponds to the fraction of the molecules whose speeds lie between zero and infinity. All molecules fall into this category, so the value of this total area is unity; that is,

$$\int_0^\infty P(v)\,dv = 1. \qquad (20\text{-}28)$$

The fraction (frac) of molecules with speeds in an interval of, say, v_1 to v_2 is then

$$\text{frac} = \int_{v_1}^{v_2} P(v)\,dv. \qquad (20\text{-}29)$$

Average, RMS, and Most Probable Speeds

In principle, we can find the **average speed** v_{avg} of the molecules in a gas with the following procedure: We *weight* each value of v in the distribution; that is, we

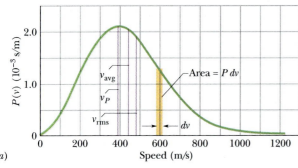

(a)

Fig. 20-7 (a) The Maxwell speed distribution for oxygen molecules at $T = 300$ K. The three characteristic speeds are marked. (b) The curves for 300 K and 80 K. Note that the molecules move more slowly at the lower temperature. Because these are probability distributions, the area under each curve has a numerical value of unity.

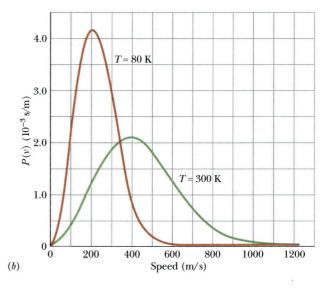

(b)

multiply it by the fraction $P(v) \, dv$ of molecules with speeds in a differential interval dv centered on v. Then we add up all these values of $v \, P(v) \, dv$. The result is v_{avg}. In practice, we do all this by evaluating

$$v_{\text{avg}} = \int_0^\infty v \, P(v) \, dv. \tag{20-30}$$

Substituting for $P(v)$ from Eq. 20-27 and using generic integral 20 from the list of integrals in Appendix E, we find

$$v_{\text{avg}} = \sqrt{\frac{8RT}{\pi M}} \qquad \text{(average speed).} \tag{20-31}$$

Similarly, we can find the average of the square of the speeds $(v^2)_{\text{avg}}$ with

$$(v^2)_{\text{avg}} = \int_0^\infty v^2 \, P(v) \, dv. \tag{20-32}$$

Substituting for $P(v)$ from Eq. 20-27 and using generic integral 16 from the list of integrals in Appendix E, we find

$$(v^2)_{\text{avg}} = \frac{3RT}{M}. \tag{20-33}$$

The square root of $(v^2)_{\text{avg}}$ is the **root-mean-square speed** v_{rms}. Thus,

$$v_{\text{rms}} = \sqrt{\frac{3RT}{M}} \qquad \text{(rms speed),} \tag{20-34}$$

which agrees with Eq. 20-22.

The **most probable speed** v_P is the speed at which $P(v)$ is maximum (see Fig. 20-7a). To calculate v_P, we set $dP/dv = 0$ (the slope of the curve in Fig. 20-7a is zero at the maximum of the curve) and then solve for v. Doing so, we find

$$v_P = \sqrt{\frac{2RT}{M}} \qquad \text{(most probable speed).} \tag{20-35}$$

A molecule is more likely to have speed v_P than any other speed, but some molecules will have speeds that are many times v_P. These molecules lie in the *high-speed tail* of a distribution curve like that in Fig. 20-7a. We should be thankful for these few, higher speed molecules because they make possible both rain and sunshine (without which we could not exist). We next see why.

Rain: The speed distribution of water molecules in, say, a pond at summertime temperatures can be represented by a curve similar to that of Fig. 20-7a. Most of the molecules do not have nearly enough kinetic energy to escape from the water through its surface. However, small numbers of very fast molecules with speeds far out in the tail of the curve can do so. It is these water molecules that evaporate, making clouds and rain a possibility.

As the fast water molecules leave the surface, carrying energy with them, the temperature of the remaining water is maintained by heat transfer from the surroundings. Other fast molecules—produced in particularly favorable collisions—quickly take the place of those that have left, and the speed distribution is maintained.

Sunshine: Let the distribution curve of Fig. 20-7a now refer to protons in the core of the Sun. The Sun's energy is supplied by a nuclear fusion process that starts with the merging of two protons. However, protons repel each other because of their electrical charges, and protons of average speed do not have enough kinetic energy to overcome the repulsion and get close enough to merge. Very fast protons with speeds in the tail of the distribution curve can do so, however, and for that reason the Sun can shine.

Sample Problem 20-5

A container is filled with oxygen gas maintained at room temperature (300 K). What fraction of the molecules have speeds in the interval 599 to 601 m/s? The molar mass M of oxygen is 0.0320 kg/mol.

SOLUTION: The Key Ideas here are

1. The speeds of the molecules are distributed over a wide range of values, with the distribution $P(v)$ of Eq. 20-27.

2. The fraction of the molecules with speeds in a differential interval dv is $P(v)\,dv$.

3. For a larger interval, the fraction is found by integrating $P(v)$ over the interval.

4. However, the interval $\Delta v = 2$ m/s here is small compared to the speed $v = 600$ m/s on which it is centered.

Thus, we can avoid the integration by approximating the fraction as

$$\text{frac} = P(v)\,\Delta v = 4\pi \left(\frac{M}{2\pi RT}\right)^{3/2} v^2 e^{-Mv^2/2RT}\,\Delta v.$$

The function $P(v)$ is plotted in Fig. 20-7a. The total area between the curve and the horizontal axis represents the total fraction of molecules (unity). The area of the thin gold strip represents the fraction we seek.

To evaluate frac in parts, we can write

$$\text{frac} = (4\pi)(A)(v^2)(e^B)(\Delta v). \tag{20-36}$$

A and B are

$$A = \left(\frac{M}{2\pi RT}\right)^{3/2} = \left(\frac{0.0320 \text{ kg/mol}}{(2\pi)(8.31 \text{ J/mol} \cdot \text{K})(300 \text{ K})}\right)^{3/2}$$
$$= 2.92 \times 10^{-9} \text{ s}^3/\text{m}^3$$

and $$B = -\frac{Mv^2}{2RT} = -\frac{(0.0320 \text{ kg/mol})(600 \text{ m/s})^2}{(2)(8.31 \text{ J/mol} \cdot \text{K})(300 \text{ K})} = -2.31.$$

Substituting A and B into Eq. 20-36 yields

$$\text{frac} = (4\pi)(A)(v^2)(e^B)(\Delta v)$$
$$= (4\pi)(2.92 \times 10^{-9} \text{ s}^3/\text{m}^3)(600 \text{ m/s})^2(e^{-2.31})(2 \text{ m/s})$$
$$= 2.62 \times 10^{-3}. \qquad \text{(Answer)}$$

Thus, at room temperature, 0.262% of the oxygen molecules will have speeds that lie in the narrow range between 599 and 601 m/s. If the gold strip of Fig. 20-7a were drawn to the scale of this problem, it would be a very thin strip indeed.

Sample Problem 20-6

The molar mass M of oxygen is 0.0320 kg/mol.

(a) What is the average speed v_{avg} of oxygen gas molecules at $T = 300$ K?

SOLUTION: The Key Idea here is that to find the average speed, we must weight speed v with the distribution function $P(v)$ of Eq. 20-27 and then integrate the resulting expression over the range of possible speeds (0 to ∞). That leads to Eq. 20-31, which gives us

$$v_{\text{avg}} = \sqrt{\frac{8RT}{\pi M}}$$

$$= \sqrt{\frac{8(8.31 \text{ J/mol} \cdot \text{K})(300 \text{ K})}{\pi(0.0320 \text{ kg/mol})}} = 445 \text{ m/s}. \quad \text{(Answer)}$$

This result is plotted in Fig. 20-7a.

(b) What is the root-mean-square speed v_{rms} at 300 K?

SOLUTION: The Key Idea here is that to find v_{rms}, we must first find $(v^2)_{\text{avg}}$ by weighting v^2 with the distribution function $P(v)$ of Eq. 20-27 and then integrating the expression over the range of possible speeds. Then we must take the square root of the result. That leads to Eq. 20-34, which gives us

$$v_{\text{rms}} = \sqrt{\frac{3RT}{M}}$$

$$= \sqrt{\frac{3(8.31 \text{ J/mol} \cdot \text{K})(300 \text{ K})}{0.0320 \text{ kg/mol}}} = 483 \text{ m/s}. \quad \text{(Answer)}$$

This result is plotted in Fig. 20-7a. It is greater than v_{avg} because the greater speed values influence the calculation more when we integrate the v^2 values than when we integrate the v values.

(c) What is the most probable speed v_P at 300 K?

SOLUTION: The Key Idea here is that v_P corresponds to the maximum of the distribution function $P(v)$, which we obtain by setting the derivative $dP/dv = 0$ and solving the result for v. That leads to Eq. 20-35, which gives us

$$v_P = \sqrt{\frac{2RT}{M}}$$

$$= \sqrt{\frac{2(8.31 \text{ J/mol} \cdot \text{K})(300 \text{ K})}{0.0320 \text{ kg/mol}}} = 395 \text{ m/s}. \quad \text{(Answer)}$$

This result is also plotted in Fig. 20-7a.

20-8 The Molar Specific Heats of an Ideal Gas

In this section, we want to derive from molecular considerations an expression for the internal energy E_{int} of an ideal gas. In other words, we want an expression for the energy associated with the random motions of the atoms or molecules in the gas. We shall then use that expression to derive the molar specific heats of an ideal gas.

Internal Energy E_{int}

Let us first assume that our ideal gas is a *monatomic gas* (which has individual atoms rather than molecules), such as helium, neon, or argon. Let us also assume that the internal energy E_{int} of our ideal gas is simply the sum of the translational kinetic energies of its atoms. (As explained by quantum theory, individual atoms do not have rotational kinetic energy.)

The average translational kinetic energy of a single atom depends only on the gas temperature and is given by Eq. 20-24 as $K_{avg} = \frac{3}{2}kT$. A sample of n moles of such a gas contains nN_A atoms. The internal energy E_{int} of the sample is then

$$E_{int} = (nN_A)K_{avg} = (nN_A)(\tfrac{3}{2}kT). \qquad (20\text{-}37)$$

Using Eq. 20-7 ($k = R/N_A$), we can rewrite this as

$$E_{int} = \tfrac{3}{2}nRT \qquad \text{(monatomic ideal gas).} \qquad (20\text{-}38)$$

Thus,

> The internal energy E_{int} of an ideal gas is a function of the gas temperature *only*; it does not depend on any other variable.

With Eq. 20-38 in hand, we are now able to derive an expression for the molar specific heat of an ideal gas. Actually, we shall derive two expressions. One is for the case in which the volume of the gas remains constant as energy is transferred to or from it as heat. The other is for the case in which the pressure of the gas remains constant as energy is transferred to or from it as heat. The symbols for these two molar specific heats are C_V and C_p, respectively. (By convention, the capital letter C is used in both cases, even though C_V and C_p represent types of specific heat and not heat capacities.)

Molar Specific Heat at Constant Volume

Figure 20-8a shows n moles of an ideal gas at pressure p and temperature T, confined to a cylinder of fixed volume V. This *initial state i* of the gas is marked on the p-V diagram of Fig. 20-8b. Suppose now that you add a small amount of energy to the gas as heat Q by slowly turning up the temperature of the thermal reservoir. The gas temperature rises a small amount to $T + \Delta T$, and its pressure rises to $p + \Delta p$, bringing the gas to *final state f*.

In such experiments, we would find that the heat Q is related to the temperature change ΔT by

$$Q = nC_V\,\Delta T \qquad \text{(constant volume),} \qquad (20\text{-}39)$$

where C_V is a constant called the **molar specific heat at constant volume.** Substituting this expression for Q into the first law of thermodynamics as given by

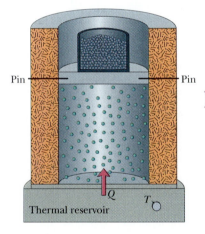

Pin ————   ———— Pin

Q T

Thermal reservoir

(a)

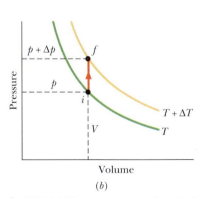

Pressure

$p + \Delta p$ f

p i

V

$T + \Delta T$

T

Volume

(b)

Fig. 20-8 (a) The temperature of an ideal gas is raised from T to $T + \Delta T$ in a constant-volume process. Heat is added, but no work is done. (b) The process on a p-V diagram.

Eq. 19-26 ($\Delta E_{int} = Q - W$) yields

$$\Delta E_{int} = nC_V \Delta T - W. \tag{20-40}$$

With the volume held constant, the gas cannot expand and thus cannot do any work. Therefore, $W = 0$, and Eq. 20-40 gives us

$$C_V = \frac{\Delta E_{int}}{n \Delta T}. \tag{20-41}$$

From Eq. 20-38 we know that $E_{int} = \frac{3}{2}nRT$, so the change in internal energy must be

$$\Delta E_{int} = \frac{3}{2}nR \Delta T. \tag{20-42}$$

Substituting this result into Eq. 20-41 yields

$$C_V = \tfrac{3}{2}R = 12.5 \text{ J/mol} \cdot \text{K} \quad \text{(monatomic gas).} \tag{20-43}$$

As Table 20-2 shows, this prediction of the kinetic theory (for ideal gases) agrees very well with experiment for real monatomic gases, the case that we have assumed. The (predicted and) experimental values of C_V for *diatomic gases* (which have molecules with two atoms) and *polyatomic gases* (which have molecules with more than two atoms) are greater than those for monatomic gases for reasons that will be suggested in Section 20-9.

We can now generalize Eq. 20-38 for the internal energy of any ideal gas by substituting C_V for $\frac{3}{2}R$; we get

$$E_{int} = nC_V T \quad \text{(any ideal gas).} \tag{20-44}$$

This equation applies not only to an ideal monatomic gas but also to diatomic and polyatomic ideal gases, provided the appropriate value of C_V is used. Just as with Eq. 20-38, we see that the internal energy of a gas depends on the temperature of the gas but not on its pressure or density.

When an ideal gas that is confined to a container undergoes a temperature change ΔT, then from either Eq. 20-41 or Eq. 20-44 we can write the resulting change in its internal energy as

$$\Delta E_{int} = nC_V \Delta T \quad \text{(ideal gas, any process).} \tag{20-45}$$

TABLE 20-2   Molar Specific Heats at Constant Volume

Molecule		Example	C_V (J/mol $\cdot$ K)
Monatomic	Ideal		$\frac{3}{2}R = 12.5$
	Real	He	12.5
		Ar	12.6
Diatomic	Ideal		$\frac{5}{2}R = 20.8$
	Real	N_2	20.7
		O_2	20.8
Polyatomic	Ideal		$3R = 24.9$
	Real	NH_4	29.0
		CO_2	29.7

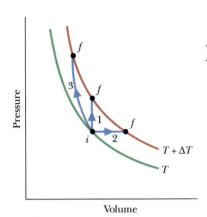

Fig. 20-9 Three paths for three different processes that take an ideal gas from an initial state i at temperature T to some final state f at temperature $T + \Delta T$. The change ΔE_{int} in the internal energy of the gas is the same for these three processes and for any others that result in the same change of temperature.

This equation tells us:

> A change in the internal energy E_{int} of a confined ideal gas depends on the change in the gas temperature only; it does *not* depend on what type of process produces the change in the temperature.

As examples, consider the three paths between the two isotherms in the p-V diagram of Fig. 20-9. Path 1 represents a constant-volume process. Path 2 represents a constant-pressure process (that we are about to examine). Path 3 represents a process in which no heat is exchanged with the system's environment (we discuss this in Section 20-11). Although the values of heat Q and work W associated with these three paths differ, as do p_f and V_f, the values of ΔE_{int} associated with the three paths are identical and are all given by Eq. 20-45, because they all involve the same temperature change ΔT. Therefore, no matter what path is actually taken between T and $T + \Delta T$, we can *always* use path 1 and Eq. 20-45 to compute ΔE_{int} easily.

Molar Specific Heat at Constant Pressure

We now assume that the temperature of the ideal gas is increased by the same small amount ΔT as previously, but that the necessary energy (heat Q) is added with the gas under constant pressure. An experiment for doing this is shown in Fig. 20-10a; the p-V diagram for the process is plotted in Fig. 20-10b. From such experiments we find that the heat Q is related to the temperature change ΔT by

$$Q = nC_p \, \Delta T \qquad \text{(constant pressure)}, \qquad (20\text{-}46)$$

where C_p is a constant called the **molar specific heat at constant pressure.** This C_p is *greater* than the molar specific heat at constant volume C_V, because energy must now be supplied not only to raise the temperature of the gas but also for the gas to do work—that is, to lift the weighted piston of Fig. 20-10a.

To relate molar specific heats C_p and C_V, we start with the first law of thermodynamics (Eq. 19-26):

$$\Delta E_{int} = Q - W. \qquad (20\text{-}47)$$

We next replace each term in Eq. 20-47. For ΔE_{int}, we substitute from Eq. 20-45. For Q, we substitute from Eq. 20-46. To replace W, we first note that since the pressure remains constant, Eq. 20-16 tells us that $W = p \, \Delta V$. Then we note that, using the ideal gas equation ($pV = nRT$), we can write

$$W = p \, \Delta V = nR \, \Delta T. \qquad (20\text{-}48)$$

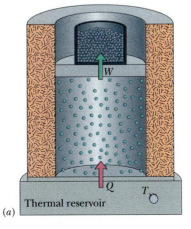

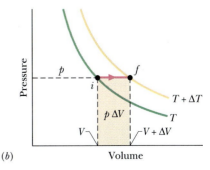

Fig. 20-10 (a) The temperature of an ideal gas is raised from T to $T + \Delta T$ in a constant-pressure process. Heat is added and work is done in lifting the loaded piston. (b) The process on a p-V diagram. The work $p \, \Delta V$ is given by the shaded area.

Making these substitutions in Eq. 20-47, and then dividing through by $n\,\Delta T$, we find

$$C_V = C_p - R$$

and then

$$C_p = C_V + R. \tag{20-49}$$

This prediction of kinetic theory agrees well with experiment, not only for monatomic gases but for gases in general, as long as their density is low enough so that we may treat them as ideal.

✔ **CHECKPOINT 4:**   The figure here shows five paths traversed by a gas on a p-V diagram. Rank the paths according to the change in internal energy of the gas, greatest first.

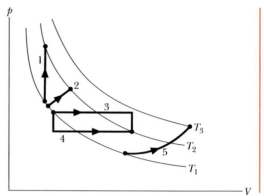

Sample Problem 20-7

A bubble of 5.00 mol of helium is submerged at a certain depth in liquid water when the water (and thus the helium) undergoes a temperature increase ΔT of 20.0 C° at constant pressure. As a result, the bubble expands. The helium is monatomic and ideal.

(a) How much energy is added as heat to the helium during the increase and expansion?

SOLUTION: One **Key Idea** here is that the heat Q is related to the temperature change ΔT by a molar specific heat of the gas. Because the pressure p is held constant during the addition of energy, we use the molar specific heat at constant pressure C_p and Eq. 20-46,

$$Q = nC_p\,\Delta T, \tag{20-50}$$

to find Q. To evaluate C_p we go to Eq. 20-49, which tells us that for any ideal gas, $C_p = C_V + R$. Then from Eq. 20-43, we know that for any *monatomic* gas (like the helium here), $C_V = \frac{3}{2}R$. Thus, Eq. 20-50 gives us

$$Q = n(C_V + R)\,\Delta T = n(\tfrac{3}{2}R + R)\,\Delta T = n(\tfrac{5}{2}R)\,\Delta T$$
$$= (5.00\ \text{mol})(2.5)(8.31\ \text{J/mol}\cdot\text{K})(20.0\ \text{C}°)$$
$$= 2077.5\ \text{J} \approx 2080\ \text{J.} \qquad\qquad \text{(Answer)}$$

(b) What is the change ΔE_{int} in the internal energy of the helium during the temperature increase?

SOLUTION: Because the bubble expands, this is not a constant-volume process. However, the helium is nonetheless confined (to the bubble). Thus, a **Key Idea** here is that the change ΔE_{int} is the same as *would occur* in a constant-volume process with the same temperature change ΔT. We can easily find the constant-volume change ΔE_{int} with Eq. 20-45:

$$\Delta E_{int} = nC_V\,\Delta T = n(\tfrac{3}{2}R)\,\Delta T$$
$$= (5.00\ \text{mol})(1.5)(8.31\ \text{J/mol}\cdot\text{K})(20.0\ \text{C}°)$$
$$= 1246.5\ \text{J} \approx 1250\ \text{J.} \qquad\qquad \text{(Answer)}$$

(c) How much work W is done by the helium as it expands against the pressure of the surrounding water during the temperature increase?

SOLUTION: One **Key Idea** here is that the work done by *any* gas expanding against the pressure from its environment is given by Eq. 20-11, which tells us to integrate $p\,dV$. When the pressure is constant (as here), we can simplify that to $W = p\,\Delta V$. When the gas is *ideal* (as here), we can use the ideal gas law (Eq. 20-5) to write $p\,\Delta V = nR\,\Delta T$. We end up with

$$W = nR\,\Delta T$$
$$= (5.00\ \text{mol})(8.31\ \text{J/mol}\cdot\text{K})(20.0\ \text{C}°)$$
$$= 831\ \text{J.} \qquad\qquad \text{(Answer)}$$

Because we happen to know Q and ΔE_{int}, we can work this problem another way: The **Key Idea** now is that we can account for the energy changes of the gas with the first law of thermodynamics, writing

$$W = Q - \Delta E_{int} = 2077.5\ \text{J} - 1246.5\ \text{J}$$
$$= 831\ \text{J.} \qquad\qquad \text{(Answer)}$$

Note that during the temperature increase, only a portion (1250 J) of the energy (2080 J) that is transferred to the helium as heat goes to increasing the internal energy of the helium and thus the temperature of the helium. The rest (831 J) is transferred out of the helium as work that the helium does during the expansion. If the water were frozen, it would not allow that expansion. Then the same temperature increase of 20.0 C° would require only 1250 J of heat, because no work would be done by the helium.

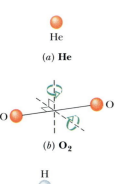

(a) **He**

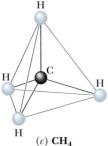

(b) **O₂**

(c) **CH₄**

Fig. 20-11 Models of molecules as used in kinetic theory: (a) helium, a typical monatomic molecule; (b) oxygen, a typical diatomic molecule; and (c) methane, a typical polyatomic molecule. The spheres represent atoms, and the lines between them represent bonds. Two rotation axes are shown for the oxygen molecule.

20-9 Degrees of Freedom and Molar Specific Heats

As Table 20-2 shows, the prediction that $C_V = \frac{3}{2}R$ agrees with experiment for monatomic gases but fails for diatomic and polyatomic gases. Let us try to explain the discrepancy by considering the possibility that molecules with more than one atom can store internal energy in forms other than translational motion.

Figure 20-11 shows common models of helium (a *monatomic* molecule, containing a single atom), oxygen (a *diatomic* molecule, containing two atoms), and methane (a *polyatomic* molecule). From such models, we would assume that all three types of molecules can have translational motions (say, moving left–right and up–down) and rotational motions (spinning about an axis like a top). In addition, we would assume that the diatomic and polyatomic molecules can have oscillatory motions, with the atoms oscillating slightly toward and away from one another, as if attached to opposite ends of a spring.

To keep account of the various ways in which energy can be stored in a gas, James Clerk Maxwell introduced the theorem of the **equipartition of energy:**

> Every kind of molecule has a certain number f of *degrees of freedom*, which are independent ways in which the molecule can store energy. Each such degree of freedom has associated with it—on average—an energy of $\frac{1}{2}kT$ per molecule (or $\frac{1}{2}RT$ per mole).

Let us apply the theorem to the translational and rotational motions of the molecules in Fig. 20-11. (We discuss oscillatory motion in the next section.) For the translational motion, superimpose an *xyz* coordinate system on any gas. The molecules will, in general, have velocity components along all three axes. Thus, gas molecules of all types have three degrees of translational freedom (three ways to move in translation) and, on average, an associated energy of $3(\frac{1}{2}kT)$ per molecule.

For the rotational motion, imagine the origin of our *xyz* coordinate system at the center of each molecule in Fig. 20-11. In a gas, each molecule should be able to rotate with an angular velocity component along each of the three axes, so each gas should have three degrees of rotational freedom and, on average, an additional energy of $3(\frac{1}{2}kT)$ per molecule. *However,* experiment shows this is true only for the polyatomic molecules. As explained by quantum theory, a monatomic gas molecule does not rotate and so has no rotational energy (a single atom cannot rotate like a top). A diatomic molecule can rotate like a top only about axes perpendicular to the line connecting the atoms (the axes are shown in Fig. 20-11b) and not about that line itself. Therefore, a diatomic molecule can have only two degrees of rotational freedom and a rotational energy of only $2(\frac{1}{2}kT)$ per molecule.

To extend our analysis of molar specific heats (C_p and C_V, in Section 20-8) to ideal diatomic and polyatomic gases, it is necessary to retrace the derivations of that analysis in detail. First, we replace Eq. 20-38 ($E_{\text{int}} = \frac{3}{2}nRT$) with $E_{\text{int}} = (f/2)nRT$, where f is the number of degrees of freedom listed in Table 20-3. Doing so leads to

TABLE 20-3 **Degrees of Freedom for Various Molecules**

| Molecule | Example | Degrees of Freedom | | | Predicted Molar Specific Heats | |
		Translational	Rotational	Total (f)	C_V (Eq. 20-51)	$C_p = C_V + R$
Monatomic	He	3	0	3	$\frac{3}{2}R$	$\frac{5}{2}R$
Diatomic	O₂	3	2	5	$\frac{5}{2}R$	$\frac{7}{2}R$
Polyatomic	CH₄	3	3	6	$3R$	$4R$

the prediction

$$C_V = \left(\frac{f}{2}\right)R = 4.16f \text{ J/mol} \cdot \text{K}, \qquad (20\text{-}51)$$

which agrees—as it must—with Eq. 20-43 for monatomic gases ($f = 3$). As Table 20-2 shows, this prediction also agrees with experiment for diatomic gases ($f = 5$), but it is too low for polyatomic gases.

Sample Problem 20-8

A cabin of volume V is filled with air (which we consider to be an ideal diatomic gas) at an initial low temperature T_1. After you light a wood stove, the air temperature increases to T_2. What is the resulting change ΔE_{int} in the internal energy of the air in the cabin?

SOLUTION: As the air temperature increases, the air pressure p cannot change but must always be equal to the air pressure outside the room. The reason is that, because the room is not air-tight, the air is not confined. As the temperature increases, air molecules leave through various openings and thus the number of moles n of air in the room decreases. Thus, one Key Idea here is that we *cannot* use Eq. 20-45 ($\Delta E_{int} = nC_V \Delta T$) to find ΔE_{int}, because it requires constant n.

A second Key Idea is that we *can* relate the internal energy E_{int} at any instant to n and the temperature T with Eq. 20-44 ($E_{int} = nC_VT$). From that equation we can then write

$$\Delta E_{int} = \Delta(nC_VT) = C_V \,\Delta(nT).$$

Next, using Eq. 20-5 ($pV = nRT$), we can replace nT with pV/R, obtaining

$$\Delta E_{int} = C_V \,\Delta\!\left(\frac{pV}{R}\right). \qquad (20\text{-}52)$$

Now, because p, V, and R are all constants, Eq. 20-52 yields

$$\Delta E_{int} = 0, \qquad \text{(Answer)}$$

even though the temperature changes.

Why does the cabin feel more comfortable at the higher temperature? There are at least two factors involved: (1) You exchange electromagnetic radiation (thermal radiation) with surfaces inside the room, and (2) you exchange energy with air molecules that collide with you. When the room temperature is increased, (1) the amount of thermal radiation emitted by the surfaces and absorbed by you is increased, and (2) the amount of energy you gain through the collisions of air molecules with you is increased.

20-10 A Hint of Quantum Theory

We can improve the agreement of kinetic theory with experiment by including the oscillations of the atoms in a gas of diatomic or polyatomic molecules. For example, the two atoms in the O_2 molecule of Fig. 20-11b can oscillate toward and away from each other, with the interconnecting bond acting like a spring. However, experiment shows that such oscillations occur only at relatively high temperatures of the gas—the motion is "turned on" only when the gas molecules have relatively large energies. Rotational motion is also subject to such "turning on," but at a lower temperature.

Figure 20-12 is of help in seeing this turning on of rotational motion and oscillatory motion. The ratio C_V/R for diatomic hydrogen gas (H_2) is plotted there against temperature, with the temperature scale logarithmic to cover several orders of magnitude. Below about 80 K, we find that $C_V/R = 1.5$. This result implies that only the three translational degrees of freedom of hydrogen are involved in the specific heat.

As the temperature increases, the value of C_V/R gradually increases to 2.5, implying that two additional degrees of freedom have become involved. Quantum theory shows that these two degrees of freedom are associated with the rotational motion of the hydrogen molecules and that this motion requires a certain minimum amount of energy. At very low temperatures (below 80 K), the molecules do not have enough energy to rotate. As the temperature increases from 80 K, first a few molecules and then more and more obtain enough energy to rotate, and C_V/R increases, until all of them are rotating and $C_V/R = 2.5$.

Fig. 20-12 A plot of C_V/R verses temperature for (diatomic) hydrogen gas. Because rotational and oscillatory motions begin at certain energies, only translation is possible at very low temperatures. As the temperature increases, rotational motion can begin. At still higher temperatures, oscillatory motion can begin.

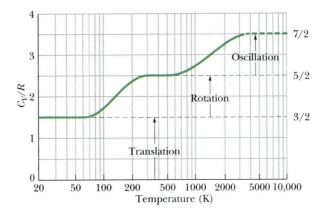

Similarly, quantum theory shows that oscillatory motion of the molecules requires a certain (higher) minimum amount of energy. This minimum amount is not met until the molecules reach a temperature of about 1000 K, as shown in Fig. 20-12. As the temperature increases beyond 1000 K, the number of molecules with enough energy to oscillate increases, and C_V/R increases, until all of them are oscillating and $C_V/R = 3.5$. (In Fig. 20-12, the plotted curve stops at 3200 K because at that temperature, the atoms of a hydrogen molecule oscillate so much that they overwhelm their bond, and the molecule then *dissociates* into two separate atoms.)

20-11 The Adiabatic Expansion of an Ideal Gas

We saw in Section 18-3 that sound waves are propagated through air and other gases as a series of compressions and expansions; these variations in the transmission medium take place so rapidly that there is no time for energy to be transferred from one part of the medium to another as heat. As we saw in Section 19-10, a process for which $Q = 0$ is an *adiabatic process*. We can ensure that $Q = 0$ either by carrying out the process very quickly (as in sound waves) or by doing it (at any rate) in a well-insulated container. Let us see what the kinetic theory has to say about adiabatic processes.

Figure 20-13a shows our usual insulated cylinder, now containing an ideal gas and resting on an insulating stand. By removing mass from the piston, we can allow the gas to expand adiabatically. As the volume increases, both the pressure and the temperature drop. We shall prove next that the relation between the pressure and the volume during such an adiabatic process is

$$pV^\gamma = \text{a constant} \qquad \text{(adiabatic process),} \qquad (20\text{-}53)$$

in which $\gamma = C_p/C_V$, the ratio of the molar specific heats for the gas. On a p-V diagram such as that in Fig. 20-13b, the process occurs along a line (called an *adiabat*) that has the equation $p = (\text{a constant})/V^\gamma$. Since the gas goes from an initial state i to a final state f, we can rewrite Eq. 20-53 as

$$p_i V_i^\gamma = p_f V_f^\gamma \qquad \text{(adiabatic process).} \qquad (20\text{-}54)$$

We can also write an equation for an adiabatic process in terms of T and V. To do so, we use the ideal gas equation ($pV = nRT$) to eliminate p from Eq. 20-53, finding

$$\left(\frac{nRT}{V}\right)V^\gamma = \text{a constant.}$$

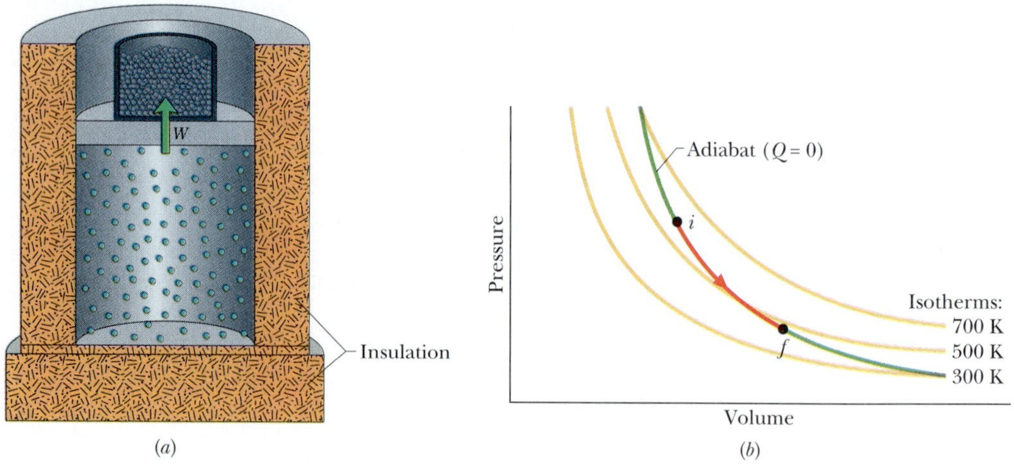

Fig. 20-13 (*a*) The volume of an ideal gas is increased by removing mass from the piston. The process is adiabatic (*Q* = 0). (*b*) The process proceeds from *i* to *f* along an adiabat on a *p-V* diagram.

Because *n* and *R* are constants, we can rewrite this in the alternative form

$$TV^{\gamma-1} = \text{a constant} \qquad \text{(adiabatic process),} \qquad (20\text{-}55)$$

in which the constant is different from that in Eq. 20-53. When the gas goes from an initial state *i* to a final state *f*, we can rewrite Eq. 20-55 as

$$T_i V_i^{\gamma-1} = T_f V_f^{\gamma-1} \qquad \text{(adiabatic process).} \qquad (20\text{-}56)$$

We can now answer the question that opens this chapter. At the top of an unopened carbonated drink, there is a gas of carbon dioxide and water vapor. Because the pressure of the gas is greater than atmospheric pressure, the gas expands out into the atmosphere when the container is opened. Thus, the gas increases its volume, but that means it must do work to push against the atmosphere. Because the expansion is so rapid, it is adiabatic and the only source of energy for the work is the internal energy of the gas. Because the internal energy decreases, the temperature of the gas must also decrease, which can cause the water vapor in the gas to condense into tiny drops, forming the fog. (Note that Eq. 20-56 also tells us that the temperature must decrease during an adiabatic expansion: Since V_f is greater than V_i, then T_f must be less than T_i.)

Proof of Eq. 20-53

Suppose that you remove some shot from the piston of Fig. 20-13*a*, allowing the ideal gas to push the piston and the remaining shot upward and thus to increase the volume by a differential amount *dV*. Since the volume change is tiny, we may assume that the pressure *p* of the gas on the piston is constant during the change. This assumption allows us to say that the work *dW* done by the gas during the volume increase is equal to *p dV*. From Eq. 19-27, the first law of thermodynamics can then be written as

$$dE_{\text{int}} = Q - p \, dV. \qquad (20\text{-}57)$$

Since the gas is thermally insulated (and thus the expansion is adiabatic), we substitute 0 for *Q*. Then we use Eq. 20-45 to substitute $nC_V \, dT$ for dE_{int}. With these substitutions, and after some rearranging, we have

$$n \, dT = -\left(\frac{p}{C_V}\right) dV. \qquad (20\text{-}58)$$

Now from the ideal gas law ($pV = nRT$) we have

$$p \, dV + V \, dp = nR \, dT. \tag{20-59}$$

Replacing R with its equal, $C_p - C_V$, in Eq. 20-59 yields

$$n \, dT = \frac{p \, dV + V \, dp}{C_p - C_V}. \tag{20-60}$$

Equating Eqs. 20-58 and 20-60 and rearranging then give

$$\frac{dp}{p} + \left(\frac{C_p}{C_V}\right) \frac{dV}{V} = 0.$$

Replacing the ratio of the molar specific heats with γ and integrating (see integral 5 in Appendix E) yield

$$\ln p + \gamma \ln V = \text{a constant}.$$

Rewriting the left side as $\ln pV^{\gamma}$ and then taking the antilog of both sides, we find

$$pV^{\gamma} = \text{a constant}, \tag{20-61}$$

which is what we set out to prove.

Free Expansions

Recall from Section 19-10 that a free expansion of a gas is an adiabatic process that involves no work done on or by the gas, and no change in the internal energy of the gas. A free expansion is thus quite different from the type of adiabatic process described by Eqs. 20-53 through 20-61, in which work is done and the internal energy changes. Those equations then do *not* apply to a free expansion, even though such an expansion is adiabatic.

Also recall that in a free expansion, a gas is in equilibrium only at its initial and final points; thus, we can plot only those points, but not the expansion itself, on a p-V diagram. In addition, because $\Delta E_{int} = 0$, the temperature of the final state must be that of the initial state. Thus, the initial and final points on a p-V diagram must be on the same isotherm, and instead of Eq. 20-56 we have

$$T_i = T_f \quad \text{(free expansion).} \tag{20-62}$$

If we next assume that the gas is ideal (so that $pV = nRT$), then because there is no change in temperature, there can be no change in the product pV. Thus, instead of Eq. 20-53 a free expansion involves the relation

$$p_i V_i = p_f V_f \quad \text{(free expansion).} \tag{20-63}$$

Sample Problem 20-9

In Sample Problem 20-2, 1 mol of oxygen (assumed to be an ideal gas) expands isothermally (at 310 K) from an initial volume of 12 L to a final volume of 19 L.

(a) What would be the final temperature if the gas had expanded adiabatically to this same final volume? Oxygen (O_2) is diatomic and here has rotation but not oscillation.

SOLUTION: The Key Ideas here are

1. When a gas expands against the pressure of its environment, it must do work.

2. When the process is adiabatic (no energy is transferred as heat), then the energy required for the work can come only from the internal energy of the gas.

3. Because the internal energy decreases, the temperature T must also decrease.

We can relate the initial and final temperatures and volumes with Eq. 20-56:

$$T_i V_i^{\gamma-1} = T_f V_f^{\gamma-1}. \tag{20-64}$$

Because the molecules are diatomic and have rotation but not oscillation, we can take the molar specific heats from Table 20-3.

Thus,

$$\gamma = \frac{C_p}{C_V} = \frac{\frac{7}{2}R}{\frac{5}{2}R} = 1.40.$$

Solving Eq. 20-64 for T_f and inserting known data then yield

$$T_f = \frac{T_i V_i^{\gamma-1}}{V_f^{\gamma-1}} = \frac{(310 \text{ K})(12 \text{ L})^{1.40-1}}{(19 \text{ L})^{1.40-1}}$$

$$= (310 \text{ K})(\tfrac{12}{19})^{0.40} = 258 \text{ K}. \qquad \text{(Answer)}$$

(b) What would be the final temperature and pressure if, instead, the gas had expanded freely to the new volume, from an initial pressure of 2.0 Pa?

SOLUTION: Here the **Key Idea** is that the temperature does not change in a free expansion:

$$T_f = T_i = 310 \text{ K}. \qquad \text{(Answer)}$$

We find the new pressure using Eq. 20-63, which gives us

$$p_f = p_i \frac{V_i}{V_f} = (2.0 \text{ Pa}) \frac{12 \text{ L}}{19 \text{ L}} = 1.3 \text{ Pa}. \qquad \text{(Answer)}$$

PROBLEM-SOLVING TACTICS

Tactic 2: *A Graphical Summary of Four Gas Processes*
In this chapter we have discussed four special processes that an ideal gas can undergo. An example of each is shown in Fig. 20-14, and some associated characteristics are given in Table 20-4, including two process names (isobaric and isochoric) that we have not used but that you might see in other courses.

✓CHECKPOINT 5:   Rank paths 1, 2, and 3 in Fig. 20-14 according to the heat transfer to the gas, greatest first.

TABLE 20-4   Four Special Processes

Path in Fig. 20-14	Constant Quantity	Process Type	Some Special Results ($\Delta E_{int} = Q - W$ and $\Delta E_{int} = nC_V \Delta T$ for all paths)
1	p	Isobaric	$Q = nC_p \Delta T$; $W = p \Delta V$
2	T	Isothermal	$Q = W = nRT \ln(V_f/V_i)$; $\Delta E_{int} = 0$
3	pV^{γ}, $TV^{\gamma-1}$	Adiabatic	$Q = 0$; $W = -\Delta E_{int}$
4	V	Isochoric	$Q = \Delta E_{int} = nC_V \Delta T$; $W = 0$

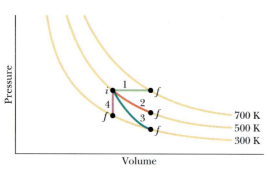

Fig. 20-14 A p-V diagram representing four special processes for an ideal gas. Table 20-4 explains the processes.

REVIEW & SUMMARY

Kinetic Theory of Gases   The *kinetic theory of gases* relates the *macroscopic* properties of gases (for example, pressure and temperature) to the *microscopic* properties of gas molecules (for example, speed and kinetic energy).

Avogadro's Number   One mole of a substance contains N_A (*Avogadro's number*) elementary units (usually atoms or molecules), where N_A is found experimentally to be

$$N_A = 6.02 \times 10^{23} \text{ mol}^{-1} \quad \text{(Avogadro's number).} \quad (20\text{-}1)$$

One molar mass M of any substance is the mass of one mole of the substance. It is related to the mass m of the individual molecules of the substance by

$$M - mN_A. \qquad (20\text{-}4)$$

The number of moles n contained in a sample of mass M_{sam}, con-

sisting of N molecules, is given by

$$n = \frac{N}{N_A} = \frac{M_{sam}}{M} = \frac{M_{sam}}{mN_A}. \qquad (20\text{-}2, 20\text{-}3)$$

Ideal Gas   An *ideal gas* is one for which the pressure p, volume V, and temperature T are related by

$$pV = nRT \qquad \text{(ideal gas law).} \qquad (20\text{-}5)$$

Here n is the number of moles of the gas present and R is a constant $(8.31 \text{ J/mol} \cdot \text{K})$ called the **gas constant.** The ideal gas law can also be written as

$$pV = NkT, \qquad (20\text{-}9)$$

where the **Boltzmann constant** k is

$$k = \frac{R}{N_A} = 1.38 \times 10^{-23} \text{ J/K}. \quad (20\text{-}7)$$

Work in an Isothermal Volume Change The work done *by* an ideal gas during an **isothermal** (constant-temperature) change from volume V_i to volume V_f is

$$W = nRT \ln \frac{V_f}{V_i} \quad \text{(ideal gas, isothermal process).} \quad (20\text{-}14)$$

Pressure, Temperature, and Molecular Speed The pressure exerted by n moles of an ideal gas, in terms of the speed of its molecules, is

$$p = \frac{nMv_{rms}^2}{3V}, \quad (20\text{-}21)$$

where $v_{rms} = \sqrt{(v^2)_{avg}}$ is the **root-mean-square speed** of the molecules of the gas. With Eq. 20-5 this gives

$$v_{rms} = \sqrt{\frac{3RT}{M}}. \quad (20\text{-}22)$$

Temperature and Kinetic Energy The average translational kinetic energy K_{avg} per molecule of an ideal gas is

$$K_{avg} = \tfrac{3}{2}kT. \quad (20\text{-}24)$$

Mean Free Path The *mean free path* λ of a gas molecule is its average path length between collisions and is given by

$$\lambda = \frac{1}{\sqrt{2}\pi d^2 N/V}, \quad (20\text{-}25)$$

where N/V is the number of molecules per unit volume and d is the molecular diameter.

Maxwell Speed Distribution The *Maxwell speed distribution* $P(v)$ is a function such that $P(v)\,dv$ gives the *fraction* of molecules with speeds in the interval dv centered on speed v:

$$P(v) = 4\pi \left(\frac{M}{2\pi RT}\right)^{3/2} v^2 e^{-Mv^2/2RT}. \quad (20\text{-}27)$$

Three measures of the distribution of speeds among the molecules of a gas are

$$v_{avg} = \sqrt{\frac{8RT}{\pi M}} \quad \text{(average speed),} \quad (20\text{-}31)$$

$$v_P = \sqrt{\frac{2RT}{M}} \quad \text{(most probable speed),} \quad (20\text{-}35)$$

and the rms speed defined above in Eq. 20-22.

Molar Specific Heats The molar specific heat C_V of a gas at constant volume is defined as

$$C_V = \frac{1}{n}\frac{Q}{\Delta T} = \frac{1}{n}\frac{\Delta E_{int}}{\Delta T}, \quad (20\text{-}39, 20\text{-}41)$$

in which Q is the energy transferred as heat to or from a sample of n moles of the gas, ΔT is the resulting temperature change of the gas, and ΔE_{int} is the resulting change in the internal energy of the gas. For an ideal monatomic gas,

$$C_V = \tfrac{3}{2}R = 12.5 \text{ J/mol}\cdot\text{K}. \quad (20\text{-}43)$$

The molar specific heat C_p of a gas at constant pressure is defined to be

$$C_p = \frac{1}{n}\frac{Q}{\Delta T}, \quad (20\text{-}46)$$

in which Q, n, and ΔT are defined as above. C_p is also given by

$$C_p = C_V + R. \quad (20\text{-}49)$$

For n moles of an ideal gas,

$$E_{int} = nC_V T \quad \text{(ideal gas).} \quad (20\text{-}44)$$

If n moles of a confined ideal gas undergo a temperature change ΔT due to *any* process, the change in the internal energy of the gas is

$$\Delta E_{int} = nC_V\,\Delta T \quad \text{(ideal gas, any process),} \quad (20\text{-}45)$$

in which the appropriate value of C_V must be substituted, according to the type of ideal gas.

Degrees of Freedom and C_V We find C_V itself by using the *equipartition of energy* theorem, which states that every *degree of freedom* of a molecule (that is, every independent way it can store energy) has associated with it—on average—an energy $\tfrac{1}{2}kT$ per molecule ($= \tfrac{1}{2}RT$ per mole). If f is the number of degrees of freedom, then $E_{int} = (f/2)nRT$ and

$$C_V = \left(\frac{f}{2}\right)R = 4.16f \text{ J/mol}\cdot\text{K}. \quad (20\text{-}51)$$

For monatomic gases $f = 3$ (three translational degrees); for diatomic gases $f = 5$ (three translational and two rotational degrees).

Adiabatic Process When an ideal gas undergoes a slow adiabatic volume change (a change for which $Q = 0$), its pressure and volume are related by

$$pV^\gamma = \text{a constant} \quad \text{(adiabatic process),} \quad (20\text{-}53)$$

in which $\gamma (= C_p/C_V)$ is the ratio of molar specific heats for the gas. For a free expansion, however, $pV = $ a constant.

QUESTIONS

1. If the temperature of an ideal gas is changed from 20°C to 40°C while the volume is unchanged, is the pressure of the gas doubled, increased but less than doubled, or increased and more than doubled?

2. In Fig. 20-15a, three isothermal processes are shown for the same gas and for the same change in volume (V_i to V_f) but at different temperatures. Rank the processes according to (a) the work done by the gas, (b) the change in the internal energy of the gas, and (c) the energy transferred as heat to the gas, greatest first.

In Fig. 20-15b, three isothermal processes are shown along a

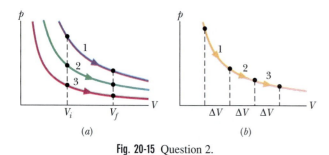

Fig. 20-15 Question 2.

single isotherm, for the same change ΔV in volume. Rank the processes according to (d) the work done by the gas, (e) the change in the internal energy of the gas, and (f) the energy transferred as heat to the gas, greatest first.

3. The volume of a gas and the number of gas molecules within that volume for four situations are (a) $2V_0$ and N_0, (b) $3V_0$ and $3N_0$, (c) $8V_0$ and $4N_0$, and (d) $3V_0$ and $9N_0$. Rank the situations according to the mean free path of the molecules, greatest first.

4. In Sample Problem 20-2, how much energy is transferred as heat during the expansion?

5. Figure 20-16 shows the initial state of an ideal gas and an isotherm through that state. Which of the paths shown result in a decrease in the temperature of the gas?

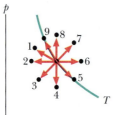

Fig. 20-16 Question 5.

6. For four situations for an ideal gas, the table gives the heat Q and either the work W done by the gas or the work W_{on} done on the gas, all in joules. Rank the four situations in terms of the temperature change of the gas, most positive first, most negative last.

	a	b	c	d
Q	−50	+35	−15	+20
W	−50	+35		
W_{on}			−40	+40

7. For a temperature increase of ΔT_1, a certain amount of an ideal gas requires 30 J when heated at constant volume and 50 J when heated at constant pressure. How much work is done by the gas in the second situation?

8. An ideal diatomic gas, with molecular rotation but not oscillation, loses energy as heat Q. Is the resulting decrease in the internal energy of the gas greater if the loss occurs in a constant-volume process or in a constant-pressure process?

9. A certain amount of energy is to be transferred as heat to 1 mol of a monatomic gas (a) at constant pressure and (b) at constant volume, and to 1 mol of a diatomic gas (c) at constant pressure and (d) at constant volume. Figure 20-17 shows four paths from an initial point to four final points on a p-V diagram. Which path goes with which process? (e) Are the molecules of the diatomic gas rotating?

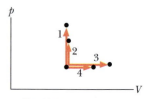

Fig. 20-17 Question 9.

10. Does the temperature of an ideal gas increase, decrease, or stay the same during (a) an isothermal expansion, (b) an expansion at constant pressure, (c) an adiabatic expansion, and (d) an increase in pressure at constant volume?

11. (a) Rank the four paths of Fig. 20-14 according to the work done by the gas, greatest first. (b) Rank paths 1, 2, and 3 according to the change in the internal energy of the gas, most positive first and most negative last.

12. In the p-V diagram of Fig. 20-18, the gas does 5 J of work along isotherm ab and 4 J along adiabat bc. What is the change in the internal energy of the gas if the gas traverses the straight path from a to c?

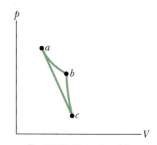

Fig. 20-18 Question 12.

EXERCISES & PROBLEMS

ssm Solution is in the Student Solutions Manual.
www Solution is available on the World Wide Web at:
    http://www.wiley.com/college/hrw
ilw Solution is available on the Interactive LearningWare.

SEC. 20-2   Avogadro's Number

1E. Find the mass in kilograms of 7.50×10^{24} atoms of arsenic, which has a molar mass of 74.9 g/mol. ssm

2E. Gold has a molar mass of 197 g/mol. (a) How many moles of gold are in a 2.50 g sample of pure gold? (b) How many atoms are in the sample?

3P. If the water molecules in 1.00 g of water were distributed uniformly over the surface of Earth, how many such molecules would there be on 1.00 cm^2 of the surface?

4P. A distinguished scientist has written: "There are enough molecules in the ink that makes one letter of this sentence to provide not only one for every inhabitant of Earth, but one for every creature if each star of our galaxy had a planet as populous as Earth." Check this statement. Assume the ink sample (molar mass = 18 g/mol) to have a mass of 1 μg, the population of Earth to be 5×10^9, and the number of stars in our galaxy to be 10^{11}.

SEC. 20-3   Ideal Gases

5E. Compute (a) the number of moles and (b) the number of molecules in 1.00 cm^3 of an ideal gas at a pressure of 100 Pa and a temperature of 220 K. ssm

6E. The best laboratory vacuum has a pressure of about 1.00×10^{-18} atm, or 1.01×10^{-13} Pa. How many gas molecules are there per cubic centimeter in such a vacuum at 293 K?

7E. Oxygen gas having a volume of 1000 cm³ at 40.0°C and 1.01×10^5 Pa expands until its volume is 1500 cm³ and its pressure is 1.06×10^5 Pa. Find (a) the number of moles of oxygen present and (b) the final temperature of the sample. ssm

8E. An automobile tire has a volume of 1.64×10^{-2} m³ and contains air at a gauge pressure (pressure above atmospheric pressure) of 165 kPa when the temperature is 0.00°C. What is the gauge pressure of the air in the tires when its temperature rises to 27.0°C and its volume increases to 1.67×10^{-2} m³? Assume atmospheric pressure is 1.01×10^5 Pa.

9E. A quantity of ideal gas at 10.0°C and 100 kPa occupies a volume of 2.50 m³. (a) How many moles of the gas are present? (b) If the pressure is now raised to 300 kPa and the temperature is raised to 30.0°C, how much volume does the gas occupy? Assume no leaks.

10E. Calculate the work done by an external agent during an isothermal compression of 1.00 mol of oxygen from a volume of 22.4 L at 0°C and 1.00 atm pressure to 16.8 L.

11P. Pressure p, volume V, and temperature T for a certain material are related by

$$p = \frac{AT - BT^2}{V},$$

where A and B are constants. Find an expression for the work done by the material if the temperature changes from T_1 to T_2 while the pressure remains constant. ssm

12P. A container encloses two ideal gases. Two moles of the first gas are present, with molar mass M_1. The second gas has molar mass $M_2 = 3M_1$, and 0.5 mol of this gas is present. What fraction of the total pressure on the container wall is attributable to the second gas? (The kinetic theory explanation of pressure leads to the experimentally discovered law of partial pressures for a mixture of gases that do not react chemically: *The total pressure exerted by the mixture is equal to the sum of the pressures that the several gases would exert separately if each were to occupy the vessel alone.*)

13P. Air that initially occupies 0.14 m³ at a gauge pressure of 103.0 kPa is expanded isothermally to a pressure of 101.3 kPa and then cooled at constant pressure until it reaches its initial volume. Compute the work done by the air. (Gauge pressure is the difference between the actual pressure and atmospheric pressure.) ssm ilw www

14P. A sample of an ideal gas is taken through the cyclic process *abca* shown in Fig. 20-19; at point *a*, $T = 200$ K. (a) How many moles of gas are in the sample? What are (b) the temperature of the gas at point *b*, (c) the temperature of the gas at point *c*, and (d) the net energy added to the gas as heat during the cycle?

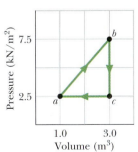
Fig. 20-19 Problem 14.

15P. An air bubble of 20 cm³ volume is at the bottom of a lake 40 m deep where the temperature is 4.0°C. The bubble rises to the surface, which is at a temperature of 20°C. Take the temperature of the bubble's air to be the same as that of the surrounding water. Just as the bubble reaches the surface, what is its volume? ssm

16P. A pipe of length $L = 25.0$ m that is open at one end contains air at atmospheric pressure. It is thrust vertically into a freshwater lake until the water rises halfway up in the pipe, as shown in Fig. 20-20. What is the depth h of the lower end of the pipe? Assume that the temperature is the same everywhere and does not change.

17P. Container A in Fig. 20-21 holds an ideal gas at a pressure of 5.0×10^5 Pa and a temperature of 300 K. It is connected by a thin tube (and a closed valve) to container B, with four times the volume of A. Container B holds the same ideal gas at a pressure of 1.0×10^5 Pa and a temperature of 400 K. The valve is opened to allow the pressures to equalize, but the temperature of each container is kept constant at its initial value. What then is the pressure in the two containers? ilw

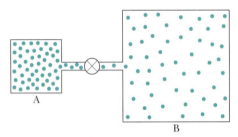
Fig. 20-20 Problem 16.

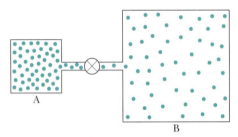

Fig. 20-21 Problem 17.

SEC. 20-4   Pressure, Temperature, and RMS Speed

18E. Calculate the rms speed of helium atoms at 1000 K. See Appendix F for the molar mass of helium atoms.

19E. The lowest possible temperature in outer space is 2.7 K. What is the root-mean-square speed of hydrogen molecules at this temperature? (The molar mass of hydrogen molecules (H_2) is given in Table 20-1.) ssm

20E. Find the rms speed of argon atoms at 313 K. See Appendix F for the molar mass of argon atoms.

21E. The temperature and pressure in the Sun's atmosphere are 2.00×10^6 K and 0.0300 Pa. Calculate the rms speed of free electrons (mass $= 9.11 \times 10^{-31}$ kg) there, assuming they are an ideal gas.

22E. (a) Compute the root-mean-square speed of a nitrogen molecule at 20.0°C. The molar mass of nitrogen molecules (N_2) is given in Table 20-1. At what temperatures will the root-mean-square speed be (b) half that value and (c) twice that value?

23P. A beam of hydrogen molecules (H_2) is directed toward a wall, at an angle of 55° with the normal to the wall. Each molecule in the beam has a speed of 1.0 km/s and a mass of 3.3×10^{-24} g. The beam strikes the wall over an area of 2.0 cm², at the rate of

10^{23} molecules per second. What is the beam's pressure on the wall? ssm

24P. At 273 K and 1.00×10^{-2} atm, the density of a gas is 1.24×10^{-5} g/cm³. (a) Find v_{rms} for the gas molecules. (b) Find the molar mass of the gas and identify the gas. (*Hint:* The gas is listed in Table 20-1.)

SEC. 20-5 Translational Kinetic Energy

25E. What is the average translational kinetic energy of nitrogen molecules at 1600 K? ssm

26E. Determine the average value of the translational kinetic energy of the molecules of an ideal gas at (a) 0.00°C and (b) 100°C. What is the translational kinetic energy per mole of an ideal gas at (c) 0.00°C and (d) 100°C?

27P. Water standing in the open at 32.0°C evaporates because of the escape of some of the surface molecules. The heat of vaporization (539 cal/g) is approximately equal to εn, where ε is the average energy of the escaping molecules and n is the number of molecules per gram. (a) Find ε. (b) What is the ratio of ε to the average kinetic energy of H_2O molecules, assuming the latter is related to temperature in the same way as it is for gases? ssm www

28P. Show that the ideal gas equation, Eq. 20-5, can be written in the alternative form $p = \rho RT/M$, where ρ is the mass density of the gas and M is the molar mass.

29P. *Avogadro's law* states that under the same conditions of temperature and pressure, equal volumes of gas contain equal numbers of molecules. Is this law equivalent to the ideal gas law? Explain. ssm

SEC. 20-6 Mean Free Path

30E. The mean free path of nitrogen molecules at 0.0°C and 1.0 atm is 0.80×10^{-5} cm. At this temperature and pressure there are 2.7×10^{19} molecules/cm³. What is the molecular diameter?

31E. At 2500 km above Earth's surface, the density of the atmosphere is about 1 molecule/cm³. (a) What mean free path is predicted by Eq. 20-25 and (b) what is its significance under these conditions? Assume a molecular diameter of 2.0×10^{-8} cm. ssm

32E. At what frequency would the wavelength of sound in air be equal to the mean free path of oxygen molecules at 1.0 atm pressure and 0.00°C? Take the diameter of an oxygen molecule to be 3.0×10^{-8} cm.

33E. What is the mean free path for 15 spherical jelly beans in a bag that is vigorously shaken? The volume of the bag is 1.0 L, and the diameter of a jelly bean is 1.0 cm. (Consider bean–bean collisions, not bean–bag collisions.) ssm

34P. At 20°C and 750 torr pressure, the mean free paths for argon gas (Ar) and nitrogen gas (N_2) are $\lambda_{Ar} = 9.9 \times 10^{-6}$ cm and $\lambda_{N_2} = 27.5 \times 10^{-6}$ cm. (a) Find the ratio of the effective diameter of argon to that of nitrogen. What is the mean free path of argon at (b) 20°C and 150 torr, and (c) −40°C and 750 torr?

35P. In a certain particle accelerator, protons travel around a circular path of diameter 23.0 m in an evacuated chamber, whose residual gas is at 295 K and 1.00×10^{-6} torr pressure. (a) Calculate the number of gas molecules per cubic centimeter at this pressure. (b) What is the mean free path of the gas molecules if the molecular diameter is 2.00×10^{-8} cm? ssm

SEC. 20-7 The Distribution of Molecular Speeds

36E. Twenty-two particles have speeds as follows (N_i represents the number of particles that have speed v_i):

N_i	2	4	6	8	2
v_i (cm/s)	1.0	2.0	3.0	4.0	5.0

(a) Compute their average speed v_{avg}. (b) Compute their root-mean-square speed v_{rms}. (c) Of the five speeds shown, which is the most probable speed v_P?

37E. The speeds of 10 molecules are 2.0, 3.0, 4.0, . . . , 11 km/s. (a) What is their average speed? (b) What is their root-mean-square speed? ssm

38E. (a) Ten particles are moving with the following speeds: four at 200 m/s, two at 500 m/s, and four at 600 m/s. Calculate their average and root-mean-square speeds. Is $v_{rms} > v_{avg}$? (b) Make up your own speed distribution for the 10 particles and show that $v_{rms} \geq v_{avg}$ for your distribution. (c) Under what condition (if any) does $v_{rms} = v_{avg}$?

39P. (a) Compute the temperatures at which the rms speed for (a) molecular hydrogen and (b) molecular oxygen is equal to the speed of escape from Earth. (c) Do the same for the speed of escape from the Moon, assuming the gravitational acceleration on its surface to be $0.16g$. (d) The temperature high in Earth's upper atmosphere is about 1000 K. Would you expect to find much hydrogen there? Much oxygen? Explain. ssm www

40P. It is found that the most probable speed of molecules in a gas when it has (uniform) temperature T_2 is the same as the rms speed of the molecules in this gas when it has (uniform) temperature T_1. Calculate T_2/T_1.

41P. A molecule of hydrogen (diameter 1.0×10^{-8} cm), traveling with the rms speed, escapes from a furnace ($T = 4000$ K) into a chamber containing atoms of *cold* argon (diameter 3.0×10^{-8} cm) at a density of 4.0×10^{19} atoms/cm³. (a) What is the speed of the hydrogen molecule? (b) If the H_2 molecule collides with an argon atom, what is the closest their centers can be, considering each as spherical? (c) What is the initial number of collisions per second experienced by the hydrogen molecule? (*Hint:* Assume that the cold argon atoms are stationary. Then the mean free path of the hydrogen molecule is given by Eq. 20-26 and not Eq. 20-25.) ssm

42P. Two containers are at the same temperature. The first contains gas with pressure p_1, molecular mass m_1, and root-mean-square speed v_{rms1}. The second contains gas with pressure $2p_1$, molecular mass m_2, and average speed $v_{avg2} = 2v_{rms1}$. Find the mass ratio m_1/m_2.

43P. Figure 20-22 shows a hypothetical speed distribution for a sample of N gas particles (note that $P(v) = 0$ for $v > 2v_0$). (a) Express a in terms of N and v_0. (b) How many of the particles have speeds between $1.5v_0$ and $2.0v_0$? (c) Express the average speed of the particles in terms of v_0. (d) Find v_{rms}. ssm

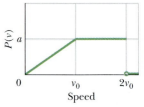

Fig. 20-22 Problem 43.

SEC. 20-8 The Molar Specific Heats of an Ideal Gas

44E. What is the internal energy of 1.0 mol of an ideal monatomic gas at 273 K?

45E. One mole of an ideal gas undergoes an isothermal expansion. Find the energy added to the gas as heat in terms of the initial and final volumes and the temperature. (*Hint:* Use the first law of thermodynamics.) ssm

46P. When 20.9 J was added as heat to a particular ideal gas, the volume of the gas changed from 50.0 cm^3 to 100 cm^3 while the pressure remained constant at 1.00 atm. (a) By how much did the internal energy of the gas change? If the quantity of gas present is 2.00×10^{-3} mol, find the molar specific heat of the gas at (b) constant pressure and (c) constant volume.

47P. A container holds a mixture of three nonreacting gases: n_1 moles of the first gas with molar specific heat at constant volume C_1, and so on. Find the molar specific heat at constant volume of the mixture, in terms of the molar specific heats and quantities of the separate gases. ssm

48P. One mole of an ideal diatomic gas goes from a to c along the diagonal path in Fig. 20-23. During the transition, (a) what is the change in internal energy of the gas, and (b) how much energy is added to the gas as heat? (c) How much heat is required if the gas goes from a to c along the indirect path abc?

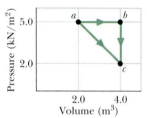

Fig. 20-23 Problem 48.

49P. The mass of a gas molecule can be computed from its specific heat at constant volume c_V. Take $c_V = 0.075$ cal/g·C° for argon and calculate (a) the mass of an argon atom and (b) the molar mass of argon. ilw

SEC. 20-9 Degrees of Freedom and Molar Specific Heats

50E. We give 70 J as heat to a diatomic gas, which then expands at constant pressure. The gas molecules rotate but do not oscillate. By how much does the internal energy of the gas increase?

51E. One mole of oxygen (O_2) is heated at constant pressure starting at 0°C. How much energy must be added to the gas as heat to double its volume? (The molecules rotate but do not oscillate.) ilw

52E. Suppose 12.0 g of oxygen (O_2) is heated at constant atmospheric pressure from 25.0°C to 125°C. (a) How many moles of oxygen are present? (See Table 20-1 for the molar mass.) (b) How much energy is transferred to the oxygen as heat? (The molecules rotate but do not oscillate.) (c) What fraction of the heat is used to raise the internal energy of the oxygen?

53P. Suppose 4.00 mol of an ideal diatomic gas, with molecular rotation but not oscillation, experienced a temperature increase of 60.0 K under constant-pressure conditions. (a) How much energy was transferred to the gas as heat? (b) How much did the internal energy of the gas increase? (c) How much work was done by the gas? (d) How much did the translational kinetic energy of the gas increase? ssm www

SEC. 20-11 The Adiabatic Expansion of an Ideal Gas

54E. (a) One liter of a gas with $\gamma = 1.3$ is at 273 K and 1.0 atm pressure. It is suddenly compressed adiabatically to half its original volume. Find its final pressure and temperature. (b) The gas is now cooled back to 273 K at constant pressure. What is its final volume?

55E. A certain gas occupies a volume of 4.3 L at a pressure of 1.2 atm and a temperature of 310 K. It is compressed adiabatically to a volume of 0.76 L. Determine (a) the final pressure and (b) the final temperature, assuming the gas to be an ideal gas for which $\gamma = 1.4$. ssm

56E. We know that for an adiabatic process $pV^\gamma = $ a constant. Evaluate "a constant" for an adiabatic process involving exactly 2.0 mol of an ideal gas passing through the state having exactly $p = 1.0$ atm and $T = 300$ K. Assume a diatomic gas whose molecules have rotation but not oscillation.

57E. Let n moles of an ideal gas expand adiabatically from an initial temperature T_1 to a final temperature T_2. Prove that the work done by the gas is $nC_V(T_1 - T_2)$, where C_V is the molar specific heat at constant volume. (*Hint:* Use the first law of thermodynamics.) ssm

58E. For adiabatic processes in an ideal gas, show that (a) the bulk modulus is given by

$$B = -V\frac{dp}{dV} = \gamma p,$$

and therefore (b) the speed of sound in the gas is

$$v_s = \sqrt{\frac{\gamma p}{\rho}} = \sqrt{\frac{\gamma RT}{M}}.$$

See Eqs. 18-2 and 18-3.

59E. Air at 0.000°C and 1.00 atm pressure has a density of 1.29×10^{-3} g/cm^3, and the speed of sound in air is 331 m/s at that temperature. Use those data to compute the ratio γ of the molar specific heats of air. (*Hint:* See Exercise 58.)

60P. (a) An ideal gas initially at pressure p_0 undergoes a free expansion until its volume is 3.00 times its initial volume. What then is its pressure? (b) The gas is next slowly and adiabatically compressed back to its original volume. The pressure after compression is $(3.00)^{1/3}p_0$. Is the gas monatomic, diatomic, or polyatomic? (c) How does the average kinetic energy per molecule in this final state compare with that in the initial state?

61P. One mole of an ideal monatomic gas traverses the cycle of Fig. 20-24. Process $1 \rightarrow 2$ occurs at constant volume, process $2 \rightarrow 3$ is adiabatic, and process $3 \rightarrow 1$ occurs at constant pressure. (a) Compute the heat Q, the change in internal energy ΔE_{int}, and the work done W, for each of the three processes and for the cycle as a whole. (b) The initial pressure at point 1 is 1.00 atm. Find the pressure and the volume at points 2 and 3. Use 1.00 atm = 1.013×10^5 Pa and $R = 8.314$ J/mol·K. ssm

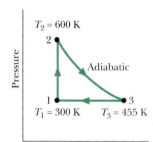

Fig. 20-24 Problem 61.

21 Entropy and the Second Law of Thermodynamics

An anonymous graffito on a wall of the Pecan Street Cafe in Austin, Texas, read: "Time is God's way of keeping things from happening all at once." Time also has direction—some things happen in a certain sequence and could never happen on their own in a reverse sequence. As an example, an accidentally dropped egg splatters in a cup. The reverse process, a splattered egg reforming into a whole egg and jumping up to an outstretched hand, will never happen on its own—but why not? Why can't that process be reversed, like a videotape run backward?

What in the world gives direction to time?

The answer is in this chapter.

21-1 Some One-Way Processes

Suppose you come indoors on a very cold day and wrap your cold hands around a warm mug of cocoa. Then your hands get warmer and the mug gets cooler. However, it never happens the other way around; that is, your cold hands never get still colder while the warm mug gets still warmer.

The system consisting of your hands and the mug is a *closed system,* one that is isolated from its environment. Here are some other one-way processes that occur in closed systems: (1) A crate sliding over an ordinary surface eventually stops—but you never see an initially stationary crate start to move all by itself. (2) If you drop a glob of putty, it falls to the floor—but an initially motionless glob of putty never leaps spontaneously into the air. (3) If you puncture a helium-filled balloon in a closed room, the helium gas spreads throughout the room—but the individual helium atoms will never clump up again into the shape of the balloon. We say that such one-way processes are **irreversible,** meaning that they cannot be reversed by means of only small changes in their environment.

The one-way character of such thermodynamic processes is so pervasive that we take it for granted. If these processes were to occur *spontaneously* (on their own) in the "wrong" direction, we would be astonished beyond belief. *Yet none of these "wrong-way" events would violate the law of conservation of energy.* In the cocoa mug example, that law would be obeyed even for a wrong-way transfer of energy as heat between hands and mug. It would be obeyed even if a stationary crate or a stationary glob of putty suddenly were to transfer some of its thermal energy to kinetic energy and begin to move. It would also be obeyed even if the helium atoms released from a balloon were, on their own, to clump together again.

Thus, changes in energy within a closed system do not set the direction of irreversible processes. Rather, that direction is set by another property that we shall discuss in this chapter—the *change in entropy* ΔS of the system. The change in entropy of a system is defined in the next section, but we can here state its central property, often called the *entropy postulate:*

> If an irreversible process occurs in a *closed* system, the entropy S of the system always increases; it never decreases.

Entropy differs from energy in that it does *not* obey a conservation law. The *energy* of a closed system is conserved; it always remains constant. For irreversible processes, the *entropy* of a closed system always increases. Because of this property, the change in entropy is sometimes called "the arrow of time." For example, we associate the egg of our opening photograph, breaking irreversibly as it drops into a cup, with the forward direction of time and with an increase in entropy. The backward direction of time (a videotape run backward) would correspond to the broken egg re-forming into a whole egg and rising into the air. This backward process would result in an entropy decrease, so it never happens.

There are two equivalent ways to define the change in entropy of a system: (1) in terms of the system's temperature and the energy it gains or loses as heat, and (2) by counting the ways in which the atoms or molecules that make up the system can be arranged. We use the first approach in the next section, and the second in Section 21-7.

21-2 Change in Entropy

Let's approach this definition of *change in entropy* by looking again at a process that we described in Sections 19-10 and 20-11: the free expansion of an ideal gas.

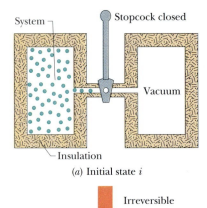

System

Stopcock closed

Vacuum

Insulation

(a) Initial state i

Irreversible process

Stopcock open

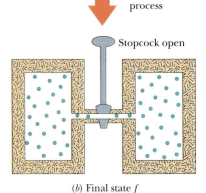

(b) Final state f

Fig. 21-1 The free expansion of an ideal gas. (a) The gas is confined to the left half of an insulated container by a closed stopcock. (b) When the stopcock is opened, the gas rushes to fill the entire container. This process is irreversible; that is, it does not occur in reverse, with the gas spontaneously collecting itself in the left half of the container.

Figure 21-1a shows the gas in its initial equilibrium state i, confined by a closed stopcock to the left half of a thermally insulated container. If we open the stopcock, the gas rushes to fill the entire container, eventually reaching the final equilibrium state f shown in Fig. 21-1b. This is an irreversible process; all the molecules of the gas will never return, by themselves, to the left half of the container.

The p-V plot of the process, in Fig. 21-2, shows the pressure and volume of the gas in its initial state i and final state f. Pressure and volume are *state properties,* properties that depend only on the state of the gas and not on how it reached that state. Other state properties are temperature and energy. We now assume that the gas has still another state property—its entropy. Furthermore, we define the **change in entropy** $S_f - S_i$ of a system during a process that takes the system from an initial state i to a final state f as

$$\Delta S = S_f - S_i = \int_i^f \frac{dQ}{T} \qquad \text{(change in entropy defined).} \qquad (21\text{-}1)$$

Here Q is the energy transferred as heat to or from the system during the process, and T is the temperature of the system in kelvins. Thus, an entropy change depends not only on the energy transferred as heat but also on the temperature at which the transfer takes place. Because T is always positive, the sign of ΔS is the same as that of Q. We see from Eq. 21-1 that the SI unit for entropy and entropy change is the joule per kelvin.

There is a problem, however, in applying Eq. 21-1 to the free expansion of Fig. 21-1. As the gas rushes to fill the entire container, the pressure, temperature, and volume of the gas fluctuate unpredictably. In other words, they do not have a sequence of well-defined equilibrium values during the intermediate stages of the change from initial equilibrium state i to final equilibrium state f. Thus, we cannot trace a pressure–volume path for the free expansion on the p-V plot of Fig. 21-2 and, more important, we cannot find a relation between Q and T that allows us to integrate as Eq. 21-1 requires.

However, if entropy is truly a state property, the difference in entropy between states i and f must depend *only on those states* and not at all on the way the system went from one state to the other. Suppose, then, that we replace the irreversible free expansion of Fig. 21-1 with a *reversible* process that connects states i and f. With a reversible process we can trace a pressure–volume path on a p-V plot, and we can find a relation between Q and T that allows us to use Eq. 21-1 to obtain the entropy change.

We saw in Section 20-11 that the temperature of an ideal gas does not change during a free expansion: $T_i = T_f = T$. Thus, points i and f in Fig. 21-2 must be on the same isotherm. A convenient replacement process is then a reversible isothermal expansion from state i to state f, which actually proceeds *along* that isotherm. Furthermore, because T is constant throughout a reversible isothermal expansion, the integral of Eq. 21-1 is greatly simplified.

Fig. 21-2 A p-V diagram showing the initial state i and the final state f of the free expansion of Fig. 21-1. The intermediate states of the gas cannot be shown because they are not equilibrium states.

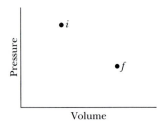

Pressure

Volume

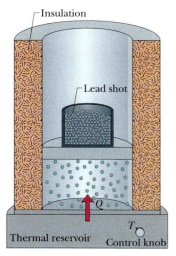

Insulation

Lead shot

Q

T

Thermal reservoir   Control knob

(a) Initial state i

Reversible process

Lead shot

T

(b) Final state f

Fig. 21-3 The isothermal expansion of an ideal gas, done in a reversible way. The gas has the same initial state i and same final state f as in the irreversible process of Figs. 21-1 and 21-2.

Figure 21-3 shows how to produce such a reversible isothermal expansion. We confine the gas to an insulated cylinder that rests on a thermal reservoir maintained at the temperature T. We begin by placing just enough lead shot on the movable piston so that the pressure and volume of the gas are those of the initial state i of Fig. 21-1a. We then remove shot slowly (piece by piece) until the pressure and volume of the gas are those of the final state f of Fig. 21-1b. The temperature of the gas does not change because the gas remains in thermal contact with the reservoir throughout the process.

The reversible isothermal expansion of Fig. 21-3 is physically quite different from the irreversible free expansion of Fig. 21-1. However, *both processes have the same initial state and the same final state and thus must have the same change in entropy*. Because we removed the lead shot slowly, the intermediate states of the gas are equilibrium states, so we can plot them on a p-V diagram (Fig. 21-4).

To apply Eq. 21-1 to the isothermal expansion, we take the constant temperature T outside the integral, obtaining

$$\Delta S = S_f - S_i = \frac{1}{T} \int_i^f dQ.$$

Because $\int dQ = Q$, where Q is the total energy transferred as heat during the process, we have

$$\Delta S = S_f - S_i = \frac{Q}{T} \qquad \text{(change in entropy, isothermal process)}. \qquad (21\text{-}2)$$

To keep the temperature T of the gas constant during the isothermal expansion of Fig. 21-3, heat Q must have been energy transferred *from* the reservoir *to* the gas. Thus, Q is positive and the entropy of the gas *increases* during the isothermal process and during the free expansion of Fig. 21-1.

To summarize:

> To find the entropy change for an irreversible process occurring in a *closed* system, replace that process with any reversible process that connects the same initial and final states. Calculate the entropy change for this reversible process with Eq. 21-1.

When the temperature change ΔT of a system is small relative to the temperature (in kelvins) before and after the process, the entropy change can be approximated as

$$\Delta S = S_f - S_i \approx \frac{Q}{T_{avg}}, \qquad (21\text{-}3)$$

where T_{avg} is the average temperature of the system in kelvins during the process.

✔CHECKPOINT 1:   Water is heated on a stove. Rank the entropy changes of the water as its temperature rises (a) from 20°C to 30°C, (b) from 30°C to 35°C, and (c) from 80°C to 85°C, greatest first.

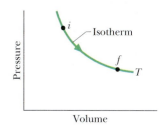

Fig. 21-4 A p-V diagram for the reversible isothermal expansion of Fig. 21-3. The intermediate states, which are now equilibrium states, are shown.

Sample Problem 21-1

One mole of nitrogen gas is confined to the left side of the container of Fig. 21-1a. You open the stopcock and the volume of the gas doubles. What is the entropy change of the gas for this irreversible process? Treat the gas as ideal.

SOLUTION: We need two **Key Ideas** here. One is that we can determine the entropy change for the irreversible process by calculating it for a reversible process that provides the same change in volume. The other is that the temperature of the gas does not change in the free expansion. Thus, the reversible process should be an isothermal expansion—namely, the one of Figs. 21-3 and 21-4.

From Table 20-4, the energy Q added as heat to the gas as it expands isothermally at temperature T from an initial volume V_i to a final volume V_f is

$$Q = nRT \ln \frac{V_f}{V_i},$$

in which n is the number of moles of gas present. From Eq. 21-2

the entropy change for this reversible process is

$$\Delta S_{rev} = \frac{Q}{T} = \frac{nRT \ln(V_f/V_i)}{T} = nR \ln \frac{V_f}{V_i}.$$

Substituting $n = 1.00$ mol and $V_f/V_i = 2$, we find

$$\Delta S_{rev} = nR \ln \frac{V_f}{V_i} = (1.00 \text{ mol})(8.31 \text{ J/mol} \cdot \text{K})(\ln 2)$$

$$= +5.76 \text{ J/K}.$$

Thus, the entropy change for the free expansion (and for all other processes that connect the initial and final states shown in Fig. 21-2) is

$$\Delta S_{irrev} = \Delta S_{rev} = +5.76 \text{ J/K}. \qquad \text{(Answer)}$$

ΔS is positive, so the entropy increases, in accordance with the entropy postulate of Section 21-1.

✔**CHECKPOINT 2:**   An ideal gas has temperature T_1 at the initial state i shown in the p-V diagram here. The gas has a higher temperature T_2 at final states a and b, which it can reach along the paths shown. Is the entropy change along the path to state a larger than, smaller than, or the same as that along the path to state b?

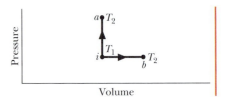

Sample Problem 21-2

Figure 21-5a shows two identical copper blocks of mass $m = 1.5$ kg: block L at temperature $T_{iL} = 60°C$ and block R at temperature $T_{iR} = 20°C$. The blocks are in a thermally insulated box and are separated by an insulating shutter. When we lift the shutter, the blocks eventually come to the equilibrium temperature $T_f = 40°C$ (Fig. 21-5b). What is the net entropy change of the two-block system during this irreversible process? The specific heat of copper is 386 J/kg · K.

SOLUTION: The **Key Idea** here is that to calculate the entropy change, we must find a reversible process that takes the system from the initial state of Fig. 21-5a to the final state of Fig. 21-5b. We can calculate the net entropy change ΔS_{rev} of the reversible process using Eq. 21-1, and then the entropy change for the irreversible process is equal to ΔS_{rev}. For such a reversible process we need a thermal reservoir whose temperature can be changed slowly (say, by turning a knob). We then take the blocks through the following two steps, illustrated in Fig. 21-6.

Step 1.   With the reservoir's temperature set at 60°C, put block L on the reservoir. (Since block and reservoir are at the same temperature, they are already in thermal equilibrium.) Then slowly lower the temperature of the reservoir and the block to 40°C. As the block's temperature changes by each increment dT during this process, energy dQ is transferred as heat *from* the block to the reservoir. Using Eq. 19-14, we can write this transferred energy as $dQ = mc \, dT$, where c is the specific heat of copper. According to Eq. 21-1, the entropy change

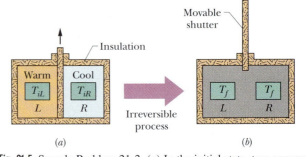

Fig. 21-5 Sample Problem 21-2. (*a*) In the initial state, two copper blocks L and R, identical except for their temperatures, are in an insulating box and are separated by an insulating shutter. (*b*) When the shutter is removed, the blocks exchange energy as heat and come to a final state, both with the same temperature T_f.

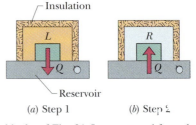

Fig. 21-6 The blocks of Fig. 21-5 can proceed from their initial state to their final state in a reversible way if we use a reservoir with a controllable temperature (*a*) to extract heat reversibly from block L and (*b*) to add heat reversibly to block R.

ΔS_L of block L during the full temperature change from initial temperature T_{iL} (= 60°C = 333 K) to final temperature T_f (= 40°C = 313 K) is

$$\Delta S_L = \int_i^f \frac{dQ}{T} = \int_{T_{iL}}^{T_f} \frac{mc\ dT}{T} = mc \int_{T_{iL}}^{T_f} \frac{dT}{T}$$

$$= mc \ln \frac{T_f}{T_{iL}}.$$

Inserting the given data yields

$$\Delta S_L = (1.5 \text{ kg})(386 \text{ J/kg} \cdot \text{K}) \ln \frac{313 \text{ K}}{333 \text{ K}}$$

$$= -35.86 \text{ J/K}.$$

Step 2.   With the reservoir's temperature now set at 20°C, put block R on the reservoir. Then slowly raise the temperature of the reservoir and the block to 40°C. With the same reasoning used to find ΔS_L, you can show that the entropy change ΔS_R of

block R during this process is

$$\Delta S_R = (1.5 \text{ kg})(386 \text{ J/kg} \cdot \text{K}) \ln \frac{313 \text{ K}}{293 \text{ K}}$$

$$= +38.23 \text{ J/K}.$$

The net entropy change ΔS_{rev} of the two-block system undergoing this two-step reversible process is then

$$\Delta S_{rev} = \Delta S_L + \Delta S_R$$

$$= -35.86 \text{ J/K} + 38.23 \text{ J/K} = 2.4 \text{ J/K}.$$

Thus, the net entropy change ΔS_{irrev} for the two-block system undergoing the actual irreversible process is

$$\Delta S_{irrev} = \Delta S_{rev} = 2.4 \text{ J/K}. \qquad \text{(Answer)}$$

This result is positive, in accordance with the entropy postulate of Section 21-1.

Entropy as a State Function

We have assumed that entropy, like pressure, energy, and temperature, is a property of the state of a system and is independent of how that state is reached. That entropy is indeed a *state function* (as state properties are usually called) can only be deduced by experiment. However, we can prove it is a state function for the special and important case in which an ideal gas is taken through a reversible process.

To make the process reversible, it is done slowly in a series of small steps, with the gas in an equilibrium state at the end of each step. For each small step, the energy transferred as heat to or from the gas is dQ, the work done by the gas is dW, and the change in internal energy is dE_{int}. These are related by the first law of thermodynamics in differential form (Eq. 19-27):

$$dE_{int} = dQ - dW.$$

Because the steps are reversible, with the gas in equilibrium states, we can use Eq. 19-24 to replace dW with $p\ dV$ and Eq. 20-45 to replace dE_{int} with $nC_V\ dT$. Solving for dQ then leads to

$$dQ = p\ dV + nC_V\ dT.$$

Using the ideal gas law, we replace p in this equation with nRT/V. Then we divide each term in the resulting equation by T, obtaining

$$\frac{dQ}{T} = nR\frac{dV}{V} + nC_V\frac{dT}{T}.$$

Now let us integrate each term of this equation between an arbitrary initial state i and an arbitrary final state f to get

$$\int_i^f \frac{dQ}{T} = \int_i^f nR\frac{dV}{V} + \int_i^f nC_V\frac{dT}{T}.$$

The quantity on the left is the entropy change ΔS (= $S_f - S_i$) defined by Eq. 21-1. Substituting this and integrating the quantities on the right yield

$$\Delta S = S_f - S_i = nR \ln \frac{V_f}{V_i} + nC_V \ln \frac{T_f}{T_i}. \qquad (21\text{-}4)$$

Note that we did not have to specify a particular reversible process when we integrated. Therefore, the integration must hold for all reversible processes that take the gas from state i to state f. Thus, the change in entropy ΔS between the initial and final states of an ideal gas depends only on properties of the initial state (V_i and T_i) and properties of the final state (V_f and T_f); ΔS does not depend on how the gas changes between the two states.

21-3 The Second Law of Thermodynamics

Here is a puzzle. We saw in Sample Problem 21-1 that if we cause the reversible process of Fig. 21-3 to proceed from (a) to (b) in that figure, the change in entropy of the gas—which we take as our system—is positive. However, because the process is reversible, we can just as easily make it proceed from (b) to (a), simply by slowly adding lead shot to the piston of Fig. 21-3b until the original volume of the gas is restored. In this reverse process, energy must be extracted as heat *from the gas* to keep its temperature from rising. Hence Q is negative and so, from Eq. 21-2, the entropy of the gas must decrease.

Doesn't this decrease in the entropy of the gas violate the entropy postulate of Section 21-1, which states that entropy always increases? No, because that postulate holds only for *irreversible* processes occurring in closed systems. The procedure suggested here does not meet these requirements. The process is *not* irreversible and (because energy is transferred as heat from the gas to the reservoir) the system—which is the gas alone—is *not* closed.

However, if we include the reservoir, along with the gas, as part of the system, then we do have a closed system. Let's check the change in entropy of the enlarged system *gas + reservoir* for the process that takes it from (b) to (a) in Fig. 21-3. During this reversible process, energy is transferred as heat from the gas to the reservoir—that is, from one part of the enlarged system to another. Let $|Q|$ represent the absolute value (or magnitude) of this heat. With Eq. 21-2, we can then calculate separately the entropy changes for the gas (which loses $|Q|$) and the reservoir (which gains $|Q|$). We get

$$\Delta S_{gas} = -\frac{|Q|}{T}$$

and

$$\Delta S_{res} = +\frac{|Q|}{T}.$$

The entropy change of the closed system is the sum of these two quantities, *which is zero*.

With this result, we can modify the entropy postulate of Section 21-1 to include both reversible and irreversible processes:

> If a process occurs in a *closed* system, the entropy of the system increases for irreversible processes and remains constant for reversible processes. It never decreases.

Although entropy may decrease in part of a closed system, there will always be an equal or larger entropy increase in another part of the system, so that the entropy of the system as a whole never decreases. This fact is one form of the **second law of thermodynamics** and can be written as

$$\Delta S \geq 0 \qquad \text{(second law of thermodynamics),} \qquad (21\text{-}5)$$

where the greater-than sign applies to irreversible processes, and the equals sign to reversible processes. Equation 21-5 applies only to closed systems.

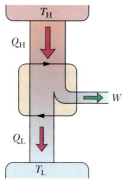

Fig. 21-7 The elements of an engine. The two black arrowheads on the central loop suggest the working substance operating in a cycle, as if on a p-V plot. Energy $|Q_H|$ is transferred as heat from the high-temperature reservoir at temperature T_H to the working substance. Energy $|Q_L|$ is transferred as heat from the working substance to the low-temperature reservoir at temperature T_L. Work W is done by the engine (actually by the working substance) on something in the environment.

In the real world almost all processes are irreversible to some extent because of friction, turbulence, and other factors, so the entropy of real closed systems undergoing real processes always increases. Processes in which the system's entropy remains constant are always idealizations.

21-4 Entropy in the Real World: Engines

A **heat engine,** or more simply, an **engine,** is a device that extracts energy from its environment in the form of heat and does useful work. At the heart of every engine is a *working substance*. In a steam engine, the working substance is water, in both its vapor and its liquid form. In an automobile engine the working substance is a gasoline–air mixture. If an engine is to do work on a sustained basis, the working substance must operate in a *cycle*; that is, the working substance must pass through a closed series of thermodynamic processes, called *strokes*, returning again and again to each state in its cycle. Let us see what the laws of thermodynamics can tell us about the operation of engines.

A Carnot Engine

We have seen that we can learn much about real gases by analyzing an ideal gas, which obeys the simple law $pV = nRT$. This is a useful plan because, although an ideal gas does not exist, any real gas approaches ideal behavior as closely as you wish if its density is low enough. In much the same spirit we choose to study real engines by analyzing the behavior of an **ideal engine.**

> In an ideal engine, all processes are reversible and no wasteful energy transfers occur due to, say, friction and turbulence.

We shall focus on a particular ideal engine called a **Carnot engine** after the French scientist and engineer N. L. Sadi Carnot (pronounced "car-no"), who first proposed the engine's concept in 1824. This ideal engine turns out to be the best (in principle) at using energy as heat to do useful work. Surprisingly, Carnot was able to analyze the performance of this engine before the first law of thermodynamics and the concept of entropy had been discovered.

Figure 21-7 shows schematically the operation of a Carnot engine. During each cycle of the engine, the working substance absorbs energy $|Q_H|$ as heat from a thermal reservoir at constant temperature T_H and discharges energy $|Q_L|$ as heat to a second thermal reservoir at a constant lower temperature T_L.

Figure 21-8 shows a p-V plot of the *Carnot cycle*—the cycle followed by the working substance. As indicated by the arrows, the cycle is traversed in the clockwise direction. Imagine the working substance to be a gas, confined to an insulating cylinder with a weighted, movable piston. The cylinder may be placed at will on either of the two thermal reservoirs, as in Fig. 21-3, or on an insulating slab. Figure 21-8 shows that, if we place the cylinder in contact with the high-temperature reservoir at temperature T_H, heat $|Q_H|$ is transferred *to* the working substance *from* this reservoir as the gas undergoes an isothermal *expansion* from volume V_a to volume V_b. Similarly, with the working substance in contact with the low-temperature reservoir at temperature T_L, heat $|Q_L|$ is transferred *from* the working substance *to* the low-temperature reservoir, as the gas undergoes an isothermal *compression* from volume V_c to volume V_d.

In the engine of Fig. 21-7, we assume that heat transfers to or from the working substance can take place *only* during the isothermal processes *ab* and *cd* of Fig. 21-8. Therefore, processes *bc* and *da* in that figure, which connect the two isotherms at temperatures T_H and T_L, must be (reversible) adiabatic processes; that

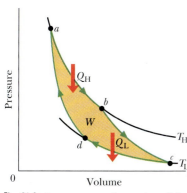

Fig. 21-8 A pressure–volume plot of the cycle followed by the working substance of the Carnot engine in Fig. 21-7. The cycle consists of two isotherm (*ab* and *cd*) and two adiabatic processes (*bc* and *da*). The shaded area enclosed by the cycle is equal to the work W per cycle done by the Carnot engine.

is, they must be processes in which no energy is transferred as heat. To ensure this, during processes *bc* and *da* the cylinder is placed on an insulating slab as the volume of the working substance is changed.

During the consecutive processes *ab* and *bc* of Fig. 21-8, the working substance is expanding and thus doing positive work as it raises the weighted piston. This work is represented in Fig. 21-8 by the area under curve *abc*. During the consecutive processes *cd* and *da*, the working substance is being compressed, which means that it is doing negative work on its environment or, equivalently, that its environment is doing work on it as the loaded piston descends. This work is represented by the area under curve *cda*. The *net work per cycle,* which is represented by *W* in both Figs. 21-7 and 21-8, is the difference between these two areas and is a positive quantity equal to the area enclosed by cycle *abcda* in Fig. 21-8. This work *W* is performed on some outside object, such as a load to be lifted.

Equation 21-1 ($\Delta S = \int dQ/T$) tells us that any energy transfer as heat must involve a change in entropy. To illustrate the entropy changes for a Carnot engine, we can plot the Carnot cycle on a temperature-entropy (*T-S*) diagram as shown in Fig. 21-9. The lettered points *a*, *b*, *c*, and *d* in Fig. 21-9 correspond to the lettered points in the *p-V* diagram in Fig. 21-8. The two horizontal lines in Fig. 21-9 correspond to the two isothermal processes of the Carnot cycle (because the temperature is constant). Process *ab* is the isothermal expansion of the cycle. As the working substance (reversibly) absorbs energy $|Q_H|$ as heat at constant temperature T_H during the expansion, its entropy increases. Similarly, during the isothermal compression *cd*, the working substance (reversibly) loses energy $|Q_L|$ as heat at constant temperature T_L, and its entropy decreases.

The two vertical lines in Fig. 21-9 correspond to the two adiabatic processes of the Carnot cycle. Because no energy is transferred as heat during the two processes, the entropy of the working substance is constant during them.

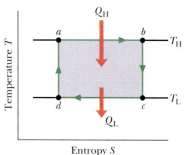

Fig. 21-9 The Carnot cycle of Fig. 21-8 plotted on a temperature–entropy diagram. During processes *ab* and *cd* the temperature remains constant. During processes *bc* and *da* the entropy remains constant.

The Work: To calculate the net work done by a Carnot engine during a cycle, let us apply Eq. 19-26, the first law of thermodynamics ($\Delta E_{int} = Q - W$), to the working substance. That substance must return again and again to any arbitrarily selected state in the cycle. Thus, if *X* represents any state property of the working substance, such as pressure, temperature, volume, internal energy, or entropy, we must have $\Delta X = 0$ for every cycle. It follows that $\Delta E_{int} = 0$ for a complete cycle of the working substance. Recalling that *Q* in Eq. 19-26 is the *net* heat transfer per cycle and *W* is the *net* work, we can write the first law of thermodynamics for the Carnot cycle as

$$W = |Q_H| - |Q_L|. \tag{21-6}$$

Entropy Changes: In a Carnot engine, there are *two* (and only two) reversible energy transfers as heat, and thus two changes in the entropy of the working substance— one at temperature T_H and one at T_L. The net entropy change per cycle is then

$$\Delta S = \Delta S_H + \Delta S_L = \frac{|Q_H|}{T_H} - \frac{|Q_L|}{T_L}. \tag{21-7}$$

Here ΔS_H is positive because energy $|Q_H|$ is *added to* the working substance as heat (an increase in entropy) and ΔS_L is negative because energy $|Q_L|$ is *removed from* the working substance as heat (a decrease in entropy). Because entropy is a state function, we must have $\Delta S = 0$ for a complete cycle. Putting $\Delta S = 0$ in Eq. 21-7 requires that

$$\frac{|Q_H|}{T_H} = \frac{|Q_L|}{T_L}. \tag{21-8}$$

Note that, because $T_H > T_L$, we must have $|Q_H| > |Q_L|$; that is, more energy is extracted as heat from the high-temperature reservoir than is delivered to the low-temperature reservoir.

We shall now use Eqs. 21-6 and 21-8 to derive an expression for the efficiency of a Carnot engine.

Efficiency of a Carnot Engine

The purpose of any engine is to transform as much of the extracted energy Q_H into work as possible. We measure its success in doing so by its **thermal efficiency** ε, defined as the work the engine does per cycle ("energy we get") divided by the energy it absorbs as heat per cycle ("energy we pay for"):

$$\varepsilon = \frac{\text{energy we get}}{\text{energy we pay for}} = \frac{|W|}{|Q_H|} \qquad \text{(efficiency, any engine).} \qquad (21\text{-}9)$$

For a Carnot engine we can substitute for W from Eq. 21-6 to write Eq. 21-9 as

$$\varepsilon_C = \frac{|Q_H| - |Q_L|}{|Q_H|} = 1 - \frac{|Q_L|}{|Q_H|}. \qquad (21\text{-}10)$$

Using Eq. 21-8 we can write this as

$$\varepsilon_C = 1 - \frac{T_L}{T_H} \qquad \text{(efficiency, Carnot engine),} \qquad (21\text{-}11)$$

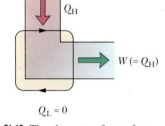

Fig. 21-10 The elements of a perfect engine—that is, one that converts heat Q_H from a high-temperature reservoir directly to work W with 100% efficiency.

where the temperatures T_L and T_H are in kelvins. Because $T_L < T_H$, the Carnot engine necessarily has a thermal efficiency less than unity—that is, less than 100%. This is indicated in Fig. 21-7, which shows that only part of the energy extracted as heat from the high-temperature reservoir is available to do work, and the rest is delivered to the low-temperature reservoir. We will show in Section 21-6 that no real engine can have a thermal efficiency greater than that calculated from Eq. 21-11.

Inventors continually try to improve engine efficiency by reducing the energy $|Q_L|$ that is "thrown away" during each cycle. The inventor's dream is to produce the *perfect engine*, diagrammed in Fig. 21-10, in which $|Q_L|$ is reduced to zero and $|Q_H|$ is converted completely into work. Such an engine on an ocean liner, for example, could extract energy as heat from the water and use it to drive the propellers, with no fuel cost. An automobile, fitted with such an engine, could extract energy as heat from the surrounding air and use it to drive the car, again with no fuel cost. Alas, a perfect engine is only a dream: Inspection of Eq. 21-11 shows that we can achieve 100% engine efficiency (that is, $\varepsilon = 1$) only if $T_L = 0$ or $T_H \to \infty$, requirements that are impossible to meet. Instead, decades of practical engineering experience have led to the following alternative version of the second law of thermodynamics:

> No series of processes is possible whose sole result is the transfer of energy as heat from a thermal reservoir and the complete conversion of this energy to work.

In short, *there are no perfect engines.*

To summarize: The thermal efficiency given by Eq. 21-11 applies only to Carnot engines. Real engines, in which the processes that form the engine cycle are not reversible, have lower efficiencies. If your car were powered by a Carnot engine, it

Fig. 21-11 The North Anna nuclear power plant near Charlottesville, Virginia, which generates electric energy at the rate of 900 MW. At the same time, by design, it discards energy into the nearby river at the rate of 2100 MW. This plant—and all others like it—throws away more energy than it delivers in useful form. It is a real counterpart to the ideal engine of Fig. 21-7.

would have an efficiency of about 55% according to Eq. 21-11; its actual efficiency is probably about 25%. A nuclear power plant (Fig. 21-11), taken in its entirety, is an engine. It extracts energy as heat from a reactor core, does work by means of a turbine, and discharges energy as heat to a nearby river. If the power plant operated as a Carnot engine, its efficiency would be about 40%; its actual efficiency is about 30%. In designing engines of any type, there is simply no way to beat the efficiency limitation imposed by Eq. 21-11.

Stirling Engine

Equation 21-11 does not apply to all ideal engines, but only to engines that can be represented as in Fig. 21-8—that is, to Carnot engines. For example, Fig. 21-12 shows the operating cycle of an ideal **Stirling engine.** Comparison with the Carnot cycle of Fig. 21-8 shows that each engine has isothermal heat transfers at temperatures T_H and T_L. However, the two isotherms of the Stirling engine cycle are connected, not by adiabatic processes as for the Carnot engine, but by constant-volume processes (Fig. 21-12). To increase the temperature of a gas at constant volume reversibly from T_L to T_H (process da of Fig. 21-12) requires a transfer of energy as heat to the working substance from a thermal reservoir whose temperature can be varied smoothly between those limits. Also, a reverse transfer is required in process bc. Thus, reversible heat transfers (and corresponding entropy changes) occur in all four of the processes that form the cycle of a Stirling engine, not just two processes as in a Carnot engine. Thus, the derivation that led to Eq. 21-11 does not apply to an ideal Stirling engine. More important, the efficiency of an ideal Stirling engine is lower than that of a Carnot engine operating between the same two temperatures. Real Stirling engines have even lower efficiencies.

The Stirling engine was developed in 1816 by Robert Stirling. This engine, long neglected, is now being developed for use in automobiles and spacecraft. A Stirling engine delivering 5000 hp (3.7 MW) has been built.

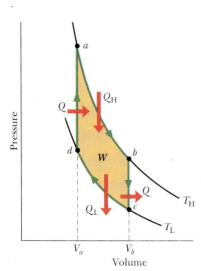

Fig. 21-12 A p-V plot for the working substance of an ideal Stirling engine, assumed for convenience to be an ideal gas.

✓**CHECKPOINT 3:**   Three Carnot engines operate between reservoir temperatures of (a) 400 and 500 K, (b) 600 and 800 K, and (c) 400 and 600 K. Rank the engines according to their thermal efficiencies, greatest first.

Sample Problem 21-3

Imagine a Carnot engine that operates between the temperatures $T_H = 850$ K and $T_L = 300$ K. The engine performs 1200 J of work each cycle, which takes 0.25 s.

(a) What is the efficiency of this engine?

SOLUTION: The **Key Idea** here is that the efficiency ε of a Carnot engine depends only on the ratio T_L/T_H of the temperatures (in kelvins) of the thermal reservoirs to which it is connected. Thus, from Eq. 21-11, we have

$$\varepsilon = 1 - \frac{T_L}{T_{II}} = 1 - \frac{300 \text{ K}}{850 \text{ K}} = 0.647 \approx 65\%. \quad \text{(Answer)}$$

(b) What is the average power of this engine?

SOLUTION: Here the Key Idea is that the average power P of an engine is the ratio of the work W it does per cycle to the time t that each cycle takes. For this Carnot engine, we find

$$P = \frac{W}{t} = \frac{1200 \text{ J}}{0.25 \text{ s}} = 4800 \text{ W} = 4.8 \text{ kW}. \quad \text{(Answer)}$$

(c) How much energy $|Q_H|$ is extracted as heat from the high-temperature reservoir every cycle?

SOLUTION: Now the Key Idea is that for any engine, including a Carnot engine, the efficiency ε is the ratio of the work W that is done per cycle to the energy $|Q_H|$ that is extracted as heat from the high-temperature reservoir per cycle ($\varepsilon = W/|Q_H|$). Thus,

$$|Q_H| = \frac{W}{\varepsilon} = \frac{1200 \text{ J}}{0.647} = 1855 \text{ J}. \quad \text{(Answer)}$$

(d) How much energy $|Q_L|$ is delivered as heat to the low-temperature reservoir every cycle?

SOLUTION: The Key Idea here is that for a Carnot engine, the work W done per cycle is equal to the difference in the energy transfers as heat: $|Q_H| - |Q_L|$, as in Eq. 21-6. Thus, we have

$$|Q_L| = |Q_H| - W$$
$$= 1855 \text{ J} - 1200 \text{ J} = 655 \text{ J}. \quad \text{(Answer)}$$

(e) What is the entropy change of the working substance for the energy transfer to it from the high-temperature reservoir? From it to the low-temperature reservoir?

SOLUTION: The Key Idea here is that the entropy change ΔS during a transfer of energy Q as heat at constant temperature T is given by Eq. 21-2 ($\Delta S = Q/T$). Thus, for the *positive* transfer of energy Q_H from the high-temperature reservoir at T_H, the change in the entropy of the working substance is

$$\Delta S_H = \frac{Q_H}{T_H} = \frac{1855 \text{ J}}{850 \text{ K}} = +2.18 \text{ J/K}. \quad \text{(Answer)}$$

Similarly, for the *negative* transfer of energy Q_L to the low-temperature reservoir at T_L, we have

$$\Delta S_L = \frac{Q_L}{T_L} = \frac{-655 \text{ J}}{300 \text{ K}} = -2.18 \text{ J/K}. \quad \text{(Answer)}$$

Note that the net entropy change of the working substance for one cycle is zero, as we discussed in deriving Eq. 21-8..

Sample Problem 21-4

An inventor claims to have constructed an engine that has an efficiency of 75% when operated between the boiling and freezing points of water. Is this possible?

SOLUTION: The Key Idea here is that the efficiency of a real engine (with its irreversible processes and wasteful energy transfers) must be less than the efficiency of a Carnot engine operating between the same two temperatures. From Eq. 21-11, we find that the effi-

ciency of a Carnot engine operating between the boiling and freezing points of water is

$$\varepsilon = 1 - \frac{T_L}{T_H} = 1 - \frac{(0 + 273) \text{ K}}{(100 + 273) \text{ K}} = 0.268 \approx 27\%.$$

Thus, the claimed efficiency of 75% for a real engine operating between the given temperatures is impossible.

PROBLEM-SOLVING TACTICS

Tactic 1: *The Language of Thermodynamics*
A rich, but sometimes misleading, language is used in scientific and engineering studies of thermodynamics. You may see statements that say heat is added, absorbed, subtracted, extracted, rejected, discharged, discarded, withdrawn, delivered, gained, lost, transferred, or expelled, or that it flows from one body to another (as if it were a liquid). You may also see statements that describe a body as *having* heat (as if heat can be held or possessed), or that its heat is increased or decreased. You should always keep in mind what is meant by the term *heat:*

▶ Heat is energy that is transferred from one body to another body owing to a difference in the temperatures of the bodies.

When we identify one of the bodies as being our system of interest, any such transfer of energy into the system is positive heat Q, and any such transfer out of the system is negative heat Q.

The term *work* also requires close attention. You may see statements that say work is produced or generated, or combined with heat or changed from heat. Here is what is meant by the term *work:*

▶ Work is energy that is transferred from one body to another body owing to a force that acts between them.

When we identify one of the bodies as being our system of interest, any such transfer of energy out of the system is either positive work W done *by* the system or negative work W done *on* the system. Any such transfer of energy into the system is negative work done *by* the system or positive work done *on* the system. (The preposition that is used is important.) Obviously, this can be confusing—whenever you see the term *work,* you should read carefully to determine the intent.

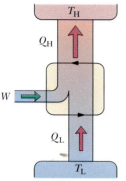

Fig. 21-13 The elements of a refrigerator. The two black arrowheads on the central loop suggest the working substance operating in a cycle, as if on a p-V plot. Energy Q_L is transferred as heat to the working substance from the low-temperature reservoir. Energy Q_H is transferred as heat to the high-temperature reservoir from the working substance. Work W is done on the refrigerator (on the working substance) by something in the environment.

21-5 Entropy in the Real World: Refrigerators

A **refrigerator** is a device that uses work to transfer energy from a low-temperature reservoir to a high-temperature reservoir as it continuously repeats a set series of thermodynamic processes. In a household refrigerator, for example, work is done by an electrical compressor to transfer energy from the food storage compartment (a low-temperature reservoir) to the room (a high-temperature reservoir).

Air conditioners and heat pumps are also refrigerators. The differences are only in the nature of the high- and low-temperature reservoirs. For an air conditioner, the low-temperature reservoir is the room that is to be cooled, and the high-temperature reservoir is the (presumably warmer) outdoors. A heat pump is an air conditioner that can be operated in reverse to heat a room; the room is the high-temperature reservoir and heat is transferred to it from the (presumably cooler) outdoors.

Let us consider an *ideal refrigerator:*

> In an ideal refrigerator, all processes are reversible and no wasteful energy transfers occur due to, say, friction and turbulence.

Figure 21-13 shows the basic elements of an ideal refrigerator that operates in the reverse of the Carnot engine of Fig. 21-7. In other words, all the energy transfers, as either heat or work, are reversed from those of a Carnot engine. We can call such an ideal refrigerator a **Carnot refrigerator.**

The designer of a refrigerator would like to extract as much energy $|Q_L|$ as possible from the low-temperature reservoir (what we want) for the least amount of work $|W|$ (what we pay for). A measure of the efficiency of a refrigerator, then, is

$$K = \frac{\text{what we want}}{\text{what we pay for}} = \frac{|Q_L|}{|W|} \quad \begin{array}{c}\text{(coefficient of performance,}\\\text{any refrigerator),}\end{array} \quad (21\text{-}12)$$

where K is called the *coefficient of performance.* For a Carnot refrigerator, the first law of thermodynamics gives $|W| = |Q_H| - |Q_L|$, where $|Q_H|$ is the magnitude of the energy transferred as heat to the high-temperature reservoir. Equation 21-12 then becomes

$$K_C = \frac{|Q_L|}{|Q_H| - |Q_L|}. \quad (21\text{-}13)$$

Because a Carnot refrigerator is a Carnot engine operating in reverse, we can combine Eq. 21-8 with Eq. 21-13; after some algebra we find

$$K_C = \frac{T_L}{T_H - T_L} \quad \begin{array}{c}\text{(coefficient of performance,}\\\text{Carnot refrigerator).}\end{array} \quad (21\text{-}14)$$

For typical room air conditioners, $K \approx 2.5$. For household refrigerators, $K \approx 5$. Perversely, the value of K is higher the closer the temperatures of the two reservoirs are to each other. That is why heat pumps are more effective in temperate climates than in climates where the outside temperature varies widely.

It would be nice to own a refrigerator that did not require some input of work— that is, one that would run without being plugged in. Figure 21-14 represents another "inventor's dream," a *perfect refrigerator* that transfers heat Q from a cold reservoir to a warm reservoir without the need for work. Because the unit operates in cycles, the entropy of the working substance does not change during a complete cycle. The

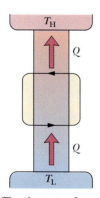

Fig. 21-14 The elements of a perfect refrigerator—that is, one that transfers energy from a low-temperature reservoir to a high-temperature reservoir without any input of work.

entropies of the two reservoirs, however, do change: The entropy change for the cold reservoir is $-|Q|/T_L$, and that for the warm reservoir is $+|Q|/T_H$. Thus, the net entropy change for the entire system is

$$\Delta S = -\frac{|Q|}{T_L} + \frac{|Q|}{T_H}.$$

Because $T_H > T_L$, the right side of this equation is negative and thus the net change in entropy per cycle for the closed system *refrigerator + reservoirs* is also negative. Because such a decrease in entropy violates the second law of thermodynamics (Eq. 21-5), a perfect refrigerator does not exist. (If you want your refrigerator to operate, you must plug it in.)

This result leads us to another (equivalent) formulation of the second law of thermodynamics:

> No series of processes is possible whose sole result is the transfer of energy as heat from a reservoir at a given temperature to a reservoir at a higher temperature.

In short, *there are no perfect refrigerators.*

✓CHECKPOINT 4:   You wish to increase the coefficient of performance of an ideal refrigerator. You can do so by (a) running the cold chamber at a slightly higher temperature, (b) running the cold chamber at a slightly lower temperature, (c) moving the unit to a slightly warmer room, or (d) moving it to a slightly cooler room. The magnitudes of the temperature changes are to be the same in all four cases. List the changes according to the resulting coefficients of performance, greatest first.

21-6 The Efficiencies of Real Engines

Let ε_C be the efficiency of a Carnot engine operating between two given temperatures. In this section we prove that no real engine operating between those temperatures can have an efficiency greater than ε_C. If it could, the engine would violate the second law of thermodynamics.

Let us assume that an inventor, working in her garage, has constructed an engine X, which she claims has an efficiency ε_X that is greater than ε_C:

$$\varepsilon_X > \varepsilon_C \qquad \text{(a claim)}. \qquad (21\text{-}15)$$

Let us couple engine X to a Carnot refrigerator, as in Fig. 21-15a. We adjust the strokes of the Carnot refrigerator so that the work it requires per cycle is just equal

Fig. 21-15 (a) Engine X drives a Carnot refrigerator. (b) If, as claimed, engine X is more efficient than a Carnot engine, then the combination shown in (a) is equivalent to the perfect refrigerator shown here. This violates the second law of thermodynamics, so we conclude that engine X *cannot* be more efficient than a Carnot engine.

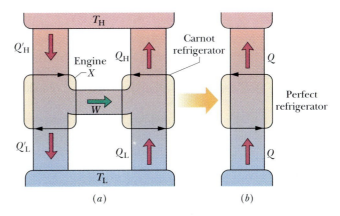

to that provided by engine X. Thus, no (external) work is performed on or by the combination *engine + refrigerator* of Fig. 21-15a, which we take as our system.

If Eq. 21-15 is true, from the definition of efficiency (Eq. 21-9), we must have

$$\frac{|W|}{|Q'_H|} > \frac{|W|}{|Q_H|},$$

where the prime refers to engine X and the right side of the inequality is the efficiency of the Carnot refrigerator when it operates as an engine. This inequality requires that

$$|Q_H| > |Q'_H|. \qquad (21\text{-}16)$$

Because the work done by engine X is equal to the work done on the Carnot refrigerator, we have, from the first law of thermodynamics (see Eq. 21-6),

$$|Q_H| - |Q_L| = |Q'_H| - |Q'_L|,$$

which we can write as

$$|Q_H| - |Q'_H| = |Q_L| - |Q'_L| = Q. \qquad (21\text{-}17)$$

Because of Eq. 21-16, the quantity Q in Eq. 21-17 must be positive.

Comparison of Eq. 21-17 with Fig. 21-15 shows that the net effect of engine X and the Carnot refrigerator, working in combination, is to transfer energy Q as heat from a low-temperature reservoir to a high-temperature reservoir without the requirement of work. Thus, the combination acts like the perfect refrigerator of Fig. 21-14, whose existence is a violation of the second law of thermodynamics.

Something must be wrong with one or more of our assumptions, and it can only be Eq. 21-15. We conclude that *no real engine can have an efficiency greater than that of a Carnot engine when both engines work between the same two temperatures.* At most, it can have an efficiency equal to that of a Carnot engine. In that case, engine X *is* a Carnot engine.

21-7 A Statistical View of Entropy

In Chapter 20 we saw that the macroscopic properties of gases can be explained in terms of their microscopic, or molecular, behavior. For one example, recall that we were able to account for the pressure exerted by a gas on the walls of its container in terms of the momentum transferred to those walls by rebounding gas molecules. Such explanations are part of a study called **statistical mechanics.**

Here we shall focus our attention on a single problem, involving the distribution of gas molecules between the two halves of an insulated box. This problem is reasonably simple to analyze, and it allows us to use statistical mechanics to calculate the entropy change for the free expansion of an ideal gas. You will see in Sample Problem 21-6 that statistical mechanics leads to the same entropy change we obtained in Sample Problem 21-1 using thermodynamics.

Figure 21-16 shows a box that contains six identical (and thus indistinguishable) molecules of a gas. At any instant, a given molecule will be in either the left or the right half of the box; because the two halves have equal volumes, the molecule has the same likelihood, or probability, of being in either half.

Table 21-1 shows four of the seven possible *configurations* of the six molecules, each configuration labeled with a Roman numeral. For example, in configuration I, all six molecules are in the left half of the box ($n_1 = 6$) and none are in the right half ($n_2 = 0$). The three configurations not shown are V with a (2, 4) split, VI with a (1, 5) split, and VII with a (0, 6) split. We see that, in general, a given configuration can be achieved in a number of different ways. We call these different

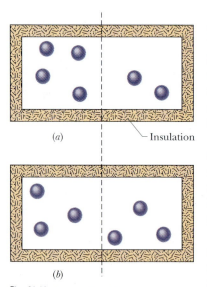

Fig. 21-16 An insulated box contains six gas molecules. Each molecule has the same probability of being in the left half of the box as in the right half. The arrangement in (a) corresponds to configuration III in Table 21-1, and that in (b) corresponds to configuration IV.

Table 21-1 Six Molecules in a Box

Configuration			Multiplicity W (number of microstates)	Calculation of W (Eq. 21-18)	Entropy 10^{-23} J/K (Eq. 21-19)
Label	n_1	n_2			
I	6	0	1	$6!/(6!\,0!) = 1$	0
II	5	1	6	$6!/(5!\,1!) = 6$	2.47
III	4	2	15	$6!/(4!\,2!) = 15$	3.74
IV	3	3	20	$6!/(3!\,3!) = 20$	4.13

Total number
 of microstates = 64

arrangements of the molecules *microstates*. Let us see how to calculate the number of microstates that correspond to a given configuration.

Suppose we have N molecules, distributed with n_1 molecules in one half of the box and n_2 in the other. (Thus $n_1 + n_2 = N$.) Let us imagine that we distribute the molecules "by hand," one at a time. If $N = 6$, we can select the first molecule in six independent ways; that is, we can pick any one of the six molecules. We can pick the second molecule in five ways, by picking any one of the remaining five molecules; and so on. The total number of ways in which we can select all six molecules is the product of these independent ways, or $6 \times 5 \times 4 \times 3 \times 2 \times 1 = 720$. In mathematical shorthand we write this product as $6! = 720$, where $6!$ is pronounced "six factorial." Your hand calculator can probably calculate factorials. For later use you will need to know that $0! = 1$. (Check this on your calculator.)

However, because the molecules are indistinguishable, these 720 arrangements are not all different. In the case that $n_1 = 4$ and $n_2 = 2$ (which is configuration III in Table 21-1), for example, the order in which you put four molecules in one half of the box does not matter, because after you have put all four in, there is no way that you can tell the order in which you did so. The number of ways in which you can order the four molecules is $4!$ or 24. Similarly, the number of ways in which you can order two molecules for the other half of the box is simply $2!$ or 2. To get the number of *different* arrangements that lead to the (4, 2) split of configuration III, we must divide 720 by 24 and also by 2. We call the resulting quantity, which is the number of microstates that correspond to a given configuration, the *multiplicity* W of that configuration. Thus, for configuration III,

$$W_{\text{III}} = \frac{6!}{4!\,2!} = \frac{720}{24 \times 2} = 15.$$

Thus, Table 21-1 tells us there are 15 independent microstates that correspond to configuration III. Note that, as the table also tells us, the total number of microstates for six molecules distributed over the seven configurations is 64.

Extrapolating from six molecules to the general case of N molecules, we have

$$W = \frac{N!}{n_1!\,n_2!} \qquad \text{(multiplicity of configuration).} \qquad (21\text{-}18)$$

You should verify that Eq. 21-18 gives the multiplicities for all the configurations listed in Table 21-1.

The basic assumption of statistical mechanics is:

▶ All microstates are equally probable.

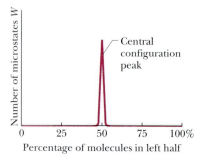

Fig. 21-17 For a *large* number of molecules in a box, a plot of the number of microstates that require various percentages of the molecules to be in the left half of the box. Nearly all the microstates correspond to an approximately equal sharing of the molecules between the two halves of the box; those microstates form the *central configuration peak* on the plot. For $N \approx 10^{22}$, the central configuration peak is much too narrow to be drawn on this plot.

In other words, if we were to take a great many snapshots of the six molecules as they jostle around in the box of Fig. 21-16 and then count the number of times each microstate occurred, we would find that all 64 microstates would occur equally often. In other words, the system will spend, on average, the same amount of time in each of the 64 microstates.

Because the microstates are equally probable, but different configurations have different numbers of microstates, the configurations are *not* equally probable. In Table 21-1 configuration IV, with 20 microstates, is the *most probable configuration,* with a probability of 20/64 = 0.313. This result means that the system is in configuration IV 31.3% of the time. Configurations I and VII, in which all the molecules are in one half of the box, are the least probable, each with a probability of 1/64 = 0.016 or 1.6%. It is not surprising that the most probable configuration is the one in which the molecules are evenly divided between the two halves of the box, because that is what we expect at thermal equilibrium. However, it *is* surprising that there is *any* probability, however small, of finding all six molecules clustered in half of the box, with the other half empty. In Sample Problem 21-5 we show that this state can occur because six molecules is an extremely small number.

For large values of N there are extremely large numbers of microstates, but nearly all the microstates belong to the configuration in which the molecules are divided equally between the two halves of the box, as Fig. 21-17 indicates. Even though the measured temperature and pressure of the gas remain constant, the gas is churning away endlessly as its molecules "visit" all probable microstates with equal probability. However, because so few microstates lie outside the very narrow central configuration peak of Fig. 21-17, we might as well assume that the gas molecules are always divided equally between the two halves of the box. As we shall see, this is the configuration with the greatest entropy.

Sample Problem 21-5

Suppose that there are 100 indistinguishable molecules in the box of Fig. 21-16. How many microstates are associated with the configuration $n_1 = 50$ and $n_2 = 50$? How many are associated with the configuration $n_1 = 100$ and $n_2 = 0$? Interpret the results in terms of the relative probabilities of the two configurations.

SOLUTION: The Key Idea here is that the multiplicity W of a configuration of indistinguishable molecules in a closed box is the number of independent microstates with that configuration, as given by Eq. 21-18. For the (n_1, n_2) configuration (50, 50), that equation yields

$$W = \frac{N!}{n_1! \, n_2!} = \frac{100!}{50! \, 50!}$$
$$= \frac{9.33 \times 10^{157}}{(3.04 \times 10^{64})(3.04 \times 10^{64})}$$
$$= 1.01 \times 10^{29}. \qquad \text{(Answer)}$$

Similarly, for the configuration of (100, 0), we have

$$W = \frac{N!}{n_1! \, n_2!} = \frac{100!}{100! \, 0!} = \frac{1}{0!} = \frac{1}{1} = 1. \qquad \text{(Answer)}$$

Thus, a 50–50 distribution is more likely than a 100–0 distribution by the enormous factor of about 1×10^{29}. If you could count, at one per nanosecond, the number of microstates that correspond to the 50–50 distribution, it would take you about 3×10^{12} years, which is about 750 times longer than the age of the universe. Even 100 molecules is *still* a very small number. Imagine what these calculated probabilities would be like for a mole of molecules, say about $N = 10^{24}$. Thus, you need never worry about suddenly finding all the air molecules clustering in one corner of your room!

Probability and Entropy

In 1877, Austrian physicist Ludwig Boltzmann (the Boltzmann of Boltzmann's constant k) derived a relationship between the entropy S of a configuration of a gas and the multiplicity W of that configuration. That relationship is

$$S = k \ln W \qquad \text{(Boltzmann's entropy equation).} \qquad (21\text{-}19)$$

This famous formula is engraved on Boltzmann's tombstone.

It is natural that S and W should be related by a logarithmic function. The total entropy of two systems is the *sum* of their separate entropies. The probability of occurrence of two independent systems is the *product* of their separate probabilities. Because $\ln ab = \ln a + \ln b$, the logarithm seems the logical way to connect these quantities.

Table 21-1 displays the entropies of the configurations of the six-molecule system of Fig. 21-16, computed using Eq. 21-19. Configuration IV, which has the greatest multiplicity, also has the greatest entropy.

When you use Eq. 21-18 to calculate W, your calculator may signal "OVERFLOW" if you try to find the factorial of a number greater than a few hundred. Fortunately, there is a very good approximation, known as **Stirling's approximation,** not for $N!$ but for $\ln N!$, which is exactly what is needed in Eq. 21-19. Stirling's approximation is

$$\ln N! \approx N(\ln N) - N \qquad \text{(Stirling's approximation).} \qquad (21\text{-}20)$$

The Stirling of this approximation is not the Stirling of the Stirling engine.

CHECKPOINT 5: A box contains one mole of a gas. Consider two configurations: (a) each half of the box contains half the molecules, and (b) each third of the box contains one-third of the molecules. Which configuration has more microstates?

Sample Problem 21-6

In Sample Problem 21-1 we showed that when n moles of an ideal gas doubles its volume in a free expansion, the entropy increase from the initial state i to the final state f is $S_f - S_i = nR \ln 2$. Derive this result with statistical mechanics.

SOLUTION: One **Key Idea** here is that we can relate the entropy S of any given configuration of the molecules in the gas to the multiplicity W of microstates for that configuration, using Eq. 21-19 ($S = k \ln W$). We are interested in two configurations: the final configuration f (with the molecules occupying the full volume of their container in Fig. 21-1b) and the initial configuration i (with the molecules occupying the left half of the container).

A second **Key Idea** is that, because the molecules are in a closed container, we can calculate the multiplicity W of their microstates with Eq. 21-18. Here we have N molecules in the n moles of the gas. Initially, with the molecules all in the left half of the container, their (n_1, n_2) configuration is $(N, 0)$. Then, Eq. 21-18 gives their multiplicity as

$$W_i = \frac{N!}{N!\ 0!} = 1.$$

Finally, with the molecules spread through the full volume, their (n_1, n_2) configuration is $(N/2, N/2)$. Then, Eq. 21-18 gives their multiplicity as

$$W_f = \frac{N!}{(N/2)!\ (N/2)!}.$$

From Eq. 21-19, the initial and final entropies are

$$S_i = k \ln W_i = k \ln 1 = 0$$

and

$$S_f = k \ln W_f = k \ln(N!) - 2k \ln[(N/2)!]. \qquad (21\text{-}21)$$

In writing Eq. 21-21, we have used the relation

$$\ln \frac{a}{b^2} = \ln a - 2 \ln b.$$

Now, applying Eq. 21-20 to evaluate Eq. 21-21, we find that

$$\begin{aligned} S_f &= k \ln(N!) - 2k \ln[(N/2)!] \\ &= k[N(\ln N) - N] - 2k[(N/2) \ln(N/2) - (N/2)] \\ &= k[N(\ln N) - N - N \ln(N/2) + N] \\ &= k[N(\ln N) - N(\ln N - \ln 2)] = Nk \ln 2. \qquad (21\text{-}22) \end{aligned}$$

From Section 20-3 we can substitute nR for Nk, where R is the universal gas constant. Equation 21-22 then becomes

$$S_f = nR \ln 2.$$

The change in entropy from the initial state to the final is thus

$$\begin{aligned} S_f - S_i &= nR \ln 2 - 0 \\ &= nR \ln 2, \qquad \text{(Answer)} \end{aligned}$$

which is what we set out to show. In Sample Problem 21-1 we calculated this entropy increase for a free expansion with thermodynamics by finding an equivalent reversible process and calculating the entropy change for *that* process in terms of temperature and heat transfer. In this sample problem, we calculate the same increase in entropy with statistical mechanics using the fact that the system consists of molecules.

REVIEW & SUMMARY

One-Way Processes   An **irreversible process** is one that cannot be reversed by means of small changes in the environment. The direction in which an irreversible process proceeds is set by the *change in entropy* ΔS of the system undergoing the process. Entropy S is a *state property* (or *state function*) of the system; that is, it depends only on the state of the system and not on how the system reached that state. The *entropy postulate* states (in part): *If an irreversible process occurs in a closed system, the entropy of the system always increases.*

Calculating Entropy Change   The **entropy change** ΔS for an irreversible process that takes a system from an initial state i to a final state f is exactly equal to the entropy change ΔS for *any reversible process* that takes the system between those same two states. We can compute the latter (but not the former) with

$$\Delta S = S_f - S_i = \int_i^f \frac{dQ}{T}. \tag{21-1}$$

Here Q is the energy transferred as heat to or from the system during the process, and T is the temperature of the system in kelvins during the process.

For a reversible isothermal process, Eq. 21-1 reduces to

$$\Delta S = S_f - S_i = \frac{Q}{T}. \tag{21-2}$$

When the temperature change ΔT of a system is small relative to the temperature (in kelvins) before and after the process, the entropy change can be approximated as

$$\Delta S = S_f - S_i \approx \frac{Q}{T_{avg}}, \tag{21-3}$$

where T_{avg} is the system's average temperature during the process.

When an ideal gas changes reversibly from an initial state with temperature T_i and volume V_i to a final state with temperature T_f and volume V_f, the change ΔS in the entropy of the gas is

$$\Delta S = S_f - S_i = nR \ln \frac{V_f}{V_i} + nC_V \ln \frac{T_f}{T_i}. \tag{21-4}$$

The Second Law of Thermodynamics   This law, which is an extension of the entropy postulate, states: *If a process occurs in a closed system, the entropy of the system increases for irreversible processes and remains constant for reversible processes. It never decreases.* In equation form,

$$\Delta S \geq 0. \tag{21-5}$$

Engines   An **engine** is a device that, operating in a cycle, extracts energy $|Q_H|$ as heat from a high-temperature reservoir and does a certain amount of work $|W|$. The *efficiency* ε of any engine is defined as

$$\varepsilon = \frac{\text{energy we get}}{\text{energy we pay for}} = \frac{|W|}{|Q_H|}. \tag{21-9}$$

In an **ideal engine,** all processes are reversible and no wasteful energy transfers occur due to, say, friction and turbulence. A **Carnot engine** is an ideal engine that follows the cycle of Fig. 21-8.

Its efficiency is

$$\varepsilon_C = 1 - \frac{|Q_L|}{|Q_H|} = 1 - \frac{T_L}{T_H}, \tag{21-10, 21-11}$$

in which T_H and T_L are the temperatures of the high- and low-temperature reservoirs, respectively. Real engines always have an efficiency lower than that given by Eq. 21-11. Ideal engines that are not Carnot engines also have efficiencies lower than that given by Eq. 21-11.

A *perfect engine* is an imaginary engine in which energy extracted as heat from the high-temperature reservoir is converted completely to work. Such would violate the second law of thermodynamics, which can be restated as follows: No series of processes is possible whose sole result is the absorption of energy as heat from a thermal reservoir and the complete conversion of this energy to work.

Refrigerators   A refrigerator is a device that, operating in a cycle, has work W done on it as it extracts energy $|Q_L|$ as heat from a low-temperature reservoir. The coefficient of performance K of a refrigerator is defined as

$$K = \frac{\text{what we want}}{\text{what we pay for}} = \frac{|Q_L|}{|W|}. \tag{21-12}$$

A **Carnot refrigerator** is a Carnot engine operating in reverse. For a Carnot refrigerator, Eq. 21-12 becomes

$$K_C = \frac{|Q_L|}{|Q_H| - |Q_L|} = \frac{T_L}{T_H - T_L}. \tag{21-13, 21-14}$$

A *perfect refrigerator* is an imaginary refrigerator in which energy extracted as heat from the low-temperature reservoir is converted completely to heat discharged to the high-temperature reservoir, without any need for work. Such would violate the second law of thermodynamics, which can be restated as follows: No series of processes is possible whose sole result is the transfer of energy as heat from a reservoir at a given temperature to a reservoir at a higher temperature.

Entropy from a Statistical View   The entropy of a system can be defined in terms of the possible distributions of its molecules. For identical molecules, each possible distribution of molecules is called a **microstate** of the system. All equivalent microstates are grouped into a **configuration** of the system. The number of microstates in a configuration is the **multiplicity** W of the configuration.

For a system of N molecules that may be distributed between the two halves of a box, the multiplicity is given by

$$W = \frac{N!}{n_1! \, n_2!}, \tag{21-18}$$

in which n_1 is the number of molecules in one half of the box and n_2 is the number in the other half. A basic assumption of *statistical mechanics* is that all the microstates are equally probable. Thus, configurations with a large multiplicity occur most often. When N is very large (say, $N = 10^{22}$ molecules or more), the molecules are nearly always in the configuration in which $n_1 = n_2$.

The multiplicity W of a configuration of a system and the entropy S of the system in that configuration are related by Boltzmann's entropy equation:

$$S = k \ln W, \qquad (21\text{-}19)$$

where $k = 1.38 \times 10^{-23}$ J/K is the Boltzmann constant.

When N is very large (the usual case), we can approximate $\ln N!$ with *Stirling's approximation:*

$$\ln N! \approx N(\ln N) - N. \qquad (21\text{-}20)$$

QUESTIONS

1. A gas, confined to an insulated cylinder, is compressed adiabatically to half its volume. Does the entropy of the gas increase, decrease, or remain unchanged during this process?

2. In four experiments, blocks A and B, starting at different initial temperatures, were brought together in an insulating box (as in Sample Problem 21-2) and allowed to reach a common final temperature. The entropy changes for the blocks in the four experiments had the following values (in Joules per kelvin), but not necessarily in the order given. Determine which values for A go with which values for B.

Block		Values		
A	8	5	3	9
B	-3	-8	-5	-2

3. Point i in Fig. 21-18 represents the initial state of an ideal gas at temperature T. Taking algebraic signs into account, rank the entropy changes that the gas undergoes as it moves, successively and reversibly, from point i to points a, b, c, and d, greatest first.

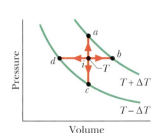

Fig. 21-18 Question 3.

4. An ideal gas, in contact with a controllable thermal reservoir, can be taken from initial state i to final state f along the four reversible paths in Fig. 21-19. Rank the paths according to the magnitudes of the resulting entropy changes of (a) the gas, (b) the reservoir, and (c) the gas–reservoir system, greatest first.

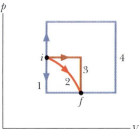

Fig. 21-19 Question 4.

5. You allow a gas to expand freely from volume V to volume $2V$. Later you allow that gas to expand freely from volume $2V$ to volume $3V$. Is the net entropy change for these two expansions greater than, less than, or equal to the entropy change that would occur if you allowed the gas to expand freely from volume V directly to volume $3V$?

6. Three Carnot engines operate between temperature limits of (a) 400 and 500 K, (b) 500 and 600 K, and (c) 400 and 600 K. Each engine extracts the same amount of energy per cycle from the high-temperature reservoir. Rank the magnitudes of the work done by the engines per cycle, greatest first.

7. Does the entropy per cycle increase, decrease, or remain the same for (a) a Carnot engine, (b) a real engine, and (c) a perfect engine (which is, of course, impossible to build)?

8. If you leave the door of your kitchen refrigerator open for several hours, does the temperature of the kitchen increase, decrease, or remain the same? Assume that the kitchen is closed and well insulated.

9. Does the entropy per cycle increase, decrease, or remain the same for (a) a Carnot refrigerator, (b) a real refrigerator, and (c) a perfect refrigerator (which is, of course, impossible to build)?

10. A box contains 100 atoms in a configuration, with 50 atoms in each half of the box. Suppose that you could count the different microstates associated with this configuration at the rate of 100 billion states per second, using a supercomputer. Without written calculation, guess how much computing time you would need: a day, a year, or much more than a year.

11. Figure 21-20 shows a snapshot at time $t = 0$ of molecules a and b in a box (similar to that of Fig. 21-16). The molecules have the same mass and speed v, and collisions between the molecules and the walls are elastic. What is the probability that snapshots taken at times (a) $t = 0.10L/v$ and (b) $t = 10L/v$ will show that a is in the left side of the box and b is in the right side of the box? (c) What is the probability that at some later time only half of the kinetic energy of the molecules will be in the right side of the box?

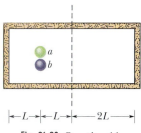

Fig. 21-20 Question 11.

EXERCISES & PROBLEMS

SEC. 21-2 Change in Entropy

1E. A 2.50 mol sample of an ideal gas expands reversibly and isothermally at 360 K until its volume is doubled. What is the increase in entropy of the gas? **ilw**

2E. How much heat is required for a reversible isothermal expansion of an ideal gas at 132°C if the entropy of the gas increases by 46.0 J/K?

3E. Four moles of an ideal gas undergo a reversible isothermal expansion from volume V_1 to volume $V_2 = 2V_1$ at temperature $T = 400$ K. Find (a) the work done by the gas and (b) the entropy change of the gas. (c) If the expansion is reversible and adiabatic instead of isothermal, what is the entropy change of the gas? **ssm**

4E. An ideal gas undergoes a reversible isothermal expansion at 77.0°C, increasing its volume from 1.30 L to 3.40 L. The entropy change of the gas is 22.0 J/K. How many moles of gas are present?

5E. Find (a) the energy absorbed as heat and (b) the change in entropy of a 2.00 kg block of copper whose temperature is increased reversibly from 25°C to 100°C. The specific heat of copper is 386 J/kg · K. **ilw**

6E. An ideal monatomic gas at initial temperature T_0 (in kelvins) expands from initial volume V_0 to volume $2V_0$ by each of the five

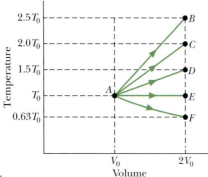

Fig. 21-21 Exercise 6.

processes indicated in the *T-V* diagram of Fig. 21-21. In which process is the expansion (a) isothermal, (b) isobaric (constant pressure), and (c) adiabatic? Explain your answers. (d) In which processes does the entropy of the gas decrease?

7E. (a) What is the entropy change of a 12.0 g ice cube that melts completely in a bucket of water whose temperature is just above the freezing point of water? (b) What is the entropy change of a 5.00 g spoonful of water that evaporates completely on a hot plate whose temperature is slightly above the boiling point of water?

8P. A 2.0 mol sample of an ideal monatomic gas undergoes the reversible process shown in Fig. 21-22. (a) How much energy is absorbed as heat by the gas? (b) What is the change in the internal energy of the gas? (c) How much work is done by the gas?

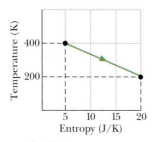

Fig. 21-22 Problem 8.

9P. In an experiment, 200 g of aluminum (with a specific heat of 900 J/kg · K) at 100°C is mixed with 50.0 g of water at 20.0°C, with the mixture thermally isolated. (a) What is the equilibrium temperature? What are the entropy changes of (b) the aluminum, (c) the water, and (d) the aluminum–water system? **ssm** **www**

10P. In the irreversible process of Fig. 21-5, let the initial temp-

atures of identical blocks *L* and *R* be 305.5 and 294.5 K, respectively, and let 215 J be the energy transfer between the blocks required to reach equilibrium. Then for the reversible processes of Fig. 21-6, what are the entropy changes of (a) block *L*, (b) its reservoir, (c) block *R*, (d) its reservoir, (e) the two-block system, and (f) the system of the two blocks and the two reservoirs?

11P. Use the reversible apparatus of Fig. 21-6 to show that, if the process of Fig. 21-5 happened in reverse, the entropy of the system would decrease, a violation of the second law of thermodynamics.

12P. An ideal diatomic gas, whose molecules are rotating but not oscillating, is taken through the cycle in Fig. 21-23. Determine for all three processes, in terms of p_1, V_1, T_1, and *R*: (a) p_2, p_3, and T_3 and (b) *W*, *Q*, ΔE_{int}, and ΔS per mole.

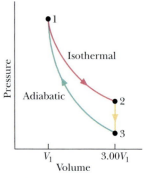

Fig. 21-23 Problem 12.

13P. A 50.0 g block of copper whose temperature is 400 K is placed in an insulating box with a 100 g block of lead whose temperature is 200 K. (a) What is the equilibrium temperature of the two-block system? (b) What is the change in the internal energy of the two-block system between the initial state and the equilibrium state? (c) What is the change in the entropy of the two-block system? (See Table 19-3.) **ilw**

14P. One mole of a monatomic ideal gas is taken from an initial pressure *p* and volume *V* to a final pressure 2*p* and volume 2*V* by two different processes: (I) It expands isothermally until its volume is doubled, and then its pressure is increased at constant volume to the final pressure. (II) It is compressed isothermally until its pressure is doubled, and then its volume is increased at constant pressure to the final volume. (a) Show the path of each process on a *p-V* diagram. For each process calculate, in terms of *p* and *V*, (b) the energy absorbed by the gas as heat in each part of the process, (c) the work done by the gas in each part of the process, (d) the change in internal energy of the gas, $E_{int,f} - E_{int,i}$, and (e) the change in entropy of the gas, $S_f - S_i$.

15P. A 10 g ice cube at $-10°C$ is placed in a lake whose temperature is 15°C. Calculate the change in entropy of the cube–lake system as the ice cube comes to thermal equilibrium with the lake. The specific heat of ice is 2220 J/kg · K. (*Hint:* Will the ice cube affect the temperature of the lake?) **ssm**

16P. An 8.0 g ice cube at $-10°C$ is put into a Thermos flask containing 100 cm³ of water at 20°C. By how much has the entropy of the cube–water system changed when a final equilibrium state is reached? The specific heat of ice is 2220 J/kg · K.

17P. A mixture of 1773 g of water and 227 g of ice is in an initial equilibrium state at 0.00°C. The mixture is then, in a reversible process, brought to a second equilibrium state where the water–ice ratio, by mass, is 1:1 at 0.00°C. (a) Calculate the entropy change of the system during this process. (The heat of fusion for water is 333 kJ/kg.) (b) The system is then returned to the initial equilibrium state in an irreversible process (say, by using a Bunsen burner). Calculate the entropy change of the system during this process. (c) Are your answers consistent with the second law of thermodynamics? **ssm**

18P. A cylinder contains n moles of a monatomic ideal gas. If the gas undergoes a reversible isothermal expansion from initial volume V_i to final volume V_f along path I in Fig. 21-24, its change in entropy is $\Delta S = nR \ln (V_f/V_i)$. (See Sample Problem 21-1.) Now consider path II in Fig. 21-24, which takes the gas from the same initial state i to state x by a reversible adiabatic expansion, and then from that state x to the same final state f by a reversible constant-volume process. (a) Describe how you would carry out the two reversible processes for path II. (b) Show that the temperature of the gas in state x is

$$T_x = T_i(V_i/V_f)^{2/3}.$$

(c) What are the energy Q_I transferred as heat along path I and the energy Q_{II} transferred as heat along path II? Are they equal? (d) What is the entropy change ΔS for path II? Is the entropy change for path I equal to it? (e) Evaluate T_x, Q_I, Q_{II}, and ΔS for $n = 1$, $T_i = 500$ K, and $V_f/V_i = 2$.

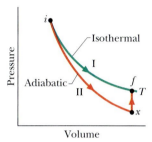

Fig. 21-24 Problem 18.

19P. One mole of an ideal monatomic gas is taken through the cycle in Fig. 21-25. (a) How much work is done by the gas in going from state a to state c along path abc? What are the changes in internal energy and entropy in going (b) from b to c and (c) through one complete cycle? Express all answers in terms of the pressure p_0, volume V_0, and temperature T_0 of state a. **ssm**

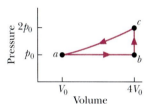

Fig. 21-25 Problem 19.

20P. One mole of an ideal monatomic gas, at an initial pressure of 5.00 kPa and initial temperature of 600 K, expands from initial volume $V_i = 1.00$ m³ to final volume $V_f = 2.00$ m³. During the expansion, the pressure p and volume V of the gas are related by $p = 5.00 \exp[(V_i - V)/a]$, where p is in kilopascals, V_i and V are in cubic meters, and $a = 1.00$ m³. What are (a) the final pressure and (b) the final temperature of the gas? (c) How much work is done by the gas during the expansion? (d) What is the change in entropy of the gas during the expansion? (*Hint:* Use two simple reversible processes to find the entropy change.)

SEC. 21-4   Entropy in the Real World: Engines

21E. A Carnot engine absorbs 52 kJ as heat and exhausts 36 kJ as heat in each cycle. Calculate (a) the engine's efficiency and (b) the work done per cycle in kilojoules.

22E. A Carnot engine whose low-temperature reservoir is at 17°C has an efficiency of 40%. By how much should the temperature of the high-temperature reservoir be increased to increase the efficiency to 50%?

23E. A Carnot engine operates between 235°C and 115°C, absorbing 6.30×10^4 J per cycle at the higher temperature. (a) What is the efficiency of the engine? (b) How much work per cycle is this engine capable of performing? **ssm**

24E. In a hypothetical nuclear fusion reactor, the fuel is deuterium gas at a temperature of about 7×10^8 K. If this gas could be used to operate a Carnot engine with $T_L = 100°C$, what would be the engine's efficiency?

25E. A Carnot engine has an efficiency of 22.0%. It operates between constant-temperature reservoirs differing in temperature by 75.0 C°. What are the temperatures of the two reservoirs? **ssm**

26P. A Carnot engine has a power of 500 W. It operates between constant-temperature reservoirs at 100°C and 60.0°C. What are (a) the rate of heat input and (b) the rate of exhaust heat output, in kilojoules per second?

27P. One mole of a monatomic ideal gas is taken through the reversible cycle shown in Fig. 21-26. Process bc is an adiabatic expansion, with $p_b = 10.0$ atm and $V_b = 1.00 \times 10^{-3}$ m³. Find (a) the energy added to the gas as heat, (b) the energy leaving the gas as heat, (c) the net work done by the gas, and (d) the efficiency of the cycle. **ssm** **ilw**

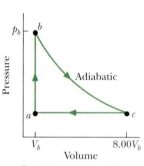

Fig. 21-26 Problem 27.

28P. Show that the area enclosed by the Carnot cycle on the temperature-entropy plot of Fig. 21-9 represents the net energy transfer per cycle as heat to the working substance.

29P. One mole of an ideal monatomic gas is taken through the cycle shown in Fig. 21-27. Assume that $p = 2p_0$, $V = 2V_0$, $p_0 = 1.01 \times 10^5$ Pa, and $V_0 = 0.0225$ m³. Calculate (a) the work done during the cycle, (b) the energy added as heat during stroke abc, and (c) the efficiency of the cycle. (d) What is the efficiency of a Carnot engine operating between the highest and lowest temperatures that occur in the cycle? How does this compare to the efficiency calculated in (c)?

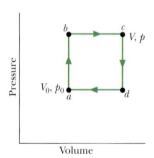

Fig. 21-27 Problem 29.

30P. In the first stage of a two-stage Carnot engine, energy Q_1 is absorbed as heat at temperature T_1, work W_1 is done, and energy Q_2 is expelled as heat at a lower temperature T_2. The second stage absorbs that energy Q_2, does work W_2, and expels energy Q_3 at a still lower temperature T_3. Prove that the efficiency of the two-stage engine is $(T_1 - T_3)/T_1$.

31P. Suppose that a deep shaft were drilled in Earth's crust near one of the poles, where the surface temperature is $-40°C$, to a depth where the temperature is 800°C. (a) What is the theoretical limit to the efficiency of an engine operating between these temperatures? (b) If all the energy released as heat into the low-temperature reservoir were used to melt ice that was initially at $-40°C$, at what rate could liquid water at 0°C be produced by a 100 MW power plant (treat it as an engine)? The specific heat of ice is 2220 J/kg · K; water's heat of fusion is 333 kJ/kg. (Note that the engine can operate only between 0°C and 800°C in this case. Energy exhausted at $-40°C$ cannot be used to raise the temperature of anything above $-40°C$.) **ssm** **www**

32P. One mole of an ideal gas is used as the working substance of an engine that operates on the cycle shown in Fig. 21-28. *BC* and *DA* are reversible adiabatic processes. (a) Is the gas monatomic, diatomic, or polyatomic? (b) What is the efficiency of the engine?

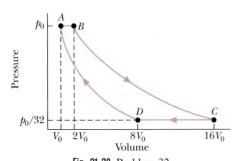

Fig. 21-28 Problem 32.

33P. The operation of a gasoline internal combustion engine is represented by the cycle in Fig. 21-29. Assume the gasoline–air intake mixture is an ideal gas and use a compression ratio of 4:1 ($V_4 = 4V_1$). Assume that $p_2 = 3p_1$. (a) Determine the pressure and temperature at each of the vertex points of the *p-V* diagram in terms of p_1, T_1, and the ratio γ of the molar specific heats of the gas. (b) What is the efficiency of the cycle? ssm

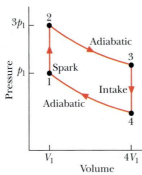

Fig. 21-29 Problem 33.

SEC. 21-5 Entropy in the Real World: Refrigerators

34E. A Carnot refrigerator does 200 J of work to remove 600 J from its cold compartment. (a) What is the refrigerator's coefficient of performance? (b) How much energy per cycle is exhausted to the kitchen as heat?

35E. A Carnot air conditioner takes energy from the thermal energy of a room at 70°F and transfers it to the outdoors, which is at 96°F. For each joule of electric energy required to operate the air conditioner, how many joules are removed from the room? ssm

36E. The electric motor of a heat pump transfers energy as heat from the outdoors, which is at −5.0°C, to a room, which is at 17°C. If the heat pump were a Carnot heat pump (a Carnot engine working in reverse), how many joules of heat would be transferred to the thermal energy of the room for each joule of electric energy consumed?

37E. A heat pump is used to heat a building. The outside temperature is −5.0°C, and the temperature inside the building is to be maintained at 22°C. The pump's coefficient of performance is 3.8, and the heat pump delivers 7.54 MJ as heat to the building each hour. If the heat pump is a Carnot engine working in reverse, at what rate must work be done to run the heat pump? ssm

38E. How much work must be done by a Carnot refrigerator to transfer 1.0 J as heat (a) from a reservoir at 7.0°C to one at 27°C, (b) from a reservoir at −73°C to one at 27°C, (c) from a reservoir

at −173°C to one at 27°C, and (d) from a reservoir at −223°C to one at 27°C?

39P. An air conditioner operating between 93°F and 70°F is rated at 4000 Btu/h cooling capacity. Its coefficient of performance is 27% of that of a Carnot refrigerator operating between the same two temperatures. What horsepower is required of the air conditioner motor? ilw

40P. The motor in a refrigerator has a power of 200 W. If the freezing compartment is at 270 K and the outside air is at 300 K, and assuming the efficiency of a Carnot refrigerator, what is the maximum amount of energy that can be extracted as heat from the freezing compartment in 10.0 min?

41P. A Carnot engine works between temperatures T_1 and T_2. It drives a Carnot refrigerator that works between temperatures T_3 and T_4 (Fig. 21-30). Find the ratio Q_3/Q_1 in terms of T_1, T_2, T_3, and T_4. ssm

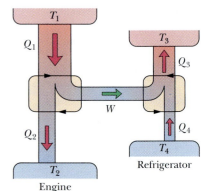

Fig. 21-30 Problem 41.

SEC. 21-7 A Statistical View of Entropy

42E. Construct a table like Table 21-1 for eight molecules.

43E. Show that for *N* molecules in a box, the number of possible microstates is 2^N when microstates are defined by whether a given molecule is in the left half of the box or the right half. Check this for the situation of Table 21-1.

44P. A box contains *N* gas molecules, equally divided between its two halves. For $N = 50$: (a) What is the multiplicity of this central configuration? (b) What is the total number of microstates for the system? (*Hint:* See Exercise 43.) (c) What percentage of the time does the system spend in its central configuration? (d) Repeat (a) through (c) for $N = 100$. (e) Repeat (a) through (c) for $N = 200$. (f) As *N* increases, you will find that the system spends *less* time (not more) in its central configuration. Explain why this is so.

45P. A box contains *N* gas molecules. Consider the box to be divided into three equal parts. (a) By extension of Eq. 21-18, write a formula for the multiplicity of any given configuration. (b) Consider two configurations: configuration *A* with equal numbers of molecules in all three thirds of the box, and configuration *B* with equal numbers of molecules in both halves of the box. What is the ratio W_A/W_B of the multiplicity of configuration *A* to that of configuration *B*? (c) Evaluate W_A/W_B for $N = 100$. (Because 100 is not evenly divisible by 3, put 34 molecules into one of the three box parts and 33 in each of the other parts for configuration *A*.) ssm www

The International System of Units (SI)*

1. The SI Base Units

Quantity	Name	Symbol	Definition
length	meter	m	". . . the length of the path traveled by light in vacuum in 1/299,792,458 of a second." (1983)
mass	kilogram	kg	". . . this prototype [a certain platinum–iridium cylinder] shall henceforth be considered to be the unit of mass." (1889)
time	second	s	". . . the duration of 9,192,631,770 periods of the radiation corresponding to the transition between the two hyperfine levels of the ground state of the cesium-133 atom." (1967)
electric current	ampere	A	". . . that constant current which, if maintained in two straight parallel conductors of infinite length, of negligible circular cross section, and placed 1 meter apart in vacuum, would produce between these conductors a force equal to 2×10^{-7} newton per meter of length." (1946)
thermodynamic temperature	kelvin	K	". . . the fraction 1/273.16 of the thermodynamic temperature of the triple point of water." (1967)
amount of substance	mole	mol	". . . the amount of substance of a system which contains as many elementary entities as there are atoms in 0.012 kilogram of carbon-12." (1971)
luminous intensity	candela	cd	". . . the luminous intensity, in the perpendicular direction, of a surface of 1/600,000 square meter of a blackbody at the temperature of freezing platinum under a pressure of 101.325 newtons per square meter." (1967)

*Adapted from "The International System of Units (SI)," National Bureau of Standards Special Publication 330, 1972 edition. The definitions above were adopted by the General Conference of Weights and Measures, an international body, on the dates shown. In this book we do not use the candela.

2.   Some SI Derived Units

Quantity	Name of Unit	Symbol	
area	square meter	m^2	
volume	cubic meter	m^3	
frequency	hertz	Hz	s^{-1}
mass density (density)	kilogram per cubic meter	kg/m^3	
speed, velocity	meter per second	m/s	
angular velocity	radian per second	rad/s	
acceleration	meter per second per second	m/s^2	
angular acceleration	radian per second per second	rad/s^2	
force	newton	N	$kg \cdot m/s^2$
pressure	pascal	Pa	N/m^2
work, energy, quantity of heat	joule	J	$N \cdot m$
power	watt	W	J/s
quantity of electric charge	coulomb	C	$A \cdot s$
potential difference, electromotive force	volt	V	W/A
electric field strength	volt per meter (or newton per coulomb)	V/m	N/C
electric resistance	ohm	Ω	V/A
capacitance	farad	F	$A \cdot s/V$
magnetic flux	weber	Wb	$V \cdot s$
inductance	henry	H	$V \cdot s/A$
magnetic flux density	tesla	T	Wb/m^2
magnetic field strength	ampere per meter	A/m	
entropy	joule per kelvin	J/K	
specific heat	joule per kilogram kelvin	$J/(kg \cdot K)$	
thermal conductivity	watt per meter kelvin	$W/(m \cdot K)$	
radiant intensity	watt per steradian	W/sr	

3.   The SI Supplementary Units

Quantity	Name of Unit	Symbol
plane angle	radian	rad
solid angle	steradian	sr

APPENDIX B
Some Fundamental Constants of Physics*

Constant	Symbol	Computational Value	Best (1998) Value Value[a]	Best (1998) Value Uncertainty[b]
Speed of light in a vacuum	c	3.00×10^8 m/s	2.997 924 58	exact
Elementary charge	e	1.60×10^{-19} C	1.602 176 462	0.039
Gravitational constant	G	6.67×10^{-11} m³/s²·kg	6.673	1500
Universal gas constant	R	8.31 J/mol·K	8.314 472	1.7
Avogadro constant	N_A	6.02×10^{23} mol⁻¹	6.022 141 99	0.079
Boltzmann constant	k	1.38×10^{-23} J/K	1.380 650 3	1.7
Stefan–Boltzmann constant	σ	5.67×10^{-8} W/m²·K⁴	5.670 400	7.0
Molar volume of ideal gas at STP[d]	V_m	2.27×10^{-2} m³/mol	2.271 098 1	1.7
Permittivity constant	ϵ_0	8.85×10^{-12} F/m	8.854 187 817 62	exact
Permeability constant	μ_0	1.26×10^{-6} H/m	1.256 637 061 43	exact
Planck constant	h	6.63×10^{-34} J·s	6.626 068 76	0.078
Electron mass[c]	m_e	9.11×10^{-31} kg	9.109 381 88	0.079
		5.49×10^{-4} u	5.485 799 110	0.0021
Proton mass[c]	m_p	1.67×10^{-27} kg	1.672 621 58	0.079
		1.0073 u	1.007 276 466 88	1.3×10^{-4}
Ratio of proton mass to electron mass	m_p/m_e	1840	1836.152 667 5	0.0021
Electron charge-to-mass ratio	e/m_e	1.76×10^{11} C/kg	1.758 820 174	0.040
Neutron mass[c]	m_n	1.68×10^{-27} kg	1.674 927 16	0.079
		1.0087 u	1.008 664 915 78	5.4×10^{-4}
Hydrogen atom mass[c]	m_{1H}	1.0078 u	1.007 825 031 6	0.0005
Deuterium atom mass[c]	m_{2H}	2.0141 u	2.014 101 777 9	0.0005
Helium atom mass[c]	m_{4He}	4.0026 u	4.002 603 2	0.067
Muon mass	m_μ	1.88×10^{-28} kg	1.883 531 09	0.084
Electron magnetic moment	μ_e	9.28×10^{-24} J/T	9.284 763 62	0.040
Proton magnetic moment	μ_p	1.41×10^{-26} J/T	1.410 606 663	0.041
Bohr magneton	μ_B	9.27×10^{-24} J/T	9.274 008 99	0.040
Nuclear magneton	μ_N	5.05×10^{-27} J/T	5.050 783 17	0.040
Bohr radius	r_B	5.29×10^{-11} m	5.291 772 083	0.0037
Rydberg constant	R	1.10×10^7 m⁻¹	1.097 373 156 854 8	7.6×10^{-6}
Electron Compton wavelength	λ_C	2.43×10^{-12} m	2.426 310 215	0.0073

[a]Values given in this column should be given the same unit and power of 10 as the computational value.

[b]Parts per million.

[c]Masses given in u are in unified atomic mass units, where 1 u = 1.660 538 73 $\times 10^{-27}$ kg.

[d]STP means standard temperature and pressure: 0°C and 1.0 atm (0.1 MPa).

*The values in this table were selected from the 1998 CODATA recommended values (www.physics.nist.gov).

APPENDIX C
Some Astronomical Data

Some Distances from Earth

To the Moon*	3.82×10^8 m	To the center of our galaxy	2.2×10^{20} m
To the Sun*	1.50×10^{11} m	To the Andromeda Galaxy	2.1×10^{22} m
To the nearest star (Proxima Centauri)	4.04×10^{16} m	To the edge of the observable universe	$\sim 10^{26}$ m

*Mean distance.

The Sun, Earth, and the Moon

Property	Unit	Sun	Earth	Moon
Mass	kg	1.99×10^{30}	5.98×10^{24}	7.36×10^{22}
Mean radius	m	6.96×10^8	6.37×10^6	1.74×10^6
Mean density	kg/m^3	1410	5520	3340
Free-fall acceleration at the surface	m/s^2	274	9.81	1.67
Escape velocity	km/s	618	11.2	2.38
Period of rotation[a]	—	37 d at poles[b] 26 d at equator[b]	23 h 56 min	27.3 d
Radiation power[c]	W	3.90×10^{26}		

[a]Measured with respect to the distant stars.
[b]The Sun, a ball of gas, does not rotate as a rigid body.
[c]Just outside Earth's atmosphere solar energy is received, assuming normal incidence, at the rate of 1340 W/m^2.

Some Properties of the Planets

	Mercury	Venus	Earth	Mars	Jupiter	Saturn	Uranus	Neptune	Pluto
Mean distance from Sun, 10^6 km	57.9	108	150	228	778	1430	2870	4500	5900
Period of revolution, y	0.241	0.615	1.00	1.88	11.9	29.5	84.0	165	248
Period of rotation,[a] d	58.7	-243^b	0.997	1.03	0.409	0.426	-0.451^b	0.658	6.39
Orbital speed, km/s	47.9	35.0	29.8	24.1	13.1	9.64	6.81	5.43	4.74
Inclination of axis to orbit	<28°	≈3°	23.4°	25.0°	3.08°	26.7°	97.9°	29.6°	57.5°
Inclination of orbit to Earth's orbit	7.00°	3.39°		1.85°	1.30°	2.49°	0.77°	1.77°	17.2°
Eccentricity of orbit	0.206	0.0068	0.0167	0.0934	0.0485	0.0556	0.0472	0.0086	0.250
Equatorial diameter, km	4880	12 100	12 800	6790	143 000	120 000	51 800	49 500	2300
Mass (Earth = 1)	0.0558	0.815	1.000	0.107	318	95.1	14.5	17.2	0.002
Density (water = 1)	5.60	5.20	5.52	3.95	1.31	0.704	1.21	1.67	2.03
Surface value of g,[c] m/s^2	3.78	8.60	9.78	3.72	22.9	9.05	7.77	11.0	0.5
Escape velocity,[c] km/s	4.3	10.3	11.2	5.0	59.5	35.6	21.2	23.6	1.1
Known satellites	0	0	1	2	16 + ring	18 + rings	17 + rings	8 + rings	1

[a]Measured with respect to the distant stars.
[b]Venus and Uranus rotate opposite their orbital motion.
[c]Gravitational acceleration measured at the planet's equator.

APPENDIX D
Conversion Factors

Conversion factors may be read directly from these tables. For example, 1 degree = 2.778×10^{-3} revolutions, so $16.7° = 16.7 \times 2.778 \times 10^{-3}$ rev. The SI units are fully capitalized. Adapted in part from G. Shortley and D. Williams, *Elements of Physics*, 1971, Prentice-Hall, Englewood Cliffs, NJ.

Plane Angle

	°	′	″	RADIAN	rev
1 degree =	1	60	3600	1.745×10^{-2}	2.778×10^{-3}
1 minute =	1.667×10^{-2}	1	60	2.909×10^{-4}	4.630×10^{-5}
1 second =	2.778×10^{-4}	1.667×10^{-2}	1	4.848×10^{-6}	7.716×10^{-7}
1 RADIAN =	57.30	3438	2.063×10^{5}	1	0.1592
1 revolution =	360	2.16×10^{4}	1.296×10^{6}	6.283	1

Solid Angle

1 sphere = 4π steradians = 12.57 steradians

Length

	cm	METER	km	in.	ft	mi
1 centimeter =	1	10^{-2}	10^{-5}	0.3937	3.281×10^{-2}	6.214×10^{-6}
1 METER =	100	1	10^{-3}	39.37	3.281	6.214×10^{-4}
1 kilometer =	10^{5}	1000	1	3.937×10^{4}	3281	0.6214
1 inch =	2.540	2.540×10^{-2}	2.540×10^{-5}	1	8.333×10^{-2}	1.578×10^{-5}
1 foot =	30.48	0.3048	3.048×10^{-4}	12	1	1.894×10^{-4}
1 mile =	1.609×10^{5}	1609	1.609	6.336×10^{4}	5280	1

1 angström = 10^{-10} m
1 nautical mile = 1852 m
= 1.151 miles = 6076 ft

1 fermi = 10^{-15} m
1 light-year = 9.460×10^{12} km
1 parsec = 3.084×10^{13} km

1 fathom = 6 ft
1 Bohr radius = 5.292×10^{-11} m
1 yard = 3 ft

1 rod = 16.5 ft
1 mil = 10^{-3} in.
1 nm = 10^{-9} m

Area

	METER^2	cm^2	ft^2	in.^{2}
1 SQUARE METER =	1	10^{4}	10.76	1550
1 square centimeter =	10^{-4}	1	1.076×10^{-3}	0.1550
1 square foot =	9.290×10^{-2}	929.0	1	144
1 square inch =	6.452×10^{-4}	6.452	6.944×10^{-3}	1

1 square mile = 2.788×10^{7} ft^2 = 640 acres
1 barn = 10^{-28} m^2

1 acre = 43 560 ft^2
1 hectare = 10^{4} m^2 = 2.471 acres

Volume

	METER^3	cm^3	L	ft^3	in.^{3}
1 CUBIC METER = 1		10^6	1000	35.31	6.102×10^4
1 cubic centimeter = 10^{-6}		1	1.000×10^{-3}	3.531×10^{-5}	6.102×10^{-2}
1 liter = 1.000×10^{-3}		1000	1	3.531×10^{-2}	61.02
1 cubic foot = 2.832×10^{-2}		2.832×10^4	28.32	1	1728
1 cubic inch = 1.639×10^{-5}		16.39	1.639×10^{-2}	5.787×10^{-4}	1

1 U.S. fluid gallon = 4 U.S. fluid quarts = 8 U.S. pints = 128 U.S. fluid ounces = 231 in.^{3}
1 British imperial gallon = 277.4 in.^{3} = 1.201 U.S. fluid gallons

Mass

Quantities in the colored areas are not mass units but are often used as such. When we write, for example, 1 kg ''='' 2.205 lb, this means that a kilogram is a *mass* that *weighs* 2.205 pounds at a location where g has the standard value of 9.80665 m/s^2.

	g	KILOGRAM	slug	u	oz	lb	ton
1 gram = 1		0.001	6.852×10^{-5}	6.022×10^{23}	3.527×10^{-2}	2.205×10^{-3}	1.102×10^{-6}
1 KILOGRAM = 1000		1	6.852×10^{-2}	6.022×10^{26}	35.27	2.205	1.102×10^{-3}
1 slug = 1.459×10^4		14.59	1	8.786×10^{27}	514.8	32.17	1.609×10^{-2}
1 atomic mass unit = 1.661×10^{-24}		1.661×10^{-27}	1.138×10^{-28}	1	5.857×10^{-26}	3.662×10^{-27}	1.830×10^{-30}
1 ounce = 28.35		2.835×10^{-2}	1.943×10^{-3}	1.718×10^{25}	1	6.250×10^{-2}	3.125×10^{-5}
1 pound = 453.6		0.4536	3.108×10^{-2}	2.732×10^{26}	16	1	0.0005
1 ton = 9.072×10^5		907.2	62.16	5.463×10^{29}	3.2×10^4	2000	1

1 metric ton = 1000 kg

Density

Quantities in the colored areas are weight densities and, as such, are dimensionally different from mass densities. See note for mass table.

	slug/ft^3	KILOGRAM/ METER^3	g/cm^3	lb/ft^3	lb/in.^{3}
1 slug per foot^3 = 1		515.4	0.5154	32.17	1.862×10^{-2}
1 KILOGRAM per METER^3 = 1.940×10^{-3}		1	0.001	6.243×10^{-2}	3.613×10^{-5}
1 gram per centimeter^3 = 1.940		1000	1	62.43	3.613×10^{-2}
1 pound per foot^3 = 3.108×10^{-2}		16.02	16.02×10^{-2}	1	5.787×10^{-4}
1 pound per inch^3 = 53.71		2.768×10^4	27.68	1728	1

Time

	y	d	h	min	SECOND
1 year = 1		365.25	8.766×10^3	5.259×10^5	3.156×10^7
1 day = 2.738×10^{-3}		1	24	1440	8.640×10^4
1 hour = 1.141×10^{-4}		4.167×10^{-2}	1	60	3600
1 minute = 1.901×10^{-6}		6.944×10^{-4}	1.667×10^{-2}	1	60
1 SECOND = 3.169×10^{-8}		1.157×10^{-5}	2.778×10^{-4}	1.667×10^{-2}	1

Speed

	ft/s	km/h	METER/SECOND	mi/h	cm/s
1 foot per second = 1		1.097	0.3048	0.6818	30.48
1 kilometer per hour = 0.9113		1	0.2778	0.6214	27.78
1 METER per SECOND = 3.281		3.6	1	2.237	100
1 mile per hour = 1.467		1.609	0.4470	1	44.70
1 centimeter per second = 3.281×10^{-2}		3.6×10^{-2}	0.01	2.237×10^{-2}	1

1 knot = 1 nautical mi/h = 1.688 ft/s   1 mi/min = 88.00 ft/s = 60.00 mi/h

Force

Force units in the colored areas are now little used. To clarify: 1 gram-force (= 1 gf) is the force of gravity that would act on an object whose mass is 1 gram at a location where g has the standard value of 9.80665 m/s^2.

	dyne	NEWTON	lb	pdl	gf	kgf
1 dyne = 1		10^{-5}	2.248×10^{-6}	7.233×10^{-5}	1.020×10^{-3}	1.020×10^{-6}
1 NEWTON = 10^5		1	0.2248	7.233	102.0	0.1020
1 pound = 4.448×10^5		4.448	1	32.17	453.6	0.4536
1 poundal = 1.383×10^4		0.1383	3.108×10^{-2}	1	14.10	1.410×10^2
1 gram-force = 980.7		9.807×10^{-3}	2.205×10^{-3}	7.093×10^{-2}	1	0.001
1 kilogram-force = 9.807×10^5		9.807	2.205	70.93	1000	1

1 ton = 2000 lb

Pressure

	atm	dyne/cm^2	inch of water	cm Hg	PASCAL	lb/in.^{2}	lb/ft^2
1 atmosphere = 1		1.013×10^6	406.8	76	1.013×10^5	14.70	2116
1 dyne per centimeter^2 = 9.869×10^{-7}		1	4.015×10^{-4}	7.501×10^{-5}	0.1	1.405×10^{-5}	2.089×10^{-3}
1 inch of water^a at 4°C = 2.458×10^{-3}		2491	1	0.1868	249.1	3.613×10^{-2}	5.202
1 centimeter of mercury^a at 0°C = 1.316×10^{-2}		1.333×10^4	5.353	1	1333	0.1934	27.85
1 PASCAL = 9.869×10^{-6}		10	4.015×10^{-3}	7.501×10^{-4}	1	1.450×10^{-4}	2.089×10^{-2}
1 pound per inch^2 = 6.805×10^{-2}		6.895×10^4	27.68	5.171	6.895×10^3	1	144
1 pound per foot^2 = 4.725×10^{-4}		478.8	0.1922	3.591×10^{-2}	47.88	6.944×10^{-3}	1

^{a}Where the acceleration of gravity has the standard value of 9.80665 m/s^2.

1 bar = 10^6 dyne/cm^2 = 0.1 MPa   1 millibar = 10^3 dyne/cm^2 = 10^2 Pa   1 torr = 1 mm Hg

Energy, Work, Heat

Quantities in the colored areas are not energy units but are included for convenience. They arise from the relativistic mass–energy equivalence formula $E = mc^2$ and represent the energy released if a kilogram or unified atomic mass unit (u) is completely converted to energy (bottom two rows) or the mass that would be completely converted to one unit of energy (rightmost two columns).

	Btu	erg	ft · lb	hp · h	JOULE	cal	kW · h	eV	MeV	kg	u
1 British thermal unit =	1	1.055×10^{10}	777.9	3.929×10^{-4}	1055	252.0	2.930×10^{-4}	6.585×10^{21}	6.585×10^{15}	1.174×10^{-14}	7.070×10^{12}
1 erg =	9.481×10^{-11}	1	7.376×10^{-8}	3.725×10^{-14}	10^{-7}	2.389×10^{-8}	2.778×10^{-14}	6.242×10^{11}	6.242×10^{5}	1.113×10^{-24}	670.2
1 foot-pound =	1.285×10^{-3}	1.356×10^{7}	1	5.051×10^{-7}	1.356	0.3238	3.766×10^{-7}	8.464×10^{18}	8.464×10^{12}	1.509×10^{-17}	9.037×10^{9}
1 horsepower-hour =	2545	2.685×10^{13}	1.980×10^{6}	1	2.685×10^{6}	6.413×10^{5}	0.7457	1.676×10^{25}	1.676×10^{19}	2.988×10^{-11}	1.799×10^{16}
1 JOULE =	9.481×10^{-4}	10^{7}	0.7376	3.725×10^{-7}	1	0.2389	2.778×10^{-7}	6.242×10^{18}	6.242×10^{12}	1.113×10^{-17}	6.702×10^{9}
1 calorie =	3.969×10^{-3}	4.186×10^{7}	3.088	1.560×10^{-6}	4.186	1	1.163×10^{-6}	2.613×10^{19}	2.613×10^{13}	4.660×10^{-17}	2.806×10^{10}
1 kilowatt-hour =	3413	3.600×10^{13}	2.655×10^{6}	1.341	3.600×10^{6}	8.600×10^{5}	1	2.247×10^{25}	2.247×10^{19}	4.007×10^{-11}	2.413×10^{16}
1 electron-volt =	1.519×10^{-22}	1.602×10^{-12}	1.182×10^{-19}	5.967×10^{-26}	1.602×10^{-19}	3.827×10^{-20}	4.450×10^{-26}	1	10^{-6}	1.783×10^{-36}	1.074×10^{-9}
1 million electron-volts =	1.519×10^{-16}	1.602×10^{-6}	1.182×10^{-13}	5.967×10^{-20}	1.602×10^{-13}	3.827×10^{-14}	4.450×10^{-20}	10^{-6}	1	1.783×10^{-30}	1.074×10^{-3}
1 kilogram =	8.521×10^{13}	8.987×10^{23}	6.629×10^{16}	3.348×10^{10}	8.987×10^{16}	2.146×10^{16}	2.497×10^{10}	5.610×10^{35}	5.610×10^{29}	1	6.022×10^{26}
1 unified atomic mass unit =	1.415×10^{-13}	1.492×10^{-3}	1.101×10^{-10}	5.559×10^{-17}	1.492×10^{-10}	3.564×10^{-11}	4.146×10^{-17}	9.320×10^{8}	932.0	1.661×10^{-27}	1

Power

	Btu/h	ft · lb/s	hp	cal/s	kW	WATT
1 British thermal unit per hour =	1	0.2161	3.929×10^{-4}	6.998×10^{-2}	2.930×10^{-4}	0.2930
1 foot-pound per second =	4.628	1	1.818×10^{-3}	0.3239	1.356×10^{-3}	1.356
1 horsepower =	2545	550	1	178.1	0.7457	745.7
1 calorie per second =	14.29	3.088	5.615×10^{-3}	1	4.186×10^{-3}	4.186
1 kilowatt =	3413	737.6	1.341	238.9	1	1000
1 WATT =	3.413	0.7376	1.341×10^{-3}	0.2389	0.001	1

Magnetic Field

	gauss	TESLA	milligauss
1 gauss =	1	10^{-4}	1000
1 TESLA =	10^{4}	1	10^{7}
1 milligauss =	0.001	10^{-7}	1

1 tesla = 1 weber/meter^2

Magnetic Flux

	maxwell	WEBER
1 maxwell =	1	10^{-8}
1 WEBER =	10^{8}	1

APPENDIX E
Mathematical Formulas

Geometry

Circle of radius r: circumference $= 2\pi r$; area $= \pi r^2$.

Sphere of radius r: area $= 4\pi r^2$; volume $= \frac{4}{3}\pi r^3$.

Right circular cylinder of radius r and height h:
area $= 2\pi r^2 + 2\pi rh$; volume $= \pi r^2 h$.

Triangle of base a and altitude h: area $= \frac{1}{2}ah$.

Quadratic Formula

If $ax^2 + bx + c = 0$, then $x = \dfrac{-b \pm \sqrt{b^2 - 4ac}}{2a}$.

Trigonometric Functions of Angle θ

$\sin\theta = \dfrac{y}{r}$ $\cos\theta = \dfrac{x}{r}$

$\tan\theta = \dfrac{y}{x}$ $\cot\theta = \dfrac{x}{y}$

$\sec\theta = \dfrac{r}{x}$ $\csc\theta = \dfrac{r}{y}$

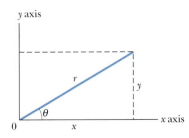

Pythagorean Theorem

In this right triangle,
$$a^2 + b^2 = c^2$$

Triangles

Angles are A, B, C

Opposite sides are a, b, c

Angles $A + B + C = 180°$

$\dfrac{\sin A}{a} = \dfrac{\sin B}{b} = \dfrac{\sin C}{c}$

$c^2 = a^2 + b^2 - 2ab\cos C$

Exterior angle $D = A + C$

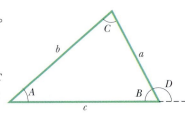

Mathematical Signs and Symbols

$=$ equals

$\approx$ equals approximately

$\sim$ is the order of magnitude of

$\neq$ is not equal to

$\equiv$ is identical to, is defined as

$>$ is greater than ($\gg$ is much greater than)

$<$ is less than ($\ll$ is much less than)

$\geq$ is greater than or equal to (or, is no less than)

$\leq$ is less than or equal to (or, is no more than)

$\pm$ plus or minus

$\propto$ is proportional to

Σ the sum of

x_{avg} the average value of x

Trigonometric Identities

$\sin(90° - \theta) = \cos\theta$

$\cos(90° - \theta) = \sin\theta$

$\sin\theta/\cos\theta = \tan\theta$

$\sin^2\theta + \cos^2\theta = 1$

$\sec^2\theta - \tan^2\theta = 1$

$\csc^2\theta - \cot^2\theta = 1$

$\sin 2\theta = 2\sin\theta\cos\theta$

$\cos 2\theta = \cos^2\theta - \sin^2\theta = 2\cos^2\theta - 1 = 1 - 2\sin^2\theta$

$\sin(\alpha \pm \beta) = \sin\alpha\cos\beta \pm \cos\alpha\sin\beta$

$\cos(\alpha \pm \beta) = \cos\alpha\cos\beta \mp \sin\alpha\sin\beta$

$\tan(\alpha \pm \beta) = \dfrac{\tan\alpha \pm \tan\beta}{1 \mp \tan\alpha\tan\beta}$

$\sin\alpha \pm \sin\beta = 2\sin\frac{1}{2}(\alpha \pm \beta)\cos\frac{1}{2}(\alpha \mp \beta)$

$\cos\alpha + \cos\beta = 2\cos\frac{1}{2}(\alpha + \beta)\cos\frac{1}{2}(\alpha - \beta)$

$\cos\alpha - \cos\beta = -2\sin\frac{1}{2}(\alpha + \beta)\sin\frac{1}{2}(\alpha - \beta)$

Binomial Theorem

$$(1 + x)^n = 1 + \frac{nx}{1!} + \frac{n(n-1)x^2}{2!} + \cdots \qquad (x^2 < 1)$$

Exponential Expansion

$$e^x = 1 + x + \frac{x^2}{2!} + \frac{x^3}{3!} + \cdots$$

Logarithmic Expansion

$$\ln(1 + x) = x - \tfrac{1}{2}x^2 + \tfrac{1}{3}x^3 - \cdots \qquad (|x| < 1)$$

Trigonometric Expansions
(θ in radians)

$$\sin \theta = \theta - \frac{\theta^3}{3!} + \frac{\theta^5}{5!} - \cdots$$

$$\cos \theta = 1 - \frac{\theta^2}{2!} + \frac{\theta^4}{4!} - \cdots$$

$$\tan \theta = \theta + \frac{\theta^3}{3} + \frac{2\theta^5}{15} + \cdots$$

Cramer's Rule

Two simultaneous equations in unknowns x and y,

$$a_1 x + b_1 y = c_1 \quad \text{and} \quad a_2 x + b_2 y = c_2,$$

have the solutions

$$x = \frac{\begin{vmatrix} c_1 & b_1 \\ c_2 & b_2 \end{vmatrix}}{\begin{vmatrix} a_1 & b_1 \\ a_2 & b_2 \end{vmatrix}} = \frac{c_1 b_2 - c_2 b_1}{a_1 b_2 - a_2 b_1}$$

and

$$y = \frac{\begin{vmatrix} a_1 & c_1 \\ a_2 & c_2 \end{vmatrix}}{\begin{vmatrix} a_1 & b_1 \\ a_2 & b_2 \end{vmatrix}} = \frac{a_1 c_2 - a_2 c_1}{a_1 b_2 - a_2 b_1}.$$

Products of Vectors

Let $\hat{i}$, $\hat{j}$, and $\hat{k}$ be unit vectors in the x, y, and z directions. Then

$$\hat{i} \cdot \hat{i} = \hat{j} \cdot \hat{j} = \hat{k} \cdot \hat{k} = 1, \qquad \hat{i} \cdot \hat{j} = \hat{j} \cdot \hat{k} = \hat{k} \cdot \hat{i} = 0,$$

$$\hat{i} \times \hat{i} = \hat{j} \times \hat{j} = \hat{k} \times \hat{k} = 0,$$

$$\hat{i} \times \hat{j} = \hat{k}, \qquad \hat{j} \times \hat{k} = \hat{i}, \qquad \hat{k} \times \hat{i} = \hat{j}.$$

Any vector $\vec{a}$ with components a_x, a_y, and a_z along the x, y, and z axes can be written as

$$\vec{a} = a_x \hat{i} + a_y \hat{j} + a_z \hat{k}.$$

Let $\vec{a}$, $\vec{b}$, and $\vec{c}$ be arbitrary vectors with magnitudes a, b, and c. Then

$$\vec{a} \times (\vec{b} + \vec{c}) = (\vec{a} \times \vec{b}) + (\vec{a} \times \vec{c})$$

$$(s\vec{a}) \times \vec{b} = \vec{a} \times (s\vec{b}) = s(\vec{a} \times \vec{b}) \quad (s = \text{a scalar}).$$

Let θ be the smaller of the two angles between $\vec{a}$ and $\vec{b}$. Then

$$\vec{a} \cdot \vec{b} = \vec{b} \cdot \vec{a} = a_x b_x + a_y b_y + a_z b_z = ab \cos \theta$$

$$\vec{a} \times \vec{b} = -\vec{b} \times \vec{a} = \begin{vmatrix} \hat{i} & \hat{j} & \hat{k} \\ a_x & a_y & a_z \\ b_x & b_y & b_z \end{vmatrix}$$

$$= \hat{i} \begin{vmatrix} a_y & a_z \\ b_y & b_z \end{vmatrix} - \hat{j} \begin{vmatrix} a_x & a_z \\ b_x & b_z \end{vmatrix} + \hat{k} \begin{vmatrix} a_x & a_y \\ b_x & b_y \end{vmatrix}$$

$$= (a_y b_z - b_y a_z)\hat{i} + (a_z b_x - b_z a_x)\hat{j} + (a_x b_y - b_x a_y)\hat{k}$$

$$|\vec{a} \times \vec{b}| = ab \sin \theta$$

$$\vec{a} \cdot (\vec{b} \times \vec{c}) = \vec{b} \cdot (\vec{c} \times \vec{a}) = \vec{c} \cdot (\vec{a} \times \vec{b})$$

$$\vec{a} \times (\vec{b} \times \vec{c}) = (\vec{a} \cdot \vec{c})\vec{b} - (\vec{a} \cdot \vec{b})\vec{c}$$

Derivatives and Integrals

In what follows, the letters u and v stand for any functions of x, and a and m are constants. To each of the indefinite integrals should be added an arbitrary constant of integration. The *Handbook of Chemistry and Physics* (CRC Press Inc.) gives a more extensive tabulation.

1. $\dfrac{dx}{dx} = 1$

2. $\dfrac{d}{dx}(au) = a\dfrac{du}{dx}$

3. $\dfrac{d}{dx}(u + v) = \dfrac{du}{dx} + \dfrac{dv}{dx}$

4. $\dfrac{d}{dx}x^m = mx^{m-1}$

5. $\dfrac{d}{dx}\ln x = \dfrac{1}{x}$

6. $\dfrac{d}{dx}(uv) = u\dfrac{dv}{dx} + v\dfrac{du}{dx}$

7. $\dfrac{d}{dx}e^x = e^x$

8. $\dfrac{d}{dx}\sin x = \cos x$

9. $\dfrac{d}{dx}\cos x = -\sin x$

10. $\dfrac{d}{dx}\tan x = \sec^2 x$

11. $\dfrac{d}{dx}\cot x = -\csc^2 x$

12. $\dfrac{d}{dx}\sec x = \tan x \sec x$

13. $\dfrac{d}{dx}\csc x = -\cot x \csc x$

14. $\dfrac{d}{dx}e^u = e^u\dfrac{du}{dx}$

15. $\dfrac{d}{dx}\sin u = \cos u\dfrac{du}{dx}$

16. $\dfrac{d}{dx}\cos u = -\sin u\dfrac{du}{dx}$

1. $\displaystyle\int dx = x$

2. $\displaystyle\int au\, dx = a\int u\, dx$

3. $\displaystyle\int (u + v)\, dx = \int u\, dx + \int v\, dx$

4. $\displaystyle\int x^m\, dx = \dfrac{x^{m+1}}{m + 1}\quad (m \neq -1)$

5. $\displaystyle\int \dfrac{dx}{x} = \ln |x|$

6. $\displaystyle\int u\dfrac{dv}{dx}\, dx = uv - \int v\dfrac{du}{dx}\, dx$

7. $\displaystyle\int e^x\, dx = e^x$

8. $\displaystyle\int \sin x\, dx = -\cos x$

9. $\displaystyle\int \cos x\, dx = \sin x$

10. $\displaystyle\int \tan x\, dx = \ln |\sec x|$

11. $\displaystyle\int \sin^2 x\, dx = \tfrac{1}{2}x - \tfrac{1}{4}\sin 2x$

12. $\displaystyle\int e^{-ax}\, dx = -\dfrac{1}{a}e^{-ax}$

13. $\displaystyle\int xe^{-ax}\, dx = -\dfrac{1}{a^2}(ax + 1)e^{-ax}$

14. $\displaystyle\int x^2e^{-ax}\, dx = -\dfrac{1}{a^3}(a^2x^2 + 2ax + 2)e^{-ax}$

15. $\displaystyle\int_0^\infty x^ne^{-ax}\, dx = \dfrac{n!}{a^{n+1}}$

16. $\displaystyle\int_0^\infty x^{2n}e^{-ax^2}\, dx = \dfrac{1 \cdot 3 \cdot 5 \cdots (2n - 1)}{2^{n+1}a^n}\sqrt{\dfrac{\pi}{a}}$

17. $\displaystyle\int \dfrac{dx}{\sqrt{x^2 + a^2}} = \ln(x + \sqrt{x^2 + a^2})$

18. $\displaystyle\int \dfrac{x\, dx}{(x^2 + a^2)^{3/2}} = -\dfrac{1}{(x^2 + a^2)^{1/2}}$

19. $\displaystyle\int \dfrac{dx}{(x^2 + a^2)^{3/2}} = \dfrac{x}{a^2(x^2 + a^2)^{1/2}}$

20. $\displaystyle\int_0^\infty x^{2n+1}e^{-ax^2}\, dx = \dfrac{n!}{2a^{n+1}}\quad (a > 0)$

21. $\displaystyle\int \dfrac{x\, dx}{x + d} = x - d\ln(x + d)$

APPENDIX F
Properties of the Elements

All physical properties are for a pressure of 1 atm unless otherwise specified.

Element	Symbol	Atomic Number Z	Molar Mass, g/mol	Density, g/cm³ at 20°C	Melting Point, °C	Boiling Point, °C	Specific Heat, J/(g·°C) at 25°C
Actinium	Ac	89	(227)	10.06	1323	(3473)	0.092
Aluminum	Al	13	26.9815	2.699	660	2450	0.900
Americium	Am	95	(243)	13.67	1541	—	—
Antimony	Sb	51	121.75	6.691	630.5	1380	0.205
Argon	Ar	18	39.948	1.6626×10^{-3}	−189.4	−185.8	0.523
Arsenic	As	33	74.9216	5.78	817 (28 atm)	613	0.331
Astatine	At	85	(210)	—	(302)	—	—
Barium	Ba	56	137.34	3.594	729	1640	0.205
Berkelium	Bk	97	(247)	14.79	—	—	—
Beryllium	Be	4	9.0122	1.848	1287	2770	1.83
Bismuth	Bi	83	208.980	9.747	271.37	1560	0.122
Bohrium	Bh	107	262.12	—	—	—	—
Boron	B	5	10.811	2.34	2030	—	1.11
Bromine	Br	35	79.909	3.12 (liquid)	−7.2	58	0.293
Cadmium	Cd	48	112.40	8.65	321.03	765	0.226
Calcium	Ca	20	40.08	1.55	838	1440	0.624
Californium	Cf	98	(251)	—	—	—	—
Carbon	C	6	12.01115	2.26	3727	4830	0.691
Cerium	Ce	58	140.12	6.768	804	3470	0.188
Cesium	Cs	55	132.905	1.873	28.40	690	0.243
Chlorine	Cl	17	35.453	3.214×10^{-3} (0°C)	−101	−34.7	0.486
Chromium	Cr	24	51.996	7.19	1857	2665	0.448
Cobalt	Co	27	58.9332	8.85	1495	2900	0.423
Copper	Cu	29	63.54	8.96	1083.40	2595	0.385
Curium	Cm	96	(247)	13.3	—	—	—
Dubnium	Db	105	262.114	—	—	—	—
Dysprosium	Dy	66	162.50	8.55	1409	2330	0.172
Einsteinium	Es	99	(254)	—	—	—	—
Erbium	Er	68	167.26	9.15	1522	2630	0.167
Europium	Eu	63	151.96	5.243	817	1490	0.163
Fermium	Fm	100	(237)	—	—	—	—
Fluorine	F	9	18.9984	1.696×10^{-3} (0°C)	−219.6	−188.2	0.753
Francium	Fr	87	(223)	—	(27)	—	—
Gadolinium	Gd	64	157.25	7.90	1312	2730	0.234
Gallium	Ga	31	69.72	5.907	29.75	2237	0.377

Element	Symbol	Atomic Number Z	Molar Mass, g/mol	Density, g/cm^3 at 20°C	Melting Point, °C	Boiling Point, °C	Specific Heat, J/(g · °C) at 25°C
Germanium	Ge	32	72.59	5.323	937.25	2830	0.322
Gold	Au	79	196.967	19.32	1064.43	2970	0.131
Hafnium	Hf	72	178.49	13.31	2227	5400	0.144
Hassium	Hs	108	(265)	—	—	—	—
Helium	He	2	4.0026	0.1664×10^{-3}	−269.7	−268.9	5.23
Holmium	Ho	67	164.930	8.79	1470	2330	0.165
Hydrogen	H	1	1.00797	0.08375×10^{-3}	−259.19	−252.7	14.4
Indium	In	49	114.82	7.31	156.634	2000	0.233
Iodine	I	53	126.9044	4.93	113.7	183	0.218
Iridium	Ir	77	192.2	22.5	2447	(5300)	0.130
Iron	Fe	26	55.847	7.874	1536.5	3000	0.447
Krypton	Kr	36	83.80	3.488×10^{-3}	−157.37	−152	0.247
Lanthanum	La	57	138.91	6.189	920	3470	0.195
Lawrencium	Lr	103	(257)	—	—	—	—
Lead	Pb	82	207.19	11.35	327.45	1725	0.129
Lithium	Li	3	6.939	0.534	180.55	1300	3.58
Lutetium	Lu	71	174.97	9.849	1663	1930	0.155
Magnesium	Mg	12	24.312	1.738	650	1107	1.03
Manganese	Mn	25	54.9380	7.44	1244	2150	0.481
Meitnerium	Mt	109	(266)	—	—	—	—
Mendelevium	Md	101	(256)	—	—	—	—
Mercury	Hg	80	200.59	13.55	−38.87	357	0.138
Molybdenum	Mo	42	95.94	10.22	2617	5560	0.251
Neodymium	Nd	60	144.24	7.007	1016	3180	0.188
Neon	Ne	10	20.183	0.8387×10^{-3}	−248.597	−246.0	1.03
Neptunium	Np	93	(237)	20.25	637	—	1.26
Nickel	Ni	28	58.71	8.902	1453	2730	0.444
Niobium	Nb	41	92.906	8.57	2468	4927	0.264
Nitrogen	N	7	14.0067	1.1649×10^{-3}	−210	−195.8	1.03
Nobelium	No	102	(255)	—	—	—	—
Osmium	Os	76	190.2	22.59	3027	5500	0.130
Oxygen	O	8	15.9994	1.3318×10^{-3}	−218.80	−183.0	0.913
Palladium	Pd	46	106.4	12.02	1552	3980	0.243
Phosphorus	P	15	30.9738	1.83	44.25	280	0.741
Platinum	Pt	78	195.09	21.45	1769	4530	0.134
Plutonium	Pu	94	(244)	19.8	640	3235	0.130
Polonium	Po	84	(210)	9.32	254	—	—
Potassium	K	19	39.102	0.862	63.20	760	0.758
Praseodymium	Pr	59	140.907	6.773	931	3020	0.197
Promethium	Pm	61	(145)	7.22	(1027)	—	—
Protactinium	Pa	91	(231)	15.37 (estimated)	(1230)	—	—
Radium	Ra	88	(226)	5.0	700	—	—
Radon	Rn	86	(222)	9.96×10^{-3} (0°C)	(−71)	−61.8	0.092
Rhenium	Re	75	186.2	21.02	3180	5900	0.134

Element	Symbol	Atomic Number Z	Molar Mass, g/mol	Density, g/cm³ at 20°C	Melting Point, °C	Boiling Point, °C	Specific Heat, J/(g · °C) at 25°C
Rhodium	Rh	45	102.905	12.41	1963	4500	0.243
Rubidium	Rb	37	85.47	1.532	39.49	688	0.364
Ruthenium	Ru	44	101.107	12.37	2250	4900	0.239
Rutherfordium	Rf	104	261.11	—	—	—	—
Samarium	Sm	62	150.35	7.52	1072	1630	0.197
Scandium	Sc	21	44.956	2.99	1539	2730	0.569
Seaborgium	Sg	106	263.118	—	—	—	—
Selenium	Se	34	78.96	4.79	221	685	0.318
Silicon	Si	14	28.086	2.33	1412	2680	0.712
Silver	Ag	47	107.870	10.49	960.8	2210	0.234
Sodium	Na	11	22.9898	0.9712	97.85	892	1.23
Strontium	Sr	38	87.62	2.54	768	1380	0.737
Sulfur	S	16	32.064	2.07	119.0	444.6	0.707
Tantalum	Ta	73	180.948	16.6	3014	5425	0.138
Technetium	Tc	43	(99)	11.46	2200	—	0.209
Tellurium	Te	52	127.60	6.24	449.5	990	0.201
Terbium	Tb	65	158.924	8.229	1357	2530	0.180
Thallium	Tl	81	204.37	11.85	304	1457	0.130
Thorium	Th	90	(232)	11.72	1755	(3850)	0.117
Thulium	Tm	69	168.934	9.32	1545	1720	0.159
Tin	Sn	50	118.69	7.2984	231.868	2270	0.226
Titanium	Ti	22	47.90	4.54	1670	3260	0.523
Tungsten	W	74	183.85	19.3	3380	5930	0.134
Un-named	Uun	110	(269)	—	—	—	—
Un-named	Uuu	111	(272)	—	—	—	—
Un-named	Uub	112	(264)	—	—	—	—
Un-named	Uut	113	—	—	—	—	—
Un-named	Unq	114	(285)	—	—	—	—
Un-named	Uup	115	—	—	—	—	—
Un-named	Uuh	116	(289)	—	—	—	—
Un-named	Uus	117	—	—	—	—	—
Un-named	Uuo	118	(293)	—	—	—	—
Uranium	U	92	(238)	18.95	1132	3818	0.117
Vanadium	V	23	50.942	6.11	1902	3400	0.490
Xenon	Xe	54	131.30	5.495×10^{-3}	−111.79	−108	0.159
Ytterbium	Yb	70	173.04	6.965	824	1530	0.155
Yttrium	Y	39	88.905	4.469	1526	3030	0.297
Zinc	Zn	30	65.37	7.133	419.58	906	0.389
Zirconium	Zr	40	91.22	6.506	1852	3580	0.276

The values in parentheses in the column of molar masses are the mass numbers of the longest-lived isotopes of those elements that are radioactive. Melting points and boiling points in parentheses are uncertain.

The data for gases are valid only when these are in their usual molecular state, such as H_2, He, O_2, Ne, etc. The specific heats of the gases are the values at constant pressure.

Source: Adapted from J. Emsley, *The Elements,* 3rd ed., 1998, Clarendon Press, Oxford. See also www.webelements.com for the latest values and newest elements.

Periodic Table of the Elements

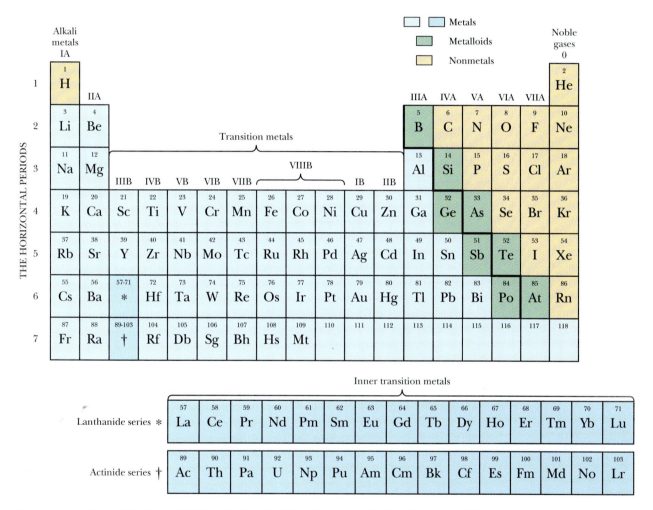

Metals

Metalloids

Nonmetals

Alkali metals
IA

Noble gases
0

Transition metals

VIIIB

Inner transition metals

THE HORIZONTAL PERIODS

	IA	IIA											IIIA	IVA	VA	VIA	VIIA	0
1	1 H																	2 He
2	3 Li	4 Be											5 B	6 C	7 N	8 O	9 F	10 Ne
3	11 Na	12 Mg	IIIB	IVB	VB	VIB	VIIB				IB	IIB	13 Al	14 Si	15 P	16 S	17 Cl	18 Ar
4	19 K	20 Ca	21 Sc	22 Ti	23 V	24 Cr	25 Mn	26 Fe	27 Co	28 Ni	29 Cu	30 Zn	31 Ga	32 Ge	33 As	34 Se	35 Br	36 Kr
5	37 Rb	38 Sr	39 Y	40 Zr	41 Nb	42 Mo	43 Tc	44 Ru	45 Rh	46 Pd	47 Ag	48 Cd	49 In	50 Sn	51 Sb	52 Te	53 I	54 Xe
6	55 Cs	56 Ba	57-71 *	72 Hf	73 Ta	74 W	75 Re	76 Os	77 Ir	78 Pt	79 Au	80 Hg	81 Tl	82 Pb	83 Bi	84 Po	85 At	86 Rn
7	87 Fr	88 Ra	89-103 †	104 Rf	105 Db	106 Sg	107 Bh	108 Hs	109 Mt	110	111	112	113	114	115	116	117	118

Lanthanide series *	57 La	58 Ce	59 Pr	60 Nd	61 Pm	62 Sm	63 Eu	64 Gd	65 Tb	66 Dy	67 Ho	68 Er	69 Tm	70 Yb	71 Lu
Actinide series †	89 Ac	90 Th	91 Pa	92 U	93 Np	94 Pu	95 Am	96 Cm	97 Bk	98 Cf	99 Es	100 Fm	101 Md	102 No	103 Lr

The names for elements 104 through 109 (Rutherfordium, Dubnium, Seaborgium, Bohrium, Hassium, and Meitnerium, respectively) were adopted by the International Union of Pure and Applied Chemistry (IUPAC) in 1997. Elements 110, 111, 112, 114, 116, and 118 have been discovered but, as of 2000, have not yet been named. See www.webelements.com for the latest information and newest elements.

ANSWERS
to Checkpoints and Odd-Numbered Questions, Exercises, and Problems

CHAPTER 1

EP **1.** (a) 10^9; (b) 10^{-4}; (c) 9.1×10^5 **3.** (a) 160 rods; (b) 40 chains **5.** (a) 4.00×10^4 km; (b) 5.10×10^8 km²; (c) 1.08×10^{12} km³ **7.** 1.9×10^{22} cm³ **9.** 1.1×10^3 acre-feet **11.** (a) 0.98 ft/ns; (b) 0.30 mm/ps **13.** C, D, A, B, E; the important criterion is the constancy of the daily variation, not its magnitude **15.** 0.12 AU/min **17.** 2.1 h **19.** 9.0×10^{49} **21.** (a) 10^3 kg; (b) 158 kg/s **23.** (a) 1.18×10^{-29} m³; (b) 0.282 nm **25.** (a) 60.8 W; (b) 43.3 Z **27.** 89 km **29.** $\approx 1 \times 10^{36}$

CHAPTER 2

CP **1.** b and c **2.** zero (zero displacement for entire trip) **3.** (check the derivative dx/dt) (a) 1 and 4; (b) 2 and 3 **4.** (see Tactic 5) (a) plus; (b) minus; (c) minus; (d) plus **5.** 1 and 4 ($a = d^2x/dt^2$ must be a constant) **6.** (a) plus (upward displacement on y axis); (b) minus (downward displacement on y axis); (c) $a = -g = -9.8$ m/s² **Q** **1.** (a) all tie; (b) 4, tie of 1 and 2, then 3 **3.** E **5.** a and c **7.** $x = t^2$ and $x = 8(t-2) + (1.5)(t-2)^2$ **9.** same **EP** **1.** 414 ms **3.** (a) +40 km/h; (b) 40 km/h **5.** (a) 73 km/h; (b) 68 km/h; (c) 70 km/h; (d) 0 **7.** (a) 0, −2, 0, 12 m; (b) +12 m; (c) +7 m/s **9.** 1.4 m **11.** (a) −6 m/s; (b) negative x direction; (c) 6 m/s; (d) first smaller, then zero, and then larger; (e) yes ($t = 2$ s); (f) no **13.** 100 m **15.** (a) velocity squared; (b) acceleration; (c) m²/s², m/s² **17.** 20 m/s², in the direction opposite to its initial velocity **19.** (a) 80 m/s; (b) 110 m/s; (c) 20 m/s² **21.** (a) m/s², m/s³; (b) 1.0 s; (c) 82 m; (d) −80 m; (e) 0, −12, −36, −72 m/s; (f) −6, −18, −30, −42 m/s² **23.** 0.10 m **25.** (a) 1.6 m/s; (b) 18 m/s **27.** (a) 3.1×10^6 s = 1.2 months; (b) 4.6×10^{13} m **29.** 1.62×10^{15} m/s² **31.** 2.5 s **33.** (a) 3.56 m/s²; (b) 8.43 m/s **35.** (a) 5.00 m/s; (b) 1.67 m/s²; (c) 7.50 m **37.** (a) 0.74 s; (b) −6.2 m/s² **39.** (a) 10.6 m; (b) 41.5 s **41.** (a) 29.4 m; (b) 2.45 s **43.** (a) 31 m/s; (b) 6.4 s **45.** (a) 3.2 s; (b) 1.3 s **47.** (a) 3.70 m/s; (b) 1.74 m/s; (c) 0.154 m **49.** 4.0 m/s **51.** 857 m/s², upward **53.** 1.26×10^3 m/s², upward **55.** 22 cm and 89 cm below the nozzle **57.** 1.5 s **59.** (a) 5.4 s; (b) 41 m/s **61.** (a) 76 m; (b) 4.2 s **63.** (a) 1.23 cm; (b) 4 times, 9 times, 16 times, 25 times **65.** 2.34 m

CHAPTER 3

CP **1.** (a) 7 m ($\vec{a}$ and $\vec{b}$ are in same direction); (b) 1 m ($\vec{a}$ and $\vec{b}$ are in opposite directions) **2.** c, d, f (components must be head-to-tail; $\vec{a}$ must extend from tail of one component to head of the other) **3.** (a) +, +; (b) +, −; (c) +, + (draw vector from tail of $\vec{d_1}$ to head of $\vec{d_2}$) **4.** (a) 90°; (b) 0° (vectors are parallel— same direction); (c) 180° (vectors are antiparallel—opposite directions) **5.** (a) 0° or 180°; (b) 90° **Q** **1.** $\vec{A}$ and $\vec{B}$ **3.** no, but $\vec{a}$ and $-\vec{b}$ are commutative: $\vec{a} + (-\vec{b}) = (-\vec{b}) + \vec{a}$ **5.** (a) $\vec{a}$ and $\vec{b}$ are parallel; (b) $\vec{b} = 0$; (c) $\vec{a}$ and $\vec{b}$ are perpendicular **7.** all but (e) **9.** (a) 0 (vectors are parallel); (b) 0 (vectors are antiparallel) **EP** **1.** The displacements should be (a) parallel, (b) antiparallel, (c) perpendicular **3.** (a) −2.5 m; (b) −6.9 m **5.** (a) 47.2 m; (b) 122° **7.** (a) 168 cm; (b) 32.5° above the floor **9.** (a) 6.42 m; (b) no; (c) yes; (d) yes; (e) a possible answer: (4.30 m)î + (3.70 m)ĵ + (3.00 m)k̂; (f) 7.96 m **11.** (a) 370 m; (b) 36° north of east; (c) 425 m; (d) the distance **13.** (a) (−9 m)î + (10 m)ĵ; (b) 13 m; (c) +132° **15.** (a) 4.2 m; (b) 40° east of north; (c) 8.0 m; (d) 24° north of west **17.** (a) (3.0 m)î − (2.0 m)ĵ + (5.0 m)k̂; (b) (5.0 m)î − (4.0 m)ĵ − (3.0 m)k̂; (c) (−5.0 m)î + (4.0 m)ĵ + (3.0 m)k̂ **19.** (a) 38 m; (b) 320°; (c) 130 m; (d) 1.2°; (e) 62 m; (f) 130° **21.** (a) 1.59 m; (b) 12.1 m; (c) 12.2 m; (d) 82.5° **27.** (a) Put axes along cube edges, with the origin at one corner. Diagonals are $a\hat{i} + a\hat{j} + a\hat{k}$, $a\hat{i} + a\hat{j} - a\hat{k}$, $a\hat{i} - a\hat{j} - a\hat{k}$, $a\hat{i} - a\hat{j} + a\hat{k}$; (b) 54.7°; (c) $\sqrt{3}a$ **29.** (a) 30; (b) 52 **31.** 22° **35.** (b) $a^2b \sin\phi$ **37.** (a) 3.0 m; (b) 0; (c) 3.46 m; (d) 2.00 m; (e) −5.00 m; (f) 8.66 m; (g) −6.67; (h) 4.33

CHAPTER 4

CP **1.** (a) (8î − 6ĵ) m; (b) yes, the xy plane (no z component) **2.** (draw $\vec{v}$ tangent to path, tail on path) (a) first; (b) third **3.** (take second derivative with respect to time) (1) and (3) a_x and a_y are both constant and thus $\vec{a}$ is constant; (2) and (4) a_y is constant but a_x is not, thus $\vec{a}$ is not **4.** 4 m/s³, −2 m/s, 3 m **5.** (a) v_x constant; (b) v_y initially positive, decreases to zero, and then becomes progressively more negative; (c) $a_x = 0$ throughout; (d) $a_y = -g$ throughout **6.** (a) −(4 m/s)î; (b) −(8 m/s²)ĵ **7.** (a) 0, distance not changing; (b) +70 km/h, distance increasing; (c) +80 km/h, distance decreasing **8.** (a)–(c) increase **Q** **1.** (a) (7 m)î + (1 m)ĵ + (−2 m)k̂; (b) (5 m)î + (−3 m)ĵ + (1 m)k̂; (c) (−2 m)î **3.** a, b, c **5.** (a) all tie; (b) 1 and 2 tie (the rocket is shot upward), then 3 and 4 tie (it is shot into the ground!) **7.** (a) 3, 2, 1; (b) 1, 2, 3; (c) all tie; (d) 6, 5, 4 **9.** (a) less; (b) unanswerable; (c) equal; (d) unanswerable **11.** (a) 2; (b) 3; (c) 1; (d) 2; (e) 3; (f) 1 **13.** (a) yes; (b) no; (c) yes **EP** **1.** (a) (−5.0î + 8.0ĵ) m; (b) 9.4 m; (c) 122°; (e) (8î − 8ĵ) m; (f) 11 m; (g) −45° **3.** (a) (−7.0î + 12ĵ) m; (b) xy plane **5.** 7.59 km/h, 22.5° east of north **7.** (a) (3.00î − 8.00tĵ) m/s; (b) (3.00î − 16.0ĵ) m/s; (c) 16.3 m/s; (d) −79.4° **9.** (a) (8tĵ + k̂) m/s; (b) 8ĵ m/s² **11.** (a) (6.00î − 106ĵ) m;

(b) $(19.0\hat{i} - 224\hat{j})$ m/s; (c) $(24.0\hat{i} - 336\hat{j})$ m/s^2; (d) $-85.2°$ to $+x$ **13.** (a) $(-1.5\hat{j})$ m/s; (b) $(4.5\hat{i} - 2.25\hat{j})$ m   **15.** (a) 45 m; (b) 22 m/s   **17.** (a) 62 ms; (b) 480 m/s   **19.** (a) 0.205 s; (b) 0.205 s; (c) 20.5 cm; (d) 61.5 cm   **21.** (a) 2.00 ns; (b) 2.00 mm; (c) 1.00×10^7 m/s; (d) 2.00×10^6 m/s   **23.** (a) 16.9 m; (b) 8.21 m; (c) 27.6 m; (d) 7.26 m; (e) 40.2 m; (f) 0   **25.** 4.8 cm   **29.** (a) 11 m; (b) 23 m; (c) 17 m/s; (d) 63° below horizontal   **31.** (a) 24 m/s; (b) 65° above the horizontal   **33.** (a) 10 s; (b) 897 m   **35.** the third   **37.** (a) 202 m/s; (b) 806 m; (c) 161 m/s; (d) -171 m/s   **39.** (a) yes; (b) 2.56 m   **41.** between the angles 31° and 63° above the horizontal   **43.** (a) 7.49 km/s; (b) 8.00 m/s^2 **45.** (a) 19 m/s; (b) 35 rev/min; (c) 1.7 s   **47.** (a) 0.034 m/s^2; (b) 84 min   **49.** (a) 12 s; (b) 4.1 m/s^2, down; (c) 4.1 m/s^2, up   **51.** 160 m/s   **53.** (a) 13 m/s^2, eastward; (b) 13 m/s^2, eastward   **55.** 36 s, no   **57.** 60°   **59.** 32 m/s   **61.** (a) 38 knots, 1.5° east of north; (b) 4.2 h; (c) 1.5° west of south   **63.** (a) 37° west of north; (b) 62.6 s

CHAPTER 5

CP   **1.** c, d, and e ($\vec{F}_1$ and $\vec{F}_2$ must be head-to-tail, $\vec{F}_{net}$ must be from tail of one of them to head of the other)   **2.** (a) and (b) 2 N, leftward (acceleration is zero in each situation)   **3.** (a) and (b) 1, 4, 3, 2   **4.** (a) equal; (b) greater (acceleration is upward, thus net force on body must be upward)   **5.** (a) equal; (b) greater; (c) less   **6.** (a) increase; (b) yes; (c) same; (d) yes   **7.** (a) $F \sin \theta$; (b) increase   **8.** 0 (because now $a = -g$)   **Q**   **1.** (a) 5; (b) 7; (c) $(2$ N$)\hat{i}$; (d) $-(6$ N$)\hat{j}$; (e) fourth; (f) fourth   **3.** (a) 2 and 4; (b) 2 and 4   **5.** (a) 2, 3, 4; (b) 1, 3, 4; (c) 1, $+y$; 2, $+x$; 3, fourth quadrant; 4, third quadrant   **7.** (a) less; (b) greater   **9.** (a) 20 kg; (b) 18 kg; (c) 10 kg; (d) all tie; (e) 3, 2, 1   **11.** (a) 4 or 5, choose 4; (b) 2; (c) 1; (d) 4 or 5, choose 5; (e) 3; (f) 6; (g) 3 and 6; 1, 2, and 5; (h) 3 and 6; (i) 1, 2, and 5   **EP**   **1.** (a) $F_x = 1.88$ N; (b) $F_y = 0.684$ N; (c) $(1.88\hat{i} + 0.684\hat{j})$ N   **3.** 2.9 m/s^2 **5.** $(3\hat{i} - 11\hat{j} + 4\hat{k})$ N   **7.** (a) $(-32\hat{i} - 21\hat{j})$ N; (b) 38 N; (c) 213° from $+x$ **9.** (a) 108 N; (b) 108 N; (c) 108 N   **11.** (a) 11 N; (b) 2.2 kg; (c) 0; (d) 2.2 kg   **13.** 16 N   **15.** (a) 42 N; (b) 72 N; (c) 4.9 m/s^2 **17.** (a) 0.02 m/s^2; (b) 8×10^4 km; (c) 2×10^3 m/s   **19.** 1.2×10^5 N   **21.** 1.5 mm   **23.** (a) $(285\hat{i} + 705\hat{j})$ N; (b) $285\hat{i} - 115\hat{j})$ N; (c) 307 N; (d) $-22°$ from $+x$; (e) 3.67 m/s^2; (f) $-22°$ from $+x$ **25.** (a) 0.62 m/s^2; (b) 0.13 m/s^2; (c) 2.6 m   **27.** (a) 494 N, up; (b) 494 N, down   **29.** (a) 2.2×10^{-3} N; (b) 3.7×10^{-3} N   **31.** (a) 1.1 N   **33.** 1.8×10^4 N   **35.** (a) 620 N; (b) 580 N   **37.** (a) 3260 N; (b) 2.7×10^3 kg; (c) 1.2 m/s^2 **39.** (a) 180 N; (b) 640 N   **41.** (a) 1.23 N; (b) 2.46 N; (c) 3.69 N; (d) 4.92 N; (e) 6.15 N; (f) 0.25 N   **43.** (a) 0.735 m/s^2; (b) downward; (c) 20.8 N   **45.** (a) 1.18 m; (b) 0.674 s; (c) 3.50 m/s   **47.** (a) 4.9 m/s^2; (b) 2.0 m/s^2; (c) upward; (d) 120 N   **49.** (a) 2.18 m/s^2; (b) 116 N; (c) 21.0 m/s^2 **51.** (b) $F/(m + M)$; (c) $MF/(m + M)$; (d) $F(m + 2M)/2(m + M)$ **53.** $2Ma/(a + g)$ **55.** (a) 31.3 kN; (b) 24.3 kN

CHAPTER 6

CP   **1.** (a) zero (because there is no attempt at sliding); (b) 5 N; (c) no; (d) yes; (e) 8 N   **2.** (a) same (10 N); (b) decreases;

(c) decreases (because N decreases)   **3.** greater (from Sample Problem 6-5, v_t depends on $\sqrt{R}$)   **4.** ($\vec{a}$ is directed toward center of circular path) (a) $\vec{a}$ downward, $\vec{N}$ upward; (b) $\vec{a}$ and $\vec{N}$ upward   **5.** (a) same (must still match the gravitational force on the rider); (b) increases ($N = mv^2/R$); (c) increases ($f_{s,max} = \mu_s N$)   **6.** (a) $4R_1$; (b) $4R_1$ **Q**   **1.** (a) F_1, F_2, F_3; (b) all tie   **3.** (a) same; (b) increases; (c) increases; (d) no   **5.** (a) decrease; (b) decrease; (c) increase; (d) increase; (e) increase   **7.** (a) the block's mass m; (b) equal (they are a third-law pair); (c) that on the slab is in the direction of the applied force; that on the block is in the opposite direction; (d) the slab's mass M **9.** 4, 3, then 1, 2, and 5 tie   **EP**   **1.** (a) 200 N; (b) 120 N   **3.** 0.61   **5.** (a) 190 N; (b) 0.56 m/s^2 **7.** (a) 0.13 N; (b) 0.12   **9.** (a) no; (b) $(-12\hat{i} + 5\hat{j})$ N   **13.** (a) 300 N; (b) 1.3 m/s^2 **15.** (a) 66 N; (b) 2.3 m/s^2 **17.** (b) 3.0×10^7 N   **19.** 100 N   **21.** (a) 0; (b) 3.9 m/s^2 down the incline; (c) 1.0 m/s^2 down the incline   **23.** (a) 3.5 m/s^2; (b) 0.21 N; (c) blocks move independently   **25.** 490 N   **27.** (a) 6.1 m/s^2, leftward; (b) 0.98 m/s^2, leftward   **29.** $g(\sin \theta - \sqrt{2}\mu_k \cos \theta)$ **31.** 9.9 s   **33.** 6200 N   **35.** 2.3   **37.** about 48 km/h   **39.** 21 m   **41.** $\sqrt{Mgr/m}$ **43.** (a) light; (b) 778 N; (c) 223 N   **45.** 2.2 km   **47.** (b) 8.74 N; (c) 37.9 N, radially inward; (d) 6.45 m/s

CHAPTER 7

CP   **1.** (a) decrease; (b) same; (c) negative, zero   **2.** d, c, b, a **3.** (a) same; (b) smaller   **4.** (a) positive; (b) negative; (c) zero   **5.** zero   **Q**   **1.** all tie   **3.** (a) positive; (b) zero; (c) negative; (d) negative; (e) zero; (f) positive   **5.** (a) A, B, C; (b) C, B, A; (c) C, B, A; (d) $A, 2$; $B, 3$; $C, 1$ **7.** all tie   **9.** c, d, then a and b tie, then f, e **11.** (a) $2F_1$; (b) $2W_1$ **13.** B, C, A **EP**   **1.** 1.2×10^6 m/s   **3.** (a) 3610 J; (b) 1900 J; (c) 1.1×10^{10} J   **5.** (a) 2.9×10^7 m/s; (b) 2.1×10^{-13} J   **7.** (a) 590 J; (b) 0; (c) 0; (d) 590 J   **9.** (a) 170 N; (b) 340 m; (c) -5.8×10^4 J; (d) 340 N; (e) 170 m; (f) -5.8×10^4 J   **11.** (a) 1.50 J; (b) increases   **13.** 15.3 J   **15.** (a) 98 N; (b) 4.0 cm; (c) 3.9 J; (d) -3.9 J   **17.** (a) 1.2×10^4 J; (b) -1.1×10^4 J; (c) 1100 J; (d) 5.4 m/s   **19.** (a) $-3Mgd/4$; (b) Mgd; (c) $Mgd/4$; (d) $\sqrt{gd/2}$ **21.** (a) -0.043 J; (b) -0.13 J   **23.** (a) 6.6 m/s; (b) 4.7 m   **25.** 800 J   **27.** 0, by both methods   **29.** -6 J   **31.** 490 W   **33.** (a) 0.83 J; (b) 2.5 J; (c) 4.2 J; (d) 5.0 W   **35.** 740 W   **37.** 68 kW   **39.** (a) 1.8×10^5 ft · lb; (b) 0.55 hp

CHAPTER 8

CP   **1.** no (consider round trip on the small loop)   **2.** 3, 1, 2 (see Eq. 8-6)   **3.** (a) all tie; (b) all tie   **4.** (a) CD, AB, BC (zero) (check slope magnitudes); (b) positive direction of x **5.** all tie   **Q**   **1.** (a) 12 J; (b) -2 J   **3.** (a) all tie; (b) all tie   **5.** (a) 4; (b) returns to its starting point and repeats the trip; (c) 1; (d) 1   **7.** (a) fL; (b) 0.50; (c) 1.25; (d) 2.25; (e) b, center; c, right; d, left   **9.** (a) increasing; (b) decreasing; (c) decreasing; (d) constant in AB and BC, decreasing in CD **EP**   **1.** 89 N/cm   **3.** (a) 4.31 mJ; (b) -4.31 mJ; (c) 4.31 mJ; (d) -4.31 mJ; (e) all increase   **5.** (a) mgL; (b) $-mgL$; (c) 0; (d) $-mgL$; (e) mgL; (f) 0; (g) same   **7.** (a) 184 J; (b) -184 J; (c) -184 J

9. (a) 2.08 m/s; (b) 2.08 m/s; (c) increase
11. (a) $\sqrt{2gL}$; (b) $2\sqrt{gL}$; (c) $\sqrt{2gL}$; (d) all the same
13. (a) 260 m; (b) same; (c) decrease   **15.** (a) 21.0 m/s;
(b) 21.0 m/s; (c) 21.0 m/s   **17.** (a) 0.98 J; (b) -0.98 J;
(c) 3.1 N/cm   **19.** (a) 39.2 J; (b) 39.2 J; (c) 4.00 m
21. (a) 35 cm; (b) 1.7 m/s   **23.** (a) 4.8 m/s; (b) 2.4 m/s
25. 10 cm   **27.** 1.25 cm   **31.** (a) $2\sqrt{gL}$; (b) $5mg$; (c) 71°
33. $mgL/32$ **37.** (a) $1.12(A/B)^{1/6}$; (b) repulsive; (c) attractive
39. (a) 5.6 J; (b) 3.5 J   **41.** (a) 30.1 J; (b) 30.1 J; (c) 0.22
43. (a) -2900 J; (b) 390 J; (c) 210 N   **45.** 11 kJ   **47.** 20
ft · lb   **49.** (a) 1.5 MJ; (b) 0.51 MJ; (c) 1.0 MJ; (d) 63 m/s
51. (a) 67 J; (b) 67 J; (c) 46 cm   **53.** (a) 31.0 J; (b) 5.35 m/s;
(c) conservative   **55.** (a) 44 m/s; (b) 0.036   **57.** (a) -0.90 J;
(b) 0.46 J; (c) 1.0 m/s   **59.** 1.2 m   **63.** in the center
of the flat part   **65.** (a) 216 J; (b) 1180 N; (c) 432 J;
(d) motor also supplies thermal energy to crate and belt
67. (b) $\rho(L-x)/2$; (c) $v = v_0[2(\rho L + m_f)/(\rho L + 2m_f - \rho x)]^{0.5}$;
(e) 35 m/s

CHAPTER 9

CP   1. (a) origin; (b) fourth quadrant; (c) on y axis below origin;
(d) origin; (e) third quadrant; (f) origin   **2.** (a) to (c) at the center of mass, still at the origin (their forces are internal to the system and cannot move the center of mass)   **3.** (Consider slopes
and Eq. 9-23.) (a) 1, 3, and then 2 and 4 tie (zero force); (b) 3
4. (No net external force; $\vec{P}$ conserved.) (a) 0; (b) no; (c) $-x$
5. (a) 500 km/h; (b) 2600 km/h; (c) 1600 km/h   **6.** (a) yes;
(b) no (because of net force along y)   **Q   1.** (a) to (d) at the
origin   **3.** (a) at the center of the sled; (b) $L/4$, to the right;
(c) not at all (no net external force); (d) $L/4$, to the left; (e) L;
(f) $L/2$; (g) $L/2$ **5.** (a) ac, cd, and bc; (b) bc; (c) bd and ad
7. c, d, and then a and b tie   **9.** b, c, a   **EP   1.** (a) 4600 km;
(b) $0.73R_e$ **3.** (a) 1.1 m; (b) 1.3 m; (c) shifts toward topmost
particle   **5.** (a) -0.25 m; (b) 0   **7.** 6.8×10^{-12} m from the
nitrogen atom, along axis of symmetry   **9.** (a) $H/2$; (b) $H/2$;
(c) descends to lowest point and then ascends to $H/2$;
(d) $\dfrac{HM}{m}\left(\sqrt{1 + \dfrac{m}{M}} - 1\right)$ **11.** 72 km/h   **13.** (a) 28 cm;
(b) 2.3 m/s   **15.** 53 m   **17.** (a) halfway between the containers;
(b) 26 mm toward the heavier container; (c) down; (d) -1.6×10^{-2} m/s²   **19.** 4.2 m   **21.** 24 km/h   **23.** (a) 7.5×10^4 J;
(b) 3.8×10^4 kg · m/s; (c) 38° south of east   **25.** (a) $(-4.0 \times 10^4 \hat{i})$ kg · m/s; (b) west; (c) 0   **27.** 3.0 mm/s, away from
the stone   **29.** increases by 4.4 m/s   **31.** 4400 km/h
33. (a) 7290 m/s; (b) 8200 m/s; (c) 1.271×10^{10} J;
(d) 1.275×10^{10} J   **35.** (a) 1.4×10^{-22} kg · m/s; (b) 150°;
(c) 120°; (d) 1.6×10^{-19} J   **37.** (a) 1010 m/s, 9.48° clockwise
from the $+x$ direction; (b) 3.23 MJ   **39.** 14 m/s, 135° from
the other pieces   **41.** 108 m/s   **43.** (a) 1.57×10^6 N;
(b) 1.35×10^5 kg; (c) 2.08 km/s   **45.** 2.2×10^{-3}
47. (a) 46 N; (b) none   **49.** (a) 0.2 to 0.3 MJ; (b) same amount
51. (a) 8.8 m/s; (b) 2600 J; (c) 1.6 kW   **53.** 24 W
55. (a) 860 N; (b) 2.4 m/s   **57.** (a) 2.1×10^6 kg;
(b) $\sqrt{100 + 1.5t}$ m/s; (c) $(1.5 \times 10^6)/\sqrt{100 + 1.5t}$ N;
(d) 6.7 km   **59.** 1.5 cm/s downward (the bubbles rise but the
layers descend)

CHAPTER 10

CP   1. (a) unchanged; (b) unchanged (see Eq. 10-4); (c) decrease
(see Eq. 10-8)   **2.** (a) zero; (b) positive (initial p_y down y; final
p_y up y); (c) positive direction of y **3.** (a) 10 kg · m/s;
(b) 14 kg · m/s; (c) 6 kg · m/s   **4.** (a) 4 kg · m/s; (b) 8 kg · m/s;
(c) 3 J   **5.** (a) 2 kg · m/s (conserve momentum along x);
(b) 3 kg · m/s (conserve momentum along y)   **Q   1.** all tie
3. b and c   **5.** (a) rightward; (b) rightward; (c) smaller
7. (a) one was stationary; (b) 2; (c) 5; (d) equal (pool player's
result)   **9.** (a) 2; (b) 1; (c) 3; (d) yes; (e) no   **EP   1.** 2.5 m/s
3. 3000 N   **5.** 67 m/s, in opposite direction   **7.** (a) 42 N · s;
(b) 2100 N   **9.** (a) $(7.4 \times 10^3 \hat{i} - 7.4 \times 10^3 \hat{j})$ N · s;
(b) $(-7.4 \times 10^3 \hat{i})$ N · s; (c) 2.3×10^3 N; (d) 2.1×10^4 N;
(e) $-45°$ **11.** 10 m/s   **13.** (a) 1.0 kg · m/s; (b) 250 J;
(c) 10 N; (d) 1700 N; (e) answer for (c) includes time between
pellet collisions   **15.** 41.7 cm/s   **17.** (a) 1.8 N · s, upward in
figure; (b) 180 N, downward in figure   **19.** (a) 9.0 kg · m/s;
(b) 3000 N; (c) 4500 N; (d) 20 m/s   **21.** 3.0 m/s
23. ≈ 2 mm/y   **25.** (a) 4.6 m/s; (b) 3.9 m/s; (c) 7.5 m/s
27. (a) $mR(\sqrt{2gh} + gt)$; (b) 5.06 kg   **29.** 1.18×10^4 kg
31. (a) $mv_i/(m + M)$; (b) $M/(m + M)$ **33.** 25 cm
35. (a) 1.9 m/s, to the right; (b) yes; (c) no, total kinetic energy
would have increased   **37.** (a) 99 g; (b) 1.9 m/s; (c) 0.93 m/s
39. 7.8%   **41.** (a) 1.2 kg; (b) 2.5 m/s   **43.** (a) 100 g;
(b) 1.0 m/s   **45.** (a) 1/3; (b) $4h$ **47.** (a) 4.15×10^5 m/s;
(b) 4.84×10^5 m/s   **49.** (a) 41°; (b) 4.76 m/s; (c) no
51. 120°   **53.** (a) 6.9 m/s, 30° to $+x$ direction; (b) 6.9 m/s,
$-30°$ to $+x$ direction; (c) 2.0 m/s, $-x$ direction   **57.** (a) $5mg$;
(b) $7mg$; (c) 5 m

CHAPTER 11

CP   1. (b) and (c)   **2.** (a) and (d) ($\alpha = d^2\theta/dt^2$ must be a constant)   **3.** (a) yes; (b) no; (c) yes; (d) yes   **4.** all tie   **5.** 1, 2, 4,
3 (see Eq. 11-29)   **6.** (see Eq. 11-32) 1 and 3 tie, 4, then 2 and
5 tie (zero)   **7.** (a) downward in the figure ($\tau_{net} = 0$); (b) less
(consider moment arms)   **Q   1.** (a) positive; (b) zero; (c) negative; (d) negative   **3.** finite angular displacements are not commutative   **5.** (a) c, a, then b and d tie; (b) b, then a and c tie,
then d **7.** 3, 1, 2   **9.** 90°, then 70° and 110° tie   **11.** (a) de-
crease; (b) clockwise; (c) counterclockwise   **EP   1.** (a) $a + 3bt^2 - 4ct^3$; (b) $6bt - 12ct^2$ **3.** (a) 5.5×10^{15} s; (b) 26
5. (a) 2 rad; (b) 0; (c) 130 rad/s; (d) 32 rad/s²; (e) no   **7.** 11 rad/s
9. (a) -67 rev/min²; (b) 8.3 rev   **11.** 200 rev/min   **13.** 8.0 s
15. (a) 44 rad; (b) 5.5 s, 32 s; (c) -2.1 s, 40 s   **17.** (a) 340 s;
(b) -4.5×10^{-3} rad/s²; (c) 98 s   **19.** 1.8 m/s², toward the center
21. 0.13 rad/s   **23.** (a) 3.0 rad/s; (b) 30 m/s; (c) 6.0 m/s²;
(d) 90 m/s²   **25.** (a) 3.8×10^3 rad/s; (b) 190 m/s
27. (a) 7.3×10^{-5} rad/s; (b) 350 m/s; (c) 7.3×10^{-5} rad/s;
(d) 460 m/s   **29.** 16 s   **31.** (a) -2.3×10^{-9} rad/s²; (b) 2600 y;
(c) 24 ms   **33.** 12.3 kg · m²   **35.** (a) 1100 J; (b) 9700 J
37. (a) $5md^2 + \frac{8}{3}Md^2$; (b) $(\frac{5}{2}m + \frac{4}{3}M)d^2\omega^2$ **39.** 0.097 kg · m²
41. $\frac{1}{3}M(a^2 + b^2)$ **45.** 4.6 N · m   **47.** (a) $r_1F_1 \sin \theta_1 - r_2F_2 \sin \theta_2$; (b) -3.8 N · m   **49.** (a) 28.2 rad/s²; (b) 338 N · m
51. (a) 155 kg · m²; (b) 64.4 kg   **53.** 130 N   **55.** (a) 6.00 cm/s²;
(b) 4.87 N; (c) 4.54 N; (d) 1.20 rad/s²; (e) 0.0138 kg · m²
57. (a) 1.73 m/s²; (b) 6.92 m/s²   **59.** 396 N · m
61. (a) $mL^2\omega^2/6$; (b) $L^2\omega^2/6g$ **63.** 5.42 m/s   **65.** $\sqrt{9g/4L}$

67. (a) $[(3g/H)(1 - \cos \theta)]^{0.5}$; (b) $3g(1 - \cos \theta)$; (c) $\frac{3}{2}g \sin \theta$; (d) $41.8°$ **69.** 17

CHAPTER 12

CP **1.** (a) same; (b) less **2.** less (consider the transfer of energy from rotational kinetic energy to gravitational potential energy) **3.** (draw the vectors, use right-hand rule) (a) $\pm z$; (b) $+y$; (c) $-x$ **4.** (see Eq. 12-21) (a) 1 and 3 tie, then 2 and 4 tie, then 5 (zero); (b) 2 and 3 **5.** (see Eqs. 12-23 and 12-16) (a) 3, 1; then 2 and 4 tie (zero); (b) 3 **6.** (a) all tie (same τ, same t, thus same ΔL); (b) sphere, disk, hoop (reverse order of I) **7.** (a) decreases; (b) same ($\tau_{net} = 0$, so L conserved) (c) increases
Q **1.** (a) same; (b) block; (c) block **3.** (a) $0.5L$; (b) L
5. b, then c and d tie, then a and e tie (zero) **7.** a, then b and c tie, then e, d (zero) **9.** (a) same; (b) increase; (c) decrease; (d) same, decrease, increase **11.** (a) 1, 2, 3 (zero); (b) 1 and 2 tie, then 3; (c) 1 and 3 tie, then 2 **13.** (a) 3, 1, 2; (b) 3, 1, 2
EP **1.** (a) 59.3 rad/s; (b) 9.31 rad/s²; (c) 70.7 m **3.** -3.15 J
5. 1/50 **7.** (a) 8.0°; (b) more **9.** 4.8 m **11.** (a) 63 rad/s; (b) 4.0 m **13.** (a) 8.0 J; (b) 3.0 m/s; (c) 6.9 J; (d) 1.8 m/s
15. (a) 13 cm/s²; (b) 4.4 s; (c) 55 cm/s; (d) 1.8×10^{-2} J; (e) 1.4 J; (f) 27 rev/s **19.** (a) 10 N · m, parallel to yz plane, at 53° to $+y$; (b) 22 N · m, $-x$ **21.** (a) $50\hat{k}$ N · m; (b) 90°
23. 9.8 kg · m²/s **25.** (a) 0; (b) $(8.0\hat{i} + 8.0\hat{k})$ N · m
27. (a) mvd; (b) no; (c) 0, yes **29.** (a) $-170\hat{k}$ kg · m²/s; (b) $+56\hat{k}$ N · m; (c) $+56\hat{k}$ kg · m²/s² **31.** (a) 0; (b) $8t$ N · m, in $-z$ direction; (c) $2/\sqrt{t}$ N · m, $-z$; (d) $8/t^3$ N · m, $+z$
33. (a) -1.47 N · m; (b) 20.4 rad; (c) -29.9 J; (d) 19.9 W
35. (a) $14md²$; (b) $4md²\omega$; (c) $14md²\omega$
37. $\omega_0 R_1 R_2 I_1/(I_1 R_2^2 + I_2 R_1^2)$ **39.** (a) 3.6 rev/s; (b) 3.0; (c) in moving the bricks in, the forces on them from the man transferred energy from internal energy of the man to kinetic energy
41. (a) 267 rev/min; (b) 2/3 **43.** (a) 149 kg · m²;
(b) 158 kg · m²/s; (c) 0.746 rad/s **45.** $\dfrac{m}{M + m}\left(\dfrac{v}{R}\right)$
47. (a) $(mRv - I\omega_0)/(I + mR²)$; (b) no, energy transferred to internal energy of cockroach **49.** 3.4 rad/s **51.** (a) 0.148 rad/s; (b) 0.0123; (c) 181° **53.** the day would be longer by about 0.8 s
55. (a) 18 rad/s; (b) 0.92 **57.** (a) 0.24 kg · m²; (b) 1800 m/s
59. $\theta = \cos^{-1}\left[1 - \dfrac{6m²h}{d(2m + M)(3m + M)}\right]$

CHAPTER 13

CP **1.** c, e, f **2.** directly below the rod (torque on the apple due to $\vec{F}_g$, about the suspension, is zero) **3.** (a) no; (b) at site of $\vec{F}_1$, perpendicular to plane of figure; (c) 45 N **4.** (a) at C (to eliminate forces there from a torque equation); (b) plus; (c) minus; (d) equal **5.** d **6.** (a) equal; (b) B; (c) B Q **1.** (a) yes; (b) yes; (c) yes; (d) no **3.** a and c (forces and torques balance) **5.** $m_2 = 12$ kg, $m_3 = 3$ kg, $m_4 = 1$ kg **7.** (a) 15 N (the key is the pulley with the 10 N piñata); (b) 10 N **9.** A, then tie of B and C EP **1.** (a) 2; (b) 7 **3.** (a) $(-27\hat{i} + 2\hat{j})$ N; (b) 176° counterclockwise from $+x$ direction **5.** 7920 N
7. (a) $(mg/L)\sqrt{L² + r²}$; (b) mgr/L **9.** (a) 1160 N, down;
(b) 1740 N, up; (c) left; (d) right **11.** 74 g **13.** (a) 280 N;
(b) 880 N, 71° above the horizontal **15.** (a) 8010 N;
(b) 3.65 kN; (c) 5.66 kN **17.** 71.7 N **19.** (a) 5.0 N; (b) 30 N;

(c) 1.3 m **21.** $mg\dfrac{\sqrt{2rh - h²}}{r - h}$ **23.** (a) 192 N; (b) 96.1 N;
(c) 55.5 N **25.** (a) 6630 N; (b) 5740 N; (c) 5960 N **27.** 2.20 m
29. 0.34 **31.** (a) 211 N; (b) 534 N; (c) 320 N **33.** (a) 445 N;
(b) 0.50; (c) 315 N **35.** (a) slides at 31°; (b) tips at 34°
37. (a) 6.5×10^6 N/m²; (b) 1.1×10^{-5} m
39. (a) 867 N; (b) 143 N; (c) 0.165 **41.** (a) 51°; (b) $0.64Mg$

CHAPTER 14

CP **1.** all tie **2.** (a) 1, tie of 2 and 4, then 3; (b) line d
3. negative y direction **4.** (a) increase; (b) negative
5. (a) 2; (b) 1 **6.** (a) path 1 (decreased E (more negative) gives decreased a); (b) less (decreased a gives decreased T)
Q **1.** (a) between, closer to less massive particle; (b) no; (c) no (other than infinity) **3.** $3GM²/d²$, leftward **5.** (a) 1 and 2 tie, then 3 and 4 tie; (b) 1, 2, 3, 4 **7.** $U_i/4$ **9.** (a) all tie; (b) all tie
11. (a)–(d) zero EP **1.** 19 m **3.** 29 pN **5.** 1/2 **7.** 2.60×10^5 km **9.** 0.017 N, toward the 300 kg sphere **11.** 3.2×10^{-7} N **13.** $\dfrac{GmM}{d²}\left[1 - \dfrac{1}{8(1 - R/2d)²}\right]$ **15.** 2.6×10^6 m
17. (b) 1.9 h **21.** 4.7×10^{24} kg **23.** (a) $(3.0 \times 10^{-7}$ N/kg)m;
(b) $(3.3 \times 10^{-7}$ N/kg)m; (c) $(6.7 \times 10^{-7}$ N/kg · m)mr
25. (a) 9.83 m/s²; (b) 9.84 m/s²; (c) 9.79 m/s² **27.** (a) -1.3×10^{-4} J; (b) less; (c) positive; (d) negative **29.** (a) 0.74;
(b) 3.7 m/s²; (c) 5.0 km/s **31.** (a) 5.0×10^{-11} J; (b) -5.0×10^{-11} J **35.** (a) 1700 m/s; (b) 250 km; (c) 1400 m/s
37. (a) 82 km/s; (b) 1.8×10^4 km/s **39.** 2.5×10^4 km
41. 6.5×10^{23} kg **43.** 5×10^{10} **45.** (a) 7.82 km/s;
(b) 87.5 min **47.** (a) 6640 km; (b) 0.0136
49. (a) 1.9×10^{13} m; (b) $3.5R_P$ **53.** 0.71 y **55.** $\sqrt{GM/L}$
57. (a) 2.8 y; (b) 1.0×10^{-4} **61.** (a) no; (b) same; (c) yes
63. (a) 7.5 km/s; (b) 97 min; (c) 410 km; (d) 7.7 km/s; (e) 92 min; (f) 3.2×10^{-3} N; (g) no; (h) yes, if the satellite–Earth system is considered isolated

CHAPTER 15

CP **1.** all tie **2.** (a) all tie (the gravitational force on the penguin is the same); (b) $0.95\rho_0$, ρ_0, $1.1\rho_0$ **3.** 13 cm³/s, outward
4. (a) all tie; (b) 1, then 2 and 3 tie, 4 (wider means slower); (c) 4, 3, 2, 1 (wider and lower mean more pressure) Q **1.** e, then b and d tie, then a and c tie **3.** (a) 2; (b) 1, less; 3, equal; 4, greater **5.** (a) moves downward; (b) moves downward
7. all tie **9.** (a) downward; (b) downward; (c) same
EP **1.** 1.1×10^5 Pa or 1.1 atm **3.** 2.9×10^4 N **5.** 0.074
7. (b) 26 kN **9.** 5.4×10^4 Pa **11.** (a) 5.3×10^6 N; (b) 2.8×10^5 N; (c) 7.4×10^5 N; (d) no **13.** 7.2×10^5 N
15. $\frac{1}{4}\rho gA(h_2 - h_1)²$ **17.** 1.7 km **19.** (a) $\rho gWD²/2$;
(b) $\rho gWD³/6$; (c) $D/3$ **21.** (a) 7.9 km; (b) 16 km **23.** 4.4 mm
25. (a) 2.04×10^{-2} m³; (b) 1570 N **27.** (a) 670 kg/m³;
(b) 740 kg/m³ **29.** (a) 1.2 kg; (b) 1300 kg/m³ **31.** 57.3 cm
33. 0.126 m³ **35.** (a) 45 m²; (b) car should be over center of slab if slab is to be level **37.** (a) 9.4 N; (b) 1.6 N **39.** 8.1 m/s
41. 66 W **43.** (a) 2.5 m/s; (b) 2.6×10^5 Pa **45.** (a) 3.9 m/s;
(b) 88 kPa **47.** (a) 1.6×10^{-3} m³/s; (b) 0.90 m **49.** 116 m/s
51. (a) 6.4 m³; (b) 5.4 m/s; (c) 9.8×10^4 Pa **53.** (a) 74 N;
(b) 150 m³ **55.** (b) 2.0×10^{-2} m³/s **57.** (b) 63.3 m/s

59. (a) 180 kN; (b) 81 kN; (c) 20 kN; (d) 0; (e) 78 kPa; (f) no
61. (a) 0.050; (b) 0.41; (c) no; (d) Lay back on the surface, slowly pull your legs free, and then roll over to the shore.

CHAPTER 16

CP   **1.** (sketch x versus t) (a) $-x_m$; (b) $+x_m$; (c) 0   **2.** a (F must have form of Eq. 6-10)   **3.** (a) 5 J; (b) 2 J; (c) 5 J   **4.** all tie (in Eq. 16-29, m is included in I)   **5.** 1, 2, 3 (the ratio m/b matters; k does not)   Q   **1.** c   **3.** (a) 2; (b) positive; (c) between 0 and $+x_m$ **5.** (a) toward $-x_m$; (b) toward $+x_m$; (c) between $-x_m$ and 0; (d) between $-x_m$ and 0; (e) decreasing; (f) increasing   **7.** (a) π rad; (b) π rad; (c) $\pi/2$ rad   **9.** (a) varies; (b) varies; (c) $x = \pm x_m$; (d) more likely   **11.** b (infinite period; does not oscillate), c, a   **13.** one system: $k = 1500$ N/m, $m = 500$ kg; other system: $k = 1200$ N/m, $m = 400$ kg; the same ratio $k/m = 3$ gives resonance for both systems   EP   **1.** (a) 0.50 s; (b) 2.0 Hz; (c) 18 cm   **3.** (a) 0.500 s; (b) 2.00 Hz; (c) 12.6 rad/s; (d) 79.0 N/m; (e) 4.40 m/s; (f) 27.6 N   **5.** $f > 500$ Hz   **7.** (a) 6.28×10^5 rad/s; (b) 1.59 mm   **9.** (a) 1.0 mm; (b) 0.75 m/s; (c) 570 m/s^2 **11.** (a) 1.29×10^5 N/m; (b) 2.68 Hz   **13.** 7.2 m/s   **15.** 2.08 h   **17.** 3.1 cm   **19.** (a) 5.58 Hz; (b) 0.325 kg; (c) 0.400 m   **21.** (a) 2.2 Hz; (b) 56 cm/s; (c) 0.10 kg; (d) 20.0 cm below y_i **23.** (a) $0.183A$; (b) same direction   **29.** (a) $(n+1)k/n$; (b) $(n+1)k$; (c) $\sqrt{(n+1)/n}\,f$; (d) $\sqrt{n+1}\,f$ **31.** 37 mJ   **33.** (a) 2.25 Hz; (b) 125 J; (c) 250 J; (d) 86.6 cm   **35.** (a) 130 N/m; (b) 0.62 s; (c) 1.6 Hz; (d) 5.0 cm; (e) 0.51 m/s   **37.** (a) $\frac{3}{4}$; (b) $\frac{1}{4}$; (c) $x_m/\sqrt{2}$ **39.** (a) 16.7 cm; (b) 1.23%   **41.** (a) 39.5 rad/s; (b) 34.2 rad/s; (c) 124 rad/s^2 **43.** 99 cm
45. 5.6 cm   **47.** (a) $2\pi\sqrt{\dfrac{L^2 + 12d^2}{12gd}}$; (b) increases for $d < L/\sqrt{12}$, decreases for $d > L/\sqrt{12}$; (c) increases; (d) no change
49. (a) 0.205 kg · m^2; (b) 47.7 cm; (c) 1.50 s   **53.** $2\pi\sqrt{m/3k}$
55. (a) 0.35 Hz; (b) 0.39 Hz; (c) 0   **57.** (b) smaller   **59.** 0.39
61. (a) 14.3 s; (b) 5.27   **63.** (a) $F_m/b\omega$; (b) F_m/b

CHAPTER 17

CP   **1.** a, 2; b, 3; c, 1 (compare with phase in Eq. 17-2, then see Eq. 17-5)   **2.** (a) 2, 3, 1 (see Eq. 17-12); (b) 3, then 1 and 2 tie (find amplitude of dy/dt)   **3.** (a) same (independent of f); (b) decrease ($\lambda = v/f$); (c) increase; (d) increase   **4.** (a) increase; (b) increase; (c) increase   **5.** 0.20 and 0.80 tie, then 0.60, 0.45   **6.** (a) 1; (b) 3; (c) 2   **7.** (a) 75 Hz; (b) 525 Hz   Q   **1.** 7d   **3.** (a) $\pi/2$ rad and 0.25 wavelength; (b) π rad and 0.5 wavelength; (c) $3\pi/2$ rad and 0.75 wavelength; (d) 2π rad and 1.0 wavelength; (e) $3T/4$; (f) $T/2$ **5.** (a) 4; (b) 4; (c) 3   **7.** a and d tie, then b and c tie   **9.** d   **11.** (a) decrease; (b) disappears   EP   **1.** (a) 3.49 m^{-1}; (b) 31.5 m/s   **3.** (a) 0.68 s; (b) 1.47 Hz; (c) 2.06 m/s   **7.** (a) $y(x, t) = 2.0 \sin 2\pi(0.10x - 400t)$, with x and y in cm and t in s; (b) 50 m/s; (c) 40 m/s   **9.** (a) 11.7 cm; (b) π rad   **11.** 129 m/s   **13.** (a) 15 m/s; (b) 0.036 N   **15.** $y(x, t) = 0.12 \sin(141x + 628t)$, with y in mm, x in m, and t in s   **17.** (a) $2\pi y_m/\lambda$; (b) no   **19.** (a) 5.0 cm; (b) 40 cm; (c) 12 m/s; (d) 0.033 s; (e) 9.4 m/s; (f) $5.0 \sin(16x + 190t + 0.93)$, with x in m, y in cm, and t in s   **21.** 2.63 m from the end of the wire from which the later pulse originates   **25.** (a) 3.77 m/s; (b) 12.3 N; (c) zero; (d) 46.3 W; (e) zero; (f) zero; (g) ± 0.50 cm   **27.** $1.4y_m$ **29.** 5.0 cm   **31.** (a) $0.83y_1$; (b) 37°   **33.** (a) 140

m/s; (b) 60 cm; (c) 240 Hz   **35.** (a) 82.0 m/s; (b) 16.8 m; (c) 4.88 Hz   **37.** 7.91 Hz, 15.8 Hz, 23.7 Hz   **39.** (a) 105 Hz; (b) 158 m/s   **41.** (a) 0.25 cm; (b) 120 cm/s; (c) 3.0 cm; (d) zero   **43.** (a) 50 Hz; (b) $y = 0.50 \sin[\pi(x \pm 100t)]$, with x in m, y in cm, and t in s   **45.** (a) 1.3 m; (b) $y = 0.002 \sin(9.4x) \cos(3800t)$, with x and y in m and t in s   **47.** (a) 2.0 Hz; (b) 200 cm; (c) 400 cm/s; (d) 50 cm, 150 cm, 250 cm, etc.; (e) 0, 100 cm, 200 cm, etc.   **51.** (a) 323 Hz; (b) eight

CHAPTER 18

CP   **1.** beginning to decrease (example: mentally move the curves of Fig. 18-7 rightward past the point at $x = 42$ m)   **2.** (a) 0, fully constructive; (b) 4λ, fully constructive   **3.** (a) 1 and 2 tie, then 3 (see Eq. 18-28); (b) 3, then 1 and 2 tie (see Eq. 18-26)   **4.** second (see Eqs. 18-39 and 18-41)   **5.** loosen   **6.** a, greater; b, less; c, can't tell; d, can't tell; e, greater; f, less   **7.** (measure speeds relative to the air) (a) 222 m/s; (b) 222 m/s   Q   **1.** pulse along path 2   **3.** (a) 2.0 wavelengths; (b) 1.5 wavelengths; (c) fully constructive, fully destructive   **5.** (a) exactly out of phase; (b) exactly out of phase   **7.** (a) one; (b) nine   **9.** (a) increase; (b) decrease   **11.** all odd harmonics   **13.** d, e, b, c, a EP   **1.** divide the time by 3   **3.** (a) 79 m, 41 m; (b) 89 m   **5.** 1900 km   **7.** 40.7 m   **9.** (a) 0.0762 mm; (b) 0.333 mm   **11.** (a) 1.50 Pa; (b) 158 Hz; (c) 2.22 m; (d) 350 m/s   **13.** (a) $343(1 + 2m)$ Hz, with m being an integer from 0 to 28; (b) $686m$ Hz, with m being an integer from 1 to 29   **15.** (a) 143 Hz, 429 Hz, 715 Hz; (b) 286 Hz, 572 Hz, 858 Hz   **17.** 15.0 mW   **19.** 36.8 nm   **21.** (a) 1000; (b) 32   **23.** (a) 59.7; (b) 2.81×10^{-4} **25.** (b) 5.76×10^{-17} J/m^3 **27.** (b) length^2 **29.** (a) 5200 Hz; (b) amplitude$_{SAD}$/amplitude$_{SBD}$ = 2   **31.** (a) 57.2 cm; (b) 42.9 cm   **33.** (a) 405 m/s; (b) 596 N; (c) 44.0 cm; (d) 37.3 cm   **35.** (a) 1129, 1506, and 1882 Hz   **37.** 12.4 m   **39.** (a) node; (c) 22 s   **41.** 45.3 N   **43.** 387 Hz   **45.** 0.02   **47.** 17.5 kHz   **49.** (a) 526 Hz; (b) 555 Hz   **51.** (a) 1.02 kHz; (b) 1.04 kHz   **53.** 155 Hz   **55.** (a) 485.8 Hz; (b) 500.0 Hz; (c) 486.2 Hz; (d) 500.0 Hz   **57.** (a) 598 Hz; (b) 608 Hz; (c) 589 Hz   **59.** (a) 42°; (b) 11 s

CHAPTER 19

CP   **1.** (a) all tie; (b) 50°X, 50°Y, 50°W   **2.** (a) 2 and 3 tie, then 1, then 4; (b) 3, 2, then 1 and 4 tie (from Eqs. 19-9 and 19-10, assume that change in area is proportional to initial area)   **3.** A (see Eq. 19-14)   **4.** c and e (maximize area enclosed by a clockwise cycle)   **5.** (a) all tie (ΔE_{int} depends on i and f, not on path); (b) 4, 3, 2, 1 (compare areas under curves); (c) 4, 3, 2, 1 (see Eq. 19-26)   **6.** (a) zero (closed cycle); (b) negative (W_{net} is negative; see Eq. 19-26)   **7.** b and d tie, then a, c (P_{cond} identical; see Eq. 19-32)   Q   **1.** 25 S°, 25 U°, 25 R°   **3.** A and B tie, then C, D **5.** (a) both clockwise; (b) both clockwise   **7.** c, a, b   **9.** sphere, hemisphere, cube   **11.** (a) at freezing point; (b) no liquid freezes; (c) ice partly melts   EP   **1.** 0.05 kPa, nitrogen   **3.** 348 K   **5.** (a) $-40°$; (b) 575°; (c) Celsius and Kelvin cannot give the same reading   **7.** (a) Dimensions are inverse time.   **9.** $-92.1°$X   **11.** 960 μm   **13.** 2.731 cm   **15.** 29 cm^3 **17.** 0.26 cm^3 **19.** 360°C   **23.** 0.68 s/h, fast   **25.** 7.5 cm   **27.** (a) 523 J/kg · K; (b) 26.2 J/mol · K; (c) 0.600 mole   **29.** 42.7 kJ   **31.** 1.9 times as great   **33.** (a) 33.9 Btu; (b) 172 F°

35. 160 s   **37.** 2.8 days   **39.** 742 kJ   **41.** 82 cal   **43.** 33 g
45. (a) 0°C; (b) 2.5°C   **47.** 8.72 g   **49.** A: 120 J, B: 75 J,
C: 30 J   **51.** -30 J   **53.** (a) 6.0 cal; (b) -43 cal; (c) 40 cal;
(d) 18 cal, 18 cal   **55.** (a) 0.13 m; (b) 2.3 km   **57.** 1660 J/s
59. (a) 16 J/s; (b) 0.048 g/s   **61.** 0.50 min   **63.** (a) 17 kW/m^2;
(b) 18 W/m^2 **65.** 0.40 cm/h   **67.** (a) 90 W; (b) 230 W;
(c) 330 W

53. (a) 6980 J; (b) 4990 J; (c) 1990 J; (d) 2990 J
55. (a) 14 atm; (b) 620 K   **59.** 1.40   **61.** (a) In joules, in the
order Q, ΔE_{int}, W: $1 \rightarrow 2$: 3740, 3740, 0; $2 \rightarrow 3$: 0, -1810,
1810; $3 \rightarrow 1$: -3220, -1930, -1290; Cycle: 520, 0, 520;
(b) $V_2 = 0.0246$ m^3, $p_2 = 2.00$ atm, $V_3 = 0.0373$ m^3, $p_3 =$
1.00 atm

CHAPTER 20

CP   **1.** all but c **2.** (a) all tie; (b) 3, 2, 1   **3.** gas A **4.** 5
(greatest change in T), then tie of 1, 2, 3, and 4   **5.** 1, 2, 3
($Q_3 = 0$, Q_2 goes into work W_2, but Q_1 goes into greater work
W_1 and increases gas temperature)   **Q**   **1.** increased but less
than doubled   **3.** tie of a and c, then b, then d   **5.** 1–4   **7.** 20 J
9. (a) 3; (b) 1; (c) 4; (d) 2; (e) yes   **11.** (a) 1, 2, 3, 4; (b) 1, 2, 3
EP   **1.** 0.933 kg   **3.** 6560   **5.** (a) 5.47×10^{-8} mol; (b) $3.29 \times$
10^{16} **7.** (a) 0.0388 mol; (b) 220°C   **9.** (a) 106; (b) 0.892 m^3
11. $A(T_2 - T_1) - B(T_2^2 - T_1^2)$ **13.** 5600 J   **15.** 100 cm^3
17. 2.0×10^5 Pa   **19.** 180 m/s   **21.** 9.53×10^6 m/s
23. 1.9 kPa   **25.** 3.3×10^{-20} J   **27.** (a) 6.75×10^{-20} J;
(b) 10.7   **31.** (a) 6×10^9 km   **33.** 15 cm   **35.** (a) 3.27×10^{10};
(b) 172 m   **37.** (a) 6.5 km/s; (b) 7.1 km/s   **39.** (a) $1.0 \times$
10^4 K; (b) 1.6×10^5 K; (c) 440 K, 7000 K; (d) hydrogen, no;
oxygen, yes   **41.** (a) 7.0 km/s; (b) 2.0×10^{-8} cm;
(c) 3.5×10^{10} collisions/s   **43.** (a) $\frac{2}{3}v_0$; (b) $N/3$; (c) $122v_0$;
(d) $1.31v_0$ **45.** $RT \ln(V_f/V_i)$ **47.** $(n_1C_1 + n_2C_2 + n_3C_3)/(n_1 +$
$n_2 + n_3)$ **49.** (a) 6.6×10^{-26} kg; (b) 40 g/mol   **51.** 8000 J

CHAPTER 21

CP   **1.** a, b, c   **2.** smaller (Q is smaller)   **3.** c, b, a   **4.** a, d,
c, b   **5.** b   **Q**   **1.** unchanged   **3.** b, a, c, d **5.** equal
7. (a) same; (b) increase; (c) decrease   **9.** (a) same; (b) increase;
(c) decrease   **11.** (a) 0; (b) 0.25; (c) 0.50   **EP**   **1.** 14.4 J/K
3. (a) 9220 J; (b) 23.0 J/K; (c) 0   **5.** (a) 5.79×10^4 J;
(b) 173 J/K   **7.** (a) 14.6 J/K; (b) 30.2 J/K   **9.** (a) 57.0°C;
(b) -22.1 J/K; (c) $+24.9$ J/K; (d) $+2.8$ J/K   **13.** (a) 320 K;
(b) 0; (c) $+1.72$ J/K   **15.** $+0.75$ J/K   **17.** (a) -943 J/K;
(b) $+943$ J/K; (c) yes   **19.** (a) $3p_0V_0$; (b) $\Delta E_{int} = 6RT_0$,
$\Delta S = \frac{3}{2}R \ln 2$; (c) both are zero   **21.** (a) 31%; (b) 16 kJ
23. (a) 23.6%; (b) 1.49×10^4 J   **25.** 266 K and 341 K
27. (a) 1470 J; (b) 554 J; (c) 918 J; (d) 62.4%
29. (a) 2270 J; (b) 14 800 J; (c) 15.4%; (d) 75.0%, greater
31. (a) 78%; (b) 81 kg/s   **33.** (a) $T_2 = 3T_1$, $T_3 = 3T_1/4^{\gamma-1}$,
$T_4 = T_1/4^{\gamma-1}$, $p_2 = 3p_1$, $p_3 = 3p_1/4^\gamma$, $p_4 = p_1/4^\gamma$;
(b) $1 - 4^{1-\gamma}$ **35.** 21 J   **37.** 440 W   **39.** 0.25 hp
41. $[1 - (T_2/T_1)]/[1 - (T_4/T_3)]$ **45.** (a) $W = N!/(n_1!\ n_2!\ n_3!)$;
(b) $[(N/2)!\ (N/2)!]/[(N/3)!\ (N/3)!\ (N/3)!]$; (c) 4.2×10^{16}

PHOTO CREDITS

CHAPTER 18
Page 398: Stephen Dalton/Animals Animals. Page 399: Howard Sochurak/The Stock Market. Page 407: Ben Rose/The Image Bank. Page 409: Bob Gruen/Star File. Page 410: John Eastcott/ Yva Momativk/DRK Photo. Page 419: U.S. Navy photo by Ensign John Gay.

CHAPTER 19
Page 425: ©Dr. Mosato Ono, Tamagawa University. Reproduced with permission. Page 431: AP/Wide World Photos. Page 445: Courtesy Daedalus Enterprises, Inc. Page 453: Kim Westerskov/ Tony Stone Images/New York, Inc.

CHAPTER 20
Page 454: Tom Branch.

CHAPTER 21
Page 482: Steven Dalton/Photo Researchers. Page 492: Richard Ustinich/The Image Bank.

INDEX

Figures are noted by page numbers in *italics;* tables are indicated by t following the page number.

Problem Solving Tactics